BIRDS IN WALES

The Dee at Llangollen

Birds in Wales

by ROGER LOVEGROVE, GRAHAM WILLIAMS
and IOLO WILLIAMS

Illustrated by
Kim Atkinson, Phil Bristow, Terence Lambert, Darren Rees,
Philip Snow and Owen Williams

T & A D POYSER
London

First published in 1994 by T & A D Poyser Ltd
24–28 Oval Road, London, NW1 7DX

United States Edition published by
Academic Press Inc., San Diego, CA 92101

This book is printed on acid-free paper

Text set in 9/10pt Linotron Ehrhardt
Typeset by Phoenix Photosetting, Chatham, Kent
Printed and bound in Great Britain by
The Bath Press, Bath, Avon

A CIP record for this book is available from the British Library

ISBN 0-85661-069-0

Contents

I Meg, Dai a'r teulu am eich cefnogaeth.

To Judith and Charlotte with love.

For the Lady Mary
A Myrddin sy'n gwau hud a lledrith

Preface

It seemed a good idea in April 1981 when two of us talked with Trevor Poyser about the merits of setting out on a project to produce, for the first time, a *Birds of Wales*, as a companion volume to that which Valerie Thom was planning in Scotland. We knew then that we were taking on a fairly big commitment and agreed with Trevor that it would be over-optimistic to aim to complete it in less than five years. Little did we realise just how much of a commitment it would turn out to be! We slowly grasped the astuteness of Valerie Thom in negotiating a Sabbatical year to produce *Birds in Scotland* as we struggled to match our working commitments with the RSPB to the ebb tide of time in researching the book. Deadline passed deadline and how Trevor Poyser and later Andrew Richford remained so patient is a constant wonder to us. In 1989 we were joined on the permanent staff of RSPB in Wales by Iolo Williams, who with the enthusiasm of relative youth did not resist the invitation to become the third author. He has not only brought an added dimension of Welshness to the book but has worked like a Trojan and we believe it is true to say that without his participation and the share that he took on it is almost certain that the book would not have been completed.

We have thoroughly enjoyed writing this book and, despite the staccato progress which has marked its gestation, never for a moment have we tired of it or wished it were not there. Rather we have ached to have the time to put into the research which the subject fully deserves. Despite our full-time involvement in birds and bird conservation in Wales we have, apart from anything else, learned so much ourselves through the writing of it. Even at this final stage of its preparation we would wish for six months free from other commitments to delve more deeply for historical nuggets which we are sure are still to be found.

This is the first attempt there has been to produce an authoritative *Birds in Wales*, which has meant that we have had to search as far back into the annals of written records as we are able. We are not clear whether our job has been made easier or more difficult by the fact that Wales has been very poorly served by way of historical ornithological information. Before the flush of growing interest in birds which manifested itself in a modest tide of written records, county lists and avifaunas for several "fashionable" areas of Wales around the end of the 19th and early 20th centuries, there is desperately little by way of written accounts to tell us the bird populations of any areas of the Principality. Even A. G. More in 1865, attempting the first comprehensive collation of information on the distribution of birds during the nesting season in Great Britain, could find few correspondents in Wales able to supply him with information and his coverage of birds in the Principality is barely even a veneer. Accordingly the one or two local lists produced in the middle of the 19th century, by men such as Doddridge-Knight and Lewis Dillwyn are invaluable records – would that there were more of them, whatever the doubts or queries we might have about the accuracy of one or two of the inclusions.

However, it is not only for the 19th century that such a paucity of information and written record exists in Wales. Although, with the rise of concern for nature conservation and the environment in general, we have some high-quality data on the numbers and distribution of many species within threatened groups or communities, great gaps in our knowledge are still evident. Thus we have good indices for breeding seabird populations since the 1970s, accurate data for many birds of prey, notably the rarer ones, and regular figures for intertidal populations of wildfowl and wading birds but at the same time we know desperately little about the fluctuations of many common passerine species. No more than seven Common Bird Census plots have been worked for as long as ten years in Wales and even these few do not lend themselves easily to comparisons, one with another, as they cover different types of habitat. It has not been possible for us to quote even one example of density for a range of some of our commoner species, for example, Swift, Jay, Yellowhammer. The comparison between the level of information known about our commonest species in Wales and most areas in England is a sobering one, mitigated only in part by the frequent plea that Wales is thinly populated (certainly by field birdwatchers!). The status of many of our commoner species is actually most poorly known in those areas which have the highest human populations!

Regular county bird reports (or annotated summaries) began first in Montgomeryshire in 1947 (now defunct even in summary since 1981–82) followed by the *Cambrian Bird Report* in 1953–54 (covering Anglesey, Caernarfonshire, Denbighshire and Merioneth) and the *Glamorgan Bird Report* in 1962. At the time of writing (1993), annual reports are now published covering all the (old) Welsh Counties except, sadly, Montgomeryshire. County avifaunas have been produced at one time or another for all counties except Anglesey – one of the most exciting bird areas of Wales – Montgomeryshire and Denbighshire (although Hope Jones and Lawton Roberts produced a list of Denbighshire birds in 1982). We have certainly felt handicapped in producing this volume, by the absence of these accounts.

It is inevitable that we have no illusions about the fact that this account of the birds of Wales, albeit as full as we can make it, reveals the enormous gaps that exist in our knowledge of bird occurrences and populations in Wales; we can do little about the past and that will remain substantially a tantalising mystery, but we can apply ourselves to the future and it is our hope that others will seek to update this volume in the not-too-distant future and be able to bridge many of the gaps which we are, through necessity, leaving unfilled. In this respect we hope that the book will act as a catalyst for improved bird recording throughout Wales, especially in those areas which are currently most poorly covered.

Acknowledgements

It is self-evident that no attempt could be made to produce a book such as this without enormous help from a great number of people: amateur birdwatchers, county recorders, professional ornithologists, archivists, museum staff, foresters and a host of others. To all those who have given help in innumerable respects we record a deep debt of gratitude, which is difficult to express adequately. Wales is a small country and it is therefore easier to feel that one knows such people individually. In addition, the fact that all three authors work for The Royal Society for the Protection of Birds (RSPB) means that part of our daily work involves us in contact with many individuals throughout Wales and we have the privilege to number most of those who are quoted below among our personal friends. We are deeply grateful for all they have contributed.

Our first thanks are to our publisher, initially Trevor Poyser himself and latterly Andrew Richford, for help, advice, encouragement and a seemingly bottomless well of patience. It has been a pleasure working with both.

County bird recorders have tolerated innumerable requests from us and have responded with kindness and generosity to an almost endless barrage of questions and queries: Peter Davis (Cardiganshire), Jack Donovan and Graham Rees (Pembrokeshire), Brayton Holt (Montgomeryshire), Peter Jennings (Radnorshire), Martin Peers (Breconshire), Peter Rathbone (Denbighshire and Flintshire), Dilwyn Roberts (Carmarthenshire), Reg Thorpe (Merioneth), Tom Gravett and Alan Davies (Caernarfonshire and Anglesey), Steve Moon (Mid and South Glamorgan), Brian Gregory (Monmouthshire), Harold Grenfell (West Glamorgan).

Many other individuals have helped with specific requests for information or clarification: Ian Armstrong (Heddon-on-the-Wall), Dr Richard Arnold (Bangor) who has unselfishly made manuscript notes and diaries available to us, Raymond Bark-Jones (Formby), John Barnes (Waunfawr), Michael Betts (Skokholm), Ron Birch (Saughall), Nigel Brown (Bangor) has been unstinting with his time on innumerable occasions, Rob Cockbain (Liverpool), Ken Croft (Anglesey), Rob Evans (Llanfairfechan), Peter Fraser (Bristol), Andrew Gouldstone (Holywell), Bob Haycock (Stackpole), Steve Howe (Cardiff), Mark Hughes (Penmaenmawr), Merfyn Hughes (Llanfairfechan), Rob Hume (RSPB, Sandy), Clive Hurford (Cardiff), (late) Gordon Ireson (Bridgnorth), Richard Knight (Rhayader), Peter Lansdown (Cardiff), John Lawton Roberts (Llangollen), J. C. Wyn Lewis (Llanrhidian), Harold McSweeny (Aberedw), Alastair Moralee (Holyhead), Sally Moralee (Holyhead), Tony Mercer (Brynsiencyn), Julian Moulton (Rhyl), (late) John Mullins (Ruislip), Keith Naylor (Nottingham), (late) Desmond Nethersole-Thompson (Sutherland), John O'Sullivan (RSPB, Sandy), Jack Parry (Anglesey), Tony Pickup (Rhandirmwyn), David Saunders (Milford Haven), Brig. Clive Simson (Aston Rowant), Dick Squires (Eglwysfach), Clive Stephenson (Minffordd), Ken Stott (Guilsfield), Dr Derek Thomas (Gower), Mike Walker (Llanwddyn), Colin Wells (Dee Estuary), Paul Whalley (Llanfair P.G.), Dr David Worrell (ex Flatholm), Ivor Wynmclean (Caergeiliog).

We have a special debt to those who have read and commented on drafts of individual species accounts, many of them people making important contributions to the accuracy and fullness of the texts. While the authors hold themselves wholly responsible for all the statements made throughout the book, we gratefully acknowledge the collective wisdom and experience of all these people: Ted Abraham, Duncan Brown, Nigel Brown, Ken Croft, Peter Davis, Peter Dare, Jack Donovan, Rob Evans, Dr Peter Ferns, Julian Friese, Tony Fox, Ken Jones, Owen Leyshon, Alastair Moralee, Steve Moon, Ian Morgan, Steve Parr, Martin Peers, Peter Rathbone, Ivor Rees, Graham Rees, Dilwyn Roberts, David Saunders, Peter Schofield, Mike Shrubb, Barry Stewart, Anna Sutcliffe, Steve Sutcliffe, Stephanie Tyler, John Underhill-Day, Eddie Urbanski.

Help has been given to us in a wide spectrum of fields by many others. Ben Averis, Mick Green, David Thomas and Christie Wild put in many hours of work in the earlier years in extracting records from written sources and museum collections. Andy Irwin latterly gave much help in extracting wildfowl data. Stephen Howe in the Geology Department at the National Museum of Wales has repeatedly pointed us in the direction of many important archaeological sources for some of the earlier records of individual species in Wales. Other Museum staff have also been of major assistance, mainly through their oological collections of Welsh material: Dr Colin Harrison, British Museum at Tring; Prof. Peter Morgan and Dr Stephen Green, National Museum of Wales; Prof. W. G. Hale, Liverpool Polytechnic; Dr C. A. Parsons, Jourdain Society Collection at Gloucester Museum; David M. Niles, Delaware Museum of National History; R. D. James, Royal Ontario Museum; Philip Nagle, Smithsonian Institute; Lloyd Kiff, Western Federation of Vertebrate Zoology; John Bull, American Museum of Natural History; Raymond Paynter, Agassiz Museum; Gillian King, Oxford Museum; Janet Hamber, Santa Barbara Museum of Natural History; Prof. Geoffrey Mathews, Wildfowl and Wetlands Trust (WWT), Slimbridge. Other very fruitful sources have been the Alexander Library, Edward Grey Institute (EGI) Oxford (Coward, Oldham, Jourdain, Whitaker and J. H. Owen diaries) and the Merseyside County Archives (Boyd diaries).

We are also extremely grateful to the Seabirds at Sea Team of the Joint Nature Conservation Committee

(JNCC), especially Mark Tasker and Paul Walsh, who kindly made available to us the complete set of their data for Welsh seabird breeding colonies.

We wish to make particular mention of Lynn Giddings and Ian Dawson in the RSPB Library at Sandy, who have never once failed to trace references for us, however vague or tenuous the information we have given them. Their co-operation over the years has been priceless. Special thanks are also due to Dr Glyn Tegai Hughes and Gerallt Jones, wardens at Gregynog for permission to use the excellent library facilities at the Hall.

Special mention must also be made of the direct contributions from Peter Davis (Red Kite) and Graham Rees (Offshore Waters) who have been responsible for the two sections in their fields of specialism.

Covering up our own inadequacies with word processors over the years has been a succession of skilled operators on the support staff of the RSPB in its Wales office in Newtown. They have willingly(?) been coerced into weekend and evening typing, editing and re-editing. Our grateful thanks are due to Sally Graham, Nicola Rowland, Pauline Warren, Ros Tudor, Diane Owen and Margaret Anthony. Ffion Worthington kindly volunteered herself into the thankless task of typing and checking the reference section and our colleague Richard Farmer spent much time perfecting the computerised production of the histograms and graphs. Mary Lovegrove kindly undertook the task of indexing.

Jeremy Greenwood, Director of the British Trust for Ornithology (BTO), is in our debt for graciously agreeing to allow us to have complete print-outs of Welsh ringing data, Birds of Estuary Enquiry (BoEE) data and Welsh Common Bird Census (CBC) results. We are greatly appreciative of that substantial assistance and extend that warm appreciation to other BTO staff who produced the actual material for us: Stephen Baillie, David Gibbons, John Marchant, Will Peach and Ray Waters with additional help from Chris Mead. At the Wildfowl and Wetlands Trust our thanks are especially due to Simon Delaney and Carl Mitchell.

We express our thanks to the artists elsewhere but make particular mention of Jeremy Jones, who put in much time in preparing all the maps for us.

The artists

Early in the planning of this book, we recognised that it would give the opportunity to place on record the work of some of the best bird illustrators and artists living in Wales at the present time.

Wales is well blessed in that it can boast a considerable number of the modern generation of bird artists who are in the top tier of their field not only in the UK, but also internationally. We are grateful to the following six artists whose illustrations enhance this volume:

Kim Atkinson. Born in 1962 and brought up on Ynys Enlli (Bardsey Island) from 1963-69, she attended Falmouth Art College and Cheltenham College of Art and then finally the Royal College of Art, where she specialised in natural history illustration. In 1987, after college, she returned to Ynys Enlli, where she now lives. (*pp. 36, 40, 46, 48, 49, 51, 53, 58, 62, 209, 214, 223, 226.*)

Phil Bristow. Born in Cardiff in 1960, where he still lives, Phil is a full-time local government officer by profession. His life-long interest in birds has led to active involvement in ornithology in south Wales, where he is county recorder for South Glamorgan. Although never having received formal art training, he has contributed vignettes to many local bird reports and other similar publications. (*pp. 182, 185, 190, 192, 194, 197, 200, 201, 206, 207.*)

Terence Lambert. Born in 1951, Terence went to Guildford College of Art and has been a professional artist and illustrator for 20 years, during which time he has illustrated several books and countless journals. His finished paintings are held by collectors all over the world. He lives and works in mid Wales. (*pp. 133, 139, 142, 146, 297, 299, 304, 308, 311, 313, 316, 322, 327.*)

Darren Rees. Born in Hampshire in 1961, Darren moved to mid Wales in 1986, working with the RSPB. His credits include several books and various painting awards. In 1993 his own book *Bird Impressions* was published and his finished paintings are now acclaimed in many countries. (*pp. 3, 13, 14, 17,18, 19, 24, 27, 29, 33, 148, 151, 153, 155, 158, 159, 161, 162, 163, 164, 166, 169, 172, 175, 178, 180, 330, 332, 335, 339, 342, 344, 346, 347, 349.*)

Philip Snow. Born in 1947, Philip trained in illustration at Manchester Polytechnic. He has lived in north Wales for 20 years and is well established as a wildlife and landscape painter. His work is published and exhibited worldwide in books, magazines and prints and in many Royal, public and private collections. (*pp. 1, 43, 54, 55, 109, 114, 117, 119, 125, 128, 227, 229, 232, 236, 238, 240, 243, 244, 246, 249, 255, 258, 261, 263, 266, 271, 272, 276, 278, 281, 283, 284, 285, 287, 291, 292, 371.*)

Owen Williams. Born in 1956, Owen has been a professional wildlife artist since 1985 when he started selling his wildfowl paintings in the USA. In recent years his experience as a painter of wildfowl has become increasingly recognised by publishers, with his work featured in several books including *The Illustrated Encyclopaedia of Birds*. He lives with his wife and young family in west Wales and is a fluent Welsh speaker. (*pp. 63, 66, 69, 74, 77, 79, 83, 87, 91, 95, 98, 102, 105.*)

Snowdonia from Malltraeth, Anglesey

1: The Welsh Counties – administrative boundaries

One of the first problems confronting us was the question of which administrative (i.e. county) boundaries to use in Wales. Such an issue may appear at first consideration to be slightly fanciful. However it does present us as authors with a serious dilemma. The 13 "old" counties of Wales, established for some four hundred years, are those which are shown inside the back cover. In 1974 the "old" counties were summarily cast aside in favour of a slimmed-down set of eight super-counties – Gwynedd, Powys, Clwyd, Gwent, Dyfed and the three fragmented Glamorgans. The generation of Welsh youth now emerging from schools and colleges knows only these new names and is unfamiliar with names such as Monmouthshire, Cardiganshire, Flintshire or Denbighshire. However, the geographical units for bird recording in Wales have remained substantially unchanged, linked as they are into well-rooted traditions and frequently strongly tied to clubs, Trusts and other organisations firmly based on the "old" counties. In this respect, our task in transposing all the old records to a revised county basis which bore no relationship either to the former system or the on-going recording units, would have been both unscalable and incomprehensible to most readers. Wales is a small country and to try to discuss the "county" distribution of bird species from an area such as Powys, which stretches from the summit of the Berwyn Mountains above Bala to the Swansea Valley, would have been to demean the book itself. Thus we believe we have had no option but to adhere to the "old" county boundary system, consistent with the areas on which bird recording in Wales has been based historically and on which it substantially continues to be collated. We apologise to any readers who may find this unfamiliar but hope that the maps on the inside of the covers will help make the transition easy.

As the book is prepared, another twist to the story arises as the Welsh Office now proposes the discarding of the 1974 counties in favour of (probably) some 23 "unitary authorities". Thus the administrative boundaries of Wales will be redrawn once again with the great likelihood that, certainly in most of rural Wales, we shall see the genuine resurrection of full county names such as Caernarfonshire, Pembrokeshire, Flintshire and Anglesey!

As a footnote to this explanation, it is perhaps relevant to mention one or two changes in recording which have tried to accommodate the 1974 changes to some degree. Since 1985, the *Glamorgan Bird Report* ceased to cover the erstwhile county of Glamorgan and became the *Mid and South Glamorgan Bird Report*, leaving West Glamorgan to be covered by an extended *Gower Bird Report*. For the same reason, *An Atlas of Breeding Birds in West Glamorgan*, published in 1992, covered the transient county of West Glamorgan.

At the same time (1985) in north Wales, the *Cambrian Bird Report* ceased to include records for the old county of Denbighshire which had already been pooling records with Flintshire to produce the *Clwyd Bird Report*. The Cambrian Bird Report simultaneously metamorphosed into the *Gwynedd Bird Report* covering the (former) counties of Anglesey, Caernarfonshire and Merioneth.

Bird Rock (Craig y Aderyn)

2: Myth and history

Throughout the process of researching for this book it has been easy to become diverted and sidetracked into fascinating glimpses of the past and to touch on the "visible" parts of stories – factual, legendary and sometimes mythical – which are woven into the tapestry of Welsh ornithology. No factual seam is richer in this respect than that which runs through the amazing history of the protection of the remnant pairs of Red Kites in Wales (and therewith the entire British population) from the final decade of the 19th century up to the equivalent decade of the current one. It is the story of the longest running protection of a single bird species anywhere in the world. It is in the end a story – so far – of success: the saving and eventual recovery of a species snatched from the very edge of the abyss of extinction in Britain (see Peter Davis's account on page 109). At the same time the history is also a cornucopia for those who enjoy the salty mixtures of human intrigue, deviousness, loyalty, disloyalty, secrecy, conflict, double standards or unswerving devotion to duty. These elements are not for the current volume, however, but are dealt with elsewhere. Other gems from the past, less factually-based in some cases, have caught our eye and one or two are worth relating.

We were particularly attracted to the account of the strange arrival of a Kiwi on Anglesey which was recorded in the *Zoologist* of 1853 by Josiah Spode from Rugeley in Staffordshire. He recounts the details of a bird shot that same year in a marsh on Anglesey

> "I was startled with the resemblance to ... *Apteryx*, which I had seen in the gardens of the Zoological Society. I saw an absence alike of wings and tail: the feathers of hard, rough texture, the colour being an uniform dark brown: the head round the base of the beak and the eyes thinly covered with bristly hairs. Nothing I had ever seen bore a resemblance to the creature before me, except the *Apteryx* and I hastened to consult an eminent ornithologist ... He assured me that it belongs to the genus *Apteryx* ... How this creature could arrive at Anglesey seems a mystery unless it had escaped from a wrecked vessel..."

Mr Spode was rewarded with the combined advice of several eminent ornithologists in succeeding volumes of the *Zoologist*, amongst which was this contribution from Edmund Brown.

> "As *Apteryx* is confined to a very limited region on the opposite side of our planet, not aquatic in its habits, totally incapable of flight and consequently one of the last birds ever likely to be met with in Britain ... I fear it will prove as difficult to account for the presence of *Apteryx* in Anglesey as it is for a physiologist to explain the growth of Mr Pickard-Cambridge's spider which increased to more than ten times its original size without material sustenance."

We have, sadly, chosen not to include this record of a Kiwi within our volume, thereby depriving Welsh avifauna of what would have been an exciting addition to its list.

Two events on Grassholm are notable for one or two features. In the first incident, Thomas Henry Thomas relates (in *Report and Transactions of the Cardiff Naturalists' Society* 1890–91) how he, together with colleagues J. J. Neale and T. W. Proger ventured to Grassholm on a camping expedition in May 1890 where one of them had previously discovered Gannets nesting. His account is a minor classic of its time, encapsulating the thrill of discovery with a constant reflection of their evident delight at the privilege of being encamped on a tiny island in the midst of such a host of seabirds.

> "We were in the midst of a metropolis, thousands of white wings and breasts before us, on the grey and orange rocks, all ringed about by the azure white-flecked sea and sky; our ears were filled with a wild concert. Every rocky ledge and terrace has its rows of Puffins, and among them Guillemots and Razorbills; clinging under the ledges above the sea were pearly Kittiwakes and there among the rocks a Herring Gull ... and Black-backed gulls. High above a Peregrine falcon soared."

However, their greatest thrill was evidently the Gannet colony, then a modest (but increasing) two hundred pairs or so; this was their real discovery.

Imagine their amazement on being woken in their tent on Whit Monday by the crack of rifle fire to discover "an attack was made upon the settlement of Gannets by a company from on board HMS Sir Richard Fletcher, followed by a landing and general battle upon shore, terminating in the slaughter of ... many of the birds".

Seldom had such a national mayhem been created by an incident involving birds than that which followed the events nine miles out to sea on that peaceful May morning. The daily papers were alive with the story; questions were asked in the House by Mr Webster, MP for St Pancras, as the government with customary disdain, wriggled and twisted to protect its officers. Eventually, a private prosecution was taken successfully at Haverfordwest Sessions by the RSPCA, who still describe it as one of the most important they had fought to that time. Thomas Thomas concludes his piece with the note that "a large case containing Gannets, Puffins, Guillemots etc., killed by the party is now set up in the Cardiff Museum with a background painted to represent the Gannet settlement".

The Royal Navy had an opportunity to atone for this crime 39 years later. On 9 June 1929 Mr Sturt, the then owner of Grassholm visited the island with R. M. Lockley, and they were alarmed to see smoke rising above it. The RSPB *Watcher's Report* for the combined years of 1929 and 1930 records the graphic account of the incident and its eventual resolution:

> "On landing (we) found that some party had been there, left a good deal of picnic litter, and a fire burning. Grassholm is covered entirely by a deep (1 to 3 ft) bed of dead grass of the consistency of peat and this was flaming and burning as peat does, the fire being in the centre and as yet small, and not too near the peaty area where the Gannets nest.
>
> "The next day was misty, and it was not until the evening that, from Skokholm, it became clear enough to see that Grassholm was still afire, by the dense smoke that poured out. As some of the Fleet is anchored in Dale Roads I sent a letter to the Admiral asking if it is possible for him to help. The letter we delivered by boat to the flagship 'Renown'.
>
> "Within three hours of delivering my appeal to the Admiral, a destroyer had been sent out to Grassholm and landed blue-jackets to fight the fire."

On 13 June, 1929, the Vice-Admiral Commanding Battle Cruiser Squadron *Renown*, at Milford Haven, reported:

> "'Wallace' proceeded to Grassholm on Wednesday 11th June, and reported at 6 p.m. that the fire consisted of smouldering peat about two feet deep and 200 yards in circumference. The Commanding Officer of 'Wallace' anticipated being able to extinguish the fire before dark, but by 8 p.m. the swell had increased considerably and it was necessary to re-embark the men. At the time that the men were re-embarked, a trench had been dug about three-quarters of the way round the fire and the Commanding Officer considered that the bird sanctuary was then out of danger.
>
> On Thursday, 12th June, a party from H.M.S. 'Vortigern' was landed on Grassholm and this party finally extinguished the fire on the western and southern slopes. The fire on the summit was also subdued by this party, but the officer in charge considered it would probably break out again as the island was excessively hot."

On 19 June Admiral Pound added: "On the evening of 13th June a volunteer party arrived from St. David's and helped the men of H.M.S. 'Velox' to complete the work of extinguishing the last remains of the fire. The party remained on the Island for the night".

Stories of birds in vast numbers always have a special appeal. Peter Conder, then warden of Skokholm Bird Observatory (later Director of the RSPB), recorded (*British Birds*, vol. 47 p. 349) the emigration of huge numbers of passerines from the Dale and Marloes area of Pembrokeshire in advance of bitter weather in January 1952. With starlings alone he estimated "millions of birds in an area of about two square miles". Forty-eight years earlier O. V. Aplin (*Zoologist* 1905 pp. 17–173) described – second-hand

> "A great Immigration of birds at Pwllheli on the night of 17th March 1904. The wind was north-east and the night fine at Pwllheli, but at St Tudwal's the weather was what they call 'misty rain', i.e. bordering on fog. The men in the quarry on the Gimblet Rock (Careg yr Imbril), at the entrance of Pwllheli harbour, were working extra time loading vessels, and flares were burning which lit up the whole place. This island-like rock juts out to some extent from the coast-line, and from its height is a very noticeable feature in a long stretch of low coast. Suddenly the workmen were startled by what some have termed a 'flow of birds' and others a 'shower' descending on the rock. Thousands of birds dropped on the quarry, the rock, the wharves, and the vessels close to, in a dying state. In a short time the ground was thickly covered with birds, most of them dead or in a dying condition, whilst a cloud of birds hovered in a helpless condition a few yards up in the air. At daybreak the seashore was found strewn with hundreds of birds, evidently drowned at sea, and washed ashore by the tide".

This account states that it was an inky dark night, and notices a theory that the birds struck the Rock, which would be between the sea and the place where the flares were burning. Another account said that a shower of birds suddenly fell on the workmen. "Thousands of birds covered the ground in a few minutes – some dead, some half-dead. The vessels at anchor close to the wharf appeared to be instantly covered by birds from stem to stern, every available space on the riggings, stays, crosstrees, and yards being occupied." Dead birds were found in every direction.

> "At daybreak the birds on the vessels flew away. Upon inspection there were thousands of dead birds in the quarry, on the top of the Rock, and on the stretch of land that reaches from the Gimblet Rock in the direction of the South Beach. Many of the birds were injured about the head, and there is no doubt but that, having lost their bearings in the thick dark night, and being attracted by the light of the flares, numbers of them struck against the precipitous seaward face of the Rock and other obstacles, while others more fortunate settled down within the influence of the light. On the morfa* thousands stood on the rigging and the ropes of vessels, and they say it was a very grand sight to see the glittering colours in the light ... There were birds of every description. Most of them that were on the vessels flew away at daybreak."

The following species were mentioned in the papers: Starling, Thrush, Blackbird, Snipe, Woodcock, Robin, Curlew. At St. Tudwal's Lighthouse the same night they had lots of Starlings, Blackbirds and Redwings about the light, but saw no other birds; there were, however, more Blackbirds than the keeper ever saw at one time previously.

How much more exciting discoveries must have been two hundred years or so ago – and how the climate off the Pembrokeshire coast has evidently deteriorated in the interim. Robin Fenton in his *Historical Tour Through Pembrokeshire* 1811 leaves us a seductively alluring account of the seabirds on the Bishops and Clerks – a series of rocky islets north-east of Ramsey.

> "On these rocks an infinite number of seabirds breed, whose eggs are so thickly deposited all over the surface of them that if one egg on the summit be stirred in its irregular rotation it is known to carry with it hundreds of others. Though the birds are perpetually hovering around the rock yet no regular incubation is performed and the eggs are chiefly hatched by the sun here felt at the season in an almost tropical degree."

Thomas Pennant in his *Tours of Wales* quotes "a very uncommon wreck of seafowl" on Llŷn in 1776 when "the beach near Criccieth for miles together was covered with dead birds, especially those kinds which annually visit the rocks in summer, such as Puffins, Kittiwakes, Razorbills and Guillemots, Gannets, wild geese, Barnacles, Brent Geese, Scoters and Tufted Ducks". Such a gathering of "seafowl" as reported to Pennant by The Revd Hugh Davies clearly needed divine inspiration as one cannot imagine the circumstances which could bring together such a spectrum of species, at whatever time of the year, along the Welsh coast, even if viewed from the distance of Beaumaris on Anglesey where the good Revd Davies exercised his living. Such stories make good reading even if they do little more than add to the wealth of myths.

* Morfa is a Welsh term for saltmarsh.

The Dinas Hill, Tywi Valley

3: Sources and references

A full list of references is given at the back of the book. However, and in addition to the individual references quoted in the text, we have used a standard set of references in the preparation of each species account. These are listed below and, as they are taken as the "basic" source material for all species, they are not normally referenced in text in the interests of readability.

An Atlas of Breeding Birds in West Glamorgan. Thomas *et al*. 1992
Bardsey Bird Report 1985–1991
Birds of Bardsey. Roberts 1985
Birds of Brecknock. Ingram & Salmon 1957
Birds of Breconshire. Massey 1976
Birds of Breconshire. Peers & Shrubb 1990
Breconshire Birds 1978–1991
Birds of Caernarfonshire. Hope Jones & Dare 1976
Birds of Cardiganshire. Ingram *et al*. 1966
Birds of Denbighshire. Hope Jones & Roberts 1982
Birds of Flintshire. Birch *et al*. 1968
Birds of Glamorgan. Heathcote *et al*. 1967
Birds of Gwent. Ferns *et al*. 1977
Birds of Merioneth. Hope Jones 1974
Birds of Monmouthshire. Ingram & Salmon 1937
Birds of Pembrokeshire & its islands. Mathew 1894
Birds of Pembrokeshire. Lockley *et al*. 1949
Birds of Radnorshire and Powys. Peers 1985
Birds of Skokholm. Betts 1992
Cambrian Bird Report 1952/53–1984
Carmarthenshire Bird Report 1982–1991
Ceredigion Bird Report 1982/83–1991
Clwyd Bird Report 1975–1989
Dyfed Bird Report 1967–1981
Glamorgan Bird Report 1962–1991
Gower Bird Report 1967–1991
Gwent Atlas of Breeding Birds Tyler *et al*. 1987
Gwynedd Bird Reports 1985–1991
Handbook of the Birds of Europe, the Middle East and North Africa vols. 1-6, Cramp et al. 1977–1992, OUP (referred to in text as BWP)
Vertebrate Fauna of North Wales. H. E. Forrest 1907
Handbook to the Fauna of North Wales. H. E. Forrest 1919
Handlist of the Birds of Carmarthenshire. Ingram & Salmon 1954
Handlist of the Birds of Radnorshire. Ingram & Salmon 1955
Montgomeryshire Field Society Report 1947–1981
Pembrokeshire Bird Report 1981-1991
Skokholm & Skomer annual reports
The Atlas of Breeding Birds of Britain and Ireland. Sharrock 1976 (referred to in text as *Breeding Atlas*)
The Atlas of Wintering Birds in Britain and Ireland. Lack 1986 (referred to in text as *Winter Atlas*)
The New Atlas of Breeding Birds in Britain & Ireland: 1988–1991. D. W. Gibbons, J. B. Reid & R. A. Chapman, Poyser 1993 (referred to in text as the *New Atlas*)
Welsh Bird Reports 1978–1991

Grassholm

4: Principal sites mentioned in species accounts

Individual sites mentioned in the texts of species accounts are accompanied by the county name in parentheses. However, it is inevitable that the most important bird sites recur repeatedly throughout the book. In order to avoid the tiresome and unnecessary repetition of their location, these sites are therefore used without their appropriate county names or other identifying descriptions. The sites treated in this way are listed fully below, with their county names.

- Bardsey (Llŷn, Caernarfonshire)
- Berwyn Mountains (Montgomeryshire/Denbighshire/Merioneth)
- Blackpill (Swansea Bay, Glamorgan)
- Burry Inlet (Gower, Glamorgan/Carmarthenshire)
- Caldey Is. (Pembrokeshire)
- Cambrian Mountains (Cardiganshire/Breconshire/Radnorshire/Carmarthenshire)
- Cardigan Bay (Cardiganshire/Pembrokeshire/Merioneth/Caernarfonshire)
- Carmarthen Bay (Carmarthenshire)
- Cleddau estuary (Milford Haven, Pembrokeshire)
- Cors Caron (Cardiganshire)
- Cors Fochno (Cardiganshire)
- Dee estuary (Flintshire)
- Dyfi estuary (Cardiganshire/Merioneth)
- Gower (Glamorgan)
- Grassholm (Pembrokeshire)
- Great Orme and Little Orme (Caernarfonshire)
- Gwent Levels (Monmouthshire)
- Kenfig NNR (Glamorgan)
- Lavernock Point (Glamorgan)
- Llangorse Lake (Breconshire)
- Llŷn (Caernarfonshire)
- Mynydd Hiraethog (Denbighshire)
- Newborough Warren NNR (Anglesey)
- Oxwich Marsh NNR (Gower, Glamorgan)
- Point Lynas (Anglesey)
- Point of Air (Flintshire)
- Puffin Is. (Anglesey)
- Ramsey Is. (Pembrokeshire)
- Severn estuary (Monmouthshire/Glamorgan)
- Skokholm Is. (Pembrokeshire)
- Skomer Is. (Pembrokeshire)
- Snowdonia (Caernarfonshire)
- South Stack (Holyhead Island, Anglesey)
- St. Margaret's Is. (Pembrokeshire)
- Strumble Head (Pembrokeshire)
- Traeth Bach (Merioneth)
- Traeth Lafan (Caernarfonshire)
- Witchett Pool (Carmarthenshire)
- Ynys-hir (Cardiganshire)

Cardiff Bay

5: Physical characteristics and bird habitats

GEOLOGY

Not only does Wales have an immense variety of rocks of different ages, a circumstance entirely disproportionate to its size, but it also includes the classical region for Lower Palaeozoic stratigraphy where names such as Cambrian, Ordovician and Silurian were first defined, then subsequently used throughout the world (Bowen 1981).

Four major structural provinces are recognised:

1. Pre-Cambrian basement outcrops in Anglesey, Llŷn, the Padarn Ridge (Caernarfonshire) and in the St David's, Roch-Trefgarn area and Johnston in Pembrokeshire. Radiometric dating shows that the Johnston Diorite was intruded 643 million years ago.
2. Lower Palaeozoic (Cambrian, Ordovician and Silurian) volcanic rocks extend over much of north and mid-Wales, including Snowdonia and the Harlech Dome.
3. The Variscan fold-belt of south Wales, which includes the Coal Measures and Carboniferous Limestone outcrops adjoining older rocks, in both south and north Wales.
4. Mesozoic and Cenozoic rocks of comparatively recent times, represented notably by Triassic New Red Sandstone in the Vale of Clwyd and the Vale of Glamorgan.

RELIEF

Wales is a land of peaks and plateaux dissected by river valleys (Lewin 1980). Craggy peaks are to be found especially in the north-west mainland. Here Snowdon rises to 1085 m together with the Glyders (Glyder Fawr 999 m), the Carneddau 1062 m, Moel Siabod 872 m, Arenig Fawr 854 m, and further south the Rhinogs (Y Llethr 754 m). These mountains, developed on ancient Cambrian and pre-Cambrian rocks, are bordered on the south by the escarpment of Cadair Idris 892 m, the Arans (Aran Fawddwy 907 m) and Berwyn (Cadair Berwyn 827 m). The Hiraethog moors and then the limestone hills of Denbighshire and Flintshire are beyond to the east. In mid-Wales the plateaux are developed on Silurian and Ordovician sedimentary rocks and are more in the nature of rolling uplands – Pumlumon 752 m and its extensive continuation to the south and Radnor Forest 660 m. In south-west Wales only Mynydd Preseli 536 m is of significant height but south Wales has a variety of north-facing escarpments: Mynydd Eppynt 475 m, the Black Mountains on the English border, the Brecon Beacons 886 m, Fforest Fawr and Mynydd Du.

Striking though many of these peaks are, with the characteristic U-shaped valleys and steep ridges of a glaciated landscape, much of Wales consists of rolling upland rather than rocky outcrop. Many steep valleys do have their cwm, cliff or scree, and in some areas like Mynydd Preseli there are dramatic sky-line tors, but it is the plateau and valley forms that are ubiquitous. Even the mountains gain their grandeur from the local scale of their relief, for they are not large by world standards.

Few parts of Wales are really flat but the low relief of Anglesey, the coastal plateaux of Pembrokeshire and South Glamorgan, and parts of the Welsh Marches where river valleys penetrate into the upland plateaux are exceptions to this.

The largest Welsh rivers drain east and west from the main north–south watershed which forms the backbone of Wales.

In terms of landform Howe and Thomas (1963) offered the following simple classification of landscapes in Wales:

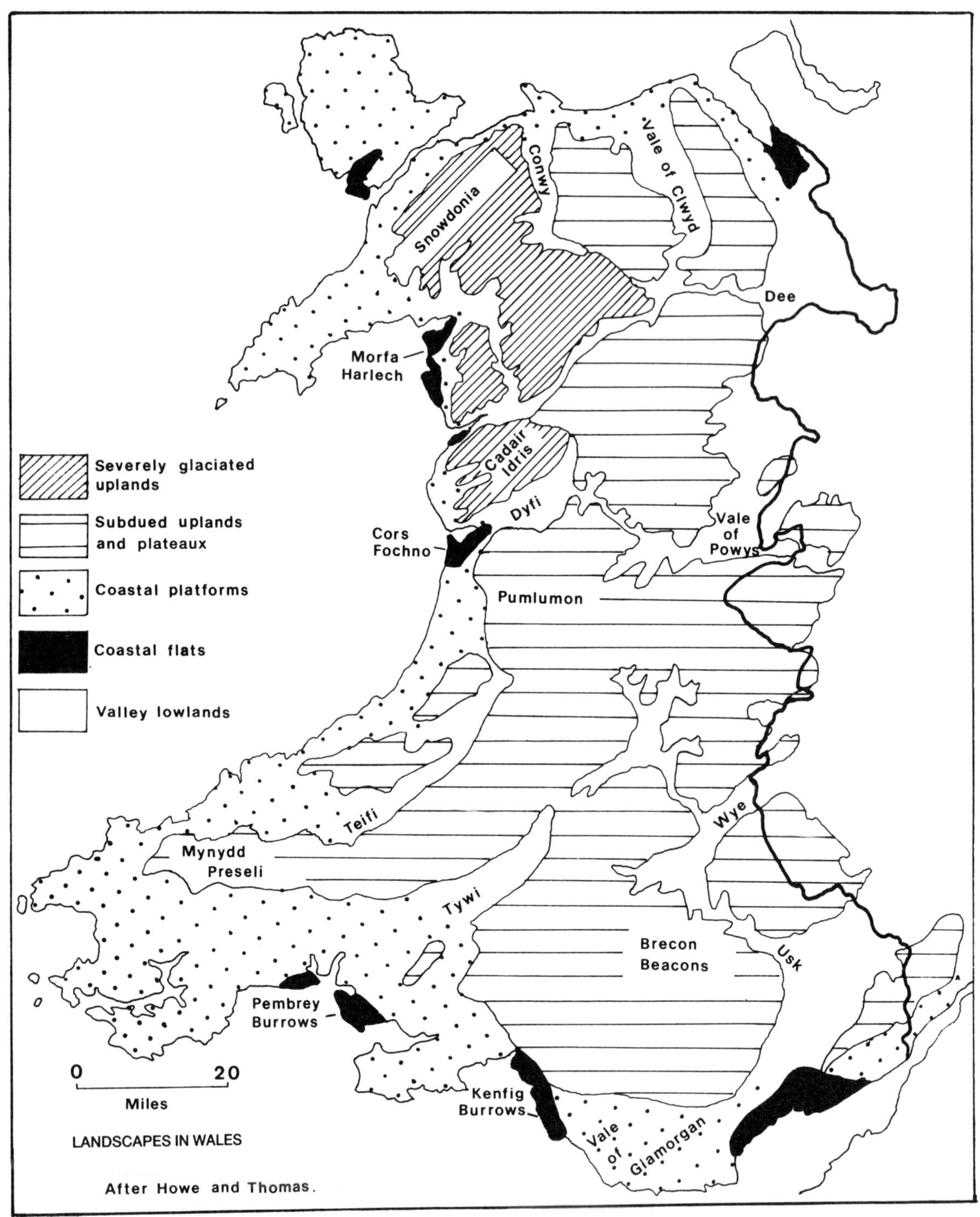

Landscapes in Wales

1. Uplands – generally between about 230 m and 1070 m –
a) Severely glaciated uplands, e.g. Snowdonia and Cadair Idris with features such as cirques, arêtes, ribbon lakes and great near-vertical crags found in profusion.
b) Subdued uplands and plateaux, e.g. Mynydd Hiraethog, Brecon Beacons. These regions, although most were certainly glaciated to a marked degree do not show the striking results of glaciation in such profusion as do the regions of group a). Expanses of moorland are common.

2. Lowlands – generally below 215 m.
a) Coastal platforms, e.g. Gower and Anglesey – mainly peneplanes of marine erosion which have been uplifted to a moderate degree, forming plateaux usually 60–180 m above sea level.
b) Coastal flats, nearly always below 15 m and formed as a result of marine and stream deposition during recent (Quaternary) times.
c) Valley lowlands, too large to be included as part of their surrounding highland blocks and regarded as separate regions in their own right, both physically and culturally, e.g. Vale of Clwyd and Vale of Conwy.

CLIMATE

The character of the Welsh climate owes much to an exposure to moist westerly air streams whereas continental air masses, which generally are less frequent, usually bring drier, harder weather to eastern Wales, sometimes including snow in winter and warm spells in summer (Jones & Taylor 1983).

The mountains of Wales enhance the west–east contrast in climate. They present an unbroken barrier to the moist, warm air from the Atlantic, a barrier about 500 m high and without any major gap between Conwy and the Rhondda. That part of the Atlantic west of Britain is warm by latitudinal standards as a result of the North Atlantic Drift; the Welsh mountains stand squarely in the path of westerly wind flow causing it to rise and cool, forming low clouds and shedding precipitation.

It is not surprising therefore that the area of highest rainfall is concentrated along the mountainous backbone with an extensive area that has more than 200 cm of rain per annum with part of Snowdonia receiving more than 400 cm. Low-lying coasts and the rain-shadow area to the east of the mountain spine have less than 120 cm of rain per annum, some areas such as the Vale of Clwyd having under 70 cm.

The effect of the sea is illustrated by temperature maps in which the effect of topography (i.e. the general fall in temperature with altitude) is removed by the conventional correcting of temperatures to mean sea level. Corrected mean daily temperatures in January for the south-west tip of Pembrokeshire and the Gower are nearly 2.5°C higher than in the Welsh borderland. In July the temperature gradient is reversed but the contrast is confined to the Cardigan Bay area with south-coast temperatures comparable to the eastern counties of Wales. The temperature distribution suggests in general that distance from the coast is at least as strong a factor as latitude.

As shown by Jones and Taylor (1983) the climatic environment of Wales is truly a product of the juxtaposition of sea and mountains. There are three broad climatic zones: the coast, which has relatively low rainfall and mild winters with few frosts; the north-south watershed with high rainfall and lower temperatures than the rest of Wales; and the area east of the watershed, which has relatively low rainfall and, generally, a greater temperature range than the coast.

BIRD HABITATS

Wales comprises a mosaic of habitat types, each with its characteristic bird assemblage. The following is a brief account of the main types with an indication of the threats to them and their important birds. For a more complete treatment the reader is referred to Fuller (1982).

The total land surface of Wales comprises nearly 2.1 million hectares, of which some 82% is devoted to agriculture. The remainder includes forestry and woodland (12%) (Welsh Office 1990).

Broad divisions of habitat are as follows:

Montane

The montane areas of Wales above 611 m cover only a small proportion of the 560 000 hectares in Britain as a whole. Most of the montane areas of significance for breeding birds are in Scotland (Dotterel, Ptarmigan, Snow Bunting, etc.) and the nature conservation importance of high ground in Wales is mainly botanical (Arctic-alpine communities rather than ornithological although the occurrence of passage and, occasional, breeding Dotterel is of interest.

Upland heaths and grassland

These form the largest extent of semi-natural vegetation in Wales, comprising in the 1960s nearly 30% of the total land area of the Principality. This habitat includes the heather-dominated moorland along the Berwyn/Llandegla Moor axis extending to Radnor Forest and the Black Mountains in the south and Mynydd Hiraethog in the north. Outcrops of this dry heath include the Rhinogau and Migneint in the north, and Preseli mountains in the south-west, where the maritime climate imposes sharp lapse rates of temperature and a severe climate at quite low altitudes.

This habitat has been severely fragmented in recent decades through afforestation by non-native conifers, agricultural improvements and widespread overgrazing by sheep. Important breeding species include Red Grouse, Hen Harrier and Merlin. Golden Plover and Dunlin are characteristic breeding species of the limited amount of upland mire which occurs on Mynydd

Hiraethog, Elenydd and Migneint amongst other areas.

An important sub-division is the "ffridd", a characteristic habitat at the periphery of the Welsh uplands occupying the zone between the fenced-in bye land of the valley bottoms and the open hill land above. This zone is dominated by bracken, interspersed with scatterings of trees. Bracken covers about 6% of the land surface of Wales which is a reasonable measure of the extent of the "ffridd" habitat. This is a particularly important habitat for Whinchat, Yellowhammer, Tree Pipit and Cuckoo, among others.

Broadleaved woods

The 1980 Forestry Commission Census showed that the total area of broadleaved woodland amounted to 3.3% of the total land area of Wales. The principal species in broadleaved high forest was oak (44% of the total). The relict "hanging oakwoods" of the valley sides, which are a feature of much of Wales, have survived through being uneconomic to clear for agricultural purposes and because of the value of sheltered grazing for stock in the winter. They have a very characteristic bird assemblage, with such species as Red Kite, Pied Flycatcher, Redstart, Wood Warbler and Tree Pipit. The main threat to this habitat is lack of regeneration through overgrazing.

Plantations

Conifer plantations, mainly of non-native species, cover more than 8% of the land area of Wales.

In addition to common species such as Goldcrest and Coal Tit, the breeding assemblage of mature plantations includes Siskin, Crossbill, Goshawk and (recently) Merlin. When plantations are clear-felled or are in the early stages of growth, Nightjar is a characteristic species.

Rivers and streams

The high rainfall of the Welsh hills gives rise to an extensive network of rivers. Within the Welsh Water Region there is a total of 4647 km (2881 miles) of rivers and this excludes that part of Wales which falls within the Severn–Trent catchment. Water quality was defined as Good and Fair along 94% of river length in the Welsh Region in 1990 and Poor and Bad along the remaining 6%. This classification broadly related to the potential use of the waters, especially in the support of fisheries; acidification, which is occurring in many of the upland rivers of Wales is having a marked effect on breeding numbers of Dipper in some areas.

Open waters

The natural lakes and reservoirs of upland Wales are generally too acidic to attract any numbers of waterfowl but the eutrophic lakes and reservoirs of lowland Wales have significant breeding populations of several species, e.g. Shoveler, Pochard, Gadwall. The many shallow lakes of Anglesey are particularly important in this respect but there is concern as to their future through siltation and over-enrichment by phosphates.

Swamp, fen and carr

This habitat mainly comprises reedbeds of which there are more than 200 in Wales, chiefly in Anglesey, Caernarfonshire, Glamorgan and Carmarthenshire. The total area of reedbeds in Wales is only between 450 and 500 ha and many reedbeds are under threat through natural succession to willow and alder scrub and from pollution. It is a very important habitat for Reed Warbler and scarcer species such as Marsh Harrier, Bearded Tit and Bittern (and Cetti's Warbler in south Wales where scrub forms a part of the vegetation mosaic).

Lowland heath

The maritime heath developed along the western seaboard of Wales is now represented only by a few comparatively small fragments such as at South Stack on Anglesey, a few sites on the Llŷn peninsula and parts of Pembrokeshire. Cultivation up to the edge of the sea cliffs has removed most of the former extent of this habitat. Maritime heaths are important feeding areas for Chough and the gorse component of the heath provides breeding sites for Stonechat and Linnet.

Wet lowland grassland

This is now a scarce habitat in Wales, as elsewhere in Britain, since the flood meadows of the major river systems have largely been drained as part of agricultural improvement schemes in the 1960s and 1970s. These have had a major adverse effect on breeding waders and wintering wildfowl, e.g. White-fronted Geese.

Arable land

Arable land (which includes grass leys etc.) is a relatively small and declining component of agricultural land in Wales. In 1989 it amounted to 225 000 ha compared with a total of more than 5 million hectares in England. Characteristic species include Corn Bunting, Lapwing and Grey Partridge, which have all declined substantially in recent years.

Permanent grass

Permanent grass, with just over 900 000 ha in Wales (Welsh Office 1990) amounted to more than 53% of the agricultural land in Wales in 1989. Grasslands are important for a wide variety of species including Barn Owl,

Corncrake, Lapwing and Curlew. Locally, grasslands may be threatened by afforestation, chemical pollution and urbanisation but more significant problems for bird conservation arise through overgrazing and the recent tendency for early silage cuts which seriously affect the productivity of ground-nesting species.

Intertidal flats and saltmarshes

There are 35 significant estuaries around the coast of Wales covering about 700–750 km^2, but this includes the English parts of the two largest estuaries of the Dee and Severn. These two alone represent about 4% of the land area of Wales at about 500 km^2. Excluding these the remainder of about 200 km^2 represents about 1% of the surface of Wales, a large proportion of this being accounted for by a few relatively large estuaries such as the Burry Inlet and Milford Haven.

Beaches

Sand and shingle beaches are the main breeding habitat of Little Tern and Ringed Plover. Sadly, Little Terns are now reduced to only one site in Wales and the main problem comes from recreational disturbance.

Sea-cliffs, rocks and islets

This forms a particularly important habitat in Wales where internationally significant populations of seabirds breed, e.g. Manx Shearwater and Gannet, especially on the Pembrokeshire islands. Fortunately the most important sites are protected as nature reserves and threats are minor and localised although there can be problems, e.g. through the presence of rats.

Coastal waters

The shallow inshore waters of the Irish Sea, St. George's Channel and Bristol Channel (generally less than 40 m in depth) provide important feeding grounds for many of the seabirds that breed round the Welsh coast although some species range widely in the feeding forays from their breeding colonies. In winter large numbers of birds of foreign origin, including many sea-ducks such as Scoter, congregate in favoured areas such as Conwy Bay and Carmarthen Bay.

Broadwater from Aber Dysynni

6: Bird recording in Wales

There are disappointingly few contemporary accounts of birds from mediaeval Wales, with a small number of honourable exceptions.

The earliest references to birds in Wales relate to the laws of Hywel Dda (Hywel the Good), codifying Welsh laws and customs in about the year 945. References are all related to the important mediaeval sport of falconry and the place of the falconer in the King's court.

The first to leave actual written records of birds, though these were generally quite incidental to their main tasks, were the various chroniclers who travelled through Wales. Giraldus Cambrensis (Gerald the Welshman), born at Manorbier Castle, Pembrokeshire, in 1147 was one of the greatest scholars of his day. In 1188 he accompanied Archbishop Baldwin through Wales preaching the Third Crusade (during which the prelate was to die in Palestine). Giraldus wrote down his impressions which included a few ornithological passages – he mentions the high esteem placed on the falcons which bred in Pembrokeshire and records a golden oriole heard in Caernarfonshire although this observation is open to doubt.

John Leland, who catalogued the antiquities of Wales between 1536 and 1539, was one of many early travellers to mention eagles nesting. However, not one of them actually encountered a breeding site themselves despite some most graphic accounts. George Owen in his *Description of Pembrokeshire* (1603) devoted four pages to "Of abundance of foule that the County yeeldeth, and of the feverall fortes thereof". This provides a fascinating glimpse of birds breeding in Pembrokeshire at that time, including Cranes (long since extinct as breeding birds), and an insight into visitors to the county such as Woodcock arriving in numbers far exceeding those of modern times.

John Ray and Francis Willughby visited Wales in 1658 and 1662 and are described by W. M. Condry in his book *Snowdonia* (1966) as "the two most celebrated English naturalists of their time". Towards the end of the same century Edward Lloyd, possibly the greatest of all naturalists to have worked in Wales, made repeated visits to the northern uplands. He is best remembered for his botanical discoveries, one of which – the Snowdon lily *Lloydia serotina* – is named after him. We are the poorer for the loss by fire in 1810 of his unpublished manuscript notes including those on birds (Saunders 1974).

One of the most prolific correspondents of Gilbert White, the curate of Selborne, was Thomas Pennant, a Welshman born in 1726 at Downing Hall near Mostyn, Flintshire. Although one of the most eminent of the great 18th century naturalists, author of masterpieces like *A Tour in Scotland* (1771), *British Zoology* (1776) and *A History of British Quadrupeds* (1793), Pennant's contribution to the natural history of Wales is disappointing. Ornithological information in his *Tour of Wales* (1784) is sparse, relating chiefly to the mountains of Snowdonia and seabird stations in north Wales.

The first comprehensive accounts of birds in the Principality appeared in the 19th century. These accounts from the early 19th century were largely annotated lists or accounts of the birds of a particular neighbourhood. At this time the recording activities of most ornithologists were restricted to the areas around their homes in the tradition of Gilbert White, due to the problems of travelling. However, some Victorian natural historians managed to travel quite widely, and accounts of the birds encountered on these journeys were also published. The information that follows is largely based on a paper by Christine Johnson (1983) that appeared in *Nature in Wales* with updates where relevant.

The first of the Victorian local bird lists for Wales was compiled by Lewis Weston Dillwyn, Member of Parliament for Glamorgan and later Mayor of Swansea. In common with many of his contemporaries he was a proficient all-round natural historian and in 1848 he published the first floral and faunal list for Swansea. The family tradition in natural history was continued by his grandson, Sir John Talbot Dillwyn-Llewelyn, and although not primarily an ornithologist, his unpublished records include bird lists for Swansea for the years 1900 and 1925.

The Reverend E. Dodderidge-Knight was not only an enthusiastic naturalist but, like many of the Victorian ornithologists, a renowned shot. His attitudes to ornithology were similar to many Victorian gentlemen natural historians in that most of his observations were made along the barrel of a shotgun. At this time the need for conservation had not been recognised, and it was often necessary to provide the specimen before a record of an unusual bird was accepted. However, his records are valuable and appear in a typical account of the period entitled *Notes on less common birds killed in the neighbourhood of Newton Nottage* published in 1853. Later lists for parts of Carmarthenshire were published in the *Zoologist* in 1865 by Thomas Dix.

Another Glamorgan rector, The Revd Digby Seys Whitlock Nicholl of Llantwit Major, compiled the earliest comprehensive list of the birds of a Welsh county. This was probably one of the earliest attempts at the compilation of a county avifauna, and it took the form of a series of notes published in the *Zoologist* in 1889.

The earliest list for Pembrokeshire is a catalogue produced by the taxidermist John Tracy in 1851. More detailed accounts were produced by Thomas Dix and published in the *Zoologist* in 1866 and 1869.

In the latter part of the 19th century, members of the Cardiff Naturalists' Society made important contributions to Welsh ornithology by both recording birds and promoting their protection through legislation. In the *Transactions* of the Society, notes on the birds of Grassholm and Skomer were recorded in the 1890s by Robert Drane and J. J. Neale.

In north and central Wales at this time, there was a dearth of ornithologists and no flourishing natural history societies developed. One of the earliest lists was produced in 1858 by William Davies for Llandeilo Fawr and the surrounding area, and by John Cordeaux in 1866 for North Wales generally. However, the most important contribution to our knowledge of the birds of mid-Wales in the 19th century was made by Prof. J. H. Salter, the holder of the Chair of Botany at the University College of Wales, Aberystwyth. From 1895 to 1904 he published his observations on the birds of mid-Wales, followed by a list of the birds of Aberystwyth.

Throughout the 20th century local bird lists continued to appear in scientific and natural history journals, together with records of more unusual birds. Many of these accounts were still based on personal records for particular areas, although there was a growing tendency to base them on detailed and systematic censuses of an area and to discuss the occurrence of birds in relation to the available habitats.

The first comprehensive Welsh county avifauna was published in 1882 (revised in 1899) as *The Birds of Breconshire* by E. Cambridge Phillips and based on a series of papers that appeared in the *Zoologist*. This was followed by the *Birds of West Cheshire, Denbighshire and Flintshire* by Dr W. H. Dobie, a Chester physician, published by the Chester Society of Natural Science and Literature (1893); the districts covered in this well researched account also included a small part of Caernarfonshire east of the Conwy and of east Merioneth.

The Birds of Pembrokeshire based on the records of the Revd M. A. Matthew came soon after in 1894 and in 1905 *Handbook of the Natural History of Carmarthenshire* by T. W. Barker was published.

An early example of the modern systematic county avifauna was produced by a committee of the Cardiff Naturalists' Society in 1900. *The Birds of Glamorgan* was based not only on contemporary records, but also included the observations of local gamekeepers, sportsmen and taxidermists. The committee later carried out the first revision of this important work in 1925.

In the first decade of the 20th century the first area avifauna for the whole of north Wales appeared in H. E. Forrest's *Vertebrate Fauna of North Wales* in 1907, followed by his *Handbook* in 1919. These two works covered the counties of Denbighshire, Flintshire, Caernarfonshire, Anglesey, Merioneth and Montgomeryshire and were based on a painstaking collation of published information, correspondence with other active naturalists and personal observations. Like himself, domiciled in Shrewsbury, many of the active field naturalists of this time were visitors from neighbouring English counties (e.g. T. A. Coward and S. G. Cummings from Cheshire).

The production of Welsh county avifaunas from the early 1930s to the mid-1960s was primarily due to the efforts of two members of the Cardiff Naturalists' Society – G. C. S. Ingram and H. Morrey Salmon. They were inspired by the work of H. E. Forrest, and aspired to produce a similar work for south Wales. Their efforts eventually resulted in the production of seven county avifaunas: Glamorgan (1936), Monmouthshire (1939), Pembrokeshire (with R. M. Lockley 1949), Carmarthenshire (1954), Radnorshire (1955), Breconshire (1957), and Cardiganshire (with W. M. Condry 1966). Some of these important systematic lists have subsequently been revised, *The Birds of Breconshire* by M. E. Massey in 1976 and by M. Peers and M. Shrubb in 1990 and *The Birds of Monmouthshire* by P. N. Humphreys in 1963. This latter volume has now been superseded by *The Birds of Gwent* published in 1977 by the Gwent Ornithological Society. In 1967, the centenary year of the Cardiff Naturalists' Society, *The Birds of Glamorgan* was revised by A. Heathcote, D. Griffin, and H. Morrey Salmon, and in 1974 a supplement was produced by H. Morrey Salmon. *The Birds of Radnorshire* was revised by Martin Peers in 1985.

The remaining county avifaunas have been produced primarily as a result of the efforts of Peter Hope Jones. The county list for Merioneth was published in 1974, followed by *Birds of Caernarfonshire* which was produced in 1976 in conjunction with Peter Dare. This list for Caernarfonshire was the first comprehensive account of the birds recorded in the county since Forrest. The county avifauna list for Denbighshire was published by

Peter Hope Jones and John Roberts, in *Nature in Wales* (1983). *The Birds of Flintshire* was published by the Flintshire Ornithological Society in 1968.

Sadly, there has been no update on the Montgomeryshire county avifauna since Forrest nor, comprehensively of Anglesey, although there are useful updates contained in a number of journals (e.g. Whalley 1954). These are gaps that remain to be filled.

In recent years there has been a welcome trend in the gathering of information on the breeding status of even the most common birds, stimulated by the publication of *The Atlas of Breeding Birds in Britain and Ireland* by the BTO in 1976. The Atlas inspired several local bird societies to embark on county tetrad (2 km × 2 km) surveys and atlases of all breeding species. In Wales this has already led to the publication of *The Gwent Atlas of Breeding Birds* in 1987 covering the years 1981 to 1985 (Gwent Ornithological Society) and *An Atlas of Breeding Birds in West Glamorgan* in 1992 covering the years 1984 to 1989 (Gower Ornithological Society). The results of surveys covering 203 out of 445 tetrads in Breconshire for the years 1988 to 1990 form, likewise, an important part of the updated *Birds of Breconshire* produced by Peers and Shrubb (1990) as does the comprehensive survey of tetrads in Pembrokeshire (1984–1988) in the revised Pembrokeshire avifauna in course of preparation by J. Donovan and G. H. Rees. It is a pity that the initiative taken in these south Wales counties has not, so far, been emulated by similar surveys in more northerly counties.

The longest periods of systematic bird recording and observation in Wales relate to the islands of Bardsey and Skokholm. There has been regular recording of all migrant and breeding birds at Bardsey, mainly within a period from March to November, from the opening of the Bird Observatory in 1954 to the present day (with the exception of the period 1971 to July 1973 when the Observatory was forced to close due to lack of an island boat and ferry service). Roberts (1985) has published a valuable account of the 276 species recorded on Bardsey to the end of 1984. In 1933 Ronald Lockley established the first British bird observatory on Skokholm which continued until ringing programmes ended in 1976, with a gap during the years of the Second World War. Systematic records have, however, continued to the present time and Betts (1992) has produced a similar volume to the Bardsey account, analysing the status of the 262 species that have occurred on Skokholm. The Welsh bird list owes a great deal to the systematic observation carried out on Skokholm and Bardsey; many first, and frequently only, records of rare vagrants have come from those two locations.

A milestone in the annals of bird recording in Wales was the commencement of a combined annual report of records for the whole of Wales produced by Peter Hope Jones and Peter Davis for the years 1967–1977 inclusive. Published annually in *Nature in Wales* these marked the first time that records had been compiled for the country as one entity and successfully achieved a two-fold object of bringing some cohesion to bird recording in Wales and providing a vehicle for the publication of records from those counties which had no local report when the *Welsh Bird Report* began. In parallel with this a Welsh Records Advisory Group (WRAG) was established to assess the records of birds, which though not rare enough to be included in the national list of species considered by the British Birds Rarities Committee, were nevertheless felt to be rare in a Welsh context.

The *Welsh Bird Report* eventually foundered as a result of a change of policy by the then editors of *Nature in Wales*. In fact *Nature in Wales* itself foundered shortly afterwards. Happily, the Welsh Ornithological Society (WOS) founded in 1987 as a result of discussions between WRAG and county bird recorders in Wales, has been able to continue the work started in 1967. WOS has now published *Welsh Bird Reports* for the years 1978–87, 1988 and 1989 under the editorship of Clive Hurford, and 1990 and 1991 under the editorship of Mike Shrubb.

Berwyn Mountains

7: The impact of agriculture on birds

Wales is a predominantly rural country, notwithstanding the highly industrialised areas of the south Wales valleys and the urban centres they have spawned on the coast – Newport, Cardiff, Swansea. Agriculture has therefore been the primary industry throughout most of Wales both before and since the Industrial Revolution. It is overwhelmingly a stock-rearing country with landform, soils and climate only poorly suited to the growing and ripening of grain crops, despite the earlier need for crop diversity when self-sufficiency and subsistence were the foundations of family farms.

With the growth of demand for food from the urban areas of Britain, the opening up of easier transport routes and consequent opportunities for two-way trade and later, developing international markets, the pattern of agriculture in Wales has changed. Following the agricultural recession of the 1920s, and particularly with the imperative to modernise food production after the Second World War, Wales has experienced rapid intensification of stocking levels, concurrent with a shift away from the growing of arable crops. The mixed stock-rearing regime of cattle and sheep has given way increasingly to a monoculture of sheep, most notably with Britain's entry into the EEC and the Common Agricultural Policy in 1972.

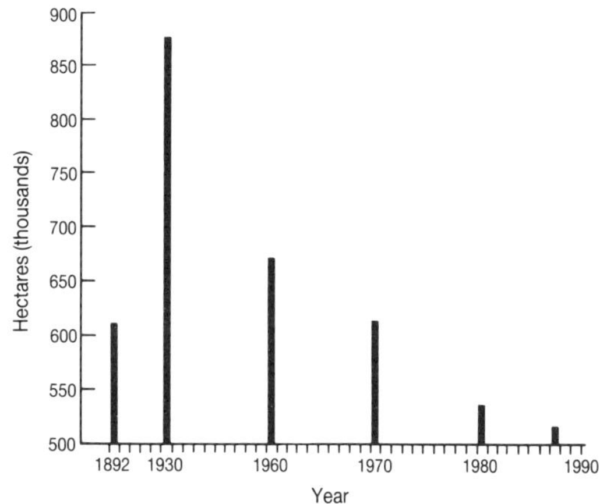

Rough grazing and common rough grazing in Wales at intervals since 1892

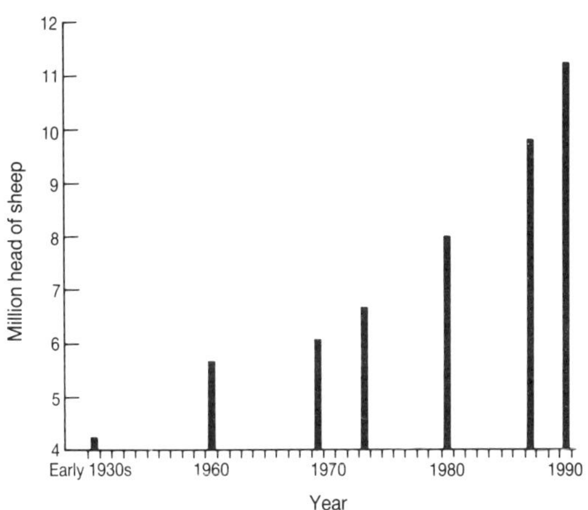

Sheep numbers in Wales since 1930 (source WOAD June Census Statistics)

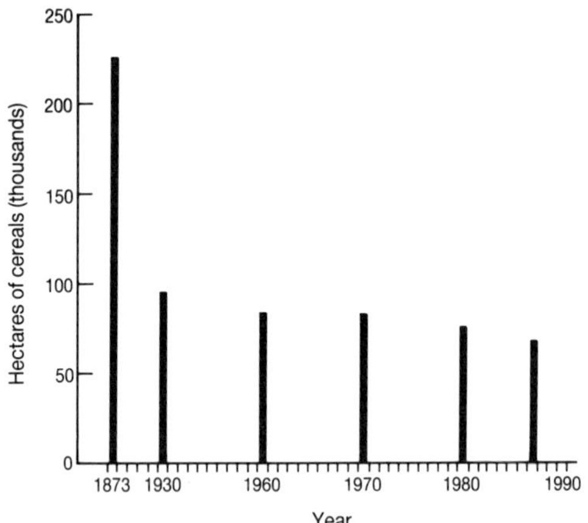

Area of cereals in Wales, 1873–present

These changes have had an enormous impact on farmland and moorland birds species (as indeed they have on a wide spectrum of other wildlife). Agricultural diversity and extensification provides the variety of habitats which encourages a breadth of species: conversely, monoculture and intensification lead to a reduction of species and a narrowing of the variations in farmland habitats. The accompanying histograms show, dramatically, three of the most telling changes that have affected bird populations, particularly those on farmland and sheep walks, in the 20th century. The species accounts, which form the main body of this book, reflect these effects and also refer to the significance of the high stocking levels on the state of Welsh woodlands and the make up of their bird communities.

The intensification of sheep grazing with the concomitant pasture "improvement" and moorland reclamation, together with the decline of cereals and other arable crops and not forgetting the decline in cattle numbers, has been to the serious detriment of birds such as Red Grouse, Golden Plover, Lapwing, Snipe, Grey Partridge, Skylark, Linnet and Yellowhammer. In any situation of change, there are winners as well as losers and here the modern rise in Carrion Crows to previously unimaginable levels and to a lesser extent Magpies and Ravens, can be attributed to the same factor – in this case an abundance of sheep carrion and the widespread provision of concentrated foods for lambing flocks in spring. Perversely, from the point of view of birds, these corvids are probably helping to deliver the *coup de grâce* to the breeding populations of many of the ground-nesting species mentioned above.

Bardsey

8: The offshore waters of Wales

by Graham Rees

Birdwatchers have paid considerable attention to the offshore waters of Wales in recent years. Mostly this has taken the form of sea-watching from prominent headlands, favoured localities including Point Lynas and South Stack on Anglesey, the island of Bardsey and Skokholm, Strumble Head and the Gower coast. Many other points have also been used and it may be that places like Carmel Head and St. David's Head could prove to be as good as the more established vantage points. To a lesser extent, passenger ferries and smaller craft have been used and offer considerable potential for gaining a greater understanding of how birds use the offshore waters. The following account attempts to outline our current understanding of this aspect of the birds of Wales. No doubt some of the views put forward will need to be modified following further observation.

It may be stretching literalness to regard the land between the Dee and the Severn as an isthmus to peninsular Wales, nevertheless the Principality has north, west and south coasts. Off the north and for most of the length of the west coast, lies the Irish Sea. The extreme south of the west coast is bounded by the "outflow" of the Irish Sea – St. George's Channel – and by the open Atlantic. The south coast is flanked by the Bristol Channel and Severn estuary. The nature of these different waters and the effect of Ireland almost landlocking the Irish Sea should be kept in mind when reviewing bird movement.

Normally there is constancy in the bird populations using the offshore waters during January and February. The shorelines and inshore zones are exploited by the commoner gulls. Divers, grebes and Common Scoters feed in the tide races and in the shallows above underwater shelves. Kittiwakes, Guillemots, Razorbills, Fulmars and a few Gannets utilise the whole sea area. Other species occur in very small numbers and in sporadic fashion. Winter gales occasionally briefly drive the more pelagic species inshore and at such times it becomes apparent that the waters that are out of range of land-based observers support species and numbers that would otherwise go undetected. Piecing together observations after the gales, it is clear that a few Manx Shearwaters winter offshore, perhaps late-hatched birds which lacked the stimulus or guidance to journey across the Atlantic to the species' principal wintering areas off the coasts of South America. A few Great Skuas, Little Gulls and Little Auks briefly appear inshore and more rarely a winter Leach's Petrel or Grey Phalarope. More significantly, large numbers of Kittiwakes, Guillemots, Razorbills and Fulmars appear and the sudden increase of Gannets indicates that the sparse inshore population is heavily outweighed by those fishing out of sight of land. Immature Gannets of all ages, including first-year birds, are among these, like the Manx Shearwaters showing that not all individuals conform to the general post-breeding migration pattern of the species.

Many birds which breed locally begin to arrive in Welsh waters from about the middle of March. Manx Shearwaters and Gannets become common inshore, Puffins arrive from the Atlantic and Storm Petrels quarter the offshore zones again. Terns and skuas pass through in April and May: only small numbers are seen off the South Wales and Pembrokeshire coasts, but numbers seen increase progressively northwards off Cardiganshire and

Caernarfonshire/Anglesey, suggesting that the majority move in off the sea across the south-east of Ireland. A few skuas and terns are still to be found passing into June, many showing immature plumage.

The tempo of seabirds passing offshore increases throughout June and July, as adults forage to meet the increasing food demands of their nestlings. The bulk of the breeding Manx Shearwaters in Wales and all of the Gannets, are to be found on the islands off the Pembrokeshire coast. Each day streams of both species can be seen radiating from the islands in the morning and returning in the evening in processions of many thousands. Many go out to the open sea from the islands, particularly to the south-west, towards Jones Bank and La Badie Bank. Others move into the Bristol Channel, sometimes penetrating up the Severn estuary as far as Aust. Large numbers use Cardigan Bay and some penetrate even farther beyond Llŷn and Anglesey, although the extent of these more distant movements is difficult to establish. Ringing has shown that there is considerable interchange of Manx shearwaters between the various Irish Sea breeding colonies.

In July, Guillemots and Razorbills gather offshore with their young in favoured fishing areas and prepare for moult, and the first appearances of Arctic Skuas, Mediterranean Shearwaters and Sooty Shearwaters heralds the start of a prolonged autumn passage. This passage continues to December, various species peaking at different times while the process as a whole is fairly continuous. Sooty Shearwaters, Arctic and Great Skuas, Sandwich and Common/Arctic Terns pass mainly in the period between late July and early October. Pomarine Skuas, auks, Kittiwakes and wildfowl come later, mainly in September to November. Common Scoters, divers and Gannets pass throughout the whole period, each usually peaking in numbers more than once. The overall impression of this movement is that in "normal" weather conditions, birds generally pass north to south through the Irish Sea and thence out of Welsh waters. Moderate to strong winds with a westerly component result in more birds being visible from the shore as they are deflected eastwards by the wind.

Gales alter the pattern completely and often result in displaced birds passing through inshore waters in spectacular numbers and increased variety. When south to south-west gales rage in the Western Approaches for a day or two, the birds are pushed downwind towards the Welsh coast. It is very likely that many of these birds are blown from great distances, some of them having previously passed down the west coast of Ireland or beyond, before being displaced by the gale. Naturally enough it is difficult to observe the actual mechanics of this displacement, especially from the shore. However, over a period of years, observations from headlands and occasionally from the unsteady deck of a ferry, suggest that although birds will turn downwind they do not normally maintain this direction willingly. Instead, they more often beat into the wind, maintaining air speed but, being captured in the air stream, actually progress backwards relative to the ground. When the depression causing the gale progresses to the north-east, the wind in Welsh waters veers around to the west or north-west before moderating. During this phase the displaced birds are able to beat out to sea again by flying obliquely across the wind and the day-long processions occur that are visible from the headlands and which delight the sea-watchers so much.

The south to south-west gales that cause so many birds to appear in the Irish Sea also drive smaller numbers onto the south Wales coast, from where they can move out to sea again or travel along the north Devon and Cornish coasts once the wind goes around or moderates.

It seems likely also that prolonged west or north-west winds push birds through the North Channel and into the "pocket" of Liverpool Bay. Perhaps south-east gales also cause birds to seek comparative shelter by entering the Irish Sea through the North Channel. In either event, Liverpool Bay becomes their ultimate lee shore and this has been demonstrated on many occasions by the congregation of Leach's Petrels in that area. Post-gale recovery of such birds may well be variable. It is likely that some of these petrels make their way out to sea again through the North channel while there is evidence to suggest that they also move southwards through the southern Irish Sea. Probably, however, this movement normally takes place out of sight of land, with sightings of only a small proportion of the total being seen past the North Wales coast, off Bardsey or Strumble Head. Occasionally, the weather or the condition of the birds, or a combination of both, results in greater numbers passing closer inshore – e.g. 181 passed Point Lynas in six hours on 1 October 1978.

In summary, when the weather systems affecting Welsh waters are not severe, there is a general spring passage northwards and autumn passage southwards. The onset of gales modifies this pattern inasmuch as species which would normally be found well out to sea become concentrated into a narrow inshore zone.

A limited account of this nature can communicate very little of the complexity or detail of individual species' patterns or of the variety of species that do occur, or indeed of the excitement of spectacle or discovery available to the patient birdwatcher. A little of the flavour of these aspects is offered in the following specimen sea-watch accounts extracted from the Strumble Head record.

3 September 1983

"A SSW gale had been blowing for the previous 24 hours, gusting to force 10 at times. At dawn the wind was round to WNW and had begun to moderate slowly; it was down to force 6 by the late afternoon.

"Shortly after first light, Manx shearwaters began to stream by on their way out to sea. The rate of passage seemed not to diminish all day and at least 40 000 birds were estimated to have passed. Fulmars and gannets were also passing in good numbers but kittiwakes in only moderate numbers. Visible among this throng were the flitting forms of storm petrels, mostly beating seawards but frequently pausing to form little flocks at some transient upwelling of planktonic food source. A very difficult species to detect, a nominal 100 were logged as having passed, although many more than this probably did so.

"Very few Sandwich terns passed but a total of 277 common/arctic terns was logged and at least 500 lesser black-backed gulls, mostly juveniles.

"A trickle of waders passed during the day, which seemed rather odd as they could so easily have waited for calmer conditions. Most of the waders were too far out for certain identification but those within range included dunlin, knot, turnstone, whimbrel and bar-tailed godwit.

"With all this activity claiming attention, it is perhaps not surprising that counting was not too precise and this was compounded by the counter-attraction of skuas passing close inshore. Only two Pomarine skuas were seen, a juvenile and an adult with full 'tail regalia'. A total of 103 arctic skuas exhibited a wide variety of plumages which made for intriguing viewing. 198 bonxies passed, some of which displayed great interest in the little groups of feeding storm petrels. The skuas would close in and alight among the small fluttering forms but were seemingly more interested in whatever was attracting the petrels rather than in the birds themselves.

"Occasional Mediterranean shearwaters were picked out among the stream of passing Manx, a total of 13 by the end of the watch. Despite the skuas, terns and petrels claiming so much attention the shearwater procession was kept under surveillance and eventually a great shearwater was discovered among them, to be followed shortly afterwards by two more together. A few sooty shearwaters began to appear and as the day progressed they passed in increasing numbers. At first they came by singly with occasional twos and threes but by the afternoon were in groups of 6 or 7 and by the end of the watch up to 14 at a time were passing. A total of 397 was logged this day."

25 September 1985

"Wind S force 3-4, hazy, visibility moderate – not classically 'good' sea-watching conditions. Sea-watching though is full of surprises. All through the day, rafts of kittiwakes formed up on the water just offshore. Periodically there must have been a spate of small prey coming to the surface for the birds would rise and begin to feed furiously. The set of the tide was diagonally away from the shore to the NE and the feeding birds followed this track until the food ran out upon which they returned inshore to form up on the water to await the next upwelling.

"The feeding groups attracted other species as they passed by, many pausing to exploit the transient food source. Consequently there was a different mixture of birds each time there was a feeding flurry, only the kittiwakes remained a constant feature. The following were noted at or passing through the feeding flocks: 2 Mediterranean shearwaters, a pomarine skua, 2 great skuas, a juvenile long-tailed skua and 4 arctic skuas, 81 Sandwich terns, 76 common/arctic terns, 3 black terns and a total of 5 juvenile Sabine's gulls."

24 October 1984

"Classic W gale following a SW gale the previous day.

"There was a day-long procession of some 35 000 mixed guillemots and razorbills and of 18 000 kittiwakes. Additionally, 10 Mediterranean and 265 Manx shearwaters were noted (quite late for so many Manx) and, surprisingly, 10 fulmars – a species which is normally absent inshore during October as the birds stay out to sea to moult.

"As ever, the skuas provided great interest with 64 bonxies, 22 arctics, 21 pomarines (several with complete tails) and a juvenile long-tailed.

"4 common/arctic terns, 2 little gulls and a juvenile Sabine's gull were picked out in the stream of kittiwakes and 3 late puffins and 6 little auks among the other auks.

"A leavening of other species added an interesting element of variety, notably a great crested grebe, 5 red-throated divers, 10 common scoter, 2 teal, 4 grey plover, a flock of 17 dunlin and 4 grey phalarope.

"Two latish storm petrels seen in the early afternoon provided a good control for identifying a solitary following Leach's petrel."

13 November 1987

"A SW gale veering round to NW had not previously occurred all autumn, so coming this late in November, it added an air of expectancy. We were not to be disappointed.

"Mixed lines of guillemots and razorbills passed during the cold 7 hour watch. Sample 10 minute counts were made every hour and, when extrapolated, suggested a total of 26,000 during the watch. With all this auk traffic going by, it was a little surprising that we only managed to find 2 little auks.

"Although a little late in the season for shearwaters and skuas, by the end of the day we had managed to pick up 4 Mediterranean shearwaters, a Leach's petrel, 8 bonxies and a single arctic skua. More typical of the time of year was a total of 7,000 kittiwakes, 76 little gulls, 2 black-throated divers, 20 red-throated divers and 19 great northern divers. Several of the divers were still in full or partial summer plumage.

"An enjoyable aspect was also supplied by a few passing wildfowl, with occasional wigeon, a shelduck, small parties of light-breasted brent geese (total of 24 birds) and 5 scaup. Surprisingly, only 35 common scoter were seen but in line astern of one male common scoter were 4 male surf scoters, a dramatic climax to a memorable sea-watch."

17 October 1991

"A SW gale had been blowing for 24 hours but had gone round to the NW by dawn. Guillemots and razorbills streamed by, in line after line, whirring rapidly seawards. Two little auks were seen close inshore but with mountainous seas running many more may well have passed further out, completely undetected.

"Manx shearwaters were passing in surprising numbers for so late in the year, at least 150 during the day, and were accompanied by a total of 10 Mediterranean and 5 sooty shearwaters. Wildfowl were prominent in the form of white-fronted geese, a total of 94 passing in five flocks. Over 50 common scoter and a tufted duck were also recorded.

"A stream of kittiwakes was closely scanned and revealed an accompanying little gull and seven common/arctic terns. Along the same line came the skuas, a total of 50 great and 5 arctics but also four juvenile long-tailed skuas, yet the bird that impressed the most was the frequent passage of pomarine skuas, either singly or in groups of up to five, totalling 97 by the end of the day."

18 October 1991

"The all-important wind had veered to the north overnight but had not abated. The sea revealed another day-long procession of birds heading out to sea, with a similar mix to the previous day although there were few notable differences. Among the wildfowl were two long-tailed ducks and two scaup and, a day later than the passing shearwaters, seven Leach's petrels. However, the pomarine skua continued to dominate the scene, passing frequently in groups of up to 14 birds, a total of 130 being logged by the end of the day. Most were juveniles but among the adults were several with their impressive tail 'spoons' intact."

There is evidently a great deal of interest to see and to discover regarding birds in the offshore waters of Wales. Rare species such as Black-browed Albatross, Little Shearwater, Wilson's Petrel and Gull-billed Tern have all been recorded in the region in recent years. Other challenging species, perhaps birds like Bulwer's Petrel and Bridled Tern, are potential new discoveries and this should give bird watchers the incentive for further watching.

Many questions about seabird passage off the Welsh coast arising from existing observations remain to be investigated. For instance, do Fulmars actually become flightless during their post-breeding moult? Why are feeding Storm Petrels sparsely distributed in the southern Irish Sea but frequently encountered immediately south of St. George's Channel and inside the Bristol Channel? Why does fog "ground" Manx Shearwaters at sea but apparently not affect Gannets?

The brief accounts of specimen sea-watches given above concern the highlights from several years of watching. They do not tell of other long hours expended when very little was seen. Yet it is often during these quiet periods that questions are posed that provide the spur for further watching. Almost inevitably, one of the questions is: "why not go out to sea to seek the birds, instead of waiting passively for them to pass within sight of land?" Realisation of this aim would involve overcoming problems of logistics and cost, as well as enduring the vagaries of *mal de mer*, but it would seem to be the way ahead over the next few years. Who knows what revelations might then unfold?

Llangorse Lake

9: Background to the species accounts

PERIOD COVERED

We have collated what relevant and verifiable records we have been able to find, reaching back in a few cases to archaeological remains many thousands of years ago. More usually – with a few exceptions of records from Roman and mediaeval middens, occasional written references (e.g. Giraldus Cambrensis) and a handful of 18th century accounts – the earliest accounts for most species date from the middle of the 19th century or later. At the current end of the time scale, individual species accounts are taken up to the most recent available data. In the case of the *Welsh Bird Report* this is 1991 and for county bird reports usually 1990 or 1991 (Denbighshire and Flintshire 1989). Other more recent data are incorporated where they exist, e.g. RSPB/Countryside Council for Wales (CCW) monitoring and survey results up to 1992; Pembrokeshire islands seabird data up to 1992; Wales Raptor Study Group data also up to 1992. Other individual records of particular interest have been added more recently, where they have come to hand (e.g. numbers of breeding Short-eared Owls on Skomer in 1993).

ARRANGEMENT OF MATERIAL

The sequence and nomenclature followed is that given by Voous in his *List of Recent Holarctic Bird Species* (1977) as incorporated in the *Checklist of Birds of Britain and Ireland* (British Ornithologists' Union 1992). The English names of species are, however, those in present usage rather than the new names for some species advocated in the *Checklist*.

INCLUSION OF RECORDS

Records of rare species are included only if they have been accepted by the British Birds Rarities Committee (BBRC). Every record accepted by that Committee from the years 1958 to 1991 is included. As a matter of principle only accepted records of species considered by BBRC are included in the accounts that follow. It therefore follows that records under consideration by the Committee and not published by the time of the Report on Rare Birds in Great Britain in 1991 (*British Birds* 85, 510–552) are not referred to. These include some notable records that would constitute firsts for Wales if accepted (e.g. South Polar Skua and Soft-plumaged Petrel at Strumble Head).

SOURCES

A "standard list" of references (see Ch. 3 Sources and references) has been used as a basic bank of source material for all species. This fairly lengthy list represents both the most reliable ornithological archive for Wales and – through county bird reports and the updating of county avifaunas – the ongoing published account of birds in the Principality year-by-year. In a country such as Wales where many counties are sparsely populated and birdwatchers and ornithologists are often very thin on the ground, it is inevitable that the fortunes of county bird reports are frequently heavily dependent on one or two individuals. Great credit is due to those who ensure the annual production of these reports and it is hardly surprising that dependence on so few people has meant that there are periodic ebbs and flows in the regularity of some county reports. It is encouraging to see that the situation at the time of writing, in early 1993, is probably more secure than at any time in the past. The most serious omission at present is the total absence of a report for Montgomeryshire; in Denbighshire and Flintshire, no report has yet appeared for the years since 1989.

The production of breeding bird atlases for some of the south Wales counties in recent years has provided important data. These data are invaluable but at the same time we are only too well aware that the absence of any comparable work in mid or north Wales gives an unbalanced feel to many of the species accounts in this book, particularly passerines. We are particularly grateful to the authors of the *Status and Atlas of Birds in Pembrokeshire* for making their manuscripts available to us prior to publication. We are equally grateful to David Gibbons who generously allowed us to have pre-publication copies of the distribution maps from the *New Atlas of Breeding Birds in Britain and Ireland*.

Computer print-outs of Welsh ringing recovery data up to June 1992 have been provided by the BTO (one or two subsequent recoveries of note have also been included in the species accounts). The BTO has also provided copies of all CBC data from Welsh Common Bird Census sites up to and including 1991.

In addition to these all-embracing references, information on particular groups has been gleaned from other sources as indicated below.

Wildfowl and waders

Much use is made of the invaluable series of data on wader numbers on estuaries provided by the British Trust for Ornithology's Birds of Estuaries Enquiry covering the period 1969 to date. In particular, reference is frequently made in the species accounts to average winter peaks of numbers for five-year periods. It should be noted that the quoted years refer to the BoEE "counting year" and that, for example, the year 1991 refers to the period which begins in July 1991 and finishes in June 1992. In the same context, "winter" comprises the months of November to March inclusive.

The Wildfowl and Wetlands Trust wildfowl counts, alongside the BoEE data, provide important baseline information on the numbers and distribution of wintering wildfowl. Unfortunately, coverage is by no means complete in Wales although most of the major inland waters are covered. Accordingly it is not yet possible to give fully accurate wintering populations for some wildfowl species, as will be evident from the texts.

Birds of prey

The Welsh Raptor Study Group was established in 1991 to monitor the populations of diurnal birds of prey through a network of volunteer fieldworkers. It has become the principal source of data for this group and also covers one or two other species, notably Raven, Chough, Barn Owl and Short-eared Owl. The account of the Red Kite written by the Kite recorder, Peter Davis, has drawn on a range of historical and archive material, notably the confidential document collated by Col. H. Morrey Salmon in 1971. Since that date Peter Davis himself has kept the detailed records of Red Kite in Wales on behalf of the Countryside Council for Wales (formerly Nature Conservancy Council) and the RSPB.

Seabirds

We are reasonably fortunate in Wales that records of some of the seabird colonies around the coast have been documented, albeit sporadically, for many years by a number of travellers and, later, amateur and then professional ornithologists. As an example, the origin and growth of the Gannet colony on Grassholm is fairly accurately known (see p 51). The periodic counts, or estimates, from some of the offshore islands have been particularly valuable but the main sources for recent years have been the co-ordinated census of all coastal breeding seabirds in 1969–70 (Operation Seafarer), summarised in *The Seabirds of Britain and Ireland* (Cramp *et al.* 1974) and the subsequent Seabird Colony Register (SCR), 1985–87. This Register was established in 1984 on the joint initiative of the Seabird Group and the (then) Nature Conservancy Council. Now administered through the Joint Nature Conservation Committee (as the Seabirds Team), its aim has been to produce a single source of information on breeding seabirds in Britain and Ireland, and in 1985–87 it took on the major task of repeating the Operation Seafarer survey. Between the years of Operation Seafarer and the establishment of the Seabird Colony Register (SCR), annual monitoring was carried out at selected sites (Stowe 1982) and has now been brought under the SCR. Owing to the increasing fragility and reducing status of populations of all tern species in Wales, the RSPB has kept accurate figures for these species annually since the opening of an office in Wales in 1971.

Seabirds at sea are amongst the most difficult of groups to assess and although much still remains to be learned of the movements, moulting areas and feeding zones of many individual species, a considerable understanding of the patterns of occurrence of some of the principal species has been accumulated in the past decade or so by the Seabirds Team.

This expanding knowledge has also been supplemented by the efforts of sea-watching enthusiasts who have added a great amount to the understanding of inshore seabird occurrences and movements in the past 10–15 years; principal Welsh sites in this respect have been Strumble Head (Pembrokeshire) (see summary account by Graham Rees on p 29) and Point Lynas (Anglesey).

Passerines

Data on many of our commoner passerines are lamentably lacking in Wales. Since its inception in 1961 the Common Birds Census has been carried out at over sixty sites in Wales. However, most of these have run for only very short periods of time and do not easily lend themselves to comparative analysis. The number of CBC sites which have run for ten years or more is only seven (discounting island ones on Skokholm and Bardsey and ones which terminated more than ten years ago). These sites cover both woodland and farmland plots but both the geographical spread and the range of habitat types covered is so great that comparison of findings is very difficult to make. For example the three long-running farmland sites range from a plot on the Gwent Levels, to urban-fringe farmland at Swansea and a hill farm in Merioneth surrounded by the conifer forest of Coed Y Brenin. Another problem is geographical imbalance, the majority of the up-to-date and long-running sites again being in south Wales (mainly Monmouthshire).

County bird reports, sadly, have proved of very little use in helping to define the populations of most of our breeding passerines. In many instances the most meaningful data have been those produced by a series of survey and monitoring programmes initiated by the RSPB since the mid-1970s. The general scope of these surveys is referred to elsewhere in this book; several of the more important surveys and some of the research programmes have been carried out jointly with other bodies, including the National Rivers Authority, Local Authorities, Welsh Water, Forestry Commission, British Coal (Opencast Executive), (former) Nature Conservancy Council and much of the work in the 1970s and 1980s was implemented under successive Manpower Services Commission initiatives. Since 1990 a more structured approach to the monitoring of key species and groups has been instituted jointly by RSPB and the Countryside Council for Wales.

DEFINITIONS

Categories of status for non-breeding species are defined as follows:

Vagrant	1–10 records this century
Rare	11–50 records this century
Scarce	1–5 birds/records per annum in recent years
Uncommon	6–50 birds/records per annum in recent years
Fairly common	51–250 birds/records per annum in recent years
Common	251–1000 birds/records per annum in recent years
Abundant	1000+ birds/records per annum in recent years

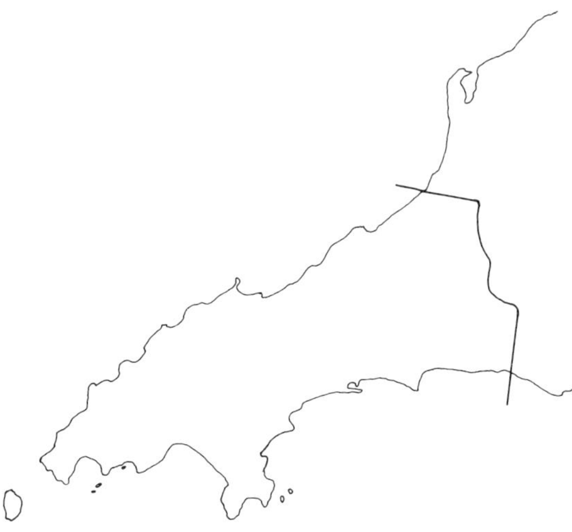

Area of Llŷn covered by RSPB Surveys 1986–87

The species accounts

Red-throated Diver *Gavia stellata*
Trochydd Gyddfgoch

Winter visitor and passage migrant, regularly recorded offshore, sometimes in large concentrations. Rare inland.

The winter distribution of the Red-throated Diver along the west coast of Britain is distinctly patchy, and numbers fluctuate widely in response to weather conditions and other factors which may affect food supply. Although its British breeding range is confined to the Scottish lochs, in winter it is most commonly found in shallow inshore waters and sandy bays where it feeds on crustaceans. It is the most numerous and widely distributed of all the divers wintering in Wales, and is found offshore between the months of July and May inclusive. It is concentrated mainly off the north-west coast, although occasionally large numbers are encountered elsewhere, especially following stormy weather.

Even during the late 19th century, observers noted that this was the most numerous of the three divers. At that time, it was apparently quite common along the west coast of Merioneth and the mouth of the Dyfi, but less common off the north Wales coast. Birds were also recorded inland on a few occasions, especially on Llyn Tegid (Merioneth), and on 28 December 1907, a Red-throated Diver was shot at Pistyll Rhaeadr (Denbighshire). Several of the early records refer to individuals in full summer plumage seen in late April or early May. At times, large concentrations were recorded, such as the 50 birds offshore between Llanddulas and Old Colwyn (Denbighshire) on 31 March 1907 and 20 near the west coast of Pembrokeshire from 21 to 25 April 1930.

Throughout the early and mid-20th century most of the records refer to single birds or small groups seen around the coasts of west and north Wales, with the odd record from inland waters. Even during the early years of this century, however, large numbers were recorded on passage from some of the better watched headlands, such as the Great Orme (Caernarfonshire).

From the mid-1970s onwards, the wintering population of Red-throated Divers has apparently increased markedly, particularly on the Merioneth and north Cardiganshire coasts. Since 1977, over 300 birds have been seen annually off the coast between Sarn Cynfelyn and the mouth of the Dyfi (Fox & Roderick 1989), with a maximum of 427 birds offshore in the winter of 1991–92 between Aberdyfi and Black Rock Sands (Thorpe 1992). These birds appear in late August, and hence include moulting birds, with maximum numbers occurring in December. There is a great deal of movement within north Cardigan Bay, depending on the weather, with wintering flocks often encountered as far north as Morfa Harlech, and even Criccieth (Caernarfonshire), where 40 were counted on 23 January 1987. This increase has also been noted around the north Wales coast, with 150 at Llanfairfechan (Caernarfonshire) in March 1993 and up to 40 in Red Wharf Bay (Anglesey) in February 1987. Previously, the highest numbers had been recorded following severe gales (e.g. over 100 sheltering in the Dee estuary on 28 September 1958).

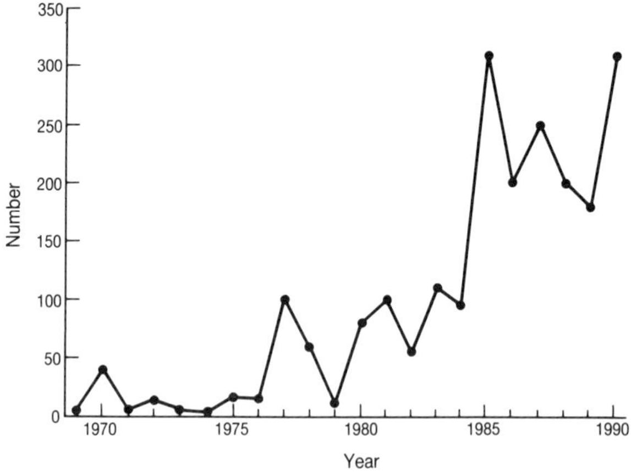

Maximum annual counts of Red-throated Diver off Borth – Ynys Las

Its status in south Wales is not properly understood although large numbers are occasionally seen offshore at sites such as Strumble Head, Worm's Head (Glamorgan) and Carmarthen Bay (e.g. 70 at the last site in December 1980). The first record for Monmouthshire was not until

16 April 1964, reflecting the scarcity of this species in the Severn estuary. Records from the inland counties are also scarce, although there have been several records of single birds from Llangorse Lake, one record from Radnorshire, and two from Montgomeryshire.

There is considerable turnover in the number of Red-throated Divers off the Welsh coast during passage. The largest numbers pass Point Lynas and Strumble Head in November and December with dramatic spring passage at the former site in late April and early May (Fox & Roderick 1989). By contrast, autumn movements on Bardsey are small, possibly suggesting that birds are passing the island well offshore. This species is recorded as being a scarce passage migrant to all the major islands of north and west Wales, with even a record of two from Grassholm (Pembrokeshire) on 18 November 1968.

Black-throated Diver *Gavia arctica*
Trochydd Gyddfddu

Winter visitor and passage migrant, recorded in small numbers annually.

This is the rarest of the common divers around the coast of Britain, although it is probably under-recorded due to the difficulties of identification in conditions that are anything less than ideal. It is less gregarious than the Red-throated Diver and is usually found in small groups only on passage or following hard weather. In Wales, it is generally very scarce and records usually come from traditional Red-throated Diver areas. The Welsh wintering population is probably made up of breeding birds from Scotland with possibly a few immigrants from Scandinavia. Inland records are very rare.

Forrest (1907) described this species as "the rarest of the three divers", and stated that in north Wales it had been recorded from the Dyfi estuary (Merioneth) and on occasion, off the Great Orme. Mathew (1894) also described it as a scarce visitor to Pembrokeshire, although it was later said to be almost as common as the Great Northern Diver in the county (Lockley 1961). During the last century, a bird was shot at East Moors in Cardiff, but no exact date is given, and a bird was shot inland on Llangorse Lake on 27 February 1892.

The first county record for Caernarfonshire was on 30 April 1910 when a bird was seen in Llandudno Bay, and there was a further sighting, of an oiled bird in the same area in 1921. Most of the sightings at around that time came from the Llandudno/Great Orme area, with 20 sightings from Caernarfonshire up to 1973. In Carmarthenshire, the first record was of a bird shot at the mouth of the Afon Taf on 16 November 1942, but since then the species has become a regular visitor to the county, mainly to the Burry Inlet with up to four recorded most winters. Records for Glamorgan generally come from this site, Eglwys Nunydd Reservoir and Kenfig Pool.

One unusual record concerns a bird picked up dead in a garden in Newport (Monmouthshire) on 5 February 1915; the next sighting for the county was not until 1962. During the 1940s and early 1950s, Black-throated Divers were often recorded from the Dee estuary, and from the 1960s they became much more common around the Gower peninsula, and have been recorded annually in this area ever since. The species remains a scarce winter visitor to the coasts of Merioneth, Cardiganshire and Pembrokeshire, but since 1968, there have been six records from Breconshire, all from Llangorse Lake or Talybont Reservoir. Between 4 May and at least 3 July 1989, a first-year bird was present at Nant-y-Moch Reservoir (Cardiganshire), a most unusual record for a predominantly marine winter visitor.

In 1979 and 1986, and to a lesser extent 1987, unusually high numbers of this species were recorded around the Welsh coast, presumably involving birds moving in response to hard weather in the early part of these years (Fox & Roderick 1989). In particular, 1979 was a year of considerable influxes of this species, especially inland, due to severe weather (Chandler 1981). Up to ten individuals were present in Holyhead Harbour (Anglesey) in March 1979 and up to six in St. Bride's Bay (Pembrokeshire) from January to March 1986. In 1987, on 31 January three were seen at Llanfairfechan, eight were present in Red Wharf Bay (Anglesey) on 22 February, and four were recorded at Morfa Dinlle (both Caernarfonshire) on 29 March.

Black-throated Divers are rarely recorded from the islands, with most records in March–April or October–December. There had been 14 records of this species from Bardsey up to 1985, the most notable being three together on 13 March 1984 and one bird which stayed around the island throughout January 1963.

Great Northern Diver *Gavia immer*
Trochydd Mawr

Winter visitor and passage migrant, recorded offshore in small numbers annually. Recorded from inland waters more frequently than other diver species.

The largest of the three divers that visit the Welsh coasts each winter, the Great Northern Diver winters in deeper water and farther offshore than the other two species, and is therefore probably under-recorded. However, it is also found on inland waters more often than the other two species, and during severe weather it often seeks shelter in shallow bays and harbours.

Its winter distribution in Wales shows a strong north-westerly and south-westerly bias, being most common around the coasts of Anglesey, Caernarfonshire, Pembrokeshire and Glamorgan. It is nowhere near as numerous as the Red-throated Diver and the numbers recorded on passage are relatively small. It is likely that the Welsh wintering population originates from Iceland with possibly some Greenland birds and even some from northern Canada.

At the turn of the century, this species' winter distribution was very similar to that of today, with regular sightings around Holyhead and South Stack (Anglesey), Llŷn (Caernarfonshire) and the Merioneth, Pembrokeshire and Glamorgan coasts. Most of the records refer to single birds, but Mathew (1894) stated that the species was numerous at times in Milford Haven. Inland records came from the meres on the Flintshire/Cheshire border, Llandrillo Weir in Denbighshire and Llangorse Lake. Again, most records were of single birds but three were present on Llangorse Lake in the winter of 1881.

The species appears to have become a much scarcer visitor during the first half of this century, although this may be due to a lack of observers and the Great Northern Diver's habit of fishing a few kilometres offshore. Scattered sightings were still submitted from many coastal areas and occasionally inland, e.g. one on Roath Park Lake (Glamorgan) in March 1929 and one on Ynys-y-Fro Reservoir (Monmouthshire) on 23 October 1938. Between 1930 and 1974, there were only nine records for Merioneth, and between 1900 and 1960, only five for Carmarthenshire.

Over the past 30 years, the number of records has increased, and this species is now a regular visitor to many favoured areas. Birds are recorded most years, albeit generally singletons, from Holyhead Bay on Anglesey (five in December 1992), the Llŷn coastline, north Cardigan Bay, around Pembrokeshire and the Gower coast. The largest numbers recorded were seven or eight in Conwy Bay (Caernarfonshire) in March 1956 and eight in the Dee estuary in December 1961 (Dickinson & Howells 1962), but more recently three were recorded at Llanfairfechan (Caernarfonshire) on 13 February 1987, and up to six were present in Carmarthen Bay between 17 and 25 February 1985.

Inland, this species has been recorded recently from Llyn Trawsfynydd (Merioneth), Lake Vyrnwy (Montgomeryshire), Llangorse Lake, Talybont Reservoir and Llandegfedd Reservoir (Monmouthshire). Of 11 records for Monmouthshire up to 1975, six were birds recorded at Llandegfedd between 1973 and 1975.

Sea-watches off Point Lynas (Anglesey) have shown that the ratio of Great Northern to Red-throated Divers recorded on passage is 1:9 (Fox & Roderick 1989), the former generally recorded as singletons. Up to 26 in a day have been recorded at Strumble Head during autumn passage (peak during September–October) and it is most frequently recorded from Bardsey during this time, although small numbers are seen in spring (March–May). Three passage birds were seen at Llanfairfechan (Caernarfonshire) in April 1992. In Pembrokeshire some of the species' most favoured haunts are the sheltered bays between the offshore islands and the mainland, where it is most frequently recorded in September and October, and occasionally in the spring. One bird spent the entire year in the Cleddau estuary (Pembrokeshire) in 1991.

White-billed Diver *Gavia adamsii*

Trochydd Pigwen

A vagrant.

There is only one fully authenticated record of this Arctic breeding species:

24 February–19 May 1991 – One first-winter bird in Holyhead Harbour and adjoining bays, Anglesey.

One was claimed to have been seen off Harlech (Merioneth) on 14 January 1925 but there are insufficient details to warrant acceptance of the record.

Pied-billed Grebe *Podilymbus podiceps*

Gwyach Ylfinfraith

A trans-Atlantic vagrant.

It has been recorded from two localities in Wales:

13 November–30 December 1984 – One, Aber Ogwen (Caernarfonshire)
31 January–23 April 1987 – One, Kenfig Pool
31 October 1987–1 April 1988 – One, Kenfig Pool

This is seemingly an unlikely vagrant to cross the Atlantic, but it is typical for this species to remain in one locality for long periods.

Little Grebe *Tachybaptus ruficollis*

Gwyach Fach

Rather uncommon breeding resident on well vegetated lakes, marshes and water courses; scarcer in the land-locked counties.

The only factor limiting the distribution of this species in Britain is its requirement for luxuriant vegetation both on lake bottoms and as emergent plants. It occupies small farm ponds and urban lakes in addition to larger lakes, gravel pits, marshes and slow-moving streams and canals. Little Grebes spend a great deal of their time hidden amongst dense vegetation, but because they are the most vocal of European grebes, breeding pairs are unlikely to go undetected.

Much of Wales is unsuitable for the Little Grebe as the upland lakes are largely oligotrophic and therefore devoid of vegetation. On some lowland waters it may be absent due to its inability to compete with Great Crested Grebes. However, on lowland waters where the habitat is ideal, Little Grebes may reach high densities so that nesting is almost colonial. Parslow (1973) stated that there was evidence of an increase in the breeding population since the end of the 19th century, but that there had been little change in more recent years. In Wales, however, it is apparent that the population has declined in many counties since the turn of the century, and that it is now a rather uncommon breeder over much of the country.

At the turn of the century, the Little Grebe was reported to be a common breeding bird in the coastal north Wales counties and in Glamorgan and Monmouthshire. At that time, it bred up to altitudes of 390 m on the moors above Llanuwchllyn (Merioneth), and breeding pairs were recorded on many lowland waters, including Roath Park Lake in Cardiff (Glamorgan). However, the population declined in most areas during the 1930s, possibly as a result of the severe weather during the winter of 1928–29. Little Grebes declined further as a breeding species in Wales following the hard winters of 1939–40, 1947–48 and 1962–63, and by the mid-1960s, the species was recorded as a scarce breeder in most Welsh counties, with the population probably numbering no more than 50 breeding pairs.

Parslow (1973) noted that Pembrokeshire was one of only two counties in Britain where this species did not breed, and during that time it was probably absent from much of Snowdonia and the Cambrian Mountains. It became extinct on Llangorse Lake during the early 1970s, about the same time that the Great Crested Grebe population crashed. A combination of greatly increased disturbance and an associated change in the aquatic vegetation are the most probable causes for this decline (M. Shrubb, pers. comm.). Since about 1980, however, several counties have experienced a slight increase in populations, none more so than Pembrokeshire where there are now 8–12 breeding pairs, although it is evident that sporadic breeding had occurred previously at several sites in the county. One of the principal reasons for the recent increase is thought to be the colonisation of newly created farm ponds, and this is also thought to have contributed to recent increases in Caernarfonshire and Anglesey (N. Brown, pers. comm.). In Carmarthenshire also, the creation of artificial lakes and ponds has benefited this species and the county population is now thought to be between 20 and 22 breeding pairs (I. Morgan, pers. comm.). The creation of farm ponds has not resulted in an increase in Montgomeryshire where the population of no more than ten pairs is confined mainly to small, natural lakes such as Gregynog and Gungrog Flash.

Breeding Little Grebes are still absent from much of Llŷn (0–3 breeding pairs in 1986), Snowdonia, the Cambrian Mountains and the majority of the Welsh uplands, and few Welsh sites individually support three or more breeding pairs. During the 1950s, up to eight pairs were recorded on Llyn Hilyn (Radnorshire), but the maximum in recent years has been three. Garth Lake (Breconshire) and Oxwich Marsh NNR also support up to three pairs, but nowhere in Wales does the breeding density approach that of some of the lowland lakes in England where this species is almost colonial. Anglesey supports much suitable habitat; in 1986, the island's population was thought to be between 15 and 25 pairs (RSPB 1986).

Following a series of mild winters and the creation of new, shallow pools in some areas, the Welsh breeding population has continued to increase slowly throughout the 1980s and probably stands at 120–150 pairs today.

Little Grebes are much more catholic in their choice of habitat in winter. Although they do not form large post-breeding flocks like the Great Crested Grebe, they often gather into small groups and roost together during the winter months. They also undergo widespread dispersal and, in Wales, many birds which probably bred a few miles inland gather together on western estuaries such as the Cleddau (Pembrokeshire), the Teifi and Dyfi (Cardiganshire), the Mawddach (Merioneth) and the Ogwen (Caernarfonshire).

This post-breeding migration was identified by Forrest (1907), who stated that pairs which, at that time, bred at high altitudes in north Wales, descended to the lowlands during the winter. At the turn of the century, small wintering flocks gathered on Llyn Tegid (Merioneth) and many of the estuaries where the species is still found today.

Vinicombe (1982) showed that only 43% of the August 1976 population remained on Anglesey's many shallow lakes the following January, 11% having moved to the coasts and estuaries, with most of the missing 5% presumed to have moved elsewhere. On the island, small parties are often recorded on several lakes, with larger groups occasionally recorded both here (e.g. 27 on Llyn Cefni, 30 July 1989) and in sheltered bays (e.g. 50 on the Inland Sea on 18 November 1989).

The Dee estuary regularly supports small flocks of Little Grebes (maximum 24 at Sealand on 1 January 1979), but the largest concentrations are found in north-west Wales, with 30+ regularly seen on the Glaslyn estuary in Caernarfonshire (42 in November 1989) and smaller numbers on the Mawddach (18 in January 1989) and Broadwater in Merioneth (22 in December 1987). Similar numbers are also recorded on the Cleddau in Pembrokeshire (29 in January 1978), although the largest flock recorded in the county was 40 on the millpond at Pembroke on 25 December 1925. On 20 November 1983 20 were recorded on the Afon Leri (Cardiganshire) and 25–30 are often seen on the Dyfi at the point where salt and freshwater mix.

On 2 February 1965, 32 were seen on Cadoxton Ponds in Glamorgan, but the most regular site in the county for this species is Eglwys Nunydd Reservoir (18 on 6 August 1966). Most records for the other Welsh counties involve small flocks of no more than half a dozen birds, but larger numbers are occasionally recorded. On 11 January 1970, 29 were present on the Gwendraeth estuary (Carmarthenshire), and ten were recorded at Llanwenarth (Monmouthshire) in January 1972. A small flock of 6–10 birds winters each year on the Wye at Glasbury (Radnorshire), and similar numbers have been recorded at Llyn Hilyn in the same county (10+ here on 14 April 1970). Small numbers are also recorded from some of the low-lying lakes in Breconshire, the maximum recorded to date being a flock of 13 at Talybont Reservoir in November 1988.

Little Grebes have been recorded on some of the

smaller offshore islands, almost all records referring to single birds. Most sightings from Bardsey involve birds seeking shelter in the protected bays or attracted by the lighthouse, the maximum number being three on 7 October 1965. The occurrence of occasional birds on Skokholm and Skomer probably refers to grounded night migrants. Ringing on the Afon Leri (Cardiganshire) has shown that the same birds return to this area in successive years (A. V. Cross, pers. comm.)

Great Crested Grebe *Podiceps cristatus*

Gwyach Gopog

Resident breeding species in all counties except Pembrokeshire, breeding on shallow, vegetated lakes. Shows a more coastal distribution in winter, and recorded on passage from several counties.

Great Crested Grebes breed almost exclusively on shallow, well vegetated lakes with a large expanse of open water. Although they favour lowland lakes, they do breed up to an altitude of over 200 m. In Britain, this species demonstrates a strong south-eastern bias in its breeding distribution, its stronghold being East Anglia and the English Midlands. In Wales, it is most common on the shallow Anglesey lakes and in many of the counties along the English border.

Up until the middle of the 19th century, it was widely distributed in small numbers in England and Wales, but during this time, under-pelts became fashionable as "grebe-furs" and Great Crested Grebes were killed in their hundreds. By 1860, there were no known breeding pairs in Wales, and the first nesting attempt following the widespread massacre did not occur until after the species was given full protection in 1880. It was in 1882 that Great Crested Grebes first colonised Llangorse Lake, and the population here grew to 20 breeding pairs in 1902 with the same number still present in 1934. During the 1880s and 1890s birds also colonised the Anglesey Lakes, with three pairs present on Llyn Penrhyn in 1891 and 1892, two pairs on Llyn Maelog in 1885 and a pair each on Llyn Traffwll and Llyn Llywenan at the turn of the century (Harrisson & Hollom 1932).

In 1894, a pair reared young on a protected estate at Sant-y-nyll Pond near St. Fagan (Glamorgan), and for many years from 1915 onwards up to eight pairs bred on an estate belonging to Sir Frances Rose-Pryce at Hensol Castle (Glamorgan). The birds were allowed to breed undisturbed here until the estate was sold in the late 1920s and its subsequent use as a mental hospital resulted in the birds abandoning the lake from 1932 until 1948, when a pair bred there once more. The lake, however, was drained in 1961.

During the 1931 Great Crested Grebe Enquiry (Harrisson & Hollom 1932), 16–18 of the 21–23 pairs found in Wales were located in counties adjacent to the English border. The exceptions to this were the two pairs breeding on Llyn Tegid (Merioneth), one pair on Llyn Llywenan (Anglesey) and two pairs on Hensol Castle Lake (Glamorgan). At that time this species was much more numerous in the English Midlands, and particularly in Cheshire, and it is probable that pairs nesting in the adjacent Welsh counties originated as overspills from those centres of population.

Much of southern Britain experienced an increase in the numbers of breeding Great Crested Grebes during the 20th century, associated with climate amelioration and the creation of gravel pits in many counties. In Wales however, there appears to have been a decrease in many areas following the 1931 census, and no breeding was recorded in Merioneth between 1936 and 1971, when a pair bred at Tal-y-llyn. Similarly, no breeding was recorded in Glamorgan between 1953 and 1973, and numbers declined or disappeared from favoured sites such as Llyn Ystumllyn (Caernarfonshire) and Llangorse Lake.

However, as the number of pairs decreased in many areas, breeding was recorded for the first time from other counties. Despite rumours that a pair bred at Middleton Hall (Carmarthenshire) earlier this century, the first confirmed record for the county was in 1967 when two pairs bred at Talley Lakes after up to five birds had been present during the 1966 breeding season. The first breeding record for Cardiganshire occurred in 1949 when a pair reared four young at Rhos-rhydd, and breeding was not recorded in Monmouthshire until 1971 when a pair nested at St. Pierre Lake.

The 1965 census (Prestt & Mills 1966) revealed mixed fortunes for this species in the Welsh counties, numbers having increased since 1931 only in Anglesey (17 adult birds), Montgomeryshire (16), Radnorshire (2) and Cardiganshire (2). The most important waters were Llangorse Lake, Hanmer Mere (Flintshire) and Llyn Ebyr (Montgomeryshire), all of which held eight adult birds. Declining numbers were recorded from Breconshire (12), Glamorgan (0), and Merioneth (0). However, counts were incomplete for some counties and Shrubb (pers. comm.) believes that even during the mid 1960s, 18–20 pairs were still found on Llangorse Lake.

By 1975, a notable increase was noted in all Welsh counties except Montgomeryshire (no change), Monmouthshire (marked decrease) and Pembrokeshire where no birds were recorded (Hughes *et al.* 1979). The total Welsh population had now increased from an estimated 71 birds in 1965 to 163 in 1975, the most important

Great Crested Grebe numbers in Wales in 1931, 1965 and 1975

County	*1931* *Adult Birds*	*1965* *Adult Birds*	*1975* *Adult Birds*
Anglesey	2	23	51
Breconshire	16–20	12	39
Caernarfonshire	0	0	2
Cardiganshire	0	2	4
Carmarthenshire	0	0	6
Denbighshire	2	4	18
Flintshire	8	8	NC
Glamorgan	4	0	7
Merioneth	4	0	4
Monmouthshire	0	4	2
Montgomeryshire	6	16	22
Pembrokshire	0	0	0
Radnorshire	0	2	8
Totals	42–46	71	163

NC=No count

counties being Anglesey (51 birds), Breconshire (39), Montgomeryshire (22) and Denbighshire (18).

More recent population estimates have been given for West Glamorgan (3–4 pairs), Monmouthshire (6), Carmarthenshire (5) and Breconshire (10), and despite the marked increase in Britain in recent years, this species is still a scarce and sporadic breeder over much of Wales with few sites supporting more than two pairs. An RSPB survey on Anglesey in 1986 estimated the county total to be 20–25 pairs with the majority on lakes greater than 5 ha in area (RSPB 1986). The Merioneth population has increased from one pair during the early 1980s to five or six pairs at four sites in 1991 (R. Thorpe, pers. comm.). In England, much of the increase has been due to the creation of flooded gravel pits, a habitat which is not widely available in Wales where the population is largely dependent on natural lakes. On some waters such as Llangorse Lake, an increase in water sports, particularly speedboats, has resulted in fewer breeding pairs and poor productivity, with only two or three pairs nesting in 1992 (Francis 1992). Breeding has still not been recorded in Pembrokeshire despite the presence of pairs on seemingly suitable waters on more than one occasion, and the species' stronghold in Wales is still the Anglesey Lakes where about 30 pairs breed each year (A. Moralee, pers. comm.). The Welsh breeding population is small, and nowhere in Wales does the density reach that encountered on many waters in lowland England. Breeding birds are still absent from Pembrokeshire and much of Caernarfonshire, Denbighshire, Carmarthenshire and Monmouthshire. The total population in the Principality is probably about 100 breeding pairs.

Traeth Lafan (Caernarfonshire) supports a nationally important moult assembly of post-breeding and non-breeding birds between July and October. Birds feed on the sea off Llanfairfechan and move to form roosting rafts between here and Aber Ogwen at high tide. Moulting birds begin to arrive in early July and peak during August/September, although large concentrations sometimes persist until November. Numbers are usually between 250 and 450 birds, with a maximum of 536 in October 1976. During the 1960s, 150–250 birds were recorded and it is likely that the site had been used as a moulting ground for some time prior to this.

It was thought that the moulting birds at Traeth Lafan wintered mainly around the Merseyside coast (Dare & Schofield 1976), but recently Fox & Roderick (1989) have shown that there is strong evidence to suggest that the moulting flocks move to Cardigan Bay during the winter months. Great Crested Grebes were first recorded here in large numbers at the turn of the century, but it is only recently that the full importance of the site has been recognised. Recently, the population has exceeded the level of national importance of 100 birds (Salmon *et al.* 1988) each winter, with 500+ recorded here in 1981. Peak numbers in recent years have been recorded between December and February, although smaller numbers are recorded during stormy weather.

Elsewhere, sizeable flocks are recorded most winters from the north Wales coast between Abergele (Denbighshire) and Point of Air (Flintshire), with larger numbers seen on occasion (e.g. 251 on 12 October 1986). It is probable that birds move between the north Wales and Lancashire coasts in response to food supply and weather (Fox & Roderick 1989). Large numbers are usually seen at Traeth Lafan in winter, particularly during severe weather (e.g. 170 on 18 January 1973).

Wintering Great Crested Grebes are much more scarce off the south Wales coast from Pembrokeshire eastwards, although it is regular in small numbers in the Burry Inlet, particularly at Whiteford Point (Glamorgan). Inland, some waters hold small numbers, particularly during hard weather, but up to 30 have been recorded on Llangorse Lake (e.g. 30 on 8 March 1971 and again in November 1982). This lake also supports sizeable numbers on occasion during spring passage (44 on 2 March 1976), and larger numbers are recorded on passage from Traeth Lafan (200 on 17 March 1973), with birds sometimes staying into the spring (e.g. 104 in full breeding plumage on 5 May 1973).

Despite the fact that this species is a common winter visitor in large numbers to the north Wales coast, it is rarely recorded from Bardsey, where all seven records up to 1983 involved single birds.

Red-necked Grebe *Podiceps grisegena*

Gwyach Yddfgoch

Scarce but regular winter visitor.

This is the rarest of the five grebes which regularly winter around the shores of Wales, and like all the other predominantly coastal visitors, it prefers sheltered localities. Although sometimes subject to hard weather influxes from the Continent, it is a very scarce winter visitor to Wales.

During the 19th century, Red-necked Grebes were described as being occasional visitors to the Dee estuary and west Anglesey, and two were seen on the Dyfi in October 1899, one of which was shot. These two birds were apparently also seen farther upriver in Montgomeryshire, the only record for the county. The first sighting in Glamorgan was on 6 March 1833 when one was killed on Newton Pool but, following one more record from the last century, there were no subsequent reports until one was seen at Llanishen Reservoir on 8 February 1922, to be joined by another bird on 12 March.

Three birds were noted on the Menai Straits (Caernarfonshire) every winter between 1925 and 1929 and, at the turn of the century, several were apparently killed on the millpond at Pembroke (Pembrokeshire). There were several records throughout the first half of the 20th century, mainly from coastal areas such as the Dyfi. One interesting record during this time concerns a bird shot three kilometres upriver from Aberystwyth on the Afon Rheidol (Cardiganshire) in February 1947.

Over the past 40 years, birds have been recorded from all Welsh counties with the exception of Montgomeryshire and Radnorshire. Most records are of single birds from coastal areas, but there have been several sightings from Talybont Reservoir and Llanishen Reservoir (Glamorgan) where two were seen in February and March 1956. Three displaying near Llandanwy (Merioneth) on 12 March 1987 is an unusual record, and three were seen at Llanfairfechan (Caernarfonshire) in the same month.

Monthly totals of Red-necked Grebe sightings in Wales, 1988–91

Jan	*Feb*	*Mar*	*Apr*	*May*	*Jun*	*Jul*	*Aug*	*Sep*	*Oct*	*Nov*	*Dec*
16	10	9	2	1	0	1	0	1	6	9	15

This species is now a regular winter visitor around the coasts of Anglesey, where it is sometimes encountered on some of the inland waters, and to Conwy Bay. Elsewhere it remains a scarce winter visitor, and it has not yet been recorded from the offshore islands.

Slavonian Grebe *Podiceps auritus*

Gwyach Gorniog

A scarce and local winter visitor, chiefly to coastal waters, occasionally recorded on passage. Rare inland.

The Slavonian Grebe has a more northerly coastal winter distribution compared with Red-necked and Black-necked Grebes in Britain, although it is a much more common visitor to the Welsh coast than the former. It is a comparatively conspicuous species, which visits traditional sites each winter, often for long periods. The small Scottish breeding population probably accounts for some of the birds wintering in Wales, but others may originate from Iceland, Scandinavia and Russia.

In north Wales at the turn of the century this species was virtually restricted to the Merioneth coast where it was said to be more common off Barmouth than the Great Crested Grebe. It had, however, been recorded from Anglesey (one obtained near Holyhead in winter 1884–85) and Flintshire (one obtained on the Dee in January 1839), and one was even sighted on the River Severn at Welshpool in December 1900. In Pembrokeshire, Slavonian Grebes were reported to be regular visitors, although there are no confirmed records, and a very tame bird is reputed to have visited Penllergaer (Glamorgan) regularly for over a month. The first confirmed record for this county, however, was in January 1887 when one was killed near Wenvoe, and the first record from the neighbouring county of Monmouthshire occurred in the 1890s when a female was shot at Marshfield.

For over half a century, from approximately 1910 onwards, Slavonian Grebes appear to have been scarce visitors to the Merioneth coast, while they were recorded with increasing frequency elsewhere. A pair was recorded on Roath Park Lake in Cardiff (Glamorgan) on 24 January 1912 (Ingram & Salmon 1920), and in the summer of 1932 a pair in full breeding plumage was present on Llangorse Lake. Two birds were shot on Lake Vyrnwy (Montgomeryshire) on 6 February 1912, and another was shot there a few days later on 21 February. One was present on the Afon Ystwyth on 11 February 1947 and one on the Afon Rheidol (both Cardiganshire) during hard weather on 20 January 1963.

The species continues to be a scarce visitor to Flintshire, there are no recent records for Montgomeryshire, and it has never been sighted in Radnorshire. Elsewhere however, the number of records has increased and it is regarded as being a regular visitor to several traditional sites around the Welsh coast. In Glamorgan, the number of sightings has increased markedly since the Second World War, with Llanishen Reservoir, Lisvane Reservoir and Kenfig Pool being favoured sites during the 1960s, but more recently this species has become a regular winter visitor to the north Gower coast (ten off Whiteford in March 1987). In Carmarthenshire, a party of nine grebes, almost certainly Slavonian, was present on the Gwendraeth estuary on 3 January 1950, but recent records have mainly involved single birds.

Slavonian Grebes are now recorded each winter from the coast of Merioneth, with a remarkable gathering of 21 off Llandanwg on 12 March 1987. Fourteen were present at Morfa Harlech in December 1989, and eight were recorded near Criccieth (Caernarfonshire) on 18 October 1989. There is little doubt that this section of north Cardigan Bay has become an important wintering site for this species, as has Conwy Bay between the north Caernarfonshire coast and Anglesey. Three were recorded in this area, near Penmon on 18 February 1989 and ten were observed at Llanfairfechan in December 1991.

Another regular wintering site in Wales is on the west coast of Anglesey near Llanddwyn. Most of the records from this area during the 1960s and early 1970s related to one or two birds, but up to nine have been seen here recently. At this site, maxima invariably occur in January–February (Fox & Roderick 1989), perhaps associated with hard weather. Slavonian Grebes are now regular visitors in small numbers to the Pembrokeshire coast, usually at St. Bride's Bay (up to four) and Angle Bay (up to seven).

The most regular inland sites for this species are Llangorse Lake and Talybont Reservoir, the most recent record being two birds seen at the former site on 29 October 1989. The only record of this species from the offshore islands of Wales is a single bird seen during severe weather off Bardsey between 28 and 30 January 1963.

Black-necked Grebe *Podiceps nigricollis*
Gwyach Yddfddu

Rare winter visitor to coastal waters, occasionally recorded on passage. A regular visitor in small numbers only to Anglesey and Caernarfonshire. Has bred on Anglesey.

Black-necked Grebes frequent well vegetated, eutrophic lakes during the breeding season, often with very little open water. This habitat is rare in Britain, and particularly in Wales, although they may be under-recorded due to the nature of the lakes they frequent. Small colonies sometimes build up at favoured locations, but rarely last for more than a few years.

The first breeding record for Great Britain came from Llyn Llywenan on Anglesey when Oldham and Cummings discovered at least five pairs feeding young and one additional pair on 8 June 1904 (T. A. Coward, Diaries). It is probable that pairs bred here prior to this, and Forrest (1907) stated that "Eared Grebes" used to nest on Llyn Maelog (Anglesey). However, there is confusion as to whether this record applies to Great Crested or Black-necked Grebes. In 1905 at least eight pairs were present on Llyn Llywenan, and breeding was regularly recorded at this site until at least 1934, and probably later.

In 1904 Coward noted that one observer had seen young birds that were suspected to be Black-necked Grebes on Llyn Traffwll. The observer, however, could not confirm their identity as he was unable to get close enough to the birds.

Black-necked Grebes bred on Llyn Llywenan again in 1957, and although there are no further confirmed records from this site, breeding has been suspected on more than one occasion. An adult was present all summer on Llyn Penrhyn (Anglesey) in 1981 and 1982, and more recently, a pair was present for at least a month on the same lake in 1990. Three birds were present on a private lake near Builth Wells (Breconshire) from August to September 1932, and a pair was present on this site in July 1934 but no trace was found of a nest or young. These dates coincide with a general expansion in range by this species across Europe, attributed by Kalela (1949) to the desiccation of the Steppe Lakes in the Caspian region.

In winter, the Black-necked Grebe is a scarce visitor whose distribution is much more south-westerly than both the Slavonian and Red-necked Grebes. They are fairly conspicuous birds, favouring sheltered coastal bays and estuaries, with few Welsh inland records apart from Llangorse Lake and Llandegfedd Reservoir (Monmouthshire). As very few pairs now breed in Britain, the wintering population is presumably augmented by birds that breed farther east, on the Continent.

G. H. Caton Haigh, writing in the *Zoologist* in 1890, describes how he shot six out of eight Black-necked Grebes (or "Eared Grebes" as he called them) on the Glaslyn estuary (Merioneth) between January and March 1890. He described the species as being much less shy than other grebes and consequently easier to shoot. At that time, this species occurred quite regularly on the sea around the north Wales coast but appeared to be much scarcer in the south. Indeed, the first confirmed record for this species south of the Dyfi estuary is a bird seen inland on the Wye near Builth Wells (Breconshire) on 9 April 1905.

At the turn of the century, it was regularly recorded from the Merioneth coast, particularly around the Glaslyn estuary, and on the Dee Marshes (Flintshire). On occasion, it was also encountered inland in Flintshire, and two were killed on Hanmer Pool in 1864. However, written records after the first few years of this century are scarce, and it was not until the 1920s and 1930s that birds were recorded in most of the south Wales counties. This increase in records coincides with the growth of an extraordinary colony at Lough Funshinagh in Ireland, which numbered some 250 breeding pairs in 1930, but which petered out once the site was drained a few years later (Hutchinson 1989).

The first confirmed records for Glamorgan, Monmouthshire, Pembrokeshire, Cardiganshire and Carmarthenshire fell in 1921, 1923, 1925, 1929 and 1938 respectively, although there appears to have been a decline in the number of wintering birds seen around the Merioneth coast over the same period.

Black-necked Grebes are still scarce off much of the Welsh coast, and are regularly encountered only in two or three of the more sheltered bays. From the winter of 1953–54 onwards, birds were recorded from the Whiteford area of the Gower each year, with a maximum of eight on 8 March 1970. Numbers decreased during the late 1970s and 1980s, however, and this species is now rarely recorded there.

Prior to 1956, there are no firm records of Black-necked

Grebes having been recorded in Caernarfonshire, but it is now regularly seen offshore between the Great Orme and Penmon (Anglesey). Between three and eight birds are seen here most winters, with peak numbers generally recorded in November and February (Fox & Roderick 1989). The presence of 24 birds here on 20 February 1977 may indicate that the species is under-recorded and that this site is of much greater importance than was previously thought.

A few birds are usually recorded in Beddmanarch Bay, the Inland Sea and Holyhead Harbour (Anglesey) and around the Merioneth coast (e.g. four at Morfa Harlech on 4 March 1971) most winters, and it has been recorded every year since 1982 in Pembrokeshire. Elsewhere, it is still a scarce visitor with sightings rarely involving more than two birds. It has not been recorded in Montgomeryshire or Radnorshire, but there have been 24 records from Breconshire, the majority from Llangorse Lake.

Black-browed Albatross *Diomedea melanophris*
Albatros Aelddu

A vagrant. The only accepted record to date is in 1990:

19 August 1990 – One, Skokholm

An Albatross flying south of Bardsey on 1 May 1976 was thought to have been Black-browed.

Fulmar *Fulmarus glacialis*
Aderyn-Drycin y Graig

A numerous species, breeding in all coastal counties except Monmouthshire. Frequent offshore in most months.

It is strange for us now, at the end of the 20th century, to realise that the Fulmar – such a numerous and familiar bird around the Welsh coasts – was virtually unknown here at the turn of the century. Mathew (1894) could document only one record for Pembrokeshire, from Tenby in December 1890, when one was brought in to shore on a cod line. One was found dead at Harlech in 1890 and Hutchins, the bird stuffer from Aberystwyth, claimed in 1894 that he had handled half a dozen or so in previous years. In north Wales, Forrest (1907) could list only vagrant occurrences on the Merioneth coast and only one from the north coast – an individual shot near Rhyl around 1860.

The situation today is very different and few bird species are better documented from the time of their first colonisation of Wales than the Fulmar. The dramatic explosion of breeding numbers away from the single original British breeding site on St. Kilda has been documented in detail by Fisher (1952), who showed that they had evidently been established there for many hundreds of years. For reasons that are not wholly understood, but speculated to include a relationship with rapid increases of available fish offal at sea, evolution of special genotype and subtle oceanographic factors, the Fulmar population started to increase rapidly in the mid-19th century. In 1878, the first birds were found breeding away from St. Kilda (Foula, Shetland – probably from Suduroy in the Faeroes) and thereafter the colonisation, first of the northern islands and then the mainland was underway. As the spread of prospecting and then breeding Fulmars moved inexorably southwards at a remarkable speed (they had reached the Isle of Man by 1930) they were clearly destined to colonise Wales. There was also very early colonisation on the Saltee Is. (Co. Wexford) in 1930, which suggests that the colonisation of Wales may eventually have occurred by both routes round Ireland. From around 1930 records of Fulmars around the Welsh coast became more frequent and by the late 1930s and early 1940s prospecting was taking place at one or two sites in north Wales – 1937, 1938 and 1939 at Great Orme; 1938 and 1939 at South Stack. In south Wales birds had been ashore at Elegug Stacks as early as 1931 with up to seven sites occupied (Fisher 1952). However, it was not until the 1940s that breeding took place, when a pair with an egg was discovered on the Great Orme in May 1945 by which time prospecting birds were also being recorded at St. Tudwal's Is., Little Orme and the Pembrokeshire and Cardiganshire cliffs. Two years after the nesting on Great Orme, the first pair in Cardiganshire bred on Newquay Head and two years later again at Mathry and Skomer (although Saunders (in litt.) believes breeding may have begun there as early as 1940).

By 1964 Fisher (1966) was able to list some 45 sites in Wales where Fulmars were present during the breeding season, although nesting was only proved at just over half of these.

The increase of nesting Fulmars after the initial colonisation is further exemplified by the distribution of breeding pairs in Pembrokeshire as shown by the results of the breeding atlas (Donovan & Rees, in press). This shows only 20 unoccupied coastal tetrads compared with 75 occupied ones on this cliff-dominated coastline. The distribution here, where most of the cliffs are very suitable for breeding, is denser than on other coastlines but the pattern is otherwise similar where nesting sites are available, especially in Anglesey, Cardiganshire (now over 250 pairs) and Llŷn.

The spread still continues, with breeding for the first time on the cliffs of mainland Glamorgan (i.e. excluding Gower) at two sites in 1988. St. Margaret's Is. (Pembrokeshire) saw its first breeding pair in 1986, while the numbers on Skokholm (90 apparently occupied sites in 1988; 109 in 1990) and Skomer (592 apparently occupied sites in 1988; 742 in 1990) continue to rise, as at other sites on the mainland. The Skomer population rose by 50% between 1987 and 1992. In the Seabird Colony Register in 1987; 2472 apparently occupied sites were distributed in 123 localities round the Welsh coast. The bulk of these sites were in Pembrokeshire (1409) with other sizeable totals being in Cardiganshire (318),

Current breeding distribution of Fulmar round the Pembrokeshire coast, by tetrads

Caernarfonshire (c400), Anglesey (c300); Monmouthshire unsurprisingly remains the only coastal county with no breeding pairs. No more than 30 pairs nest in Merioneth, c50 pairs in Denbighshire and c20 pairs in Flintshire. In the last two counties, where the coasts are low and sandy, colonies are all in quarry sites just behind the coastline, e.g. Llandulas Quarry, Dyserth Quarry.

Nesting sites are eventually deserted in September and Fulmars are usually absent offshore through October and November. Their absence from the breeding cliffs is short-lived, however, and by late November or December the first birds are back prospecting round the nest sites. Passage off the western headlands is sometimes impressive, for example 5500 birds passed Strumble Head on 10 February 1988. The British breeding population is of the pale morph. In the high Arctic the great majority of the birds are of the dark or "blue" morph. Such individuals are not infrequently recorded off the Welsh coast, giving an indication of the origins of at least some of the birds which occur outside the breeding season.

Although Fulmars are supreme aerial masters at sea, irrespective of weather, occasionally birds are blown inland and have been recorded there from most counties, albeit rarely: e.g. Monmouthshire 1967, Breconshire 1986, Montgomeryshire 1976 (being chased by a Peregrine in the Tanat Valley, Berwyn Mountains). Fulmars occasionally prospect around cliffs a few kilometres inland, such as Aber Falls (Caernarfonshire) and Craig Aderyn (Merioneth) 7 km inland.

Cory's Shearwater *Calonectris diomedea*

Aderyn-Drycin Cory

A scarce passage species off the Welsh coast, chiefly recorded from west coast localities in late summer. It breeds on Mediterranean islands, the Canaries, the Salvages, Madeira group, Azores and Cape Verde Island and disperses in the Atlantic, wintering off South Africa.

The first contemporary record from Wales was one off Sker Point (Glamorgan) on 27 July 1975. Since then there have been 29 further records involving 34 individuals. Most of these are from Pembrokeshire (16 records, 20 individuals) but Anglesey also features well (6 records, 7 individuals); other records are from Monmouthshire (1), Glamorgan (5) and Caernarfonshire (2).

All but seven of the records are from August to September, the exceptions being in February, June (2), July (3) and October.

There are a handful of records of either this species or Great Shearwater but these are excluded from the analysis.

Harrison (1977) records Cory's Shearwater from Ipswichian Interglacial deposits in two Gower caves and thought it likely that it may have been present as a breeding species.

Great Shearwater *Puffinus gravis*

Aderyn-Drycin Mawr

A scarce passage species occasionally seen in inshore Welsh waters in summer when it ranges the north Atlantic from its breeding islands of the Tristan da Cunha group.

The first definite record for Wales was one seen off the Great Orme on 22 June 1912 and since then there have been a further 41 records involving 57 individuals, making it more frequent than the other large shearwater (Cory's) that occurs in Welsh waters.

All but four of the records have been in the period July to September, the others comprising June (1), October (2) and November (1).

As would be expected from the recent intensity of sea-watching, most of the records come from Strumble Head since 1980 and no less than 31 records of 46 individuals come from that county, excluding birds seen at sea from the Fishguard/Rosslare Ferry (23 seen on 15 September 1981). Other records come from Anglesey (1), Caernarfonshire (3) Merioneth (1), Cardiganshire (1), Carmarthenshire (1) and Glamorgan (4 records, 5 individuals).

Sooty Shearwater *Puffinus griseus*

Aderyn-Drycin Du

A fairly common passage migrant to western coastal areas, principally in the period late August to early October.

Most records come from Pembrokeshire where Lockley *et al.* (1949) regarded it as fairly regular during the early part of September, and Saunders (1976) assessed it as annual in its occurrence off the coast in extremely small numbers. The records for the 1950s to 1970s were predominantly from the Skokholm area and averaged only three birds per autumn, maximum six in 1970 and 1971. Regular and frequent observation from Strumble Head during the 1980s (Donovan & Rees in prep.) has shown, however, that far more than this pass southwards each autumn, an average of 81 per annum between 1983 and 1987, excluding sharp daily peaks in passage. These peaks all occurred following south-west gales when Sooty Shearwaters were presumably diverted downwind but beat back out to sea in procession when the wind veered to north of west. Much larger numbers pass Strumble Head in such circumstances, with maxima of 237 on 1 September 1985 and 397 on 3 September 1983.

Bardsey is the only other site from which Sooty Shearwaters are recorded with any regularity, seen most years in very small numbers. There was an unprecedented southerly movement in strong westerly gales from 9 to 17 September 1980: a total of 208 was counted with 24, 126 and 44 on 11, 12 and 13 September respectively.

Otherwise there is a scattering of recent records of very small numbers from coastal areas in Flintshire, Anglesey, Caernarfonshire, Merioneth, Cardiganshire, Glamorgan and Carmarthenshire.

The peak period is late August to early October but there are records from January and March and during the period May to December.

Manx Shearwater *Puffinus puffinus*
Aderyn-Drycin Manaw

Breeding on Pembrokeshire islands and also in Caernarfonshire. Large feeding movements and autumn passage are regular off western coasts and even well up into the Bristol Channel.

Although the Manx Shearwater has a breeding range which now spans the north Atlantic (there has been recent colonisation – since 1973 – of north America), it is principally restricted to breeding colonies on the north-west European coasts from Iceland and the Faeroe Is. in the north, to northern France. The world population is put at 260 000–330 000 pairs (Lloyd *et al.* 1991), of which it is estimated that some 94% (280 000+ pairs) breed in Britain and Ireland, with 140 000–150 000 on six Welsh islands.

Shearwaters were exploited for food in Pembrokeshire in the past and, until the Seabirds Preservation Act in 1869 and subsequently the Wild Birds Protection Act (1880), were also extensively caught and used as bait in lobster pots (Howells 1968).

R. M. Lockley and others studied the Manx Shearwater colony on Skokholm intensively for over 40 years, including the pioneering work on migration and homing ability. Here, there is a very large colony of Shearwaters, referred to by Mathew (1894) but obviously well established long before that. Lockley himself (1949) thought that there were probably 10 000 pairs on the island but this was clearly an underestimate. In 1969 a calculated figure gave a population of 30 000–40 000 pairs, which Brooke (1990) thought might be increasing slowly, or at least be comfortably maintaining its size. There has been no recent estimate of numbers but the belief is that the population is still increasing.

The adjacent island of Skomer supports an even larger colony, which Lockley thought might number 25 000 pairs, vying with Rhum in the Inner Hebrides as the largest single colony in the world and it was estimated at about 60 000 pairs in 1969 (Seafarer) but revised further upwards to 95 000 pairs by Corkhill (1973) a few years later. Perrins (1968) showed an 11% increase in numbers within a sample area of the colony by 1978 and by 1981 Brooke (1990) estimated a total in the region of 100 000 pairs. No estimate was made for the Seabird Colony Register survey, 1985–87, but in 1988 and 1989 calculations gave a population of c160 000 pairs, borne out by a further distribution survey in 1990 which produced a total of between 135 000 and 185 000 pairs (S. J. Sutcliffe, in litt.). As for Skokholm, it is suspected that numbers here may still be rising; recent evidence from Ramsey (see below) may also support this contention. Certainly the small colony on Middleholm, just inshore from Skomer, estimated at 100 pairs on the Seafarer survey (1969–70), had risen twentyfold to over 2000 pairs by 1983 (and was suggested as being even higher in 1991).

Lockley recalls that the *Modern Universal Traveller* of 1779 mentions Manx Shearwaters breeding on Ramsey and although their status in recent decades has been dubious because of rat infestation, they certainly breed there at present and by 1992 it was believed that there were some ten separate small colonies with a total of up to at least 150 pairs on the west side and north end (T. Sutton, pers. comm.). A few pairs nest on North Bishop, a kilometre or two north of Ramsey.

Mathew, on the evidence of Jeffrey, taxidermist of Tenby, suggested that Manx Shearwaters formerly bred on St. Margaret's Is. where individuals were found in crevices among the rocks in 1893. There was perhaps slightly stronger evidence for the claim that they bred on

Caldey Is. where several local residents believed this to be the case up to the 1880s, when evening flocks were seen gathering offshore. However, this phenomenon can still be seen today and these birds are clearly part of the Skokholm/Skomer population; there is often a strong inshore passage towards the islands at dusk. Thomas Dix (*Zoologist* 1869) supported the claim of breeding on St. Margarets Is. and there is some evidence (Sage 1956) of birds still coming ashore in the 1950s.

Cardigan Is. is another site which has been over-run with rats, in this case since the foundering of the *Herefordshire* in 1934. The rats were cleared by the (then) West Wales Naturalists' Trust (now Dyfed Wildlife Trust) in 1968 and during 1980–84, 50 chicks, preparatory to fledging, were introduced onto the island each year from Skomer in the hope of establishing a colony. Although birds have reappeared on the island since then and one deserted egg was discovered in 1984, there has not yet been any proof of successful breeding.

In north Wales the only certain colony is that on Bardsey, where the number of pairs has probably risen from about 2500 in the 1969 survey to perhaps 4500 by 1986. In 1901 Aplin had estimated a "considerable number breeding", probably to be interpreted as several hundred. Twelve years later Ticehurst (1919–20) thought only 30–40 pairs bred but it was certainly higher in 1930 (Wilson 1931) and Hope Jones put it at about 1500 pairs in 1976.

Manx Shearwaters clearly bred on the nearby St. Tudwal's Is. for many years. Like Puffin Is. and Cardigan Is., St. Tudwal's became rat-infested and the colony there (together with the Puffin colony) was wiped out between 1951 and 1956. As long ago as 1893 T. A. Coward found "25 dead Shearwaters with their necks wrung, apparently killed by rabbit catchers". Congreve, Watmough and Lewis, were among many egg collectors who regularly pillaged these two Caernarfonshire colonies up to 1936. Watmough dug out over 35 burrows to find his two eggs on St. Tudwal's in 1936 and Baldwin Young "quickly and easily collected 18 eggs" on Bardsey in 1915. In Pembrokeshire, Congreve also collected numerous eggs from Skokholm and Skomer. Coward (Diaries) quotes a report from a correspondent in 1925 asserting "there is no doubt Shearwaters are nesting on Llanddwyn Island (Anglesey)" but there is no evidence to support this.

One of the enigmas is the extent to which Shearwaters nest, or have nested, on the mainland of Wales. Rumour has strongly associated breeding with Carmel Head (Anglesey) where it is claimed that six or so pairs were present in 1966; between 1972 and 1982 birds were heard calling on various occasions, but no proof of breeding exists. At the end of the last century Coward made claim to breeding on Penrhyn Du, the headland near Abersoch (Llŷn) but that is the only reference to this site. In Cardiganshire, Moore (1984) has detailed how he heard Manx Shearwaters at Wallog, near Aberystwyth on many occasions in the summer of 1983; he further quoted longer-term residents nearby who said that Shearwaters had been present at night every year for as long as they could remember. Penderi, farther south in the same county, is another mainland site where birds have been heard irregularly. At none of these sites has there been any proof of breeding. However, in a situation where breeding numbers are apparently still rising such possibility should not be discounted. At the same time Sutcliffe (in litt.) points out that calling Manx Shearwaters are a not uncommon feature around mainland coasts with birds – possibly sub-adults – starting to call as they drift over land, particularly between June and August. D. R. Saunders (in litt.) mentions that Manx Shearwaters are not infrequently heard overhead at Marloes village (Pembrokeshire) on misty/rainy nights in summer and points to the regular occurrence of remains at one particular fox earth nearby.

Lockley's famous experiments with the navigational and homing abilities of Manx Shearwaters from Skokholm took place between 1936 and 1939. By transporting birds from their breeding burrows, Lockley was able to demonstrate the impressive ability of these birds to return to their nest burrow, sometimes at remarkable speed, from locations which, in many cases, they had certainly never been near before. Some of the most impressive recoveries included the following journeys (Lockley 1942):

Female, released at Start Point (Devon) 2 p.m. 18 June 1936; returned to nest burrow within 9¾ hours (362 km by sea 225 km cross country).

Female, released 160 km south of Faeroe Islands 31 July 1936 and female, released in Faeroes, 1 August 1936: both back in nest burrows by 10 August.

Two females released inland at Frensham Ponds (Surrey) at dusk on 8 June 1937; found in nest burrows the following night, 352 km.

The two intrepid Frensham birds were then released at Venice on 10 July 1937; one was not seen again until the following March, the other returned to Skokholm in 14 days, 5 hours. Speculation continues as to whether they followed a sea route through the Straits of Gibraltar (5920 km) or flew overland, far away from their natural environment, across the Alps and western Europe in a direct line (1488 km).

To try and shed light on that aspect, further releases took place, high in the Alps in 1939:

Two Skokholm birds returned from Lugano in 10 and 15 days.

One from Berne in 13 days.

Nine other birds released in Switzerland were not found again on Skokholm.

In the 1950s Matthews (1953) took the navigational experiments to a further degree of refinement with successive releases of individual Skokholm birds at inland sites in Britain, under different sky conditions. Homing speeds were often impressive; several birds released at Cambridge, for example, reached Skokholm (380 km) the same night. Matthews also released birds at coastal locations, his most renowned bird being one released at Boston (Massachusetts) which found its way back to Skokholm (4880 km) in 12½ days

– famously, 10 hours before the postal news of the details of its release arrived at Skokholm! At the time (1952) it was assumed that no Manx Shearwater ever visited that part of the Atlantic; we now know that they are regular in that area but nonetheless the mystery of that impressive flight cannot be diminished.

Brooke (1990) believes that all breeding adults feed within 400 km of the breeding islands. This encompasses the whole of the Irish Sea and extensive areas of the Western Approaches, the English Channel and the Bay of Biscay. Throughout the season when the colonies are tenanted (March–September) daily feeding movements off the coasts of Cardigan Bay are large, with many thousands of birds often passing by headlands such as Strumble Head, St. Govan's (Pembrokeshire) and Aberdaron (Caernarfonshire) and similar movements up the Bristol Channel. In summer the Manx Shearwater is the most abundant species in the western, stratified part of the Irish Sea, most assumed to be from the Welsh colonies.

At either end of the breeding season such movements are further confused by the passage of birds from the northern colonies such as Copeland (Co. Down) and Rhum (Highland Region).

Once the young Manx Shearwaters leave the nesting colonies they head southwards immediately. This is the season (September) when many of the newly fledged young are at the mercy of equinoctial gales and storm-blown individuals not infrequently occur in inland counties. The successful young head south, sometimes at impressive rates. One fledgling ringed on Skokholm before leaving the island was recovered in the Canary Is. six days later. Another remarkable youngster was found on the coast of Brazil 16 days after having been ringed on Skokholm as a fledgling – a journey of 9600 km at an average daily rate of some 740 km immediately after leaving its nest burrow. The fact that British Shearwaters winter off the Atlantic coast of Brazil and northern Argentina has been well demonstrated by a large number of ringing recoveries. (It is the only European bird species to migrate regularly to South America.) The recoveries also confirm a southerly route down the eastern Atlantic seaboard before veering westwards off west Africa. The return journey is almost certainly farther west, taking at least some of the birds through the rich fishing grounds off Newfoundland. Brooke (1990) has suggested that some of these birds have been responsible for the recent establishment of breeding colonies in Newfoundland and Massachusetts.

Ringing recoveries of birds from Wales have been very numerous due to the vast numbers ringed over the years and much is now understood about their winter quarters in the south Atlantic.

The most distant ringing recovery of any British bird is a Manx Shearwater ringed as a chick on Skokholm on 9 September 1960, which was found, long dead, near Venus Bay, Great Australian Bight, South Australia on 22 November 1961.

Mediterranean Shearwater *Puffinus yelkouan*

Aderyn-Drycin Môr y Canoldir

The Mediterranean Shearwater was separated as a distinct species only in 1991. Until then it had been regarded as a race (the Balearic Shearwater) of the Manx Shearwater, P. puffinus.

It breeds in a very restricted area of the western Mediterranean on the Balearic Isles and after the breeding season it disperses more widely into the Mediterranean before entering the Atlantic. Some birds move north into the Irish Sea and the English Channel with a few reaching as far north as the coasts of Scandinavia. In the course of these movements, birds have been seen with increasing frequency in recent years off the coasts of Wales, particularly off Strumble Head where there is regular sea-watching. Here there is fairly strong passage in autumn (mainly September and October). In addition to the autumn movement, however, records at Strumble Head have occurred in all months except February, March and May (but only single records or two to three in January, April, June and December). The maximum number seen off the Pembrokeshire coast in any one year, to date, is 274. Individuals of this species are also seen off some of the coasts of north Wales, e.g. Cemlyn (Anglesey), Little Orme and Bardsey, all in autumn 1989.

Little Shearwater *Puffinus assimilis*

Aderyn-Drycin Bach

A vagrant to Wales. It breeds on the islands of the Atlantic, Pacific and Indian Oceans south of 32°N. The North Atlantic population apparently rarely wanders very far from breeding islands.

There has been a marked increase in records in the 1980s after a long gap from the first record in 1951:

7, 9 May 1951	–	One off Aberdaron (Caernarfonshire)
26 June–10 July 1981	–	Male, Skomer
21 June–25 July 1982	–	Same individual as above, Skomer
27 September 1982	–	One, Strumble Head
16 October 1983	–	One, Strumble Head
21 September 1984	–	One, Strumble Head
12 September 1987	–	Four, South Stack
15 September 1987	–	Two, South Stack
7 October 1988	–	One, Bardsey
5 November 1989	–	One, Strumble Head

The 1981–82 records from Skomer are of particular interest and have been fully described by James & Rawlings (1986). They refer to one male, which held territory in a burrow in a boulder slope, regularly calling to advertise its presence. When trapped in 1981 the bird was identified as belonging to the Madeiran race. In 1982 the same bird, confirmed by its ring number, returned to a location about 5 m from its previous one. It was not seen or heard in 1983 despite a complete search of the island.

This sequence of records marked the first, and so far only, prospecting Little Shearwater in the United Kingdom.

Wilson's Petrel *Oceanites oceanicus*

Pedryn Wilson

A vagrant.

Recorded twice in Wales:

12 September 1980	–	One, St. George's Channel (off Pembrokeshire)
3 September 1986	–	One, Strumble Head

It breeds in Antarctica and islands of the southern oceans.

Storm Petrel *Hydrobates pelagicus*

Pedryn Drycin

Breeds regularly on several island sites off the west coast, mainly in very small colonies. Skokholm holds the only population over 1000 pairs. Infrequent offshore, most likely to be seen in autumn.

The Storm Petrel, smallest of our seabirds with a body length of no more than 15.5 cm, is a truly pelagic species which approaches land only for the purpose of breeding. Because of its small size, it is very vulnerable to depredation by other avian species such as gulls or (on Skomer and Skokholm) Little Owls during the time it is on shore. For this reason it is entirely nocturnal during its land visits; it cannot exist where mammalian predators such as rats or cats are present and accordingly is confined to small offshore islands. Partly because of its nocturnal habit but also owing to the fact that the colonies are placed in a wide variety of rock cavities, stone walls and burrows

in soft ground, numbers are particularly difficult to census. In Wales the earliest birds are to be found on the breeding islands in the second half of April but most arrive in mid-May; the last birds have normally departed the colonies by the end of September.

The Storm Petrel is restricted as a breeding species to the eastern Atlantic and Mediterranean. Colonies occur from Vestmannaeyjar (Iceland) in the north to Iberia and the Canaries in the south and in the Mediterranean it breeds sparingly as far east as Sicily and Malta. In Wales there are only six to eight islands where it breeds regularly:

Caernarfonshire; Bardsey	–	probably no more than 10 pairs
Pembrokeshire; Skomer	–	100–200 pairs
Skokholm	–	6000+ pairs
Ramsey	–	probable breeding re-established around 1980, at least 3–5 apparently occupied burrows
Middleholm	–	probably breeds but very few pairs
North Bishop	–	fewer than 100 pairs
Green Scar	–	proved 1928 and 1930: current status unknown
Stack Rocks (St. Brides Bay)	–	present in the past; current status unknown
St. Margaret's Is.	–	present in the past; current status unknown. Rats present.

None of these colonies has been censused regularly and, in some cases, very rarely, if at all. Even on Skokholm the only two serious attempts to estimate the population have been in 1967–69 by D. A. Scott, using a trapping and recapture technique (calculated 6200 pairs) and 1991–92 by M. Betts with the use of an image intensifier, based on the sites located by Scott (ringing of birds on Skokholm ceased in 1976 thus Scott's methodology could not be repeated); Betts believes that his preliminary findings indicate a likely decline since 1967–69, but this is at present unquantified. Lockley's estimate of 600 pairs in 1939 was presumably an underestimate. All the other colonies are extremely small, some probably not involving more than a handful of birds. Although "first dates of known breeding" are quoted for some of the island colonies, e.g. Skokholm 1931 (Lockley), it is to be presumed that most of the main ones, at least, are long established. The small Bardsey colony was first recorded in 1926 (and egg collected) by Bark-Jones, not in 1953 (Norris) as widely quoted. There is little known about the long-term trends in the populations of this elusive species and it is quite possible that the exact breeding distribution in Wales is not even known accurately. The North Bishop colony has only recently (1985) been discovered and it is quite possible that pairs breed on some of the other islets off the west coast of Pembrokeshire. Breeding took place in the past on Ramsey but the accidental introduction of rats in the early 19th century is believed to have caused the demise of the species as a breeding bird on the island, although, in the light of the evidence above, it is just possible that a few pairs hung on until rediscovery in 1980.

There has never been conclusive proof of breeding on Grassholm over the years, despite several attempts to determine it; it seems strange if breeding has not occurred there and some evidence was forthcoming from the Seabird Colony Register surveys in 1991 when birds were shown to be present at the beginning of July. At another site in north Wales, Bark Jones (1954) recorded the remains of Storm Petrels on Ynys Gwylan Fawr (Caernarfonshire) in late spring. Elsewhere on the mainland Welsh coast, Storm Petrels are occasionally recorded on summer nights and the possibility of breeding has been inferred. Such a likelihood, however, seems remote. Storm Petrels are great wanderers and there is much night-time visiting to other colonies or prospecting for new sites, some of it established by ringing and retrapping, e.g. Skokholm to Bardsey, Skokholm to Irishtearaght (Co. Kerry); Channel Islands and Brittany to Skokholm. They also respond well to tape-lure calls even at places well removed from breeding colonies. In Wales tape-luring has taken place on Bardsey and the Deer Park at Martin's Haven (Pembrokeshire), and on Bardsey up to 44 non-breeding adults were attracted to mist-net sites in late July and early August 1979.

Away from the breeding colonies Storm Petrels are sparsely distributed in the southern Irish Sea but are more numerous south of St. George's Channel and in the Bristol Channel, particularly in the Celtic Sea frontal region south-west of the Smalls where they often associate with Common Dolphins.

Passage off the coastal headlands in the west is difficult to quantify for a bird which seems to be more numerous further offshore than can normally be picked up from the headlands. Off Bardsey it is very rare in spring, and autumn only normally produces 10–20 each season; 125–150 in south-westerly gales from 3 to 5 September 1984 was exceptional. At Strumble Head ones and twos are sporadically seen in spring and summer and the average autumn total recorded is about 20 birds although numbers in excess of 100 have been recorded in early September following south-westerly gales. Occasionally, storm-driven birds are picked up on the shore line, especially after onshore gales in early September, and have been recorded from time to time in all coastal counties. During the large wreck of Leach's Petrels in October 1952 (see next species) only nine Storm Petrels were recorded.

Occasional examples exist of storm-driven birds well inland, e.g. Breconshire, Radnorshire, Monmouthshire. Once away from British waters after the breeding season, the Storm Petrels move far away to the south Atlantic; adults are believed to winter in the waters off south and south-west Africa, with the young birds possibly staying farther north. Ringing recoveries (95 recoveries of Welsh-ringed birds to date) confirm the movement of individual birds from colony to colony but, unsurprisingly, do not shed light on wintering areas.

Leach's Petrel *Oceanodroma leucorhoa*
Pedryn Gynffon-fforchog

A scarce autumn visitor, commonest off the north Wales coast.

Early records of Leach's Petrel in Wales go back as far as the middle of the 19th century. In the present century, there have been sporadic records from scattered parts of the coasts of Wales. Most records in the past have come from the north Wales coast, especially Denbighshire and Flintshire, where wind-blown birds tend to gather in Liverpool Bay after north-westerly gales in October or November and, less frequently, in late September or early December. It is usually a distinctly scarce autumn visitor, although it may well be that it is somewhat more numerous than currently believed, on the basis of the increasing number of sightings which have been made in recent years from coastal headlands such as Point Lynas and Strumble Head with the upsurge in the popularity of sea-watching. For example, 109 passed Strumble Head on 13 September 1988 (in which year 304 birds were recorded around the Welsh coasts, principally in Flintshire and Pembrokeshire) and up to 181 were seen off Point Lynas on 1 October 1978. In most years, however, 1–10 are more usual in Pembrokeshire, with similar numbers off Anglesey. The numbers in given years are variable, depending upon the extent to which they are pushed into the North Channel of the Irish Sea by adverse winds, instead of passing down the west coast of Ireland.

The species was recorded off the Dee estuary (Flintshire) before 1899, in which year the remarkable total of 89 was reported as having been shot close inshore. In more recent years, up to 51 (20 October 1983) have been seen off Prestatyn in a single day but in other years, much dependent on the force and direction of autumn gales, records may be few or non-existent. Elsewhere along the north Wales coast individual Leach's Petrels maybe seen offshore at the same time of year, usually singly. On Bardsey it is uncommon in September and October with at least one old record (1913) for December; 32 passing south on 12 September 1988 is a maximum but more usually one to five are recorded each autumn. All other coastal counties can produce scattered autumn records, even as far up the Channel as Monmouthshire where December gales in 1989 (see below) produced a surprising 30 individuals with a further 22 in Glamorgan.

Even in the inland counties there are miscellaneous records of storm-blown birds being found. At least one or two records have been reported from all Welsh counties. Some of these relate to years when there have been notable "wrecks" of Leach's Petrels, for the species is recognised as being prone to such occasional disasters. Although the causes of such "wrecks" are not properly understood (but are popularly attributed to prolonged storms at sea) they can evidently cause extensive mortality. There was a large "wreck" of adults in 1891 producing Welsh records in a few coastal counties and as far inland as Radnorshire. The most famous "wreck" was that of late October and early November 1952 when 6700 birds, including adults and recently fledged juveniles, were found all across Britain; 900 were recorded in Wales, including 29 inland, and presumably many more than this went unrecorded (Boyd 1954). Earlier, a cluster of records for 1929 (Carmarthenshire, Cardiganshire, Glamorgan, Monmouthshire (several) and Merioneth) suggests some degree of "wreck" that year. In December 1989 another "wreck" of birds occurred, principally in south Wales where 191 birds were seen (Monmouthshire to Cardiganshire) between 17th and 24th of the month; at the same time one was found inland in Breconshire and three in north Wales (Powell 1989).

There is always speculation about the possibility of Leach's Petrels breeding, undiscovered, on offshore islands in the Irish Sea. Like Storm Petrels, these birds are great wanderers and are known to be night-time visitors to such island sites during the breeding season. One was recorded on St. Margaret's Is. in May 1902 (Walpole-Bond 1904). On Skokholm individual Leach's Petrels have been found in the Storm Petrel colony in June or July in several years – 1966, 1976, 1977 (two birds regularly for a month), 1978, 1980 and 1989. Such appearances excite much speculation although there is no real suspicion that breeding has ever been attempted. The nearest colonies remain St. Kilda and the Flannan Is. off the western coast of Scotland.

Gannet *Sula bassana*

Hugan

A common bird offshore throughout the year, most plentiful during the extended breeding season, March–October. One major breeding colony on Grassholm.

The gannetry on Grassholm, the only one in Wales, 14 km off the Pembrokeshire coast, is of relatively recent origin. It was certainly not in existence at the beginning of the last century when Grassholm was a huge puffinry and it is suggested that the Gannet colony here was initiated by the demise – at the hands of human exploitation – of the long standing colony on Lundy Island off the north Devon coast; certainly the timings were coincident even though there is no other firm evidence to support the theory. The colony was certainly well established by 1860 and quite possibly as early as 1820. H. T. Thomas (1890–91) gives a vivid description of a prolonged visit to the island in 1890 when he and three colleagues stayed on the island and studied the Gannet colony which, at the time, numbered around 200 pairs. Their visit ended after the infamous incident in which a naval party landed, shot innumerable birds and destroyed all the Gannet eggs. This incident led to questions in the House and eventually a successful prosecution taken at Haverfordwest by the RSPCA. Since then, the colony has grown steadily, although in uneven steps rather than in gradual and consistent annual increments. For example, between 1914 and 1922 Salmon and Lockley calculated the growth at around 16%; over the following two years it was at 42%, falling to 10% per annum in the next decade. Nelson (1978) has suggested that the explanation for the irregular but substantial increases can only be attributed to immigration from other overflowing colonies such as that on Little Skellig (Co. Kerry) but also probably from colonies farther north.

As shown on the map below, the original colonisation was on the western, windward, side of the island in the vicinity of the West Tump. The West Tump itself has remained one of the principal "club" areas where the non-breeding birds congregate in large numbers during the breeding season. The colony has gradually extended as numbers have increased and has spread along the remainder of the western side and inexorably up to the central ridge of the island and latterly onto the leeward slopes. The eventual limit of the expansion is conjecture and obviously depends on whether the numbers continue the increase which has now taken place unabated since the first half of the 19th century.

Counts of breeding Gannets on Grassholm (other intermediate counts are omitted)

c1820?	Possible establishment of colony	
1883	20 nests	Lort Phillips
1886	250 nests	M. D. Propert
1895	200+ nests	T. H. Thomas
1903	250–300 nests	J. H. Gurney
1922	800–1000 pairs	Capt V. Hewitt
1924	1 800–2000 pairs	C. M. Ackland and H. M. Salmon
1933	4750 nests	H. M. Salmon and R. M. Lockley
1939	5875 nests	J. Fisher and R. M. Lockley
1949	c9500–± 13%	Skokholm Bird Report
1956	10 550 pairs	R. M. Lockley
1964	15 500 pairs	J. H. Barrett and M. P. Harris
1969	16 128 pairs	Seafarer Survey
1975	18 300–18 400 pairs	C. S. Lloyd
1978	20 200–20 300 pairs ± 5%	Tickell *et al.*
1984	28 500 pairs	M. Alexander
1987	28 535 pairs	R. Prytherch

Grassholm is the only site where Gannets are ever known to have bred in Wales. Prospecting adults have lingered on and around other cliffs from time to time, e.g. Great Orme 1908 and 1941, but none has ever made an attempt to breed.

At the end of the breeding season, the young Gannets wander at random for the first few weeks before making their way southwards to winter off the west African coasts of Morocco and Senegal, where they remain to spend most of the following two to three years of immaturity. A small number of the juveniles remain in Welsh waters through the winter. Breeding and non-breeding adults disperse in the Irish Sea, Western Approaches and the North Atlantic when they leave the colony (September–October) to feed mainly within the continental shelf, but their absence from Grassholm is fairly short-lived and the first birds are again prospecting the nest sites during clement weather as early as January. Many recoveries of birds ringed as fledglings on Grassholm and recovered before they are breeding age, indicate the strongly southerly movement with many reports from Morocco, Senegal, Spanish West Africa and Mauretania. Other winter recoveries of adults – and a few sub-adults – from North Sea coasts as far north as Norway and the Faeroe Islands reflect the pelagic wanderings of these individuals outside the breeding season. Passage birds from the northern colonies are numerous in the Irish Sea and off the Welsh coast in autumn and passage is frequently very strong off headlands such as Strumble and Aberdaron (Caernarfonshire). A young bird ringed on the Lofoton Isles (Norway) was recovered in October the same year at Rhossili (Gower).

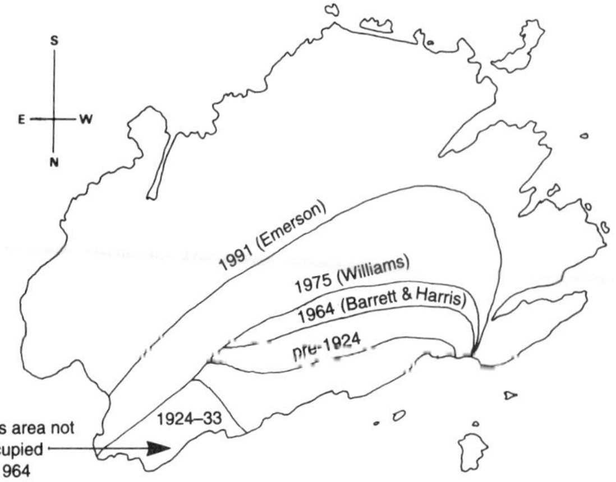

Spread of Grassholm gannetry 1924–91

All Welsh counties have produced vagrant inland records, even as far east as Radnorshire and Montgomeryshire: an adult near Knighton was shot by a farmer when it sought company with farmyard geese in June 1911; an adult was found walking across the dam at Lake Vyrnwy (Montgomeryshire) on 3 July 1982.

The gannetry on Grassholm is arguably the most spectacular bird-watching experience in Wales. Although well out to sea, it is frequently visited from mid-June onwards, and landing is possible in calm weather. The sight, sound and smell of this immense colony is an unforgettable experience, either from land or, even more so, from circumnavigation of the island; it is now the second largest colony of this species in the world (after St. Kilda, 72 km west of the Outer Hebrides) but is also the largest colony on one site (that on St. Kilda is distributed over three separate adjacent islets). Many of the Gannets that fly over and around the island can be seen to be carrying discarded fishing net and other flotsam which they build into their nest mounds; some have coloured netting tangled around their legs or neck, and others clearly die from such causes as can frequently be seen from corpses washed up on mainland shores. Young birds on the nests are often maimed, or trapped on the nest mounds by the same materials.

Cormorant *Phalacrocorax carbo*

Mulfran

Breeding resident in western counties from Pembrokeshire to Anglesey. Frequently seen offshore in other counties. A regular visitor to inland freshwaters, mainly between late summer and early spring.

Britain and Ireland support something like 14% of the breeding Cormorant population of Europe (Lloyd *et al.* 1991) which is currently believed to be about 83 000–85 000. The surveys for the Seabird Colony Register (1985-87) estimated the British and Irish population at around 11 700; in Wales the survey found c1700 pairs, an overall increase of some 19% since the original national survey (Operation Seafarer) in 1969–70, and some 15% of the British and Irish population.

Numbers of breeding Cormorants in Wales: seabird surveys 1969–70 and 1985–87 (adapted from Lloyd et al. *1991)*

	1969–70	*1985–87*
Glamorgan (Gower)	10	0
Cardiganshire and Pembrokeshire	550	650
Merioneth, Caernarfonshire and Anglesey	908	1063
	c1500	c1700

This gradual increase in numbers can probably be attributed mainly to a reduction in human persecution. Cormorants have long been regarded as pests of commercial fisheries and have been widely shot on that account. In particular, their habit of visiting stocked lakes, reservoirs or other put-and-take fisheries has led them into direct conflict with anglers who see them as competitors. Under the Wildlife and Countryside Act 1981 Cormorants can still be controlled under licence for the purpose of preventing serious damage to fisheries. However, in Wales in the 1980s the then Welsh Water Authority actively discouraged the killing of Cormorants by adopting a policy that effectively precluded control of any avian fish predators throughout Wales; this policy has to date been continued by the Welsh region of the National Rivers Authority since its inception in 1989. A small number of Cormorants is drowned in gill nets set inshore (D. Thomas, pers. comm.) but the problem is not yet of serious proportions despite a steady increase in the use of monofilament gill nets.

Cormorants do not breed in those counties with low coastlines – Flintshire, Denbighshire, Monmouthshire, Carmarthenshire; in Glamorgan a handful of pairs bred at Thurba Head on Gower up until the 1970s but ceased to do so after 1971. Breeding has been suspected from time to time in the Wharley Point and Telpyn Point areas of Carmarthenshire but not conclusively proved. The principal breeding areas around the Welsh coast are in western counties as shown on the accompanying map.

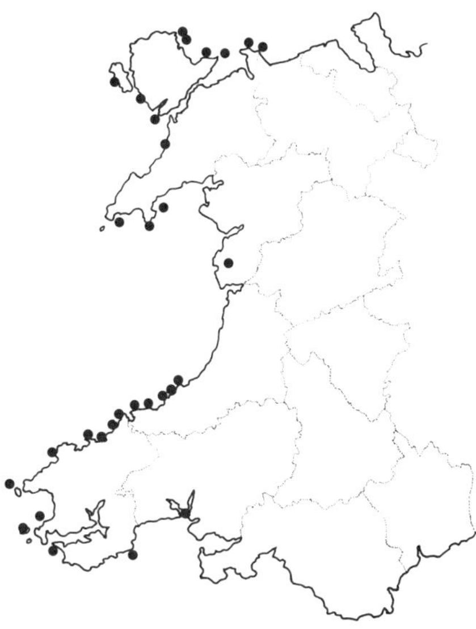

Principle Cormorant breeding colonies around the Welsh coast

Numbers vary at all these colonies from year to year, possibly in response to the availability of food, as in some seasons some adults appear not to breed. The largest colonies in Wales, those usually totalling over 100 nests, are found on St. Margaret's Is. (Pembrokeshire) – maximum 322 pairs (1973), Penderi (Cardiganshire) – maximum 176 pairs (1978 and 1982), Ynys yr Adar (Anglesey) – 116 pairs (1974), Puffin Is. (Anglesey) – 370 pairs (1986), Little Orme (Caernarfonshire) – 308 pairs (1986). The long-established colony on St. Margaret's Is. has been studied by S. J. Sutcliffe over a 25-year period from 1967 to 1992 (see figure); as long ago as 1930 Bertram Lloyd estimated a minimum of 100 pairs there.

Although it is smaller in size, the colony on Craig yr Aderyn (Merioneth) is undoubtedly the best known in Wales. Situated some 7 km inland from Tywyn (Merioneth), the rock with its colony of 30–70 nests (recent maximum 67 pairs in 1990) is a well-known and spectacular feature, presumably tenanted by the Cormorants since the sea lapped at its base in pre-historic times. Thomas Pennant in 1778 recognised Craig yr Aderyn as "The Rock of Birds, so called for the numbers of Cormorants, rock pigeons and hawks which breed on

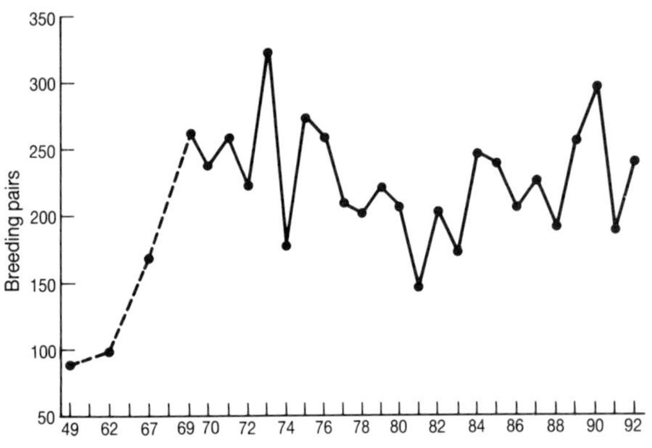

Year (1949–1992)

Numbers of breeding Cormorants, St. Margaret's Island, Pembrokeshire

it". Other inland breeding in Wales is unrecorded in modern times, although Mathew (1894) claimed breeding in trees in Pembrokeshire in the 19th century.

Outside the breeding season Cormorants are numerous all round the coasts of Wales, often abundant in estuaries and other inshore waters and frequently resorting to inland lakes, reservoirs and rivers. Notable movements are sometimes visible offshore, e.g. 154 off Aberystwyth in August 1983 and loose lines of Cormorants regularly pass southwards along the Irish Sea coast in autumn, presumably from more northerly breeding colonies. Extensive ringing, mainly on St. Margaret's Is. (Pembrokeshire) and Puffin Is., shows that adults and immature birds disperse, mainly southwards and eastwards, although an increasing number of birds now seem to be going northwards (S. J. Sutcliffe, in litt.). Some of these dispersive movements are overland. Notable recoveries of young birds in the months after fledging have been from Holland (4), Rhône delta (southern France)(1) and Italy (1). A Norwegian bird was recorded in Glamorgan in its first winter. Roosts of 100 birds or more exist in several counties, e.g. Morfa Harlech (Merioneth, maximum 235, September 1986), Whiteford Point lighthouse (Gower, maximum 113, July 1981), Friog cliffs (Merioneth, maximum 120), Wharley Point (Carmarthenshire, maximum 283, August 1974), and even in Monmouthshire (maximum 100, Piercefield, January 1988). In the Dwyryd estuary (Merioneth) a large pylon roost rose to 248 in September 1987. Of the birds that remain locally during winter, some are shown (S. J. Sutcliffe, pers. comm.) to be strongly site-faithful on inland waters.

The subspecies *P. c. sinensis* occupies a huge Eurasian range and breeds in inland and coastal locations throughout central and southern Europe as far west as the Netherlands. Because much of the population is necessarily migratory – the amount of movement depending on the severity of freezing in different winters – individuals of this "white-headed" race are not infrequent around British coasts in winter, some eventually reaching Welsh waters. Published records of individuals of this race in the

Sub-species *P. c. sinensis* (adult)

decade 1981–1990 occur for Anglesey (2), Caernarfonshire (6), Cardiganshire (2), Carmarthenshire (2), Glamorgan (19), Monmouthshire (10). The great majority (38) of these records have been between February and April with other single records in January, June and October.

Shag *Phalacrocorax aristotelis*

Mulfran Werdd

A regular breeding species on rocky coasts from Gower northwards to Little Orme.

The Shag is a marine species, confined to the north-east Atlantic with a range extending from Iceland and northern Scandinavia to the south of Portugal. (A separate subspecies (*P. a. desmarestii*) inhabits the Mediterranean and Black Sea, while another subspecies (*P.a. riggenbachi*), now exceedingly rare, is found on islets off north Morocco). Shags are usually restricted to the open sea rocky coasts, shunning large sandy bays and estuaries and other brackish waters except in severe weather and being a very rare straggler into freshwaters. The breeding population in Britain and Ireland, as shown by the Seabird Colony Register surveys (1985–87) is around 47 300 pairs – some 38% of the European total. In Wales there were only a modest 770 pairs or so (Lloyd *et al.* 1991) in that survey. However, it should be noted that, as a more fecund species than many seabirds, the number of breeding pairs fluctuates widely from year to year: 1986 appears to have been a year of particularly high numbers.

As a bird strongly associated with rocky coasts, the Shag is exclusively restricted, for breeding, to the islands, headlands and inlets of the western counties of Wales. A maximum of five pairs nest on the Worm's Head at the western end of the Gower peninsula but as the coastline of Carmarthen Bay is unsuitable for the species, the next nesting area westwards is on St. Margaret's Is. off Tenby. It is a less colonial species than the Cormorant but at least 32 loose colonies are scattered round the Pembrokeshire coast of which those on Grassholm (10–12 pairs), St. Margaret's Is. (10–20 pairs – but down to a maximum of

3 pairs in 1992) and Middleholm (20–30 pairs) are the largest. In addition, 60 pairs nest in four loose colonies in Cardiganshire. The coast of Merioneth does not offer the rocky aspect which Shags require, neither does much of the south coast of Llŷn so that the western half of that peninsula and its islands together with Carreg y Llam on the north coast are the main strongholds on the Llŷn peninsula. Anglesey has a number of scattered colonies with main concentrations in the south-west on Llanddwyn Is., Pen y Parc, Bodorgan and Ynys yr Adar and the south-east on Puffin Is. The most easterly nesting site on the north Wales coast is on Little Orme, now with only sporadic individual pairs.

Numbers of pairs of shags breeding in Wales, 1969–70 and 1985–87 (adapted from Lloyd et al. *1991)*

	1969–70	*1985–87*
Glamorgan (Gower)	2	1
Cardiganshire and Pembrokeshire	198	152
Caernarfonshire and Anglesey	349	613
	c549	c766

Numbers of pairs of shags breeding at principal Welsh colonies 1969 and 1986 (adapted from Lloyd et al. *1991)*

Colony	*1969*	*1986*
Grassholm	27	7
Skomer Is.	20	3 (breeding birds moved to Middleholm)
Middleholm	22	?
Cardigan Is.	6	26
St. Margaret's Is.	18	18
Bardsey	28	23
Ynys Gwylan Fawr	65	125
Carreg y Llam (Llŷn)	5	27
St. Tudwal's Is.	11	123
Bodorgan Head (Anglesey)	43	33
Ynys yr Adar (Anglesey)	51	28
Puffin Is.	70	113

Shags have increased markedly this century, a trend which has been particularly well recorded since 1969–70, when Operation Seafarer was the first attempt at achieving a comprehensive estimate of all British breeding seabirds. Between then and 1985–87 (the second comprehensive survey) the Welsh breeding population rose by some 39% (Lloyd *et al.* 1991) although on the western side of the Irish Sea the rise was as much as 249% in the same period. The reasons for this long-term rise are unclear, although increased legal protection in the present century both in UK and more recently in Eire (where bounties for Cormorants were removed in 1976 – many Shags were shot in mistake for Cormorants and 3527 Cormorants were known to have been shot for bounties between 1973 and 1976) and (poorly understood) links with food availability are suggested. (Lloyd *et al.* 1991). Shags feed in deeper water than Cormorants and prey mainly on free-swimming fish rather than bottom feeders.

The Shag's normal breeding season is between March and July. On the Ynys yr Adar rocks off Llanddwyn Is. in south-west Anglesey it has long been recognised (Tong 1967) that some of the Shags regularly display an abnormal breeding cycle with first eggs being laid as early as 20 November (1966) and 21 January (1967); L. S. and U. M. Venables found as many as 22 nests, 16 with sitting birds on 27 February 1968.

After the end of the breeding season Shags, especially the young ones, disperse along adjacent coasts and are seen more regularly away from the breeding areas. Even then they occur relatively infrequently off the low coastlines of Glamorgan, Carmarthenshire, Denbighshire and Flintshire. First-year recoveries of birds from Middleholm demonstrate the wide extent to which these young birds disperse: Glamorgan coast (several), Cornwall (several), Aberystwyth, Llŷn (2), Staffordshire, Northampton and Finisterre. It is only rarely that individual Shags are blown inland to turn up unexpectedly in Welsh counties as far east as Radnorshire, Montgomeryshire or Monmouthshire. Shags are to be seen in Welsh coastal waters throughout the year notwithstanding the dispersal of some young birds at the end of the breeding season, as mentioned. In autumn, birds from the Anglesey and Caernarfonshire colonies and presumably other colonies further north, pass southwards through the Irish Sea as has been shown by ringing, and this movement can be seen from western headlands. The dispersal of young is random however and further ringing recoveries indicate that some at least move northwards (nine recoveries from Scotland). Some 30 Welsh birds have been recovered from northern England and 70 from southern England. However, the great bulk of recoveries of Welsh ringed birds (646) have later been recovered off Welsh coasts.

Bittern *Botaurus stellaris*

Aderyn y Bwn

Scarce, but regular winter visitor chiefly to Anglesey, Cardiganshire, Carmarthenshire, Glamorgan and Pembrokeshire. Has bred, and birds occasionally summer in suitable areas.

Bitterns are secretive birds which require large, wet

reedbeds for breeding. Were it not for the deep, resonant booming "song" of the male, it would be very difficult to locate but suitable localities are now so few in Wales that a potential nesting pair is unlikely to be overlooked.

Prior to the 19th century, it is likely that the Bittern was a widespread resident in the marshes of lowland Wales but persecution, either accidental in the course of shooting wildfowl or for taxidermy and egg collections, together with the extensive drainage of the larger wetlands resulted in it becoming extinct before 1900 as a breeding species. Owen (1603) recorded it as breeding in Pembrokeshire, and the Bittern is often alluded to in early Welsh poetry and in the laws of Hywel Dda (940 AD). By the middle of the last century however it was confined to the larger remaining marshes of Glamorgan (Margam and Crymlyn), Anglesey (although Forrest could find no proof of breeding), and probably Cardiganshire (Cors Caron and Cors Fochno). On Gower, it was called "Bumbagus", a term specific to the Oxwich area which would indicate that folk in the area were very familiar with the bird prior to the 20th century.

There were some records from elsewhere and booming birds were heard in suitable locations on several occasions, particularly at the turn of the century. A pair may have bred in a marsh between Bala and Dolgellau (Merioneth) during the 1880s, although the evidence is inconclusive, and in 1899 a female shot on Anglesey on 13 August was thought to have been breeding. Whether these were genuine breeding records or not, it is apparent that the Bittern had become extinct as a regular breeding species in Wales long before 1900.

During the early years of the 20th century Bitterns recolonised East Anglia from the continent (Riviere 1930), and as this small population slowly increased and spread, booming was once again heard on Anglesey in 1947 (Day & Wilson 1978). No more "boomers" were recorded in Wales until 1955 when two were present on Anglesey, and one or two were heard on the island annually until 1966 in which year booming was recorded at five sites. The population reached 5–6 presumed breeding pairs (although up to 11 "boomers" were present) until the mid–70s. The first and only confirmed breeding attempt was in 1968 when a nest containing five young was found at Llyn Traffwll on 3 June (I. Wynmclean, pers. comm.). The maximum number of breeding records was in 1973 when up to six pairs were present on Anglesey and one pair was present at Oxwich Marsh NNR. Booming males were also present at the latter site in 1969–74 (two in 1972) and again in 1979. A pair was present at the Witchett Pool in 1973, but one of the birds was accidentally shot in April.

Nationally, the breeding population collapsed during the 1960s and 1970s, but breeding probably continued sporadically on Anglesey until the mid-1980s, the principal sites being Llyn Llywenan, Bodgylched, Penrhyn, Dinam, Garreg-lwyd, Traffwll, Maelog, Padrig and Coron. It is not known whether breeding occurred at Oxwich Marsh NNR during this period but certainly this reedbed and several of the Anglesey ones are now probably too dry to support nesting Bitterns and most other Welsh reedbeds are probably too small (G. Tyler, pers. comm.).

Bitterns are recorded away from the main breeding areas outside the nesting season, the number each year being dependent on the severity of the weather (Bibby 1981). In hard winters when large numbers often perish, there is an apparent movement of birds westwards into Wales from north-west Europe and eastern England. At such times, they are seen in a wider variety of wetland habitats than during the breeding season and are therefore recorded more frequently, with birds noted almost annually from several Welsh counties.

During the last century, this species was encountered more commonly than it is at present, particularly on the Anglesey marshes, Dyfi estuary and Cors Caron. During the severe weather of 1842 a taxidermist in Pembrokeshire obtained no less than ten which had all been shot in one week in the county, and Hutchings, the taxidermist at Aberystwyth, received 12 birds in the winter of 1925. It was recorded in all Welsh counties, although the number of birds and frequency of sightings decreased markedly from about 1910 onwards until the Anglesey breeding population became established.

Two unusual records are of birds killed by trains; one on Cors Caron in February 1954, and one killed by a night train between Newtown and Llanidloes (Montgomeryshire) on 6 January 1951. An alleged albino Bittern seen in Pembrokeshire near Solva in 1886, is thought to have been a Spoonbill and a bird caught near Penybont (Radnorshire) was kept in captivity for almost three years before it died.

Influxes are still recorded during hard winters with unusually high numbers seen in the Principality during the cold weather of early 1982. At that time, at least 27 birds were reported from south-east and south-west Wales. Today, Bitterns are reported in most winters from favoured sites such as Anglesey, Teifi Marshes (Cardiganshire/Pembrokeshire), Oxwich Marsh NNR, Kenfig NNR, the Witchett Pool and Llangorse Lake. Between 1988 and 1991, 44 sightings of Bitterns were reported in Wales, with a maximum of 14 in 1989.

Evidence of the influx of Continental birds into Wales is provided by the recovery of a bird ringed at Kalmthout (Belgium) on 23 April 1957 at Kenfig (Glamorgan) on 2 October 1957, a distance of 568 km west.

American Bittern *Botaurus lentiginosus*
Aderyn-bwn America

A trans-Atlantic vagrant to Wales.

It breeds in North America and winters in the southern USA to Central America. It was formerly more frequent, with about 50 British records before 1958 and only 10 between 1958 and 1991, two of which were from Wales.

There are six Welsh records in all:

About 1851	– One shot in Anglesey
October 1872	– One shot at St. David's (Pembrokeshire)

11 December 1905	– Male shot near Dale (Pembrokeshire)
19 October 1946	– One seen, Ramsey
12–16 September 1962	– One, Bardsey (found dying on the 15th, skin preserved in National Museum of Wales)
29 October 1981 – 7 January 1982	– One, Magor Reserve (Monmouthshire) enjoyed by many hundreds of birdwatchers

---

Little Bittern *Ixobrychus minutus*

Aderyn-bwn Leiaf

A rare visitor to Wales.

It breeds in most of Europe (except Scandinavia, Britain and Ireland) and eastwards to Sinkiang, also central and southern Africa and Australia. The European population winters in tropical Africa.

There are 33 Welsh records of which, depressingly, 24 refer to dead birds, most of them shot during the 19th or early 20th centuries. All records are of single birds except for two shot near Arenig Fach (Merioneth) in 1867 or 1868. There are 13 records in the period 1958–88, nine of which refer to live birds. Of the nine, four (all in Glamorgan) have involved stays of two weeks or more in freshwater marshes, where this skulking species has presumably found suitable feeding conditions and sufficient cover.

The dated occurrences cover the period 28 January to 4 October, with 11 of the total of 17 concentrated in April, May and June. These spring records are mostly overshooting migrants to the south coast of Wales, Glamorgan and Pembrokeshire being particularly favoured. The only other recent record is from Flintshire. Historical records come from Breconshire, Merioneth, Anglesey, Cardiganshire and Carmarthenshire as well as from Pembrokeshire, Glamorgan and Flintshire.

Night Heron *Nycticorax nycticorax*

Crëyr y Nos

A rare visitor to Wales, with seven records in the 19th century and 22 records to date in the 20th century.

Predictably, all seven early records involved shot specimens but slightly more heartening is the fact that two out of three roosting in a tree within 30 yards of St. David's Cathedral (Pembrokeshire) in 1876 escaped that fate, as did one out of two near Cydweli (Carmarthenshire) a few years later.

The Night Heron breeds in Iberia, north-west Africa, France and Netherlands eastwards to Japan, also southern Africa and the Americas. The European population winters in Africa south of the Sahara. Fifteen of the recent records from Wales are in the period mid-March to June, typical of overshooting migrants.

It was considered that an immature bird that frequented the Menai Bridge (Anglesey) in February and March 1960, and possibly earlier, may have been an escape from Edinburgh Zoo where there was a breeding colony in an aviary with no roof since 1951. There is no real evidence either way on this.

Post 1900 records come from Glamorgan (4), Caernarfonshire (2), Anglesey (2), Cardiganshire (1), Pembrokeshire (8), Carmarthenshire (1) and Monmouthshire (1), with a series of records (3) from the Teifi Marshes (Pembrokeshire/Cardiganshire). Most of the records are of single birds but there were two at Oxwich on 6 May 1990, with two on the Teifi Marshes on 20 March and 29 April 1990 and two and one dead on 6 April 1990. There was a substantial influx into south-west Wales in March–April 1990.

Squacco Heron *Ardeola ralloides*

Crëyr Melyn

The Squacco Heron is a vagrant to Wales.

It breeds in Iberia and north-west Africa eastwards to south-west Asia, also east, central and southern Africa. It winters in Africa south of the Sahara.

There are only five records for Wales:

August 1828	– One shot at Furnace (Denbighshire); specimen went to British Museum
3 May 1867	– One shot at Clyro (Radnorshire)
about 1875	– One shot at Garthmyl (Montgomeryshire)
17–30 May 1954	– One, possibly present three days earlier at Nottage Court, Porthcawl (Glamorgan)
11 June 1988	– One seen at Cemlyn Bay (Anglesey)

Numbers in much of southern Europe have declined drastically, with fewer than 250 pairs in the whole of Spain in 1988 so there is increasingly less chance of further records from Wales.

Cattle Egret *Bubulcus ibis*

Crëyr y Gwartheg

A vagrant to Wales.

There are no Welsh records of this small, stocky heron until 1980–81 when in common with other parts of Britain there was a scattering of sightings. The closest breeding areas are southern France and Iberia where breeding numbers are increasing so occurrences in Britain can be expected to follow suit. The species is not a true migrant but is a known dispersive wanderer.

The series of Welsh records is as follows:

11 December 1980 – 17 January 1981	– One, Haroldston Chins, Redstone Cross and Stackpole area, Pembrokeshire (presumed same individual)
15 January – 24 May 1981	– One, Llyn Alaw (Anglesey)
25 January – 13 March 1981	– One, Aber (Caernarfonshire)
1 March 1981	– One juvenile found dead at Llandenny (Monmouthshire)
about 5 –14 April 1981	– One, Crundale (Pembrokeshire), a different bird from the previous Pembrokeshire individual

The records of single birds reported from Merioneth and Anglesey in 1971 were not submitted to the British Birds Rarities Committee for assessment and are not therefore included in the above list.

Little Egret *Egretta garzetta*

Crëyr bach

The Little Egret is an uncommon but increasing visitor to Wales with about 70 records of 78 individuals up to 1991, principally in the period from 1970 onwards, reflecting an increase and spread in its breeding range northwards in France.

It breeds from Iberia and north-west Africa through the Mediterranean, eastwards to southern Asia and Australia, also southern Africa, and the bulk of the European population winters in Africa south of the Sahara.

Three-quarters of the records are concentrated in the period April to early July which is as one would expect of a south European species overshooting its breeding range. There are also a few autumn and winter records including some long-stay individuals in favoured estuarine habitats e.g. one at different localities on Anglesey from 5 August 1961 until found dead on 31 December 1961 and two in Pembrokeshire from October 1969 until mid-December when one disappeared, believed shot, the other remaining in the Sandy Haven/Gann estuary area until 10 April 1971.

An unprecedented influx of Little Egrets occurred in Britain and Ireland in the latter half of 1989, probably originating from the recent year-round population increase in Brittany (Cambridge & Parr 1992). In Wales the most notable feature of the influx was the series of records from the Burry Inlet from 6 August until at least 29 September, with a maximum of six birds at Wernffrwd and Penclawdd (Glamorgan) from 11–13 August.

There was also a good summer in 1990 at the Burry Inlet with a maximum of five birds at Penclacwydd (Carmarthenshire) on 13 August.

Again, as one would expect, the majority of the records are from the south coast counties with three, quarters of the total from Glamorgan, Carmarthenshire and Pembrokeshire, the remainder from Monmouthshire and the west coast counties of Cardiganshire, Merioneth, Caernarfonshire, Anglesey and Flintshire.

Great White Egret *Egretta alba*

Crëyr Mawr Gwyn

A vagrant to Wales with five records.

It breeds in small numbers in the Netherlands and from Hungary eastwards to Japan and south to Australia, Africa south of the Sahara, southern North, Central and South America. The Welsh occurrences are probably connected either with the colonisation of the Netherlands which took place during the 1970s or trans-Atlantic crossings as suspected particularly for the Penmaenpool bird.

The five records to date are:

29 May 1981	– One, Gronant (Flintshire)
21 October–10 November 1982	– One, Penmaenpool (Merioneth)
7 August 1984	– One, Minfford, Bangor (Caernarfonshire)
12 April 1988	– One, near St. Davids (Pembrokeshire)
17 October 1990	– One, Bardsey

Grey Heron *Ardea cinerea*

Crëyr Glas

Resident breeding species occurring widely around estuaries, lakes and inland watercourses; the population is increasing in some counties.

The Grey Heron is widely distributed in Wales, many heronries having been used for well over one hundred years. Breeding was formerly recorded from sea cliffs in addition to the usual sites in tall trees, and there is little doubt that, prior to the extensive drainage of wetlands and felling of woodlands adjacent to watercourses, it was a far more common species in the Principality. During the 20th century, the population appears to have been reasonably stable in most counties (Stafford 1971), short-term declines occurring only after severe winters. Despite the birds' large size and the conspicuousness of most heronries, isolated pairs and groups of two or three nests can be difficult to locate, especially when such heronries are newly established. However, the Welsh population has been well monitored this century, particularly as a result of national censuses undertaken in 1928, 1954, 1964 and 1985. Data are also collected each year, in some counties, for the BTO's census of heronries, started in 1928 by *British Birds* (Nicholson 1929).

In his *Tours of Wales* published in 1778 and 1781, Pennant mentioned a heronry amongst the nests of several other birds on the rocks at South Stack (Anglesey). He wrote:

> "The part of the Head fronting the sea is either an immense precipice or hollowed into most magnificent caves. Birds of various kinds breed in the rocks; among them are Peregrine Falcons, Shags, Herons, Razorbills and Guillemots. Their eggs are sought after for food, and are gathered by means of a man, who is lowered down by a rope held by one or more persons".

Herons evidently still nested here in 1838 when Eyton wrote: "Numbers of nests may be seen on the precipitous rocks in the neighbourhood of the South Stack lighthouse. When the young are nearly fledged, if a noise be made under the nests by striking the oars against the side of the boat, they will often spring out and fall into the sea." Cliff nesting was also recorded from the Llŷn Peninsula and the St. Bride's Bay area of Pembrokeshire during the last century.

In 1768, there is said to have been an "extensive Cranery" at Llanegryn in Merioneth, a site which is still occupied to this day, although numbers have fluctuated greatly over the years. At the end of the 19th century only three heronries were known in the county of Glamorgan, at Margam, Penrice and Hensol. Others were occupied prior to this, however, as Lord Aberdare recalls how, as a boy (c1800), he visited the nest of a Red Kite in a wood near Penrhiwceiber tenanted by Herons and Rooks. Davies (1858) referred to a "splendid heronry containing scores of herons" at the western end of the Deer Park at Dinefwr (Carmarthenshire), and prior to 1870 there was a large heronry in the same county at Llwynywermud near Llandovery, but the birds moved to Cilycwm where about 50 pairs nested in 1882. Cambridge Phillips (1899) stated that in Breconshire herons bred sparsely in scattered pairs throughout the county with no regular heronry. He did contradict himself however by mentioning that he was aware of four small heronries in the south of the county. Forrest (1907) believed that this species was most numerous in the western counties of north Wales, and he knew of no active heronries in Montgomeryshire at that time. The heronry at Llanllyr in the Aeron Valley (Cardiganshire) was said to be the largest in the county in the 1840s, and it is even mentioned in an old folk story published in 1845, which states "The boys and girls shall play in the hay fields until the herons shall leave their nests at Llanllyr".

Several Welsh heronries have been consistently occupied for over one hundred years, whereas others have been prematurely deserted as a result of tree felling, disturbance or persecution. Once heronry trees are felled, the birds tend to nest as isolated pairs for two or three years before coalescing once more, often close to the original site. This pattern often results in confusion over the number of occupied heronries within a county over a series of years, and some "new" sites are, in fact, the result of the loss of the traditional heronry. At Hensol in Glamorgan, for example, a large heronry in existence since at least 1872 was felled during the First World War. The birds dispersed to nearby sites but returned to a wood near the original location during the 1930s. This attachment of Herons to traditional sites was seemingly exemplified in 1974 when a pair nested at Allt Trefenty near Llandovery (Carmarthenshire), a site which had been extinct since about 1870!

The first national census of heronries was undertaken in 1928 (Nicholson 1929), and although coverage for all Welsh counties was not comprehensive, the results provided good baseline data on the number of heronries in the Principality. The corrected results (Burton 1957) revealed a total of 327–343 breeding pairs at 46 occupied heronries, the most important county being Breconshire with 40–49 pairs at six sites, closely followed by Caernarfonshire with 44 pairs at seven sites. Numbers in some counties had shown a sharp decline, none more so than Merioneth where the population had decreased from 50 nests at six sites in 1907, to 18–22 nests at three sites in 1928. The average size of the colonies varied greatly between counties, the smallest recorded average being three (Radnorshire), and the largest ten (Cardiganshire and Denbighshire). The Radnorshire total of eight pairs at three sites was misleading, however, as it included one site that was wholly within Shropshire but omitted another included under Herefordshire.

During the 1920s and 1930s, Herons suffered heavy persecution in many areas, particularly where they were thought to be depleting fish stocks. Scores of birds were said to have been shot on the rivers Wye and Usk in Radnorshire, Breconshire and Monmouthshire during this time, and all the birds at a small heronry near Cwm Taf (Breconshire) were shot in 1936. Persecution was undoubtedly widespread throughout Wales at this time, and during the two World Wars, several heronries were lost as timber was badly needed for the construction industry. Whereas Herons were able to withstand the latter onslaught by moving to suitable sites nearby, persecution sometimes resulted in local extinctions.

By 1954, the Welsh Heron population had increased to 483–494 breeding pairs, although the number of heronries had declined to only 39. Burton (1956) had given a total of 422–433 breeding pairs, but he had included the Monmouthshire total with the English counties, and eleven breeding pairs had been omitted from the Breconshire total. In his additions and correction to the 1954 Heronry census, Burton (1957) added three breeding pairs at one Denbighshire heronry and four breeding pairs at a Caernarfonshire colony, giving a Welsh total of 490–501 breeding pairs at 41 colonies. The most important Welsh county at that time was Carmarthenshire with 81–84 breeding pairs, closely followed by Breconshire and Caernarfonshire with 72 and 69 respectively.

Carmarthenshire also boasted the largest heronry in Wales during the survey, at Aberglasnant, with 42 occupied nests, although there were also large heronries at Picton Wood (Pembrokeshire) and Faynol Park (Caernarfonshire) with 40 and 37 occupied nests respectively.

In Carmarthenshire and Glamorgan a surprising number of heronries were found in the vicinity of old mansions, often sites which had been used for many centuries. The tall trees containing nests may have been retained by the gentry as status symbols of good family fortunes, and in the more distant past, as a provider of traditional game for the sport of falconry (Roberts 1983). Leland in his *Itinerary in Wales* (about 1536–39), observed: "There is very good hawkyne for herons on Vendraith Vetian" (River Gwendraeth Fach). During the 1954 survey, the Gwendraeth estuary was still one of five heronry sites in the county, the others located in the mid-Tywi valley, the Upper Tywi valley, the Taf estuary and the Cothi valley.

The Carmarthenshire population showed a gradual decline following the 1954 census, and the severe winters of 1962 and 1963 resulted in a marked decline throughout Britain (Stafford 1971) which was reflected in the county. After the 1962–63 winter, the number of occupied nests in Radnorshire also fell from 16 to only four (Stafford 1969), with similar declines recorded in other Welsh counties. The national census in 1964 revealed that although the total number of heronries in Wales had remained fairly constant since 1954, the number of occupied nests had decreased by over 50% to only 208 (Stafford 1979). Carmarthenshire showed the most marked decline, with only 17 nests at three heronries compared with 83 at three sites in 1954. Merioneth was the only county where the population was found to be stable, although no heronries were found in Flintshire in either the 1954 or 1964 censuses.

During the early 1960s, residues of organochlorine pesticides were discovered in the eggs and corpses of several species, with higher concentrations in Herons than any other bird (Cramp & Olney 1967). Although this probably had little direct effect on the decline caused by two successive severe winters, a sample census between 1969 and 1973 showed that numbers in Wales increased by 41% over this period (Reynolds 1974).

A great deal of data was collated annually from several counties during the 1970s and 80s, showing that by 1977, the Welsh Heron population had increased to a minimum of 711 pairs (Marquiss & Reynolds 1986). By 1985, a fourth national census revealed that the population had increased further to 750 pairs, and numbers were now higher than those found by any previous national survey. Carmarthenshire held a total of 172 occupied nests and substantial numbers were also found in Cardiganshire (108) and Glamorgan (92). The largest single heronry was Alltygaer (Carmarthenshire) with 55 occupied nests. The data showed that smaller heronries had grown in size whereas the larger heronries had become smaller. The results also demonstrated that the population had not suffered a set-back as a result of severe weather during the 1981 winter, unlike the impact of the prolonged frost of 1962–63.

Since the mid-1980s, the Heron population appears to

Results of National Heron Censuses in Wales in 1928, 1954, 1964 and 1985

	1928		*1954*		*1964*		*1985*	
	Occupied heronries	*Breeding pairs*	*Occupied heronries*	*Breeding pairs*	*Occupied heronries*	*Breeding pairs*	*Occupied heronries*	*Breeding pairs*
Anglesey	2	9	4	28–29	6	20	8	26
Breconshire	6	46–49	7	72	2	15	6	37
Caernarfonshire	7	47	6	69	8	63	7	43
Cardiganshire	3	30–32	5	40–47	5	37	9	108
Carmarthenshire	5	39–40	3	81–84	3	17	15	172
Denbighshire	3	30	3	19	0	0	2	16
Glamorgan	3	29	2	26	2	12	8	92
Merioneth	3	18–22	2	7	3	7	7	53
Monmouthshire	3	27–28	3	50	5	14	5	90
Montgomeryshire	3	22–23	2	30	0	0	6	51
Pembrokshire	4	22–26	2	43	3	16	4	43
Radnorshire	3	8	2	25	1	7	3	19
Wales totals	45	327–343	41	490–501	38	208	80	750

have remained stable in most counties, Pembrokeshire being the only one where declines have been noted at several colonies during the late 1980s. Carmarthenshire is still the most important county in terms of the number of occupied nests, with well over a hundred in most years. The total Welsh population fluctuates from one year to another, but is probably between 700 and 800 pairs at present.

Herons, notably the youngsters, disperse widely in late summer and autumn, and it is during this period that they are most frequently recorded from the offshore islands. A pair actually bred on Ramsey in 1931, 1932, 1946 and 1948, and a pair with one juvenile was seen fishing the rock pools there in 1973. Forrest (1907) had recorded birds on the Skerries off north Anglesey and on Bardsey and St. Tudwal's Is. Today, post-breeding influxes are recorded from most islands, with maxima of up to seven seen together on Bardsey, and up to ten on Skokholm. One bird was recorded on Grassholm on 1 September 1954, and small parties of juveniles have been seen flying due west out to sea on several occasions.

The winter distribution of the Grey Heron is not inconsistent with that of the breeding season. Individual birds return repeatedly to specific feeding sites, particularly flooded, marshy fields, estuaries and creeks in lowland Wales. Very little has been written about the habits of Welsh herons in winter, early ornithologists concentrating on their activity during the breeding season. Forrest (1907) merely stated that they were widely distributed in the lowlands, and on occasion on some of the higher Snowdonia lakes. Birds are often found along the edges of shallow pools and marshy ground in the uplands, especially at frog-spawning times in early spring.

Although generally found in small groups of two or three birds, larger gatherings do occur at favoured sites. On 8 November 1987, 66 were present at Aber Ogwen (Caernarfonshire) and up to 16 have been recorded at Llandegfedd Reservoir and Llanfihangel Gobion (Monmouthshire), the Mawddach estuary (Merioneth), Llanrhidian (Glamorgan) and the Dee estuary. At the last site, and on the Dyfi estuary, Herons can be seen feeding on small mammals and birds such as the Water Rail, especially during exceptionally high tides. A roost of up to 23 has been noted at Llangorse Lake in the autumn, and up to 15 here and eight at Glasbury, in the same county, during the winter.

Ringing recoveries demonstrate that birds move into Wales during late autumn and early winter from the northern and western European seaboards. The longest distance recorded involved a bird ringed at Raudoy in Norway on 2 June 1979 and recovered on 16 February 1981 near Abergavenny (Monmouthshire), a distance of 1067 km south-south-west. Five recoveries of Welsh-ringed birds demonstrate small movements east and north-east into England.

Purple Heron *Ardea purpurea*

Crëyr Porffor

A scarce visitor to Wales.

This is very much a Glamorgan speciality with no less than 19 of a Welsh total of 34 records coming from that county, a reflection in part at least of the species liking for freshwater marshes with reedbeds, several of which occur along the Glamorgan coast.

The species breeds from the Netherlands, Iberia and north-west Africa eastwards to Manchuria and Indonesia, also southern Africa. The European population winters in Africa, mostly south of the Sahara.

Of the Welsh records, 80% are in spring, from April to early July, the remainder being in autumn and early winter. This is as one would expect with a species which overshoots its breeding grounds on the continent. At the most favoured site, the extensive reedbeds of Oxwich Marsh NNR on the Gower, individuals have stayed for the summer on several occasions, with two in at least two years. Breeding would be quite possible here albeit difficult to prove.

There has been a substantial increase in sightings since 1970 with six before that year and 28 since. The only record from an inland county is an historical one from Breconshire where one was shot out of three seen near Talybont on Usk some time before 1882. A wintering record from Witchett Pool, a first winter bird from 8 December 1985 to 17 January 1986, is a notable one, being the only British record over that particular period.

Black Stork *Ciconia nigra*

Ciconia Du

A vagrant.

This species, which breeds from Iberia and France eastwards to Manchuria, has been recorded five times in Wales:

3 May 1989	– One, Cwm Eigiau, Carneddau (Caernarfonshire), presumed same Foel Fras, Carneddau on 7 May
30 August – 6 September 1990	– One, Upper Teme valley (Radnorshire/Shropshire), possibly since 9 August
27-28 April 1991	– One, Skokholm
22 June 1991	– One near Carmel Head (Anglesey)
29 July 1991	– One, Skomer, same bird Marloes (Pembrokeshire) on 2 August

---

White Stork *Ciconia ciconia*

Ciconia Gwyn

A rare visitor to Wales but there has been a dramatic increase in sightings in the past 19 years.

It breeds discontinuously from Iberia and north-west Africa through central Europe to Iran, Turkistan and Manchuria; the European population winters in Africa south of the Sahara.

The first record for Wales was in May 1900 when one was shot out of three seen at Bishopston (Glamorgan). There was then a lengthy gap until the second record, from Bronant (Cardiganshire) on 9 April 1971 but one or more sightings in 13 of the next 20 years. The total number of records now stands at 28, all of single birds except the 1900 occurrence and two at Roath Park Lake (Glamorgan) on 19 November 1982. Of the 28 records, 18 are in the period 28 March to 22 June, presumably of birds overshooting their breeding areas farther south.

The distribution of records is relatively even throughout Wales, covering all counties except Breconshire.

Two ringing recoveries provide an indication of the origin of Welsh birds:

> One ringed at Frostrip (Denmark) on 16 June 1976 was seen near Ruabon (Denbighshire) on 11 March 1978 and subsequently a movement of 890 km west-south-west). One ringed at Odeborg (West Germany) on 17 July 1973 was found injured at Pendine (Carmarthenshire) on 24 September 1973 (a movement of 825 km west-south-west).

Some of the records may be the result of a re-introduction scheme in the Netherlands.

Glossy Ibis *Plegadis falcinellus*

Crymanbig Ddu

A vagrant to Wales with only 11 records, three of them dating back to the 19th century.

It breeds very discontinuously from the Balkans to southern Asia, Indonesia and Australia, southern Africa and the Caribbean; with numbers declining in Europe it is likely that records from Wales will be even more infrequent.

The records are:

1806	– "Several" obtained in Anglesey
Autumn 1834	– One shot, Slebech (Pembrokeshire)
19 April 1858	– One shot near Laugharne (Carmarthenshire)
February 1900	– One shot near Llangendeirne (Carmarthenshire)
11 October 1902	– Adult male shot between Newport and Caerleon (Monmouthshire)
1 November 1910	– One shot above Ferryside (Carmarthenshire)
23 October 1917	– One picked up, Tenby (Pembrokeshire)
2 October 1945	– Adult male shot, Malltraeth (Anglesey)
16 September to at least 6 October 1959	– Juvenile, Shotton Marsh (Flintshire)
20 August 1976	One, Gann estuary (Pembrokeshire)
Late September to 13 October 1986	– One, Skew Bridge, Cardiff (Glamorgan)

Spoonbill *Platalea leucorodia*

Llwybig

A scarce visitor to Wales, recorded mainly from estuaries and coastal lagoons and occasionally overwintering and summering.

It breeds in central and southern Europe, through to south-east Asia, and north Africa. The Dutch population winters principally on the Atlantic coast of Africa; the eastern Europe population principally on the Mediterranean and the northern tropics of Africa.

The Spoonbill is mentioned under the name "Shoveler" by George Owen in 1603 as nesting on high trees in Pembrokeshire. This name for the Spoonbill is the one generally used by the earlier naturalists and although the species is most frequently associated at present with reedbeds or ground nesting sites it is recorded also as nesting in trees such as willows and poplars. There have been no further indications of Spoonbills breeding in Wales.

In the 19th century it was described as not infrequent in coastal areas of Cardiganshire and Pembrokeshire, particularly the Dyfi estuary and Milford Haven. At the former there was a remarkable record of a flock of 14 on 16 May 1893, settled in the river opposite Glandyfi Castle where they were watched "running about restlessly on a sandy spit, and wading off to a mud-bank, shovelling up the ooze with their bills". No fewer than 11 were shot in Milford Haven, 1854-55. In trying to assess the historical records there is one danger that comes to mind in that "Spoonbill" is also a name used on occasion for the Shoveler!

There are about ten records from Wales in the first half of the 20th century but a marked increase thereafter with 52 records in the period 1950–1991. The peak arrival time is mid-May to the end of June, and October; at least four of the autumn birds have overwintered. Most of the records are of single birds but up to 4 together have been noted in the period since 1950. The contemporary records come from all coastal counties except Denbighshire, particularly from Glamorgan, Pembrokeshire and Anglesey.

There is interesting evidence of the origin of several of the birds through colour-ringing of young in the Netherlands:

24 August to about 6 September 1974, one to 15 September	– Two marked as young in July – 1974 present in Teifi estuary (Pembrokeshire/ Cardiganshire)
17-18 April 1977	– One of three adults at Wernffrwd (Glamorgan) had been colour-ringed in 1974
12 October 1988 to end of year at least	– One of two birds in the Gann estu ary area (Pem–brokeshire) had been colour ringed on 11 June 1988

Mute Swan *Cygnus olor*

Alarch Dôf

Resident breeding species, now increasing in most counties.

Historically, the Mute Swan is an indigenous species which was brought into semi-domestication in mediaeval times. Over the past few centuries, however, it has been reverting to the wild state, although many pairs are still closely associated with man and man-made waters. Today, Mute Swans breed over the whole of lowland Britain and Ireland, their distribution limited only by the fact that they rarely nest above 300 m and that waters must be large enough for their take-off runs. Their large size, white plumage and huge nest make it one of the easiest species to survey.

Much of Wales is unsuitable for this swan, recent censuses confirming the expected absence of breeding pairs over most of the Cambrian Mountains and Snowdonia. Elsewhere, it is confined to the larger lakes, ponds, coastal wetlands and canals, although some nests are located in small, open ox-bow lakes and slow-moving sections of large rivers. Many of the Welsh breeders nest in close association with man on artificial ponds and castle moats.

At the turn of the century, Forrest (1907) knew of no wild swans in north Wales other than the few pairs on Whixall Moss on the Flintshire/Cheshire border. Others, he stated, bred in an almost wild state near Pwllheli (Caernarfonshire), and Mathew (1894) mentioned only semi-domesticated birds breeding in Pembrokeshire. Prior to this, Richard Fenton in his *tours in Wales* 1804–1813 quoted the eminent naturalist Edward Llwyd (1660–1709) who described two small lakes in Pembrey (Carmarthenshire) known as the Swan Pool. During Llwyd's time, up to 80 Mute Swans were kept here by the Lord of the Manor, and it appears to have been a haven for other wildfowl also. Unfortunately the pond was drained and filled for the construction of the Pembrey airfield immediately prior to the Second World War.

Encouraged by the introduction of birds onto ornamental lakes and ponds, the Welsh Mute Swan population increased during the first half of the present century, but no national census was undertaken until 1955–56 (Campbell 1960). Since then, national surveys have been carried out in 1978 (Ogilvie 1981), 1983 (Ogilvie 1986) and 1990 (Delany *et al.* 1992). Unfortunately Eltringham's 1961 census did not cover any of the Welsh counties.

In 1955–56, 228 breeding adults were located with a further 459 non-breeding birds, giving a total of 687 individuals (Campbell 1960). Monmouthshire was the most important county with 17 breeding pairs, but nowhere in Wales did the density of nesting pairs reach the levels found in some areas of south-east England. The majority of Welsh nests were located on canals or small pools, although tidal estuarine reedbeds also held significant numbers in many western counties.

A number of observers reported a decline in local populations following the severe winter of 1962–63, and the results of the 1978 survey confirmed that in Britain as a whole, the Mute Swan population had decreased by about 15% (Ogilvie 1981). Coverage in 1955–56 and 1978 varied greatly from one county to the next, therefore the figures presented in the table are based on a revision of Campbell's figures for 1955–56.

In 1978, the Welsh population consisted of 158 breeding adults and 228 non-breeding birds, a total of 386 individuals. Ogilvie's revised estimate, allowing for poor coverage in several counties, was 586 individuals. Anglesey, Caernarfonshire, Montgomeryshire and Radnorshire were the only Welsh counties showing an increase in the population since 1955–56, with Monmouthshire showing a marked decrease. However, the number of actual breeding pairs had increased in this county as it had in Cardiganshire, Radnorshire and Glamorgan. The most important areas for this species were found to be the Usk Valley (Breconshire and Monmouthshire) and the Severn Valley (Montgomeryshire), with Monmouthshire and Glamorgan holding the highest number of breeding adults. It should be noted however, that the breeding population fluctuates widely from one year to the next.

The next national census was carried out in spring 1983 as a contribution to a programme of monitoring the effects of lead poisoning from anglers' weights on the Mute Swan population. Again, full coverage was not achieved, therefore extrapolation of the data was necessary. The results showed that over the country as a whole numbers had increased by 7% since the 1978 survey, but remained 8% below the 1955–56 level (Ogilvie 1986). In Wales, the total number of birds recorded was 606, although a revised estimate of 700 individuals was quoted. Monmouthshire again held the highest number of birds (95), with 76 in Anglesey and Carmarthenshire. In Anglesey, an RSPB survey estimated the breeding population in 1986 to be between 11 and 19 pairs (RSPB 1986).

Totals of Mute Swans found in each old (pre-1974) county of Wales in 1955–56, 1978 and 1983 (the 1990 survey used different county boundaries for some areas of Wales)

County	*1955–56*	*1978*	*1983*
Anglesey	16	71	76
Caernarfonshire	47	75	66
Denbighshire	24	8	24
Flintshire	104	0	14
Merioneth	45	20	39
Cardiganshire	4	41	33
Pembrokeshire	50	10	36
Carmarthenshire	43	20	76
Radnorshire	4	16	8
Breconshire	60	0	65
Glamorgan	87	51	32
Monmouthshire	176	53	95
Montgomeryshire	27	33	42
Wales totals			
Unadjusted	687	398	606
Estimated	780	586	700

The results of the 1990 survey showed that the Welsh population had continued to increase during the 1980s with 95 breeding pairs located and an estimated total of 840 individual birds (Delany *et al.* 1992). The number of breeding birds however, was less than that recorded during the 1955–56 survey, and the increase was in the number of summering birds. The population in Monmouthshire had increased to 130 birds, with relatively high numbers in eastern Clwyd (Flintshire and parts of Denbighshire – 91 birds), Breconshire (84) and Pembrokeshire (75 birds). The highest densities of territorial pairs were recorded on the lower Wye (Breconshire and Radnorshire) and the River Severn (Montgomeryshire). In many areas, numbers are increasing due to the creation of artificial lakes and ponds, and nine pairs are now recorded in the Llanelli area (Carmarthenshire) compared with only two regular pairs in the 1970s.

The surveys revealed that in addition to the relatively small numbers of breeding pairs, several areas of Wales supported summer flocks of non-breeding birds. During the 1950s and 1960s, one of the most important sites was Llyn Coron on Anglesey where the summering herd reached a maximum of 54 on 21 July 1965 (Griffiths 1967). This herd regularly numbered over 40 individuals but the birds did not return after 1970, possibly due to a change in their food supply (L.S.V. & U.M. Venables 1972).

Large flocks of non-breeding birds formerly gathered on Roath Park Lake in Glamorgan (maximum c150), but numbers declined after the mid-1960s. Some of the birds moved to the nearby Lisvane Reservoir (44 on 8 August 1962), but here too, numbers dwindled. Sizeable flocks gather to moult at Llangorse Lake (up to 80) and Glasbury (up to 78) in Breconshire although recent records suggest that these flocks are largely the same birds moving between the sites in response to disturbance at

Llangorse (M. Shrubb, pers. comm.). Pwllheli Harbour in Caernarfonshire supports up to 50 birds on occasion, although this site is currently under threat from marina development. During the 1990 survey, flocks of over 30 non-breeders were recorded on the estuaries of the Rivers Dee (Flintshire), Cleddau (Pembrokeshire), Ogmore (Glamorgan) and Glaslyn (Caernarfonshire), and in July and August 1992, 64 birds were present at Aber Ogwen and 60 in Caernarfon harbour (both Caernarfonshire). The former site is now an important summering site for non-breeding birds with over 50 Mute Swans recorded here most years (N. Brown, pers. comm.).

Most wintering flocks of Mute Swan are small, with few exceeding 100. Breeding and winter distributions are almost identical, with many breeding pairs remaining on their territories throughout the year, while immature and non-breeding birds join flocks in traditional sites. In Wales, wintering herds favour low-lying lakes, slow-moving rivers and sheltered estuaries.

There are few early records of wintering flocks of Mute Swan, although Mathew (1894) mentioned that the summer flock of about 100 individuals at Stackpole Court (Pembrokeshire) dispersed to nearby creeks and estuaries during the winter months. A sizeable flock wintered on Llangorse Lake during severe weather around the 1850s, but birds were rarely recorded here in large numbers at the turn of the century. The number of wintering birds in Wales at that time reflected the small number of breeding pairs, and many met with the predictable fate described in the account of a party of six wild Mute Swans which appeared near Penarth in 1890: "They were fired at and dispersed, but one by one they all fell to the guns of their pursuers" (Heathcote *et al.* 1967).

As the breeding population gradually increased, so the size of many winter flocks also increased. Flocks of 40–50 birds were found at Llangorse Lake during the first half of this century, but many were culled by fishermen, and by the 1950s only small numbers remained. This flock has since increased however and now numbers 20–25, with a further 50 at Glasbury. Up to 60 are now recorded each winter on Barry Boating Lake and up to 30 at Port Talbot Docks (both Glamorgan).

Flocks of between 20 and 50 birds are found along many of the western estuaries, particularly the Cleddau (Pembrokeshire), the Teifi estuary (Pembrokeshire/Cardiganshire), the Mawddach and Glaslyn estuaries (Merioneth), Foryd Bay (Caernarfonshire) and many of the Anglesey lakes. The largest flocks have been recorded on the Cleddau (50), the Glaslyn Marshes (44 in February 1992), Pwllheli harbour (32 in December 1987), Aberdyfi (21 on 6 November 1931) and at one inland site on the River Severn near Newtown (Montgomeryshire). Here, 35 Mute Swans were recorded with a flock of 22 Whoopers in December 1990.

Mute Swans are rare vagrants to the Pembrokeshire islands and Bardsey, and are usually recorded between July and November (maximum of four on Bardsey on 10 July 1957).

The majority of ringing records concerning Mute Swans recovered in Wales involve relatively short distances, principally birds ringed in Cheshire or the English Midlands, with no seasonal trend to the recoveries. However, one bird ringed at Malahide, near Dublin in 1984 was recovered on the Menai Straits (Caernarfonshire) in 1985, a movement of 134 km east.

Bewick's Swan *Cygnus columbianus bewickii*
Alarch Bewick

Regular winter visitor in small numbers to Breconshire, Carmarthenshire, Monmouthshire and Radnorshire. Elsewhere recorded in larger flocks on spring passage.

The Bewick's Swan is the smallest of the three British swans, and whereas the distribution of the Whooper Swan in winter is mainly north-western, this species is found predominantly in the south of the country. They favour shallow freshwater lakes and slow-moving rivers adjacent to grasslands liable to flooding where they readily mix with other swans. In Wales, the most suitable habitat for this species is found on Anglesey, and this island was formerly the Bewick's Swan's stronghold in the Principality. Now however, small flocks are found elsewhere, particularly in low-lying fields adjacent to water-courses which are liable to flooding during the winter months.

During the 19th century, Bewick's Swans appear to have been not infrequent visitors to some coastal areas, especially during severe winters. Forrest (1907) notes that a flock of 32 was present for several days on one of Sir Pryse Pryse's lakes, at Lodge Park, Tre'r Ddol (Cardiganshire) in January 1893. Fourteen were present on Llyn Conach (Cardiganshire) for two weeks in January 1894, and up to 40 were present at times in the same county on the Dyfi estuary. A flock of about 50 was present in Pembrokeshire during the hard winter of 1890, although most historical records from this and many other Welsh counties were of small groups shot by wildfowlers.

Throughout the first half of the current century, the Bewick's Swan appears to have been a regular visitor in small numbers to a few estuaries, lowland bogs and shallow lakes. On occasions, large flocks have been encountered on the Dee estuary (60 at Sealand on 21 February 1979), but these birds rarely stay for any length of time. A small flock was also present on Llyn Ystumllyn (Caernarfonshire) until its destruction through drainage in the 1950s. At that time, Anglesey was the stronghold for this species in Wales, and totals of 40 or more birds were recorded from several sites (e.g. 40 at Malltraeth on 26 January 1951). Slightly smaller herds were recorded from the Dyfi estuary and Cors Caron, with 40–50 at the former site on 24 February 1923.

Until recently, a small flock visited Kenfig NNR each winter, the birds often mixing with Whooper Swans and moving between this site and Eglwys Nunydd Reservoir. They were first recorded here in 1919, and the herd reached a maximum of 35 on 17 January 1939. In Carmarthenshire, the first record of Bewick's Swans was on 16 January 1935 when two were seen on the Witchett Pool. Following the Second World War numbers increased rapidly, and a small flock became established on the River Tywi near Dryslwyn where over 30 were recorded most years.

Prior to 1962, there were only two records of Bewick's Swans for Monmouthshire but during the 1960s, a rapidly increasing flock became established on the Gwent Levels and the Usk Valley/Llandegfedd Reservoir. Twenty were present here during the early 1960s, but by 1973-74, the population had increased to 70 with 46 on Llandegfedd Reservoir alone on 17 November 1990. This marked increase over a relatively short time is probably due to an overspill of birds from the large flocks at the Wildfowl and Wetlands Trust reserve at Slimbridge.

In Breconshire, birds are recorded most winters, but there is no regular site. In the adjacent county of Radnorshire, however, a small flock has become established on the River Wye between Boughrood and the English border. The herd usually numbers about 30, and these birds are occasionally recorded from other sites, particularly Pwll Patti (20 here on 23 November 1981).

Bewick's Swans no longer visit Anglesey in large numbers, and in recent years, flocks have rarely exceeded half a dozen birds, although 17 were seen on Llyn Alaw on 27 October 1991. Wetlands such as the Severn/Vrynwy confluence and Hem Flash in Montgomeryshire, which formerly attracted small herds in some winters, have long been drained or canalised, and now rarely flood during the winter months.

In recent years, a strong spring and, to a lesser extent, autumn passage of Bewick's Swans has been identified, particularly in north Wales. These are thought to be birds making their way between Ireland and their northern breeding grounds, and flocks of several hundred have been recorded on occasion. On 4 November 1968, about 320 were seen flying over Brynsiencyn (Anglesey), and flocks of up to 200 are regularly recorded over the Dyfi estuary in late February and early March (R. Squires, pers. comm.). This species is a rare vagrant to the offshore islands.

Two birds ringed at the Wildfowl and Wetlands Trust Reserve, Slimbridge, were seen at Pwll Patti (Radnorshire) on 29 January 1984. Most interesting, however, is the record of a bird ringed at Slimbridge in 1982 that was later recorded for two winters at Welney (Cambridgeshire), one winter in Germany, a spring record in Holland *en route* to Siberia, and finally seen on 10 December 1987 at Llyn Alaw on Anglesey!

Whooper Swan *Cygnus cygnus*

Alarch y Gogledd

Regular winter visitor in small numbers to several Welsh counties, notably Anglesey, Caernarfonshire, Merioneth and Montgomeryshire.

The winter distribution of this species in Britain is predominantly a northern and western one. Whooper Swans are found on a wide range of different waters, from lowland pools and estuaries to oligotrophic lakes and ponds where they feed on a variety of emergent and submergent plants, grass, winter cereal and, in estuaries, on eel grass. Most of the British and Irish wintering birds probably originate from the Icelandic breeding population, although Swedish birds have also been recorded here. In Wales, this species was formerly a rare winter visitor, particularly in the southern counties, but following an increase in numbers during the 1950s and 1960s, it is now recorded annually in most Welsh counties. Even at regularly used sites, however, flocks rarely exceed 60.

During the 19th century, Whooper Swans were regularly recorded only in the northernmost Welsh counties. Forrest (1907) described it as a not uncommon visitor on flat coasts and estuaries in north Wales, with a marked influx of birds in the winter of 1890–91. At that time, flocks were present on several of the Anglesey lakes, Penrhyndeudraeth (Caernarfonshire) and Llyn Tegid (Merioneth), and a herd of 20 was seen on Afon Dysynni (Merioneth). A herd of about 40 was recorded on the Dee estuary (Flintshire) on 6 March 1890, and during hard winters, small flocks were often seen on Afon Conwy (Caernarfonshire), Llyn Tegid (Merioneth) and some of the mountain lakes on the Berwyns (Denbighshire).

It was recorded on only two occasions before 1900 in Cardiganshire and Pembrokeshire. In the former county, three were shot out of a flock of six on Cors Caron in

1855, and a cygnet was shot from a herd of 19 on the Dyfi estuary on 23 November 1897. The only record from Glamorgan during this time was of four birds seen on East Mud, Cardiff during severe weather.

Throughout the 1950s and 1960s, the numbers and range of wintering Whooper Swans in Wales increased, and a large flock of over 50 individuals (maximum 65 in 1952–53) was present on Llyn Ystumllyn (Caernarfonshire) during the early 1950s. Numbers increased on many of the Anglesey lakes and in Snowdonia the wintering population regularly moved between the larger, oligotrophic lakes and the Glaslyn Marshes. On Llyn Cwellyn 36 were recorded on 19 January 1963, and 25 were seen off the coast near Conwy on 24 November 1962. From at least 1950, a small group became established on Cors Caron, peaking at 50 in 1974–75 and 39 in 1978–79, although since then, numbers have declined considerably. Up to 16 were present on several waters in Breconshire during the severe winter of 1962–63.

The first county records of Whooper Swans in Carmarthenshire, Monmouthshire and Radnorshire all came in the 1950s and 1960s, but more recently small numbers have been recorded most years. In counties such as Flintshire, Denbighshire, Pembrokeshire, Glamorgan and Monmouthshire, small flocks are recorded from several sites some years, with larger numbers occasionally recorded during prolonged cold weather. The largest herds in Wales, however, are found in Anglesey and Caernarfonshire, with regular counts of over 50 individuals. On Anglesey, Whooper Swans regularly visit Llyn Alaw (58 in December 1987), Llyn Cefni (36 on 19 December 1989) and Llanerchymedd (57 on 31 March 1989), although numbers over the past two years have been very low. A large herd returns each winter to the Glaslyn Marshes on the Caernarfonshire/Merioneth border (79 in November 1992), where this species often mixes with Mute Swans. An international Whooper Swan census in January 1991, however, found low numbers in Wales with a total of only 45 birds recorded in five groups, with 40 of the birds found on Anglesey (Kirby *et al.* 1992).

Maximum winter counts of Whooper Swans on the Glaslyn Marshes, 1982–92

1982	*1983*	*1984*	*1985*	*1986*	*1987*	*1988*	*1989*	*1990*	*1991*	*1992*
36	36	23	12	62	80	59	59	56	69	79

Elsewhere in Wales, large flocks have been recorded in flight (e.g. 42 over Ruabon, Denbighshire, in 1970), but the only other sizeable wintering herd is found in the Severn Valley near Caersws (Montgomeryshire). In 1989–90, the flock numbered 25 birds, and as with the Glaslyn herd, was associated with large numbers of Mute Swans. This has recently led to conflict with local farmers as the birds are reputed to be damaging autumn-sown crops.

This species is a rare visitor to the offshore islands of Wales, although on occasion it is recorded in large herds. Sixteen were seen over Bardsey on 12 October 1961 and 47 were seen over Skokholm on 19 February 1956.

A bird ringed in Iceland has been recovered near Harlech (Merioneth), and a swan ringed on the Dee estuary in 1984 was recovered at Graenhlid in Iceland in 1987, both involving distances of c1500 km.

Bean Goose *Anser fabalis*

Gŵydd y Llafur

Rare and irregular winter visitor.

Between 200 and 300 Bean Geese winter in Britain each year, with up to twice as many in severe winters. It is a rare and irregular winter visitor to Wales, and is not recorded every year. Its status has changed considerably over the past one hundred years since Forrest (1907) described it as being the most frequent of the wild grey geese away from the Dee estuary. The earliest record of the species' occurrence in Wales, however, comes from excavated remains from Bacon Hole, Deep Slade, on Gower (Glamorgan). This suggests that Bean Geese may have been commoner in this part of Wales during the interglacial period.

Several observers comment that there has in the past been a great deal of confusion between this and other grey geese, particularly Pink-footed Geese. However, it is apparent that at the turn of the century, this species was much more common than it is today. It is described as being plentiful around Rhosneigr (Anglesey), and occurring most winters on the Menai Straits. At that time, Bean Geese were recorded in small flocks almost every winter on the Dee estuary (e.g. 15 on 22 January 1908) and near Penrhyndeudraeth (Merioneth) where it is reputed to have been the most frequently recorded wild goose. The Severn estuary also appears to have been a favoured haunt during this time, with large numbers recorded between Cardiff and Llanwern (Monmouthshire) between September and November 1899.

Occasionally, cold weather influxes were recorded in some areas; a large flock on the Mawddach estuary (Merioneth) in the winter of 1881 and seven or eight near Stone Hall (Pembrokeshire) during the 1880 winter being two examples.

The only county records for Caernarfonshire (one shot at Afonwen on 30 October 1898 and one preserved at Broom Hall on the Llŷn peninsula, 20 October 1898), Radnorshire (one obtained at Dolyhir on 23 December 1903) and Breconshire (two killed on the River Usk near Talybont in January 1890 and seven at Pencelli in January 1894) all come from the turn of the century. In January 1893, one was also obtained from Cors Fochno, and most subsequent records for Cardiganshire have come from the Dyfi estuary (e.g. one at Ynyshir from 8 to 11 November 1982, one at Cors Fochno on 23 January 1955). Bean Geese were supposedly common in Carmarthenshire during the 19th century, but the first confirmed record was of at least one, possibly two birds with the flock of White-fronted Geese at Dryslwyn on 28 January 1973. Most recent records from this county have again involved a few birds among the same flock of Whitefronts (e.g. five on 30 January 1976).

One unusual record involved a party of five Bean Geese flying westwards off Prestatyn (Flintshire) on 29 July 1921 (*British Birds* Vol. 15, pp. 141–142). This is a very early date for a species which normally arrives in Britain during the latter part of September or October, and may have involved feral birds.

There are very few records of Bean Geese in Wales after about 1920 and many of these will refer to escaped feral birds. More recent sightings generally involve fewer than three birds together although nine were seen at East Hook (Pembrokeshire) on 2 November 1978 and 14 at Blackpill (Glamorgan) on 7 November 1982.

Pink-footed Goose *Anser brachyrhynchus*
Gŵydd droed-binc

A scarce and irregular winter visitor, chiefly to north Wales. Subject to hard weather influxes from the Ribble Estuary.

The population of Pink-footed Geese wintering in Britain has increased from 48 000 to 172 000 during 1960-87 (Fox *et al.* 1989), with the majority in Scotland and north-west England. Despite this increase, which is thought to be mainly due to a higher survival rate, no site in Wales holds important numbers. The majority of records come from north Wales, and it is thought that many records from south and mid-Wales involve birds that have escaped from wildfowl collections.

Since this is a goose which exploits arable farmland, it is probable that it was much commoner in Wales in the 19th century when this type of agriculture was at its peak in the Principality. In the late 19th and early 20th centuries, large flocks of Pink-footed Geese were recorded on the Dee estuary. Over 400 were recorded on the marshes there on 1 December 1903, and by 1917, numbers had increased to over 1000. Forrest (1907) records that the birds generally departed in late March, but that the date often depended on the weather. At this time, it was by far the most common goose on the estuary, but by the 1930s, the flocks had moved to the potato fields of south Lancashire. Farrar (1938) thought that this was due to constant disturbance caused by the spread of industrialisation in the Dee Valley. Whatever the reason, the number of Pink-footed Geese on the estuary declined dramatically, and by the 1950s the species was present only in small numbers.

At the turn of the century, this species was found on the Dyfi estuary and the Menai Straits on occasion, and in 1901 and 1902, small flocks were recorded inland at Llandrillo (Merioneth) where, on 15 January 1901, one was shot from a flock of nine by a gamekeeper who "was after cormorants, and when he picked it up, he was much puzzled" (Bye-gones 1901).

Historical records reveal much confusion between this species and the Bean Goose, making the authenticity of some older sightings dubious. In Carmarthenshire and Glamorgan during the last century, it is said to have been recorded on several occasions, but few definite records exist. A large flock was recorded at Llanwern (Monmouthshire) between October and December 1899, and two were shot here in 1898. A female was obtained at Peterstone Wentlooge (Monmouthshire) in December 1935 when three large flocks were recorded in this area. Since that time, however, most records from both Glamorgan and Monmouthshire have involved mainly single birds, a notable exception being a flock of 12 at Peterstone on 2 November 1975.

During the 1950s and 1960s, a small flock regularly visited the lower Clwyd Valley (Flintshire/Denbighshire), but by the early 1970s, their numbers had dwindled. During severe weather, however, large numbers are still recorded in the area, and in 1979 1200 were present at Sealand on 23 February and 350 at Bodelwyddan on 25 March. In the same year, up to 500 were seen over the Gresford/Ruabon (Denbighshire) area, with smaller flocks at Rhos-on-Sea and inland at Pentrefoelas. The displacement of birds from the Ribble that year resulted in a maximum count of 316 on the Dyfi estuary on 2 February. Other, smaller flocks were recorded from various parts of Cardiganshire (e.g. 60 over Ffair Rhos on 26 January), and 16 were seen at Marloes Mere (Pembrokeshire) on 8 January and c60 at Llanferian in the same county during February. Another cold weather movement in late December 1981 was very localised to Cardiganshire. During this period, a maximum of 42 was recorded at Ynyslas on 27 December and 20 passed north over Aberporth on 23 December.

Hard weather movements of birds into Wales have been recorded in many other years, notably in 1979 and 1981. During January to March 1979, it is estimated that in excess of 3500 birds were present in north Wales, the main influx occurring in late January. The December 1981 influx was concentrated on Anglesey where in excess of 500 birds were present, with small flocks elsewhere in Wales.

North Wales experienced large cold weather influxes of Pink-footed Geese during the early months of 1987. About 300 birds, probably of this species, were recorded flying over the Menai Bridge on 26 February and a flock of 29 was seen on Llyn Traffwll (Anglesey) on 21 February. In the nearby county of Caernarfonshire, 150 were seen at Llandudno on 27 February and 63 were present in Aber Ogwen on 18 January. Eleven were recorded farther down the coast at Tonfannau (Merioneth) on 22 January. A smaller influx of birds occurred during 1989 when a flock of about 100 were seen at Gwalchmai (Anglesey), and one bird was recorded inland at Llyn Tegid (Merioneth) in December 1989; on 22 January 1991, 150 were seen at Llyn Traffwll.

Records from the three inland Welsh counties are scarce, with many of the earlier records involving birds being shot. The earliest Montgomeryshire record is of a bird which was shot at Bryntanat near Llanymynech on 18 February 1895, with all subsequent records involving fewer than six birds associating with other goose species. The only Radnorshire record is of a bird shot near Llanelwedd in January 1909, but several records exist for

Breconshire. The largest number seen together in the county is 20, on 5 December 1945 (six of which were shot) and near Glasbury on 9 February 1977.

Pink-footed Geese are rare vagrants to the Welsh islands, two being seen briefly on Skokholm (Pembrokeshire) on 27 May 1988. One seen on Skomer on 25 April 1991 was the first record for the island.

Several recoveries exist of Pink-footed Geese ringed in Iceland and recovered in Denbighshire, Flintshire and Cardiganshire. The greatest distance travelled involved a bird ringed in Iceland on 1 August 1953 and recovered in Cardiganshire on 14 February 1954, a distance of 1581 km south-south-east.

White-fronted Goose *Anser albifrons*
Gŵydd Dalcen-wen

Scarce winter visitor; regular flocks of A. a. albifrons *occur in Anglesey and Carmarthenshire only, whilst Cardiganshire and Montgomeryshire support the only regular flocks of* A. a. flavirostris. *Rare elsewhere, except occasionally during hard weather.*

Two distinct subspecies of White-fronted Goose winter in Wales, with very little overlap between the two. The Greenland White-fronted Goose (*A. a. flavirostris*) has been intensively studied in Wales in recent years, and is found each winter on the Dyfi estuary and various upland sites in Montgomeryshire. The European White-fronted Goose (*A. a. albifrons*), which breeds in the tundra belt of Eurasia, is found in Anglesey and Carmarthenshire, although scattered records of both races are reported from other areas on occasion. Since it was not until 1948 that the Greenland race was described for science (Delgety & Scott 1948), many of the early published records of this goose obviously do not distinguish between the two subspecies.

The European White-fronted Goose is more numerous in Wales than the Greenland subspecies, but both forms have declined over the past century. During this time, however, individual populations have fluctuated markedly, with many former haunts no longer suitable, and some new sites being occupied.

Anglesey

Both races of White-fronted Goose have been recorded in Anglesey, although numbers have probably never been high. At the turn of the century, it was described as being plentiful at times near Rhosneigr and one was shot from a flock of nine at Llyn Maelog on 21 October 1879. There are few subsequent records until the 1950s when Tunnicliffe (C.F. Tunnicliffe Sketchbooks, No. 2) describes seeing a flock of about 400 on Malltraeth Marsh on 2 March 1951. In the same year, he also records that two birds of the Greenland race were shot on the same marsh, and suggested that large numbers of this subspecies gathered here each winter. Prater (1981) stated that up to 200 were present on the island until the 1970s since when the European race has become most prevalent. The largest flock of Greenland White-fronted Geese recorded on the island in recent years is one of 24 on Llyn Traffwll on 11 December 1987, although flocks of up to 40 of the European race are regularly recorded from several sites including Llyn Coron, Llyn Bodgylched and Llyn Alaw (e.g. 45 on 14 December 1987).

Cardiganshire

Forrest (1907) described this species as common on the Dyfi estuary, and Salter recorded small flocks here during the latter part of the 19th century (J. H. Salter, Diaries). Salter also recorded birds on some of the nearby uplands around Pumlumon, but there are very few records from the Dyfi until the 1940s and 1950s. The majority were probably of the Greenland race, but it is not until 1954 that large numbers of this race were positively identified by W. A. Cadman in a flock of 155 White-fronted Geese of both races on Cors Fochno (Fox & Stroud 1985). A maximum of 220 Greenlands was recorded here on 3 March 1951, and an injured bird summered on the estuary in 1955. Numbers of the European race on the estuary peaked at about 200 between January and March, with exceptional numbers of up to 600 recorded some winters. By the 1960s, however, the population had declined to approximately 100 birds each winter (Ogilvie 1968), and by the 1970s, the European birds had become no more than sporadic vagrants.

A flock of White-fronted Geese wintered on Cors Caron for at least a century, and Salter recorded a flock of about 30 here during the 1890s. The fact that the birds arrived in October and left in mid-April suggests that

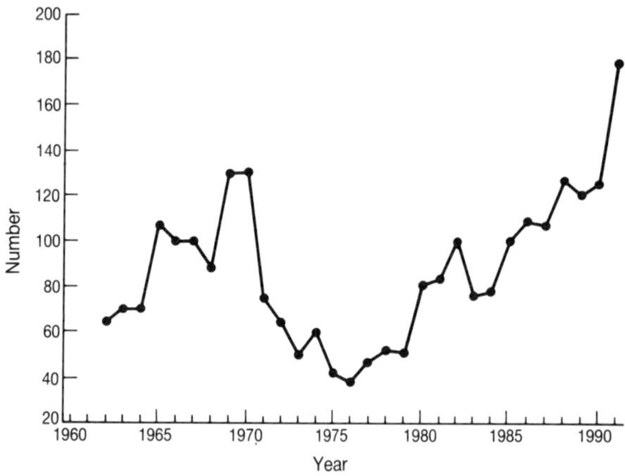

Numbers of Greenland White-fronted Geese wintering on the Dyfi estuary

they were Greenland birds, but this was not confirmed until 30 January 1947 when a bird ringed in Ikamiut, west Greenland, on 21 July 1946 was shot here (Salomonsen 1947). At this time, geese were also reported from sites around Tregaron, and in January 1940 Salter estimated the population to be about 250–300 birds. Numbers increased gradually to a maximum of 550–600, with flocks often moving to the Teifi Pools and other upland areas when disturbed. The severe winter of 1962–63, however, proved disastrous for this population as geese died of starvation because their feeding areas were frozen for weeks at a time. Many geese sought refuge in fields nearer the coast, but indiscriminate shooting and predation resulted in very high mortality. Geese near Llanon were so weak that men were killing them with sticks, and of the 600 of so which arrived in the autumn, only a third survived until the thaw (Fox & Stroud 1985). The following winter, a maximum of 117 was counted on 20 November 1963, but numbers declined subsequently and the geese ceased to winter here completely after 1967–68. Since that time, small numbers have occasionally been recorded at Cors Caron, with a maximum of 39 in December 1982, but very few since then.

It is not known whether the Cors Caron flock moved to Ireland, or whether it moved north to the Dyfi estuary, where this race was recorded regularly from the late 1960s onwards. The flock was known to move to nearby uplands on the Cardiganshire/Montgomeryshire border, but by the winter of 1975–76, they were recorded feeding mainly on the Dyfi estuary itself. A voluntary ban on shooting since 1972 in association with the declaration of the area as a National Nature Reserve probably contributed to a gradual increase in the population from 36 in January 1976 to 179 in December 1991.

Montgomeryshire

During the first two decades of the 20th century, the White-fronted Goose was regarded as an irregular visitor to this country and Forrest (1907) quoted only two records, a bird shot near the Breidden on 23 March 1901 and seven at Lake Vyrnwy on 26 April 1905. Salter frequently saw White-fronted Geese at Bugeilyn, high on the Pumlumon massif (e.g. 40 on 6 February 1936), but these were undoubtedly Greenland birds from Cardiganshire. By the late 1940s, a flock of European birds on the Severn had increased to over 500 birds, and over the next decade, reached a maximum of 1400 birds in 1958 (Montgomeryshire Field Society Reports, 1950–60). This flock was very mobile, moving mainly between Leighton and Hem Moor to feed, and onto hill lakes to the north to roost at night. Numbers generally reached a peak of 1000+ in March, but a gradual decline throughout the 1960s and 1970s resulted in a flock of fewer than 50 by the early 1980s (e.g. 45 at Leighton on 16 February 1982). The decline is thought to be due to increased disturbance (Ogilvie 1968) and loss of habitat, although the major factor is probably the fact that better feeding areas exist in the Netherlands and Germany where there have been large increases in recent years (A. Fox, pers. comm.). By the late 1980s, the regular flock had disappeared from Montgomeryshire although scattered birds are still seen on occasion. The increase during the 1940s may have been due to the displacement of birds from the Mersey (Atkinson-Willes 1963), and it is possible that during the decline in the 1960s and 1970s, the Severn valley birds moved to join the Tywi flock.

It had long been known that the flock of European birds in the Severn Valley often roosted on the upland lakes above Caersws and at Llyn Hir, but it was not until 1980-81 that a small flock of the Greenland race was identified here. On 15 October 1982, a flock of 16 was seen on Llyn Gwyddior and subsequent sightings have confirmed that the flock moved between several upland sites in the county. Llyn Gwyddior formerly held over 100 birds each winter (N. Hill-Trevor, pers. comm.), but it has recently been lost to afforestation. The presence of these birds may explain small influxes into the Dyfi and Cors Caron flocks during cold weather (Fox & Stroud 1985).

Up to 40 birds wintered on the uplands during the 1970s and early 1980s, moving between Llanbrynmair Moor, Llyn Hir/Gwyddior and the lakes above Caersws. In March 1984, the flock numbered 22–24, and on 13 November 1988, 26 were seen at Llyn Hir with one European bird. Since then, however, it appears that numbers have declined markedly with only nine seen briefly in October 1990. From the time of year when birds have been seen recently, it also appears that birds arrive in late October/early November but move on soon afterwards, possibly to the Dyfi estuary.

Carmarthenshire

Until the 1930s, European White-fronted Geese were common and regular winter visitors to the Tywyn Burrows/Gwendraeth estuary, with up to 500 and 50–100 at the two sites respectively. However, the planting of Pembrey Forest followed by the construction of an airfield prior to the Second World War resulted in rapidly dwindling numbers (D. H. V. Roberts, in litt.).

Since the late 1920s the Twyi meadows, centred mainly at Dryslwyn, have attracted large flocks of the European race. This flock increased during the time that numbers were declining at Pembrey, and approximately 500 were present in December 1953. Over 1000 were present during February counts between 1964 and 1973, the maximum recorded being 2500 on 31 January 1971. Since then numbers have decreased, with 255 present in 1979–80. However, 1100 recorded here on 3 January 1979 was probably the result of hard weather movement from the flock at Slimbridge, on the Severn estuary. Today, numbers rarely reach 100, the main reason for the decline probably being the fact that the geese have found better feeding sites in Germany and the Netherlands. Flocks of over 50 birds seen elsewhere in the county were doubtless birds from the Dryslwyn group.

Greenland White-fronted Geese are rare visitors to the county and most records involved birds from the Cors Caron flock, often amongst the Tywi Valley birds (e.g. six on 31 January 1971).

Glamorgan

Large flocks of over 2000 European birds formerly visited the Margam Moors in winter. This flock declined when the railway line was built across the main feeding areas, and disappeared altogether with the construction of the large steelworks and other associated industries. Peak numbers recorded were c2500 in 1939–40 and c2000 in 1940–41 and 1941–42. In 1946–47, numbers were down to fewer than 500 because of an increase in the level of disturbance and heavy shooting. By the 1950s, the flock had disappeared, as had a smaller flock of up to 300 birds on the Burry Inlet. Up to 100 birds were recorded on Eglwys Nunydd between 1949 and 1956, but these were probably offshoots from the main flock at Margam.

Since that time, numbers visiting the county have been small, and most were probably stragglers from the Carmarthenshire flock. Large numbers were recorded during the severe winter of 1963, however, with c250 in the Thaw Valley, c300 on the Ogmore estuary and c150 at Eglwys Nunydd. Both races are now scarce and irregular visitors to the county, and in fact the first confirmed record of a Greenland bird was at Whiteford on 23 February 1977.

Monmouthshire

European White-fronted Geese are regular winter visitors to this county, probably involving local movements of birds from Slimbridge. Flocks of up to 80 are seen on occasion with cl000 recorded over Abergavenny on 7 March 1969. A small flock of up to 50 birds was formerly recorded annually at Caldicot and Undy, but an increase in the level of disturbance has resulted in a decrease in the number of geese. Most records of European birds come from the Gwent Levels and Usk Valley, and the only record of a Greenland White-fronted Goose was a single bird at Llandegfedd in March 1974.

Flintshire

Although Dobie (1893) described this species as an occasional visitor to the Dee marshes, Forrest (1907) stated that it was common here at times, being particularly plentiful in February 1904. Few subsequent records exist, and most records are from Burton Marsh. The largest number recorded here was 400 on 30 January 1950, although good numbers were also present in 1939–40. Elsewhere, odd birds are occasionally recorded from the Clwyd estuary, but this species is generally a scarce visitor.

Denbighshire

An occasional winter visitor to the Clwyd Valley, it appears that the White-fronted Goose has never been a regular visitor to this county in large numbers although small flocks attached to the Mersey population were formerly recorded in the Pulford/Rossett area. Over the past 20 years, records have generally involved fewer than ten birds although a flock of 60 was seen over Ruthin on 26 January 1979 and a flock of 12 was seen over the same town on 20 December 1980.

Caernarfonshire

Described as being the only grey goose on the Conwy estuary and the most numerous in the Porthmadoc area at the turn of the century, this species is no longer recorded regularly from Caernarfonshire. North *et al.* (1947) stated that the species was seen regularly at Llyn Ystumllyn, and it appears that between about 1958–60 and 1965–68, a flock of up to 30 Greenland birds frequented the marsh. There have been no records since that time, and the occasional birds encountered at Foryd Bay, Glaslyn Marshes and the Conwy estuary in recent years are probably from the Anglesey flocks.

Merioneth

There are a few unconfirmed records from the Traeth Mawr area at the turn of the century, and Forrest stated that several were obtained at Llwyngwril at that time. Since then, there have been scattered records mainly from the Dysynni and Mawddach valleys, which were probably wandering Dyfi birds.

Pembrokeshire

Mathew (1894) described this species as an occasional visitor, particularly during severe weather and it is true to say that the only time it has been recorded in some numbers was during the 1963 winter. Most of these were probably displaced Cors Caron birds, and several flocks of between 30 and 130 were recorded in the county. Since then, both races have been recorded on occasion, with a maximum of 22 Greenland White-fronted Geese on

Llys y fran Reservoir in February 1976. There appears to be a fairly regular passage of Greenland birds through the county in October and November, probably involving birds migrating to Ireland.

Radnorshire

White-fronted Geese are very scarce visitors to Radnorshire, the maximum number recorded being a flock (probably of this species) of 70 over Gladestry in April 1954. Several parties were seen in 1969, 28 over Stanner Rocks the maximum number, and there has not yet been a confirmed record of the Greenland race in the county.

Breconshire

Numerous records exist of 'grey geese' being seen in the county, and many will undoubtedly involve this species. The maximum recorded is 40 over Mynydd Llangynidr on 5 November 1974, and the only confirmed record of a Greenland White-front was between 19 and 26 November 1969 when one was present on Llangorse Lake. This species was recorded in eight winters between 1969 and 1989.

White-fronted Geese are regularly recorded from the offshore islands, the majority of records coming in October. Interestingly, there are more January records from the Pembrokeshire islands than for Bardsey, although the reverse is true for March. At times, large flocks are recorded offshore (e.g. 64 off Bardsey on 17 October 1991 and 99 on Skomer in 1963). One Greenland White-fronted Goose present on the island of Grassholm between 5 and 14 October 1982 attached itself throughout its stay to *Larus* gulls.

There are several records of European White-fronts ringed at Slimbridge and recovered in Wales, and several birds recovered here had also been ringed in the Netherlands. The only record of a Greenland White-fronted Goose recovered in this country involves a bird ringed at Ikamiut on 21 July 1946 and shot on Cors Caron (Cardiganshire) on 30 January 1947, a distance of 3142 km south-east. A Greenland White-fronted Goose fitted with a neck collar as a gosling on the Wexford Slobs (Ireland) in autumn 1990 was resighted on the Dyfi estuary on 30 November 1991. It subsequently spent the whole winter with the flock, but it has not been seen since.

Lesser White-fronted Goose *Anser erythropus*

Gŵydd Dalcen-wen Leiaf

A vagrant to Wales. It breeds from northern Scandinavia to north-east Siberia and the western population winters mainly in the south Caspian Sea and Transcaucasia.

There have been two Welsh records:

11 February 1955	– Adult female shot, Dyfi Estuary
7 March 1971	– Adult seen, Dryslwyn Meadows (Carmarthenshire)

The first bird was flying with a flock of White-fronted Geese, with which there were also a few Pink-footed Geese.

Excavations at Coygan Cave, Laugharne (Carmarthenshire) found bones identified as this species by Don Bramwell (E. Walker, in litt.).

Greylag Goose *Anser anser*

Gŵydd Wyllt

Rare winter visitor, occasionally recorded on passage. Its true status is clouded by the presence of several resident feral populations.

The Greylag Goose is our only native breeding goose, although the truly wild populations are now restricted to the more remote parts of north Scotland and the Western Isles. Since the 1930s – and in Wales since the 1960s – feral stocks have now become established so that its breeding range has extended as far as south-east England and parts of Ireland. Whereas the breeding habitat of the truly wild population in Scotland is confined to heather moorland adjacent to lochs, feral populations have taken to nesting on lowland lakes, in urban parks and alongside slow-moving rivers.

The earliest recorded breeding of this species in Wales was in the early years of the 20th century when three or four birds nested each year on an island in the middle of Roath Park Lake (Ingram & Salmon 1920). No further records exist until Greylag Geese were released into Anglesey under the supervision of the Wildfowlers' Association for Great Britain and Ireland (WAGBI) for sporting purposes during the 1960s. Some of the original eggs and young came from the feral population in the Loch Insh area of Scotland where conflicts arose because much of the pasture had been converted to arable land. The habitat on Anglesey proved ideal for this species and the population expanded rapidly. In 1986, the population was estimated to be around 50–80 pairs (RSPB 1986) and in 1987 breeding was reported from Llyn Alaw, Llyn Traffwll, Llyn Maelog and the Valley Lakes and pairs even attempted to breed on islands off Rhosneigr. Isolated pairs nested during the late 1960s and early 1970s in Denbighshire (near Holt and Allington 1970–72) and Montgomeryshire (1968), but none of these became established. The *New Atlas* showed that by 1990, several new feral populations had become established. Breeding has been recorded from several sites on Anglesey, the Vale of Clwyd, eastern Montgomeryshire, near Criccieth (Caernarfonshire) and the Three Estuaries (Carmarthenshire). The total Welsh breeding population is now probably between 200 and 300 pairs.

Greylag Geese formerly concentrated on British estuaries in winter but over the past twenty years they have moved onto farmland where they feed on grass and waste grain. They are very mobile and gregarious and roost usually on estuaries, lakes and reservoirs. The large flocks on Anglesey can sometimes be seen feeding amongst farm stock, often close to human habitation.

Whereas all the Greylag Geese breeding in Wales are of feral origin, most of the earlier wintering records were of truly wild birds. Some of the more recent records may include migrants from Iceland, but they are more likely to be from the resident breeding population. Forrest (1907) described the species as being the rarest of the grey geese to visit north Wales at the turn of the century. At that time, birds were recorded almost annually on the Dyfi estuary, particularly during exceptionally cold winters. Large flocks were seen here in 1893, and several were shot in the hard winters of 1895 and 1906–7. The Conwy estuary (Caernarfonshire) was also a regularly used site, and a large flock seen below Talycafn on 7 February 1910 was thought to be Greylag Geese.

It was formerly a rare visitor to the Dee estuary, and Farrar (1938) records only three occasions when this species was seen here, the largest number being six on 11 December 1931. Single birds were recorded from many Welsh counties during the early years of the 20th century, but flocks of ten or more were exceptional. The largest number seen together at that time was a flock of about 50 seen on the Pembrey-Kidwelly marshes (Carmarthenshire) on 27 November 1922, an area where this species was apparently a regular visitor. Despite several authors' claims that the species was seen during the last century in Glamorgan and Breconshire, the first confirmed records for these two counties were on 26 December 1925 (Kenfig Sands) and 11 March 1972 respectively. Likewise, the earliest Monmouthshire sighting was on 5 September 1967 when a single bird was observed at Llandegfedd Reservoir. Many of these more recent sightings, however, probably relate to escaped birds. Certainly the individuals seen with the flock of White-fronted Geese in Montgomeryshire during the 1950s were of domestic origin (Montgomery Field Society Reports, 1952–57).

By far the largest concentrations of Greylag Geese wintering in Wales today are found on Anglesey where the resident breeding population now numbers over 150 pairs. In winter, flocks of over 500 are occasionally seen on some lakes, particularly Llyn Alaw where 817 were present on 14 October 1987. Llyn Traffwll (358 in the winter of 1987–88) and Llyn Coron (176 in 1986–87) also hold large numbers in some winters, and it is likely that the large flocks recorded occasionally off the north Caernarfonshire coast are Anglesey birds (e.g. 170 off Llandudno on 24 February 1987 and 120 flying over Foryd Bay on 18 September 1987).

Greylag Geese are recorded more often from Bardsey than from the Pembrokeshire islands of Skomer and Skokholm, as would be expected of a species whose distribution in Wales is markedly northern. Here, birds have been recorded in the months of April–July, October, November and January, suggesting that some may have been from wild stock.

All but one ringing recovery of Welsh Greylag Geese involve birds ringed and recovered in Anglesey, suggesting that the breeding population is very sedentary. The one exception is a bird ringed in Cheshire and recovered in Anglesey.

Canada Goose *Branta canadensis*

Gŵydd Canada

A locally common breeding resident in most counties, with feral populations continuing to expand.

The Canada Goose is a north American species introduced into Britain during the 17th century. Although a feral population was well established in England at the turn of the century, it was still scarce in Wales until the 1950s and 1960s when birds were released in several parts of the country. Since then, the species has increased to such an extent that in some areas it is regarded as an agricultural pest.

Canada Geese were first introduced into Montgomeryshire in 1908, but birds from nearby English counties were seen here prior to this, although no breeding was recorded. The initial introduction was at Leighton Park near Welshpool, but by the middle of the century breeding pairs were also recorded from the nearby Powis Castle Estate where five pairs were recorded on five ponds. The breeding population for the county at that time was estimated to be 11–12 pairs on ten nesting waters (Blurton Jones 1956), with a total number of 68–84 birds. Over the next 20 years the population increased rapidly and by the early 1970s over 400 birds were present in the county with extensive breeding in the lower eastern half. An accurate breeding estimate is difficult as many Canada Geese form non-breeding flocks during the summer months and some breeding pairs are now known to use small, secluded waters (e.g. Bryngwyn Pool).

There have been several introductions of Canada Geese into Anglesey, the majority by the late Richard Palethorpe, who released birds onto the Valley Lakes where breeding is now recorded annually. In June 1956, Canada Geese were introduced onto Llyn Llywenan from Kedleston in Derbyshire (C. F. Tunnicliffe Sketchbooks No. 8), and here too pairs are now found nesting most years. Between 1960 and 1965 a pair bred in the western dunes of Newborough Warren and five broods were recorded on Llyn Rhos Ddu in 1964 (Hope Jones 1965). During a BTO/WT survey in May and June 1980, 110–120 adults on seven waters produced a minimum of 90 goslings in seven broods. Since then, numbers have increased further and in 1986, the population was estimated to be around 30–50 breeding pairs (RSPB 1986).

In north-east Wales, birds have been recorded since the last century, but breeding was first recorded in Denbighshire, near Ruthin, in 1981. The species is now firmly established on several pools around Wrexham and in the Vale of Clwyd, with sporadic records elsewhere from this and the adjoining county of Flintshire.

A small feral population has been established near Afonwen on the Llŷn peninsula since about 1965. This population has increased rapidly to its current level of about 150 birds and is posing problems to the nearby holiday camp at Pwllheli whose ponds it occasionally invades in large numbers. An RSPB survey in 1986 estimated the breeding population east of Pwllheli to be about 15 pairs (RSPB 1986). In Merioneth, small populations now frequent the upper Dee near Cynwyd, Llyn Trawsfynydd and the Dyfi estuary on the border with Cardiganshire. In south-west Wales, breeding is confined to the Aberystwyth area (Cardiganshire), the lower Teifi and Boulston in Pembrokeshire, some originating from introduced flocks. During the late 1970s, only single birds were seen on the Dyfi, but by 1984, a small flock had built up at Ynys-hir, and breeding was first proved in that year. In 1989, a pair bred on Cors Fochno (Fox *et al.* 1990), and by 1990, breeding was recorded at three other sites in Cardiganshire. The Teifi group also appeared during the late 1970s, but breeding was not recorded until 1984. Sporadic breeding has occurred elsewhere in south Wales, with populations often failing to become established.

In Monmouthshire, Canada Geese were introduced by the Newport Wildfowling and Gun Club in 1960 and 1962, and breeding took place at Newbridge-on-Usk during 1964-69 and at Undy Pool in 1968. Since then however, the population has declined, probably due to heavy shooting. Breeding has only recently been recorded from Carmarthenshire and Glamorgan (Kenfig NNR), and with much suitable habitat available in these two counties, the population will almost certainly increase further.

Canada Geese first bred in Radnorshire in 1959 when three goslings were seen in flight. The only regular site currently supporting breeding Canada Geese is Penybont Lake, where between one and three pairs have bred annually since 1969. In Breconshire, a pair bred at Dderw Pool between 1966 and 1976, and pairs have also nested near Glasbury. Only Llangorse Lake is used regularly, however, with up to six pairs breeding most years.

Lovegrove *et al.* (1980) calculated the Welsh Canada Goose population to be in the region of 900 birds, but since then breeding has been recorded in several new areas, and the population has increased markedly in some traditional areas. The current population probably exceeds 2000 birds.

Canada Geese become much more mobile in winter, moving to and from feeding areas and roost sites. Adverse weather forces birds to seek the more sheltered, milder sites and large influxes are sometimes experienced in some areas. Birds gather in large, noisy flocks and are therefore much easier to count than during the breeding season when pairs disperse to lakes, rivers and small pools.

Anglesey and Montgomeryshire have the largest number of Canada Geese in Wales, wintering flocks in the former sometimes reaching 400–500 birds. Favoured sites on the island include Llyn Alaw (202 in winter 1984–85), Llyn Traffwll (170 on 9 August 1989) and Llyn Coron (126 in 1971–72), although many other waters support flocks of over 50 birds. In Montgomeryshire, a flock of several hundred birds is found each year on the River Severn, moving between Caersws and Welshpool (e.g. 420 at Y Gaer on 11 September 1989). Because of their habit of feeding on winter-grown crops near the water's edge, the geese have become exceedingly unpopular with farmers.

Elsewhere in Wales, small flocks of up to 50 birds are recorded from many of the areas where breeding pairs have become firmly established, and a flock of over 100 individuals is present on the Llŷn (e.g. 170 at Penychain on 22 December 1989). Elsewhere, up to 300 are recorded each winter at Llangorse Lake, and there is evidence to prove that some of these are birds from Herefordshire. Large influxes have been recorded from some areas during cold winters (e.g. 63 on the Wye near Glasbury (Radnorshire) in January 1963 and 35 at Eglwys Nunydd in Glamorganshire on 23 February 1963), and many of these may be birds that have flown across from English counties. In Pembrokeshire, Canada Geese gather on the upper Cleddau to moult and are rarely seen away from here during the winter. This flock and those on the Dyfi (Cardiganshire) and Teifi (Cardiganshire/Pembrokeshire) estuaries now number over 100 birds each winter.

Canada Geese were formerly rare vagrants to the offshore islands but recently, records have become much more frequent as the mainland breeding population has increased.

Several records involving Welsh-ringed or Welsh-controlled birds have been reported, many involving movements within the Principality. However, several birds, have been ringed in Wales and recovered on the Beauly Firth in Scotland, a well-known moulting area. Another interesting record was of a bird ringed in Pembrokeshire and found with a broken wing on Bardsey on 13 June 1960.

Barnacle Goose *Branta leucopsis*

Gŵydd Wyran

A small flock regularly wintered in Pembrokeshire until the late 1980s. Elsewhere it is a rare winter visitor, with the majority of records relating to escapes from captivity.

Of the three separate breeding populations of Barnacle Geese in the world, two winter exclusively in Britain and Ireland. Whereas the population breeding in the Svalbard island group winters exclusively on the Solway Firth, the

Greenland population winters on the north and west coasts of Scotland and Ireland. The small flock that wintered until recently in Pembrokeshire probably consisted of wanderers from the Greenland population, whereas most other Welsh records almost certainly involve birds that have escaped from wildfowl collections.

A large collection of Barnacle Geese bones dating back to 20 000 BC was unearthed at Little Hoyle Cave (Pembrokeshire) and this is thought to represent a breeding population (Dr S. H. R. Aldhouse-Green, pers. comm.).

One of the earliest written records of this species in Wales was in 1776 when several Barnacle and Brent Geese were picked up dead on the coast near Criccieth after severe storms. Up to the early 1870s, Barnacle Geese were present in their thousands on the Dee estuary (Williams 1977), but numbers declined rapidly until, by 1900, records were very scarce. Dobie (1893) described how a large flock of Barnacle Geese infested the wheat fields on Llandrillo Bay (Caernarfonshire) one severe winter about 1816. By the late 19th century, however, the only record from this area was of a small group on the River Conwy in winter 1894.

There is reported to have been a small number of Barnacle Geese on the Dyfi estuary during the 1854–55 winter, but there are only three further records for Cardiganshire between 1884 and 1965. Mathew (1894) described it as an infrequent winter visitor to Pembrokeshire, often in the company of the more common Brent Geese. The first record for Carmarthenshire was in January 1931 when birds were seen on the Tywi near Ferryside. A flock of about 30 was seen on 15 December 1945 on the Gwendraeth estuary, and six were still present here on 8 January 1946. This species is mentioned in the county list of the adjacent county of Glamorgan in the last century, but the first published record was on 9 December 1933 when five were seen on the Burry Inlet. In October and November 1968, a flock of c50 settled on Ramsey, but it is not known whether they wintered there.

Most recent records probably involve birds that have escaped from wildfowl collections. Feral birds are now seen regularly on Anglesey and on the Dyfi estuary, and birds seen at Llyn Trawsfynydd and near Tywyn had probably escaped from a private collection at Maentwrog (Merioneth). Other records often involve single birds or small flocks seen in association with other geese such as Brent or White-fronted Geese.

The one major exception to this is a wintering flock which has appeared in Pembrokeshire in recent years. A single bird was present on Marloes Mere in 1979, but by 1981, the number had increased to 26, and to 80 by 1987. The flock moved between Marloes Mere and Skomer, where a maximum of 136 was recorded in 1989. In 1990, however, no more than eight were recorded on the island, and since then, the flock has not returned. Kirby *et al.* (1990) suggest that the Skomer flock may be the same as the one which spends some time on Bittell Reservoir in Worcestershire. The Pembrokeshire flock generally arrived in small numbers in early January, around the time that birds moved on from Bittell Reservoir, and built up throughout the month before departing in April. Other scattered records from Pembrokeshire and birds recorded in Carmarthenshire (c50 on the upper Taf Estuary in mid-January 1989), Radnorshire (60 at Bronydd between 29 December 1988 and 20 February 1989) and at Llangorse Lake are probably from the Marloes Mere/Skomer flock.

A bird ringed at Orsted Dal in Greenland on 12 July 1963 was recovered at Pontrhydfendigaid (Cardiganshire), 2377 km south-south-east, on 13 November 1965.

Brent Goose *Branta bernicla*

Gŵydd Ddu

A regular wintering population is found in the Burry Inlet. Elsewhere, small numbers are recorded in winter and on passage.

The Brent is the smallest of the geese wintering in Wales and its distribution is exclusively coastal with a marked preference for large estuaries and intertidal mudflats. Both the dark-bellied *B. b. bernicla*, which breeds in Arctic Europe, and the light-bellied race *B. b. hrota*, which breeds in Greenland and Spitzbergen, are found here in winter, but the latter is a scarce visitor in very small numbers. In 1990 and 1991, where the race was specifically determined for flocks and individuals seen away from the Burry Inlet, the ratio was 75% dark-bellied to 25% light-bellied geese.

The Burry Inlet is the most important wintering site for dark-bellied Brent Geese in Wales, regularly holding over 1000 individuals (maximum 1410 in November 1988). Birds arrive here in September and depart in March, reaching a peak in November and December. Small numbers of light-bellied birds are also present most years, but the numbers rarely exceed three or four individuals. Historically, this species has always been a regular winter visitor to the Burry Inlet, but records show that the numbers have never been as high as they are at present.

At the turn of the century, the Brent Goose was described as being an uncommon visitor to Glamorgan. Numbers had obviously increased by the 1930s when this species was noted on the Burry in very small numbers. At this time, there was a sharp decline in the British population of Brent Geese due in part to a disease of eel grass, their favoured food plant. By the 1960s, however, the dark-bellied race was a regular visitor to the area with flocks of 30–40 seen most winters. Numbers continued to increase throughout the 1970s, thanks to the removal of shooting and particularly since 1975, in tandem with a ten-fold increase in the world population (Kirby *et al.* 1990).

Large flocks of Brent Geese were formerly found on the Dee estuary. Dobie (1893) saw about 200 here on 7 February 1888, and Coward (1910) stated that it was found in considerable numbers on the estuary at the turn of the century. Since then, however, numbers have declined markedly and following the population crash of the 1930s, most observations have been of one or two birds, with one exceptional record of 39 seen flying up the estuary on 17 January 1960. Most of the published records do not distinguish between the two races, but of those that do, the majority are of the dark-bellied form.

Forrest (1907) stated that this species frequented flat shores of estuaries, particularly in north-west Wales, and that although it was generally an irregular visitor, it was occasionally encountered in large flocks. The Conwy estuary (Caernarfonshire) was a favoured haunt at that time, although birds had all but disappeared from here by 1924. Forrest mentions one remarkable record of a bird at Valley (Anglesey) on 3 August 1912 (presumably a pricked bird) and another of a single bird which stayed for 5½ months between January and June 1924 at Deganwy on the Conwy estuary. Small flocks were also seen on the Dyfi at this time, with a maximum of 30 recorded in two different flocks by Salter (1900) on 9 January 1893.

Several authors noted that small numbers frequented the Teifi estuary (Pembrokeshire/Cardiganshire) at the turn of the century, and two geese were shot here on 2 February 1912. Mathew (1894) described the species as a winter visitor to Pembrokeshire, sometimes in large numbers, particularly to Goodwick and Milford Haven. Large flocks have been recorded from sites usually not associated with this species (e.g. 70 at Goldcliff in Monmouthshire in February 1929) and flocks are sometimes recorded offshore from the Pembrokeshire islands (e.g. 16 over Skokholm on 10 April 1936).

Today, Brent Geese are still found in small numbers most winters on the Teifi and Dyfi estuaries or along the open coast between them, and also on occasion at Angle Bay (Pembrokeshire). Small flocks of the light-bellied form regularly visit Foryd Bay in Caernarfonshire (e.g. nine on 1 November 1987) and Beaumaris on Anglesey (e.g. seven on 20 March 1987). The Anglesey birds appear to be very mobile, often turning up at other sites on the island and along the Menai Straits, and in January 1992, at least 70 birds were present in the Menai Straits and Inland Sea (N. Brown, pers. comm.). A few are recorded along the estuaries of the Merioneth coast most winters (e.g. 18 at Broadwater on 23–24 September 1989), and along the Monmouthshire coast, all of these usually being light-bellied Brent Geese. In 1992, one bird summered in Foryd Bay (Caernarfonshire).

Inland records are scarce, but this species has been seen in all three inland Welsh counties. In Montgomeryshire, Brent Geese were recorded among a flock of White-fronted Geese at Leighton in 1958 and 1964, and the only record for Radnorshire is of a bird killed "many years ago near The Skreen". Brent Geese were supposedly shot on several occasions in Breconshire during the last century, including a few birds among a flock of Canada Geese on Llangorse Lake in 1891. A single bird was seen at the same site between 23 November 1986 and 10 January 1987 and in 1991, a total of four birds were recorded, including two together at Glasbury between 18 and 24 February.

This species is occasionally seen passing offshore from the Pembrokeshire islands during the winter, but most records from here and from Bardsey occur between September and December. The majority of Brent Geese recorded on passage from Pembrokeshire are of the light-bellied form, the largest group seen to date being 48 passing through Jack Sound on 8 October 1949.

Red-breasted Goose *Branta ruficollis*

Gŵydd Frongoch

A vagrant to Wales.

It breeds in Siberia and winters principally around the Caspian, Iraq and Bulgaria.

There are two records:

18 January 1935 – Immature male shot in Milford Haven (Pembrokeshire); it had been seen for about three weeks feeding on grass or stubble fields in the neighbourhood

4 March 1950 – One seen on Camlad Meadows (Montgomeryshire/ Shropshire) in company with 650 White-fronted Geese and two Barnacle Geese

---

Ruddy Shelduck *Tadorna ferruginea*

Hwyaden Goch yr Eithin

A vagrant, recorded satisfactorily only once in Wales.

This species breeds in North Africa and south-east Europe eastwards through central USSR to Amurland and China. Rogers (1982) has reviewed the recent occurrences of Ruddy Shelducks in Britain and concluded that no record during the previous 50 years could with certainty be regarded as other than an escape from captivity. These records included a party of three at Aberaeron (Cardiganshire) in the winters of 1974–75 and 1975–76 and of eight at Llyn Alaw (Anglesey) on 26 July 1979.

The only record that appears to be of definitely wild birds is from Pembrokeshire where Mathew (1894) recorded that one was shot out of a small flock near St. David's in July 1892. The wild origin of these birds is given particular credence by the well-documented fact that an unprecedented invasion took place in summer 1892 with records from Ireland and Britain (flocks up to 20), Iceland, Greenland, Norway and Sweden.

Sheldduck *Tadorna tadorna*

Hwyaden yr Eithin

A locally common breeding resident in coastal areas, but rare passage migrant to inland counties.

The Shelduck differs from most other Welsh breeding ducks in that it nests adjacent to tidal waters. This species decreased locally in the 19th century, but there are no records to confirm that the decline was significant over the country as a whole. Subsequent increases during this century are due to protection both in this country and at the Shelduck's moulting grounds in the Heligoland Bight. Against this background of general expansion, however, increased coastal disturbance, particularly through recreation, has led to local reductions in the population.

The distribution of breeding Shelduck in north Wales is very similar today to what it was at the turn of the century. Forrest (1907) described it as being common on the coasts and estuaries of all the coastal counties of north Wales. He discovered two colonies of 40–50 birds each near Aberffraw (Anglesey), and noted that they were also common around Arthog and Penmaenpool (Merioneth) and on the south side of the Dyfi estuary. At that time, a few pairs nested a mile to the west of Glandyfi in a woodland, the female having to lead her brood across a road, through a hedge, a fence, down two banks and across the railway before reaching the water!

Coward and Oldham (1905) found this species numerous on Anglesey, with breeding colonies at Pen-y-Parc, Aberffraw, Malltraeth, Cymyran and the mouth of the Alaw. Shelduck also bred in good numbers on Ynys Llanddwyn at that time, and between 1930 and 1933 a pair probably bred on the Skerries.

There is little doubt that in many of these areas, Shelduck have declined since the turn of the century due to habitat loss and an increase in public pressure on the coastline, but most of the old haunts mentioned by Forrest and others are still used today. In 1986, an RSPB survey estimated the Anglesey population to be around 40–70 pairs, the majority in the dunes around Aberffraw and Newborough (RSPB 1986). Few pairs breed on the Llŷn peninsula due to its rocky coastline, but where this is broken by sheltered inlets and bays in places such as Abersoch and Pwllheli, Shelduck are known to nest. In Merioneth, as on Llŷn, good numbers can be found only on estuaries, such as the Dysynni, Mawddach and Dwyryd. On the Dee estuary, over a thousand adults were recorded in the 1960s (1081 adults and 33 young on 2 July 1961), but by the early 1990s, the number of adults in late summer had increased to 4000, numbers augmented by a pre-moult-migration build up. Elsewhere in Denbighshire, flocks of over 100 adults and juveniles are found only on the Clwyd estuary.

Shelduck have never been common in Cardiganshire, and today breeding is confined mainly to the Dyfi (10–20 successful pairs most years) and Teifi. This species was formerly a common breeder in the Ynyslas Dunes, but the population is now much reduced due to disturbance. In Pembrokeshire, breeding is restricted to the Cleddau and Teifi estuaries, Caldey Is. and occasionally Bosherston and Skomer. It breeds in moderate numbers in the extensive dune systems of Carmarthen Bay, the Taf-Tywi-Gwendraeth estuaries and also on the north shore of the Burry Inlet and the Loughor. In 1869, Dix recorded this species as being "numerous in the sand hills near Laugharne Marsh", an area still used. However, there has been a steady decrease in the breeding population of Shelduck in Carmarthenshire and Glamorgan as a result of increased coastal disturbance.

Sir Richard Colt Hoare (1793–1810) mentions seeing several "Burrow Ducks" (Sheldrakes) on a visit to Barry Island (Glamorgan) on 22 May 1803 and told how an acquaintance had reared and domesticated a brood of this species. At the turn of the century, they were recorded as being numerous in rabbit burrows along the coast of the Bristol Channel, and even bred along Penarth Road in Cardiff. At this time, it was common on Sully Is. (c12 pairs), Kenfig, Margam and Baglan Burrows, but numbers had declined markedly by the 1960s. At that time, they were numerous only on the Whiteford Dunes and Taff-Ely estuary, the latter an area currently under threat from a barrage scheme. The highest breeding density in Wales is found on the island of Flatholm where 80-100 pairs are thought to have bred in 1991 (Worral, in litt.). The offspring of this population, the most important in the Severn estuary, are almost certainly taken to Cardiff Bay for brood rearing (Fox & Salmon in press). The other stronghold in the county is the south side of the Burry Inlet where numbers have increased over the past 20 years and by the early 1990s, an estimated 20–25 pairs bred annually.

In Monmouthshire, the Shelduck's favoured breeding haunts are the Gwent Levels, but it also occurs around the lower Wye Valley and Caerwent. Recently there has been an increase in the number of young seen around the Monmouthshire coast although here, too, its breeding habitats are under constant threat. This is one of the few counties in Wales where Shelduck have bred inland, a phenomenon which has become increasingly common in many parts of England (Linton & Fox 1991). A pair reared seven young at Llandegfedd Reservoir in 1991 at a time when birds were noted spreading inland up the Usk and Wye valleys. Two isolated instances of inland breeding also occurred in south-west Wales in 1984, and birds have bred at the Cardiff reservoirs.

Lovegrove *et al.* (1980) estimated the Welsh breeding

population of Shelduck to be about 220 pairs, making it the second most numerous breeding wildfowl in the Principality. However, provisional results from the 1992 Wildfowl and Wetlands Trust national Shelduck survey show that their estimates for many areas were too low (S. Delany, pers. comm.). The data from this survey show a total of 1317 occupied territories located in spring (25 April – 17 May) and 1194 juveniles recorded in the summer (27 June to 2 August). Overall coverage was good although non-estuarine parts of the coasts of Denbighshire and Flintshire were not surveyed in 1992. Unfortunately, the results were not separated into 'old' counties, but they show that the highest numbers of territories (776) and juveniles (609) were recorded in Gwynedd (Anglesey, Caernarfonshire and Merioneth) with the highest totals of adults in summer (3368) recorded in Clwyd (Denbighshire and Flintshire), the majority on the Dee estuary.

Provisional results in Wales of the 1992 Wildfowl & Wetlands Trust national Shelduck survey

	Spring		*Summer*	
	Adults	*Territories*	*Adults*	*Juveniles*
Gwynedd	1940	776	232	609
Clwyd	1016	203	3368	213
Dyfed	692	177	193	241
Powys	0	0	0	0
West Glamorgan	131	0	27	14
Mid Glamorgan	26	0	4	15
South Glamorgan	171	71	12	23
Gwent	327	90	441	79
Total	4303	1317	4277	1194

The Shelduck's winter distribution in Wales is very similar to that of the breeding season – almost exclusively coastal, and mainly in sheltered bays and estuaries. Welsh Shelducks migrate to the Wadden Sea after breeding where they moult before returning to our shores in the autumn. Allen and Rutter (1956) described how adult Shelduck from the Dee rest on the Mersey before continuing their journey eastwards to moult. Young Shelduck, however, do not migrate with their parents but stay mainly around the coast of Britain.

It is usually during this migration that birds are recorded inland. Sightings of Shelduck away from the sea are generally recorded annually from every county, although certain sites appear to be more favoured than others. In Merioneth, Llyn Tegid and Llyn Trawsfynydd, and Lake Vyrnwy in Montgomeryshire appear to be frequented almost annually, as does Llangorse Lake where 16 were recorded on 3 December 1973.

The most important wintering site for Shelduck in Wales, and the second most important in Britain is the Dee estuary whose invertebrate-rich muds provide a home for over 3000 birds most winters. Numbers here have increased over the past 20 years thanks to the creation of sanctuary areas and, possibly, an increase in their main food item, the snail *Hydrobia ulvae*. The population generally peaks in the autumn when about 6000 are present, then there is a decrease in numbers with c3000 on the estuary in winter. Immediately prior to moult migration, numbers increase once more to over 4000 individuals.

Another important wintering area for this species is the Severn estuary, although most of the birds are found on the English side. However, the mud flats of the Taff-Ely estuary and the Uskmouth–Chepstow stretch of coastline are also important sites. The Severn supported an average maximum of over 2500 birds over the five winters 1985-89 (Kirby *et al.* 1990), with numbers generally reaching a peak prior to the moult migration in July.

The Burry Inlet also supports large numbers of Shelduck, with up to 2000 birds present most winters (e.g. 1779 in January 1990). Numbers decreased here during the 1950s but have since recovered and the population has now been stable for many years. The nearby Cleddau estuary in Pembrokeshire regularly supports over 1000 birds (e.g. 1419 in January 1985), the sheltered mud-flats providing ideal wintering conditions for this, and many other species of wildfowl.

Other sites holding significant numbers most years include Traeth Lafan in Caernarfonshire (490 in March 1990), the Dwyryd/Glaslyn estuary on the Merioneth/Caernarfonshire border (267 in January 1988), the Cefni estuary in Anglesey (524 in January 1989), the Dyfi estuary (598 on 22 February 1988) and the Mawddach estuary in Merioneth (220 in March 1989).

One site which held large numbers in the past has recently been damaged by the tipping of thousands of tonnes of spoil. The Conwy estuary (Caernarfonshire) formerly held up to 500 birds each winter, but spoil from the construction of the Conwy tunnel was dumped on one of the Shelduck's most important feeding grounds near Glan Conwy. During the early 1990s, however, numbers have increased once more, and 404 were counted off Glan Conwy on 10 May 1992 (D. Elliott, pers. comm.). The Taff-Ely estuary (Glamorgan) is currently under threat from the government-backed scheme to barrage Cardiff Bay. Up to 200 birds were regularly recorded here during the 1960s and 1970s.

There are several records of Shelduck ringed in other parts of Britain and Ireland and recovered in Wales, one of a bird ringed in the Netherlands and seven of birds ringed on their moulting grounds on the River Weser in Germany.

Mandarin *Aix galericulata*

Hwyaden gribog

Scarce breeding resident, recorded, but not breeding, in the majority of Welsh counties.

A native of Russia, China and Japan, the Mandarin was first imported into Britain in 1747, with breeding first recorded in England in 1834. A feral population did not become established, however, until the 20th century, and it was not admitted onto the British list until 1971. The most important breeding population is centred around

Windsor Great Park in south-east England, but the species is slowly spreading westwards as far as Wales. More arboreal than most native breeding ducks, the Mandarin nests in holes in trees and the preferred habitat is open, mature broadleaved woodland adjacent to secluded streams or ponds, a habitat which is found throughout Wales.

A small breeding population has become established in Montgomeryshire, and breeding was also confirmed in Glamorgan in 1985, both of which can be linked to nearby waterfowl collections or bird farms. The Montgomeryshire population is well established along the Severn, Vyrnwy and some of their tributaries, with at least one pair breeding on a nearby lake. They were first recorded near Llandinam in the early 1980s and have since expanded farther downstream and along the lower Vyrnwy. The sections of river where pairs are most frequently observed during the breeding season are well-wooded, containing old, hollow oak, alder and willow, although no nest has yet been found.

Breeding was first confirmed in Glamorgan in 1985 when a pair nested near Kenfig Hill, and breeding was suspected at Port Talbot in 1990. Pairs were recorded at three sites in Denbighshire – Afon Alyn (April 1974), Gresford (7 January 1975) and Trefalyn (17 July 1976), before the first breeding attempt was reported at a small, wooded pool near Abergele. The gamekeeper who found the nest does not believe that the birds had been released at the site, but unfortunately the nest and its eggs were predated. A female was present in woodland at the RSPB's Ynys-hir reserve (Cardiganshire) in April 1988 and may have bred, although no male was seen until October.

The *New Atlas* recorded breeding at one site near Wrexham (Denbighshire), the Severn Valley (Montgomeryshire) and Glan Conwy (Denbighshire), with scattered records of birds during the breeding season from four other counties. On 1 April 1990, a pair was located on the River Wye at Doldowlod on the Radnorshire/Breconshire border, the male seen displaying to his mate for over an hour. The wooded river valleys which cover much of Wales should be ideal for this species, and it is likely that the Mandarin will spread farther west in years to come. A colony of up to 13 birds had become established on the river Clwyd at Bodfari (Denbighshire) by 1992, although these birds resulted from recent introductions (J. Moulton, pers. comm.). At present, the British population is estimated at 850–1000 breeding pairs (Davies 1985), but the Welsh population is unlikely to exceed 30 pairs. Ironically, whereas the British Mandarin is expanding and colonising new areas each year, in its native land in the Far East, the species is declining rapidly.

In winter, Mandarins flock together, particularly on larger lakes and ponds. Feeding is usually at night, when the main flocks disperse to their feeding grounds which may be by lakes and rivers, or in adjoining woodlands. In Britain, there are only relatively local movements, yet in the Far East, the species is dispersive and migratory. The presence of a single male on the sea at Aber Ogwen (Caernarfonshire) on 8 November 1987, however, indicates that individuals will sometimes move farther than usual.

The small population on the River Severn (Montgomeryshire) is known to abandon its breeding haunts and move a few kilometres to nearby pools, particularly the Lymore Pools near Montgomery. The Vyrnwy population likewise gather at Llyn Du near Meifod, a shallow lake with well-wooded fringes. Little is known about dispersal from its other breeding sites in Wales.

Interestingly, during the fieldwork for the *Winter Atlas* (1981–84), no records were submitted from Montgomeryshire, but records of between two and six birds came from Kenfig Hill and Roath Park Lake (both Glamorgan). Birds are occasionally recorded from outside their breeding sites, but whether these may, in some instances, be new releases is not known. Four were seen at Millin Pill (Pembrokeshire) on 13 February 1972, and one at Westfield Pill in the same county between 26 and 28 June 1990. Mandarins have been noted on at least three occasions from Carmarthenshire, including six which visited a fish farm near Ammanford in November 1990.

Wigeon *Anas penelope*

Chwiwell

Scarce breeding resident, common passage migrant and locally abundant winter visitor, chiefly to coastal counties.

The Wigeon is a subarctic duck which colonised Britain during the 19th century. Following the first recorded breeding attempt in Sutherland in 1834, the numbers and distribution of breeding Wigeon expanded rapidly until the 1950s, since when there have been a few local reversals. In Wales, nesting attempts have always been irregular, and in some cases are thought to have involved birds which escaped from waterfowl collections or "pricked" individuals which were unable to migrate as a result of being injured during the shooting season. Its preferred nesting habitat in Britain is upland lakes and pools, although ox-bow lakes along slow-moving rivers and similar boggy areas are also frequented. At present, the majority of the British breeding population is confined to the uplands of Scotland and northern England.

Despite Parslow's claim that sporadic breeding was recorded in north Wales during the early 1970s and that breeding records in Wales were probably more frequent 20–30 years earlier, very few definite breeding records exist. The first recorded breeding was on Llyn Mynyllod, near Bala (Merioneth) in 1898, a pair successfully rearing their brood on marshy ground close to the edge of this small upland lake. It is possible that breeding took place here in 1902 also as two pairs were seen on the lake over a period of 11 days in late April, and a young Wigeon was shot at Llyn Mynyllod on 30 September 1904. A pair bred successfully at this site again in 1934, but there have been no breeding records from this or any other site in north Wales since.

A pair summered at one locality in Caernarfonshire for several years prior to 1946 and one bird was present on Dowrog Common (Pembrokeshire) in June 1924, but no breeding has ever been proved for either county. Similarly, a pair was present on Llandegla Moor (Denbighshire) in May 1980, but again no breeding was recorded. More recently, in 1987, a pair summered on Broadwater (Merioneth) and an unusually high number of summering birds was recorded on Anglesey in 1978. During fieldwork for the *Breeding Atlas*, breeding was reported from the Gwent Levels, but this has since been discounted for lack of hard evidence (Lovegrove *et al.* 1980). It is unlikely that Wigeon have nested in Wales in recent years, although summering (probably injured) birds are seen on occasion. Certainly no site is used regularly for breeding despite the fact that several seemingly suitable sites exist.

In Wales, wintering Wigeon are predominantly coastal, with the major concentrations on estuaries. However, a move to inland habitats has taken place this century (Owen & Williams 1976) and, as a result, their diet has become much more varied than the traditional eel grass and algae of intertidal areas. Wigeon's winter movements are little understood, but it is known that birds wintering in southern Britain come predominantly from Scandinavia and Siberia, with Icelandic birds wintering mainly in Scotland and Ireland (Owen & Mitchell 1988). There is some movement south and west in Britain as the winter progresses, therefore the highest numbers at the most important Welsh sites are usually recorded in January. During severe winters, several Welsh sites experience an influx of Wigeon seeking refuge from extreme weather conditions farther east.

Forrest (1907) described Wigeon as being abundant on flat coasts, especially in the west, sometimes appearing in vast flocks during the winter. It also visited many inland waters during Forrest's time, but seldom in large numbers, and this is a fairly accurate description of Wigeon in north Wales today. The greatest concentrations in Wales are found on the Dyfi estuary, where numbers have increased in recent years, possibly due to the protection of key feeding and roosting areas as nature reserves. The wintering population at this site is normally over 3000 but numbers increase markedly during cold spells (e.g. 6483 in December 1991). The most important feeding areas for these birds are the salt marshes within the RSPB's Ynys-hir reserve.

The Burry Inlet is another regionally important site for wintering Wigeon, with over 2000 counted in most winters. The many estuaries and freshwater lakes of Anglesey provide ideal wintering conditions for Wigeon, and several sites on the island regularly support over a thousand individuals. In January 1987, over 2000 birds were recorded on the Braint Estuary and a further 1108 on Llyn Alaw. During that month, the Anglesey population at the eight most important sites was reckoned to be over 4900, and in January 1979, the Anglesey and north Caernarfonshire total was calculated at between 6000 and 6500 birds. There is considerable movement between these two areas, and sites in the latter county which regularly support over 1000 Wigeon include Foryd Bay, Traeth Lafan, the Glaslyn/Dwyryd estuary and formerly the Conwy estuary. The Conwy has lost much of its saltmarsh in recent years due to development and no longer holds large flocks of Wigeon.

The importance of the Dee estuary for wintering Wigeon has increased after a long period of decline before sanctuary areas were secured during the 1980s. In January 1949, 5000 were recorded at Burton Marsh on the Cheshire side of the estuary, but numbers subsequently decreased, with an average peak winter count of only 445 during 1972–75. Since that time, with the removal of shooting disturbance from some areas, the population has increased once more and 4300 were recorded here in November 1990.

Along the west coast of Wales, large concentrations are encountered on most of the larger estuaries, with over 500 at Broadwater (Merioneth) and the Teifi estuary (Cardiganshire). In January 1985, 3253 were observed on the Cleddau estuary (Pembrokeshire), and over 500 were recorded at Marloes Mere on 26 January 1985. In the same county, Angle Bay also supports over 1000 Wigeon in some winters. Many sites experienced unusually large influxes of this species during the 1980–81 winter, when 2000 were recorded at Coedbach on the Gwendraeth estuary (Carmarthenshire).

Wigeon were recorded as being the most numerous duck in the autumn at Orielton decoy (Pembrokeshire), with 4150 taken during eight seasons in the late 19th century (1877–85). Up to 12000 were recorded here during this period, but by the late 1930s, numbers had declined to a maximum of 3000 (Mackworth–Praed 1941). Large numbers were formerly recorded on occasion on the Taf estuary at Laugharne (Carmarthenshire), over 1000 being seen on 7 November 1950. Up to 1000 were formerly recorded on the east mud flats in Cardiff but the importance of this site together with the Ogmore estuary (600 in 1962–63), Hensol Lake (up to 200 pre-1930) and Kenfig NNR (up to 600) has diminished, mainly because of industrial and recreational pressures although some of these sites still hold large flocks during cold weather (e.g. 3540 at Kenfig Pool in January 1987).

Inland, concentrations of over 500 are rare except on the Anglesey lakes, but Llandegfedd Reservoir (Monmouthshire) supported 523 Wigeon in January 1988, and several hundred are recorded during cold winters at Cors Caron. During the severe winter of 1980–81, 1200 were counted on the Tywi at Dryslwyn (Carmarthenshire) and over 200 were sometimes recorded at Talybont Reservoir but numbers here and at nearby Llangorse Lake have declined in recent years, probably due to disturbance (Tuite *et al.* 1983). At times, they are fairly numerous on Llan Bwch-llyn (Radnorshire) and on the Wye near Glasbury where 130 were recorded on 26 January 1985.

Wigeon are usually recorded in all months except July on Bardsey with a predominance of birds migrating south offshore in September and October (maximum 52 on 27 October 1980). Large numbers are occasionally recorded on some of the Pembrokeshire islands, with up to 100 recorded on Skokholm on occasion between August and April. Small parties are seen offshore at Strumble Head (Pembrokeshire) during the autumn but little visible movement is seen during the rapid spring departure which is presumably nocturnal.

Foreign-ringed Wigeon recovered in Wales

Birds ringed as far afield as Denmark, Finland, Iceland, Switzerland and Russia have been recovered in Wales, the greatest distance travelled being a bird ringed at Novoladoga on Lake Ladoga in Russia and recovered at Cosheston in Pembrokeshire, 2522 km west-south-west. A bird ringed at Orielton on 31 October 1938 and shot in Portugal on 12 January 1939 demonstrates that Wigeon may also show south-westerly movements in winter.

American Wigeon *Anas americana*

Chwiwell America

A vagrant from North America.

There are eight dated records for Wales:

21 and 23 June 1910	– Male, Llyn Llywenan (Anglesey)
19 October – 2 November 1975	– Immature male, Kenfig Pool
30 – 31 January and 11 February 1977	– Male, Llyn Bodgylched (Anglesey)
21 January – 25 February and 1 March 1979	– Male, Llyn Bodgylched (Anglesey)
2 November 1981	– Female, Ynys-hir
29 October – 9 November 1985	– First winter or female, Kenfig
26 April 1990	– Male, Inner Marsh Farm (Flintshire/ Cheshire)
30 November – 26 December 1990	– Male, Inner Marsh Farm (Flintshire/ Cheshire)

The recent increase in records from Wales mirrors the situation for Great Britain as a whole where there has been a slight increase in numbers since about 1968. All the Welsh records are at times consistent with trans-Atlantic vagrancy except for the first where Forrest (1919) considered that the bird could possibly have been an escapee from the wildfowl collection at Woburn.

Gadwall *Anas strepera*

Hwyaden Lwyd

Regular breeding bird in small numbers on Anglesey, and has bred in at least two other counties. Elsewhere a regular winter visitor to most counties in small numbers.

The Gadwall is a scarce breeding bird over much of Britain, with an estimated population of only 600 pairs (Fox 1988). As it favours shallow, lowland lakes and marshland surrounded by luxuriant vegetation, much of Wales is unsuitable for this species. Its British stronghold is East Anglia where breeding was first proved in the mid-19th century and where the population has now risen to well over 150 pairs. These originated from wing-clipped birds introduced at Dersingham (Norfolk) in 1850, and it is thought that many of the English and Welsh breeding records stem from similar liberations.

The first record of birds summering in a suitable location in Wales was a pair seen on a coastal marsh (probably Llyn Ystumllyn) in Caernarfonshire in June 1924. It is not known whether this pair attempted to breed. It was not until 1969 that breeding was proved in the Principality when a pair with seven young was seen at Margam (Glamorgan). This, however, was an isolated record and no subsequent breeding has been recorded at this site. Since 1975, breeding has been proved at three sites in Anglesey, with two pairs in one area that year, and this has formed the core of a small population on the island. Its shallow, reed-fringed lakes, ponds and marshes are ideal for this species, and it is hoped that the small population will expand further, particularly since one of the regular sites, Llyn Penrhyn, is now an RSPB reserve.

Breeding was suspected, but not proved, at Llandeilo Graban (Radnorshire) in 1971 and a nest with three eggs was found in the same county, at Rhosgoch, on 19 May 1973 (the late G. Ireson, pers. comm.). A pair was present on Lisvane Reservoir (Glamorgan) for 11 days in April 1968, and a female was seen with three ducklings at Eglwys Nunydd Reservoir in 1971. Breeding has been suspected at Oxwich Marsh NNR in recent years, and in 1986–87, breeding was suspected at two sites on the Llŷn (RSPB 1988). However, the only recent proven breeding record outside Anglesey was a pair which nested at Llwyngwair, near Nevern (Pembrokeshire) in 1981, although, again, these were probably introduced birds. Certainly, no birds were present at the site prior to 1981, and no breeding has taken place since.

Breeding has been suspected at one suitable site in Monmouthshire, where pairs are often present in late spring and on one or two dates later in the season, and one to two pairs have bred on the Afonwen Pools (Caernarfonshire) since the late 1980s. However, it is only on Anglesey that regular breeding occurs and an RSPB survey in 1986 confirmed breeding at two sites in the county with suspected breeding at a another eight lakes (RSPB 1986). The small population on the island now constitutes some 15 pairs, mainly concentrated on the Valley Lakes and Llyn Llywenan. Elsewhere, sporadic records of summering birds and isolated breeding records have not led to regular breeding despite the fact that many seemingly suitable sites exist in the lowlands. The total Welsh population is unlikely to exceed 20 pairs at present.

The Gadwall's habitat requirements in winter are very similar to those during the breeding season, as it is predominantly a freshwater duck, preferring shallow lakes to marine habitats. Approximately 80% of wintering Gadwalls recorded by the National Wildfowl Counts are inland (Owen *et al.* 1986), the vast majority in south-east England. The small Welsh breeding population is augmented in winter by a few birds from Iceland, the Netherlands, Scandinavia and central Europe (Fox & Mitchell 1988). Numbers in Britain have increased dramatically since 1960, but as the expansion appears to be related to the proliferation of shallow inland waters such as gravel pits (Fox & Salmon 1989), the increase has not been as marked in Wales which does not have many artificial, nutrient-rich lowland waters.

This species was formerly a rare winter visitor to Wales, with Forrest (1919) only recording single sightings in north Wales during the late 19th century and Mathew (1894) mentioning Gadwall as an occasional winter visitor to Pembrokeshire. Salmon & Ingram (1957) only lists one record for Breconshire, that of a male killed on the River Usk at Scethrog in August 1885, although six were seen on a small pool at Llandeilo Hill (Radnorshire) in 1880. More recently, however, numbers have increased and the Gadwall is a regular winter visitor to most Welsh counties, albeit usually in very small numbers.

No Welsh site is internationally important for this species, and only a handful regularly support over 20 individuals. Most of these are on Anglesey, with Llyn Llygeirian (31 in 1980–81), the Valley Lakes (25 in 1985–86) and Llyn Alaw (28 in 1986–87) being the most important, but small winter gatherings are also recorded at Cemlyn Lagoon. The maximum count recorded on Anglesey to date was 36 near South Stack in the north-west corner of the island on 11 November 1987.

Elsewhere, concentrations of over twenty birds are rare, but with the number of breeding Gadwall increasing, the wintering population is also on the increase. In 1969, 25+ individuals were recorded at Kenfig and 28 at Margam (both Glamorgan) in January 1969. Small numbers are also seen regularly at Marloes Mere (15 on 3 March 1985) and Bosherston Pools (12 on 19 November 1983) in Pembrokeshire. The maximum number recorded for this county was 15 at St. Ishmaels on 31 December 1981. Eglwys Nunydd (Glamorgan) has a few Gadwall most winters (nine on 7 January 1968), and seven were recorded on Rhyl's marine lake (Denbighshire) in December 1978. The latter site has been an important refuge for several duck species, particularly during severe weather, but its importance has diminished as pressure from human recreational activities has increased.

Prior to 1937, only one Gadwall sighting was recorded for Cardiganshire, but this species is now an annual visitor to the county in small numbers, mainly to the Dyfi estuary and Cors Caron. In Carmarthenshire, Gadwall regularly winter on the Witchett Pool, with up to 24 recorded, and it is also a regular visitor to the Dinefwr Oxbows near Llandeilo. The ponds at the WWT's reserve at Penclacwydd and the Machynys Ponds near Llanelli are also frequented by this species. In Merioneth, the Gadwall is very uncommon and is not always recorded annually, and in Breconshire records were formerly very scarce, but more recently it has been recorded regularly on Llangorse Lake with a maximum of 17 in winter 1990–91.

Records for Montgomeryshire and Radnorshire are few, not only because these counties are poorly watched, but also because much of the habitat is unsuitable. Three birds were seen at Leighton (Montgomeryshire) on 17 December 1981, however, and occasional birds are recorded from some of the shallower Radnorshire pools (e.g. one at Llyn Hilyn on 21 August 1979). Gadwall are much more regular visitors to low-lying pools and lakes in Monmouthshire, with seven recorded at Peterstone on 22 June 1974 and eight at nearby Magor in October 1969.

Sightings of Gadwall at most of the offshore islands are mainly of single birds or two at the most. However, four were seen at Pen Cristin on Bardsey in 1958 and a total of seven birds had been recorded on Skomer by 1991.

There are several records of birds ringed in southern England and recovered in Wales, and two of birds ringed on the Continent. One involves a Spanish bird recovered in Builth Wells (Breconshire), the other a bird ringed at Kolmenasva in Estonia on 21 June 1977 and recovered on the Dee estuary six months later, 1769 km west-south-west.

Teal *Anas crecca*

Corhwyaden

Small numbers breed in most counties, otherwise a widespread and abundant winter visitor and passage migrant.

The Teal is the second most widespread, but not the second most numerous breeding duck in Wales. In many parts of the Principality, this small duck is a rare breeder, confined mainly to rushy moorland, heath pools and bogs. Small numbers breed in most counties, although it is particularly scarce in south Wales. Its exact distribution is complicated by the fact that breeding is less easy to prove than for the Mallard because broods are much more secretive, preferring to skulk in aquatic vegetation rather than venture into open water. Also some records of Teal possibly breeding may refer to late passage birds. It is therefore an extremely difficult bird to study or census accurately during the breeding season and in Wales, even in some of the better surveyed areas, definite breeding is not often proved.

The earliest evidence of Teal in Wales comes from the excavations of Roman drains at Caerleon (Monmouthshire) during the first and third centuries. Teal remains were also found in Roman excavations at Caerwent in the same county, perhaps indicating that this species was more widespread during this period.

Forrest (1907) describes the Teal as an uncommon breeder in all the north Wales counties, and as a bird which was frequently found nesting at high altitude on remote hill-tarns. He quotes a pair which bred in Cwm Eigiau at an altitude of over 490 m, and another pair at nearby Ffynnon Llugwy, nearly 550 m above sea level.

At the turn of the century, there is little doubt that the Teal was a more common breeder in some areas as several suitable lakes, bogs and wetlands have since been drained. Forrest described how Teal used to nest at Afon Ganol, a marsh that extended from Conwy to Rhos on the Caernarfonshire/Denbighshire border, which was doubtless the old course of the River Conwy. Mathew (1894) found no evidence of breeding in Pembrokeshire towards the end of the last century, but it is difficult to believe that Teal did not nest on the rushy moorland of the Preseli Mountains. In Glamorgan, Teal used to breed at Eglwys Nunydd until the habitat was destroyed with the construction of the reservoir. Breeding was also recorded in the dunes around Kenfig Pool until 1946, and on the moors above Margam. The latter have since suffered blanket afforestation, rendering the habitat unsuitable.

Parslow (1973) recorded no evidence of marked change in the number and distribution of breeding Teal nationally, but suggested that there may have been a slight recent decrease. In Caernarfonshire between 1968 and 1972, breeding was proved in only five areas, all of which were in south Llŷn and Snowdonia. In this county, Llyn Ystumllyn formerly held more pairs than it does at present, but the area has lost much of its wildfowl importance since it was partially drained in 1955. Even on Anglesey where small rushy pools and lakes are common, Teal are only sporadically proved to have bred. In 1986, an RSPB survey found Teal present in a total of 21 one-kilometre squares, with breeding proved or suspected in 13 of these, and a survey on the Llŷn in 1986–87 estimated a population of only 2–10 pairs (RSPB 1988).

A few pairs breed regularly in Denbighshire, confined mainly to the uplands of Mynydd Hiraethog and Ruabon. An RSPB survey of Mynydd Hiraethog between 1975 and 1978 found only four pairs, and in Flintshire, breeding has only ever been proved on one occasion, near Cilcain in 1955. Teal breed annually on the uplands and coastal marshes of Merioneth, with three pairs found nesting on the Migneint by an RSPB survey between 1975 and 1978, and the county total at present is thought to be around 20–25 pairs (R. Thorpe, pers. comm.). In Montgomeryshire, at least six pairs have been found breeding on the Berwyn Mountains (Lovegrove *et al.* 1980), and pairs nest most years at Llyn Hir and Clywedog Reservoir. Small numbers are scattered throughout the mawn pools and larger bogs of Radnorshire but the largest numbers in Wales are found in the old county of Cardiganshire. Here, over 40 pairs nest, mainly on lowland bogs such as Cors Caron (10–20 pairs in the 1970s) and Cors Fochno, and on the Elan and Pumlumon uplands. Fox (1986a) found between 13 and 18 pairs breeding on Cors Fochno during 1980–83, and 12 pairs were found in 1975–76 on the extensive moors west of the Elan Valley.

Elsewhere in south Wales the Teal is mainly a sporadic breeder, with regular breeding only suspected from a few localities. In Breconshire, Teal have apparently bred very rarely with only a handful of records in recent years. It formerly bred regularly on Mynydd Eppynt and near Onllwyn, but no breeding has been recorded here in recent years. Few Teal breed in Carmarthenshire, Pembrokeshire or Glamorganshire, with only six records of possible breeding for the latter during fieldwork for the *Breeding Atlas*, and no confirmed breeding during the

West Glamorgan Atlas, although it is suspected that pairs nest most years at Oxwich Marsh NNR and Crymlyn Bog. In Monmouthshire a breeding pair on the River Usk in 1982 and 1985 were the first confirmed records since 1943. All records for this county during the compilation of the *Gwent Atlas* came from lowland sites, with two confirmed and six probable breeding pairs seen. During fieldwork for the *Pembrokeshire Atlas* (1984–88), breeding was confirmed only once on the Preseli Mountains although a pair reared young at Trefayog in 1989 and in the past sporadic breeding has been recorded at several sites, including Skomer where a pair raised two young in 1992.

An accurate estimate of the Welsh Teal breeding population is difficult because of their very secretive nature during the nesting season, but Lovegrove *et al.* (1980) calculated the Welsh population to be fewer than 100 pairs in most years. This estimate was probably too low, and although the population has undoubtedly declined in some areas because of habitat loss, in Cardiganshire numbers may have increased after the creation of new flight ponds. The current Welsh population is likely to be between 125 and 150 breeding pairs.

The paucity of passage records from Wales indicates that the main movement is probably nocturnal, but small numbers are seen passing off Strumble Head and the offshore islands. The largest numbers are recorded from Skokholm in March although large flocks are seen on Bardsey and Skomer in October, with 400 recorded at the latter on 30 October 1981.

Although large flocks of Teal do occur in Wales, winter wildfowl counts have revealed that over 40% of wintering Teal in Britain are found in flocks of fewer than 200. The distribution of Teal in the Principality in winter is in many ways the reverse of its breeding distribution as the birds favour low-lying lakes, marshes and estuaries. However, there are many inland areas of significance, particularly where there is adequate protection from disturbance and where inland flooding occurs regularly. In the past decade, there has been a marked increase in wintering Teal, particularly during 1989–90 when numbers were at their highest level since the cold winter of 1984–85. This is somewhat surprising given the mildness of the winter and shows that cold weather may not be the only important factor causing the immigration of large numbers into Britain.

The two most important wintering areas for Teal in Wales are the Dee estuary and the Cleddau complex (Pembrokeshire). The former held 9825 individuals in November 1989, and together with other estuaries in the north-west of England, supported a very large proportion of the British wintering population. This species has benefited from the establishment of reserves on the Dee estuary, the average peak winter counts increasing from 381 between 1972 and 1975 to 5180 between 1986 and 1991. The importance of the Cleddau estuary has also increased in recent years and in January 1989 a maximum of 3243 Teal was recorded.

Elsewhere in Wales, concentrations of over 500 are regularly seen only at the Glaslyn/Dwyryd estuaries (Caernarfonshire/Merioneth), Llyn Alaw (Anglesey), the Dyfi estuary and the WWT's Penclacwydd reserve (Carmarthenshire). Over 500 have been recorded in the past on the Teifi estuary (Cardiganshire/Pembrokeshire), with a maximum of 1000 in January 1981, Dryslwyn (Carmarthenshire), with a maximum of 794 on 13 December 1980 and the Taff-Ely estuary (Glamorgan) with a maximum of 500 in February 1963. The large numbers in 1963 were undoubtedly a result of immigration from the Continent during the particularly cold spell, and in very hard winters, Teal will readily emigrate farther south to milder countries (Ridgill & Fox 1990).

Orielton (Pembrokeshire) and Kenfig NNR formerly supported several hundred in most winters with a maximum of 2300 at the former site in 1935 (Mackworth-Praed 1941), and many of the Anglesey lakes often support several hundred individuals. Farther inland, large concentrations are rarely seen but over 200 are recorded at Cors Caron (Cardiganshire) most winters, with a maximum of 380 recorded in December 1981.

Recoveries of Teal ringed in Wales

Many Teal ringed on the Continent have been recovered in Wales, most recoveries involving birds ringed in the Netherlands, Sweden and Denmark. Birds ringed in Wales in winter have subsequently been found much farther afield with several recoveries from Russia, Poland,

Norway and Italy. The greatest distance recorded so far was a bird ringed at Orielton (Pembrokeshire) on 3 December 1934 and recovered in August the following year 4165 km east-north-east at Maslovo in Russia.

Green-winged Teal *Anas crecca carolinensis*
Corhwyaden Asgell-Werdd

This North American subspecies is is a vagrant.

It has been noted in Wales as follows:

11 December 1968	– Two males, Cors Caron
30 January 1972	– One male, Teifi reedbeds (Cardiganshire/ Pembrokeshire)
22–27 April 1977	– One male, Gann estuary (Pembrokeshire)
12–23 November 1981	– One male, Kenfig Pool
16 January – 28 February 1983	– One male, Oxwich
19 November–13 December 1984	– One male, Llyn Alaw (Anglesey)
10 December 1984	– One male, Pentwd Meadows (Cardiganshire)
1–2 March 1986	– One male, Gann estuary (Pembrokeshire)
2–3 December 1991	– One male, Llyn Traffwll (Anglesey)

It is not surprising that all the records refer to males because the duck cannot be distinguished from the European Teal.

Mallard *Anas platyrhynchos*
Hwyaden Wyllt

Common and widespread breeding resident, abundant passage migrant and winter visitor to all counties.

The Mallard is by far the most widespread and numerous breeding wildfowl species in Wales. Despite this, it is by no means common in all areas and its true distribution is modified by the supplementing of wild stocks by releases from wildfowlers as well as the escape of domestic Mallard from town parks and other collections. Decreases have occurred in some areas as a result of changes in land-use, particularly drainage, but local increases have also taken place through the creation of new waters.

The earliest evidence of Mallard in Wales comes from remains which were excavated together with those of several other bird species, from Merlin's Cave in the Wye Valley (Monmouthshire) in 1901. Associated mammal remains indicated a late Devensian age which means that Mallard frequented this area of the Welsh Marches some 10 000 years ago.

It is apparent that in recent centuries the Mallard has always been the most numerous breeding duck in Wales. Several observers recorded it as being common during the late 19th century, although Mathew (1894) stated that in Pembrokeshire, the Mallard had suffered a marked decline since the mid-1800s. Cambridge Phillips, on the other hand, declared that wild duck were on the increase in Breconshire at the turn of the century probably due to the release of captive-bred birds into the wild. At that time, Mallard were also numerous in suitable localities over the six north Wales counties, not only in the low-lying wet areas, but also near open water on high moorland. Forrest (1907) records seeing small "flappers" on marshy ground near a lake, over 500 m above sea level on the Berwyn Mountains where breeding is still recorded today. He also stated that the Mallard was particularly abundant as a breeding species on Anglesey, with even one record of a nest on a sea-cliff near Bull Bay.

Anglesey is still by far the most important county in Wales for breeding Mallard, its abundance of small lakes and marshes providing ideal habitat for most breeding wildfowl. Lovegrove *et al.* (1980) showed that the average density of breeding Mallard on the island was between two to three pairs per km^2, giving a population of 1500 and 2000 breeding pairs. Elsewhere in Wales, similar breeding densities are found only in areas such as the Gwent Levels in Monmouthshire. The *Gwent Atlas* also records high densities in the river valleys, particularly the River Usk and its tributaries. The *Atlas* remarks on this species' absence from the coalfield valleys in the west of the county, probably due to industrial pollution and human pressures, particularly shooting.

In other Welsh counties, the Mallard is described as a ubiquitous breeder, its only absolute requirement being the presence of either fresh or brackish water, from large lakes to small ponds, rivers and streams. Since 1990, about 50 pairs have been recorded breeding at the WWT's Penclacwydd reserve on the north coast of the Burry Inlet, although none of these birds was introduced. It has been recorded breeding on many of the small offshore islands, including Ramsey, Skomer, Skokholm and Bardsey. Despite this, it is not common in all areas, a thorough survey of open water sites in Glamorgan in 1978 revealed only 55 pairs and the *West Glamorgan Breeding Atlas* confirmed breeding in only 42 one-kilometre squares. Lovegrove *et al.* (1980) speculated that the total for the county was unlikely to exceed 100 wild pairs at that time.

RSPB surveys of the River Wye in 1977 and the Vyrnwy and Welsh section of the Severn in 1978 produced totals of approximately 150 and 295 pairs respectively. Other than those surveys, there is little definitive data on Mallard numbers and breeding densities in Wales. The estimate of 4000–5000 wild breeding pairs quoted by Lovegrove *et al.* (1980) is probably low, and the total number is likely to be between 7000–8000 pairs, with the main concentration in Anglesey.

The winter population of Mallard in Britain and

Ireland numbers about 700 000 birds (Owen *et al.* 1986) and consists of native birds augmented by an influx of winter visitors from Scandinavia and Iceland. Although no one site in Wales comes anywhere near the 20 000 required for international recognition, both the Dee and Severn estuaries regularly support over 3000 birds.

In winter, as in the breeding season, the Mallard is highly adaptable and can be found on virtually every kind of wetland, but it will feed on arable or pasture land well away from water. In Wales it is often absent from, or present only in very small numbers in the most mountainous areas, particularly the high peaks of Snowdonia, but even on the highest lakes, Mallard can be found in all but the severest winter weather. In hard weather, it prefers to "sit out" short freezing spells and as there is little emigration by British breeders, a large proportion of the Welsh winter population consists of birds which bred locally.

Anglesey is again one of the most important Welsh counties with counts of up to 1000 birds at Llyn Alaw and Llyn Traffwll. Other freshwater lakes on the island such as the Cefni Reservoir occasionally experience an influx with over 700 recorded in November 1976. The Dyfi estuary normally holds peak numbers of over 1000 individuals, with a maximum of 1777 on 4 November 1986, and Traeth Lafan and the Glaslyn/Dwyryd estuary on the Caernarfonshire/Merioneth border regularly hold upwards of 500 individuals. Along the south Wales coast, numbers on the Taff-Ely estuary (Glamorgan) and the Burry Inlet occasionally reach 400+, and an exceptional total of 1174 was recorded on the Gwendraeth estuary (Carmarthenshire) on 20 October 1981. Large numbers are also recorded occasionally on the Gwent Levels. In December 1973, the 760 Mallard counted at Peterstone Wentlooge, represented a quarter of all Mallard in Welsh estuaries that month and 30% of the Severn estuary total.

Numbers have increased significantly on the Dee estuary since the establishment of sanctuary areas during the 1970s and 1980s. The average peak winter counts between 1972 and 1975 was only 730 whereas by 1983–86, this had increased to 3670. Since the maximum figure of c5000 was attained during the early 1980s numbers have, however, declined slightly with a series of mild winters. Wildfowl counts have revealed that the British Mallard population has shown a statistically significant increase since 1960, a trend which has been confirmed on the Dee.

Other inland lakes and some offshore islands occasionally support substantial numbers of wintering Mallard, Talybont Reservoir being one of the most important with over 500 individuals recorded regularly. Formerly there was some movement between this site and the nearby Llangorse Lake, but the latter's importance for wintering wildfowl has diminished greatly in recent years as water-based recreation, particularly water-skiing, has increased (Francis 1992). Over 200 birds are seen in most winters on Llan Bwch-llyn and at Glasbury (Radnorshire), and Cors Caron normally has maxima of 250–400. Dryslwyn in the lower Tywi valley is another important site for several species of wildfowl, and 800 Mallard were seen here in September–October 1979. Skomer occasionally supports up to 700 in the autumn, but the 1500 present on 23 October 1981 probably included some passage migrants.

Boyd and Ogilvie (1961) stated that there was little evidence of long-distance movements of British Mallards, but there have been numerous recoveries of foreign-ringed birds. Several birds ringed in the Netherlands have been recovered in Wales, as have birds from Belgium, Denmark and Sweden. The longest distance travelled however, involved a bird ringed at Kandalaksha Reserve in Murmansk (Russia) on 28 July 1969 and recovered at Whitland (Carmarthenshire), some 2663 km south-west on 24 December 1971. Welsh-ringed birds have also been recovered far afield, in countries such as Denmark, Sweden, Finland, the Netherlands and Russia, the farthest travelled being an individual ringed on Anglesey on 4 March 1965 and recovered at Arkhangelst (USSR), 2849 km east-north-east on 20 August 1966.

Black Duck *Anas rubripes*

Hwyaden Ddu

A trans-Atlantic vagrant. Up to 1991 there has only been one individual in Wales.

A long-staying adult male was present at Aber in Caernarfonshire from 11 February 1979 until at least 29 January 1985. During at least part of this period it was paired with a female Mallard and produced a number of full-grown hybrid young. It is characteristic for an individual to stay in the same locality for a long period once it has crossed the Atlantic; a female on Tresco, Scilly, stayed even longer than the Aber bird, from October 1976 to 1983.

Pintail *Anas acuta*

Hwyaden Lostfain

Rare breeder, passage migrant and locally abundant winter visitor to some counties.

The Pintail is much more catholic in its choice of breeding habitat than many other species of wildfowl, ranging from upland lakes to lowland fens. However, as it is a dabbling duck, the waters must be shallow at their margins, a habitat which is scarce over much of Wales. It is still a rare breeder over the whole of Britain, but its true distribution is often difficult to ascertain as feral populations have grown from released birds in some parts of England and south Wales.

Pintail first bred in Britain in Scotland in 1869 (Inverness-shire), England in 1910 (Kent) and Ireland in 1917 (County Roscommon). The population expanded slowly until the 1950s but until recently, no breeding is known to have occurred in Wales. Sporadic summering birds have been recorded on occasion (e.g. one to two birds on Skomer in April 1984) but breeding was not proved until 1988 when two broods were seen on Skomer, and in 1989, four broods were recorded there and another pair nested near Pont ar Sais in the Gwili valley (Carmarthenshire). There is, however, some doubt as to the origin of those birds as feral Pintail are known to have been released on the Severn estuary at about that time. In 1990, 1991 and 1992, it is thought that two pairs of Pintail bred on Skomer.

In winter, Pintail are found virtually exclusively in the most low-lying areas, particularly the larger estuaries. It is not known whether the small British population remains here during the winter, but from September onwards large influxes occur from western Siberia and, to a lesser extent, Scandinavia and Iceland. Peak numbers formerly occurred in December, but recently Pintail have been most numerous during October followed by a second peak in December or January.

In the winters between 1985 and 1990, the Dee surpassed the Mersey as the most important wintering site for this species and in October 1989, 11 945 on the Dee represented almost 50% of all Pintail counted in Britain at that time. During the first decade of this century, flocks of a few hundred individuals were the maxima recorded on the Dee, but a massive increase took place until the early 1950s when this site held about half the British wintering population with flocks of at least 5000 and possibly many more in 1952 and 1953 (Williams 1977). Numbers gradually declined from the late 1950s onwards until the population increased once more during the 1980s, probably due to the establishment of reserves and sanctuary areas on the estuary.

The Burry Inlet (Carmarthenshire/Glamorgan) also supports important numbers each winter with a maximum of over 3000 recorded in February 1991. The National Wildfowl Count average for this site over the five-year period 1985–90 was 1878 birds, making it the fifth most important site for wintering Pintail in Britain. Within the Burry, Pintail show a preference for the north shore near Penclacwydd (Carmarthenshire).

Elsewhere in Wales, over 300 individuals overwinter most years in the Rhymney estuary (Glamorgan) and Malltraeth (Anglesey). The former usually supports highest numbers in February (427 in February 1989) and the latter in January (350 in January 1988), but this is not always the case. Numbers wintering on the Dyfi estuary have declined in recent years, but influxes often occur during hard weather (326 here on 20 December 1981). Moderate numbers regularly visit the Glaslyn/Dwyryd estuary (Caernarfonshire/Merioneth), with a maximum of 204 recorded during the 1990–91 winter.

Apart from the flocks on the Dee estuary, no historical reports of large numbers of wintering Pintail exist. Forrest (1907) documents small numbers around the Anglesey, Caernarfonshire and Merioneth coasts, with flocks of fewer than ten most often encountered. Since the early years of this century, Pintail visiting coastal estuaries have increased but large flocks inland are rare, although up to 100 were recorded at Orielton (Pembrokeshire) during the 1930s (Mackworth-Praed 1941). However, Pintail are known to favour shallow floodwater pools in some areas of Britain, and 42 were recorded at one such site, the Severn/Vyrnwy confluence (Montgomeryshire) on 23 January 1981 and birds from the Dee and Mersey estuaries formerly visited the Dee flood meadows below Aldford (Cheshire/Denbighshire) until flood alleviation schemes were carried out in the 1970s.

Welsh-ringed birds have been recovered in the Netherlands and Iceland, and birds from Estonia and France have been recovered here. The longest distance involved a bird ringed at Kolmenasva in Estonia on 29 June 1977 and recovered at Connah's Quay on the Dee estuary on 6 November 1977, a distance of 1769 km miles) west-south-west.

Garganey *Anas querquedula*

Hwyaden Addfain

Scarce breeding summer visitor, recorded in small numbers on passage.

The Garganey is unique amongst British waterfowl in being a summer visitor from wintering grounds in north Africa. Britain is on the extreme western edge of its range and the numbers breeding here each summer are largely determined by the variable number of arrivals in early spring. This species is a bird of shallow fresh and brackish water, so much of Wales is unsuitable for it, although some favourable habitat exists in the more low-lying counties of Anglesey and Monmouthshire. Since they are also difficult birds to find, remaining hidden amongst rank vegetation during the nesting season, many isolated breeding attempts in Wales probably go unrecorded.

Forrest (1907) knew of only six records of this species in north Wales prior to 1900, the first being a bird "shot some years before 1873". Several other observers confirmed that this species was a scarce passage migrant and summer visitor to Wales at that time, with records of more than one bird from the same locality very unusual. Most of the early records came from Anglesey, and it is quite possible that Garganey may have bred on one of the many shallow pools in the early part of the 20th century.

Parslow (1973) stated that the first confirmed breeding record for Wales came in 1936, but does not mention where the birds nested or whether they were successful. There were no subsequent breeding records until 1940 when a pair with a brood of ducklings was seen on Kenfig Pool (Glamorgan), and in 1946 a pair probably bred at Llyn Ystumllyn (Caernarfonshire). In 1951, a pair may have bred at Cefni Marshes on Anglesey, and the following year at least two pairs were confirmed breeding at Llyn Llywenan. Also in 1952, five pairs were seen on the Valley Lakes and one on Llyn Maelog, but breeding was not proved. The next confirmed breeding record from this county was not until 1980 when a nest with eggs was discovered at Llyn Dinam on 13 May (the late G. Ireson, pers. comm.).

In Flintshire, a pair was believed to have nested at Shotton Pools in 1959 and 1960 but this was never confirmed. In the same county, a pair was present at Padeswood Pool in April 1974 and again in May 1978, but it is not thought that they bred on either occasion. Nowadays, a pair or two breed occasionally on the English side of the Dee marshes and these birds are regularly seen on the Flintshire side of the border.

During the fieldwork for the *Breeding Atlas*, a pair bred at Cors Caron in 1968, and breeding was suspected at Kenfig NNR and on the Gwent Levels, where breeding probably occurred in 1969, and possibly also in 1970. Since then, records of pairs on suitable waters during the breeding season in Wales have decreased slightly, and the only likely breeding record during the 1980s is of a pair at Kenfig NNR in 1984. In 1992, at least one pair probably bred at Llyn Llywenan where two pairs were present throughout the summer.

Spring and autumn birds have been recorded from every Welsh county, generally in small numbers. The earlier records are mainly from Anglesey (e.g. two to three birds on Llŷn Penrhyn in July and August 1925), but records also came from Llyn Ystumllyn (Caernarfonshire), the Dee estuary and Cors Caron. Nowadays, most records come from Anglesey, Pembrokeshire (e.g. a pair at Trefeiddon on 2 April 1983), Glamorgan (especially Kenfig NNR), Carmarthenshire (especially Penclacwydd) and Monmouthshire. Most are of one or two individuals, often a pair, but flocks of six or more occur occasionally, e.g. eight at the Witchett Pool from 28 March to 1 April 1969, up to five pairs at Kenfig NNR between March and May 1962, and a flock of nine on the foreshore at Llanfairfechan (Caernarfonshire) on 13 March 1993. Records are usually between March and September, but one bird overwintered in Cardiff in 1929–30.

All Montgomeryshire records involve single birds only, but pairs have been seen at two sites in Radnorshire, although breeding was not suspected. Most Breconshire records are from Llangorse Lake, a total of 17 recorded here between 1962 and 1977. One bird was shot in January 1885 near Talybont-on-Usk. Records from Bardsey and the Pembrokeshire islands are few, and merely reflect the scarcity of this species as a passage migrant to the west coast of Wales.

Blue-winged Teal *Anas discors*

Corhwyaden Asgell-las

A vagrant from North America.

This species has been recorded on only four occasions in Wales:

Early 1919	– Male shot near Holland Arms (Anglesey)
17 September 1960	– Female or immature male (Skokholm)
13–16 March 1983	– Male, Cemlyn (Anglesey)
8–9 June 1983	– Pair, Shotwick Meadows (Flintshire)

There have been no fewer than 154 records in Britain and Ireland between 1958 and 1991 with a considerable increase in the period 1976–83 which may be partly a result of an apparent increase in captive breeding, providing a higher escape risk.

Shoveler *Anas clypeata*

Hwyaden Lydanbig

Resident breeding species on Anglesey, and occasionally in small numbers elsewhere in Wales. Otherwise a locally common winter visitor and passage migrant.

A duck of shallow, eutrophic waters, the Shoveler is concentrated mainly in south-east England, particularly on the East Anglian Broads, and in central Scotland. Formerly a rare breeding species, a major increase and spread, mainly during the first half of the 20th century, saw the Shoveler expand north and west into northern England, Scotland, Wales and Ireland (Yarker & Atkinson-Willes 1971). This expansion coincided with a similar increase in western Europe and may have been linked with climatic changes.

In Wales, the Shoveler has for many years been a regular breeder in Anglesey where the marshy pools and lakes with their shallow margins and waterweeds provide ideal breeding habitat. The nest is usually well hidden in the cover of tall grass or shrubs, therefore breeding is not easy to prove unless ducklings are seen. Apart from Anglesey, this species is sparsely distributed in Wales, and although it is not clear whether the population as a whole has declined over the past 30 years, losses of habitat, particularly through drainage, have led to decreased numbers locally.

Forrest (1907) records the species breeding at one site on the island where shooting was not allowed, but he stated that birds also bred elsewhere in the county. At that time, Shoveler nested at Llyn Llywenan and Llyn Penrhyn where a nest with eggs was found in 1900. Breeding was also recorded at Llyn Bodgylched in south-east Anglesey, two pairs rearing 17 young in 1911. In April 1905, at least 10 pairs were seen on Llyn Llywenan (T. A. Coward, Diaries), and a pair nested at Plas Bog in 1904. After the turn of the century, several breeding records are catalogued from the Anglesey lakes, particularly Llyn Llywenan and Llyn Bodgylched.

Shoveler continue to breed on Anglesey today, albeit in small numbers. During fieldwork for the *Breeding Atlas* (1968–72), breeding was proved in only six 10 kilometre squares, four of which were in Anglesey. During the late 1970s, breeding was recorded at Llyn Traffwll, Llyn Llywenan, Llyn Alaw and Llyn Bodgylched and pairs were present at Malltraeth Marsh in May 1976. In 1980, a survey of the Anglesey lakes produced 31 ducklings in 10 broods on three areas of water, and in 1986, breeding was confirmed or suspected at a total of 19 sites (RSPB 1986). By 1990, the number of breeding pairs had risen to approximately 25–30 (A. Moralee, pers. comm.).

Forrest (1907) does not record that Shoveler bred on the Dee estuary at the turn of the century, but it is apparent that by the 1930s, a small colony had become established there. Farrar (1938) stated that Shoveler bred in the privately owned freshwater marshes at Inner Marsh Farm. Williams (1977) describes this species as being an irregular breeder between Point of Air and Ffynongroew and on Burton Marsh in very small numbers, but there are no recent records from this area. He also stated that a few pairs bred irregularly on some of the neighbouring freshwater marshes, but makes no mention of specific sites. In 1959, however, two pairs were found nesting at Shotton Pools, and breeding has also been recorded at Llyn Helyg.

By 1946, Shoveler had been breeding in some numbers for several years at Llyn Ystumllyn in Caernarfonshire, with up to a dozen pairs breeding some years (North *et al.* 1949). Unfortunately the site's importance for wildfowl was much diminished in 1955 when a major drainage scheme reduced the area of wetland, and no breeding has been recorded since. Indeed, there was no other breeding record for the Llŷn until 1986 when an RSPB survey found a female with ducklings on a small marshy pool (RSPB 1986). Since then, sporadic breeding has occurred once more at Llyn Ystumllyn (one to two pairs), Llyn Glasfryn (one to two pairs) and the Afonwen Pools (one pair).

In nearby Merioneth, breeding was first recorded at Llyn Mynyllod in 1896 where a young bird was shot in early August following several sightings of a pair in April and May. The only other breeding record for the county is of a female with ten ducklings at Morfa Harlech on 25 June 1975.

No confirmed breeding records exist for Denbighshire although a pair was present on Ty-mawr Reservoir in April 1975 and breeding was suspected near Ruabon in 1979. Breeding has been recorded from the inland counties of Montgomeryshire and Breconshire. In the former county, a pair hatched six young at one site in 1954 and a pair bred near Montgomery in 1961, although it is thought that the latter probably involved birds which had escaped from a nearby collection. The only breeding record for Breconshire was a pair with one young at Llangorse Lake on 27 July 1973.

Although summering birds were often seen in the county, no breeding was recorded for Cardiganshire until 1949 when a female with a young brood was seen at Cors Caron. A nest with ten eggs was found at the same locality on 24 April 1955, but this was subsequently robbed. Breeding has been suspected at this site more recently and since 1985, breeding has been recorded at the RSPB's Ynys-hir reserve. In the neighbouring county of Pembrokeshire, nesting has been very sporadic throughout the 20th century. Breeding formerly occurred at Angle, Castle Martin, Dowrog and St. David's, and breeding pairs were recorded occasionally during the 1970s. In 1988, breeding was proved at Skokholm and Marloes Mere, and breeding has been recorded annually since 1989 on Skomer. In 1992, three pairs bred on Skokholm and two on Skomer.

A pair allegedly bred at Kidwelly Burrows in Carmarthenshire in 1897, but the first definite record was in 1929 when a nest with 11 eggs was found at Witchett Pool on 6 May. Between six and eight pairs nested here up until 1939, and four to five pairs were present in May 1950. Pairs have nested here since, but breeding is no longer recorded annually.

Glamorgan is now the only county apart from Anglesey and Pembrokeshire where Shoveler are thought to breed each year, albeit at only one site. At the turn of the century, several pairs bred regularly on the Margam Moors, and a nest containing 11 eggs was found here in May 1900 (Baker Gabb, Diaries). Numbers increased steadily until the early 1920s when seven to eight pairs were present. Soon after, however, the habitat was destroyed and no breeding has been recorded there since. Two pairs bred at Eglwys Nunydd for several years in the late 1940s and early 1950s, but the habitat here too has since been destroyed. A pair probably bred at Kenfig NNR in 1971 and 1986, but the only site where breeding occurs regularly in the county is Oxwich Marsh NNR. Shoveler were first recorded nesting here in 1957, since when one or sometimes two pairs have probably bred each year.

In Monmouthshire, records suggestive of breeding have occurred sporadically for the past 30 years with pairs seen at suitable sites in late spring, disappearing then reappearing on one or two dates later in the season. Summering birds are often seen on the Gwent Levels, and during the fieldwork for the *Breeding Atlas*, a pair was observed displaying on a pond near Llandevenny. The only confirmed breeding record concerns a female with newly fledged young at Undy Pool in 1966.

Lovegrove *et al.* (1980) estimated the Welsh breeding population of Shoveler to be less than ten pairs in most years. Over the past ten years, numbers have increased slowly but the Welsh population is still very small, with probably fewer than 40 pairs nesting each year. The Anglesey population, albeit small, represents a significant aggregation of pairs in a British context.

Winter gatherings of more than a few hundred Shoveler are rare in north-west Europe, with only about a dozen sites in the whole of Britain and Ireland holding more than 250 birds. The birds that winter in these islands include immigrants from as far east as western Siberia. The highest numbers were formerly recorded in February or March (Atkinson-Willes 1956) but now numbers peak in November. Recent analyses by the IWRB show that the Shoveler population is increasing in Britain and throughout Europe.

The only site in Wales that regularly supports large numbers of wintering Shoveler is the Burry Inlet. Numbers here reach a peak of over 400 during November and December, with 449 recorded in December 1988. The first birds generally arrive in late August, and by early March, most have moved back eastwards to their breeding grounds. On occasion, birds linger until the spring and 90 were still present at Penclacwydd (Carmarthenshire) on 14 April 1990.

Wentlooge on the Gwent Levels (Monmouthshire) and Pwllheli harbour (Caernarfonshire) both support over 50 individuals most winters, although the latter's importance will undoubtedly be affected when a proposed marina is built. In January 1989, 89 birds were recorded at the Wentlooge (maximum of 160 here in December 1973), and 80 birds were present in Pwllheli harbour in December 1988. Up to 100 individuals were recorded in Pwllheli during the 1950s and 1960s, but they disappeared from here during the early 1970s, only to reappear again a few years later. Elsewhere in Wales, the Shoveler is a regular winter visitor in small numbers to most counties, the most important being Anglesey, where the Valley Lakes between them support over 100 Shoveler each winter, with smaller concentrations of up to 30 individuals at Llyn Llygeirian, Llyn Coron, Cemlyn Lagoon, Llyn Cefni and Llyn Alaw. Many of the smaller lakes and pools also support populations of over ten birds, and the peak winter population on the island is probably in the region of 300 individuals.

Llyn Ystumllyn (Caernarfonshire) was formerly a favoured wintering site, but since its partial drainage, records have been scarce. Since the late 1980s, numbers have increased on the Glaslyn/Dwyryd estuary (Caernarfonshire/Merioneth) with over 40 recorded most winters and a maximum of 80 in December 1990. Up to 20 individuals are sometimes recorded in the Ogwen Estuary (Caernarfonshire), and large flocks formerly gathered on the Dee estuary. Two hundred were counted off Point of Air on 28 January 1967, but the estuary's importance diminished during the 1970s until the establishment of sanctuaries during the 1980s resulted in an increase in the population with 120 recorded here in October 1990.

In Denbighshire and the inland counties of Montgomeryshire, Radnorshire and Breconshire, smaller numbers are generally recorded. Forrest (1907) observed that the Shoveler "occurred rarely near Carno", and records of Shoveler from this county were scarce until the 1950s and 1960s when a spate of records were submitted, many from the Forden area, where 25 were recorded on 17 March 1963 (Montgomery Field Society Reports 1952, 1953, 1956, 1963). There have been a few more recent reports, particularly from Lake Vyrnwy, which is now one of the best-watched waters in the county.

The first record for Radnorshire was in 1947, but it has become more numerous recently with a maximum of seven at Pencerrig in April 1969. However, this species is most frequently recorded from the shallow, reed-fringed waters of Llyn Heilyn. In Breconshire, the species was formerly a rare winter visitor but became much more common during the 1960s and 1970s, particularly at Llangorse Lake and Talybont Reservoir, with much movement between the two. Numbers usually ranged between one and five, but between 10 and 20 were occasionally seen during February and March. Since the mid-1970s, however, the numbers of wintering and passage birds visiting this county have decreased by about 70%.

Large flocks were formerly recorded at Cors Caron, with smaller concentrations on some of the estuaries (e.g. c40 on the Dyfi 17 December 1961). More recently, numbers at both these sites have decreased, with rarely more than two or three at the former site and 5–15 the usual numbers at the latter. In Pembrokeshire, Mathew (1894) described the Shoveler as a rare winter visitor, but by the 1950s, Lockley (1961) described it as "more numerous than formerly" with at least 300 at Orielton on 28 December 1955. Marloes Mere still holds large numbers on occasions, with up to 50 recorded several times during the early 1980s, and smaller numbers are recorded in winter from the offshore islands, some of which may be birds which bred there (e.g. seven on Skokholm on 6 April 1984).

Up to 40 are recorded on the coastal marshes and estuaries of Carmarthenshire, and up to 50 have been seen at the Witchett Pool (e.g. on 31 December 1977). Farther

Foreign-ringed Shoveler recovered in Wales

along the coast in Glamorgan, the Shoveler was a regular winter visitor to Margam Moors, Morfa Pools, Hensol and Kenfig during the early 20th century. It was also formerly recorded irregularly at Roath Park Lake, but nowadays most records are from the Burry Inlet, Oxwich Marsh NNR, Kenfig NNR, the Ogmore estuary and Porthcawl. Numbers of Shoveler recorded in Monmouthshire during the winter months have increased since the 1950s, with most records coming from the Peterstone-Uskmouth area of the Gwent Levels where the shallow lagoons and marshy ground provide ideal feeding conditions.

Passage birds are usually recorded from most Welsh counties during March, with counts of over ten birds sometimes submitted from the Pembrokeshire islands. An unusually high concentration of 37 birds was recorded on Ramsey on 12 March 1972.

There are several records of Shoveler ringed in England and recovered in Wales, mainly in Pembrokeshire (Orielton) or Anglesey. Ringed birds recovered in Wales have come from Denmark (two), the Netherlands (two), Sweden (one) and the USSR (four). The farthest travelled was a bird ringed at Haademeeste in Estonia on 29 June 1978 and recovered near Milford Haven (Pembrokeshire) on 8 January 1979, a distance of 2009 km) west-south-west.

Red-crested Pochard *Netta rufina*

Hwyaden Gribgoch

A rare visitor to Wales, with about 34 individuals recorded in the past 50 years from all parts of the country but mainly from Glamorgan.

Unfortunately, it is not possible to evaluate which, if any, of these records refer to genuinely wild birds. Several of the published records in Wales have the caveat that there are considerable doubts as to the origin of the bird concerned: there is a general feeling that the possibility of escapes from wildfowl collections or birds from feral populations is more likely than the occurrence of genuine vagrants from European breeding areas.

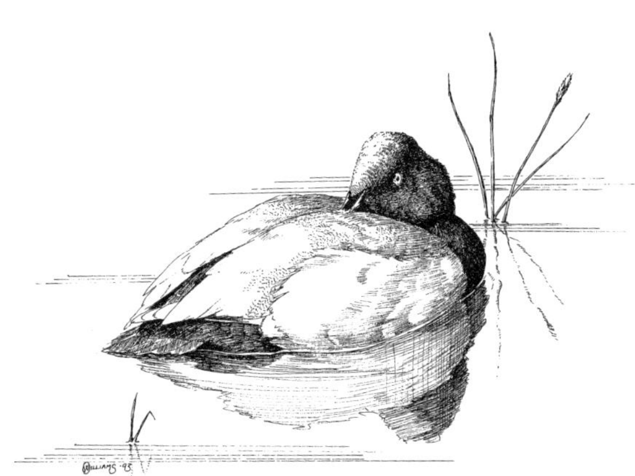

Pochard *Aythya ferina*

Hwyaden Bengoch

Resident breeding species in small numbers in several counties, otherwise a locally common winter visitor, chiefly to coastal counties, and passage migrant.

The Pochard is a scarce breeding bird in Wales, preferring eutrophic lakes and pools with dense beds of emergent vegetation. It does not nest at altitudes over 300 m therefore its distribution in Wales is confined mainly to low-lying river valleys and lakes. In Britain, it is far more widespread in the eastern half of the country, and although it was first recorded breeding over 200 years ago, its expansion has been slow. Its spread into some areas of Wales has been further hindered by the fact that Pochard rarely breed on man-made waters despite the fact that this has been one of the major reasons for its expansion in England (Fox 1991).

Forrest (1907) observed a pair of Pochard on a lake near Bala (Merioneth) on several occasions in May 1901, but suspected that breeding did not take place that year. The first recorded breeding of Pochard in this country was on Llyn Tegid (Merioneth) in 1906 where Bolam (1913) recorded that at least one, and probably two pairs bred. Forrest (1919) also stated that a female and her brood were observed on Llyn Tegid that same year, but he believed that sporadic breeding was already taking place on Anglesey. He also stated that a gamekeeper near Llanarmon Dyffryn Ceiriog (Denbighshire) had reported a breeding pair, but there are no definite records to substantiate this claim. In June 1915, two nests were found on Whixall Moss (Flintshire/Shropshire border) by Norman Gilroy. During the next 20 years Anglesey was the only Welsh county where Pochard were recorded breeding, a female with two young being seen on one occasion at Llyn Maelog (C. Oldham, Diaries).

Pochard were suspected to have bred at the Witchett Pool in 1936, but nesting was not proved until the following year, the first confirmed breeding record in south Wales. A breeding pair was reported from the same site in 1950, and in 1959 a pair bred at Orielton in Pembrokeshire. In the meantime, regular breeding was reported from Anglesey with pairs in most years on Llyn Llywenan in particular. In July 1950, a female with seven ducklings was seen on Eglwys Nunydd (Glamorgan), but breeding has not apparently taken place at this site subsequently.

By the late 1970s, several pairs were nesting each year on Anglesey, at the Witchett Pool and Oxwich Marsh NNR, and in 1983 a pair bred on the River Wye between Aberedw and Glasbury in Breconshire (H. McSweeney, pers. comm.). A pair was recorded from Llyn Du (Montgomeryshire) in late July 1964, but it is unlikely that they bred there. Regular breeding is now reported from Llyn Bwch-llyn (Radnorshire) and an RSPB survey of the Llŷn in 1986–87 revealed two possible breeding pairs (RSPB 1988).

Anglesey remains the stronghold for nesting Pochard in Wales, and breeding has now been recorded from Llyn Penrhyn, Llyn Bodgylched, Llyn Llywenan, Llyn Alaw, Llyn Dinam and Llyn Traffwll. In 1978, a minimum of seven broods was recorded from this county, and a BTO/WWT survey in 1980 revealed a total of 25 ducklings from four broods. In 1986, an RSPB survey estimated the island's population to be between 8 and 18 pairs, with breeding confirmed or suspected at 11 sites (RSPB 1986). Breeding has occurred at five sites in Carmarthenshire, including two fish farms. The Welsh breeding population is increasing slightly at present, and is probably in the region of 30 pairs, with 20–24 of these on Anglesey. Lovegrove *et al.* (1980) estimated the breeding population to be around 15 pairs in the late 1970s.

In winter, the Pochard is widely distributed, but large concentrations are unusual. It is a diving duck generally associated with freshwater and the fact that its breeding and wintering distribution is similar reflects the year-round habitat preference. It often consorts with Tufted Duck and although the number wintering in Great Britain increased considerably following the Second World War, a considerable decline has been recorded since the mid-1970s (Fox & Salmon 1988).

Forrest (1907) found this species most numerous in winter on the sea off Anglesey and on many of the lakes on the western half of the county. He also recorded Pochard on most of the Merioneth estuaries, and occasionally on the Dee. Mathew (1894), on the other hand assessed this species as a not uncommon winter visitor to Pembrokeshire, but cited only a few records.

The most important site for Pochard in Wales is the Severn estuary, with an average of 1831 recorded over the five winters 1985–90. Elsewhere, flocks of over 500 are rare with many inland lakes and pools supporting populations of 50 or less. The Rhymney estuary (Glamorgan) has recently become an important site for wintering Pochard within the Severn complex with over 500 birds here in February 1988, and in December of that year 490 were seen at the Wentlooge (Monmouthshire).

Cold weather movements are common, and a feature of the hard winter of 1962–63 was the exodus of Pochard to coastal waters such as the Taff-Ely estuary, with 600 recorded in February 1963. Such movements are also recorded periodically at Holyhead harbour (Anglesey), and on the River Wye at Glasbury (Radnorshire), but numbers rarely exceed 50. Over 100 individuals are recorded from several of the Anglesey lakes most winters, and similar numbers are often seen on Tal-y-Llyn and Llyn Trawsfynydd (Merioneth), although numbers have declined at the latter site since the late 1980s. Groups of 20–25 winter regularly on all the main Snowdonia lakes but there is much movement between them and the total population is probably less than 200.

Numbers increased markedly in Breconshire between the 1950s and 1980s, with several hundred recorded in some winters from Talybont Reservoir and Llangorse Lake, although numbers have declined over the past ten years. In November 1965, 550 Pochard were seen on Talybont Reservoir, and 440 were recorded here and a further 360 at Llangorse on 20 January 1968. This species is generally an uncommon winter visitor to the estuaries and inland waters of Cardiganshire, and 92 on the Dyfi estuary on 17 March 1976 was exceptional. In Pembrokeshire, over 100 are regularly observed at Bosherston Pools and Llys-y-fran Reservoir, with a maximum of 462 at the former site on 26 December 1981. The Pochard is common on suitable waters in Glamorgan, Roath Park Lake and Llanishen Reservoir being two of the most important sites. Several hundred formerly wintered at Kenfig NNR, and a maximum of 870 on Eglwys Nunydd in 1964 was unprecedented in the county, although 680 were also recorded here in 1966. Pochard are often seen in large flocks on new reservoirs and lakes as such water bodies often abound with charophytes, an algae whose oospores provide favoured food for this species.

Small numbers are recorded on passage from the offshore islands, mainly in October and November, and similar numbers pass off Strumble Head (Pembrokeshire) between August and November. On 2 October 1981, a passage of 70 Pochard was observed flying south off Bardsey in severe north-westerly gales. There are no recoveries of Pochard ringed in Wales, but several of birds ringed in the south of England and subsequently recovered in the Principality.

Ring-necked Duck *Aythya collaris*
Hwyaden Dorchag

The Ring-necked Duck is a rare visitor from North America, which has been recorded with increasing frequency in recent years following the first Welsh record at Bosherston Pools (Pembrokeshire) from 12 February to 8 March 1967.

Since that date there have been further records from Bosherston in March–April 1976 and winter 1977–78 and 1981. Also in Pembrokeshire a male was at Llysyfrân Reservoir on 15 September 1978. In Carmarthenshire there was a male on the River Tywi near Dryslwyn and Bethlehem in winter 1981–82 and in Breconshire a male at Talybont Reservoir on 3 February 1980.

In Montgomeryshire there was a male on pools near Welshpool each winter from January 1983 until at least January 1987. In Pembrokeshire there was a female on Skokholm on 12 October 1986. In Flintshire and Denbighshire there was one at Rhyl and Abergele in February–March 1988 and in the same year one at Upper Lliedi Reservoir from 10 January to 17 February and then at Old Castle Pond, Llanelli (Carmarthenshire) from 28 February to 21 April.

In 1989 there was a Glamorgan record from Broad Pool on 6 May and in 1990 a Cardiganshire record from Llyn Fanod and Llyn Eiddwen (a female) intermittently from 13 January to early March.

There is one outstandingly interesting ringing recovery from Breconshire: one ringed in New Brunswick Canada on 7 September 1967 was shot at Llangorse Lake on 26 December 1967, a movement of 4505 km east, providing direct evidence of trans-Atlantic vagrancy.

Ferruginous Duck *Aythya nyroca*
Hwyaden Lygadwen

A rare visitor to Wales. It is a bird of southern Europe and western Asia which occasionally wanders northwards and there is a sprinkling of records from Wales. Since this duck is commonly kept in captivity some at least of the recent records probably relate to escaped birds.

In all there have been 28 dated records this century, involving 37 individuals. These all refer to one or two birds apart from an early record from Montgomeryshire where one was shot out of a party of seven at Machynlleth on 2 April 1906 (three were seen next day on a small pool near the R. Dyfi, Cardiganshire, and one was on the same pool a week later). Thirteen of the 28 records come from Glamorgan with five from Pembrokeshire and the remainder from Montgomeryshire, Caernarfonshire, Breconshire, Monmouthshire, Merioneth and Radnorshire.

The records fall in the period from August to April inclusive, chiefly in October to December.

Tufted Duck *Aythya fuligula*
Hwyaden Gopog

Resident breeding species in small numbers in several Welsh counties and a locally common winter visitor.

Tufted Duck were first recorded breeding in Britain in 1849 when a pair nested in Yorkshire. By the 1930s they had occupied most suitable waters over much of the country, but even today, this species is absent from much of Wales. Its preference for well-vegetated lakes and ponds and sluggish reaches of river means that most of Wales is unsuitable, although small breeding populations have now been established, mainly in low-lying areas. Added to the paucity of birdwatchers in the Principality, the fact that Tufted Duck nest considerably later than most waterfowl and that they usually select areas of rank herbage means that many nesting attempts may go unrecorded.

Anglesey is the most important area for this species, and Forrest (1907) mentioned that the only confirmed breeding records known to him were from this county. However, he noted only one incidence of breeding, when a brood was seen on Llyn Maelog on 1 August 1892. This is the first known Welsh breeding record although Forrest suspected that breeding pairs went unnoticed at several other sites in the west of the county. By 1917, this species was known to breed annually on Llyn Bodgylched, and on 25 July 1925, a female with a brood of well-grown young was seen at Cors Goch (C. Oldham, Diaries).

The number of pairs on the island gradually increased until in 1976, 16 broods were counted on five different lakes: Llyn Alaw (nine), Pen-y-parc (three), Llyn Penrhyn (two), Llyn Llywenan (one) and Llyn Bodgylched (one). A WWT/BTO survey in 1980 revealed that the population had increased further with 163 ducklings recorded in 31 broods. Lovegrove *et al.* (1980) estimated the Anglesey population at this time to be in the region of 40–60 pairs but by 1986, this had increased to 70–80 pairs (RSPB 1986).

In 1906, one or two pairs nested on Llyn Tegid (Merioneth), but the ducklings were not reared to maturity. This, however, was an isolated breeding record, and no nesting was subsequently noted in the county until two pairs nested near Llyn Trawsfynydd in 1933. Since then, there have been no further confirmed records for Merioneth although several individuals have been seen in suitable habitat during the breeding season. Forrest (1907) stated that this species probably bred at Llanarmon Dyffryn Ceiriog (Denbighshire) at the turn of the century, but he gave no evidence to support this statement. Sporadic breeding subsequently occurred in this county (e.g. Mynydd Hiraethog 1974), but it has not become established as a regularly breeding species.

In Caernarfonshire, five or six pairs breed annually, the most regularly used site being Llyn Glasfryn. Breeding was first recorded in the county during the 1940s when a pair nested at Tremadoc and on the Llŷn peninsula (probably Llŷn Ystumllyn). No further breeding was recorded until 1970 when a female with seven young was seen at Llyn Glasfryn. In 1973, two broods were present on this lake, and a survey of the Llŷn in 1986–87 revealed three probable and two possible breeding pairs (RSPB 1988).

In Flintshire, a pair bred at Shotton in 1959 and regular breeding now occurs on flood prevention lagoons in the Deeside Industrial Park. Breeding has never been proved in Cardiganshire, although it has been suspected at Falcondale Lake on several occasions recently, or in Pembrokeshire although a pair behaved as though they had young at Treginnis in June 1988. In Carmarthenshire, Tufted Duck probably bred at the Witchett Pool in 1950 when a male and two females were seen well into the breeding season. In 1969, a female and three juveniles were seen at Llyn Pencarreg, and sporadic breeding also took place throughout the 1970s at Talley Lakes. The first confirmed breeding record from the Witchett Pool was in 1976, and pairs have bred here annually ever since, with 10–12 pairs present in 1981. In this county, regular breeding also occurs at Talley Lakes and recently pairs have probably bred at two sites near Llanelli.

Breeding was first confirmed in Glamorgan in May 1922 when two nests with eggs were found at Margam, and a month later a female was seen on the lake with her brood. Tufted Duck continued to nest here until the 1940s, and breeding has been proved sporadically since, most recently in 1989. At least one pair bred annually at Eglwys Nunydd until this habitat was destroyed by the construction of the reservoir in the early 1960s. From 1963 onwards, this species bred at Hensol and Talygarn lakes, where it is now firmly established. In 1966, three broods were reared at the former and five to seven at the latter, and breeding has been suspected at Oxwich Marsh NNR recently, although the only confirmed record was in 1980. Breeding was recorded at Kenfig NNR between 1978 and 1981 and although it has not been suspected since, up to 30 birds are present at this site in May and June most years.

In the adjacent county of Monmouthshire, this species was not recorded breeding until 1984 when a nest with eggs was found on Llangattock Moor. Unfortunately, the eggs were later predated when the water level fell. More recently, it has bred at Pen-y-fan Pond on at least two occasions and in 1990, a female with four ducklings was seen near Penallt (S.J. Tyler, pers. comm.).

In Montgomeryshire, breeding has been recorded each year since at least the 1950s, with four sites used regularly during the fieldwork for the *Breeding Atlas*. Pairs have bred near Forden on the Camlad, at Llyn Tarw and at Lymore Pools most years, the total number of pairs in any one year rarely exceeding ten. A pair was suspected to have bred at Llan Bwch-llyn in Radnorshire when one or two pairs were present in late April and early May. By the 1970s, breeding regularly took place at this site and at Pencerrig Lake, and until 1973, pairs bred most years at Pen-y-clawdd, near Llangunllo. Single pairs have occasionally bred elsewhere (e.g. Llyn Heilyn, 1985 and 1992, and Wern Fawr, Howey in 1981).

Breeding was first proved in Breconshire on 15 July 1930 when a female with a brood of ducklings was seen on Llangorse Lake. No subsequent breeding was recorded from the county until 1966 when a pair nested at Talybont Reservoir. In 1970, a pair nested once more at Llangorse Lake, but no breeding has been recorded at this seemingly suitable site since that time, possibly due to increased disturbance from water-based recreation. Since 1980, pairs have bred regularly on various pools in the north of the county, possibly as a result of successful breeding at Llan Bwch-llyn in Radnorshire. Garth Lake is now the only regularly used breeding site with one to three pairs annually, but breeding has also been recorded at Brechfa Pool, Caer Beris, Llanwrtyd Wells and Ffordd Fawr near Glasbury.

Lovegrove *et al.* (1980) estimated the Welsh breeding population of Tufted Duck to be approximately 80 pairs, with 40–60 of these on Anglesey. Since that time, there has been a slight increase in some counties as this species has colonised new areas. The current population is probably in the order of 100–120 breeding pairs.

In winter, the numbers of Tufted Duck in Britain are greatly augmented by immigrants from northern Europe. Flocks are generally small, and there are few lowland lakes which do not hold this species in winter. Immigrants arrive from late September onwards from Iceland, northern Scandinavia and Russia, with some of these moving on to France, particularly during severe weather (Ogilvie 1987).

The most important wintering area for this species in Wales is the Severn estuary, particularly Wentlooge (Monmouthshire) and, unusually for a species with a marked preference for freshwater habitats, the Rhymney estuary (Glamorgan). Numbers were generally low in these areas until a sharp increase occurred during the 1960s. The numbers visiting Llandegfedd Reservoir (Monmouthshire) increased from 30 in 1965 to 180 in 1968. More recently, up to 860 have been recorded at Wentlooge, and in Glamorgan 800 were seen on the Rhymney estuary in February 1988, and 454 at Llanishen Reservoir that same month. The importance of Eglwys Nunydd has increased over the past 30 years and, immediately after flooding in January 1966, 430 Tufted Duck were recorded. Flocks regularly commute between this site and Kenfig NNR where a maximum of 269 was recorded in August 1991. In February 1986, 220 were seen at Port Talbot Docks, and Ynysyfro in Monmouthshire has also held over 200 in recent years.

On Anglesey, over 200 have been recorded on Llyn Cefni (e.g. 201 in 1980–81), and flocks of over 50 are recorded on several other lakes on the island. Elsewhere in Wales, it is a common winter visitor to many freshwater lakes, but rarely in flocks of 50 or more. Tal-y-llyn and Llyn Trawsfynydd (Merioneth) have all supported over 50 birds on at least one occasion, and a flock of 100 on the Lleidi Reservoirs (Carmarthenshire) in December 1976 was exceptional. Up to 60 winter on the Menai Straits off Bangor most years, and up to 80 regularly gather on Llan Bwch-llyn in Radnorshire.

Talybont Reservoir and Llangorse Lake both hold flocks of over 100 each winter, the maximum count at the latter being 400 on 3 December 1973. Numbers have declined by about 60% since the mid-1970s due to an increase in water-based recreation (Tuite *et al.* 1983), but the November peak at Talybont can still reach a maximum of 200.

In Pembrokeshire, up to 150 formerly visited Orielton but today smaller flocks of up to 26 visit Pembroke Mill Pond. Tufted Duck are rare visitors to the Pembrokeshire islands, and before October 1979, there had only been one record from Bardsey. On 11 November 1984, however, a flock of 30 was recorded from the island.

There have been several records of Tufted Duck ringed in England and Scotland and recovered in Wales, but more interesting is the record of a duckling ringed at Nagli in Latvia (USSR) on 21 June 1984 and recovered 2034 km west-south-west in Brecon on 27 December 1985. One bird ringed at Orielton (Pembrokeshire) on 31 January 1938 was recovered in the Pechora Delta (USSR) on 7 June 1938.

Scaup *Aythya marila*

Hwyaden Benddu

Scarce and local winter visitor, also recorded on passage. Has bred once.

The Scaup breeds in the colder climates of Iceland and western Siberia, and is primarily a winter visitor to Britain. It has bred sporadically in Scotland, but most of these were isolated nesting records. In Wales, pairs had been recorded in suitable areas during early summer, but no breeding had been recorded prior to 1988. In that year, a pair bred successfully at Llyn Traffwll on Anglesey, a female with five young being seen on the lake in August. No Scaup were seen here in 1987 and none has been recorded at the site since.

Scaup are one of the most marine of all the diving ducks wintering in Wales, and inland records are generally scarce. The largest concentrations are found around the coast of north and north-west Wales, usually in February and March, and these may be passage birds. There are also smaller flocks in sheltered areas off the south Wales coast.

Forrest (1907) noted that this species was more common around the flat coastline of the Dee estuary, Anglesey and Porthmadog (Caernarfonshire), and that they were found inland generally following rough weather. Large numbers were recorded from Llyn Ystumllyn (Caernarfonshire) in January 1947, and during this period, Scaup were frequently encountered on the Snowdonia lakes. Farrar (1938) also stated that Scaup were common on the Dee, and Williams (1977) records that there was a spectacular increase in number off the mouth of the estuary between 1946 and 1950. In the latter year, up to 5000 were recorded here, mainly on the English side. By the early 1970s, however, numbers had declined dramatically and in 1973, the maximum recorded was a mere 15.

Large flocks are seen most years off the north coast of Flintshire and Denbighshire, between the Point of Air and Llanddulas. This flock increased as the Liverpool Bay flock decreased, and 196 were recorded here on 31 December 1973. In January 1974, the flock had increased to 280, but numbers decreased slightly during the 1980s with 101 at Abergele on 10 January 1988, and 150 between Rhyl and Llanddulas in January 1990, although 400 were seen off Pensarn on 20 February 1986.

At the end of the last century, this species was said to be more common than Tufted Duck and Pochard in Glamorgan. By the middle of the 20th century, small numbers were recorded most years from the Taff and Ogmore estuaries, Llanishen Reservoir and particularly Whiteford in the Burry Inlet. Seven females were seen here on 18 November 1967, and the same number was recorded again in January 1969. During the next two decades, the small flock wintering in the Rhymney Estuary and off Peterstone/Wentlooge (Monmouthshire) increased markedly from 15 on 10 January 1977 to a maximum of 72 at the former site in March 1989. The following year, however, no birds were recorded here, and only two were seen at Wentlooge.

A small flock (usually 5–15 birds) winters in Tremadog Bay each year, moving between Black Rock Sands (Caernarfonshire) and Morfa Harlech (Merioneth). A flock of 90 was present here on 5 February 1989, and 12 were present in Pwllheli harbour in February 1987. Smaller flocks are occasionally present elsewhere in north Cardigan Bay, particularly offshore at Aberdysynni and Aberdyfi (e.g. seven on 23 November 1987), and around the Anglesey coast.

Small numbers are recorded annually off the coast and in the estuaries of south-west Wales. On 17 January 1970, 36 were observed off Borth (Cardiganshire), 28 at Amroth (Pembrokeshire) on 8 February 1986, and 36 were recorded at Witchett Pool (Carmarthenshire) on 8 February 1981, a year in which flocks were frequently seen in Carmarthen Bay. On 11 February 1988, a flock of 88 was recorded off Ragwen Point (Carmarthenshire).

Inland records have increased over the past two decades, but this may merely reflect an increase in the number of birdwatchers in Wales. In Montgomeryshire, a male at Llyn Tarw on 6 November 1981 was the first county record since 1895 when a small flock was seen on the River Vyrnwy near its confluence with the Severn. In nearby Radnorshire, the first record came from the Wye near Glasbury in January 1963, but since then birds have been seen at more than three other sites. Scaup have been recorded almost annually from Breconshire since 1954, with a total of about 18 seen between 1969 and 1989. The largest number seen together was a flock of six at Llangorse Lake on 25 April 1974.

Scaup are vagrants to the Welsh islands, but small

numbers have been recorded on passage off Strumble Head (Pembrokeshire) in October and November.

Eider *Somateria mollissima*

Hwyaden Fwythblu

Generally scarce non-breeding resident and winter visitor.

Although the Eider is the commonest sea-duck around the British coast, in Wales it is rather scarce and nowhere near as abundant as the Common Scoter. Predominantly estuarine birds, Eiders are almost exclusively coastal in their distribution, with inland records being extremely rare. The British Eider population has been increasing since the end of the last century, and in Wales the trend has also been upwards, especially at the two most important sites, the Burry Inlet and Aberdysynni (Merioneth).

It was suspected that Eiders may have attempted to breed at Caldey Is. in the early 1980s and a flightless group seen at St. Govan's Head in the same county in mid-July 1983 was reputed to be a brood. There has never been any suggestion of breeding from the Burry Inlet, although male birds have been seen displaying on many occasions, but no more frequently with other Eiders than with Long-tailed Duck, Common Scoter and Red-breasted Merganser! On a few occasions, Eiders have been reported with chicks but all the evidence indicates that these were young Shelduck.

This species was formerly a very irregular visitor to the Welsh coast, but had been recorded regularly from the Burry Inlet since at least 1917, although investigations by Morrey Salmon revealed that Eider had been visiting this area since the 1870s (Harris 1959). Up to the end of 1957, the maximum number seen together was 30, but a marked increase was noted in 1958, when a maximum of 54 was recorded in September. Over the next 30 years the population increased further and during the 1980s, over 200 individuals were seen on several occasions (e.g. 203 in December 1988). Birds presumably from the Burry Inlet population regularly visit the mussel beds at Salmon Scar at the mouth of the Gwendraeth estuary (Carmarthenshire) where up to 100 have been seen. Over the past five years, however, the population at this site has declined rapidly, possibly due to tidal current changes, and now numbers between 15 and 25 birds most years (S. Moon, pers. comm.).

Eider were first recorded in Merioneth in 1951 when birds were seen off Harlech in June. Small numbers were again present here in June 1952, April 1953, and up to ten in September 1957. By the early 1970s, birds were present in small numbers throughout the year, and a flock of about 30 was seen in the Dwyryd estuary on 15 and 16 April 1972. In the summer of 1973, a raft of 50 birds was seen off Tonfannau, and this population gradually increased until the early 1980s, when up to 90 were recorded. Since then, the population has declined gradually and in 1991 the peak count was 31.

The post-war spread of wintering Eider is well documented (Taverner 1959, 1963), and it was during the late 1950s that small flocks of Eider were first recorded on a regular basis from the Dee estuary. Although most of these records were from the English side of the Dee, mainly around Hilbre Island, smaller numbers were also seen at the Point of Air. The first record for the estuary was a flock of ten seen in 1939, and during the late 1950s and throughout the next two decades, Eider were recorded almost annually. Flocks were usually small, with records of 20 or more birds rare, although 40 were seen off the Point of Air on 2 February 1976. More recently, this species has declined on the Dee, ironically probably owing to the fact that raw sewage is no longer pumped out into the estuary.

At the time that the post-war spread of Eider was at its peak during the 1950s and early 1960s, a small flock was recorded most winters off the Anglesey coast near Llanddwyn. This flock peaked in February 1958 when 23 birds were recorded, but since then flocks off the north-west coast of Anglesey rarely number more than six. Most records from the north Wales coast fall between November and March and, although large numbers are unusual, 30 were recorded in the Menai Straits on 16 April 1963 and 14 were seen at Foryd Bay (Caernarfonshire) on 31 January 1987.

Small parties of Eider are regularly recorded during the winter months off the coast of Cardiganshire, Pembrokeshire and Glamorgan, and records appear to be increasing in the Severn estuary. Unusual records include two to six birds in Aberystwyth harbour between 25 December 1962 and 25 February 1963, and a flock of about 30 off Green Bridge (Pembrokeshire) on 26 April 1984. Eider are scarce visitors to Bardsey and the Pembrokeshire islands, and inland sightings are rare. There are no records from Montgomeryshire or Radnorshire, and only two from Breconshire, a first winter male at Llangorse Lake on 2 February 1976 and two males at Traeth Mawr on 20 March 1977.

The *Winter Atlas* estimates the Welsh population to be no more than 500 birds.

King Eider *Somateria spectabilis*

Hwyaden Fwythblu'r Gogledd

A vagrant.

The first, and so far only, record of this Arctic-breeding species in Wales was a female which frequented the coast off Black Rock Sands near Criccieth (Caernarfonshire) from 28 January 1989. It remained in the north-east corner of Cardigan Bay as far south as Aberdysynni (Merioneth) until at least 23 September.

One was reported seen in the same general area at Abererch near Pwllheli (Caernarfonshire) in about 1958 but there are no details to substantiate the record.

Long-tailed Duck *Clangula hyemalis*
Hwyaden Gynffon-hir

Regular winter visitor and passage migrant, most frequently recorded offshore from counties in north-west Wales. Rare inland.

The Long-tailed Duck's breeding range extends from northern Siberia to Scandinavia, Iceland, Greenland and Canada. Its winter distribution is markedly coastal, and although occasionally found close inshore, it also occurs in large numbers well offshore and is therefore often under-recorded. Formerly a scarce winter visitor to Wales, it is now regularly recorded, mainly from the coasts and estuaries of the north-west. In the south, small numbers are often recorded from some inland reservoirs and large lakes. The paucity of inland reports in north Wales may be due to the lack of shallow waters where the birds can feed. Long-tailed Ducks usually arrive around the Welsh coast in mid-October, leaving again in early April.

Forrest (1907) stated that Long-tailed Duck were recorded from most north Wales counties, but that before 1900, the majority of records was of single birds in winter. However, he noted that several visited the estuaries off Porthmadoc in March and April 1898 and 1899, whilst two specimens obtained on the Dee were shot in October. More than half the north Wales records during the turn of the century came from the west coast of Merioneth and Caernarfonshire, especially off Porthmadog, Penrhyndeudraeth and the Dyfi estuary.

Caernarfonshire and Merioneth are still the two most important counties for this species, with small numbers recorded annually offshore. Black Rock Sands and Conwy Bay (Caernarfonshire) both support over ten birds some years, with highest counts usually in February or March. In December 1991, 48 were recorded at Black Rock Sands, and 45 in the Dwyryd/Glaslyn estuary (Merioneth) in the same month. This species has increased in Caernarfonshire where, prior to 1900, only two records existed. The most marked increase has been since the 1960s, before which time most records consisted of only one or two birds. In Merioneth, most records occur between December and February, and usually involve small flocks. Four were recorded off Morfa Harlech during the winter of 1970–71, and 12 were seen off Aberdysynni on 18 January 1984. One unusual inland record for this county is a Long-tailed Duck on Llynnau Gamallt, near Ffestiniog.

Maximum winter counts of Long-tailed Ducks in Tremadog Bay, 1982–91

1982	*1983*	*1984*	*1985*	*1986*	*1987*	*1988*	*1989*	*1990*	*1991*
40	NC	NC	5	21	14	34	9	1	45

NC, no count

Forrest (1907) recorded birds shot off the coast of Anglesey at Holyhead (two) and a female at Aberffraw. In his *Handbook of the Vertebrate Fauna of North Wales* (1919), he also mentioned an immature male shot at Malltraeth Bay on 23 November 1911. Records for the county have been very sporadic, and even more recently, sightings are usually of single birds. Exceptions to this include six seen near Newborough on 24 April 1963 and six on Llyn Coron on 30 March 1968. (C. F. Tunnicliffe Sketchbooks No. 10).

The Long-tailed Duck is a rare and irregular visitor to Denbighshire and Flintshire, recorded from the Dee estuary and offshore between Pensarn and Rhos Point. Farrar (1938) stated that it was formerly regarded as a great rarity on the Dee, but that it had, at the time of his writing, become a regular, if scarce, winter visitor. Williams (1977) records this species at Connah's Quay and Burton Marsh between September and February. Most records refer to single birds, and the increase in the number of records since the 1930s may be a reflection of the lack of observers at that time. Two other records of note from Flintshire are a bird at Shotton Pools for several days in late November 1977 and one at Llyn Helyg on 9 February 1975.

Most Cardiganshire records come from the Dyfi estuary, the earliest dating back to at least 1893 (Salter 1900). A minimum of four were seen here in 1893–94, and Salter remarked that one Captain Cosens once killed three in one shot on the Dyfi! One Long-tailed Duck was also present at Nanteos that same winter, but the next record for the county is not until 1929 when two were seen on the Dyfi estuary on 9 and 12 January. Recently, records have been annual, but still only small numbers are usually recorded (e.g. one at Aberystwyth, 28–30 January 1977; two at the Dyfi estuary, 31 December 1978), with most sightings off Ynyslas and Borth. By far the largest number seen in the county was 28 off Borth in January and February 1989.

Mathew (1894) makes mention of two birds shot off Stackpole (Pembrokeshire) and an unusual record of a summer-plumaged bird shot at Haverfordwest on 15 June 1843. On 7 December 1906, a female was shot at Milford Haven, but there are no subsequent records for many years from this county. By the 1970s, however, records were submitted almost annually, mostly from St. Brides and Carmarthen Bay but also on occasion from Bosherston, Llysyfran Reservoir or the Gann estuary. The majority of these records are of singles seen between October and February, although up to 12 were in the Saundersfoot–Amroth area between January and March 1989 and four were present at Bosherston on 30 October 1981, a time when several birds were recorded in the county.

Records for Carmarthenshire are much scarcer, the first being a bird on the Witchett Pool from 20 December 1932 until 13 January 1933. An adult male was seen flying down the Tywi near Llansteffan on 4 January 1951, and following this, records became more regular throughout the 1960s and 1970s, and by the late 1980s, sightings of Long-tailed Duck off the Carmarthenshire coast were submitted annually. On 12 May 1991, eight were seen off Pendine, a very late date for this species.

The first record for Glamorgan was a bird that stayed at Roath Park Lake, Cardiff, for a fortnight in December 1905. Following this, birds were recorded in 20 winters out of 61 up until the late 1960s, and records have increased further over the past 20 years. Early records include a pair at Margam Marshes in December 1911, one of which (the female) was subsequently shot (*British Birds* vol. 5, p. 229). Most coastal records in Glamorgan are from the estuaries of the Rhumney, Ely and Burry, as well as Worms' Head and Margam, but inland sightings have also been recorded, notably at Kenfig Pool, Eglwys Nunydd and Lisvane Reservoir. This species was particularly common during the mid-1960s, with records of up to four at Kenfig Pool and Eglwys Nunydd in 1967, three immatures at Lisvane Reservoir on 3 November 1967, and at least five birds in the county in April 1968.

The Long-tailed Duck has become a fairly regular winter visitor to Monmouthshire over the past 30 years, including some lengthy stays of between one and four months on inland waters. All but one of these involved one bird, and include a female on Llandegfedd Reservoir from 18 December 1969 until 1 March 1970 and a male at Pant-yr-Eos Reservoir between 27 March and 24 April 1973. Two birds were present on Llandegfedd Reservoir throughout December 1970.

Records from the three inland counties are much scarcer, with only two known sightings of this species for Radnorshire. The first was a female that stayed at Hindwell from 30 November 1969 until 4 January 1970, and the second a female at Llan Bwch-llyn on 27 December 1983. Forrest (1907) recorded a male shot on the River Severn above Welshpool (Montgomeryshire) in December 1874, which had been mounted and kept in Welshpool Museum. This was the only known record for the county until 1987 when one was seen on Llyn Clywedog. The first record for Breconshire was a pair at Llangorse Lake on 27 October 1893. Two immatures were seen at Llwynon Reservoir on 4 November 1928, and another bird was present here during the winter of 1954. Most subsequent records have been from Llangorse Lake, including one on 2 August 1971, but a male was also seen on Garth Lake from 7 until 19 November 1982. All except one of the records for this county are in October and November.

Five records of Long-tailed Duck were recorded on Bardsey between 1960 and 1983, all but one involving single birds, and all occurring between December and April. Records are also scarce from the Pembrokeshire islands, and mainly occur in late winter.

Common Scoter *Melanitta nigra*

Môr-hwyaden Ddu

A locally abundant visitor, recorded all year round but mainly during the winter months. The distribution is predominantly coastal, although it is regular in small numbers on inland waters on passage.

Although some Common Scoter turn up on inland lakes each year on passage, the winter distribution pattern is strongly coastal, with the bulk of the population found in a few large flocks where there are sandy substrates and shallow offshore waters. Moulting birds occur at several coastal sites, one of the most notable being Carmarthen Bay. Large numbers are present at these sites in late summer and autumn, but the main influx is in October and November, with peak numbers usually between December and February. It is likely that the birds that winter around the coast of the British Isles breed in Fennoscandia and Russia, but this has not been proved.

Carmarthen Bay has long been one of the most important sites for this species in Britain, but until fairly recently, accurate estimates of the size of this population were difficult to obtain as many of the birds stay well offshore. Early records refer to flocks of a few hundred, rarely reaching 1000 plus. It was not until early 1973 that a serious attempt was made to count the birds, when a report from the Royal Air Force unit at Tenby counted 8000 from a stationary boat four miles off Pendine and an estimated possible further 12 000 plus on the radar (Sutcliffe 1975). Shortly after this, an aerial survey of the Bay again revealed that over 10 000 birds were present, and a boat count two weeks prior to the flight estimated that over 20 000 birds were present.

The next detailed survey was not undertaken until 13 March 1974 when a boat count made in an arc five kilometres offshore between Saundersfoot and Worm's Head estimated that about 25 000 birds were present. A further count of about 16 000 in August 1974 off Pembrey confirmed the Bay's importance for this species.

Between 18 October 1975 and 31 March 1976, the RSPB carried out aerial surveys of Carmarthen Bay following pre-determined transect lines. The maximum count during these preliminary flights was relatively small, with 6700 recorded on 17 January 1975 (Lovegrove 1976). The surveys revealed that Common Scoter occurred in loose flocks over wide areas of the Bay, although the number of birds recorded was considerably fewer than had been recorded in previous years. Where sample counts of sexes were made, over 80% were found to be males.

Despite the report's recommendations that more aerial surveys should be undertaken, no detailed aeroplane counts of Carmarthen Bay were carried out until 1977, again by the RSPB. The results of these surveys, undertaken between 16 July 1976 and 21 January 1978, confirmed that the number of birds in the Bay was nowhere near the 20 000 plus recorded in the early 1970s, but over 10 000 were recorded on two occasions in July 1976 (maximum of 10 600 on 17 July). However, the survey did not set out to count all the Common Scoter in the Bay, and it was inevitable that some birds would be overlooked between transect lines, although no large flocks were missed (Lovegrove 1978b). These surveys also showed that the scoter had preferred areas within the Bay (e.g. off Cefn Sidan Sands), and that they were often encountered immediately behind the surf lines where waves disturb the benthic fauna, making it more accessible to the birds.

For the next ten years, surveys of Common Scoter in Carmarthen Bay were restricted to land-based counts at high tide, the results of which confirmed that numbers were lower than they had previously been. Between 1979 and 1981, no more than 600 were usually counted off Cefn Sidan, although 4500 were seen here on 25 June 1977, and a maximum of 12 000 was recorded here on 31 July 1978. Three thousand seen off Saundersfoot on 7 December 1980 was the highest number recorded here for several years, but since then, numbers have generally been low.

Recently, aerial surveys of Carmarthen Bay have been carried out by the RSPB and by the NCC's Seabirds at Sea Team. The latter estimated a total of 7700 birds in the Bay on 12 January 1991, whereas the results of the RSPB's surveys, following the transect lines established by Lovegrove (1976) showed that the decline of this species has continued, with a maximum count of only 2151 on 26 July 1991. These counts were not wholly satisfactory, however, as the frequency was much less than planned because of weather conditions and military activity. However, it would appear that the Bay's importance for this species may be increasing once more with a count of 11 000 off Cefn Sidan on 25 August 1992. Military disturbance in the area may reduce in the near future, to the likely benefit of the offshore flocks.

The coast of north Wales is another important area for wintering Common Scoter. Forrest (1907) described this species as being "numerous at sea all around the coast", and that in Anglesey it was often called the summer duck. Large concentrations were often reported, such as the many hundreds seen off Abergele (Denbighshire) between 31 July and 15 August 1909 and several thousands seen travelling east from Colwyn Bay towards Abergele on 22 June 1913. Regular movements of large numbers were recorded off the Great Orme in the summer of 1912 and 1913, and several thousands were present in the Menai Straits on 26 November 1933.

Recently, large numbers have been recorded from Red Wharf Bay, Anglesey (992 in December 1990) and off Llanfairfechan, Caernarfonshire (3000 in April 1992). There is a great deal of movement between the birds in Red Wharf Bay, Llanfairfechan and the large population off the Denbighshire/Flintshire coast. Flocks of several thousand are regularly recorded at the latter site, with a maximum of 10,000 seen off Pensarn on 29 February 1964. Smaller numbers are recorded here during the summer months (up to 700 in 1973).

Numbers have declined in the Dee estuary since the beginning of the century when it was described as abundant (Williams 1977). Two hundred were seen near the West Hoyle Bank on 24 April 1966, but numbers are usually much smaller. Flocks of over 1000 are regularly seen off Rhyl and Prestatyn, these birds undoubtedly being part of the larger concentration off the coasts of Denbighshire, Caernarfonshire and Anglesey. The results of an RSPB aerial survey of the north Wales coast between Point of Air and Red Wharf Bay in the winter of 1990–91 showed that the largest concentrations are usually found off Llanddulas, but that numbers vary greatly from one month to the next.

The northern half of Cardigan Bay, between Aberystwyth (Cardiganshire) and Criccieth (Caernarfonshire) has long been recognised as an important area for wintering Scoter. Aplin (1902) records seeing thousands off the Merioneth coast in October 1884 and Forrest (1907) states that many thousands were present off Barmouth (Merioneth) in May 1901. Over 1000 were recorded off Tywyn on 2 November 1937, and "fully

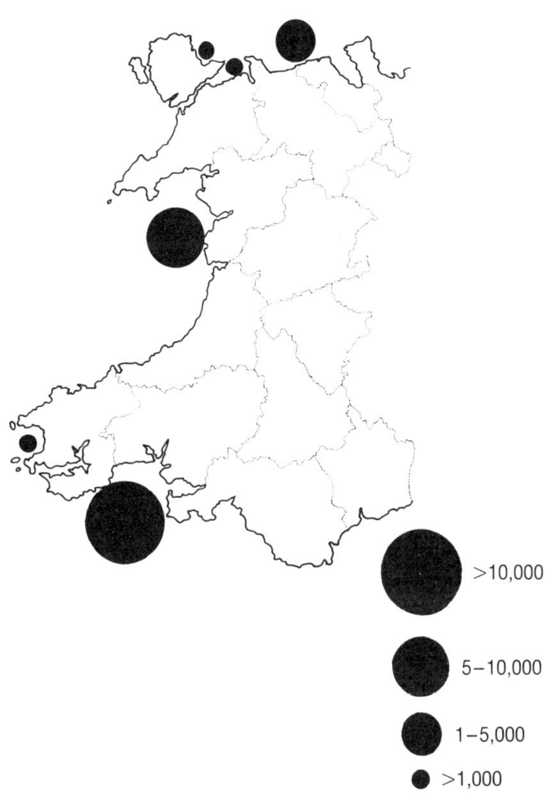

Regular wintering sites for Common Scoter around the coast of Wales

2000" were seen near Harlech (Merioneth) on 7 August 1933 (A. W. Boyd, Diaries). The fact that numbers vary greatly from one year to the next and that this population is highly mobile was confirmed by A. W. Boyd when he could find only one bird in the same area in August 1937. Large numbers have been recorded most years since then, and recently, the population appears to be increasing. On 8 April 1969, 1650 were recorded off Harlech (Merioneth), and during an aerial survey of north Cardigan Bay by the RSPB on 3 March 1991, a total of 999 birds was recorded. Numbers increased in the 1990s and on 20 November 1991, a total of 6421 was recorded during a shore-based count of the Bay between the Dyfi estuary and Black Rock Sands (R. Thorpe, pers. comm.).

Smaller numbers are sometimes seen off the Cardiganshire coast, with Salter recording flocks of 50 below Constitution Hill in February 1892 and about 100 in the same place a short time later. Most of these birds are recorded in the shallower waters to the north of the county. Small numbers are occasionally recorded in the Teifi and Dyfi estuaries, and an exceptionally large flock of 1200 was recorded off Borth in January 1961. A small autumn gathering is seen most years off Aberporth with 200–300 recorded in November and December 1991.

A flock in St. Brides Bay (Pembrokeshire), which formerly held several hundred, declined markedly throughout the 1970s. Up to 400 had been seen here in the early 1960s, but by January 1977, a maximum of only 120 was recorded. The population increased during the 1980s, however, and on 30 December 1988, 2500 were recorded here. Some of these birds often move offshore around the Pembrokeshire islands, and flocks of over 100 are sometimes recorded from Skokholm and Skomer (e.g. 145 off Skomer on 12 September 1985). Common Scoter are common winter visitors to the coast of Glamorgan (1640 off the Gower in August 1968) and Monmouthshire.

Inland records of Common Scoter are scarce, but even in Forrest's day, occasional birds had been seen several miles from the coast. He described how one had been shot at Corwen (Merioneth) on 15 January 1896, and two birds had been shot in Montgomeryshire, one at Maesmawr near Welshpool in about 1885 and one at Llansilin. The rarity of the species inland in Merioneth is reflected by the fact that the first island record for the county was not until 13 November 1972 when three birds were seen on Llyn Tegid. On the other hand, Common Scoter were frequently recorded from Montgomeryshire, particularly from Lake Vyrnwy where this species is now almost an annual visitor (e.g. 24 here in May 1990) during the overland spring passage.

The first county record for Radnorshire was an immature male seen at Rhiw Pool in August 1954, and since then there have been several records, the majority from the Elan Valley lakes. The most seen at any one time was a party of five on Caban Coch Reservoir on 28 June 1970. In Breconshire, records are much more frequent and usually involve birds seen at Llangorse although the first record was an immature killed on the Usk near Brecon on 9 January 1891. From the 1960s onwards, birds were seen more regularly in this county and a total of about 95 birds were recorded between 1969 and 1989, the majority being seen in July (48). Small parties are not uncommon, the largest being a flock of 18 males at Llangorse Lake on 21 July 1971. Today, the species is virtually an annual visitor on passage to the lake.

Common Scoter are occasionally recorded inland from most Welsh counties, but are rare in Caernarfonshire and Cardiganshire. Most inland records, and records from the Pembrokeshire islands and Bardsey, occur between June and October. Rees (1985) states that peak numbers of Common Scoter are recorded off Strumble Head (Pembrokeshire) in June and July, and probably consist largely of birds heading to Carmarthen Bay to moult. A later peak passage in November was probably birds arriving in their wintering areas and the paucity of records in spring indicated that birds were almost certainly migrating by night.

Common Scoter clearly benefit from extensive areas of unpolluted, undisturbed soft coast with shallow waters typical of north Wales and Cardigan and Carmarthen Bays. Welsh waters hold a high proportion of the total British wintering population, particularly as numbers in the Moray Firth have been declining in recent years (R. Evans, pers. comm.). Kirby *et al.* (in press) concluded that wide fluctuations in Scoter populations were due to flocks locally eating out large age-classes of bivalves and subsequently abandoning feeding sectors whilst these populations recovered to a level where bivalve size made a return worthwhile.

A male Black Scoter *M. n. americana* was at Newgale (Pembrokeshire) in November and December 1991, the only record of this subspecies in Wales.

Surf Scoter *Melanitta perspicillata*

Môr-hwyaden Yr Ewyn

A rare trans-Atlantic visitor to Wales.

The first fully substantiated record was an adult male, freshly dead, at Ginst Point (Carmarthenshire) on 15 January 1971. Prior to that there was only a doubtful record of its occurrence in Glamorgan in the 19th century on the authority of L. L. Dillwyn.

Between 1971 and 1991 there have been a total of 23 records involving at least 29 individuals, all in the winter period from October to April inclusive. Most of these refer to single birds but there were four males seen flying past Strumble Head on 13 November 1987 and two males and one female off Harlech (Merioneth) from 31 December 1988 until at least 24 January 1989.

The bulk of the records come from Pembrokeshire and Caernarfonshire and all except one come from coastal waters (the exception was an immature male at Eglwys Nunydd Reservoir/Kenfig, Glamorgan from 10 October to 12 November 1981). Several of the individuals have been long-staying, particularly at favoured localities such as the north Caernarfonshire coast and St. Brides Bay in Pembrokeshire where it is possible that the same birds have returned in successive winters.

The records show an increase in the 1980s with at least one sighting each year since 1981 but this may be a reflection of an increase in sea-watching rather than a genuine increase.

Velvet Scoter *Melanitta fusca*
Môr-hwyaden y Gogledd

A scarce and local winter visitor, regular in small numbers to some coastal counties. Occasionally recorded on passage. Rare inland.

Around the coast of Britain and Ireland, the Velvet Scoter is the rarest of the major wintering species of European sea-duck. At most sites, it is found amongst flocks of Common Scoter, and it is often difficult to distinguish between the two, especially when they are far offshore. However, they sometimes occur in distinct subgroups where they feed and display away from other duck species. Although small numbers are known to flock with Common Scoter during moult, the largest numbers are recorded during the winter months, often peaking as late as March or April.

The only sites in Wales that regularly hold over 15 individuals are north Cardigan Bay, particularly off Black Rock (Caernarfonshire) or Aberdysynni (Merioneth) and Llanfairfechan (Caernarfonshire). Forrest (1907) stated that a few Velvet Scoter had been seen at Barmouth (Merioneth), and that three immatures had been shot near Penrhyndeudraeth (Caernarfonshire) on three dates in the six years prior to 1898. He also noted that all of the birds reported to him had been quite separate from flocks of Common Scoter. At the turn of the century, one was also shot on the Dyfi estuary (Merioneth/Cardiganshire), although Forrest states, rather regretfully, that the bird was not retrieved.

Up until the 1970s, this species was still regarded as being an uncommon and irregular winter visitor to the Merioneth coast, with most records involving single birds. Five birds together off the south-west Merioneth shoreline on 1 April 1956 and 11 April 1958 were considered to be exceptional, but by the 1980s, flocks of over ten individuals were seen offshore most winters. The maximum recorded to date is 31 at Red Wharf Bay on 22 December 1976 (Vinicombe 1977). Two were present off Aberdysynni as late as 16 July 1989 and on 2 December 1991, 14 were present along with 370 Common Scoter near the sewage outlet off the Barmouth coast (R. Thorpe, pers. comm.). The only inland records for Merioneth are one bird on Llyn Tegid on 3 November 1906 and one on Llyn Trawsfynydd between 25 and 27 November 1987.

Velvet Scoter are now recorded most years amongst the larger flocks of Common Scoter off the north Wales coast. Forrest (1907) first reported this species here in 1901 when an adult was shot near Deganwy (Caernarfonshire), and four were seen off Llandudno on 30 November 1909. These were all females or immatures and were quite separate from the more numerous Common Scoter present in the Bay at the same time. On 4 August 1919, a flock of five was seen close offshore near the Great Orme, and later that year, six were present in the Conwy estuary for a few weeks. In September 1921, three more were seen off the Great Orme, and in April 1929, a party of six and another party of seven were seen in the Conwy estuary.

These two sites yielded most Velvet Scoter records for Caernarfonshire during the early half of the 20th century, but more recently, regular sightings have been submitted from the Llanfairfechan area, with a staggering 79 here in April 1992. The late G. Ireson (pers. comm.) found small numbers whenever he visited this site (e.g. seven males and nine females on 19 November 1959; eight males and 13 females in 1963), and up to three birds summered here in 1992. One interesting record was a single bird seen off Bardsey on 14 September 1965, the first time it had been recorded around the island.

In the Dee estuary, this species was recorded in most years up until the 1970s following the first record in 1921, with birds being seen during all months of the year except July and August. Now, most Flintshire and Denbighshire records come from between Llanddulas and the Point of Air. In 1924, an oiled male was washed ashore at Llanddulas along with several Common Scoter, and most records from this stretch of the north Wales coast have involved single birds. In late 1939, however, 19 were seen off Llanddulas, and the same number was again recorded off Rhos Point on 20 February 1977. In April 1944, a male was reported some miles inland at Nannerch (Flintshire), an unusual occurrence for this predominantly marine species.

Most recent records off the Anglesey coast come from Red Wharf Bay where the shallow waters and sandy substrate provide ideal feeding areas for this species (e.g. one on 22 February 1987 and one on 12 December 1987). Hope Jones (1965) also recorded Velvet Scoter offshore at Newborough, on the west coast of the island, during the early 1960s (e.g. singles on 25 February 1962 and 24 April 1963).

Most Cardiganshire records refer to birds which were seen offshore in the north of the county between Aberystwyth and the Dyfi estuary. Five were seen off Borth on 28 November 1951, and four were recorded along with several hundred Common Scoter off Aberystwyth between 19 November and 8 December 1955. Ten were seen in the same area some years earlier (no date given), and five were recorded here in early March 1956. One bird was seen at Glandyfi near the Montgomeryshire border on 26 October 1961, and since 1986, this species has been recorded annually, usually off Ynyslas or Borth (20 on 3 March 1986).

This species is now a regular winter visitor to Pembrokeshire, often recorded at Strumble Head, St. Brides Bay and Druidston (e.g. five on 25 December 1985). It was first recorded at Goodwick Bay when six were seen together on 16 November 1886, and one was picked up exhausted from the shore at Tenby in December 1889. Inland records are scarce, but birds have been seen at Slebech Mill Pool on 8 February 1939 and recently at Bosherston (one on 3 December 1983). The maximum recorded in the county was 12 off Newgale on

30 December 1988. This species is rarely recorded from the islands, although singles are sometimes recorded with other sea-duck.

In the adjacent county of Carmarthenshire, Ingram and Salmon (1954) could find no recent records of this species despite the fact that occasional birds had been reported at the turn of the century. During the past few years, sea-watching from the Amroth-Pendine cliffs and from Cefn Sidan beach has produced several sightings of up to two birds amongst the Common Scoter flocks of Carmarthen Bay.

At the end of the 19th century, the Velvet Scoter was a regular winter visitor to the Glamorgan coast, but during the first half of this century, it was recorded only three times. Since the 1950s, however, it has once again become a regular, albeit uncommon winter visitor, mainly to the west of the county. Most records are of one or two birds, usually off the Gower coast, and sightings range from September to April, with the majority in December and January. The maximum recorded was eight at Worms' Head on 15 March 1964, but three have been recorded on several occasions (e.g. off Whiteford on 12 March 1967 and near Rhossili on 11 November 1987). On 5 January 1958 one bird was seen inland on Lisvane Reservoir. Velvet Scoter rarely venture farther eastwards up the Bristol Channel to the Monmouthshire coast, with only nine records from this county to date.

This species has never been recorded in any of the three inland counties of Wales.

Goldeneye *Bucephala clangula*

Hwyaden Lygad-aur

Locally common winter visitor and passage migrant, with an increasing number of individuals summering on inland lakes.

The Goldeneye's main breeding area is the taiga of northern Europe, Asia and North America, where it generally nests in deep holes in trees. In Britain, breeding is confined to Speyside in northern Scotland where it first nested in 1970 although, recently, pairs have been seen displaying in spring in suitable habitat in northern England and Wales.

No breeding has ever been recorded in Wales, although single pairs, probably "pricked" birds, were recorded nesting in 1931 and 1932 on the Dee estuary near Burton (Cheshire), a short distance from the Welsh border. One clutch was collected and the records have been accepted despite the unusual site (Hardy 1941). More recently, displaying pairs have been recorded into May in many suitable sites in Wales and some birds have spent the summer here, although these may be injured or sick birds, or escapes from wildfowl collections. It is hoped that the provision of nest-boxes in suitable areas may encourage this species to breed in future.

Although it is most frequently recorded from estuaries and fairly sheltered coastal areas, the Goldeneye's winter distribution shows that it is also widespread inland. Small wintering flocks and individuals are widely scattered on lakes, reservoirs and even rivers, and are often seen in association with other diving ducks. In Wales, the Goldeneye appears to be on the increase, albeit on a small scale. It is never present in high numbers, with the largest offshore assemblies rarely exceeding 120.

Forrest (1907) records the Goldeneye as being more common on the west rather than the north coast of north Wales. Today, however, Goldeneye are most common in the north, particularly in the sheltered areas off Llanfairfechan (Caernarfonshire), and on occasion, large numbers are recorded in the adjacent Bangor harbour (85 in January 1989). Several of the Anglesey lakes regularly hold up to 20 birds, and the total population on the island is estimated at 150 each winter (Evans 1990). However, in March 1986, over 150 were recorded on the Inland Sea, a favoured site for this species on the island.

Its main haunt in Caernarfonshire is the Ogwen estuary and the Menai Straits, where 50–70 winter annually. The Dyfi estuary and Llyn Tegid regularly support over 30 birds, with a maximum of 45 seen at the former on 2 January 1991. The latter site is one of the inland lakes where this species is regularly encountered in late spring.

Formerly recorded as being an uncommon winter visitor to most inland and south Wales counties, the Goldeneye is now a regular winter visitor. The Cleddau estuary (Pembrokeshire) has become an important wintering site for this species, its sheltered bays and invertebrate-rich mud providing ideal feeding conditions. In 1985, 45 individuals were recorded here, and 43 in February 1989. Up to 30 have been recorded on the Tywi at Llansteffan, 38 were noted off Burry Port (both Carmarthenshire) on 31 October 1986 and similar numbers have been recorded in Breconshire at Llangorse Lake and Talybont Reservoir (20 in January 1991). Two sites in Glamorgan, Eglwys Nunydd and the Ogmore estuary, support over 30 Goldeneye each winter, with recent maxima of 31 at the latter in March 1989 and 36 at the former in December of the same year. Up to 55 have been recorded at Eglwys Nunydd in the past.

Several observers record an increase in wintering Goldeneye since the early years of the 20th century, and many also note an increase in the number of males. Forrest (1907) records that "adult males are curiously rare", with the ratio estimated as one in thirty. Today, however, males are regularly observed both inland and on the coast.

On most of the offshore islands, records of Goldeneye are rare, most being of singles. On Bardsey, this species is recorded only on passage.

Hooded Merganser *Mergus cucullatus*

Hwyaden Benwen

A vagrant from North America.

The first, of only six in Britain and Ireland, was a first-winter male killed in the Menai Straits near Bangor (Caernarfonshire) in the winter of 1830–31.

There were doubtful records from Glamorgan in 1838 and of a male in the Menai Straits in March 1911. In addition one adult male and one immature were said to have been shot near Barmouth (Merioneth) in 1864. Although the skins were genuinely of this species, there is some doubt as to their true origin and the record has not been generally accepted.

Smew *Mergus albellus*

Lleian Wen

Scarce winter visitor, not recorded annually in any county, though subject to cold weather influxes from the Continent.

Smews do not nest in Britain or Ireland, and in winter their British distribution is distinctly south-eastern. In Wales, this species is very scarce in most winters, and a true estimate of numbers, especially during cold weather influxes, is difficult to obtain owing to the fact that they are extremely mobile. Most of the birds recorded in Wales are redheads (adult females or immatures), but the ratios vary from one year to the next.

Eyton (1838) stated that Smew occurred on Welsh lakes during hard weather, but most of Forrest's records were either on estuaries or on rivers near the sea. Forrest also mentioned an influx of this species during the severe weather of January 1891 when seven were seen on the Dee near Saltney Ferry (Flintshire). Several Smew were also shot on the Dee in 1861, and records from the 19th century also refer to single birds being shot on the Cefni River (Anglesey) about 1868, a single male near Llanrwst (Denbighshire) in 1830 and 1898, an immature which stayed at Traeth Mawr (Merioneth) for a month in February 1897 and an unusual record of one obtained at Porthmadog market (Caernarfonshire) on 21 January 1881!

Since the 1940s, the Smew has become a slightly more frequent visitor to many parts of north Wales, particularly the Dee and Anglesey. Most records refer to single birds, but there have been occasional records of small flocks, particularly during large influxes such as in 1987 when six were seen on Afon Glaslyn (Caernarfonshire) on 25 January, nine at Ynysmaengwyn (Merioneth) on 17 January and six at Afonwen (Caernarfonshire) on 30 December.

In about 1882, a Smew was shot on the River Taff near Cardiff, and there are subsequent records of birds being shot at Gabalfa in 1892, near Cardiff again in December 1927, and one at Peterstone-Super-Ely in January 1940. Smew became more frequent, but still irregular, after 1920–21. In the 45 years after this date, one or more birds occurred in the county in 26 winters. An influx during the winter of 1953–54 saw three males and 10 redheads alternating between Roath Park Lake and the Cardiff reservoirs, and the following year, two males and 14 redheads were seen in the county.

The late 1960s saw small influxes of birds into Glamorgan, one particularly long-staying redhead being seen on Eglwys Nunydd Reservoir between 30 November 1966 and 25 April 1967. The following year, up to two birds were seen on Lisvane Reservoir, Kenfig Pool and Eglwys Nunydd Reservoir. Of the two most recent influxes in 1978–79 and 1984–85, birds were recorded in Glamorgan only during the latter. The former was strangely very localised, involving an influx of 13 birds all into Monmouthshire, from where there have been several recent records, e.g. two at Llandegfedd Reservoir in 1988. A male was seen at Kenfig on 10 May 1991, a very late date for the species.

Mathew (1894) records the Smew as being a not uncommon winter visitor to Pembrokeshire, and makes mention of birds being shot at Goodwick and Stackpole Court. All records during this time were of ones or twos, and it was not until 1939 that a small flock was seen in the county, when six were seen at Orielton. Since then, records have come from several sites in the county, but most often from Bosherston Pools, and again records have usually involved single birds.

Only two historical records exists for the adjacent county of Cardiganshire, a male shot at Nanteos at the turn of the century, and a bird shot on the Dyfi by Sir Pryse Pryse before 1894. In recent years, this species has become more frequent, although most records are from the River Dyfi. The earliest Carmarthenshire record is a bird shot on "Llanelli Marsh" in 1841, although at the beginning of this century, two specimens are reported to have been taken on the River Cothi. The greatest number of birds seen together was four on the Witchett Pool on 4 February 1940, and in recent years this species has remained scarce, with the few records of singletons occurring on or near the coast, often in periods of hard weather.

Forrest (1919) records examining an immature male shot near Churchstoke (Montgomeryshire) on 5 January 1909. There have been only three subsequent records from this county, and Radnorshire likewise has very few records of Smew, with no sightings prior to 1967. Since then, this species has been recorded on five occasions, most of these from the Elan Valley reservoirs.

Unlike the other two inland counties of Wales, several records of Smew exist from Breconshire. In the last century, this species was known to visit Llangorse Lake and the River Usk. The first mention of a large influx of birds came in 1971–72, however, when several Smew were recorded on the southern reservoirs. Between 1969 and

1989, 14–16 birds were recorded, the majority from Llangorse Lake and Talybont Reservoir, with much movement between the two. Interestingly, only one of these records involved an adult male.

Smew are very rare vagrants to the islands around the Welsh coast, the only Bardsey record being a single bird seen offshore on 10 November 1962.

Red-breasted Merganser *Mergus serrator*
Hwyaden Frongoch

Locally common winter visitor, particularly to the north Wales coast. Small numbers breed, chiefly around the coastline from Cardiganshire to Anglesey.

Unlike its close relative the Goosander, the Red-breasted Merganser is predominantly a coastal breeder in Wales. In Scotland, Ireland and the north of England, this species is regularly found on inland lakes and rivers, but in the Principality it has not colonised these habitats to such an extent as the Goosander. It generally nests on the ground amongst rank vegetation and sometimes even in rabbit burrows or clumps of marram grass on sand dunes.

Although in the past, pairs were recorded from suitable areas during the breeding season (e.g. a male and two females at Llyn Ebyr in Montgomeryshire on 2 April 1927), the first confirmed nesting did not take place until 1953 when R. L. Vernon found a nest with eggs by an unnamed bay in Anglesey (*British Birds* vol. 48, pp. 135–136). The pair was later seen on the sea with 14 young, but when they nested here again the following year, they were evidently unsuccessful, but a second pair nested nearby, hatching 16 ducklings. By 1963, breeding was recorded at several sites, and two nests containing six and eight eggs respectively were found at Dulas Bay (the late G. Ireson, pers. comm.). The small population on the island subsequently increased to seven pairs by 1965 and in 1957 two pairs bred in Merioneth, one near Traeth Bach and another east of Aberdyfi. By 1971, eight pairs were breeding around Traeth Bach, about six on the Mawddach estuary and up to five on the Merioneth bank of the Dyfi. Over the past two decades, breeding has been recorded inland on several Welsh lakes and rivers.

In 1958, a pair bred in Caernarfonshire, and subsequently birds quickly colonised the river systems in the north of the county. In 1973, 25 pairs were recorded, with 12 of these on the River Ogwen alone, and since then, breeding has been recorded on Afon Lledr, Llyn Padarn and Afon Dwyfor. In 1992, 8–10 pairs were recorded on the Ogwen (P. Schofield, pers. comm.) By 1957, Red-breasted Mergansers were also breeding on the Dyfi, and a female with a brood was seen a few kilometres up-river at Llanwrin (Montgomeryshire) in 1970. By the late 1970s, at least five pairs were nesting on the Cardiganshire shore of the Dyfi, and from 1976 onwards, at least one pair nested on the River Leri nearby. On 2 August 1960, 24 young were seen in the charge of one duck near Glandyfi (Cardiganshire), and 115 ducklings were counted on the upper reaches of the estuary on 4 August 1976. From 1984 or 1985 this species began nesting on the rivers Rheidol and Ystwyth, and as birds have been more frequent on the Teifi estuary in recent years, this river may also be colonised soon. In 1967 and 1970, breeding probably took place on the Glamorgan side of the Burry Inlet, where juveniles hardly able to fly were found in late summer. No breeding has been recorded here since however.

In 1981, the RSPB undertook a survey of sawbills on selected rivers in north and mid-Wales, and the results show that the number of Red-breasted Mergansers breeding inland had increased in number and distribution (RSPB 1981). Between 27 and 43 breeding pairs were recorded on nine rivers, the majority (9–20) on the Dyfi, but with good numbers (seven to nine) on the Dee. In 1985, this survey was repeated, and the results revealed that the total number of pairs (24–41) and their distribution was similar to that in 1981 (Tyler 1985). Interestingly, however, the 1985 survey showed that there was considerable segregation between Red-breasted Merganser and the Goosander, with the latter species being most common on the Wye and the Vyrnwy.

In 1990, the National Rivers Authority funded another survey of sawbills on Welsh rivers in response to fishermen's complaints that they were increasing rapidly and having an adverse effect on fish numbers. This survey indicated that the number of Goosanders had increased since 1985, but that the inland Red-breasted Merganser population had declined markedly during this time (Griffin 1990). A total of 22 adults and three juveniles were recorded on the Dysynni and the Dee (Merioneth), the Conwy (Caernarfonshire), the Ystwyth (Cardiganshire) and the Wye (Radnorshire/Breconshire). There is some speculation as to the reasons for the decline but it may be that this species is being ousted by the more successful Goosander.

Numbers of Red-breasted Mergansers found in Welsh rivers in 1981 and 1985

River	*1981 (RSPB)*	*1985 (WW)*
Tywi	2	–
Conwy	Not surveyed	1
Banwy	1	Not surveyed
Clwyd/Elwy	–	2–3
Dee/Alwen	7–9	9
Teifi	1	–
Wye	2	1
Irfon	–	1
Ystwyth	1–2	2–3
Rheidol	2	–
Dyfi	9–20	7–22
Mawddach	2–4	1
Total	27–43	24–41

Lovegrove *et al.* (1980) estimated the Welsh Red-breasted Merganser population to be in the region of 100 pairs, but since that time, the inland population has increased despite the most recent survey results (Griffin 1990). There have also been local increases along the coast of north-west Wales, therefore the current population is probably around 150 pairs.

In winter, the Welsh Red-breasted Merganser population is almost exclusively coastal. During hard weather, however, small flocks will sometimes seek refuge inland, although they rarely stay for more than a few days. As with the breeding population, the majority of wintering birds are found around the estuaries and bays of north Wales, although sizeable flocks overwinter on occasion in Pembrokeshire and the Burry Inlet.

Forrest (1907) recorded this species as a regular winter visitor to the west coast of Merioneth, with the last flocks dispersing at the beginning of April. He also stated that it was much scarcer around the coasts of Anglesey and north Wales, and that inland records were mainly from the Dee between Bala and Corwen (Merioneth).

At the turn of the 20th century, the Red-breasted Merganser was a scarce winter visitor to Wales, with sizeable flocks of over 20 birds recorded only on the Dyfi and around Penrhyndeudraeth (Merioneth). By the middle of the century, wintering numbers increased as the breeding population expanded in Scotland, northern England and eventually reached Wales. The population increased most rapidly with the onset of breeding from the late 1950s onwards and by 1970, several coastal sites regularly held more than 20 birds.

The area between the Ogwen estuary and Llanfairfechan (Caernarfonshire) is now an important moulting area for this species. Numbers build up at this site in July and generally reach a peak in August when over 200 may be present (e.g. 250 in August 1989). The population has increased greatly since July 1962 when c70 were present, and on 26 August 1973, an exceptional flock of 441 was recorded. Smaller numbers are present in more sheltered areas along the north Wales coast in winter, with over 80 occasionally recorded off Rhos Point (Denbighshire). Numbers have also increased in the Dee estuary since about 1960, with an unprecedented flock of 250 birds present in February and March 1965 (Williams 1977).

One of the most important wintering sites for this species in Wales is the Dyfi estuary where numbers reach a peak in March and April. Large flocks are sometimes seen offshore (e.g. c100 off Aberdyfi on 22 January 1968), but most often the birds seek refuge in the calmer waters of the estuary itself. Numbers here regularly reach over 100 birds, with a maximum of 166 recorded in April 1989. Sizeable flocks are also recorded on the Mawddach estuary, and on occasion, the Dysynni Valley (both Merioneth). At the former site, birds are recorded all year round, with a maximum of 63 recorded in February 1989. In May 1987, 32 birds were seen in the Dysynni Valley, but numbers here are generally much lower.

The small wintering flock on the Cleddau estuary (Pembrokeshire) has increased over the past twenty years; the largest flock recorded was one of 28 in Hook Reach on 31 January 1987. Similar numbers winter on the Burry Inlet and the Gwendraeth and Tywi estuaries (Carmarthenshire), although these populations probably coalesce to give higher numbers. More than 50 were present on the Gwendraeth estuary on 22 November 1975 and 48 at Ginst Point on 8 April 1978.

Smaller flocks are recorded in many other Welsh estuaries, and large numbers sometimes congregate in the more sheltered bays around the coast of Anglesey (e.g. 63 in Beddmanarch Bay on 29 September 1989). Inland, small flocks are reported from all counties in some years and here, too, winter records have been increasing. This species is recorded most frequently on the offshore islands of Bardsey, Skomer and Skokholm during the months of April, May, September and October, but even during these months, they are usually seen in small flocks of fewer than six birds.

There is evidence to suggest that the Welsh wintering population of Red-breasted Merganser may be augmented by migrants from Iceland and possibly Scandinavia, but there are no relevant ringing recoveries from Wales to support this. The overall increase in wintering numbers mirrors the expansion of the breeding population and is in line with the overall pattern in the trend of the British population (Owen *et al.* 1986).

Goosander *Mergus merganser*

Hwyaden Ddanheddog

Breeding resident on most Welsh rivers in mid and north Wales, the breeding population slowly expanding. In winter, numbers are augmented by birds dispersing from farther north.

The Goosander is a sawbill duck which favours fast-flowing rivers and upland lakes with wooded shores. It first bred in Britain in Perthshire in 1871, and the population has gradually spread southwards, reaching Northumberland by 1941. This species is now well established in Scotland and northern England and, during the 1970s, Goosanders also colonised Ireland.

The fast-flowing, unpolluted rivers and streams of upland Wales, with their plentiful supply of fish, are ideal habitat for breeding Goosander, but summering pairs were not recorded here until the 1950s. Prior to this, it had been recognised as a winter visitor in small numbers to freshwaters, but the numbers had gradually increased since the turn of the century. In 1952, a pair was present on the River Vyrnwy (Montgomeryshire) until late May, but breeding was not recorded. A pair was seen here in late April and early May each year for over a decade, but breeding was not proved (Lovegrove 1978a).

In July 1968, an adult female and one juvenile were seen on the Afon Dyfi, with what was presumed to be the remainder of the brood found nearby. That year, and in 1969, broods were also reported, but not confirmed, from the River Vyrnwy, but it was not until 1970 that breeding was finally proved in Wales when a female and her brood were seen on Lake Vyrnwy in Montgomeryshire. Thereafter this pair bred annually in a hollow tree near the edge of the reservoir, and during the 1970s, several other pairs bred on the streams running into the lake.

In 1972, breeding was recorded on the Elan Valley reservoirs in Radnorshire, and by 1975 at least two pairs were nesting at this well-wooded site. Two years later, three pairs nested here and another breeding pair was discovered on the upper Wye (Radnorshire). By this time, breeding had been proved as far south as the Chain Bridge on the River Usk in Monmouthshire where the first nesting in the county was recorded in 1975 and the same site was used until 1978 when floods washed away the hollow tree.

By 1977, a minimum of 10 pairs was nesting in these three counties, although it is likely that several other pairs bred unnoticed. However, even this early in their colonisation of Welsh rivers, Goosanders were being shot quite extensively on game-fishing rivers such as the Wye and the Severn (Lovegrove 1978a).

By 1980, the Goosander had increased its range to include most of the larger rivers in mid and north Wales, and in that year, breeding was first confirmed in Carmarthenshire. In July 1981, the RSPB undertook a survey of sawbills on selected Welsh rivers for the Welsh Water Authority. Because of the timing of the fieldwork, some breeding pairs were probably missed, but a total of 56–59 Goosander were recorded, with broods seen on the Rivers Alwen (Denbighshire), Banwy (Montgomeryshire) and Wye (Radnorshire and Breconshire). The survey did not cover several rivers known to support this species at that time, including the Severn, Usk and Vyrnwy.

A more thorough survey of Welsh rivers was undertaken by the RSPB in 1985, with all likely rivers surveyed in March/April to record breeding pairs, and again in July, in order to record broods. The results revealed that the population had increased to a minimum of 85 pairs, with the highest numbers on the Vyrnwy and the Wye. The overall estimate was 100 breeding pairs, with breeding recorded on 14 rivers (Tyler *et al.* 1988).

In 1987, a female with six ducklings was seen on the River Neath (Glamorgan) in July and August, the first reported breeding of the species in the county. At this time, the population was known to be expanding rapidly and in 1990, the National Rivers Authority funded another survey of Welsh rivers by the RSPB. The results indicated a spring total of 525 Goosanders, with a minimum of 82 pairs (Griffin 1990). Highest numbers were found on the Severn, Usk, Vyrnwy and Wye and birds had also colonised the Mawddach (Merioneth), Dwyfor (Caernarfonshire), Clwyd (Denbighshire), Elan (Radnorshire), Ystwyth and Rheidol (Cardiganshire), the Neath (Glamorgan) and Loughor (Carmarthenshire).

Despite an increase in the number of spring records since 1985, the number of broods was almost identical in both surveys. The main reason for this is undoubtedly an increase in the illegal shooting of this species, particularly on the Vyrnwy, Wye, Severn and Teifi. During the July counts in 1985 and 1990, very few males were recorded on Welsh rivers as these birds are known to migrate to the Tana estuary in north Norway to moult (Little & Furness 1985). The current Welsh population is likely to be in the region of 150 breeding pairs, and would be higher were it not for the heavy persecution endured on many rivers.

Comparison of spring Goosander numbers on selected Welsh rivers, 1985 and 1990

River	*1985* Total birds	*1985* Estimated pairs	*1990* Total birds	*1990* Estimated pairs
Tywi	9	4	22	4
Dyfi	8	3	25	2
Mawddach	1	0	4	0
Conwy	8	1	4	1
Dee/Alwen	30	9	34	6
Wye/Ithon	41	20	89	7
Usk	20	7	50	4
Severn/Clywedog	43	7	78	14
Vyrnwy	44	15	65	14
Banwy	16	6	42	6
Total	220	72	413	58

In winter, Goosanders favour the larger lakes and reservoirs of Wales. The few that winter around the Welsh coast occur at particularly sheltered localities. In Northumbria, Meek and Little (1977) have shown that Goosanders do not usually move far from their natal areas in winter and the same probably applies in Wales. Certainly its winter distribution closely resembles the current breeding distribution, with the largest numbers found in the uplands of mid Wales.

Unusually, Forrest (1907) describes this species as being most common around the coast of north Wales with Llyn Tegid the only regularly used inland site. The Goosander was, at that time, absent from many of the Welsh rivers and lakes, and Forrest described it as being much more common in the adjacent English county of Shropshire. At the turn of the century, it was most commonly encountered on the upper Dee, but even here it was much less common than the Red-breasted Merganser. He also mentioned Llyn Ogwen (Caernarfonshire) and Lake Vyrnwy (Montgomeryshire) as being favoured haunts of this species, and the latter site has remained in regular use ever since.

The Goosander remained a scarce and irregular visitor to all Welsh counties until the 1950s when the range expansion in Scotland and northern England resulted in an increased number of birds visiting Wales during the winter months. By the early 1970s, large flocks were regularly encountered on several inland lakes, particularly Lake Vyrnwy, the Elan Valley lakes (Radnorshire), with

40+ present here on 10 March 1973, and Llyn Brenig (Denbighshire) where 22 were present in November 1977. At Lake Vyrnwy, the winter roosts had built up to a total of over 40 birds by 1985, many of which were probably birds which bred locally. At Overton-on-Dee (Flintshire), an exceptional number of c40 birds was present in January 1980, but this total has never been approached since.

A count of Goosanders made on the evening of 6 February 1985 at all the known wintering sites in Wales revealed Hanmer Mere in Flintshire to be the most important, with 63 birds present. Llandegfedd Reservoir (Monmouthshire) held 28 birds, while Lake Vyrnwy and Talybont Reservoir (Breconshire) both held 11 birds. A total of 141 birds were recorded at 11 sites, with a further five sites showing negative results.

The fluctuating numbers at several of the Welsh lakes and reservoirs from one year to the next and from one week to another suggests that this species may be more mobile than previously thought and it is known that many areas in Wales see cold weather influxes of Goosanders. In recent years, the most important site for this species has been Talybont Reservoir, with 84 birds present here in November 1990. Caban Coch Reservoir in Radnorshire also supports large numbers outside the breeding season (e.g. 68 in September 1990), as does Llyn Tegid in Merioneth (44 in January 1990) and some of the Monmouthshire reservoirs. Llandegfedd generally holds the largest number, but in 1990 this site was surpassed by Garnlydan which held 33 birds in December 1990.

Over the past few years, as the resident Welsh Goosander population has increased, several new sites now hold more than 20 birds some winters, including some stretches of the major Welsh rivers, particularly the Rivers Vyrnwy, Severn and Usk. Parties of over 20 have also been recorded on surprisingly small waters, and recently Bryn Bach Lake, an artificial lake situated on a reclaimed coal tip in Monmouthshire has held over 25 individuals. The total Welsh wintering population is likely to be around 250–300 birds, and is increasing in parallel with the British breeding population (Owen *et al.* 1986).

Goosanders are scarce visitors to the offshore islands, with small flocks sometimes recorded during periods of severe weather. One Welsh-ringed Goosander has been recovered in Cumbria and two birds ringed in Northumbria have been recovered in Wales.

Ruddy Duck *Oxyura jamaicensis*

Hwyaden Goch

Scarce, but increasing breeding resident in several counties. Locally common on Anglesey, but elsewhere a scarce winter visitor.

The Ruddy Duck escaped into the wild in Britain from several waterfowl collections from the 1950s onwards. Since then, it has spread rapidly across much of the English Midlands and into some lowland areas of Wales, particularly Anglesey. It is a North American species, which has no ecological rival in Britain, therefore its rapid expansion is likely to continue.

It was first recorded breeding in Somerset in 1960, and over the next decade breeding was recorded from a further five counties, chiefly in the English Midlands. The first recorded summering pair in Wales was in Montgomeryshire in 1974 when a non-breeding pair was seen at Lymore Pools. By 1976, up to five pairs were present at this site, and breeding occurred in both 1976 and 1977 when young birds were seen (Lovegrove *et al.* 1980). Since then, breeding has been proved or suspected at several small lakes and pools in Montgomeryshire, including Gungrog Flash, Lymore Pools and Mellington Pool.

Breeding was first proved on Anglesey at Llyn Llywenan in 1978, following the appearance of a pair at this site in 1977. Over the next 15 years, this species became firmly established in this county with breeding recorded at several sites. In 1980, a nest containing six eggs was found at Llyn Dinam (the late G. Ireson, pers. comm.), and during a wildfowl breeding census in 1987, breeding was recorded at Llyn Traffwll, Cefni Reservoir, the Valley Lakes and Cors Erddreiniog. In 1986, the island's population was estimated at 25–35 pairs (RSPB 1986), and it is still the stronghold for this species in Wales, its shallow, well-vegetated lakes and pools providing ideal breeding habitat. By 1991, breeding had been recorded at 13 sites on the island and the total population was thought to be about 30 breeding pairs (B. Hughes, pers. comm.).

In the mid-1980s, breeding was recorded for the first time from several other Welsh counties. The first breeding record for Carmarthenshire was in 1985 when a pair and young were seen at the Witchett Pool. In 1986, a pair bred at Kenfig NNR, but it is not thought that breeding has taken place here since, and in the same year, breeding was suspected at one site on the Llŷn (RSPB 1988). The first breeding attempt at Llan Bwch-llyn in Radnorshire was in 1987 and a pair nested at Llyn Heilyn in 1992. Despite several earlier records of summering birds and the presence of a thriving population on the nearby Cheshire meres, breeding was not recorded in Flintshire until 1990, when a female and young ducklings were seen on Llyn Helyg (B. Corran, pers. comm.). In Caernarfonshire, breeding had been recorded at two sites on the Llŷn, Llyn Glasfryn near Pwllheli and the Afonwen Pools, by the early 1990s.

Summer records are increasing from several other Welsh counties, but breeding has not yet been recorded. However, this species is expanding its range rapidly, and with little competition from any native species and a wealth of suitable breeding sites in lowland Wales it is likely that this increase in range and numbers will continue. However, as the Ruddy Duck has expanded its range into southern Europe, it has hybridised with a closely related species, the globally threatened White-headed Duck, in Spain. This has put the small population of White-headed Duck under threat, and Spanish conservationists are calling for Britain to eradicate its population of Ruddy Duck as it is believed that the European, and hence Spanish, population originated here.

Lovegrove *et al.* (1980) recorded up to five breeding

pairs in Montgomeryshire and one on Anglesey, giving a maximum Welsh total of six pairs at that time. By the early 1990s, the Welsh breeding population had increased to approximately 70 pairs, with half of these in Anglesey.

The peak winter count of Ruddy Ducks in Britain in 1989–90 (2829) exceeded that of the previous year by over 400 birds, confirming that the population is continuing to increase rapidly. Since the colonisation of Britain, this species has rapidly established migration patterns in many areas, with movements taking place almost exclusively at night. Large numbers gather on shallow waters following the late summer moult, and although the largest flocks are in England, substantial numbers have been recorded on some of the Anglesey lakes.

In September 1989, 97 birds were seen on Llyn Traffwll, and in the following month, 118 were observed on Llyn Penrhyn, the first time such a large flock had been recorded in Wales. Ruddy Duck flock together at several other sites in Anglesey, with 24 recorded at Llyn Alaw, 16 at Llyn Cefni and 18 at Llyn Coron in 1988–89, and 17 on the Inland Sea on 1 February 1987. During severe weather, small numbers will disperse farther afield, as witnessed by the fact that over 100 birds were present on Anglesey in the winter of 1981–82, when the breeding population on the island consisted of fewer than ten pairs.

Elsewhere in Wales, such large flocks have never been recorded, but smaller numbers are seen on some of the Montgomeryshire lakes and pools. Lymore Pools and Gungrog Flash both hold over ten birds most autumns, and small numbers have also been recorded at Leighton, in the Severn valley near Welshpool. Wintering birds have now been seen in all the Welsh counties, the earliest on record being a single bird at Kenfig in December 1969, two years before this species was formally admitted to the British list.

Denbighshire and Flintshire both support small populations of wintering Ruddy Duck, the most important sites being shallow pools such as Hanmer Mere (six on 9 March 1976) and Gresford Flash near Wrexham (four on 10 October 1980). The first record for Carmarthenshire was a female, which visited a pinioned male for three days at Llanpumpsaint in March 1977, but more recently, small flocks have been recorded from the Witchett Pool. In Glamorgan, Ruddy Duck are regularly seen at Lisvane and Llanishen reservoirs, Kenfig NNR and Hensol Lake, many as a result of cold weather dispersal from the English Midlands and Avon (S. Moon, pers. comm.). The first records for Cardiganshire, Pembrokeshire and Radnorshire were on 27 January 1982, 4 February 1979 and 21 November 1981 respectively, the latter being a female at Llan Bwch-llyn where a few pairs are now known to nest.

The rapid expansion of the Ruddy Duck throughout southern England and Wales is probably the most impressive by any bird species since the colonisation of Britain by the Collared Dove. As much suitable habitat exists for this species, the increase will undoubtedly continue, unless serious attempts at eradication are successful.

Honey Buzzard *Pernis apivorus*

Bod y Mêl

First confirmed breeding in 1992. Formerly a rare summer visitor, numbers have increased throughout the 1980s and it is now recorded annually.

The past status of the Honey Buzzard in Britain is not well documented, but it seems likely that it has never been numerous. Its arrival in May and the timing of the breeding cycle appear to be geared towards the maximum production of its favoured prey, bee and wasp larvae, although it also consumes other large insects and a wide variety of mammalian, reptilian and amphibian species. In a cold, wet climate such as that experienced in Wales one would expect insect larvae to form a lesser part of the diet, but no data are yet available. Formerly it was a scarce summer visitor; there are several records at the turn of the century, but as breeding numbers declined in England, so the frequency of sightings diminished in Wales. Recently, however, the breeding population has increased in number and distribution in Britain, and over the past ten years it has become an annual visitor in small numbers to Wales.

Records of Honey Buzzards in Wales go back as far as the early 19th century, and it would appear that a few pairs may have nested in the Principality during this period, although there are no confirmed breeding records. In May 1827 an egg collector called J. D. Salmon noted the presence of this species at Capel Illtyd (Breconshire), but he does not mention whether this refers to eggs, a specimen or a sighting. The next record, from Capel Curig (Caernarfonshire), involves an egg in the collection of the late A. F. Griffith of Brighton, reputed to be that of a Honey Buzzard collected in 1835. This record remains unconfirmed, and Forrest (1907) makes no mention of it in his book. He does, however, dispute the sighting of a pair near Tywyn (Merioneth) on 20 August 1859, saying that they were probably Conmmon Buzzards. He lists only four records for north Wales, all involving single birds which had been shot. Two were reported from Dolwyddelan (Caernarfonshire), one shot around 1882, the other some years earlier, one shot near Llangedwyn (Denbighshire) around 1875, and one obtained near Kerry (Montgomeryshire) on 21 May 1906. The following year a male was shot on 15 October at Abergele in Denbighshire.

In south Wales records of birds came from several counties during the 1870s and 1880s. In Breconshire singles were shot at Ffrwdgrech and Danypark near Crickhowell in 1871 or 1872, and in Radnorshire one was shot near Penybont and examined by Salter in 1904. One bird was shot at Cefn Mably (Glamorgan) in 1880, and another in the same county at Ruperra in June 1876. Intriguingly, a second bird was shot about 3 km away at Machen (Monmouthshire) in 1876 suggesting that these may have been a breeding pair. The only early record for Pembrokeshire was a bird seen over Creselly in 1851, and on 24 September 1908 a female was shot near Cardigan. There are no further confirmed records from these counties for more than half a century.

Witherby *et al.* (1940) noted that a pair of Honey Buzzards bred in one locality near the Welsh border between 1928 and 1932, but do not state whether the nest was in England or Wales. F. C. R. Jourdain in his diaries also mentions this pair, saying that one of the adults was shot by a keeper in late summer 1932 and that the bird was subsequently exhibited in a taxidermist's shop in Shrewsbury. The bird was said to have been "obtained locally" suggesting that the nest was over the border in Shropshire. Jourdain also quotes a letter he received from C. H. Gowland, who stated that Honey Buzzards had bred "for the past four years" in north Wales. However, it is likely that these were the same birds that nested near the Welsh border.

The first record for Carmarthenshire was a bird seen over Llangadog in October 1921, and a bird of prey seen digging out a wasp's nest at Abergwili on 3 October 1951 was almost certainly this species. One seen at the Dinas and Gwenffrwd RSPB reserves on 1 and 5 June and 3 July 1968 is thought to have summered in the Doethie and Pysgotwr Valleys. The only recent record for Breconshire was one seen at Llangynidr on 15 October 1971, and until 1980 sightings throughout Wales were very scarce. Since that time, however, it has become increasingly common as the British breeding population has increased, and it has been recorded every year since 1980 except for 1984. Influxes were experienced in 1981 and 1991, when seven and nine sightings respectively were reported. The majority of these came from Caernarfonshire, the Pembrokeshire islands and Glamorgan, but birds were recorded throughout Wales. All sightings refer to single birds except for a remarkable record of three Honey Buzzards flying out to sea over Strumble Head (Pembrokeshire) on 18 October 1981.

In 1991 displaying birds were recorded from two unnamed sites in Wales, and although further investigation revealed that one of these held only a single bird, a pair was located at the second site (Williams, in press). The pair built a nest but did not lay, but in 1992 they returned to the same site. It became evident in June that eggs had been laid but by the end of the month the nest had failed. On inspection the nest was found to contain several small egg fragments, but it was not possible to estimate the clutch size. Interestingly, the pair did not abandon the area but continued to display until August. The site which held a solitary bird in 1991 was occupied briefly in 1992 and displaying birds were also seen at two other sites. With this increase in records over the past few years, there is a strong possibility that other pairs have gone unnoticed and with the population increasing elsewhere in Britain, it is quite possible that Honey Buzzards will colonise several Welsh counties in the near future.

Black Kite *Milvus migrans*

Barcud Du

A vagrant.

There have been six records:

19 October 1976	– One, Llangorse Lake
17–18 April 1979	– One, Overton (Glamorgan)
19 May 1979	– One, Cardiff (Glamorgan)
26 April 1985	– One, Lake Vyrnwy (Montgomeryshire) and presumably the same bird at Dyfi Forest (Merioneth)
24 April 1987	– One, Cefn Coed (Anglesey)
20 May 1990	– One, Skokholm

The Black Kite is migratory in its European range, which includes breeding populations as close as France. It is notable that up to 1966 there had been only five records of this species in Britain whereas the total now stands, up to 1991, at 154.

Red Kite *Milvus milvus*

Barcud Coch

(Contributed by P.E. Davis : Kite recorder)

Formerly widespread, but almost eradicated between the late 18th and early 20th centuries. Now an increasing resident breeder in six counties of central Wales; rather a scarce and erratic visitor elsewhere.

Kites have a long history in Wales. Bones of a kite species, dated to about 120 000 years ago, were found in a Gower bone-cave. It is unlikely that any survived the intervening glaciation, but the Red Kite was certainly established in Wales in Roman times because its remains were excavated at Caerleon (Monmouthshire). Kites nest in trees but hunt over open ground, so their numbers must have increased with forest clearance, in the past two or three millenia. Welsh place-names containing the "barcud" or "beri" elements occur widely in east and central Wales, attesting to the bird's former abundance.

Red Kites remained widespread over much of Wales

into the early decades of the 19th century, though precious few records exist before about 1830. In the words of Morrey Salmon, "it was gone . . . before anyone thought to record its disappearance". As a frequent predator of young domestic fowls and game birds, this was among the least tolerated birds of prey, and one of the easiest to exterminate, by gun, trap, poison, and nest destruction.

There seem to be no early records of kites in Anglesey, Caernarfonshire, or in south and west Pembrokeshire, and it may be the bird was absent or uncommon in the exposed western extremities of Wales. It had probably gone from Flintshire and most of Denbighshire by 1800, though there was one at Ruthin in 1827, and the kite is listed among the birds of Llanrwst, on the Denbighshire side of the Conwy, in 1830. In Montgomeryshire, kites were still "quite common" around Newtown in the 1830s, and some persisted there, probably 20 years later. They were seen in the Berwyn hills between Llangynog and Bala in 1796. Elsewhere in Merioneth, an early reference mentions kites in the Dysynni Valley in 1768, and they survived here and in the upper Dyfi Valley at Dinas Mawddwy until at least 1855.

For Cardiganshire there are several records in the Rheidol and Ystwyth Valleys between 1796 and about 1870, when the last pair at Devils Bridge was shot. Farther south, kites were nesting at Llanarth about 1818, and they occurred all along the Teifi Valley until about 1850, and much later in the upper reaches. They were evidently known on the Cardiganshire – Pembrokeshire border through the first half of the 19th century, though "already rare" in the Preseli district by 1830, and not otherwise recorded as resident in Pembrokeshire.

In Radnorshire, kites were noted at Cwm Elan in 1798, and they remained common in that district until "killed off severely" by the Nant Gwyllt keepers about 1865–70. Farther east, they bred at Llanelwedd near Builth until 1862, and in the Radnor Forest until about 1870. They seem to have lasted a little longer in the hills of south-east Radnorshire. In 1902 a retired keeper on the Maesllwch estate claimed to have killed 30 kites in as many years, and to have destroyed several nests. In Breconshire, kites occurred over Brecon town in the early 19th century, and a few pairs persisted later in the Wye and Usk catchments, thanks largely to early protection given by several landowners, at the instigation of E. Cambridge Phillips.

Monmouthshire has several records for the lower Usk Valley in the early 19th century, where the bird was "locally well known" in 1834. There may have been a nest at Nantyderi near Usk as late as 1870. Glamorgan is comparatively well provided with early records. Kites were reported to breed in Llechwedd Wood, Cardiff, in 1796 and farther up the Taff at Penrhiwceiber around 1800. They were "very common" at Newtown Nottage about 1820 and "common as buzzards" in the Cynon Valley below Aberdare about 1830, though they had gone from that district before 1850. Near Swansea, they were becoming "ever less common, though still not infrequently seen" by 1848. The last recorded nest was near Cardiff in 1853, the female and eggs collected. In Carmarthenshire, birds occurred at Ferryside in 1857 and 1859, where "they used to breed many years ago in the Iscoed Wood". At Llandeilo in 1858, kites were "formerly plentiful, now seldom or never seen", though at nearby Llanfynydd they were said to be numerous until the 1860s and they bred at Llanddeusant until at least 1888. A few pairs survived later in the upper Tywi and Cothi catchments.

By the 1880s, kites were confined to north Carmarthenshire, north-east Cardiganshire, north Breconshire and probably small parts of west and south-east Radnorshire. By the 1890s, this was the only part of Britain where kites remained, and they survived here fortuitously, and only because the impact from human persecution was less severe than elsewhere.

The subsequent history of the Red Kite in Wales is well documented, largely due to the dedication of two individuals, J. H. Salter and H. Morrey Salmon. Salter taught botany at Aberystwyth from 1891 to 1908, and lived there again from 1923 to 1942. He interviewed landowners, keepers, farmers, and others throughout central Wales, meticulously recording all he could learn about kites. He also searched widely for the birds himself, investigating each report that came his way, and noted all the observations of his fellow naturalists. It was he who made representations to the British Ornithologists Club about the serious threat posed by egg and skin collectors to the last remnants of the British kite population, and proposed the establishment of a national protection scheme, which has operated in one guise or another ever since 1903. Morrey Salmon belonged to a later generation. A Cardiff businessman, ornithologist, photographer, and soldier in both World Wars, he knew the kite country well in the 1920s, but became closely involved in kite affairs only in 1949, when he joined the Kite Committee and soon became responsible for its records. Over the next 20 years he compiled a great digest of all the known records up to 1970, which forms the basis of any account of kites in Wales. Sadly he remained unaware of the details of many of Salter's records, which have only lately been extracted from the diaries. The earlier record has also been supplemented since 1970 by data from egg collections, and from the personal diaries of other naturalists and collectors.

Thanks to these records, we can trace the history of kites in Wales over the past one hundred years with some precision. From 1891 to about 1950, although the full picture was probably never known in any one year, most kite localities and probably the great majority of pairs are recorded intermittently, even if the details of breeding success remain obscure. The chief deficiency probably falls during the two World Wars, 1915–18 and 1940–45, when travel restrictions and other preoccupations affected fieldwork. The known population data for 1891 to 1950 are tabulated by five-year periods in the table. The population continued to contract in geographical extent and probably in absolute numbers through most of this period, and seems to have reached its nadir in the 1930s, not in the first decade of the 20th century as has generally been supposed.

The few remaining pairs in the upper Teifi Valley in Cardiganshire were exterminated by about 1900 and breeding in this county became confined to the upper Tywi Valley, although it was very intermittent after 1904.

In Breconshire, two pairs in the Gwenddwr/Llyswen area of the middle Wye persisted until 1897, one of them possibly until 1902. Farther west, at least eight pairs along the upper Usk and its tributaries in the early 1890s contracted to three or four before 1900, and the last of them, near Penpont west of Brecon, was not reported after 1909. Two or three pairs remained in the far north-west of Breconshire through the early decades of the 20th century, with the occasional breeding record from the adjacent part of Radnorshire. The main surviving population in the 1890s and early 1900s lay in the upper Tywi and Cothi catchments in Carmarthenshire and sometimes Cardiganshire, with probably up to a dozen pairs, though half of these were outliers, known only from very occasional records scattered over 20 years. A surprising outlier seems to have nested in the lower Dyfi Valley in Merioneth about 1905–6.

Salter's diaries reveal the staggering wastage of kites by keepers, farmers, egg and skin collectors through the 1880s and 90s, and continuing, perhaps to a lesser degree, even after the introduction of a partial wardening scheme in 1903 and its attendant publicity. Almost every recorded nest was pillaged of eggs or young, or the birds were shot. Professional egg and skin collectors, already at work in Breconshire, discovered the upper Tywi pairs in 1893, and as soon as a monetary value was placed upon the bird and its eggs, local men joined in with enthusiasm. The young birds had long been a saleable commodity, widely kept in captivity. It was only the partial discipline imposed upon keepers and tenants by some landowners, at the instigation of men like Cambridge Phillips and Salter, that averted the final destruction of the species. The wardening scheme, which employed local men and rewarded farmers and keepers, was not without its problems. Some nests failed owing to the activities of over-zealous watchers, others were deliberately destroyed to spite the watchers, and a few watchers stole the eggs they were supposed to guard. In the background would be a steady attrition of full-grown kites from shooting and trapping, and from poison baits, widely used to control other pests.

For a year or two around 1910 the total population may have been no more than about ten pairs, confined to the upper Tywi and Cothi Valleys and the headwaters of the western tributaries of the upper Wye. Somewhat improved breeding success following protection and publicity by the conservation lobby then allowed the population to stabilise and even increase a little, though not permitting much extension of the range. The First World War gave a respite, with most keepers and collectors occupied elsewhere, but pressures resumed in the 1920s and despite a few comparatively successful years, and the brief appearance of an outlier in north-east Cardiganshire, the population continued to decline. The upper Wye component disappeared by the early 1930s.

Matters were complicated in the late 1920s through the 1930s and into the 1940s by the appearance of a considerable number of wintering kites, and several unconfirmed reports of nests, in the Radnor Forest and the east Radnor Hills. This was taken to represent an expansion of the native population into a new area, but seems in fact to have been a totally unrelated phenomenon. In 1927–28 and again on a smaller scale in the mid-1930s, the Liverpool egg dealer C. H. Gowland arranged for Red Kite eggs supplied by his associates in Spain to be placed in buzzard nests in the area near Builth Wells and Rhayader (Radnorshire). A good many young are said to have been reared, though no systematic check was maintained. Importations were brought to an end at the outbreak of the Spanish civil war in 1936. There is little doubt that the records in north and east Radnorshire derived largely if not entirely from these introductions. A few Spanish birds may have attempted to breed, though possibly most were inhibited from mating with other kites due to imprinting on Buzzard foster-parents. Some were certainly shot or poisoned. Whatever the reason, the east Radnorshire records petered out during the Second World War, and they are not heard of again.

It is now virtually certain that no Spanish blood entered the Welsh population. Recent studies of Red Kite DNA, undertaken by Celia May, Jon Wetton and David Parkin of Nottingham University, have shown that all Welsh females (and presumably males also) until very recent times were descended from one single female that survived the population bottleneck. Another female, almost certainly of German stock, joined the Welsh population unannounced within the last 20 years (1970s), so that there are now two matrilines, though the original one still covered about 85% of the population in 1992. The figures given in the table, which excludes the Radnorshire records, indicate that the generic bottleneck most probably occurred during the early 1930s (Davis 1993). In 1931–35 only seven broods of kites, containing 10 young, are known to have been reared in Wales. Only two breeding pairs or sites were involved. One of these produced five broods with seven young, and looks to be the likeliest channel for the sole surviving matriline. It was a close-run thing.

The Welsh Red Kites pulled back from the brink of extinction in the late 1930s, with again an outlying pair appearing in the Teifi Valley in Cardiganshire. With another respite from persecution and improved production of young during the Second World War, this northward expansion was consolidated. Kites also began to reoccupy the headwater valleys of the Wye and by 1950, there were at least a dozen breeding pairs.

From about 1950 the organisation of kite monitoring moved onto a higher plane. It was co-ordinated at first by H. R. H. Vaughan who chaired the the Kite Committee until 1970 and then administered jointly by Nature Conservancy (later NCC, now CCW) and the RSPB. A serious and largely successful effort to locate every pair every year became the norm. Despite a number of setbacks, such as the destruction of rabbits (a favourite prey) by myxomatosis in 1955, the impact of two severe winters in the 1960s, and possibly some reduction of fertility due to pesticide contamination, the population still increased. Numbers rose slowly and erratically until the mid-1980s with very indifferent breeding success, but then accelerated as new pairs occupied more productive lowland sites and became successful in rearing young. The recent increase in breeding numbers and in the production of young is demonstrated in the graph.

The number of known breeding pairs in Wales

Welsh Red Kite population in five-year periods, 1891 to 1950

	1891–95	*1896–1900*	*1901–05*	*1906–10*	*1911–15*	*1916–20*	*1921–25*	*1926–30*	*1931–35*	*1936–40*	*1941–45*	*1946–50*
Territories with pairs	21	20	17	15	15	13	15	13	11	10	10	16
Successful pairs known	–	–	5	6	8	9	9	5	2	4	6	9
Known broods reared	–	–	6	9	12	15	14	12	7	7	13	14

exceeded 20 by 1967, 30 in 1978, 40 in 1985, 50 in 1988, 60 in 1990, 70 in 1991, and 80 in 1992 (when they were breeding in six of the historic counties of Wales). With immatures and non-breeders, the population in summer 1992 probably reached 380 birds. In 1992 there was also the first breeding by continental Red Kites reintroduced into southern England and northern Scotland by the NCC and RSPB since 1989. Four pairs bred successfully in England and one in Scotland that year. In 1993, the Welsh total finally passed 100 nesting pairs although breeding success was poor due to prolonged wet weather, with only 59 successful pairs producing 80 young.

Kites have remained rather scarce and irregular visitors to the rest of Wales, outside the breeding area, though becoming more frequent as the population grew. They have occurred in all counties and on at least one of the smaller islands (Bardsey) in recent years. Ringing and tagging has shown that juveniles and immatures are highly mobile, and many of them circulate widely and erratically throughout the breeding area and beyond it, especially in the first autumn and winter. Outside Wales, there have been ringing recoveries and sightings of immature wing-tagged Welsh kites in south-west Scotland, Yorkshire, Staffordshire, Worcestershire, Gloucester, Somerset, Cornwall, Wiltshire, Hampshire, Oxfordshire, Berkshire and Kent and one on a gas rig in the North Sea, 40 km off the coast of Norfolk. Those immatures remaining nearer home show a markedly westward tendency in winter, and this is echoed by a small proportion of the adults, mostly from higher and less productive territories. The majority of adults, however, remain in or near their breeding ranges all year. Many Welsh kites roost communally in winter; spectacular roosts of up to 60 individuals are on record, with larger roosts consisting mainly of immatures.

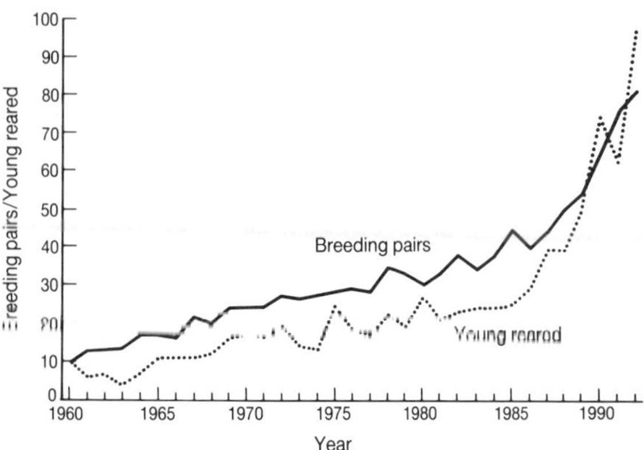

Breeding performance of Red Kite in Wales, 1960–92

Some Welsh kites begin to breed in their second year, the majority in their third or fourth year, a few not until they are five or older. They tend to settle at no great distance from their birthplace, mostly within about 12 km and more to the west than to the east of it. Fewer than half the fledglings survive to breed, but having bred, the normal life expectancy is more than ten years and the oldest recorded Welsh kite to date lived for almost 24 years. One foreign-ringed kite has been found in Wales, a juvenile from north-west Germany found dead near Rhayader in 1971.

White-tailed Eagle *Haliaeetus albicilla*

Eryr y Môr

Formerly a rare visitor but there have been no records in Wales since 1910.

This species breeds in south-west Greenland, Iceland and northern Eurasia and it bred in the British Isles until 1916 when the last confirmed breeding took place in Scotland (on the Isle of Skye). Human persecution was the cause of this demise and a successful reintroduction scheme initiated by the Nature Conservancy Council in 1975 has re-established the White-tailed Eagle as a breeding species in Scotland.

There is no evidence that White-tailed Eagles ever bred in Wales although they were known to have nested as close as the Lake District, the Isle of Man and Lundy Island. The tantalisingly vague accounts of Eagles possibly nesting in Wales more likely refer to Golden Eagles although it is impossible to be certain either way. What is certain is that the White-tailed was more regularly recorded than the Golden Eagle in the 19th and early years of the 20th centuries.

Between 1818 and 1910 there were some 17 records of White-tailed Eagles in Wales, mostly confirmed by specimens in collections of mounted birds. These covered all counties except Radnorshire and Monmouthshire.

The most recent Welsh record refers to a bird which had frequented the Llŷn peninsula in November 1910. It met the usual fate of large raptors at that time as it was shot near Abersoch (Caernarfonshire) on 29 November. In this particular instance, however, it was only winged and efforts were made to heal the injury, with what results is not recorded.

Marsh Harrier *Circus aeruginosus*

Bod y Gwerni

Scarce but increasing passage migrant and winter visitor, and a sporadic breeder.

The Marsh Harrier breeds almost exclusively in extensive *Phragmites* reedbeds, this specialised habitat inevitably restricting both numbers and distribution in Britain. Formerly it was widespread if local, but the drainage of fens and marshes together with unremitting persecution, drove the species to extinction as a breeding bird in Britain by 1900. After sporadic attempts at recolonisation from 1911, breeding occurred annually in Norfolk from 1927, but numbers remained low until a gradual increase and spread from the mid-1940s. The first proved breeding in Wales this century was in 1945, on Anglesey. From 1958 (when there were 15 nests), the British population declined markedly to a single pair in 1971, probably due to the effects of pesticides. Since 1972, numbers have risen rapidly (nearly 90 nesting attempts in 1990), and this revival has led to an increase in the number of birds seen in suitable habitat in Wales during the breeding season, with nesting probably confirmed once more in 1991 and possibly 1992.

Several authors commented that the Marsh Harrier was formerly a common resident over much of the Welsh lowlands prior to extensive drainage during the 18th and 19th centuries. George Montagu (1802), found this species to be "the most common of the falcon tribe about the sandy flats on the coast of Carmarthenshire", an indication of the wealth of wetland habitat in this area at the time. Dillwyn (1848) stated that this species was formerly commonly encountered on Crymlyn Bog (Glamorgan), with a pair possibly breeding here as late as 1882, and Forrest (1907) mentions west Anglesey and the wetlands near Llandudno (Caernarfonshire) as former breeding sites. However, there does seem to have been much confusion as to the identification of this species amongst many early naturalists, hence breeding records from the Berwyn Mountains (Denbighshire) in 1869 and 1877, near Rhayader (Radnorshire) in c1850 and on the Brecon Beacons (Breconshire) during the last century are all likely to refer to the Hen Harrier. However, Salter (1900) mentioned that it was formerly numerous on Cors Caron and Cors Fochno.

F. C. R. Jourdain received information from C. H. Gowland that Marsh Harriers had bred in a big marsh on Anglesey in 1935, and that he (Gowland) would be returning to the site the following year to obtain the eggs. It was not until 1945, however, that breeding was once again proved in Wales when a nest was found near Llyn Tai-hirion on Anglesey (Colling & Brown 1946). This pair successfully reared one young, and it is possible that another pair bred, probably in 1946, in Caernarfonshire. Although not named, it is likely that the site involved was Llyn Ystumllyn near Criccieth, but the pair was unsuccessful, probably due to egg collectors.

In 1969 and 1970 pairs summered at suitable sites on Anglesey, but breeding was not recorded again until 1973 when a pair reared three young at Llyn Bodgylched. Breeding is said to have occurred here again the following year, but this has not been confirmed, and despite the presence of single birds, and sometimes pairs, at suitable breeding habitat throughout Wales, the next probable nesting attempt was not until 1991. That year a pair is reported to have reared two young at a secret location on Anglesey, and breeding is said to have occurred again in 1992. Certainly, there is much suitable habitat for this species on the island and at other former haunts such as Oxwich Marsh NNR and Crymlyn Bog, and with the population increasing elsewhere in Britain (Underhill-Day 1984), it can be only a matter of time before Marsh Harriers are once again seen breeding regularly in Wales.

One unusual record concerns an adult male which accompanied a pair of kites at a site in Carmarthenshire for six to eight weeks in May and June 1987.

On passage, Marsh Harriers are most often recorded from former breeding haunts such as Oxwich Marsh NNR and Crymlyn Bog, the Anglesey lakes, Dyfi estuary and Cors Caron, and potential breeding sites such as Rhosgoch Common (Radnorshire) and Dowrog Common (Pembrokeshire). These were also favoured sites at the turn of the century, as were the dune systems of Anglesey, Carmarthenshire and Glamorgan, at locations such as Newborough, Pembrey and Whiteford, which are still used today. The number of birds seen has fluctuated in line with the British breeding population, and today Marsh Harriers are regularly recorded from several counties in addition to the offshore islands (particularly Skomer). Birds are usually seen on spring passage in April or May, and again in later summer in August. Between 1988 and 1991, some 15–20 birds were recorded annually, the majority (32) in May.

Monthly totals of Marsh Harrier sightings in Wales, 1988–91

Jan	*Feb*	*Mar*	*Apr*	*May*	*Jun*	*Jul*	*Aug*	*Sep*	*Oct*	*Nov*	*Dec*
3	0	1	17	32	5	5	6	8	2	3	1

Most British breeding Marsh Harriers move south to overwinter in the Mediterranean and north-west Africa. In 1937, overwintering was recorded in Britain for the first time since the turn of the century, and the number of overwintering individuals (mostly females) both in Wales and elsewhere remains low.

Forrest (1907) recorded this species as a winter visitor to the Dyfi estuary and Borth Bog and mentions one which frequented a bog on Anglesey for eight years up until 1902. Whether these were true wintering birds, passage migrants or possibly even Hen Harriers is not known, but records of wintering Marsh Harriers in Wales this century prior to the 1970s are very scarce. During the fieldwork for the *Winter Atlas*, there were a surprising number of records from south Wales, with two or more birds recorded at Cors Caron and the Three Estuaries (Carmarthenshire). Single birds were noted at several coastal and inland sites, particularly around the Gower. All three north Wales records during this period referred to single birds seen on Anglesey. During 1988–91, there were only four sightings of single wintering birds, three of which may have involved the same individual. These were from the Ffrwd Fen Reserve (Carmarthenshire), Crymlyn Bog (Glamorgan) and Blackrock (Monmouthshire) in January 1988, and another record from Crymlyn Bog on 9 December 1989.

Hen Harrier *Circus cyaneus*

Bod Tinwen

Small numbers breed on the uplands of north and mid-Wales, otherwise a winter visitor and passage migrant in small numbers, chiefly to coastal counties.

In Britain, Hen Harriers are confined mainly to rolling, heather-dominated moorland during the breeding season, although more recently they have nested increasingly in newly afforested areas. Formerly far more widespread than at present, it is difficult to assess its true distribution prior to the 1950s as it was often confused with Montagu's Harriers, and on occasion even Marsh Harriers. There is no doubt, however, that Hen Harriers have suffered severe persecution at the hands of gamekeepers, and that this persecution persists today, albeit on a much smaller scale in Wales.

Several observers noted that Hen Harriers were fairly common on the Berwyn Moors during the first half of the 19th century. Eyton (1838) had seen them near Corwen (Merioneth) and wrote "it is remarkable with what regularity they return to the same beat at the same time for many days together, which propensity often tends to their destruction". Here he referred to the birds' tendency to return to traditional sites during early spring, where more often than not they were met with the shotguns and pole-traps of gamekeepers.

During the same period Dillwyn (1848) stated that Hen Harriers were formerly abundant in the Swansea area, but that at the time of his writing they had become comparatively rare. Similarly Mathew (1894) described how this species bred on moors throughout the county of Pembrokeshire some 50 years earlier, and despite the strong possibility that many of these may have been Montagu's Harriers, other observers refer to breeding Hen Harriers in this county, and an egg in the collection of the late A. F. Griffiths of Brighton taken in Pembrokeshire is reputed to be from this species. In spring 1868 a pair was killed at Troed-yr-Aur in Cardiganshire, and it is possible that sporadic breeding occurred in the county throughout the late 19th century. Hen Harriers bred on the uplands above the Elan Valley (Radnorshire) during this time, the last record until recently being a pair near Rhayader in 1903.

The decline in the numbers and distribution of breeding Hen Harriers in Britain throughout the second half of the 19th century was mirrored in Wales (Watson 1977), and by the 1890s it was very scarce even on the extensive heather moorlands of north Wales. Dobie (1893) made surprisingly few mentions of this species despite his contacts at the Pale and Rug Estates on the Berwyns. Bolam (1913) remarked that the species had "maintained a footing here (the Berwyns) up until the middle of the last century, and has occurred occasionally since".

What was probably a pair of Hen Harriers nested near Nefyn (Caernarfonshire) in 1902 and a pair built a nest but did not lay eggs in Cwm Bychan (Merioneth) in April 1909. A pair reported in Carmarthenshire near Newcastle Emlyn during the summer of 1865 was probably Montagu's Harrier, as was a pair of harriers which bred near Nantgwyllt (Breconshire) for several years. This pair was eventually trapped and the eggs were taken, and although identified as Hen Harriers at that time, several naturalists have subsequently cast doubt upon this identification.

There is little doubt that this species had ceased to breed regularly in Wales by about 1910, and that until the late 1950s nesting occurred only sporadically. A pair is reported to have nested in Caernarfonshire sometime between 1910 and 1940, and in 1924 a pair successfully reared two young on Anglesey, the first confirmed breeding record for this county (Aspden 1939). A pair nested at the same site until 1926, and there have been rumours of pairs breeding on the island subsequently, but there are no confirmed records.

In Britain, Hen Harriers survived only on Orkney and the Western Isles until 1939, when a pair bred on the Scottish mainland once more (Campbell 1957). The population increased in numbers and distribution, aided by the afforestation of extensive areas of moorland where pairs could nest unmolested. It was not until 1958, however, that Wales was recolonised, when a pair bred on the Merioneth side of the Berwyns. Unfortunately, the contents of the nest disappeared, but the recolonisation of Wales had begun and the pair returned here in 1959 and 1960. By 1962 five pairs were located, all on the Berwyn Mountains, and in 1963 a pair bred on Mynydd Hiraethog.

The population remained at this low level until 1977, when 27–30 pairs were recorded, one as far south as Llynnoedd Ieuan in Cardiganshire (RSPB 1977). Since then the population has remained fairly stable, although fluctuations occur from one year to the next, influenced mainly by the changing vole population. The Welsh Hen Harrier population did not rely on young plantations for recolonisation as did many of the Scottish birds, and nests are rarely found outside heather moorlands.

Recently, a joint study by the RSPB and CCW to monitor breeding pairs and wing-tag fledglings has shown that the population has remained stable between 1986 and 1992, with 20–30 pairs returning to the nesting grounds each year and a maximum of 28 pairs rearing 55 young in 1992.

Nest failures are attributable mainly to natural phenomena such as the weather and depredation by foxes, but unfortunately some nests and pairs are still destroyed illegally by gamekeepers.

Hen Harriers are regularly recorded from most of the Welsh islands, with between one and three recorded most years from Bardsey, usually in October and November. An influx of birds occurred in 1982, however, although the exact number is not known as some individuals remained on the island for long periods. Birds are recorded on spring passage in small numbers from several other areas, particularly Fedw Fawr and South Stack on Anglesey.

In winter, Hen Harriers generally move away from the bleak uplands to the lower and milder coastal areas. The extent of this movement depends on several factors, including weather and prey availability, and during mild winters when vole numbers are high, many birds remain on the uplands for prolonged periods. As during the breeding season, Hen Harriers require large, open areas over which to hunt, therefore salt marshes and lowland bogs and marshes are often favoured.

During the 19th century, Hen Harriers leaving the Welsh uplands for low-lying marshes and bogs often met with the same fate as breeding birds, for having escaped the attentions of gamekeepers, they then faced the guns of wildfowlers and rough shooters. Nevertheless small numbers were present on the Anglesey marshes every winter, and many of the areas favoured in those days are still used today.

Winter records were few and far between around the turn of the century and the next few decades, and indeed it was not until the 1950s when the breeding population was increasing that birds were encountered more frequently in winter. Most records until then had come from estuaries such as the Dyfi and the Dee, and the lowland bog at Cors Caron.

During the winter, Hen Harriers often roost communally, and although most Welsh records refer to single birds, several sites support communal roosts. A survey of winter roosts in Britain and Ireland in 1983–84 and 1984–85 revealed a maximum of 14 harriers roosting at six sites in Wales on 27 November 1983 (Clarke & Watson 1990). These counts were by no means comprehensive, and were restricted to the better known and better monitored sites. Numbers of birds attending roosts are often influenced by the weather, hard winters driving more birds down off the moors. In 1963, however, Hen Harriers deserted Cors Caron, a favoured roost site, in January and did not return until March when conditions had improved.

Winter roosts of up to half a dozen birds are found at Malltraeth (Anglesey), the Dee estuary, Cors Caron and Cors Fochno and Dowrog-Tretio (Pembrokeshire). Elsewhere, and particularly inland, most records refer to single birds, but several are recorded together on occasion. Up to five birds have wintered on the Gwendraeth estuary and Tywyn Burrows (Carmarthenshire) in recent years and up to 14 have been seen at Dowrog-Tretio, and five were present at Llanrhidian Marsh (Glamorgan) on 5 and 6 January 1986. Perhaps the most remarkable sighting however, was of ten birds flying south along the western side of Skirrid Fawr (Monmouthshire) on 16 September 1974. In the inland counties, birds are recorded from many upland areas, and in Breconshire 63 sightings were recorded between 1969 and 1989.

On occasion influxes of birds do occur, as at Newborough Warren (Anglesey) in 1962–63. Whether these birds had come across from the Continent is not known, but the wing-tagging work undertaken to date shows that both Welsh and Scottish birds often overwinter in the Principality.

Ringing recoveries and the recent wing-tagging scheme show that Welsh birds generally move south and east, with birds recovered in the south-east of England and on the Continent. The longest distance recorded involves a bird ringed as a fledgling in Montgomeryshire on 17 June 1980 and recovered in Monte Aldura in Spain, 1053 km south on 20 October 1980.

Montagu's Harrier *Circus pygargus*

Bod Montagu

Rare summer visitor and passage migrant which formerly bred in small numbers.

The Montagu's Harrier was the last of the British breeding harriers to be separately identified, Colonel Montagu finally removing the previous confusion between this species and the Hen Harrier in 1802. Despite this, many breeding reports of both species from the 19th century are regarded with some suspicion by modern ornithologists, thus clouding the former true distribution of these two harriers.

Exclusively a summer migrant, the Montagu's Harrier's breeding habitats in Britain until recently ranged from lowland heaths, reedbeds and acid bogs to coastal dunes and young plantations. During the last 20 years, however, most nests have been in arable crops. It has probably never been common in Britain, numbers this century reaching a peak of 30 pairs during the mid-1950s (Underhill-Day, pers. comm.). In Wales, the population peaked at some six to ten pairs during the late 1950s, but by the mid-1960s it had become extinct.

Over the past 150 years, breeding or suspected breeding has been recorded from seven Welsh counties, the majority of records coming from Pembrokeshire and Anglesey.

Pembrokeshire

The earliest Pembrokeshire breeding record is of a nest containing one egg and three downy young found at Leweston on 2 July 1854. Unfortunately both adults were shot and mounted together with the three chicks, and the egg sent to a local collection. During the 19th century it was described as a rare summer visitor by Mathew (1894), and the only subsequent breeding record until the 1920s was of a female and four eggs taken in 1914.

Lockley (1961) stated that this species had bred in the county since 1924, and possibly for 20 years prior to this. The egg collection of the late J. H. Howell of Solva, now housed in the National Museum of Wales, certainly bears witness to the fact that pairs bred at least sporadically throughout the 1920s and 1930s. He collected clutches in the county in 1926 (two nests), 1929, 1931, 1932, 1933, 1936 and 1939. There were no subsequent nesting records until the 1950s, although there is no reason to believe that breeding did not take place during this 15-year period.

In 1954, D. Richard Evans (pers. comm.) found two nests, each containing three young, in marshland at Dowrog Common. The females, although serviced by the same male, succeeded in fledging all six chicks. Breeding was sporadically recorded or strongly suspected from this site and at least two others until 1962. The following year adults were seen but breeding was not thought to have occurred on Dowrog Common. Throughout the 1960s adult birds were seen in suitable habitat during the summer months, and breeding may have taken place again in 1968 and 1970 (J. W. Donovan, pers. comm.). During the late 1980s a small increase in the British population led to adult birds being seen once more on former haunts in Pembrokeshire, but no breeding has been recorded.

Anglesey

On 23 June 1945 a pair of Montagu's Harriers was found nesting at Newborough Warren on Anglesey, the first breeding record for north Wales since 1900 (Colling & Brown 1946). The nest was situated amongst rushes and contained three young and one infertile egg, the pair eventually succeeding in fledging all three chicks despite the fact that the female appeared to take no part in supplying food. The habitat here became increasingly ideal for this species as a large area of coastal dunes had been planted with young conifers.

In 1948 two pairs bred, but the following year two clutches out of three were robbed by egg collectors. Inevitably, a rare breeding species such as the Montagu's Harrier drew the unwanted attention of egg collectors, and the birds also ran the risk of disturbance from photographers, birdwatchers and tourists at such a public site. It was because of this that special protection measures were taken during the breeding season by the (then) Nature Conservancy, including a scheme to incubate eggs artificially, leaving the female harrier to sit on dummy eggs (Hope Jones & Colling 1984). Despite these valiant efforts, at least 11 clutches are known to have been taken by collectors between 1949 and 1964, one collector taking clutches of four and three on 12 June 1952 and 16 June 1957 respectively.

Between 1949 and 1964 a minimum of 30 clutches were laid, with the most successful year in terms of the number of pairs, clutches and young fledged being 1957, when five pairs laid six clutches and fledged seven young. The last breeding attempt occurred in 1964, when one pair laid four eggs and reared two young. At that time, the British population had decreased to 13 pairs, and the habitat at Newborough was no longer ideal as the young conifers had become too tall for ground-nesting harriers.

Caernarfonshire

The only possible breeding record for this county involves a report that a pair frequented a mountain in Caernarfonshire for many years and "no doubt nested" (Forrest 1919). There is an unconfirmed report of a pair having bred in the county in 1951, but this may refer to the record of a successful nest in the adjacent county of Merioneth. The birds breeding at Newborough Forest on Anglesey were often seen hunting in the Foryd Bay area between 1960 and 1964, but breeding was never suspected.

Merioneth

Forrest (1907) quotes the only breeding record known to him as a pair laying four eggs on a moor near Bala in 1900. Unfortunately both adults were shot by a keeper. It is half a century before the next recorded breeding attempt in the county, when a pair raised three young at Mochras. There have been no subsequent breeding attempts.

Montgomeryshire

The Montgomeryshire Field Society Report for 1961 records that breeding probably occurred at two localities in the county, and that an adult female was present at a third site. However, this was at the time when Hen Harriers were recolonising the moors in the north of the county and these records almost certainly refer to this species. In 1963 a nest was reported to have been found in reeds at an unnamed locality in the county, but this was in fact at Ynys-hir in Cardiganshire.

Cardiganshire

The first breeding record from this county involved a nest near Aberaeron in 1892 which failed due to the female being shot. A pair is known to have bred on at least two occasions between 1900 and 1904 at Llyn-y-gwaith, and a pair was reported from Myherin in 1947, although it is unlikely that breeding occurred. W. Condry (pers. comm.) reports that W. A. Cadman saw a pair at Gernos Mountain during the summer of 1955, and that when he visited the site on 2 August he saw two ringtails which could have been fledged young. Condry also suspected that this species nested sporadically in the Cors Fochno/Dyfi area during the 1950s, but there are no confirmed breeding records.

There have only been two more confirmed breeding records, a pair laying five eggs and rearing three young at Cors Caron in 1953, and a pair laying four eggs and rearing one young near Ynys-hir in 1963. A pair returned to the latter site in 1964, but did not stay for long.

Throughout the summer of 1967, a pair, including a melanistic female, was seen on Cors Caron, but breeding was not suspected.

Carmarthenshire

Ingram and Morrey Salmon (1954) wrote that this species was reported to have bred in the county, but they could find no definite information of an actual nest. A pair reported as Hen Harriers which bred near Newcastle Emlyn in 1865 was almost certainly this species, but there are no further records of pairs until 1947 when a male and female were seen together at Llanfynydd. In 1966 a pair is assumed to have bred at Pembrey Burrows and may have nested here again in 1967. The following year breeding probably occurred at Wharley Point, where a female was seen carrying nesting material or food into undergrowth in a cliff gully. The last record of possible breeding came in 1969 when a pair summered on Mynydd Llanfihangel Rhosycorn. Since then records of birds have become increasingly scarce as the British breeding population decreased.

Other than the breeding records outlined above, pairs have been reported from other counties on occasion. During the late 19th century, a pair, originally identified as Hen Harriers, bred for several years near Nantgwyllt (Breconshire). The adult birds were eventually trapped and were later identified as being this species. A pair was seen at Wentwood (Monmouthshire) in May and June 1968 and a pair was seen over cereal fields on the Breconshire/Herefordshire border in June 1992, although it is not believed that either pair attempted to nest. In 1993, a displaying male was present for several weeks at a moorland site in Denbighshire.

Montagu's Harriers have been recorded on passage in all Welsh counties, numbers having decreased since the 1960s following the demise of the breeding population in the Principality. Two interesting records are those of a melanistic male shot at Margam (Glamorgan) on 19 August 1939, and a juvenile which overwintered on Skomer in 1988–89. This species was frequently recorded from the Pembrokeshire islands up until the early 1970s, but it has since become an irregular but annual passage migrant. It is a rare vagrant to Bardsey, with all four records up to 1985 involving females or immatures. Between 1988 and 1991, there were six records of Montagu's Harriers in Wales, half of these in May.

All ringing recoveries relating to this species in Wales involve birds either ringed or recovered on Anglesey. The greatest distance recorded is a nestling ringed at Newborough on 9 July 1949 and shot at St. Jean Pied-de-Port in the French Pyrenees on 27 September 1949, 1133 km south-east. A female found injured at Bodorgan by the artist Charles Tunnicliffe on 18 August 1949 had been ringed as a nestling on Dartmoor the previous year. It was nursed back to health and later released on the island. Another female, ringed as a nestling on Anglesey in 1951, was found dead near a nest containing four eggs in Kirkcudbrightshire in June 1953.

Goshawk *Accipiter gentilis*

Gwalch Marth

An increasing breeding resident in all Welsh counties.

Goshawks inhabit extensive woodlands; coniferous, broadleaved or mixed. Despite their large size, they are easily overlooked owing to their secretive nature and their habit of hunting in or near woodlands. The nest, however, is large and fairly conspicuous and it is prior to breeding that the adult birds are most often seen as they display high above their territory.

The history of the Goshawk in Wales is clouded as one of the local names of the Peregrine – "Goose Hawk" – was often abbreviated to Goshawk. In the Domesday Book, reference is made to "4 eyries of hawks" in a forest "10 leagues long and 3 leagues wide" near Soughton in Flintshire. Whether these were Goshawks is not known, but in 1484, a warrant was issued to take for falconry "such goshawks, tarcells, fawcons, laneretts and other hawks as can be gotten within the principality of Wales" (Linnard 1982). Lewis Glyn Cothi (1447–1486), who lived in the parish of Llanybydder (Carmarthenshire), also mentions the Goshawk in one of his Welsh poems. The evidence, therefore, suggests that at this time Goshawks were widespread in the deciduous forests that covered much of Wales.

Some, if not all of these birds may have been introduced from the Continent and released into the wild in order to perpetuate and increase the supply needed for falconry (Hollom 1957). By the end of the 18th century however, the loss of extensive woodlands, persecution by gamekeepers and the theft of young had limited this species to a few scattered localities within the Principality, with the last recorded nest found at the head of the Neath Valley about 1800 (Dillwyn 1848). The Goshawk probably became extinct in north Wales at around the same time although it was still mentioned under a list of birds found around Llanrwst (Caernarfonshire) in the *Faunula Grustensis* published in 1830.

There are no further records from north Wales during the 19th century apart from a bird shot on the Mostyn Estate mentioned by Forrest (1907). In south Wales also the Goshawk had become a scarce visitor by this time, with most records probably referring to escaped falconers' birds. Inevitably, nearly all records during this period refer to birds being shot, such as the individual allegedly taken near Brecon in 1887.

Throughout the first half of the 20th century records were

few and breeding was never proved or even suspected, although a pair was reportedly seen in the south-east corner of Cardiganshire on 30 April 1903. Its scarcity in Wales during this period is reflected by the fact that the species is not even mentioned in the two books on the birds of Pembrokeshire (Mathew 1894 and Lockley *et al.* 1949).

The number of records increased during the 1950s and particularly during the late 1960s, although an accurate summary of the recolonisation of Wales is difficult due to the secrecy attached to this species. The first confirmed breeding record came from Carmarthenshire in 1969 when an adult with a fully fledged youngster was seen in the Pembrey Forest on 10 August. This site was used until at least 1974, and by the end of the 1970s breeding had also been confirmed in Monmouthshire, Radnorshire and Breconshire. Over the next ten years numbers increased rapidly as Goshawks colonised the mature coniferous woodlands of upland Wales and the mixed lowland woods.

By the 1980s, high densities were recorded in some areas (Anon 1989, 1990), but despite the small population this species was persecuted over much of its range. In 1984, at least one bird was shot at a breeding site in Radnorshire and recently birds have been shot or eggs and young stolen from nests in Monmouthshire, Carmarthenshire, Radnorshire and Denbighshire. It is because of the continuing threats from gamekeepers, egg collectors and falconers that specific sites are not mentioned here.

The speed with which this species recolonised Wales is a remarkable story. There is much evidence to suggest that several birds were released at suitable sites by falconers during the 1960s and 70s (N. Fox, pers. comm.), but the rapid expansion thereafter was possible only due to thousands of hectares of suitable habitat and an abundance of avian and mammalian prey. Recent studies in Wales show that mature trees, such as Larch and Douglas Fir are usually favoured for nesting, and that the diet is very varied, but that corvids, pigeons, thrushes and Grey Squirrels are most often taken (Anon 1990; Lindley & Jenkins 1991). Lindley and Jenkins (1991) found pigeons (45.5%) and corvids (25.4%) to be by far the most prominent categories of prey taken by number in six study sites in north Wales conifer forests.

In 1990, 46 active nests were found in Wales, and by the following year this number had increased to 60. In 1992, 81 nests were located with pairs found in all Welsh counties. Numbers continue to increase and the total Welsh population in the early 1990s is likely to be about 150 pairs.

Adult Goshawks are sedentary and generally inhabit the same woodlands all year round although immatures often disperse widely. Immatures have been recorded from Bardsey in recent years, the first sighting being on 27 December 1963. The many ringing records relating to birds ringed or recovered in Wales confirm the species' sedentary nature, although one bird ringed at a confidential site in Wales on 23 May 1985 was recovered 97 km away in Gloucestershire on 15 October 1985.

Sparrowhawk *Accipiter nisus*

Gwalch Glas

Common and widespread breeding resident throughout Wales, reported to be increasing in many counties.

Most typically found in coniferous or mixed woodland, the Sparrowhawk is much more tolerant of man than its relative, the Goshawk, and is often found nesting in parks and gardens, even in the centre of towns and cities. Formerly numerous throughout Britain despite heavy persecution by gamekeepers and other landowners, the population crashed over much of the country during the 1950s and 1960s as a result of contamination by organochlorine pesticides. Following restrictions on the use of these chemicals, however, the Sparrowhawk is again found throughout most of the country, second only to the Buzzard as the most common diurnal raptor in Wales.

There is little doubt that prior to the deforestation of large areas of Wales during the Middle Ages, the Sparrowhawk was a common bird in this country. It is mentioned several times in the Domesday Book, and at that time it was probably absent only from the highest peaks of Snowdonia. The clearance of woodlands from large areas of Wales probably did not greatly affect Sparrowhawk numbers as open farmland with small copses and hedgerow trees would have provided ideal nesting habitat and an abundance of prey, but its fortunes changed dramatically during the game preservation era of the late 18th, 19th and early 20th centuries.

Between 1874 and 1902, 738 Sparrowhawks were killed on Lord Penrhyn's estate at Betws-y-Coed (Caernarfonshire), a level of persecution attained on other estates throughout the Principality. During this period, the population declined in many parts of Wales, but it was still regarded as a common breeding bird in all counties. Mathew (1894) stated that in Pembrokeshire "Teal, Snipe and caged birds" were all taken by Sparrowhawks, and despite the fact that all birds and nests were destroyed on sight, "we never seemed to make any impression on their numbers".

In north Wales, Forrest (1907) found it to be most common in the eastern counties, especially Montgomeryshire, and it was apparently quite scarce on Anglesey and on the Llŷn at that time due to the limited amount of woodlands. It was also scarce in parts of Monmouthshire and Breconshire at the turn of the century due to persistent persecution, although it appears to have escaped the attentions of gamekeepers in Merioneth. A measure of how persecution suppressed the population throughout much of Wales is the fact that numbers increased during both World Wars when most gamekeepers were in the services.

The population remained fairly stable until the late 1950s when numbers declined markedly, especially in south Wales. By the early 1960s it had become evident that there was a severe reduction in the population of Sparrowhawks and research was undertaken in an attempt to identify the cause. It was soon established that the primary reason was the widespread use of chlorinated hydrocarbons as insecticides and seed dressings, their effects being greatest in the arable areas of south and east England (Cramp 1963). The decline was so severe in many areas that this species, until then the only unprotected raptor, was given full legal protection in Britain in 1963.

Information on the extent of the decline in Wales is confusing. Prestt (1965) undertook an analysis of small avian predators in Britain and found that in Wales between 1953 and 1963, Sparrowhawks had declined most markedly in the northern counties. However, Hope Jones (1974, 1976) stated that there was no evidence to suggest that organochlorine pesticides had any effect on the Sparrowhawk populations of Caernarfonshire and Merioneth, and it is doubtful whether Sparrowhawks declined in Cardiganshire during this time (P. E. Davis, pers. comm.). During the early 1960s, gamekeepers and foresters around the edge of the Berwyn moors described it as very scarce, therefore it is likely that organochlorine pesticides had an adverse effect on the population in parts of north Wales during this period.

Cramp (1963) noted that Pembrokeshire and Flintshire were the only two counties in England or Wales where a decline had not been noted, whereas Lockley (1961) believed that there had been a decrease in the former county. Other authors noted marked declines during this period in the counties of Carmarthenshire, Glamorgan and Monnmouthshire. One would expect the population to have decreased most markedly in the more arable counties of south and east Wales, and it is known that by 1963 Sparrowhawks were very scarce in Monmouthshire, where fewer than 10 nests were found each year despite intensive fieldwork. This is one of the most intensively cultivated Welsh counties and furthermore is bordered by Gloucestershire and Herefordshire, two counties with important orchard areas which would have been heavily contaminated by DDT (Shrubb 1985).

Despite the fact that the population had decreased in Wales as a whole, the Sparrowhawk was still regarded as a widespread, if scarce, breeding species in most counties during the late 1960s. Numbers increased in south Wales following the restrictions imposed on the use of toxic chemicals such as DDT, and in many upland areas the recovery was aided by thousands of acres of suitable habitat in the form of maturing conifer plantations. In Breconshire, the Sparrowhawk is now believed to be more common than at any time during the past 50 years, with an estimated county population of about 200 pairs, while the Pembrokeshire population in the mid-1980s was estimated to be between 500 and 1000 pairs (Donovan & Rees, in press) although this is probably too high. The *West Glamorgan Atlas* found probable or confirmed breeding in 53 tetrads and RSPB surveys on Anglesey in 1986 recorded Sparrowhawk in 31 of the 762 one-kilometre squares surveyed, although breeding was proved in only three of these.

A study of Goshawks in a coniferous woodland in north Wales in 1991, also revealed a population of 23 occupied Sparrowhawk territories in an area approximately 55 km^2 (P. Lindley, pers. comm.). A study in Breconshire during 1986–90 recorded a density of 11.3 pairs/100 km^2. The remarkably high densities recorded during the former survey may be owing to an influx of Crossbills and Siskins occurring that year, but it also reflects the fact that the study was undertaken in optimum habitat. Breeding has been recorded from most Welsh towns and cities, indeed this species is now more likely to be found in Welsh urban areas than the Kestrel.

The *New Atlas* shows that the Sparrowhawk is common throughout most of Wales, absent only from the treeless uplands and parts of the Llŷn. The total Welsh population is likely to be between 3000 and 4000 pairs.

Breeding has been recorded from several offshore islands including Bardsey, where a pair nested in 1954 and 1955. One pair formerly bred on Ramsey and Caldey Is. and numbers recorded on the islands have increased in recent years in parallel with the mainland breeding population.

During the winter, adult Sparrowhawks have individual home ranges, usually centred on nest sites (Marquiss & Newton 1982; Newton 1986). Although birds generally roost solitarily, more than ten birds have been recorded going into the same wood in one evening. Cambridge Phillips (1899) once found a female Sparrowhawk roosting next to his bantams in a bushy holly tree on a cold winter night in Breconshire, presumably benefiting from the warmth of the fowls' bodies!

Adult Sparrowhawks are generally resident although the young disperse from their natal areas soon after becoming independent. Forrest (1907) noted that in Montgomeryshire males migrated on the approach of winter, but this probably refers to the dispersal of young birds. Ringing recoveries show mainly local movements of Welsh-born birds, although there are three records of birds recovered in Ireland. The farthest distance recorded however was by a bird ringed on 4 November 1986 on Bardsey and recovered on 4 May 1987 on Fair Isle, a distance of 777 km north-north-east. There is an interesting record from Bardsey of a Sparrowhawk attracted to the lighthouse, possibly by the presence of small passerines, on the night of 9 October 1956.

Common Buzzard *Buteo buteo*

Bwncath

Common and widespread breeding resident throughout most of Wales, though less plentiful in north-west Wales and rare on Anglesey.

The Buzzard is a familiar bird throughout most of Wales due to its affinity for woodland and farmland habitat, its catholic feeding behaviour (including sheep carrion) and its ability to nest close to human habitation. It is widespread throughout the mountains of Snowdonia and the wooded valleys of central Wales, but is less common on moorland blocks where a scarcity of trees and crags allow for few nest sites. Buzzards are easy to locate as they soar on warm thermals to advertise breeding territories, and the noisy, newly fledged young make confirmation of successful breeding a simple task. Persecuted by game-keepers for many years, the Welsh population has now fully recovered in most areas, and densities in central parts of the Principality are higher than anywhere else in Britain.

Buzzards must have been very common in Britain during the Middle Ages as one of the 16th century Tudor Vermin Acts forbidding egg stealing stated that taking those of the Buzzard would still be allowed (Moore 1957). The population stayed at a healthy level until the arrival of the era of game preservation in the 19th century.

By 1800 this species could still be found breeding throughout Wales, but some 50 years later it had disappeared from Anglesey, Denbighshire and Flintshire. Towards the end of the last century, Forrest (1907) could find no evidence of breeding in any of the eastern counties of north Wales, but it is unlikely that it ever became extinct in Montgomeryshire, Radnorshire or Breconshire. Certainly it had declined dramatically by the turn of the century, and its north Wales strongholds were the mountains of Caernarfonshire and Merioneth, where birds were able to escape persecution. It was almost exclusively cliff nesting in the north of the Principality during this period, although there are few records of pairs nesting on sea-cliffs.

In Pembrokeshire, Mathew (1894) knew of only six pairs on coastal cliffs and offshore islands and in Glamorgan, the Buzzard's distribution was much restricted from the pre-coal-mining days when 25 kites and Buzzards were seen in the air together between Mountain Ash and Aberdare. Buzzards survived in the heavily industrialised Rhondda Valley due to the absence of persecution, with two pairs nesting there from 1898 until at least 1908.

Forrest (1907) knew of one nest in south Wales which had been occupied by Red Kites, Ravens and Buzzards in successive years. He also quoted several instances where the normally placid Buzzard had attacked human intruders to the nest. F. C. R. Jourdain (Diaries) estimated the total Cardiganshire breeding population in 1906 to be nine to ten pairs, and could find only four sites in the north of the county in 1902.

The Welsh population reached its lowest ebb immediately prior to the First World War, with breeding pairs confined to the remotest and most inaccessible sites. During the war years, however, numbers recovered at a remarkable rate, and in 1915 Tomkinson was able to find ten nests in one day in the Abergwesyn area of Breconshire (Tomkinson collection held at Slimbridge). Walpole Bond (1914) stated that he had periodically visited over 60 eyries in Breconshire, Radnorshire, Carmarthenshire and Cardiganshire, and that he knew of another 50 nests in Wales. He put the Welsh population at this period at 250 pairs. Oldham, writing in his diaries in 1925, told how impressed he was to see an abundance of Buzzards both inland and on the coast of Pembrokeshire, and Gilbert and Brook (1931) described how they were able to find over 40 nests with ease in the Builth Wells area (Breconshire), whereas in 1902, despite intensive searching, they were able to find only two. During this period the Buzzard returned to much of its old range in the counties of Denbighshire, Flintshire and Monmouthshire where it had become extinct, but it was still absent from the Llŷn and much of north-west Wales (Bark Jones 1954).

By the onset of the Second World War, the Buzzard population had recovered to its highest level for over 150 years and numbers continued to increase for another 15 years. Two egg collectors, J. C. Robson and E. H. R. Pye-Smith found 18 nests "in a short spell" in the Llanbrynmair area (Montgomeryshire) in 1951. In 1953 Sir William Taylor, a former Director-General of the Forestry Commission, organised a survey of Buzzards on all F. C. land in England and Wales (Moore 1957). An estimate of abundance was obtained for each forest, and the results indicated that at that time Breconshire's forests held some of the highest breeding densities recorded in Britain, although numbers were apparently still very low in Anglesey and Monmouthshire.

In 1954 the BTO organised a National Buzzard Survey in order to record its status before it could be seriously affected by myxomatosis. The results showed this species to be most common in central and south-west Wales, scarce in Flintshire and absent from Anglesey (Moore 1957). The survey also highlighted the correlation between low numbers or absence of breeding Buzzards and high numbers of gamekeepers, the latter being most common in Wales in the two counties of Flintshire and Anglesey. Indeed, game preservation probably suppressed the population in other parts of the Principality, particularly in south-east Wales, notably the Vale of Glamorgan.

Myxomatosis appears to have had little influence on breeding numbers except possibly in the lowlands of Radnorshire, Breconshire, Pembrokeshire and Monmouthshire. Indeed, on the Gower peninsula (Glamorgan) the number of nesting pairs actually rose from 15 in 1954 to 18 in 1956 (Tubbs 1974). Over most of upland Wales, its effects were probably minimised by the fact that Buzzards are capable of exploiting a wide variety of prey including sheep carrion, small mammals and invertebrates. However, few of the many pairs in Breconshire bred in 1955, and in the following winter large numbers of birds were killed by farmers, who wrongly feared that lambs and poultry would be eaten now that rabbits were scarce. In Pembrokeshire, the reduction in rabbit numbers caused widespread breeding failures among Buzzards, with brood sizes of three being extremely rare after the mid-1950s. The highest density of breeding Buzzards recorded in the British Isles during the 1954 survey was on Skomer, where eight pairs nested on 722 acres, 7.08 pairs per square mile (Davis & Saunders 1965). Over the following two years, however, numbers dropped markedly due to the spread of myxomatosis, and by 1956 only two pairs were present.

During the late 1950s and early 1960s several species of raptors declined in Britain due to the effects of organochlorine pesticides. Tubbs (1972) found four instances where Buzzard clutches had been broken in Welsh nests up to 1969, and one egg from Skomer in 1963 was found to contain small amounts of DDE and Dieldrin (Davis & Saunders 1965). Over most of Wales, however, the population appears to have been unaffected, probably due to its predominantly mammalian diet, and no less than 120 breeding pairs were recorded in Cardiganshire in 1961.

Even during this period, Buzzards continued to be heavily persecuted in some parts of Wales, most being illegally shot or trapped (Tubbs 1974). Two out of 13 birds found dead on an estate in Radnorshire in 1971 however, were found to contain up to 500 ppm of strychnine, a poison widely misused in parts of central Wales until quite recently.

By the 1960s Buzzards were breeding sporadically on Anglesey once more, with a minimum of three pairs breeding in 1967. The population has remained at about this level ever since, although an influx of birds did occur in 1989, some of these staying to breed on the island. Elsewhere in Wales, the Buzzard is common everywhere, except on the Llŷn Peninsula (Caernarfonshire) and in Flintshire. Data held by the RSPB show that persecution is still affecting the population in some areas. Poisoning by sheep farmers, although strictly illegal, is inadvertent, the main targets being foxes and corvids, but the main threats to this species today are from game rearing interests, particularly in Anglesey and parts of eastern Wales.

One interesting development since the turn of the century in many inland areas away from the mountains of central and north-west Wales has been the way in which Buzzards have relinquished their traditional cliff nesting sites in favour of tree nests. This was undoubtedly due to the fact that during the game preserving era breeding pairs could survive only on the most inaccessible cliffs. In Snowdonia, however, traditional cliff sites are still used today while new pairs have moved into deciduous and coniferous woodlands (P. J. Dare, pers. comm.).

Recent work has shown that in some parts of Wales Buzzards achieve remarkably high breeding densities, although breeding success in such areas is generally low. Newton *et al.* (1982) found higher densities in mid-Wales than anywhere else in Britain, with average densities of 41 territories per 100 km^2 in farmland, but each pair produced only 0.6 young per year. M. Shrubb (pers. comm.) found 26–27 pairs per 100 km^2 in 150 km^2 in Breconshire between 1987 and 1990 although here also, breeding success was low, with some pairs failing each year. P. J. Dare (pers. comm.) estimates the Welsh Buzzard population, based on these studies, his own and the BTO's Buzzard Survey of 1983, to be 3600–4000 pairs, some 25% of the total British population.

Buzzards breed on sea-cliffs on several of the Pembrokeshire islands, including Skomer (three to four pairs) and Ramsey (up to four pairs). This species no longer breeds on Skokholm, and has never bred on Bardsey, where it is a scarce, but annual visitor.

The Buzzard's winter distribution is very similar to its breeding distribution, reflecting its year-round territorial behaviour and affinity for well-wooded farmland. Ringing recoveries have confirmed the fact that British Buzzards are rather sedentary (Picozzi & Weir 1976; Davis & Davis 1992), with adults usually remaining on their territories throughout the year. During severe weather, however, Buzzards may have to move elsewhere to look for food. Large gatherings of up to 45 birds in one field have been reported from several parts of mid and south-west Wales in the late 1980s, the Buzzards often seen feeding on invertebrates in association with Red Kites. These examples of communal feeding do not appear to depend on specific weather conditions or farm practices, and probably mainly involve dispersing juveniles (P.J. Dare, pers. comm.).

Most Welsh ringing recoveries show local movements only, but two records involve distances of over 200 km. A nestling ringed in Cardiganshire on 17 June 1976 was recovered in Hampshire in December, a distance of 211 km south-east, and a bird ringed in Breconshire on 21 June 1976 was recovered in Cumberland, 289 km north-east on 25 August 1977. A nestling at Strata Florida (Cardiganshire) in 1971 was recovered in south Merioneth 19 years later in 1990, the second oldest British Buzzard on record.

Rough-legged Buzzard *Buteo lagopus*

Bod Bacsiog

A scarce visitor in autumn and winter.

This species has a circumpolar breeding distribution across Eurasia and North America, reaching as far south as Scandinavia. The Fenno-Scandian population winters in Denmark and southern Sweden but in some years it appears that a more westerly autumn migration results in large influxes into Britain (Scott 1968 and 1978). The bulk of the wintering records come from the coastal counties of east and south-east England but there are a handful of sightings from Wales during the years when records are more numerous than usual (e.g. three in 1966–67 and four in 1974–75).

The Rough-legged Buzzard's present status as a sporadic visitor in very small numbers is also a fair reflection of its status ever since the 19th century. Most of the early records refer to birds that were shot or trapped, e.g. in Merioneth two were shot near Bala in 1866, six were killed on the moors near Corwen in the years preceding 1891, two more in 1892 and another in 1895.

Records cover every county in Wales and mostly refer to individuals seen on a single date. There are, however, some long-staying birds such as one on Skomer from 22 October 1962 until 2 March 1963, one on Bardsey from 4 November 1974 to 27 March 1975 and one near Sennybridge (Breconshire) from the end of November 1968 to 25 April 1969.

In recent years it has been very rare with only one record for the period 1988–91 inclusive.

Golden Eagle *Aquila chrysaetos*

Eryr Euraid

Vagrant. Bred in north Wales until the 18th century.

The eagle is frequently mentioned in early Welsh literature and represented a symbol of power, perhaps reflecting memories of the Eagle Standards of the Roman legions. It is a bird of myth and legend, although it is difficult to establish whether many of the early writings refer to present-day Wales, since the "Welsh" cultural area previously extended from south-west England into southern Scotland.

W. Boyd Dawkins in his book *Cave Hunting* mentions finding the remains of eagles in a cave at Perthi-Chwaren near Llandegla (Denbighshire), but as with many of these early records, the Golden Eagle is not distinguished from the White-tailed Eagle. Golden Eagle remains were found at Cat Hole Cave on the Gower (Glamorgan) in 1864 and at Coygan Cave in Carmathenshire (E. Walker, in litt.) and the books of Aneurin and Taliesin, written in the Middle Ages, referred to the eagles' habit of scavenging on the battle-field. Thomas Johnson, the botanist, when collecting plants on Carnedd Llewelyn (Caernarfonshire) in 1639, recorded that the local guide refused to take him to some of the higher cliffs for fear of the eagles which nested there, and there are several other records of Golden Eagles nesting in Snowdonia during this period. Bolam (1913) wrote that in the time of Willoughby, who died in 1672, Golden Eagles were recorded breeding annually upon the high rocks of Snowdon, but a century later, Pennant stated that while Golden Eagles occasionally migrated into Caernarfon-shire, they rarely bred. He also noted that the Welsh name for Snowdonia is "Eryri", possibly derived from. "Eryr", the word for "eagle". Evans in his *History of Wales*, written about 1800, speaks of this species as still found in Snowdonia, though it is probable that it became extinct towards the middle of the 18th century and that records after this time referred to White-tailed Eagles or Buzzards.

During the 16th century, there is some evidence of Golden Eagles nesting in Denbighshire. Forrest (1907) quotes Leland (1540) who wrote of Castell Dinas Bran above Llangollen: "There bredith on the Rock Side that the Castelle stondith on every yere an Egle, and the Egle doth sorely assaut him that destroith the nest by going down in one basket and having another over his hedde to defend the sore stripe of the Egle." The presence of cliffs and crags throughout the rockier parts of north Wales bearing the name "eryr" bears witness to the fact that this species was once quite widespread, although probably not common, in this part of the Principality.

Golden Eagles were recorded from south Wales until the 18th century but the lack of confirmed sightings suggests that they had long been extinct as a breeding species here and the later records referred to wandering birds from farther north. There are also very few places named after this species in the south, even among the higher peaks of the Brecon Beacons, which would have been suitable during the time when these areas were much less accessible and before the invention of the gun. As with all birds of prey, persecution gradually took its toll and eventually led to its extinction.

Cambridge Phillips (1900) knew of only two records for south Wales, one involving a three-or four-year-old bird killed at Penpont near Brecon in about 1859, and the other an older record of an eagle killed near Llansanwr in Glamorgan in 1776. This bird was "shot in the act of killing a lamb, but its wing being only broken, it nearly killed a dog before it was despatched". A pair was reported from the Llanberis Pass (Caernarfonshire) in September 1870, but this probably referred to Buzzards, and in his diaries, Salter makes mention of a gamekeeper who saw an eagle at Bird Rock (Merioneth) towards the end of the last century.

During the present century, the majority of records have come from north Wales. On 2nd July 1909, one was watched for almost an hour being chased by Buzzards and Ravens over Cwm Bychan (Merioneth). The other records from this county are of single birds near Llanuwchllyn in September 1923 and June 1967, and an immature bird near Trawsfynydd on 25 November 1965. The first Carmarthenshire record came from Llanelli, where one was said to have been shot on 6 October 1908, but a bird seen here in the mid 1970s and widely reported as a Golden Eagle, proved to be an immature Buzzard. Unsubstantiated claims of Golden Eagles came from Tre'r-ddol (Cardiganshire) on 21 July 1965 and near Tretwr (Breconshire) in 1938, the former attracting a great deal of media attention.

In 1915, two birds which had been kept in captivity on the island were released on Skomer. One was shot soon after on the mainland near Marloes but the other lived on Skomer for 13 years and was seen quite far afield before it moved to Ramsey in 1928. A "mate" was provided from London Zoo in June 1929 but this bird, a female, was found drowned on the mainland coast opposite the island on 9 November the same year. The remaining bird survived until March 1932 when it was shot by the farmer on Ramsey who alleged it was killing lambs, the final irony in this sad tale being the fact that this bird was also a female!

The only recent record of this species occurring in Wales is an immature bird seen over Cors Caron on 28 January 1990, although there have been several unconfirmed reports over the past five years from several locations in north and mid Wales.

Osprey *Pandion haliaetus*

Gwalch y Pysgod

An annual spring and autumn passage migrant which has increased in parallel with the Scottish breeding population. Single birds oversummered in 1986 and 1992, and several have stayed for long periods.

There is no evidence to suggest that Ospreys have ever bred in Wales; indeed up to the late 1970s, it was regarded as a very rare passage migrant with usually one or at most two records each year. In Britain the small breeding population of some 80 pairs is confined to Scotland, and the number of sightings in Wales has increased in parallel with the increase in the Scottish population.

During the 19th century, Ospreys appear to have been fairly regular visitors to Llangorse Lake, although the first county record was a bird shot on the Wye in 1859. Most of the earlier records, however, refer to individuals being pole-trapped (e.g. at Clyro in Radnorshire on 13 April 1867) or shot (e.g. near Llandudno in Caernarfonshire in 1828). At that time this species still nested in Scotland and one assumes that it was more commonly encountered in Wales than during the first half of the 20th century when Ospreys no longer bred in Britain. On rare occasions, birds were allowed to go unmolested, and two were seen for several weeks on the mouth of the Ogmore (Glamorgan) in 1887.

The Scottish breeding population became extinct during the early years of the 20th century, and during the period between 1920 and 1960, the Osprey was a scarce visitor to Wales. Between 1938 and 1960, none was recorded in Breconshire, and there are no records from Glamorgan between 1928 and 1952. One unusual record from this period came from the Glaslyn estuary (Caernarfonshire) where an Osprey was seen to take a Lapwing on 7 June 1947. By the 1980s, however, the Osprey had become an annual spring and autumn visitor, with birds staying for long periods on occasion. Single birds summered at Llyn Brianne and Burry Inlet (both Carmarthenshire) in 1986 and 1992 respectively, and long-staying individuals have also been recorded from Llyn Brenig (Denbighshire), the Dyfi estuary (Merioneth/Cardiganshire) and the Elan Valley lakes (Radnorshire/Breconshire). In early April 1986, a male displayed on several occasions at the Upper Lleidi Reservoir (Carmarthenshire). The bird was seen to "fly in tight undulations with fish in its talons and calling simultaneously" (I. K. Morgan, pers. comm.).

Favoured passage sites include large lakes, such as Talybont Reservoir and Llangorse Lake, Llyn Brianne and Llyn Brenig, estuaries such as the Burry Inlet, and the Dyfi and river valleys such as the Usk (Monmouthshire), Wye (Radnorshire), Conwy (Caernarfonshire) and the Tywi (Carmarthenshire).

Sightings from some of these favoured sites over the years hint at distinct migration routes through Wales. Birds which appear on autumn passage in Monmouthshire are often first recorded at Talybont Reservoir or Llangorse Lake, and in spring Ospreys seem to move from Llyn Brianne (Carmarthenshire) through the Elan and Wye valleys (Breconshire). Most passage birds are recorded in April–May and August–September, extreme dates between 1988 and 1991 being 9 March and 25 October.

Monthly totals of Osprey sightings in Wales, 1988–91

Mar	*Apr*	*May*	*Jun*	*Jul*	*Aug*	*Sept*	*Oct*
3	13	23	7	16	25	21	6

The number of records from Wales has increased over the past ten years, and of 28 birds recorded from Breconshire between 1969 and 1989, ten were seen in August. This increase has occurred in parallel with the expansion in range and numbers of the Scottish breeding population, particularly during the 1980s. In 1990, a total of 36 birds were recorded in Wales, although some of these may involve the same individuals seen at two distinct sites.

On 9 September 1990, a juvenile Osprey ringed at an undisclosed site in Scotland on 10 July was recovered at Oxwich Marsh NNR, 661 km south.

Lesser Kestrel *Falco naumanni*

Cudyll Coch Lleiaf

A vagrant.

There is only a single record in Wales:

7 November 1973 – One, Vale of Neath (Glamorgan)

Lesser Kestrels breed as close as southern France and Spain, most of the European population migrates in winter to Africa south of the Sahara.

Kestrel *Falco tinnunculus*

Cudyll Coch

Rather scarce breeding resident throughout Wales. Partial and passage migrant in small numbers to some counties.

The Kestrel is one of the most ubiquitous birds of prey, breeding in a variety of habitats, from moorland and farmland to sea-cliffs and even city centres. It is the most widespread and numerous raptor in Britain, although now scarce or absent in some parts of north-west Scotland. In Wales, however, it is outnumbered by the Buzzard and Sparrowhawk, and is present only in fairly low densities over much of the country. Numbers are not easy to assess as the population in the uplands fluctuates according to the number of Field Voles, the Kestrel's main prey item in Britain (Village 1990).

During the game rearing era of the 19th and early 20th centuries, the Kestrel was heavily persecuted along with all other birds of prey. Between 1874 and 1902, 1988 Kestrels were killed on Lord Penrhyn's estate near Betws-y-Coed (Caernarfonshire), and several authors make note of the fact that this species faced the attentions of the gamekeeper's gun due to its habit of preying on Pheasant chicks in rearing fields. Mathew (1894) recorded how Kestrels would carry off young Pheasants during prolonged bad weather in early summer when other prey was difficult to catch, an occurrence also recorded by Cambridge Phillips (1899). Despite the prolonged persecution, Kestrels were described as "common" throughout the Principality at the beginning of the 20th century and were valued by some landowners for their ability to catch large numbers of harmful rodents and insects.

At the turn of the century, it was recorded as common on the sea-cliffs of the Llŷn (Aplin 1910), and Ingram and Salmon (1920) remarked that it frequently visited Roath Park in the centre of Cardiff. It was heavily persecuted in parts of Glamorgan until the First World War, but here again this appears to have had little effect on the overall population. Local increases were recorded up to the 1950s in parts of Glamorgan, Monmouthshire and Cardiganshire, as a result of large-scale afforestation, with high densities in some areas during peak vole years. In 1971 six nests were found in 40 ha of newly afforested moorland near Trawsfynydd (Merioneth). By this time, Kestrels had colonised many towns and cities in Wales, taking advantage of an abundance of small birds, mammals and insects and in 1973 a pair bred successfully on St. John's Church in the middle of Wrexham (Denbighshire).

Prestt (1965) found that the Kestrel population in north Wales had shown a slight increase since the 1950s, but that in south Wales the reverse was true. However, he calculated that numbers in south Wales at that time were higher than anywhere else in Britain, and that the severe declines in eastern England caused by organochlorine pesticides had not affected the Welsh population.

More recently, several studies and surveys have calculated Kestrel breeding densities for different areas of Wales. In 1967 a Nature Conservancy Council survey found a minimum of 24 nests on Anglesey, and an RSPB survey in the county in 1986 recorded this species in 44 of the 762 km^2 surveyed (RSPB 1986). The breeding population of Kestrels in 926 km^2 of predominantly mountainous terrain in Snowdonia (Caernarfonshire) between 1979 and 1982 was estimated at 33–64 pairs (Dare 1986a). This gave an average density of 3.6–6.9 pairs per 100 km^2, near the lower end of the recorded range for this species in Britain (Village 1990). Breeding was restricted to below 450 m and Dare suggested that the low density may have been due in part to competition for food with the more abundant Buzzard. Density estimates for Breconshire also reveal low numbers, with a maximum of 5.6 pairs per 100 km^2 recorded in an area of 360 km^2 between 1985 and 1990. The county total was estimated at 70–100 pairs, a surprisingly low number considering the amount of rough grazing in the county. The *West Glamorgan Atlas* recorded Kestrels in 63% of the 148 tetrads with breeding confirmed in 34 of these.

The *New Atlas* reveals that there have been local decreases in many of the Welsh uplands and parts of south Wales since the early 1970s. This decline is thought to be connected to the impact of an increase in sheep stocking rates on vole-rich habitats, as Kestrels are very scarce in extensive areas of improved grasslands (M. Shrubb, pers. comm.). In Pembrokeshire, where the total population is thought to be no more than 50 pairs, this species is now confined to the less intensively farmed areas including the offshore islands, the coastal strips, the Preseli mountains and large towns. The decline in new forest plantings may also have contributed to the decrease, restocked areas being far less attractive to this species than new young plantations. The current Welsh population, based on the work of Dare (1986a) and Shrubb (pers. comm.) is likely to be about 800-1000 pairs.

Kestrels nest on most of the larger offshore islands, although breeding has not been recorded on Bardsey since 1970. Birds are recorded from here and the Pembrokeshire islands every month but are particularly common in August and September. On Bardsey, small numbers are occasionally seen moving south out to sea on passage.

During the winter months, Snow (1968) showed that upland Kestrels in Britain migrated more than lowland ones. Indeed the greatest concentrations in Britain are found in eastern England, where resident birds are augmented by immigrants from the Continent, chiefly the northern Baltic and the Low Countries (Mead 1973). The distribution map in the *Wintering Atlas* shows that numbers over most of Wales are much smaller than during the breeding season.

Forrest (1907) believed that Kestrels on the west coast of Anglesey were partially migratory as few were seen in the area from September onwards. Dispersal immediately following the breeding season has also been recorded from Radnorshire and Breconshire, and may be common elsewhere but few data are available. At times, these post-breeding movements result in large concentrations of birds sometimes attracted to a particular food source, as on Ramsey on 11 September 1972 when 18 were seen feeding on beetles. Up to 30 gathered on Skomer in September 1986, 20 were seen over Twmbarlwm (Monmouthshire) in July 1971, and at least 15 were seen hunting near the Rhymney estuary in the same county on 30 August 1972.

Ringing recoveries provide evidence of migration, particularly amongst first-year birds, the longest distance recorded involving a bird ringed as a nestling near Sennybridge (Breconshire) on 22 June 1963 and recovered in Malaga (Spain) five months later, 1635 km to the south. A nestling ringed on Bardsey on 29 June 1961 was recovered in Spain in September of that year, and one of its siblings ringed on the same date was recovered dead on a beach near Abersoch (Caernarfonshire) in September 1968. Evidence that some birds come into Wales from the Continent was provided by a bird recovered at Newport (Pembrokeshire) on 30 October 1975, which had been ringed the previous year in the Netherlands. Shrubb (pers. comm.) analysed British Kestrel ringing recoveries and found that whereas it is probable that many Welsh birds disperse into the English lowlands in their first autumn, longer distance movements were much more southerly orientated.

Red-footed Falcon *Falco vespertinus*

Cudyll Troedgoch

A rare visitor.

It breeds from Austria eastwards to central Siberia and winters in southern Africa.

There have been 11 records of this species since the first, a bird which was killed at Wrexham (Denbighshire) in May 1868. Nine of the records refer to the period from

1972 onwards. The figure of 12 compares with a total of 480 from Britain and Ireland up to 1991, most of which are from the south and east counties of England.

Spring passage takes place across the Mediterranean from Algeria eastwards and therefore quite often good numbers move north across the western half of Europe. Not surprisingly the majority of the Welsh records (10) are from the period May to June, the remaining two relating to October.

In addition to the Denbighshire record referred to above sightings are from Glamorgan (5), Pembrokeshire (4), Anglesey (1) and Cardiganshire (1).

Merlin *Falco columbarius*

Cudyll Bach

Scarce breeding resident, passage migrant and winter visitor.

The Merlin is a bird of open moorland during the breeding season, although in Wales pairs formerly bred on coastal dunes, sea-cliffs and dry lowland bogs. It is an elusive falcon and nests are often difficult to locate, although its habit of returning to the same area of moorland each year does facilitate the location of breeding pairs. It has suffered a long, steady decline throughout Wales since the turn of the century although there may have been a slight recovery since the 1980s as over half the Welsh population now breeds in old corvid nests on the edges of upland conifer plantations, often in traditional territories where pairs had apparently been absent for several years.

During the last century, the Merlin was a ubiquitous breeder, widespread throughout Wales even in lowland areas. It was considered by Forrest (1907) to be common on Anglesey, and Coward and Oldham (1905) described it as numerous on the north Anglesey coast. At that time pairs bred regularly on heather-clad headlands at Rhosneigr, Red Wharf Bay, Holyhead Mountain and Carmel Head, and in 1904 four pairs were found between Amlwch and Point Lynas, and five nests were located between Bull Bay and Cemmaes in 1905. Breeding was not confined to the coastal cliffs, however, as pairs were also found in dry bogs, such as Cors-y-Bol and Cors Erddreiniog, and on sand dunes, such as Newborough Warren NNR and Aberffraw. Nesting pairs were also recorded around the coast of Llŷn, particularly near Aberdaron, these areas being well worked by egg collectors.

Even in 1896 this species had apparently declined in Merioneth, possibly due to persecution, although C. V. Stoney found up to three nests "without great difficulty" each year on the Rhinogs, and on 29 May 1914 he took a clutch of three from the sand dunes at Mochras. Salter (1895) recorded that Merlins almost certainly bred on Cors Caron and Cors Fochno, and in Pembrokeshire a nest was found amongst gorse and heather at Penberi in 1894 and breeding was confirmed at Goodwick, near St. David's and on the Preseli Mountains.

It was still believed to be breeding in the Cenarth area (Carmarthenshire) in 1865, but there are no confirmed records. In 1879 a pair possibly bred on duneland near Cydweli, and in 1898 a clutch of five was taken from the mouth of the Ogmore (Glamorgan). In this county, coastal nests were generally left unmolested, whereas inland, as on most other Welsh moors, this harmless falcon was heavily persecuted by gamekeepers. Nests were located in the adjacent county of Monmouthshire in the 1890s, mainly from the uplands around Abergavenny, where nestlings were reputed to be taken for falconry.

Up until the First World War, this species was common on several of the north Wales and mid-Wales grouse moors, particularly (it would appear from the collections of several egg collectors) Mynydd Hiraethog. W. M. Congreve in particular was an avid collector of Merlin eggs from the uplands of Denbighshire, often relying on local gamekeepers to collect the clutches for him. Although there are few records from the Berwyn Mountains during this period, there is little doubt that this bird was common here and throughout the heather uplands of Snowdonia. On Cadair Idris (Merioneth) young Merlins were known as 'Stone Falcons', doubtless due to their habit of perching on rocky outcrops and boulders while waiting for the adults to return with prey. It is only in the county of Breconshire that this bird of prey was recorded as being uncommon, the only nest cited by Cambridge Phillips (1899) being one from which a clutch of four was taken from Grouse Hill on 29 May 1888.

By the 1920s, the Glamorgan Merlin population was beginning to decline, principally on the coast, where the level of disturbance from day trippers proved to be too much for the species. Prior to the First World War, a minimum of six pairs nested in coastal dunes, and at least two of these territories had been vacated by the mid-1920s. The last few coastal pairs in the county disappeared some 20 years later, driven away by military exercises during the Second World War. By 1939 there were thought to be only three pairs remaining in the uplands of Monmouthshire, and a decline was also noted in Pembrokeshire at this time.

In Carmarthenshire, breeding was confined to a few pairs on the northern moors by the 1950s, and in Cardiganshire the population had decreased from up to ten known pairs prior to 1940 to fewer than six in the 1950s. Here, however, numbers recovered slightly during the early 1960s, and a pair bred once more on coastal cliffs in 1952 and 1961. Throughout the 1930s and 1940s O. H. Wells found several breeding pairs in the Abergwesyn area of Breconshire, with six pairs here in 1942, the majority in the nests of old Carrion Crows, although one was located in an old Buzzard nest on a crag.

During the 1950s Merlins were still recorded breeding at several sites on Anglesey, including Newborough Warren NNR (two nests here in 1954), Holyhead Mountain, Carmel Head and Porth Wen. Coastal breeding pairs were noted on the Llŷn and at one coastal site between Bangor and Conwy at this time, and pairs were located most years from the better known traditional upland sites by egg collectors and naturalists alike.

By the mid-1960s breeding records were scarce away from the core upland areas, and even here declines were evident. Scattered records of coastal breeding were reported from Carmarthenshire and Caernarfonshire throughout the 1960s and 1970s, but regular breeding at most of the Anglesey sites ceased during this period although a pair was suspected to have bred on the island in 1986 (RSPB 1986). In 1960 a pair may have bred on the Freni Fawr (Glamorgan), and in the same year it is likely that a pair bred in an old corvid nest on Dowrog Common (Pembrokeshire) and breeding was suspected on Ramsey until at least 1974. By the 1970s it was reported to be scarce in Breconshire and Radnorshire, although RSPB surveys in the late 1970s found a minimum of 12 pairs in Breconshire and up to ten pairs in Radnorshire where an egg analysed by the Institute for Terrestrial Ecology was found to contain the highest concentrations of PCBs ever discovered in a Merlin egg up to that time (24.3 ppm). In Flintshire breeding was confined to one known site on the Clwydian Range and even in core areas such as Ruabon Moor (Denbighshire), a decline had been noted by the late 1970s.

The *Breeding Atlas* showed the Welsh Merlin population to be confined almost exclusively to the uplands of north, mid and south-west Wales, with breeding pairs also recorded in Anglesey and Carmarthenshire. Organochlorine pesticides were implicated as a cause for the decline during the 1960s (Newton 1973), but it was evident that factors such as habitat loss due to afforestation and agricultural improvement were also important.

Williams (1981) estimated the Welsh breeding population between 1968 and 1975 to be at least 150 breeding pairs based on the number of territories occupied at least once during that period, with the majority of north Wales birds nesting in heather and most south Wales pairs nesting in trees. This figure however was probably too high, and information collated from fieldworkers in 1983 and 1984 revealed that the Welsh population had declined to an estimated 40-45 pairs concentrated on the Berwyn and Migneint (Merioneth) Moors. A sharp decrease was also recorded on Ruabon Moor (Denbighshire), where the population fell from eight pairs in the mid-1970s to only one in 1982, a decline where human predation, fire, poor weather and pesticides were all implicated (Roberts & Green 1983). In addition to this, it is likely that the loss of moorland edge habitat such as the ffridd to agricultural improvement had a significant effect on the Welsh upland Merlin population as these are important hunting grounds for this species (S. J. Parr, pers. comm.).

During the mid-1980s Merlins were discovered breeding in old corvid nests in mature conifer plantations adjacent to open moorland in the Cambrian Mountains. Two such nests were found in 1984 and 1985, following the discovery of the first nest in 1982. By 1989 Parr (1991) calculated the total breeding population within this upland area to be 12 pairs, all but one of which were nesting in conifer plantations. At this time four pairs were known to nest on plantation edges on Mynydd Hiraethog (Denbighshire), and this change in nesting behaviour was widely reported from most of the mid and south Wales uplands, as well as some of the north Wales moors. It is likely that the Welsh population never reached the reported low ebb of fewer than 50 pairs during the early 1980s, but that several tree-nesting pairs went undetected during this period.

Interestingly, the strongholds of heather nesting Merlins are the grouse moors of the Berwyn Mountains, although even here some pairs are now known to be nesting in coniferous woodlands. It is now thought that whereas there has been a genuine and marked decline in many Welsh counties over the past 30 years, the population may have been stable over much of mid-Wales, the apparent disappearance of breeding birds explained by the switch from traditional moorland trees to the edges of conifer plantations (Parr 1991). At present, the total Welsh population is believed to be at least 70 pairs, more than 50% of which nest in coniferous woodlands.

Merlins are recorded on passage from all Welsh counties and offshore islands, with peak numbers from the latter generally falling between March to April and September to November. On Bardsey one or two are recorded daily in spring and autumn, some birds residing on the island for long periods. One individual trapped here in October 1984 was of the Icelandic race *F.c. subaesalon*. Merlins have twice been recorded on Grassholm, on 27 September 1972 and 18 October 1978.

Merlins generally move down onto the lowlands, and often to coastal areas during the winter months, although some will stay on high ground for long periods, depending on the weather and availability of prey. They are generally solitary, although on occasion they will hunt in pairs or in association with other birds of prey. It is apparent that areas favoured by Merlins in winter today were also frequented by this species at the turn of the century. Towards the end of the 19th century this species was frequently caught in birdcatchers' nets in the Cardiff district (Glamorgan) when it attempted to seize the call-bird, and Mathews (1894) records one instance where a Merlin caught almost uninjured in a rat trap was tamed within a period of a few days.

British Merlins rarely migrate long distances, therefore the majority of birds seen in Wales during the winter will also have bred here. More than ten birds are thought to be present on Anglesey in some winters; up to four are regularly present on the Dee estuary, and similar numbers are sometimes recorded from the Burry Inlet. Other favoured sites include the lowland bogs of Cors Caron and Cors Fochno, where birds have occasionally been seen hunting in association with Hen Harriers.

Elsewhere inland a few birds are generally present on

lower ground, and in north Breconshire the presence of Merlins is thought to be related to the occasional occurrence of substantial wintering flocks of passerines, such as thrushes. Such flocks often attract several merlins, generally solitary hunters. Three Merlins were seen harassing a flock of Starlings at roost near Tywyn Point (Carmarthenshire) in October 1975, and three were present here again in November and December 1976. This association with Starling roosts may be widespread in Wales as reports of single birds hunting such flocks have also been reported from the Cambrian Mountains and Aberystwyth pier (Cardiganshire).

There are several records of Welsh ringed Merlins recovered in southern England, but the longest distance recorded involves a bird ringed as a nestling on the Migneint (Merioneth) on 27 June 1983 and recovered on 10 December 1983 at Brizambourg in France, 827 km south-east.

Hobby *Falco subbuteo*

Hebog yr Ehedydd

Rare breeding resident in at least four counties. Elsewhere a rare passage migrant, although a few summer on occasion.

The Hobby is a summer visitor which is confined mainly to the southern half of England, where it is typically found on dry heaths, downlands and low-lying farmland in broad river valleys. Outside the main breeding range, and where it occurs in south-east Wales, mixed farmland is its usual habitat. Here, pairs are often difficult to locate, with the result that the actual number breeding may be under-estimated. This century, sporadic breeding has occurred in Wales since the early 1960s, with regular breeding in Monmouthshire probably since the late 1970s. The population is still very small, although it is increasing its range westward.

It is unlikely that the Hobby was ever common in Wales. During the 19th century Dillwyn (1848) records it as a rare visitor to the Swansea area of Glamorgan, and Dobie (1893) stated that it occurred only occasionally in north-east Wales. During this period, however, Hobbies did breed sporadically in parts of south and mid-Wales, Walpole-Bond (1903) recorded it as still breeding in one or two localities in mid-Wales at the turn of the century. A pair bred at Llanmaes (Glamorgan) in 1891, although the single egg was taken by the late A. F. Griffith of Brighton. Hobbies nested in this county again near Margam in 1897 and near Bedlinog in 1892, and a pair may have bred near St. Fagans in 1913. Most early records of breeding, or possible breeding, are from south-east Wales, and in 1910 Hobbies bred at Wentwood in Monmouthshire.

Breeding was not confined exclusively to south Wales and in 1912 a pair nested in a solitary tree on open moorland near Dolgellau (Merioneth). The bird was shot off the nest and positively identified as this species, although its location is more typical of a Merlin. Forrest (1907) knew of no confirmed breeding records for north Wales, although pairs had been reported to him from Tywyn (Merioneth) and Llanllugan (Montgomeryshire). A pair was reported from the latter county again in 1920, but breeding was not suspected.

During the 1930s breeding was recorded once more from three southern counties. In 1934 an immature female was shot near Llyswen (Breconshire), the gamekeeper claiming that a pair had bred in a nearby wood. In the same year an adult female was shot on 8 August near Usk (Monmouthshire), with the adult male and juveniles also probably meeting the same fate. Breeding was again recorded in Monmouthshire, near Chepstow, in 1938, but once more the adult female was shot. A male shot in Glamorgan on 16 August 1936 was almost certainly from a breeding pair, and it is likely from these few scattered records that many other breeding attempts went unrecorded in south-east Wales.

In 1961 and 1962 a pair nested in south Radnorshire, the first breeding record for the county, and in 1966 breeding was recorded once more in Monmouthshire. It was not until the 1980s, however, that it was discovered that a small breeding population had become established in the latter county. In 1984 a pair reared three young, and the following year breeding was again recorded. Subsequent analysis of Hobby sightings for the county revealed that pairs had probably bred in nine different areas since 1980 and possibly earlier, with a maximum of four pairs in any one year. Since then, breeding has been proved annually, with up to five nests found each year. However, nests are remarkably difficult to locate and it is possible that the total county population could number more than ten pairs (S. Roberts, pers. comm.).

Breeding has been suspected at three sites in the upper Severn valley in Montgomeryshire since the late 1970s, and more recently in Breconshire, but no nest has ever been located. Since 1990, breeding has occurred once more in east Radnorshire, and breeding has also been recorded from Pembrokeshire following the discovery of a summering pair in 1989. The scattered nature of these breeding records, the occurrence of pairs in suitable habitat elsewhere during the nesting season and the birds' secretive behaviour suggest that the total Welsh breeding population could be as high as 20 pairs, and probably still increasing.

Hobbies have been recorded on passage, albeit in small numbers, from all Welsh counties. Between the late 1930s and 1970, the number of sightings of passage birds decreased markedly, but following an increase in the range of the British population and the establishment of a small Welsh breeding population, numbers have increased once more. On occasion an influx of birds is recorded, as in 1989 when several sightings were reported from every Welsh county. In Breconshire, about 20 birds were recorded on passage between 1969 and 1990, with most seen in August (9), May (7) and September (7). Llangorse Lake produces the majority of records for this county as Hobbies hunt the large flocks of Sandmartins that gather here to roost. In 1990–91, spring passage in Wales fell between 7 March and 17 June, and the autumn passage between 5 July and 8 October. However, some

July records may involve wandering non-breeders rather than true passage birds.

Hobbies are rare visitors to the offshore islands, and are always recorded singly. With an increase in the number recorded on passage, this species is now seen almost annually on Skomer, and in 1972 at least one bird was resident on Ramsey throughout August and September. The latest record from the islands was a bird seen on Bardsey on 1 November 1961.

Gyr Falcon *Falco rusticolus*

Hebog y Gogledd

A vagrant to Wales.

This species has a circumpolar Arctic breeding distribution and is migratory in high latitudes only. With a marked decline in southern parts of its breeding range in recent years there is less likelihood of birds reaching Wales and there are only two post-1960 records.

There are five 19th century records, all, sadly, of birds killed, from Pembrokeshire (1), Denbighshire (2) and Breconshire (2).

Seven 20th century records are from Anglesey (3), Carmarthenshire (2), Pembrokeshire (1) and Breconshire (1) and only two of these refer to live birds.

Peregrine *Falco peregrinus*

Hebog Tramor

The Peregrine is a resident breeding species in all counties, increasing in numbers to a current breeding population (1991) of 280 pairs.

The average population in 1930–39 was estimated at 135 pairs (62 pairs coastal, 73 pairs inland) out of a total of 800 pairs in Great Britain, but there was a decline in the coastal breeding population in the Second World War through control measures taken to protect message-carrying homing pigeons. Numbers rapidly recovered after the end of the war only to crash in dramatic fashion in the late 1950s/early 1960s due to contamination by persistent organochlorine pesticides. Recovery was very slow until the 1970s but was complete by the time of the 1981 census although the coastal population was lagging behind the inland population. The 1991 census showed that the population, both inland and coastal, has reached unprecedented levels.

The Peregrine has long been esteemed as a very special Welsh bird. Falconry, which would have referred in part at least to the use of Peregrines, featured in Welsh medieval law (the laws of Hywel Dda, c940 AD. As early as the 12th century the Peregrine rated a comment in the Description of Wales by Giraldus Cambrensis in his famous *Journey through Wales*, relating to Pembrokeshire.

> "I must not forget to tell you about the falcons of this region. They are remarkable for their good breeding, and they lord it over the river birds and those in the open fields. When Henry II, King of the English, spent some time in this neighbourhood making preparations for his journey to Ireland, he occasionally went hawking. By chance he saw a noble falcon perched on a rock. He approached it sideways and then loosed at it a huge and carefully bred Norwegian hawk which he had on his left wrist. At first the falcon seemed slower in its flight. Then it lost its temper and in its turn became the aggressor. It soared to a great height, swooped fiercely down and gave the hawk a mighty blow in the chest with its sharp talons, striking it dead at the King's feet. As a result from this year onwards Henry II always sent to this region at nesting-time for some of the falcons which breed on the sea-cliffs. Nowhere in the whole of his kingdom could he find more noble or more agile birds."

Henry II arrived in Pembroke on 21 September 1171 on his way to Ireland.

Particular sites became famous for their Peregrines in the 16th and 17th centuries. Henry Thomas Payne, in his *Collectanea et Statutia Menevensia* includes reference to a lease on Ramsey Island granted to one Thomas ap Rice of Scotsborough in 1606 by Bishop Anthony Rudd of St. David's. The lease reserved to the bishop and his successors "all hawks of what kind soever that breed or hereafter shall breed in or upon the said Island or any part or parcel thereof". In his reference to the lease Payne added a note in which he said that Ramsey had long been celebrated for a breed of that species of hawk called *Falco peregrinus* or Haggard Falcon. The interest seems to have continued, for Browne Willis, writing in 1715, tells that until eight or nine years previously there was an eyrie of excellent hawks upon the island, which were highly valued until the old falcon was stolen (Howells 1968).

North Wales also provides some early references to the Peregrine. Forrest (1919) quotes a reference which stated that in 1210 the Bishop of Bangor had to pay a fine of 200 hawks (i.e. Falcons) to King John, and was supposed to have procured the birds from Pembrokeshire. It is more probable, however, that he would have got them in his own province. Pennant refers to the Peregrines breeding on the Great Orme: "This kind was in the days of falconry so esteemed, that the great minister Burleigh sent a letter of thanks to an ancestor of Sir Roger Mostyn, for a present of a cast of Hawks from this place". John Price in his *Llandudno Guide* page 104 (published about 1875) stated that the Peregrine has never ceased to breed on the Orme's Head since one of the Mostyns presented James I of England (1603–1625) with a cast which is said to have been worth £1000.

Another traditional breeding site referred to by Pennant in the 1770s is South Stack, Holyhead. It is quite likely that this was also the site referred to by John Ray in his Third Itinerary: 21 May 1662 "an airy of Falcons at a place we did not set down the name of, near Holyhead". Thus several sites occupied at the present day, can be traced back in history for hundreds of years; tenacity to favourite breeding sites is a characteristic of the species.

The number of Peregrines breeding in Wales has shown marked fluctuations in recent years as will be discussed later. In historical times, perhaps back as far as the Roman era, it is reasonable to suppose that they may have been less numerous because most of Wales was forest-covered. The greater extent of open non-wooded country created by man during the last 2000 years favours the Peregrine and the bird may have increased markedly as the great forests of ancient times were destroyed. Ratcliffe's main conclusion (1980) was that the British Peregrine population probably did not change markedly in size from, say, the Middle Ages up to about 1800, after which there were slow declines locally. These resulted entirely from human influence, acting directly through persecution.

Traditionally the main breeding strongholds for this species have long been the dramatic sea-cliffs of Pembrokeshire and Cardiganshire and the inland crags of Merioneth and Caernarfonshire. Throughout the 20th century it has, however, bred in all parts of Wales, with the exception of Flintshire where it appears that colonisation has been relatively recent. The main requirement of a breeding site is, in general, the security of a steep cliff, either inland or coastal, with commanding views over open country. Artificial sites are also used on occasion (e.g. Britannia Bridge over the Menai Straits (Anglesey/Caernarfonshire) 1945–46 and in recent years, Llanelli Copper Works Chimney Stack (Carmarthenshire) pre-1920 and a ledge on the Telecom Tower at Swansea (Glamorgan) in recent years).

The basic need is for a ledge big enough to hold a full brood of up to four large young with sufficient earth in which the bird can make a 'scrape' to hold the eggs; old nests of Ravens, usually well sheltered under overhangs of rock, are also much favoured in Wales. Of 30 nests examined by the authors in North Wales, 17 were on grassy ledges, ten were in old Ravens' nests and three were on ledges which had the remains of old Raven nests. With the recent marked increase in the breeding population nests are now occurring regularly in what would earlier have been considered marginal sites in terms of the small size of the cliff; at least one nest is on flat ground in open moorland and there has been a recent instance of the use of an old Raven nest in a tree, a habit which could well increase.

Throughout the 19th and 20th centuries the Peregrine has been the subject of much persecution by man; this has taken many forms. Bolam (1913) refers to two traditional sites in Merioneth: one "generally subjected more or less of persecution by the farmers, who says it kills their poultry" and the second where "although one or other of the birds are often killed, others always turn up to take their places, if not at once, then certainly before the following season. Of a pair which occupied the site in 1906 the female was an inveterate slayer of grouse and sorely taxed the patience of the keeper on the beat". Persecution by gamekeepers took a heavy toll on nesting pairs in the period up to the Second World War but has probably declined since then with the reduction of viable grouse moors.

Egg collecting, too, has played a significant part from the mid-19th to the mid-20th century, although it is, fortunately, not as widespread in the contemporary record. As well as being a relatively uncommon bird, the Peregrine also had the misfortune to lay particularly beautiful and variable eggs, making them a prized target for collectors. Parts of Pembrokeshire and coastal eyries of Caernarfonshire and Anglesey were subject to regular depredations. Inland sites in south and central Wales, some of which were relatively accessible, were a strong attraction to parties of egg collectors. One pair of collectors, father and son, took 20 clutches from one Breconshire cliff during 1912–38 and 15 clutches from a second haunt in the same county during 1919–35, and other clutches were certainly taken by other collectors from the same places during those years.

D. A. Ratcliffe, the doyen of British Peregrine enthusiasts whose work has done so much to elucidate population trends, estimated that the average number of pairs nesting in Wales in 1930–39 was 135 (62 coastal pairs, 73 inland pairs). This decade was regarded as the 'standard' or 'normal' period against which to assess changes in numbers during other periods since it appeared to represent a time of relative population stability.

Between 1940 and 1945 Peregrines were heavily persecuted in some coastal areas of Wales, in particular in Pembrokeshire and Caernarfonshire as a result of the Destruction of Peregrine Falcon Order made by the Secretary of State in 1940. This destruction was organised by the Air Ministry and enabled authorised persons to kill Peregrines, the intention being to protect message-carrying homing pigeons despatched from RAF aircraft. The control measures carried out in Pembrokeshire are described in detail in copies of diaries held in the National Museum of Wales in Cardiff and indicate the substantial scale of the removal of eggs and young from nests and killing of adult and immature birds. Details from the rest of Wales are sketchy.

With the cessation of persecution, it appears that the Welsh Peregrine population began to recover reasonably rapidly but a far more serious and insidious threat to its survival was not far away. The result of the organochlorine pesticide contamination on the Peregrine was a spectacular crash of an almost unparalleled nature, now widely understood and accepted through the work of Ratcliffe but causing much controversy at the time. On the south Wales coast decline was apparent first in 1956–57, but in north Wales, although breeding success was poor at inland eyries right through the 1950s, falling population was not noted until after 1957.

Perhaps paradoxically the widespread nature of the crash only came fully to light as a result of fierce agitation about Peregrines by pigeon fanciers in the south Wales mining valleys. Ivor George of Neath appeared on television to make a vigorous protest against the depredations of falcons on the pigeons of the district and this resulted in a petition to the Home Office. The Home Office, wishing to be sure of the facts, asked the Nature Conservancy for information on the status of the Peregrine and a contract to ascertain the current position was given to the British Trust for Ornithology. In fact the 1961–62 BTO survey showed a dramatic decline in the Welsh, and other UK populations.

In Wales the following results were obtained:

Year	*Number of territories known*	*Number of territories examined*	*Birds apparently absent*	*One or both present but breeding not proved*	*Nest with eggs seen bot not re-visited*	*Nested but unsuccessful*	*Successful nesting*
1961	149	90	62	16	2	1	8
1962	149	89	71	16	–	1	2

This situation was mirrored elsewhere in Great Britain, more drastically in southern England, less so farther north. By 1962 the Great Britain population overall was at 56% of the 1930 level. Ratcliffe (1965) gives the following results for a sample of 53 Welsh eyries in 1962, 1963 and 1964:

1962	15 occupied	2 successful
1963	6 occupied	1 successful
1964	7 occupied	2 successful

As in other areas of Britain, the decline probably levelled out in 1963–64, in the latter year Ratcliffe calculated that the overall population was at 44% of its normal level even though in Wales there were desperately few successful pairs at this time. It is now a matter of history that this dramatic crash, fully chronicled by Ratcliffe (1980), was caused by lethal levels of organochlorine pesticides, residues of which were identified in eggs and corpses. The detective work carried out by Ratcliffe and his colleagues in the Toxic Chemicals and Wildlife Section of Monkswood Experimental Station linking the crash of Peregrines and other birds of prey with the use of certain pesticides was important evidence in the successful move to restrict or ban the most significantly damaging ones. The story from then on is well known now and the subsequent decadal Peregrine surveys in Wales reveal a very heartening picture, showing a virtually complete recovery in the breeding population by 1981. There was, however, a striking difference between inland and coastal sites; the slowness of recovery of the coastal populations is not easily explicable but a similar situation was noted in coastal Scotland where it was found that Peregrines were contaminated by a wide range of marine pollutants.

By 1991, as shown in the map, even the coastal population was higher (125%) than the normal figure whilst the inland population has soared to no less than 276% of the pre-war level (Williams 1992). This massive increase in inland areas is partly an increase in density in what were previously regarded as "saturated" areas, and partly a filling in at marginal nesting sites which previously did not tempt Peregrines, or only occasionally when persecution was higher. It is noteworthy in this respect that Ratcliffe (1962) found the average minimum distance between nests to be 3.0 miles (total of 63 breeding territories in four inland sample areas in north Wales, north England and Scotland) covering the years 1945–61 whereas a 1989 collation in Wales by I. T. Williams found an average of only 1.66 miles for a sample of 23 breeding territories in three inland areas of Wales. Some eyries formerly regarded as alternative sites for the same pair now regularly hold a pair in each and it can only be surmised that the inland population is at, or close to, saturation point; it

Peregrine population and breeding performance in Wales in 1961, 1971, 1981 and 1991 compared with 1930–39

Year	*Number of territories visited*	*Estimated average number of pairs 1930–39*	*Number of territories occupied*	*Number of pairs rearing young*
Coastal				
1961	28	21	7 (33%)	1
1971	70	61	12 (20%)	2
1981	76	62	44 (71%)	24
1991	87	62	78 (125%)	
Inland				
1961	62	53	21 (40%)	7
1971	71	57	14 (25%)	3
1981	100	65	81 (125%)	39
1991	206	73	202 (276%)	99

is likely, however, that further new territories will still be established in less than suitable terrain where the nearest established pair is some distance away. Future field surveys will prove, or disprove, this possibility. It is therefore a far cry from the desperate days of the 1960s and 1970s when bodies such as the RSPB found it necessary to carry out protection schemes on individual eyries to ensure their success against raids by egg collectors and falconers. At the present time a number of eyries are still robbed illegally each year by pigeon fanciers, egg collectors and falconers from home and abroad but the population is, fortunately, robust enough to withstand such pressure in the absence of any unforeseen catastrophe of the future. From being at a desperately low ebb and in danger of

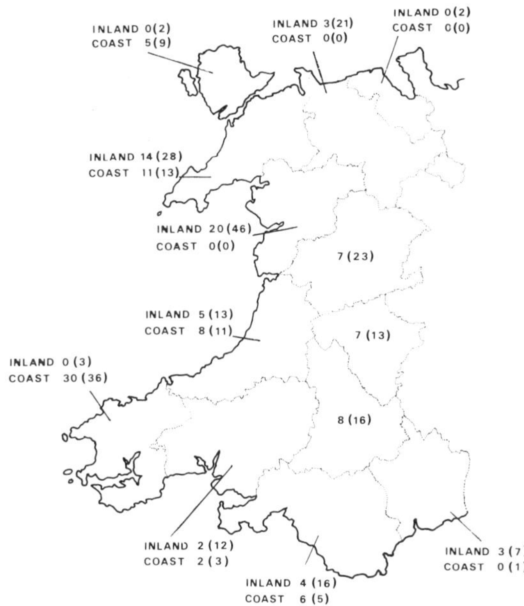

Estimated average number of pairs of Peregrine, 1930–39 compared with 1991 survey (in parentheses)

extinction in Wales some thirty years ago, this magnificent species has now re-occupied all its ancestral haunts in as dramatic recovery as the pesticide-induced decline. The map illustrates the 1991 distribution compared with 1930-39.

Many Welsh adults can be found near their breeding cliffs throughout the year and are therefore quite sedentary. There is, however, a marked influx to bird-rich coastal areas, especially estuaries, involving in particular immature birds but with some adults as well. Of 25 recoveries of young ringed in Wales only four have been recovered outside the Principality, all in southern England. By contrast there is some indication of immigration into Wales with recoveries of two birds from Cumbria, three from Scotland and two from Ireland. In addition a young bird from Norway recovered in Gwynedd is evidence that the well-documented movement from Fennoscandia into lowland Britain in winter can extend to Wales.

Red Grouse *Lagopus lagopus*

Grugiar

Breeding resident on heather uplands, declining in most areas.

A bird characteristic of heather-dominated moorland (hence the Welsh name of "Heather Hen"), the Red Grouse was once considered to be a separate species, unique to Britain and not, as is recognised today, a race of the Willow Grouse. Densities of grouse are highest on base-rich rocks, therefore Welsh grouse moors have never been as productive as those of northern England and southern Scotland, where densities can be as high as 50–60 pairs per km^2 (Jenkins & Watson 1967). The population has declined gradually throughout Britain since the Second World War following a very productive period during the 1920s and 30s. This decline has been caused principally by a deterioration in heather quality related to a reduction in the number of gamekeepers managing grouse moors and the loss of heather moorland to forestry and agriculture.

In Wales the decline has been particularly rapid as heather moorland has been lost to afforestation, agricultural improvement and overgrazing. The few grouse moors that remain no longer support enough birds to be financially viable and estates must look elsewhere for their income. The bulk of the Welsh population is now confined to the half dozen or so keepered moors, although birds are widely but very thinly distributed throughout the uplands.

Red Grouse remains were discovered at Hoyle's Mouth Cave (Pembrokeshire) in 1982 but the first written record comes from Owen (1603) who mentioned Red Grouse breeding in Wales in the county of Pembrokeshire, and Pennant mentions it breeding on Yr Eifl on the Llŷn, where small numbers were also recorded between 1773 and 1776. Little mention is made subsequently until the game preservation era of the 18th and 19th centuries when grouse shooting became an important pastime and a major source of employment to many local communities.

It is evident that the Berwyn Moors were not managed for grouse during a tour of north Wales by Arthur Aikin in 1797 as he records "kites, moor-buzzards and other birds of prey here make their nests in security; and the long heath shelters the grouse, a race that would have been extinct here but for the wide range of these wild mountains . . .". At the end of the 18th century however, the Black Mountain (Carmarthenshire) had evidently become a renowned grouse moor as Walford (1818) records that an inn at Llandybie was much frequented in the grouse shooting season.

The heyday of grouse shooting in Wales occurred between 1850 and the Second World War when grouse moors were managed intensively throughout the Welsh uplands. Some of the best moors were found in Merioneth, Denbighshire and Montgomeryshire on large estates such as Nantyr and Llymystyn, although productive moors were found in most Welsh counties. On the Rug Estate in Denbighshire an average of 198 birds was shot each year between 1835 and 1858, on the nearby Nantyr Estate, the average number of birds shot each year between 1877 and 1952 was 696 and in Montgomeryshire, the Llymystyn Estate averaged 464 Red Grouse each year between 1866 and 1951. The most productive moor however was the 3000 ha Ruabon Estate (Denbighshire) where an average of 4658 birds were shot each year between 1900 and 1913, including a Welsh record of 7142 in 1912 (J. Lawton Roberts, pers. comm.).

During the late 19th century, Red Grouse were still shot on the Preseli Mountains (Pembrokeshire) and Mynydd Eppynt (Breconshire). One of the last records

Numbers of Red Grouse shot on seven Welsh estates

Estate	*Year span*	*No. of years**	*Red Grouse (with mean no./years)*
Bryn Bach	1886–1910	25	3204 (128)
(Denbighshire)	1926–1938	13	223 (17)
Rug (Denbighshire)	1835–1858	22	4354 (198)
Nantyr (Denbighshire)	1877–1952	68	47 332 (696)
Chirk (Denbighshire)	1827–1871	35	8460 (242)
Ruabon (Denbighshire)	1900–1913	14	65 210 (4658)
Llymystyn (Mongomeryshire)	1866–1951	74	34 322 (464)
Nantgwyllt (Radnorshire)	1859–1900	42	2822 (67)

* This may not equate directly with the year span because data are missing for some years.

from the former was a brace shot in 1885, although up to three were reported here in December 1952, and on Mynydd Eppynt 30 brace were shot by two guns on the first day of one season during the late 1800s, with the last record from this area coming in 1973. During this time, grouse were plentiful on the moors around Brecon and on the Black Mountains (Breconshire/Monmouthshire), and Cambridge Phillips (1899) relates how one unusually coloured specimen shot by Rees Williams of Brecon in August 1891 on Ffriddyllt Grouse Hill fetched nearly £3, a princely sum in those days.

Even by the mid-19th century decreases were noticed in some areas, and Davies (1858) stated that grouse were not as plentiful as in former years on Mynydd Du (Carmarthenshire). By the end of the century they had disappeared from many of their former haunts in the uplands of Glamorgan, one of the main reasons here being increased urbanisation and the dumping of coal slag. In the adjacent county of Monmouthshire grouse were still plentiful on the Black Mountains during this time, and the average bag number on the Blaenavon grouse moor at the turn of the century was 410 birds, with a maximum of 165 shot by 11 guns on 12 August 1903. By 1960 numbers here had dwindled to a mere ten pairs.

During these most productive years, sporadic sightings were reported from the most unusual places. About 1898, the hard weather drove grouse across the Llŷn as far as Bardsey, and some birds remained to breed in suitable sites on Llŷn the following year. On 21 March 1905 the fresh feathers of a Red Grouse plucked by a Peregrine Falcon were found at Carmel Head on the north-west corner of Anglesey, this being the only record from the island for many years. In his diaries, T. A. Coward mentions two birds flushed near the summit of Snowdon (Caernarfonshire) on 31 December 1893, and on 11 May 1903 a clutch of 11 eggs was taken by Tompkinson from Cors Caron where small numbers were known to breed. In 1900 birds were released on Whixall Moss, a lowland bog on the Flintshire/Cheshire border, but the birds died within a short period of time, probably because the ground was too damp.

Despite its demise elsewhere, the Red Grouse continued to flourish on the well-managed grouse moors of north and mid-Wales until at least the Second World War. During the 1930s over 400 birds were shot each year on the Migneint Moor (Merioneth), and on occasion birds were so numerous that they ventured down onto farmland surrounding the village of Ysbyty Ifan to feed (J. Williams, pers. comm.). At this time, however, many upland areas were lost to afforestation and the reduction of upland gamekeepers resulted in a lack of adequate management.

By the late 1950s Red Grouse had become extinct in Pembrokeshire and were virtually absent from Carmarthenshire and Glamorgan. In Flintshire, it became confined to the higher peaks of the Clwydian Range, and in Caernarfonshire it had disappeared from the Llŷn. The *Breeding Atlas* showed that the species was still widely distributed in Wales, but it did not reveal the fact that away from the handful of keepered moors, it was present only in very small densities.

In the winter of 1991, the RSPB undertook sample surveys of all the Welsh heather moorlands in order to estimate the Red Grouse population (Williams *et al.* 1991). The results show the winter population at that time to be no more than 4800 birds, with nearly 40% of the population on the Berwyn Mountains and over 60% on the four major keepered moors. A sex ratio of 1.6 males to each female and an estimated average mortality of 52% (Hudson, pers. comm.) gave a total Welsh breeding population in 1991 of only 832–860 pairs, with a further 499–515 non-breeding males.

Red Grouse are among Wales' most hardy breeding birds, able to withstand the severest of winters on the open moor. They form flocks as the first snow of winter falls, but as the snow clears males establish territories and pair with the hens. They are very sedentary birds, with 80–90% shot within 1.5 km of where they were ringed, therefore their winter distribution is almost identical to that of the breeding season.

However, on occasion, birds will move away from the exposed high ground, and Forrest (1907) mentioned that large numbers were recorded from woodland edges near Dolwyddelan (Caernarfonshire) during hard weather. Birds will sometimes wander far from their breeding grounds and one was seen in dunes near Afon Kenfig on the south coast of Glamorgan on 11 September 1966. They are rarely found in large flocks in Wales, one notable exception being a pack of 55 in the snow on the Clwydian hills near Ruthin (Denbighshire) on 30 December 1976.

Red Grouse have been recorded from Bardsey and on 15 October 1975, a male was seen on Ramsey.

Black Grouse *Tetrao tetrix*

Grugiar Ddu

Scarce breeding resident confined to the moorlands of north and mid-Wales, declining over most of its range.

Black Grouse are birds of the interface between open heather moorland and forest, but they also require wet areas with cotton grass and sedges for feeding at certain times of the year. The males can be very conspicuous when they gather at traditional lek sites on early spring mornings, but where the population is at a low density, they can also be surprisingly elusive. Formerly much more widespread, the species disappeared from many parts of Britain during the 19th century, and today, it is confined mainly to the uplands of Scotland, northern England and Wales.

The main stronghold in Wales is the hill country of north and mid-Wales, in particular central and eastern Merioneth and southern Denbighshire. Scattered populations are also present in Montgomeryshire and Cardiganshire, whilst outlier groups are present in Caernarfonshire, Flintshire and Carmarthenshire. In all these areas, it is mainly dependent upon upland conifer plantations, although birds also occur from time to time on bracken slopes, rushy grassland and re-seeded pasture.

Fisher (1966) confirms, through the presence of fossil evidence, that Black Grouse were present in Wales in the period between the last Ice Age and the Dark Ages and bones of this species were recovered in a cave near Tenby in Pembrokeshire. There is, however, very little mention made of these birds in early Welsh literature, although Emery (1986) noted that they were present near Llanberis (Caernarfonshire) in 1693. During the 18th century, Black Grouse were said to be abundant in the Ffestiniog area of Merioneth, and Pennant mentions them on the Clwydian Range (Flintshire/Denbighshire) and in Merioneth, but there are too few references to be able to assess its true status at that time (Hope Jones 1989).

The 19th century was a period of general decline for Black Grouse in Wales despite the efforts invested in game preservation generally. The decline was particularly marked in the north-east and south-east of the country, although there was a local increase in Breconshire between 1870 and 1900. Attempts were made to introduce birds into at least seven counties between 1820 and 1920, but most of these failed. A few grouse which appeared on Anglesey during the 1920s and 1930s were almost certainly introduced (Cadman 1949), and the present occurrence of the species in parts of Denbighshire and Montgomeryshire may be due to earlier introductions.

During the 1920s and 1930s small increases were experienced in Denbighshire and Caernarfonshire, whereas the populations in Breconshire and Monmouthshire declined further. By 1940, Black Grouse had become scarce over much of Wales, with good numbers only on the uplands of Montgomeryshire, Denbighshire, Merioneth and Carmarthenshire. The reasons for the decline are not fully understood but they were probably linked to land improvement for agricultural purposes pushing farther into the uplands, the loss of heather moorland and shooting pressure on small, isolated populations. Hope Jones (1987) showed that on most estates Black Grouse formed a very small percentage of birds shot, but on the Cawdor Estate (Carmarthenshire) an average of 168 were shot each year between 1888 and 1895, with a maximum of 503 killed in 1885–86.

The period 1940–75 saw a dramatic increase in numbers and distribution of Black Grouse, an increase which was noted in all counties except Anglesey, Pembrokeshire, Glamorgan and Monmouthshire. This resurgence in numbers was closely associated with conifer afforestation in the Welsh uplands and the establishment of extensive new plantations. These new forests were always fenced off from stock grazing, thus allowing the growth of heather and bilberry in areas where they had often been suppressed by sheep grazing. During this period Black Grouse were numerous in several upland areas of Denbighshire, Merioneth, Montgomeryshire, Carmarthenshire and Breconshire.

From 1975 there was evidence of an equally dramatic

decline, with birds decreasing in many localities where they had become well known. This decrease in numbers was caused principally by the maturing conifers excluding ericaceous shrubs by shading, and therefore making such areas increasingly unsuitable for Black Grouse. Allied to this was the continued loss and degradation of heather moorland.

The decline prompted a census in spring 1986 undertaken by the RSPB and funded by the Forestry Commission. The census revealed a total of 232 males attending 91 leks, in addition to 32 non-lekking males (Grove *et al.* 1988). The total Welsh population at this time was estimated to number no more than 300 males. Subsequent research including radio telemetry revealed a great deal about this species' breeding and feeding ecology, and the RSPB was then able to advise on forestry management to benefit Black Grouse (Cayford *et al.* 1989). Sample monitoring between 1988 and 1990 showed the Welsh population to be fairly stable in areas where such work was being undertaken but a second national survey in 1992 showed that the decline had continued (Williams *et al.* 1992). The results of this census revealed a total of 210 lekking males, a decline of almost 10% since 1986. The population is now concentrated in southern Denbighshire and eastern Merioneth, with smaller populations in Caernarfonshire, Cardiganshire, Montgomeryshire and Flintshire, with only isolated birds elsewhere.

Red-legged Partridge *Alectoris rufa*

Petrisen Goesgoch

A scarce and local breeding resident over much of Wales, most common on Anglesey and along the English border. The population is maintained by releases and clouded by the occurrence of hybrids with A. chukar.

Principally a bird of agricultural land, the Red-legged Partridge was introduced into Britain on many occasions between the 17th and 20th centuries. Several attempts to introduce the species failed, but it has now become established throughout southern England and parts of Wales. The most important factor limiting distribution is their aversion to high rainfall (Howells 1963), thus rendering much of northern and western Britain unsuitable for this species.

In recent years Red-legged Partridges have hybridised with the closely related Chukar, their offspring having bred successfully in the wild. Such hybrids are often difficult to distinguish from pure-bred Red-legged Partridges and this, associated with continued releases, serves to cloud this species' true distribution and numbers.

These birds were first introduced into Wales during the latter part of the 19th century. During the 1860s it was introduced near Llanerchymedd (Anglesey), but by 1869 the population had died out. Similar introductions to Tywyn and Barmouth (Merioneth) were successful over a short period, with breeding proved near Barmouth in 1898. Releases in Glamorgan and Pembrokeshire at around the same time also appeared to fail, although several coveys were seen at Kenfig Hill (Glamorgan) in 1899–1900 and between Sker and Kenfig Dunes between 1924 and 1929, all within 10 miles of the original release site, suggesting that birds bred successfully here for several years.

Further releases have been recorded from most Welsh counties and doubtless others will have gone unrecorded. Parslow (1973) believed that numbers reached a maximum in about 1930, after which the population declined and contracted in range. Although birds were released in Montgomeryshire, Radnorshire and Breconshire, it is also believed that some of the partridges in these counties originated from the relatively healthy populations in the adjacent English counties.

A few scattered breeding records throughout the first half of the 20th century serve to indicate the species' scarcity in Wales, and by the early 1970s it was confined almost exclusively to the Welsh border counties. Since then, however, the population has increased as the decline of the Grey Partridge has resulted in widespread releases of this species, and the *New Atlas* reveals that the Red-legged Partridge is now common in parts of Anglesey and has colonised the Vale of Clwyd (Denbighshire) and the lower Dee Valley (Denbighshire/Flintshire). The species is now quite common in central and eastern Monmouthshire, with an estimated population of 250–400 pairs. In the counties of Breconshire, Radnorshire and Montgomeryshire it is confined to the low-lying agricultural land near the English border, although there are other scattered breeding records as birds are released elsewhere. A survey of 762 one-kilometre squares on Anglesey in 1986 found Red-legged Partridges present in 22 squares, although breeding was proved in only one of these (RSPB 1986).

The total Welsh population fluctuates widely as more birds are released, but at present it is likely to be approximately 500–600 pairs, over half of which are in Monmouthshire.

Red-legged Partridges are sedentary birds, although individuals do wander on occasion, one having turned up on Ramsey in recent years. The one ringing record involving this species in Wales confirms its reluctance to wander far from its natal area.

Grey Partridge *Perdix perdix*

Petrisen

A scarce and local breeding resident in much of Wales, locally common only in Monmouthshire.

In Britain, the Grey Partridge is typically found on agricultural land where hedgerows and small copses remain, although it also breeds regularly on moorland up to an altitude of 500 m. Abundant in most areas of England and Wales when game-rearing was at its zenith, throughout the 20th century there has been a gradual decline in numbers, becoming more marked from the

1940s onwards. This decline has been attributed to a combination of factors, including agricultural intensification (notably pasture improvement, overgrazing by sheep and increasing use of herbicides), cold springs and wet summers, and mammalian and avian predation.

In Wales, this species has shown a much more marked decline than in England and it is now locally common only in the low-lying agricultural land of central and eastern Monmouthshire. Small numbers persist on Anglesey, coastal areas of north-east Wales and Glamorgan, and in eastern Montgomeryshire, Radnorshire and Breconshire, but elsewhere it has virtually disappeared.

During the 19th century, the Grey Partridge was described as being abundant over most of lowland Wales. Game estates from most counties recorded good bag numbers, depending on the size of the estate and the number of shooting days. Some years were much better than others, probably due to hot, dry weather during the summer months (Ingram & Salmon 1965), with 1860 and 1879 being two of the worst years on record during the last century (Matheson 1953). Even in counties such as Pembrokeshire where Mathew (1894) considered the habitat to be unsuitable for the species, estates such as Stackpole Court were shooting an average of 789 birds a season between 1890 and 1899 (Matheson 1960). On one estate in Anglesey the annual bag number between 1875 and 1900 was often above 2000 and rarely below 1000, and in Montgomeryshire an average of 823 birds was shot on the Powis and Lymore estates between 1890 and 1899, with a maximum bag of 1713 in 1896 and 1897. Historical game records indicate that the Welsh Grey Partridge population was at its peak between 1860 and 1900, and that numbers declined gradually thereafter, with many estates having given up preserving and shooting this species by the 1920s.

Following the First World War, Grey Partridges became noticeably scarcer in parts of Merioneth, Cardiganshire and Carmarthenshire, although a decline was not noticed in counties such as Pembrokeshire and Glamorgan until the 1940s. In the former county, Ingram and Salmon (1961) laid the blame for the decline on the use of gin-traps and on the Dunraven Estate in Glamorgan, average bag numbers fell from 132 between 1927 and 1936 to only 58 between 1946 and 1953. In the Montgomeryshire Field Society Report for 1950 an extraordinary report is included of a pair of Grey Partridges that raised a brood of farmyard hen chicks from eggs laid in the Partridges' nest near Guilsfield.

By the late 1960s the species had declined markedly in all Welsh counties and had disappeared from much of west Wales. It was still widespread in the east and on Anglesey and the Llŷn during the fieldwork for the *Breeding Atlas*, and a small population still survived around Aberystwyth (Cardiganshire). Since then, however, the decline has continued and the *New Atlas* confirmed breeding only in the eastern counties, Anglesey and one record on the Llŷn. The *West Glamorgan Atlas* revealed a few breeding pairs on the Gower, and released birds appear in other areas from time to time although these small populations would not be maintained without continued introductions. On Anglesey, Grey Partridges were found in 67 of the 762 one-kilometre squares surveyed in 1986 although breeding was proved in only two of these (RSPB 1986).

The records indicate that the Grey Partridge has been declining in Wales throughout this century and its present distribution shows a clear relationship with the remaining areas of mixed farming (grass and arable). Thus the gradual decline in the early years of this century was probably related to the steady decline in arable farming which has also affected the populations of many other species of farmland birds. The steep decline following the Second World War was undoubtedly the result of the introduction and widespread use of selective herbicides on cereal crops resulting in a lack of field weeds and hence a lack of insect food for chicks (M. Shrubb, pers. comm.).

Today it is widespread, although not numerous on Anglesey where it is often found in coastal dunes (e.g. covey of 20 at Newborough Warren on 1 January 1987). In Flintshire and Denbighshire it is confined mainly to the coastal lowlands and the Vale of Clwyd, and likewise

Grey Partridge bag numbers from selected Welsh estates

Estate name	*County*	*Years*	*Average bag no.*	*Max. no. (+ year)*
Stackpole Court	Pembrokeshire	1890–1899	789	1092 (1890–1891)
Llwyngwair	Pembrokeshire	1883–1886	143	155 (1884–1885)
Gogerddan	Cardiganshire	1865–1869	605	721 (1867–1868)
Powis+Lymore	Montgomeryshire	1890–1899	823	1713 (1896–1897)
Voelas	Denbighshire	1880–1889	323	1053 (1887–1888)
Dynefwr	Carmarthenshire	1897–1903	149	199 (1902–1903)
Dunraven	Glamorgan	1927–1936	132	249 (1934–1935)

in Glamorgan where the birds are most often encountered in the dune systems around Kenfig and in the Vale of Glamorgan. The lower Severn Valley holds small numbers in Montgomeryshire, and in Radnorshire it is confined to the more fertile agricultural land in the south-east. A maximum of 50 pairs is said to be present in Breconshire, mainly confined to the Usk and Wye Valleys where there is a mixture of arable farmland, but its Welsh stronghold is in the Usk Valley and coastal areas of Monmouthshire where it is still widespread and fairly numerous in places. Within this county, it is absent only from the open moorland, large woodlands and urban areas and the population in 1985 was estimated to be between 600 and 700 pairs. The current Welsh population is estimated to be between 1000 and 2000 pairs.

The Grey Partridge is a vagrant to Bardsey where one bird was present during hard weather on 3 February 1963, with two on the island the following day. It has bred on Ramsey and Caldey and has been seen on Skomer.

In winter the Grey Partridge's distribution changes little from the breeding season. Coveys roost out in fields and in dune systems, feeding mainly on grain and weed seeds. They are not as susceptible as many other species to cold weather, and will make roost holes in soft snow to avoid the worst of the elements. However, there is some evidence to suggest that prolonged severe weather, as experienced during the 1947 and 1963 winters, does have an effect on local populations.

Quail *Coturnix coturnix*

Sofliar

A scarce and irregular breeding species, irruptive in some years; scarce passage migrant.

The Quail, the only migratory species among our gamebirds, is an enigmatic and unpredictable summer visitor to Wales, occurring sparingly, or not at all in many years but with periodic influxes as in years such as 1953, 1970 and 1989. When it does occur, it is more often heard than seen, its characteristic "whit, whit-whit" call, usually around dawn and dusk, being readily identifiable. It will occupy almost any type of herbaceous vegetation from rough grassland to cereal fields and although it is predominantly a lowland species it has been heard calling in pastures and in *Molinia* at altitudes as high as 470 m in Breconshire and Montgomeryshire. Calling does not necessarily signify breeding, however, as unmated males can be very mobile. When breeding does occur, in "good" years, there is a tendency for the birds to group together, presumably attracted by other calling males so that several pairs may then be found in the same locality; in Pembrokeshire in 1989, 50 records of individual birds came from one farm.

Judging by the frequency with which Quail appeared on the banquet menus in the Middle Ages and subsequently, it was a far more regular and numerous species then than it has been in recent times. A widespread decline has been evident throughout western Europe and although the reasons for this are not wholly clear it is thought that it is probably less attributable to agricultural change than to factors such as intensive shooting in southern Europe, particularly of passage breeding birds in spring. There are authenticated records of breeding in Wales as long ago as 1603 (G. Owen 1603) in Pembrokeshire and sometime before 1776 in Caernarfonshire (Pennant). Throughout southern Britain the Quail was certainly common until around the end of the 18th century after which time a wide-scale decline evidently took place, continuing until the second half of the following century - around 1865 (More 1865) when its numbers were at their lowest. Numbers nationally remained low until the 1940s (little recorded evidence is specific to Wales) since when there has been a modest resurgence and Quail have increasingly enjoyed years when numbers have been higher than in many past seasons.

Historically, 1870 and 1893 were widely recognised as being particularly good years, with breeding in the lowland areas of many Welsh counties. In the first of these years Forrest (1907) reported Quail as "numerous" in Denbighshire, Flintshire and the lowland areas of Merioneth and Caernarfonshire and Thomas Dix (1870) collected records of at least 350 individuals in Pembrokeshire and Cardiganshire (many of them shot) and guessed that this may have represented no more that a quarter of the real total in the counties that year. Earlier in the same century, as reported in 1904 by J. H. Salter, Professor of Botany at Aberystwyth, "seldom did a shooting season pass without 10 or 12 couples being obtained on the Gogerddan estate" near Aberystwyth. Although Quail were seen in small numbers in various counties in the first few decades of the present century – most regularly in Anglesey – it was not until 1947 that was there another notable year nationally, remarked on in Montgomeryshire but not particularly in other Welsh counties. Subsequent Quail years have occurred in 1952–53, 1964, 1970, 1983 and 1989; the increasing frequency of these is itself noteworthy. Although not particularly remarked upon in other parts of Britain, 1977 also produced a crop of records in Wales, especially in Pembrokeshire, Carmarthenshire and Cardiganshire. The number of birds reported in Wales in 1989 was the greatest for many years, Tew (1989) collated evidence of a minimum of at least 237 individuals (mostly calling males) with records from all the Welsh counties and highest numbers in Pembrokeshire (c90), Cardiganshire (45), Radnorshire (42) and Breconshire (12). Although few nests are ever found, breeding is usually proved and more frequently presumed to have occurred, in the main influx years, in one or more counties.

The reasons for the irregular nature of these irruptions is believed to be associated with particularly warm, dry spring weather in France especially when these conditions are accompanied by periods of south east winds encouraging the birds to move farther north. This is certainly a more likely explanation than that proffered by a correspondent to the *Zoologist* in 1872, who suggested that the constant firing on the continent during the Franco-Prussian war so disconcerted the birds that they were driven onwards farther north to Britain. Also the pattern of occurrences in "good" years is irregular. Whereas Pembrokeshire was the principal focus in 1989, none at all were recorded there during the 1953 summer (Moreau 1956) when counties such as Radnorshire, Denbighshire, Glamorgan, Montgomeryshire and Caernarfonshire shared the bulk of the records. After years of plenty there are sometimes one or two following seasons when a residue of summering migrants tends to return; such was particularly noticeable after the 1870 influx and slightly less so after the large 1964 one and in 1990–91 at least in Cardiganshire (P. E. Davis, in litt.)

Quail are rather late arrivals among the migratory summer visitors, often not making their presence known until the latter half of May or early June. Moreau (1956) commented on the curious fact that in years of influx the proportion of notably early records is sometimes much higher in remote, western places – notably some of the offshore islands – than elsewhere on the mainland of Britain; e.g. 1 May 1950 Skokholm, 12 May 1950 Grassholm, 20 May 1953 Skomer, 21 April 1984 Bardsey, 27 April 1988 Bardsey. Individuals have even been encountered as far offshore as the Smalls lighthouse (30 km). Return passage is most strongly marked in September with a reasonable spread of records from the island observatories concentrated in that month.

Pheasant *Phasianus colchicus*

Ffesant

A locally common breeding resident, widespread throughout Wales.

In Britain Pheasants are mainly confined to well-wooded agricultural land, parkland and large estates where the feral population is augmented by released birds each year. Despite claims by many authors (e.g. Gladstone, 1921) that Pheasants were brought to Britain by the Romans, there is no evidence to support this, and it is likely that it was brought to this country by the Normans in the 11th century (Fitter 1945). The species' spread westwards into Wales was slow, and it is not until the Tudor period that mention is made of Pheasants in the Principality, and even then in terms which indicate that they were recent arrivals.

One of the earliest written references to this species in Wales is by Owen (1603), who describes its status in Pembrokeshire at the time: "As for the Pheasant, in my memorie there was none breedinge within the shire until about XVJ years past Sir Thomas Perrot Knight procured certaine hens and cockes to be transported out of Ireland . . ." This would indicate that the first introductions to south-west Wales occurred in about 1586. Another reference to Pheasants near Llanddwywe in Merioneth occurs in an unpublished poem by 16th century Welsh writer Gruffydd Hiraethog, although here too they appear to have been recent introductions.

There were undoubtedly other introductions, but the Pheasant appears to have been an unfamiliar bird over most of Wales except within the boundaries of a few large estates. Indeed, a cock Pheasant which had strayed from a nearby estate gave rise to a story among the peasantry of Cwm Edeyrnion (Merioneth) that a strange winged viper with a long body and bright scales was on the loose (Jones 1930).

It was evidently scarce throughout most Welsh counties until towards the close of the 18th century, game bags from estates such as Gogerddan (Cardiganshire) providing evidence of the way the Pheasant became a significant percentage of all gamebirds shot by the 1840s. At Chirk in Denbighshire, the Pheasant already formed over a third of the total game shot in the decade 1827–36.

It was not until towards the end of the 19th and the beginning of the 20th century, however, that the great increase in Pheasant preservation began. During this time it often formed well over half of the total bags, although even at this time it was absent from large areas away from game-rearing estates, such as the south Wales coalfield. In Pembrokeshire during the late 1800s Mathew (1894) stated that the majority of the birds in the county were of the recently introduced Chinese *torquatus* race, and that these were often found on the uplands and wettest bogs. In north Wales, however, Pheasants were still not widespread at that time, although they were common where they had been released. Forrest (1907) quotes one record of a hybrid Pheasant and domestic fowl in the collection of Sir Pyers Mostyn at Talacre (Flintshire), which was shot sometime during the late 19th century.

In south-east Wales, despite the large numbers reared on many estates, it was still rare, and in Breconshire most releases were confined to the Usk and Wye Valleys. Throughout Wales numbers fell markedly during the First World War as there were no releases to supplement the wild stock and although gamekeepers returned to their former tasks at the end of the war, game rearing was never to be undertaken on such a large-scale in Wales again.

Between the two World Wars Pheasants showed local increases in counties such as Caernarfonshire and Cardiganshire where large-scale afforestation of the uplands allowed this species to colonise higher altitudes than it normally would. However, overall, numbers decreased and the decline became more marked during the Second World War. Whereas numbers increased over much of Britain between 1950 and 1970, the decline continued in Wales as fewer estates released birds into the wild. By the 1950s few wild Pheasants survived in Carmarthenshire and it had become rare in west Merioneth where the population was probably unable to sustain itself without releases. At that time, the small Glamorgan population was confined to the east of the county, the Vale of Glamorgan and the Gower, and in Monmouthshire, Ferns *et al.* (1976) believed the population to be smaller than was previously thought, particularly as only a handful of breeding records were submitted each year.

The *Breeding Atlas* showed the distribution of Pheasants in Wales to be patchy, with good numbers in the east of the country but large areas of Merioneth, Carmarthenshire, Radnorshire and Glamorgan with no

Numbers of pheasants shot on selected Welsh estates between 1868 and 1917

Year	*Bodorgan (Ang)*	*Voelas (Denbs)*	*Stackpole (Pembs)*	*Tytheston (Glam)*	*Powis & Lymore (Mont)*
1868–77	–	680	6840	–	8430
1878–87	3520	1010	10 080	500	10 520
1888–97	6010	3010	15 530	490	25 960
1898–07	11 700	8230	19 960	1110	46 540
1908–17	19 240	12 060	17 290	1740	29 800

breeding records. More recently, the *New Atlas* shows that the Welsh population may have declined further in areas such as Cardiganshire, Caernarfonshire, western Montgomeryshire, Radnorshire and Breconshire. Local populations, however, may have increased dramatically as shooting syndicates release Pheasants in many parts of the country and the larger upland estates now release more birds to make up for the loss of income from the decline of Grouse shooting.

Pheasants have been recorded from many of the larger offshore islands, and a small population breeds on Skomer, where birds have been seen flying to the island from the mainland across Jack Sound. Owing to the large numbers of birds released on estates throughout Wales (over 40 000 per annum on some estates), a population estimate for the Pheasant would be meaningless.

In winter, Pheasants are most often seen feeding along field and woodland edges, often in small flocks. They are sedentary birds, research by the Game Conservancy showing that 61% of a wing-tagged population were shot within 400 m of the point of release, and less than 1% dispersing more than 2 km (Bray 1968).

Golden Pheasant *Chrysolophus pictus*

Ffesant Euraid

Despite several releases, a small feral population has become established only at two sites on Anglesey.

A native of central China, Golden Pheasants were first introduced into Britain during the 18th century, but it was not until the 1880s and 1890s that feral populations became established in parts of Scotland and East Anglia. Further releases, both deliberate and accidental, led to the establishment of small populations in several southern counties of Britain, including Anglesey. Although not originally a forest species, British feral birds are generally found in coniferous or mixed woodland, often in close association with Pheasants and sometimes Lady Amherst's Pheasants.

The only well-established population in Wales is found in Anglesey although escaped birds are occasionally seen elsewhere. On Anglesey, birds are found in two distinct areas, at Pen-y-Parc woods near Beaumaris and on the Bodorgan estate near Newborough. Golden Pheasants were released at the former site during the early 1960s, probably around Herron Farm on the estate, and at the latter the birds were originally an anniversary gift, released during the late 1960s.

Very little is known about the numbers and population trends of these birds due mainly to their secretive nature. It is thought, however, that the total population of the two sites is currently about 30–35 birds and that numbers are stable mainly because the areas are well keepered and therefore potential predators are controlled and some food is provided for the birds (E. Abraham, pers. comm.). Wandering birds from these two areas are often seen elsewhere, although sightings farther afield on Anglesey and Caernarfonshire may involve escaped pheasants from private collections.

Releases in other Welsh counties have been short-lived, and as most of these have been on private estates closed to the public, little information is available. Birds are said to have been put down on the Glynllifon Estate (Caernarfonshire) over the past 20 years, but none is thought to have survived long enough to become established. During the late 1970s c30 birds were introduced at Halkyn (Flintshire), but this area is now poorly keepered and the last bird was seen here in 1988.

Single birds or small groups are occasionally seen elsewhere in Wales (e.g. a male and two females at Hirnant in Montgomeryshire August–October 1976), but these escapes from ornamental collections rarely survive for long.

There is no evidence of any migratory movement either in the wild or in the feral British populations, and although poorly studied, the Anglesey birds are thought to be sedentary throughout the year.

Lady Amherst's Pheasant *Chrysolophus amherstiae*
Ffesant Amherst

A small and declining feral population is present in Flintshire.

First introduced into England in 1828, this species is a native of the rocky mountain slopes of south-west China, Tibet and Burma. The first known attempts to release birds into the wild came around the turn of the century although many of these attempts were short-lived, partly because of interbreeding with Golden Pheasants. However, the introduced populations at Woburn (Bedfordshire) flourished, and it was from here that birds were taken to Halkyn in Flintshire.

These birds were introduced to the grounds around Halkyn Castle about 1950 at a time when this area was intensively keepered for Pheasants. The population increased and spread onto the nearby Gwynsaney estate, reaching its peak during the early 1980s. There are conflicting reports as to the size of the population, at its height local ornithologists estimating the total at no more than 40 birds (E. Abraham, pers. comm.) whereas local landowners reckoned the true number to be as high as 150 birds.

This species is particularly elusive, generally keeping to areas of dense undergrowth and therefore an accurate estimate of numbers is difficult to assess. Certainly the population has decreased rapidly over the past ten years as the area is no longer intensively keepered and many of the birds, their eggs and young fall prey to foxes and domestic cats (A. Gouldstone, pers. comm.). The introduction of c30 Golden Pheasants during the late 1970s also caused problems due to hybridisation, but those birds soon died out. There is doubt as to whether this population has ever been self-supporting as birds were fed at Halkyn Castle, and the current population is thought to consist of only four males and two females.

This species was accidentally introduced to Plas Isaf (Flintshire) around 1980 when the owner wanted to put down Pheasants. Young birds were brought in from Halkyn Castle, and amongst them were Lady Amherst's Pheasants, although the exact number is not known. The population flourished for a short time but none has been seen here since 1987 when six were recorded together.

As with the Golden Pheasant, very little is known about this species' habits although many observers believe that it is solitary in winter. The Halkyn population appears to be sedentary as birds are now rarely seen away from the grounds of Halkyn Castle and the Gwynsaney Estate.

Water Rail *Rallus aquaticus*
Rhegen y Dŵr

A scarce breeding resident, more commonly recorded as a passage migrant and winter visitor to all counties.

A bird which favours dense aquatic vegetation, the Water Rail is an extremely elusive bird which is more often heard than seen during the breeding season. Extensive drainage of marshland and swampy fields during the 18th and 19th centuries had a drastic effect on its numbers and distribution, although it is still widely distributed in small numbers throughout much of lowland Britain, thanks in part to the formation of new artificial waters such as gravel pits and canals.

The earliest record of Water Rail occurring in Wales comes from Roman times. The remains of this species, along with those of several others were discovered in Roman drains at Caerleon (Monmouthshire) dating back to the first and early third centuries.

Today, the Water Rail is a rather scarce breeder, although many breeding attempts undoubtedly go unrecorded due to its secretive nature. It is absent from most of north Wales, the Cambrian Mountains and the south Wales uplands, where suitable habitat does not exist. It also appears to be absent from much of north-east Wales despite the presence of suitable habitat, although this may again reflect the species' elusiveness.

Forrest (1907) could find no confirmation that this species had bred in north Wales by the turn of the century, although several had been shot in suitable habitat during the breeding season, particularly on the River Clwyd (Denbighshire) and on Anglesey. Even today, breeding has still not been confirmed in the county of Merioneth and Montgomeryshire, and it is a regular breeder in only about half the Welsh counties.

Cors Caron, Cors Fochno and Llangorse Lake are all traditional sites which are still used today, with some of the earliest Welsh breeding records coming from these wetlands. Oxwich Marsh NNR and Crymlyn Bog (Glamorgan) are also traditional sites for this species, although until recently Water Rail had not been confirmed breeding in Glamorgan since 1910 when a pair bred in a small marsh on Sully Is. One of the earliest records from Pembrokeshire involves a pair breeding on the island of Skokholm in 1929 and again in 1931.

Scattered breeding records have been reported from

Carmarthenshire throughout the current century, and recently breeding has been recorded regularly from a marsh east of Burry Port, where three pairs bred in 1975. The late G. Ireson (pers. comm.) found several nests in mid-Wales during the 1970s and 1980s, including one containing 13 eggs at Traeth Mawr on Mynydd Illtud (Breconshire) on 26 May 1981. It was near here also that R. Haycock (pers. comm.) found a nest at an altitude of 300 m, a rare occurrence for a predominantly lowland species. Most Radnorshire records come from Rhosgoch Common and include many confirmed breeding records (e.g. nest with eight eggs on 26 May 1973).

Water Rails appear to be scarce and irregular breeders in the counties of Caernarfonshire, Denbighshire and Flintshire, and there are no confirmed records from Merioneth and Montgomeryshire despite the presence of suitable habitat. The species' stronghold today in Wales appears to be the counties of Pembrokeshire, Glamorgan and, in particular, Anglesey, although even here confirmed breeding records are scarce. Between 1977 and 1981 breeding was strongly suspected or confirmed from eight different sites in Pembrokeshire and the county total is estimated to be a minimum of 20 pairs and it is present during the breeding season in at least ten sites on Anglesey. It also nests regularly at the RSPB's Ynys-hir reserve (Cardiganshire) and in at least two locations in Glamorgan, with up to 40 breeding pairs reputed to be present at Oxwich Marsh NNR. This may, however, be an over-estimate as accurate census figures are difficult to obtain during the breeding season. The *Gwent Atlas* found no confirmed breeding in Monmouthshire although pairs probably bred at two sites.

Despite the fact that this bird is frequently overlooked during the breeding season and that there has never been an attempt to undertake a complete survey of Water Rails in any Welsh county, the paucity of records suggest that this species is a scarce breeder in Wales, numbering probably about 100–150 breeding pairs.

In winter, resident Water Rails are joined by immigrants from the Continent, mainly in October and November. Despite its apparent weak flight, it is in fact a competent flier as ringing recoveries come from as far away as Sweden, Poland and Czechoslovakia.

Its winter distribution in Wales is similar to that of the breeding season, although there are appreciably more records from the network of wetlands along the south Wales coast. At these and other sites, three or more birds are regularly recorded, especially during severe weather or following heavy rain when the birds are forced to abandon the dense vegetation to look for food.

On passage, large numbers have been recorded during spring and autumn, especially from some of the offshore islands where the birds are attracted to the lighthouses at night. Large numbers were observed moving south at Bardsey lighthouse in November 1909, with 50 killed on the 7th alone. Fifty were seen here on 30 March 1913, and another 20 on the Skerries (Anglesey) the same night. It has been recorded in smaller numbers on the Pembrokeshire islands, and one was even seen on the Smalls, a group of rocky outcrops jutting out of the sea some 30 km offshore. Several birds found dead on a railway line near Ferryside in Carmarthenshire on 7 April 1958 is further evidence that this species passes through Wales in relatively large numbers at times, as are the number of deaths by collision with power lines at night.

In winter, Water Rail are generally as elusive and timid as they are during the breeding season, but one bird which took up residence under a bramble bush near Colwyn Bay in 1906–07 became quite tame, taking bread thrown into a nearby stream. Since the late 1980s, at least one bird has been habitually seen eating waste fish and chips dumped at a roadside lay-by near the Point of Air!

They tend to be solitary birds even during severe weather, and large numbers are exceptional. However, large concentrations have been recorded in Wales on occasion. Up to ten are regularly recorded at Llangorse Lake, with 20 here on 29 November 1979. Seven were seen in Pwllheli Harbour (Caernarfonshire) on 1 December 1989, 13 were seen at Burry Port (Carmarthenshire) on 5 October 1975 and seven were present at Llan Bwch-llyn in Radnorshire on 20 December 1971. An exceptional total of 200 was said to be present around Rosehill Marsh (Pembrokeshire/Cardiganshire) early in 1981, and several hundreds have been reported from the Dee marshes (Cheshire/Flintshire) on occasion. During exceptionally high tides on the Dyfi, 10–15 birds are often seen foraging along the waters' edge, and up to 20 on occasion at Ynys-hir (A. Fox, pers. comm.)

Several ringing recoveries show movement within Britain, but Continental ringing records show that there is a strong movement of German birds into Britain in the winter, some reaching Wales. The longest recorded distance to date involves a bird ringed at Swidwie in Poland on 19 July 1971 and recovered near Cardigan on 19 December 1973, a distance of 1285 km west.

Spotted Crake *Porzana porzana*

Rhegen Fraith

Rare vagrant and occasional breeder.

An inhabitant of marshes and bogs, the Spotted Crake is one of the most difficult species to see, and usually the only indication of its presence is its "whiplash" call, often made at night. It was formerly a local, but fairly widespread breeding bird in many parts of Wales up to the middle of the 19th century but numbers decreased with the drainage of its favoured haunts. It bred again with some regularity during the 1920s and 30s but there were then few subsequent records until the late 1970s and 1980s when breeding was recorded once more.

During the 18th and 19th century, Spotted Crake probably bred in the Dee marshes (Flintshire) before extensive drainage reduced their value for the species. E. K. Allin mentioned the nearby Buckley and Padeswood marshes in his diaries, noting that they were former haunts of the Spotted Crake, and although there are no confirmed breeding records from these sites, Forrest (1907) stated that several were shot here at the turn of the century. Dillwyn (1848) frequently encountered birds in the bogs around Swansea (Glamorgan), and it is probable that pairs bred at Crymlyn Bog and Oxwich Marsh NNR during his day. Salter believed that they bred on the Dyfi estuary, and two pairs are known to have been shot on the adjacent Cors Fochno on successive years sometime between 1890 and 1893. Forrest mentions that two young birds were shot on the Dyfi, confirming that breeding had occurred here on at least one occasion.

At the turn of the century, Spotted Crakes were evidently commoner than they are at present and many were shot at Barmouth (Merioneth) and Welshpool (Montgomeryshire) with several records of single birds from suitable habitat in Anglesey and Carmarthenshire. It formerly bred in Breconshire, notably at Onllwyn Bog and Traeth Mawr, and it also nested at Rhosgoch in Radnorshire during the last century. Cambridge Phillips recorded how Onllwyn was a favourite site for this species, and that one day he "flushed six and killed four, three of which were birds of the year". A local station master informed him that two broods had been reared that year, and it seems likely that several pairs bred here each year. No breeding was recorded in Pembrokeshire or Caernarfonshire at the turn of the century despite the presence of suitable habitat and an occasional record of birds from both these counties. During this period, Spotted Crakes were often recorded by parties shooting over trained dogs – by far the most effective way to flush such a secretive species.

Breeding was not recorded again in Wales until the 1920s and 1930s when nests were located in Flintshire, Radnorshire and Breconshire, and it is likely that pairs bred annually in at least one site in Wales for many years. In 1924, E. P. Chance took a clutch of nine eggs from a nest at Rhosgoch (Radnorshire) after having found a broken egg shell here the previous year. He searched the area intensively for three months before finding the eggs, an indication of how difficult it is to find Spotted Crakes' nests. In 1928, H. A. Gilbert found two nests at Traeth Mawr (Breconshire) and although the first was empty, the second contained a fresh clutch of nine eggs on the 21 May. The following year, a nest was found once more at Rhosgoch, and it is evident that pairs bred regularly at these two sites for several years as young were found again at Rhosgoch in 1939; in actuality, both sites were probably traditional.

Throughout the late 1920s and 1930s, breeding was either proved or suspected on many occasions in marshes on the Dee estuary. In 1926, a nest with eggs was discovered, and two years later birds were seen repeatedly at two marshes in the area. For the next decade, breeding was proved on several occasions at Puddington Bog on the Flintshire/Cheshire border, an area now known as Inner Marsh Farm. At this time, Ingram and Salmon (1937) thought it possible that it was also breeding in Monmouthshire, but there are no confirmed records.

From the 1970s onwards, the number of sightings of Spotted Crake in Wales increased, and records were received from several suitable areas during the breeding season. The first confirmed record however did not come until 1979 when a nest was found at a site in Breconshire (R. Haycock, pers. comm.). The late G. Ireson (pers. comm.) heard two males and a female here in the same year and thought that there may have been two broods. A nest and eggs probably belonging to this species was found here again in 1983, and breeding occurred on at least one occasion here again between 1984 and 1986.

Since then, it has remained an almost annual visitor to Wales although no breeding has been proved and very few sites hold calling males on a regular basis. However, this may be more a reflection of the birds' secretive nature and the lack of birdwatchers in key areas than a genuine absence of birds, and it is possible that it breeds far more regularly than the records suggest.

Presumed migrants have been recorded from all the Welsh counties, the number of records fluctuating in parallel with the number of possible breeding pairs. Spotted Crakes are rarely recorded on the offshore islands with only two records from Bardsey up to 1985 and seven from Skokholm before 1991.

Sora Rail *Porzana carolina*

Rhegen Sora

A vagrant.

This North American species has been recorded 12 times in Britain up to 1991 and two of these, widely separated in time, were from Wales:

Spring 1888 – One was caught near the low-water Pier, Cardiff and presented to the Zoological Society of London

5 August 1981 – One was caught and ringed, Bardsey

A much kinder fate for the 1981 bird! The Cardiff record was the second for Britain.

Little Crake *Porzana parva*

Rhegen Fach

A vagrant.

There have been six Welsh records of this species, which breeds from Germany eastwards to Kazakhstan and southwards to north Iran, also north Italy and, in small numbers, France and Holland.

The records are:

1839 – One was taken by hand on the River Afan and preserved at Margam (Glamorgan)

about 1894 – One picked up on the railway, Ynyslas (Cardiganshire)

1 January 1949	– One seen near the coast between New Quay and Aberaeron (Cardiganshire)
19–22 January 1967	– One seen at Llangennith Moor (Glamorgan)
18–23 April 1987	– Male, Shotton Pools (Flintshire)
30 April–2 May 1987	– Male, Upper Lliedi Reservoir (Carmarthenshire)

Baillon's Crake *Porzana pusilla*

Rhegen Baillon

A vagrant.

There have only been four Welsh records of this species which breeds from Iberia and France discontinuously eastwards through Asia to Japan and Australasia, also southern Africa.

The records are as follows:

6 November 1905	– Male caught in a ditch at Llangwstenin near Llandudno Junction (Caernarfonshire)
7 January 1932	– One found dead on golf links at Aberdyfi (Merioneth)
7–8 February 1976	– One seen, Llantwit Major (Glamorgan)
16 June 1990	– One, probably first-winter male, dead Bardsey

Corncrake *Crex crex*

Rhegen yr Ŷd

Formerly a numerous and ubiquitous summer visitor to all counties. Now almost extinct.

The demise of the Corncrake in Britain does at least have the merit of having been fairly well documented. In the days of labour-intensive farming, when all tasks were done manually, the Corncrake was an abundant and familiar bird throughout the Welsh counties. It is a bird of hay meadows and damp grassy places – often characterised by those areas with rank grasses and the presence of yellow flag. In a country where grassland is predominant, Corncrakes were abundant until the introduction of a mechanised hay harvest, exacerbated by earlier cutting (and more recently, silage taking) which coincided strongly with the bird's breeding season. Mechanised cutting gives little opportunity to avoid the nests or nestling birds and they were inexorably and literally cut out of existence. The growing predominance of sheep over the former mixed, stock-rearing, regime which had prevailed up to the 1940s signalled the final chapter of the demise (as permanently grazed pastures, often infested with rush (*Juncus* spp.), replaced less improved patterns of pastureland. Stowe *et al.* (1993) have shown that in residual areas in western Scotland which still support Corncrakes the birds are confined to the fields where hay is taken (as opposed to silage), especially where Canary grass and iris are present. They are invariably absent from short-grazed, rush-infested fields.

The first written reference to Corncrakes in Wales is the often-quoted one by G. Owen (1603), who recorded them as breeding in Pembrokeshire. In north Wales Pennant (1778) commented on their abundance on Anglesey in 1776, a statement which still applied when Forrest wrote in 1907. At these times the Corncrake was not solely a bird of lowland hay meadows and rough grazing but was also found on hills and moorland at least up to 305 m; breeding has on occasion been recorded even higher than this. Forrest's statement in relation to Anglesey (the same could have been said to apply to Llŷn) was certainly not true in parts of Wales farther south; it is useful first, however, to put the decline in Wales into a wider context. As early as the last quarter of the 19th century it was recognised that the Corncrake was becoming much rarer in the counties of south-east England; by the turn of the century it was clear that a major retraction of range was occurring as far west as south Wales. E. M. Nicholson (1951) has pointed out that the decline was not sudden or capricious but entailed a gradual withdrawal over a century or so. He suggested that the original distribution of the species in Britain was focused on the north and west and that its widespread occurrence in lowland arable England was the result of a temporary expansion of numbers and range between about 1800 and 1850. The decline after that was the reverse trend exacerbated or accelerated by machine mowing, not simply caused by it. The reasonably even withdrawal north-westwards may add credence to his claim.

Whatever the causes, the Corncrake was a rare bird in south Wales by 1938. In this year and the following one C. A. Norris carried out a detailed survey of the Corncrake for the BTO. He showed that by then it had gone from Monmouthshire and was very scarce indeed in Breconshire and Radnorshire; no records of breeding were received from any of these counties, although it was rumoured that the Corncrake still occurred in the latter two. In the coastal counties serious decreases had been noted in Glamorgan and Cardiganshire (since about 1900), Carmarthenshire (1910) and Pembrokeshire (1916). The only areas in these four counties supporting populations at that time were near Aberystwyth, near St. David's and on north-west Gower.

In north Wales the picture was different. Although no information was received from Montgomeryshire, it is now known that a relatively strong population still existed. In Merioneth there had been a steady decrease since about 1910 but Corncrakes were still breeding regularly up to the time of Norris' survey, probably becoming sporadic shortly after. Denbighshire and Flintshire maintained reasonable populations "in hilly districts". The stronghold in Wales remained on Anglesey and Llŷn, although even here the numbers were reduced.

Since Norris' survey the final demise of the Corncrake in Wales can be summarised as follows:

Monmouthshire

No regular breeding. One definite and three possible breeding attempts (1965, 1967, 1968, 1972 respectively).

Glamorgan

Regular breeding ceased by 1945. Sporadic breeding probably continued for the next 20 years. The last proved nesting was near Cowbridge in 1963, although calling birds – presumably only on passage – were heard in 1987, 1989 and 1990.

Carmarthenshire

Corncrakes could still be heard in some areas up to the early 1950s (three to four pairs annually) with occasional records occurring in most summers in the 1960s but thereafter a dearth of records until 1977. There were likely nesting attempts at three sites in 1979 but none thereafter other than occasional passage birds.

Pembrokeshire

Lockley *et al.* (1949) list birds calling from about 12 sites in 1947 and seven in 1948. Regular breeding ceased on Skokholm after 1930 and on the mainland in the early 1960s where it occurred only sporadically thereafter up to 1973. One of the last records was on a farm at Begelly where three nests were mown out; the farmer rescued some of the chicks, incubated three eggs and the birds were subsequently released. Since 1972 only occasional calling birds have indicated any possibility of breeding.

Cardiganshire

By 1966 it was considered that one or two pairs probably still attempted to breed in upland areas. Breeding was no more than sporadic after this and was last proved in 1972. There was a handful of summer records in the years following and nothing since, until an injured bird was found at Cwmystwyth in September 1990 (R. Thorpe, in litt.).

Breconshire

Five proved breeding records since 1945: 1952–54, 1959, 1965. Single calling birds were heard from 1963 to 1968. One bird summered in 1978.

Radnorshire

It had disappeared by 1938 with only rumours or occasional calling birds thereafter. No proof of breeding since pre-1938. Some summering records – 1955, 1958, 1966, 1970, 1980, 1982.

Montgomeryshire

No record of the situation prior to 1947.

1947	20 widely scattered localities
1948	Several sites; breeding proved
1949	At least six localities
1950	At least 17 localities; breeding proved
1951	At least 16 localities
1952	At least 16 localities (mainly different to those in 1951)
1953	At least 13 localities; breeding proved
1954	Six localities

1955	Four localities
1956	Three localities
1957	Six localities
1958	Three localities
1959	Two localities
1960	Two localities
1961	Two localities
1962–64	No records
1965	One pair nested
1966–67	No records
1968	Two localities
1969	Three localities; breeding proved
1970	Two localities
1971	Four localities
1972	One locality
1973	One locality; regular calling through summer
1974	No further records although birds were still calling annually at Coed y Dinas, Welshpool until 1980 (D. Howard, pers. comm.)

The spread of sites in the 1950s – some 40 different locations were involved – suggest that there was almost certainly a stronger population than the above numbers in any one year suggest.

Denbighshire

Numbers were reasonably stable up to around 1950 but thereafter decreased rapidly with the last (probable) breeding – two calling males all summer – near Denbigh in 1961 (J. M. Harrop, pers. comm). A pair bred successfully in 1992 rearing four young. A calling bird had been present near the successful 1992 site, throughout the summer of 1991.

Flintshire

Last positive breeding recorded in 1959. Several calling birds up to 1968.

Merioneth

Regular breeding is believed to have ceased as early as the 1940s but sporadic nesting has been recorded since then: 1942, 1950, 1956, 1959, 1968, 1972. The last proved breeding was near Llandderfel in 1973 although a calling bird was present in the following year and breeding is thought likely (R. Thorpe, in litt.). Interestingly, a bird turned up and was calling in the same area in June 1988 (pers. obs.).

Caernarfonshire

North *et al.* (1949) indicated that the Corncrake was still widespread in Snowdonia, including such areas as the lower Conwy valley, up to the end of the 1940s. However, it is Llŷn which was the last Welsh stronghold for Corncrakes. Many accounts in the early decades of the century testify to its abundance there, e.g. Aplin (1902) "Almost every field has a pair in it. Extremely abundant." On Bardsey the same author recorded it as "one of the noticeable land birds" and Coward (Diaries) rated it as numerous in 1905. There were still ten pairs there in 1933 but by 1952 this had reduced to only one. After that one or two pairs bred in 1953 and 1954 but none definitely since then except for a probable breeding record in 1972. On the mainland the Merseyside Naturalists' Association (1954) claimed c30 in fields around Aberdaron, although the figure seems optimistic. Hope Jones and Dare suggest that very small remnant populations continued to exist (but no proof of breeding) up to 1976. Between 1968 and 1972 breeding was proved in four sites (two pairs at one) and was probable at a fifth site. Since then there have been unconfirmed suspicions of continued breeding, at least sporadically and possibly regularly up to 1991 but at the time of writing no proof has been forthcoming.

Anglesey

In the years up to the mid-1950s Corncrakes were still fairly numerous. A. J. Mercer (pers. comm.) knew them at many sites. P. Whalley (in litt.) confirms this, quoting T.G. Walker's record (Diaries) of 36 singing males up to 1955. Mercer points to a rapid decline from then on until the final disappearance by the late 1960s with the exception of one lingering pair in the Llanfaethlu area. Thereafter they have only appeared as passage birds, mainly in autumn until the unexpected breeding of one pair, or possibly more, in the late 1980s and early 1990s. A pair is known to have hatched four young in 1992 but all four were taken by a Kestrel.

Recently, in the 1980s and early 1990s the Corncrake has been little more than a very rare passage migrant, mainly in autumn (an aberrant record was of a bird found freshly dead in Pembrokeshire on 10 December 1929). Serious attempts to retain the remnant breeding populations in Ireland are meeting with little success and the population there has fallen further by over 50% in the years 1990–92. In the Hebrides a fuller understanding of the species' requirements is leading to some success, e.g. Coll, Tiree. On Tiree the number of pairs reached an all-time high of 117 in 1993 and on Coll the population rose from 5 pairs to 13 pairs on managed land between 1991 and 1993.

In other European countries the Corncrake has suffered the same fate and it is now considered the most endangered breeding bird species in Europe. Its future in Wales and the British Isles as a whole, seems bleak although progress in the Hebrides gives some hope. Apart from the serious problems in the traditional breeding areas, Corncrakes seem to face particular problems on the long flights to and from the distant wintering areas of southern Africa. A disproportionate number – compared with other African migrants – appear to be killed on telephone wires and similar and large kills can occur at lighthouses, e.g. Bardsey.

Moorhen *Gallinula chloropus*

Iâr Ddŵr

A locally common breeding resident and winter visitor, widespread throughout Wales.

Moorhens are ubiquitous birds, inhabiting almost any freshwater habitat from large lakes to small ponds and ditches. They will readily take advantage of man-made lakes, ponds and canals, and are one of the most common riparian bird species in lowland Britain. They avoid only the highest lakes and fast-flowing rivers, and are common throughout Britain except the Scottish Highlands, the Welsh uplands and parts of south-west and northern England. This species is largely sedentary, and although it suffers setbacks in severe winters, recovery is rapid, and there is no evidence to suggest that there has been any long-term change in status nationally, although in Wales comparison between the two Breeding Atlases show that there have been local declines mainly in the east of the country.

In Wales, it is likely that the drainage of wetlands throughout the 18th and 19th centuries would have caused local decreases in Moorhen populations, although the literature suggests that at the turn of the century, the population was stable. Forrest (1907) records that it was common throughout north Wales in suitable habitat, although it was rarely encountered in the uplands. It is scarce in many areas of north and central Wales, although on occasion nests have been found at high altitudes (e.g. nest with eggs at 525 m on Llyn Du (Cardiganshire) on 6 June 1927). There were also local decreases in the south Wales coalfield valleys, probably associated with industrial pollution of the rivers, although now that most of the coal mines have closed down, Moorhens have recolonised many of these areas.

Its abundance in the Principality varies from one county to the next, depending mainly on topography and the availability of suitable waters. It is relatively scarce in parts of Caernarfonshire, Merioneth and Cardiganshire, but reaches high densities in parts of low-lying counties, such as Anglesey and Monmouthshire. Moorhens were confirmed breeding in 126 tetrads during the fieldwork for the *Gwent Atlas*, with highest densities on the drainage reens of the Levels and on small farm ponds. An ornithological survey of Anglesey in 1986 recorded this species in 253 of the 762 one-kilometre squares surveyed (RSPB 1986) and in Carmarthenshire, it was confirmed breeding in all but two 10-km squares.

In Pembrokeshire, Moorhens are most common in the lowlands of the south, with breeding confirmed in 69 one-kilometre squares and suspected in another 15. They declined throughout much of the 20th century due to the loss of habitat and pollution of watercourses but, recently, numbers have increased once more thanks to the creation of farm irrigation reservoirs associated with the potato growing areas of the county. They also breed on the islands of Skomer, Skokholm, Ramsey and Caldey, and although 16 pairs nested on Bardsey in 1987, predation by Peregrine Falcons is now thought to be causing a decline in the islands' breeding population (N. Brown, pers. comm.).

The distribution map in the *New Atlas* indicates that this species has declined in some areas of eastern Wales since the early 1970s. The absence of birds from sections of the River Monnow (Monmouthshire) and reported declines from some areas on the Rivers Teifi (Cardiganshire) and Wye in recent years has been attributed to the increase in the number of feral mink in those parts of Wales, although there is no firm evidence to support this suggestion.

Very little work has been done on this species in Wales, and in most counties no attempt has ever been made to give estimates of its breeding population. The *Gwent Atlas* has estimated a population of 1000 breeding pairs for Monmouthshire (Tyler *et al.* 1987), but this figure is probably too high. Elsewhere, the Breconshire estimate is 50–100 pairs, with 10–20 of these on Llangorse Lake, and a population of c50 pairs has been suggested for Radnorshire. The total Welsh population is likely to be 2500–3000 pairs, with the highest densities on Anglesey and the Gwent Levels.

In winter, the Moorhen is rarely found on high ground, and is therefore absent from much of central and north Wales. Whereas some pairs defend a territory throughout the year, others, mainly young birds, flock together to roost and feed. Resident British birds are joined in the winter by immigrants from north-west Europe, especially Sweden and Denmark.

During the winter months, the Moorhen does not form large gatherings like its close relative the Coot, and in Wales, wintering flocks are small despite the movements away from upland areas. Forrest was one of the first to note that some Moorhens migrate after the breeding season, citing records of birds seen on Bardsey and the Skerries (Anglesey). In Pembrokeshire, one was seen on the South Bishop lighthouse in the early hours of the morning on 9 October 1884, and again on 9 November 1975, and there have been autumn records from other offshore islands.

Although large concentrations are rare in Wales, sizeable flocks of Moorhen have been recorded from several counties. They are most common from north-east Wales and some of the southern coasts where they often favour harbours, inlets and lakes a few kilometres from the sea. In Carmarthenshire, a minimum of 45 were recorded on a pond near Burry Port on 14 January 1979, with similar numbers here again in December of the same year. An exceptional number of 91 was seen on a farm pond near Wrexham (Denbighshire) on 4 November 1972, and 29 were observed feeding in a field adjacent to Llyn Bodgylched (Anglesey) on 23 November 1972.

There are several records of Moorhen ringed in north-west Europe, particularly the Netherlands and Denmark, and recovered in Wales, the longest distance involving a bird ringed in Denmark on 12 September 1961 and shot near Brecon in November 1961, a distance of 1016 km. Others ringed in the Netherlands have been recovered on the River Wye (Breconshire), Llyn Tegid (Merioneth) and Broadhaven (Pembrokeshire).

Coot *Fulica atra*

Cwtiar

A locally common breeding resident and winter visitor, widespread throughout Wales.

Coots generally require larger waters for breeding than Moorhen, and are therefore usually found on the more open lakes and ponds. Furthermore, the Coot is a bottom-feeding vegetarian, and requires shallow waters with an abundance of submergent vegetation. Newly constructed gravel pits and industrial ponds in southern Britain have provided ideal breeding habitats, and Coots have taken advantage of such areas to show marked increases in some areas over the past two decades.

Coot bones were found in remains, presumably from the Pleistocene/Early Holocene period, excavated from Gop Cave near Dyserth (Flintshire) at the turn of the century, and more recent remains were identified from Roman drains at Caerwent (Monmouthshire) dating back to the first and early third centuries.

In Wales, large, eutrophic lakes are restricted to lowland areas, and Coots are absent from most of the upland waters. Anglesey is a stronghold for this species; the county's many lakes and open waters with rank, peripheral vegetation occasionally support high breeding densities. Even at the turn of the century, this county was the Coot's stronghold in north Wales.

At that time, Coots appeared to be increasing in some lowland areas, the first recorded breeding from the upper Dee being a nest near Llandderfel (Merioneth) in May 1894. There were also local increases elsewhere in Wales, the first recorded breeding for Monmouthshire being in 1939, and soon the Coot became a familiar sight on many lakes in lowland Wales. In 1902, up to 150 pairs bred on Llangorse Lake alone, and although 110 breeding pairs were recorded in 1979, the current estimate is 30 pairs as there appears to have been a sharp decrease during the late 1980s, probably associated with an increase in water sports (Francis 1992).

There have been increases over the past 30 years in Pembrokeshire, Carmarthenshire, Monmouthshire, Cardiganshire, western Merioneth, Radnorshire and Denbighshire, and Coots are now breeding on ponds and lakes in the centre of large towns and cities in many parts of Wales. In Pembrokeshire, the county population is estimated at c75 pairs, and in Radnorshire numbers have increased from about a dozen pairs at three main sites during the early 1970s to 45–50 pairs by the 1980s. Likewise in Monmouthshire, since the first confirmed record in 1939, the population has increased to a total of 50–100 pairs, the main site being Llandegfedd Reservoir which supports 20–25 pairs. Breeding was confined to two inland sites numbering no more than five pairs in Carmarthenshire in 1952, but by 1980, pairs were found nesting on several inland and coastal ponds, and on oxbows and ponds beside the Afon Tywi. An RSPB survey of Anglesey undertaken in 1986 recorded this species in 63 of the 762 one-kilometre squares surveyed, with breeding confirmed in 33 of these (RSPB 1986). The most important lake on the island is Llyn Alaw where an estimated 50 pairs have bred in recent years.

Today, the Coot is found on nearly all lowland eutrophic lakes in Wales, and in counties such as Breconshire it is said to be more numerous than the Moorhen. In other areas, however, it is still a relatively scarce breeding bird, and the total Welsh population is probably in the region of 1500–2000 pairs.

Ringing recoveries have demonstrated that, in winter, Coots from as far away as Russia move south and west to the milder weather on the lakes and coasts of countries such as Britain (Brown 1955). Within Britain, small-scale movements take birds to the larger lakes and reservoirs which generally remain ice-free throughout the winter months.

In Wales, several of the Coot's breeding sites are vacated over the winter, particularly during hard weather, and there are often large concentrations in a few selected sites. Some resident breeding birds will remain throughout even the most severe winters, especially in low-lying areas such as Anglesey. They rarely overwinter in large numbers on salt water in the Principality, although this was a favoured habitat at the turn of the century in north Wales. During particularly severe weather, thousands were said to have flocked to the Milford Haven Estuary (Pembrokeshire) in the winters of 1933–34 and 1946–47 and 200 were seen on Broadwater (Merioneth) on 17 December 1933.

The largest winter numbers are found on Anglesey, with several lakes regularly supporting over 200 birds. Llyn Alaw had 380 birds in 1984–85, 235 birds were present on Llyn Traffwll in 1985–86, 236 on Llyn Cefni in 1984–85 and 543 on Cemlyn Lagoon in 1986–87. Llyn Llywenan held 580 birds on 15 January 1989 and 662 were present on Llyn Maelog on 22 September that same year. Many of these birds graze on adjacent fields, and there appears to be a great deal of movement between waters.

Elsewhere, other important Welsh sites are Llandegfedd (110 in February 1991) and Ynysyfro Reservoir (225 in December 1991) in Monmouthshire, Llangorse Lake (400 in January 1990) and Talyllyn Lake in Merioneth (268 in December 1990). A few other sites regularly hold over 100 birds in most winters, with large influxes recorded on occasion (e.g. 235 on the Witchett Pool in January 1975) During the 1930s, flocks of over a thousand were seen on Kenfig NNR, but numbers here and on Lisvane Reservoir in Cardiff have been much lower in recent years. Elsewhere inland, wintering flocks rarely exceed 100 birds, and some flocks overwinter on very small pools (e.g. 87 on Fachwen Pool in Montgomeryshire on 3 January 1982).

Ringing recoveries from Wales support the suggestion that continental Coot show a southerly and westerly movement in winter. One bird ringed at Orielton (Pembrokeshire) on 14 January 1941 was recovered at Prekuln in the former USSR on 1 August 1942, a distance of 2184 km east-north-east, and another bird ringed at Orielton on 3 November 1960 was recovered at Illaunmore in County Clare, Eire on 26 December 1964. There have been several ringing recoveries indicating movements within Britain, and one involving a bird ringed in Ringvaart Haarlemmermeer in the Netherlands on 6 December 1969 and recovered at Welshpool (Montgomeryshire) later the same month.

Crane *Grus grus*

Garan

A rare visitor to Wales.

Excavations of the Roman legionary fortress of Caerleon (Monmouthshire) recovered large numbers of Crane bones, 62 bones representing at least four individuals. The morphology of the skull and other bones match skeletons of Crane in the Natural History Museum but some are as large as skeletal material of the Sarus Crane, which is not found in Europe nowadays. There is also a possibility that the bones are from the extinct *Grus primigenia*. Several of the bones have distinct knife marks (Hamilton-Dyer, in Zienkiewiez 1993).

Historically, in the Laws of Hywel Dda (10th century), it was decreed that the chief falconer of the king was to be honoured with gifts on the day that his hawk should kill one of three birds: a Bittern, a Heron or a Crane.

The Crane probably remained as a resident for long after the time of Hywel Dda. "In the bogges", says Owen in 1603 of Pembrokeshire, "breedeth the Crane, the byttur . . . and diverse others of that kynde and . . . on higher trees the heronshewe". Sadly, the Crane has long since ceased to breed, a victim of the extensive drainage of the land here, as it was in England.

The Crane was formerly esteemed for its flesh as well as for the sport it supplied. It is mentioned, along with the Heron, in the second course (consisting entirely of birds) of the menu for a special feast, as preserved in a Welsh manuscript of the early 16th century (Matheson 1932).

There is no definite information as to when the Crane ceased to breed in Wales but it was probably extinct long before the year 1700.

Within the past hundred years there have been 17 records of Cranes, commencing with birds shot near Solva (Pembrokeshire) on 28 April 1893 and at Rhosneigr (Anglesey) on 16 May 1908. There was then a long gap until 1960 but a marked increase of records from the 1970s. The 16 records from the 20th century are widely distributed from Flintshire, Caernarfonshire, Anglesey, Montgomeryshire, Cardiganshire, Pembrokeshire, Carmarthenshire, Glamorgan and Monmouthshire. They all refer to single birds except for three at South Stack on 1 May 1980 and two (an adult and first winter bird) in Cardiganshire from 16 December 1983 to 18 January 1984, then at Pembrey (Carmarthenshire) from 26 to 28 January; the adult was found shot on 28 January and the immature was last seen on the 29th.

Eight of the records are in the period April/May, and four in the period September/October. These are typical dates for a drift migrant from Scandinavia. The recent upsurge in records from Wales is consistent with the pattern in Britain as a whole.

Little Bustard *Tetrax tetrax*

Ceiliog y Waun Lleiaf

A vagrant.

There have been six definite records of Little Bustards in Wales, all sadly, referring to shot birds:

9 December 1884	–	One shot in a turnip field at Llanbabo (Anglesey)
19 November 1885	–	One shot at Gileston (Glamorgan)
November 1901	–	Female shot at Laugharne Marsh (Carmarthenshire)
5 February 1914	–	One shot at Broadway, Laugharne (Carmarthenshire)
9 September 1938	–	Male shot, Laugharne (Carmarthenshire)
23 November 1968	–	First winter bird shot at St. David's airfield (Pembrokeshire)

The species breeds in France, Iberia and north-west Africa eastwards through southern Europe to Kazakhstan and the number of records from Britain in general has decreased in recent years as Little Bustards have become less numerous on the Continent.

Great Bustard *Otis tarda*

Ceiliog y Waun

A vagrant, with no recent records.

Sadly there have been no records of this species in Wales for over eighty years and the three definite records all refer to individuals that were shot.

A few years before 1830	–	One shot near Llanrwst (Denbighshire)
Christmas week 1890	–	Female shot at Glan-rhwdw near Llanelli (Carmarthenshire) (almost certainly the same bird as one quoted as killed at Allt y Cadno in January 1891 – the two places are only slightly over a mile apart)
20 December 1902	–	One shot, Pontardawe (Glamorgan)

Barker (1905) also recorded "four or five shot in the last three or four years, some near Ferryside". Ingram and Salmon (1954) point out that while this may seem exaggerated it should not be overlooked that a number of Great Bustards were shot (including one in Glamorgan) in various parts of the British Isles in December 1902.

In the Report of the Parliamentary Commission on land in Wales (1896) there is a statement to the effect that a pair of Great Bustards was shot by Mr Thos. Gill at Llanwddyn (Montgomeryshire) a great many years ago. H. E. Forrest (1907) was informed that one had been shot on Dolargwyn Mountain near Tywyn (Merioneth) but that unfortunately it was not preserved; there are no confirmatory details.

The species breeds in Iberia, Germany, Austria, Poland and south-east Europe discontinuously eastwards through Asia and there have been recent records from England following a build-up in the east European population and hard weather influxes westwards. There is therefore still some hope of a further Welsh record!

Oystercatcher *Haematopus ostralegus*

Pioden y Môr

A common, chiefly coastal, breeding species and abundant winter visitor to all coastal areas of Wales.

The Oystercatcher is mainly a coastal bird in Wales, well distributed as a breeding species along rocky, pebbly and sandy shores, saltmarshes and the well-vegetated tops of islands; it is scarce but increasing as an inland nester along major river systems. Non-breeding flocks are present throughout the year on intertidal areas, augmented considerably from the autumn onwards by the arrival of birds from more northerly breeding grounds. Wales holds more than 6% of the East Atlantic Flyway wintering population.

At the beginning of the century the Oystercatcher was recognised as a common breeder in most coastal areas although there were some regions, e.g. Carmarthenshire, from which it was absent. In north Wales it was a common breeding species, noted by Forrest (1907) as occurring not only on flat, sandy, shingly coasts but also rocky parts especially where these had flat shelves. In Glamorgan it was recorded as common and breeding along the coast in considerable numbers i.e. Sully Is. three to four pairs, Aberthaw 12+ pairs, Ogmore River to Newton Point four to five pairs, Porthcawl to Port Talbot 50+ pairs. By 1967 the Glamorgan breeding population was reduced to a maximum of five to seven pairs, this massive decrease attributed by Heathcote *et al.* (1967) to the pressures of human disturbance and industrial development.

Although there has been a massive decline in breeding numbers in some areas, e.g. Glamorgan, and a complete demise in others, e.g. Denbighshire, overall the Welsh breeding population has shown a marked increase in the past sixty years or so, not least in Pembrokeshire where Mathew (1894) mentions breeding only on three sites compared with the comment by Lockley *et al.* (1949) that it was breeding at that time round all coasts and on all islands.

The coastal situation as regards breeding numbers in the period 1960–91 can be summarised as follows:

Dee estuary

(Cheshire/Flintshire) 40–50 pairs breed on saltmarsh and former industrial areas close to estuary.

Anglesey

200–300 pairs breed throughout the island, up to 10 miles from the coast (RSPB 1986). About 120 pairs established in 1970.

Denbighshire

None in recent years.

Caernarfonshire

Three pairs Great Orme. Several pairs breeding between Caerhun and Glan Conwy. On the Llŷn peninsula 50–75 pairs from Foryd Bay to Porthmadog. On Bardsey Island 35–50 pairs from 1953 to 1980, increasing to 60–90 pairs from 1980 onwards.

Merioneth

40 pairs between the Dwyryd and Dyfi.

Cardiganshire

20 30 pairs.

Pembrokeshire

Approximately 120 pairs on the mainland coast in the mid-1940s. The tetrad survey in the 1980s confirmed breeding in 43 mainland squares and the total population is estimated to be about 300 pairs. On Skomer the breeding population increased from 36 pairs in 1946 to 70–98 pairs in the 1980s. On Skokholm the population varies between 40 and 50 pairs.

Carmarthenshire

Very sporadic breeding only, up to two pairs.

Glamorgan

Three to seven pairs (very infrequent on Gower). Eight pairs on Flat Holm in 1991.

Monmouthshire

15 pairs in 1981–85, increasing from a single pair in 1971 (first record in the 20th century).

The total breeding population for the Welsh coast is therefore of the order of 700–900 pairs. Buxton (1961) studied the inland breeding of the Oystercatcher 1958–59 at which time he concluded that the only inland breeding in Wales occurred in Anglesey. There was, however, a pair breeding near Trawsfynydd (Merioneth) since 1947 and near Blaenau Ffestiniog (Merioneth) in 1958 or 1959; the Trawsfynydd site was occupied by a small number of pairs subsequently with 12 pairs in 1986 making Llyn Trawsfynydd the oldest and largest inland "colony" in Wales. The first confirmed breeding record well inland in Wales was not until 1974 when a nest was found close to the River Severn near Caersws (Montgomeryshire).

An inland spread such as Buxton described in southern Scotland and northern England now appears to be in progress in Wales. A 1991 survey of river shoals in Wales, primarily for Little Ringed Plover (Tyler 1992), recorded no fewer than 25–29 pairs of Oystercatchers. These were on the Severn between Newtown and Llanidloes (four to five pairs), the Vyrnwy (one to three pairs), the Dee between Chester and Bala (nine pairs), the Dyfi (ten pairs) and the Dwyryd (one pair). It would therefore seem that an inland spread is taking place along the rivers of north and west Wales.

In addition to the breeding population, a substantial number of birds remain on the coast throughout the summer; these are immature birds, which do not return to their breeding areas until their third or fourth year. Typically the summering flocks are of the order of 2000–3000 on the Burry Inlet, with small numbers elsewhere.

Numbers start to build up again in July but the main influx is in August with high population levels until February. The average January estuary counts for the years 1987–91 gave a total of 55 511 Oystercatchers, including the Dee figure (Cheshire/Flintshire). Moser (1987) showed that approximately 23% of the Oystercatcher wintering population of Britain occurred on non-estuarine coast so the figure of 55 000, which is nearly 20% of the British winter population level can be taken as very much a minimum. Just over 15 000 were counted in Wales in 1984–85 on the non-estuarine coasts (Moser & Summers 1987), so the total wintering population in Wales is of the order of 70 000 birds.

The population fluctuates from year to year and from site to site depending on the variations in main food prey, especially the cockles and mussels which other waders are incapable of opening. The Oystercatchers' skill at opening cockles and mussels has made them unpopular with fishermen. During the winter of 1963–64 there was a massive spatfall of cockles on the Burry Inlet which led to abnormally high yields in subsequent years. By the time the cockle population had decreased to its more usual level Oystercatcher numbers had increased from 8000 to 15 000. The fishermen demanded a cull of Oystercatchers to return the population to earlier levels and a total of 10 000 was shot in the autumns of 1973 and 1974. Ironically, after this cull had been carried out the cockle population plummeted due to natural fluctuations. Horwood and Goss-Custard (1977) re-analysed the data and suggested the impact of Oystercatchers had been exaggerated. Fortunately the public outrage, the shifting of opinion towards conservation and the subsequent re-analysis of the Burry Inlet data make it certain that such a cull will never be repeated (Lack 1986).

Average winter peaks for the period 1986–90 for the estuaries with more than 1000 Oystercatchers were as follows:

Dee estuary (Cheshire/Flintshire)	29 990
Burry Inlet	16 934
Traeth Lafan	3623
Carmarthen Bay	2602
Swansea Bay	2491

It is notable that the major estuaries of the Monmouth Severn (299), Taff/Ely (31) and Cleddau (502) held relatively small numbers in relation to their area, a reflection of the sparsity or absence of cockle and mussel beds.

BoEE data show that the wintering population of Oystercatchers in Wales (estuaries from the Dee (Cheshire/Flintshire) to Severn (Monmouthshire)) is increasing in recent years. This is well illustrated by the average winter peak counts:

1969–71	30 033
1972–76	38 450
1977–81	40 076
1982–86	54 906
1987–91	55 188

In inland areas, in addition to the breeding population already described the Oystercatcher is a regular, if uncommon, passage migrant chiefly in the periods March to May and in particular, July to September and especially along the valleys of the Usk, Wye and Conwy. Single birds are most usual but small parties, up to 20, are not very uncommon.

A considerable number of Oystercatchers have been ringed in Wales, mainly because of research carried out by the Ministry of Agriculture, Fisheries and Food into the movements of the birds. The map gives a summary of the recoveries. The recovery pattern indicates the main breeding areas of winter visitors to be in Northern Britain, Iceland/Faroes and Norway. Many wintering birds are site faithful and there are many cases of retraps at the same location in subsequent years. Oystercatchers are

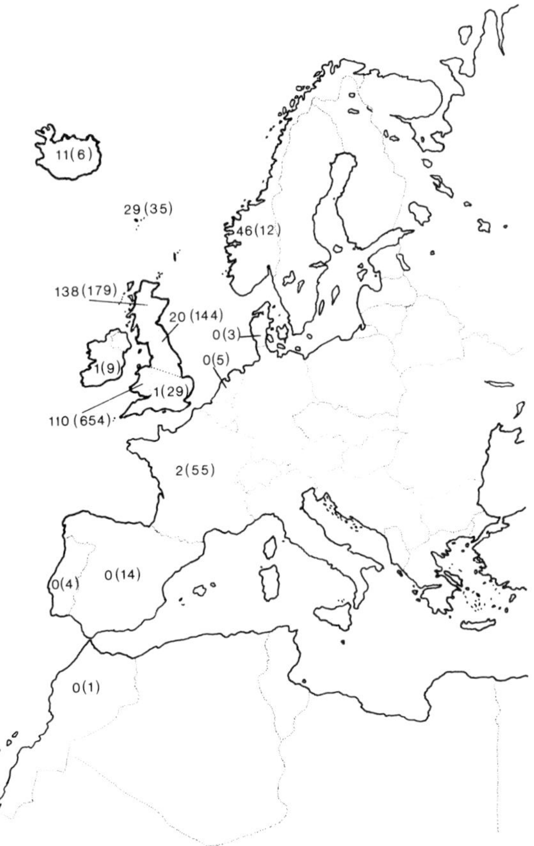

Recoveries in May to July of Oystercatchers ringed in Wales. In parentheses, recoveries for remainder of year

long-lived; ages of 20 years or more are not unusual. One individual ringed as a chick on Bardsey on 23 June 1960 was recovered there in March 1986. Recoveries from locally hatched chicks show that their wintering area lies to the south, in western France, Portugal and Spain with one recovery from Morocco.

Black-winged Stilt *Himantopus himantopus*

Hirgoes

A vagrant which has probably bred once in Wales.

This is an unmistakable species which breeds in southern Eurasia, Australia, Africa and the Americas. The European population winters mainly in Africa south of the Sahara. There is one remarkable and isolated record of probable breeding. The 1990 *Gwent Bird Report* places on record a hitherto unpublished account of the presence of a pair of Black-winged Stilts on Caldicot Moor in the early summer of one year in the 1950s (probably 1952, 1953 or 1954). The birds apparently attempted to breed but were disturbed by the flow of local people interested in seeing them. Contemporary photographs confirm the identification and show the habitat of semi-flooded pasture land.

The eight additional records are as follows:

1793	– One shot in Anglesey
23 August 1965	– One, Newborough Warren (Anglesey)
24 July 1967	– One, near Narberth (Pembrokeshire)
29 April 1984	– Two, Bardsey
5 April 1987	– One, Foryd Bay (Caernarfonshire)
5–6 April 1987	– One, West Dale (Pembrokeshire)
19 to at least 27 March 1990	– One, Penally (Pembrokeshire)
7–8 May 1990	– One, Skokholm

With a marked recent increase in sightings from the rest of Britain as well it looks as if further records can be confidently expected in Wales.

Avocet *Recurvirostra avosetta*

Cambig

A scarce visitor to Wales, recorded chiefly on spring passage but occasionally wintering.

It breeds from Iberia and north Africa through to central and south-western Asia, southern and east Africa. There are also breeding colonies on the east coast of England and the Netherlands, and a wintering flock as close to Wales as south Devon.

The seasonal distribution of records in Wales (see histogram) shows that the greatest concentration of sightings is in the period March to May when wintering birds from

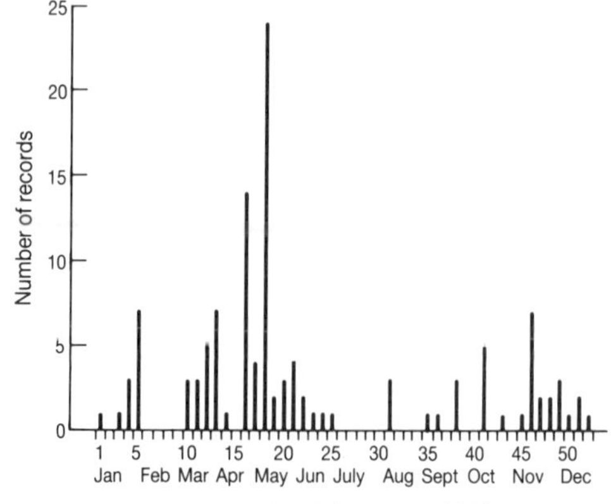

Weekly records of Avocet to 1991

Iberia and Africa are returning to breeding grounds farther north. On four occasions birds have stayed the winter in Wales (three times a single bird and once two birds). Most of the sightings are of single individuals but flocks of up to 17 have been noted.

For a species which prefers brackish or saline conditions, it is not surprising that the records are all from coastal counties, principally from the south coast where the counties of Pembrokeshire, Carmarthenshire, Glamorgan and Monmouthshire account for most of the birds seen. There are no records from Breconshire, Radnorshire, Montgomeryshire and Denbighshire.

The identification of an Avocet among late first-century food refuse from the Fortress Baths at Caerleon (Monmouthshire) is of interest as the first archaeological record of this species (Parker 1986). The bird must have presumably been caught on the nearby coast.

Stone-curlew *Burhinus oedicnemus*
Rhedwr y Moelydd

A rare visitor to Wales, with a total of only 26 acceptable records up to 1991.

It breeds in central and southern Europe, south-western and southern Asia and north Africa, wintering in western and south-western Europe, north-west and east Africa. In Britain it is a scarce breeder which has declined in recent years, confined to the chalk downland and open heaths of southern England and East Anglia.

Ten of the 23 dated records are in the period 10 April to 14 May coinciding with the peak time of spring passage, other records are from January (1), March (1), June (3), July (1), August (6) and November (1). Fourteen of the records are in the period from 1970 onwards but this apparent increase is probably due to the fact that there are now many more birdwatchers active in the field rather than a genuine increase in occurrences.

The distribution of records is quite well spread throughout Wales with a total of 17 from the four southern coastal counties and the remainder in Breconshire (1), Cardiganshire (1), Radnorshire (1), Merioneth (2), Caernarfonshire (2) and Anglesey (2).

Cream-coloured Courser *Cursorius cursor*
Rhedwr y Twyni

A vagrant which has been recorded three times in Wales.

It breeds in the desert regions of north and east Africa and south-west Asia east to Afghanistan, wintering south to northern Kenya and Arabia.

The three Welsh records are:

1793	– One shot in North Wales. Col. Montagu in his *Ornithological Dictionary* wrote: "We are assured by Mr Dickinson that a specimen of this very rare bird was shot in North Wales in the year 1793 by Mr George Kingstone of Queen's College, Oxford, a very accurate ornithologist. The bird was preserved in the collection of the late Professor Sibthorp."
2 October 1886	– One shot, Ynyslas (Cardiganshire)
23 October 1968	– One, Cefn Sidan Sands, Pembrey (Carmarthenshire)

Collared Pratincole *Glareola pratincola*
Cwtiadwennol Dorchog

A vagrant with only two definite records in Wales.

It breeds from Iberia and north-west Africa eastwards to Kazakhstan and south to Iran and Iraq, and Africa south of the Sahara. It winters in Africa south of the Sahara.

The two Welsh records are:

27 May 1973 – One, Penclawdd (Glamorgan)
6 June 1983 – One, Rhosneigr (Anglesey)

A description of a bird believed to have been this species, but not completely full, comes from Kenfig Pool (Glamorgan) on 1 May 1962. In addition a Pratincole, probably this species, was seen at Bosherston Pools (Pembrokeshire) on 13 April 1981.

Black-winged Pratincole *Glareola nordmanni*
Cwtiadwennol Aden-ddu

A vagrant. There is only one Welsh record.

2 June 1988 – One, Inner Marsh Farm (Flintshire/Cheshire)

The species breeds from Romania eastwards through the Ukraine to Kazakhstan, south to the Caucasus and winters in western and southern Africa.

Little Ringed Plover *Charadrius dubius*
Cwtiad Torchog Bach

The Little Ringed Plover is a summer visitor to Wales, breeding in small numbers on the gravel shoals of rivers in south and east Wales and on industrial sites in north-east Wales. Elsewhere it is an uncommon passage migrant.

Little Ringed Plovers were unknown as breeding birds in Britain before 1938 and, indeed, were extreme rarities even as migrants. The first record for Wales was a single bird at the Gann estuary (Pembrokeshire) on 16 May 1949 but sightings of passage birds were extremely scarce until the mid-1960s.

By 1967 at least 223 pairs of Little Ringed Plovers summered in 26 English counties and by 1973 the total had increased to at least 467 pairs (E. R. & E. D. Parrinder 1975). In Cheshire it was breeding as early as 1954 and by 1972 there were 20–30 summering pairs. In the face of this dramatic increase in adjoining areas it came as no surprise that a pair of Little Ringed Plovers bred in Flintshire in 1970. In 1971 five pairs summered at four sites in Flintshire and Denbighshire; four pairs bred successfully. One pair bred at a gravel pit in 1972 but the pit was drained in 1973 and no nesting was reported in Wales in that year, although birds were seen in the spring and autumn.

In 1974 the first pair was recorded nesting on a gravel shoal on a Welsh river, the River Severn (Montgomeryshire) with another pair in 1977 on the River Wye. In the mid-1970s a pair bred by Pendinas Reservoir (Denbighshire); this site is no longer suitable, being surrounded by conifers. During the late 1970s and early 1980s one or two pairs bred regularly on limited stretches of the Wye and Severn and on the Tywi in south-west Wales, where there were extensive shoals. Throughout Wales there were only six pairs recorded at five sites in 1984 (Parrinder 1989). In that year Little Ringed Plovers were first recorded in Monmouthshire, with pairs at two sites on dried-up banks of reservoirs. In 1985 a pair was found on a river shoal on the Usk where one to four pairs have bred annually since.

By 1988 up to ten pairs were breeding on the Tywi following the first recorded breeding in 1986; 14 pairs were recorded in 1989 and 24 in 1990. This marked increase was possibly because of low water levels in the dry springs of 1989 and 1990, allowing extensive areas of shoal to be exposed. In 1990 a pair bred on the Rheidol, the first breeding record for Cardiganshire.

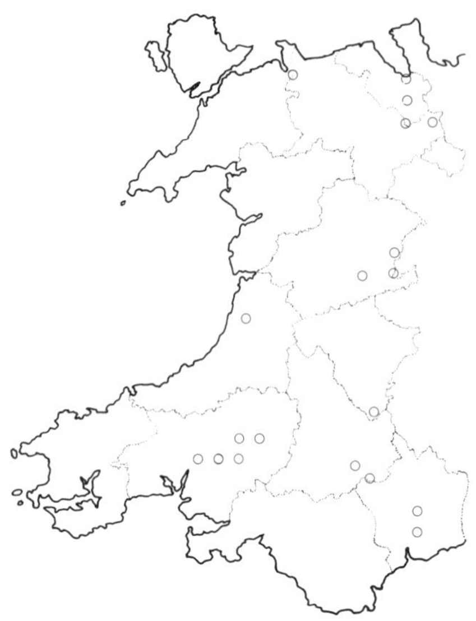

Breeding distribution of Little Ringed Plover, by 10-km squares, 1970–1992

In 1991 the RSPB carried out a survey of almost 500 km of Welsh rivers on behalf of the National Rivers Authority to assess, *inter alia*, the species' current status and distribution (Tyler 1992). The survey showed that the 1991 breeding population was at least 60 pairs, made up as follows:

R. Severn	7 pairs
R. Wye	4 pairs
R. Tywi	34 pairs
R. Cothi	4 pairs
R. Bran	4 pairs
R. Usk	1 pair
Shotton (Flintshire)	2 pairs on industrial site
Monmouthshire	1 pair on edge of upland reservoir

The highest density was recorded on the Tywi with one pair per 1.6 km. On the Severn, Usk and Wye there was a pair per 6–10 km of river surveyed. However, because pairs were not evenly distributed there were, locally, densities of one or two pairs per km. These small concentrations occurred on stretches with extensive shoals. Numbers may continue to increase on the four main rivers as there are unoccupied, but apparently suitable, shoals for more pairs. It is also possible that breeding birds will spread onto northern and western rivers although on at least some of these it may be that pebbles are too mobile or too large and angular to be suitable for the species. Pairs have bred close to the Dee estuary for more than twenty years so that colonisation of the R. Dee might have been expected already if the habitat were suitable (Tyler 1992b). In 1992 an extension of range was recorded in north Wales, a pair reared young close to the Conwy estuary (Denbighshire).

Over most of its British range Little Ringed Plovers nest at man-made sites such as gravel pits and on industrial tips. Parrinder (1989) noted that only 3% of 370 sites recorded in 1984 were "natural" with just 11 sites being on river shingle or at the edges of rivers. The river-breeding habit of the Welsh population is therefore of particular interest and is a reflection of the traditional site selection by the breeding population of continental Europe.

Away from breeding areas Little Ringed Plovers remain a scarce passage migrant in the periods March to May and July to September. There are no ringing recoveries to indicate the wintering quarters of the Welsh population.

Ringed Plover *Charadrius hiaticula*

Cwtiad Torchog

The Ringed Plover is a scarce breeder in Wales, corcentrated principally in the north-west (Llŷn peninsula and Anglesey) and the mid-section of the south coast (Carmarthenshire and Glamorgan). It is also a common passage migrant and winter visitor to all coastal areas.

At the beginning of the 20th century it was a widespread and common breeding species on all low parts of the Welsh coastline, most numerous around the northern part of Cardigan Bay, the west coast of Anglesey and the Glamorgan coast. In Glamorgan at that time it was said to have been abundant along the shore and, indeed, 25 years later it was still fairly common in suitable localities (Heathcote *et al.* 1967). Industrial and tourism developments on the Glamorgan coast and other parts of Wales have caused a gradual but substantial reduction in suitable undisturbed breeding areas.

In Pembrokeshire it was noted that Ringed Plovers were breeding at Newport, Newgale, Freshwater West and near Tenby and probably elsewhere (Lockley *et al.* 1949) and in Cardiganshire a few pairs bred on shingle beaches in the northern part of the county and a few more pairs at scattered places along the coast. The 1973 survey of the breeding population of the Ringed Plover in Britain found none in Pembrokeshire and revealed that the Cardiganshire population had reduced in range to one site

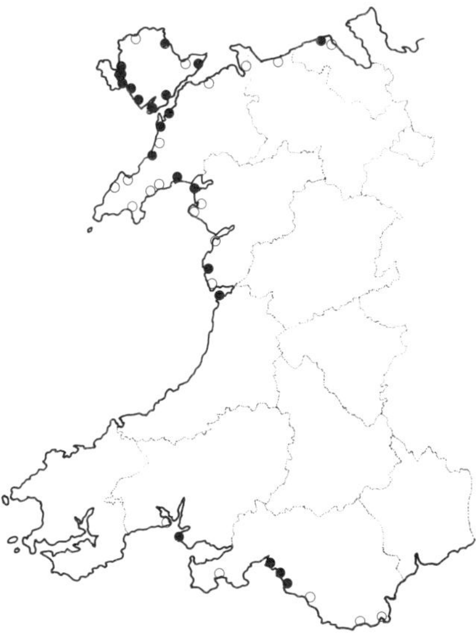

Breeding pairs of Ringed Plover in 1973 (after Prater), ○ = 1–3 pairs, ● = 4+ pairs

in the north of the county (Prater 1976a). The Dyfed Bird Report for 1972–76 does, however, give the status of Ringed Plover as a very scarce breeder on the Pembrokeshire coast with probably no more than three to five pairs annually. Four pairs bred on bare ground in the Texaco Oil Refinery in 1978 but there have been no breeding records in the county since.

The results of the 1973 survey are shown on the map. The bulk of the breeding population was located in Caernarfonshire (63 pairs on sand or shingle beaches and locally in West Llŷn and Bardsey on clifftop heath) and Anglesey but with a good concentration in Glamorgan and Carmarthenshire as well. All sites were coastal. The total breeding population was estimated at 182 pairs.

The 1984 survey found 224 pairs on the coast of Wales (Prater 1989) this was despite concerns over a decline on some beaches heavily used for recreation (as on the Gower peninsula). It was found that the populations in Cardiganshire, Caernarfonshire, Anglesey and Glamorgan had remained stable while there had apparently been substantial increases in Merioneth and Carmarthenshire. The later survey did locate three in- land nesting sites, two in Glamorgan and one in Monmouthshire but the species continues to be essentially coastal with no breeding records along rivers, unlike the situation in parts of northern Britain. Denbighshire and Flintshire were not covered for the 1984 survey so the same figure as the 1973 survey was used. However, there appears to have been an increase in the Flintshire population since 15 breeding pairs were reported in 1989 (*Clwyd Bird Report* 1989) compared with the six pairs in 1973.

Ringed Plovers are present throughout the year in most coastal areas in Wales but there are marked annual peaks in numbers in May and August. Autumn passage appears to involve a substantial number of birds from the Greenland breeding population, probably *en route* to wintering grounds in West Africa (Eades & Okill 1976). At the Dee, which is the principal passage site in Wales the average counts for the years 1986–90 were 42 in July, rising to 695 in August and declining to 248 in September, thereby showing the pronounced nature of the August passage.

In winter, numbers are much less than during the autumn migration with small numbers dispersed all round the Welsh coast. In winter Welsh estuaries from the Dee (Cheshire/Flintshire) to the Monmouthshire Severn hold more than 2% of the East Atlantic Flyway population. The average Welsh January population on estuaries for the years 1987–91 amounted to 833, but the non-estuarine coast in Wales held 776 birds in December 1984/January 1985 (Moser & Summers 1987) so the wintering population could be approximately double the estuaries total. Average winter peaks for the years 1986/90 exceeded 100 for the following estuarine sites:

Swansea Bay	254
Severn (Monmouthshire)	198
Burry Inlet	198
Carmarthen Bay	188
Dee (Cheshire/Flintshire)	158
Cleddau	107
Traeth Lafan	106

Average winter peaks show a slight decline over the past twenty years:

1969–71	1118
1972–76	1126
1977–81	889
1982–86	946
1987–91	1007

Ringing recoveries of birds ringed as young showed that the natal areas of Ringed Plovers wintering in Wales were in England, Sweden, Netherlands and Germany. A Dutch ringed bird was recaught in four different winters in the same area of Traeth Lafan, indicating how site faithful some birds are to their wintering area. Emigration of wintering birds commences in February.

There is a well-marked passage of Ringed Plovers northwards along the Welsh coast in late April to May, involving birds which are returning to breeding grounds in Iceland and Greenland. An earlier and smaller part of this heavy passage is probably attributable to Icelandic birds which normally arrive on their breeding grounds in early May (Cramp 1983). The last two weeks of May usually see a heavy passage of birds which are probably bound for Greenland and north-east Canada. Confirmation of this is provided by a recovery of a bird from Collister Pill (Monmouthshire), ringed on 19 May 1973 and recovered on Ellesmere Island, Canada on 13 July 1979. The spring passage is well recorded at most of the Welsh estuaries and, in particular, the Dee, the Severn, and to a lesser extent Burry Inlet, Gwendraeth and Dyfi.

Inland the Ringed Plover is a scarce passage migrant although, as indicated earlier, there is a modest recent trend to inland breeding in south Wales. Ringed Plovers have been recorded in all inland counties and are now seen annually in Breconshire, May and August being, not surprisingly, the peak months.

Killdeer *Charadrius vociferus*

Cwtiad Torchog Mawr

A trans-Atlantic vagrant.

There are two records of this species, which breeds in North America, West Indies and Peru and winters in the USA south to Peru:

5 February–12 March 1978	– One, Pwll near Llanelli (Carmarthenshire)
17–20 March 1982	– One, Bardsey

Winter occurrences are typical in Britain, the timing suggesting that arrival is owing to late autumn and winter storms, which sometimes result in northward movements on the American Atlantic coast.

Kentish Plover *Charadrius alexandrinus*

Cwtiad Caint

A rare visitor.

Since the first Welsh record, of two pairs at Sker (Glamorgan) in 1888 there have only been 23 records of this small plover, concentrated in the years 1951–89.

The records cover the months from April to November inclusive and are principally from the counties of Monmouthshire (4), Glamorgan (6) and Cardiganshire (7) with others from Pembrokeshire, Anglesey and Flintshire.

With the proximity of a breeding population in northern France, it is anticipated that there will continue to be sporadic sightings of this species in Wales.

Greater Sand Plover *Charadrius leschenaultii*

Cwtiad y Tywod Mwyaf

A vagrant with only one Welsh record.

16 May 1988 – One, St. Brides Wentlooge (Monmouthshire)

This was probably the same individual that had been seen at Dawlish Warren (Devon) from 27 April to 4 May 1988, the ninth British record.

This species breeds from Turkey eastwards to Mongolia and winters in south and east Africa and south Asia to Australia.

Dotterel *Charadrius morinellus*

Hutan y Mynydd

An occasional breeder, otherwise a regular but uncommon migrant in spring and autumn.

As a breeding species the Dotterel is a bird of northern latitudes, especially Fenno-Scandia, and locally in the uplands of northern Britain. In Britain it is confined almost entirely to the high montane plateaux of the Scottish Highlands where there is a population in excess of 450 pairs; there is also a small breeding population in northern England.

In 1968 a pair probably nested in north Wales and certainly did so in 1969 – "possibly the most exciting British Dotterel news of this century" (Nethersole-Thompson 1973). There has been at least one case of confirmed breeding since that time and breeding has been suspected, but not proved, on other occasions. Nethersole-Thompson (1973) referred to the suitability of many of the North Wales plateaux for breeding Dotterel with extensive areas of flat stony ground with bare soil and gravel and a good deal of *Rhacomitrium–Festuca* heath, montane *Festuca* grassland and *Festuca–Vaccinium* grass heaths favoured by the species. He suggested that "the climate, vegetation, and terrain of these mountains are so similar to those of Lakeland that the scant history of the Dotterel in north Wales is a mystery requiring explanation other than in terms of lack of suitable habitat". He further postulated that this is a district which could possibly be colonised by nesting Dotterel during a period of greater cold (Nethersole-Thompson 1973).

Dotterel winter chiefly in North Africa and use a few, but often traditional stopping-off places on passage to and from their northern breeding grounds. Green (1992) collected records of Dotterel trips in Wales between 1980 and 1989. These totalled 62, involving 317 birds, and covered every county except Flintshire. The most favoured counties were Caernarfonshire (26% of all birds), Cardiganshire (22%) and Anglesey (18%). By far the majority of records (63%) referred to May and in particular the period 1–16 May. There was one June record, no July record and three August records (from the 14th); the majority of the autumn records were in September (eight) with three in October, the latest on 20 October. The largest trips recorded in the period were 45 on Anglesey in 1980 and 32 in Cardiganshire in 1989, both in mid-May.

Many of the above records came from relatively few sites, especially at hilltop locations in Cardiganshire and Caernarfonshire. Of all records, 77% were on high plateaux and where details of habitat were available the favourite sites were ridges with short vegetation; the remaining 23% were from coastal sites or fields close to the coast. Where records prior to 1986 were readily available Green noted that it was clear that observations of Dotterel in Wales had increased during the past ten years but concluded that much of this could be attributed to an increase in observer effort (several ornithologists noted that once Dotterel were recorded from a site it has often been visited annually to check for them) although there did appear to be more records involving larger trips in recent years. There has, however, probably been an expansion of the British breeding population in the past ten years which it is hoped will give rise to further increases in sightings in Wales.

There is only one ringing recovery relating to Wales: a nestling ringed in the Highland Region of Scotland in June 1990 was recovered on passage south at Foel Fras in the Carneddau (Caernarfonshire) on 18 September 1990.

American Golden Plover *Pluvialis dominica*

Corgwtiad Aur

A vagrant.

It has been recorded twice in Wales:

26 September 1981 – One, Skokholm
3 April 1983 – One, Cemlyn (Anglesey)

American Golden Plovers breed in northern North America and winter in central south America south to Argentina.

A first summer bird seen at Cefn Sidan (Carmarthenshire) on 26 July 1987 was accepted as either this species or Pacific Golden Plover (*Pluvialis fulva*), which breeds in northern Siberia and western Alaska.

Golden Plover *Pluvialis apricaria*

Cwtiad Aur

A scarce, and declining, breeding species in Wales found chiefly on the upland plateaux of Denbighshire, Radnorshire, Cardiganshire and Breconshire.

It is restricted mainly to areas of blanket bog with cotton grass as the dominant component of the vegetation, in some areas in association with heather. It is more common as a passage migrant and winter visitor, chiefly to coastal areas, especially in south Wales.

Apart from a few pairs on Dartmoor, the Welsh population represents the southern extremity of the Golden Plover's breeding range in Britain (and Europe). The Welsh breeding population has undoubtedly declined markedly in numbers and range this century. It has never been recorded breeding in Flintshire or Pembrokeshire and only once in Glamorgan (in the 19th century) and once on Anglesey (a remarkable and isolated record, on heathland at South Stack in 1959).

In the early years of the century Forrest (1907) records that in north Wales it was most numerous as a breeder on the moors around Bala and Corwen but also noted that it bred in many other hill areas ranging from Capel Curig (Caernarfonshire) in the north to Cwm Prysor, Rhinog Fawr (Merioneth) and other moors in the west. As early as 1837 one writer describes finding four nests with eggs on the summit of Cadair Berwyn (Merioneth/ Denbighshire border). In Breconshire Phillips (1899) indicated that Golden Plovers were quite numerous on the northern hills and also bred on Mynydd Eppynt (from where they have now disappeared) whilst in Monmouthshire they used to breed on the hills around Abertillery, Tredegar and Abergavenny in the latter part of the 19th century.

Sadly, by the time of the *Breeding Atlas*, covering the years 1968–72, there had been a marked decline in the breeding range of Golden Plovers in Wales. Published records give little indication as to when the decline started, but it was certainly well under way by the 1940s if not earlier. In one area of Cardiganshire, for example, where there had been numerous pairs in 1933, it was sparse or absent since 1945.

Ratcliffe (1976) used the results of the 1968–72 Atlas Survey to make an estimate of the Welsh breeding population by applying known densities for particular habitats and areas to all 10-km squares in which nesting was certain or probable. He noted that in Wales a good deal of suitable habitat over a wide range of elevations was not occupied and that the lowest elevation at which nesting was taking place was seldom below 452 m in south Wales (formerly occasional at 305 m) and at about 365 m on Mynydd Hiraethog in north Wales. Ratcliffe's estimate for Wales was 900 pairs (30 755 pairs in Britain) but this was almost certainly very considerably too high.

During the years 1975–78 RSPB field surveys covered all the major upland areas where Golden Plover were known to breed and produced counts of 213–224 breeding pairs made up as follows:

Mynydd Hiraethog	46 pairs
Ruabon Moors (Denbighshire)	4+ pairs
Migneint/Arenig Fach (Caernarfonshire/Merioneth)	18 pairs
Berwyn Mountains	20+pairs
Llanbrynmair Moors (Montgomeryshire)	25–35 pairs
Trannon Moor (Montgomeryshire)	2 pairs
Pumlumon (Montgomeryshire/ Cardiganshire)	6–7 pairs
Kerry/Cilfaesty Hill (Radnorshire)	2 pairs
Elan/Claerwen (Radnorshire/ Breconshire)	74 pairs
Abergwesyn/Teifi Pools (Cardiganshire/Breconshire)	6+ pairs
Aberedw Hill (Radnorshire)	2 pairs
Llandeilo/Llanbedr Hill (Radnorshire)	1 pair
Black Mountain and Brecon Beacons/ Fforest Fawr (Breconshire/ Carmarthenshire)	7 pairs

It should be noted that Golden Plovers are notoriously difficult to census accurately and scattered pairs can be easily overlooked, particularly at the incubation stage. However, even allowing for this, it is unlikely that the Welsh breeding population in the late 1970s amounted to more than 250–300 pairs.

It is clear from the above survey that the prime stronghold for Golden Plovers breeding in Wales is the Elan/Claerwen catchment and adjoining areas in Cardiganshire. On the catchment, RSPB fieldwork in 1976–77 provided a figure of 74 breeding pairs and research work in 1982 (Thomas *et al.* 1983) confirmed that the breeding population was of the order of 60–80 pairs. Fieldwork over a slightly extended area and with different methodology by NCC field staff gave a population of 105 pairs in 1982. By 1990 a substantial decline had taken place as a repeat survey of the original and extended area by RSPB located only 46 pairs and this figure was confirmed by a survey by Countryside Council for Wales/Welsh Water in the following year.

The other two sites of particular importance in the 1970s have also suffered substantial declines. On Mynydd Hiraethog the 46 pairs located in an RSPB survey in 1977 had declined to about 34+ pairs by 1981 (although on a slightly less extensive area). Work by Bain (1987) which covered approximately half the moorland area of the 1981 survey found only seven pairs of Golden Plover in 1984 and four in 1985. Afforestation of Llanbrynmair Moors has made most of that upland block unsuitable for breeding waders and very few, if any, Golden Plovers are now able to breed there.

The decline in Wales is consistent with a general decrease in Britain, for reasons which are not fully understood. As in other areas pairs have disappeared from sites which still appear to remain suitable. Afforestation has, however, obliterated considerable areas of former breeding sites in Wales. For example Cefn Du Moor near Ruthin (Denbighshire) was a favourite breeding site in the 1920s and 1930s and now forms part of the Clocaenog Forest; a substantial part of the range on the northern Breconshire hills described by Phillips is now afforested; more recently, there is the loss of a prime breeding area at Llanbrynmair. However, even on remaining areas of open moor contemporary records suggest a significant decline. This may have resulted from a decline in game preservation on the moorland. Predation by Carrion Crows is the most important cause of nest loss (Ratcliffe 1976). Crows, which are effectively checked by gamekeepers, have greatly increased with the decline of keepered moors in

Numbers of Golden Plover in Breconshire, October – March

	1969–70	*1972–73*	*1975–76*	*1978–79*	*1981–82*	*1984–85*	*1989–90*	*1991–92*
Sites	5	5	4	3	4	2	1	2
Approx. Total	2510	535	835	640	475	30	30	60–100

favour of sheep farming. Similarly, controlled moor burning by keepers had a beneficial effect on Golden Plovers by creating a relative short sward in areas that would otherwise be unsuitable.

Although the total Welsh breeding population must vary from year to year, depending in part at least on prevailing weather conditions, it now seems unlikely that the average population is more than 100 pairs.

Wintering flocks of Golden Plovers are scattered thinly through most areas of lowland Wales but are particularly concentrated in coastal areas, either on saltmarshes or permanent pasture not far inland. They are very traditional in returning to favoured sites where they often associate with Lapwings. In 1977–78 co-ordinated counts throughout Britain gave the following results for Wales (Fuller & Lloyd 1981):

January 1977	18 000
November 1977	10 700
December 1977/January 1978	16 300
February 1978	5400 (count affected badly by poor coverage)

Flocks numbering more than 1000 were recorded in Anglesey, Caernarfonshire, Glamorgan, Pembrokeshire and Carmarthenshire.

The main wintering sites are in the southern part of Pembrokeshire, Carmarthenshire and Glamorgan, centred on Milford Haven (Castlemartin Range, Clynderwen and Hook) on the estuaries of the Three Rivers (Morfa Cydweli and Laugharne Burrows/Ginst Point), the Burry Inlet (Weobley saltings) and the Thaw Valley in Glamorgan (Gileston/Boverton/Llandow). Farther inland the Tywi Valley between Dryslywyn and Nantgaredig (Carmarthenshire) was a favourite haunt, with 3000 present in the January 1978 survey. In north Wales the main sites are on west Anglesey (Mona and Malltraeth) and in Caernarfonshire (Llŷn peninsula at Afon Wen, Llanengan with 2800 in January 1977, Llangwnnadl/Tudweiliog and Dinas Dinlle).

Flocks of exceptional size were recorded at Llandow on 24 November 1984 (10 000) and Weobley saltings on 12 March 1977 (6300).

There are indications that the Welsh wintering population is in decline in recent years. This is in marked contrast to the pattern for Britain and Ireland as a whole where there was a marked trough in the period 1978–83, clearly associated with the cold winters of 1978–79 and 1981–82 but where peak counts are now higher than in the early 1970s. Although the BoEE involves only a proportion of the total, it does provide some measure and the average January count has gone down from 2900 in 1971–74 to 2462 in 1987–91; the average winter peak at the main estuarine site, the Burry Inlet, went down from 2450 in 1981–85 to 1410 in 1986–90. Even more dramatically, Peers and Shrubb (1990) have shown how the wintering population of Golden Plovers in Breconshire, although never very numerous, has undergone a substantial decline in numbers and distribution. The table shows the number of sites in which wintering flocks were recorded in Breconshire from October to March and the approximate maximum number of birds involved.

Golden Plovers are badly affected by periods of prolonged severe winter weather. In Pembrokeshire, known for its equable winter climate, the county bird report noted that Golden Plovers were hard pressed in the cold spell of January/February 1985 when many birds evidently moved on for there were no major concentrations anywhere in the county during the early year. Mortality was heavy amongst those that stayed, e.g. over 100 corpses were found between Barafundle and Warren. Likewise in 1986 the hard frosts of February caused many to succumb in the Castlemartin area and 20 were found dead on Skomer. However, unlike in 1985, many groups of 10–90 stayed in the county, particularly along the banks of the Cleddau.

It is not known what proportion, if any, of the small Welsh breeding population migrates southwards for the winter but it is believed that many British birds winter in lowland habitat adjacent to nesting grounds (Ratcliffe 1976) and this is borne out by the few ringing recoveries.

In general, birds from the northern part of the breeding range (e.g. Iceland/north Scandinavia) attain full breeding plumage while those farther south, as in Wales, retain more non-breeding plumage during summer. Observations in spring show that many birds in the Welsh flocks show characteristics of birds from the northern breeding populations. There are two recoveries in Wales of birds ringed in Iceland and one of a bird ringed as a chick in Kincardine (Scotland).

Grey Plover *Pluvialis squatarola*

Cwtiad Llwyd

A widespread and common winter visitor to Welsh estuaries, and autumn and spring passage migrant.

Although it has been well distributed around the Welsh coast throughout the 20th century, there has been a marked increase in numbers in the past twenty years or so, particularly apparent since the early 1970s. The average January count for Welsh estuaries for the years 1971–74 was 730 but by 1986–90 this number had risen to 2344. This is in line with a massive increase in the British wintering population, the January index in 1989 approaching four times its 1973 level. In addition the Winter Shorebird Count, December 1984/January 1985, found 357 on non-estuarine coasts in Wales.

Average winter peaks of more than 100 Grey Plover for the years 1986–90 occurred at the following sites:

Dee estuary (Cheshire/Flintshire)	1560
Burry Inlet	743
Severn (Monmouthshire)	530
Swansea Bay	130
Cleddau	128

The Grey Plover is very much a coastal species and it is very rare inland; there have been seven records in Breconshire but none from Radnorshire and Montgomeryshire. There are no ringing recoveries to indicate the location of the breeding grounds of the birds that winter in Wales, but birds occurring in the west Palearctic are known to come from the breeding population of northern Russia.

Sociable Plover *Chettusia gregaria*

Cwtiad Heidiol

A vagrant with just one record for Wales.

20–21 October 1984 – One, south of the River Neath, Swansea (Glamorgan)

The species breeds from the Volga River eastwards through Kazakhstan and winters in Sudan, Ethiopia, Iraq and Pakistan.

Lapwing *Vanellus vanellus*

Cornchwiglen

The Lapwing is a declining breeder in all counties, with a patchy distribution. It is also a winter visitor from other parts of Britain and Europe, principally in coastal areas and especially when there is hard weather to the east.

In the 19th century and early years of the 20th century, Lapwings were described as a very common and numerous breeding species in all parts of Wales from sea level up to 750 m; it was even said to be "very abundant all over Glamorgan" (Heathcote *et al.* 1967). Sadly, it now has a very patchy and local distribution, with few colonies of more than two or three pairs. The decline started, in some areas at least, as early as the 1920s. The 1987 BTO Lapwing survey in a sample of Welsh tetrads revealed 181 pairs in 149 tetrads, giving an estimated total population of around 7500 pairs (Shrubb & Lack 1991). Marked differences in densities were apparent in different parts of Wales, with birds particularly scarce in west Wales (only 0.8 pairs per 1000 ha) and most numerous in the south (Monmouthshire and Glamorgan, 7.5 pairs per 1000 ha) and north (Anglesey, Flintshire and Denbighshire, five to seven pairs per 1000 ha). Radnorshire, Breconshire and Montgomeryshire held only 1.8–4.7 pairs per 1000 ha.

One of the preferred breeding habitats is lowland wet grassland, and surveys of the breeding waders of wet meadows carried out in 1982 and 1989 showed a drastic decline in Wales. On a sample of 22 sites surveyed in both years the numbers declined from 152 pairs to 46 pairs, a decline of 70%. Shrubb (1988) estimated a population of 630 pairs in Breconshire in 1987 (three pairs per 1000 ha) but later noted that this was an over-estimate. Results of field work in Breconshire for the *Breeding Atlas* between 1988 and 1990 suggested that there were then only about 200 pairs. From these data and others it is clear that by 1992 the declining breeding population of Lapwing is substantially less than the 7500 pairs estimated for 1987 and is probably now much less than 1000 pairs.

Most authors accept that changes in agricultural practices have brought about the Lapwing's demise. Although arable land is not a major habitat for this species in Wales it was important in parts of south Wales. Here the switch from spring-sown to autumn-sown cereals has meant that vegetation is too tall in the early spring to be suitable for Lapwings. Elsewhere, the change from cattle to sheep and the huge increase in sheep numbers and stocking densities over the last two decades has created uniform closely cropped and well-drained rye-grass swards. These swards are not favoured for breeding, nests being very conspicuous and also subject to heavy losses from trampling by stock as well as from predation. BTO nest record cards show a clear link between desertion and stock numbers. High stock numbers cause Lapwings to desert potential nesting areas or to reject them. Surveys by Bain (1987) in 1984 and 1985 on moorland and pasture on Mynydd Hiraethog showed that Lapwings were associated with the mosaic of damp, enclosed pastures, ploughed land and damp coarse grassland and avoided the dry improved pastures.

The most notable feature of the current distribution of breeding Lapwings is its comparative rarity in Pembrokeshire where there are only 70 pairs (nearly 40 of which are on Ramsey Island). In contrast, the *Gwent Atlas* gave an estimate of 1000 pairs; a survey of Llŷn (Caernarfonshire) and Anglesey in 1986–87 (RSPB 1988) gave 350–480 pairs, and a survey of most of Radnorshire and the eastern half of Montgomeryshire in 1986–87 (McFadzean & Tyler 1987) gave 245 pairs.

Throughout Wales breeding densities of Lapwings are very low when compared with other parts of Britain. For example, O'Connor and Shrubb (1986) recorded an average of 37 pairs of Lapwing per 1000 ha on intensively managed grassland of dairying or mixed arable/dairying regimes whereas Smith (1983) on certain favoured wet grassland sites recorded much higher densities (170 pairs per 1000 ha in the Ouse, 190 pairs per 1000 ha on the Nene). In the Radnorshire/Montgomeryshire study (McFadzean & Tyler 1987) the average density was 1.3 pairs per 1000 ha, although on rough pasture at Trawsfynydd (Merioneth) Lapwings occurred at a relatively high density of 40 pairs per 1000 ha in 1987 (RSPB 1988).

The Lapwing is much more numerous in winter particularly on coastal pastures and marshes. Post-breeding flocks of local breeding birds start to build up in June but it is not until October that large numbers are recorded. Ringing recoveries show that the winter influx involves, principally, continental birds: 11 birds ringed in the period May/June in Belgium, West Germany, Netherlands, Denmark, Norway, Sweden and Finland were recovered in Wales in the period September to March. An adult ringed in Flintshire in February, was recovered in the USSR, 1800 km to the east in April 1980.

The BoEE counts for the period 1986–90 showed that the average January count in Wales was 12 292, principally on the estuaries of the Dee, Burry Inlet, Carmarthen Bay and Cleddau. This figure gives, however, a far from full picture of the total present in Wales since many of the regular flocks, up to 4000 or more strong, are on favoured traditional fields near the coast, especially in Anglesey, Carmarthenshire, south Caernarfonshire, Denbighshire, Pembrokeshire and Monmouthshire. In the last county there was a huge flock of 15 000 in fields at Magor and Caldicot in December 1970. There are no definite data to support this suggestion but the general impression is that these large flocks have declined markedly in recent years; wintering counts would be very welcome to prove or disprove this suggestion.

Cold winter weather to the east drives many additional Lapwings to Welsh coastal areas, the arrival often preceding the onset of cold weather in Wales by a day or so; hard weather on the Welsh coast forces many of these farther west to Ireland (e.g. c8900 over Kenfig dunes in three hours on 27 December 1965, 6000 moving west over Abergele (Denbighshire) on 28 January 1973 and 1150 passing over the Smalls (Pembrokeshire) towards Ireland in February 1983). Wintering numbers are declining quite markedly in inland counties; for example the figures for wintering flocks in Breconshire represent less than 20% of the levels recorded in the 1960s and 1970s.

Ringing recoveries of young Welsh birds show that they move principally to the south in winter with recoveries from France (4), Spain (5) and Portugal (2); there is only one recovery from Ireland.

Knot *Calidris canutus*

Pibydd yr Aber

The Knot is an abundant winter visitor to some of the major Welsh estuaries, notably the Dee, Burry Inlet, Traeth Lafan and Severn, but relatively scarce on other estuaries. It is also an uncommon passage migrant in spring and autumn and it is rare inland.

Historical accounts (e.g. Forrest 1907 and Cardiff Naturalists Society 1900) show that the Knot was reasonably common at the turn of the century, Forrest noting for North Wales that the flocks often reached 200 or more but the Glamorgan account referred to "small numbers" on the Cardiff mud-flats. Numbers have certainly increased since that period with flocks numbering many thousands being normal at the present time in favoured intertidal areas, dating back at least as far as 1948 on the Burry Inlet and the 1930s on the Dee.

Data from the BoEE counts show that there has been a decline in numbers wintering on Welsh estuaries from the Dee (Cheshire/Flintshire) to the Monmouthshire Severn in recent years. Mean winter counts were as follows:

1969–71	30 623
1972–76	44 880
1977–81	25 462
1982–86	24 455
1987–91	25 574

The Enquiry has shown that since the late 1960s there has been a decline in the total number of Knots of Greenland and Canadian origin. Peak numbers in Britain and Ireland of 400 000 in 1971–72 declined to 230 000 in 1973–74 but have now recovered a little to 300 000 in 1989–90 and 1990–91. The Welsh numbers amount to more than 7% of the East Atlantic Flyway wintering population.

Average winter peaks on Welsh estuaries for the years 1986–90 numbered more than 500 birds for the following estuaries:

Dee (Cheshire/Flintshire)	21 158
Burry Inlet	4243
Severn (Monmouthshire)	1687
Carmarthen Bay	769
Traeth Lafan	690

Only two other estuaries (Taff/Ely and Cefni) had averages of over 100 birds in the same period, indicating how wintering flocks of this gregarious species concentrated in a very few areas. Maxima occur, in recent years, in December on the Dee and Carmarthen Bay, January at Traeth Lafan and the Burry Inlet, and February on the Severn. The average January count for the years 1987–91 in Wales, including the whole of the Dee (i.e. English and Welsh sides) as one unit, amounted to more than 17 000 birds which is nearly 8% of the total British wintering population.

Virtually all the Knot that winter in the United Kingdom are the subspecies *C. c. islandica,* which breed in north-east Canada and Greenland. There have been six recoveries of birds ringed in the period August/February in Wales and recovered in Greenland at locations between 68°N and 77°N. Recoveries in Wales show that there are staging areas in autumn in Iceland and the tip of south-west Norway. Recoveries also show that at least some birds pass through the Wash prior to moving west to Wales for the winter. Not all arrivals in Wales follow this route, however, since in at least one instance (a bird ringed in Iceland on 28 July 1972 and recovered at the Point of Air, Flintshire on 12 August 1972) it appears likely that a direct flight from Iceland to Wales was undertaken.

Recoveries in Wales of 12 birds ringed in Iceland in May and nine birds ringed in Wales and recovered in Iceland in May indicate the importance of Iceland as a staging area in spring. Recoveries of Welsh ringed birds in northern England in March (6), April (7) and May (7) indicate the route for the northward spring exodus; Morecambe Bay records peak numbers prior to departure northwards in early May as birds build up their fat reserves for the flight. A recovery of a Welsh ringed Knot in north Norway in late May (69°N) at first sight appears likely to refer to a bird from the Siberian breeding population, but studies have shown that birds in north Norway in May are in fact of Nearctic origin.

Sanderling *Calidris alba*

Pibydd y Tywod

A winter visitor, locally common on sandy beaches in Flintshire and Carmarthenshire/Glamorgan and an autumn and spring passage migrant on the coast; it is scarce inland.

The status of the Sanderling in Wales appears to have changed little in the past one hundred years or so although the indications are that it is more numerous now in the winter than it was at the turn of the century. In the three main wintering areas recorded now it was noted as "occasional" in Carmarthenshire (Barker 1905), "not very common"' in Glamorgan (1900) and "some few spend the winter months here" in North Wales (Forrest 1907). As a passage migrant the statement by Forrest (1907) that it "is met with generally in flocks or small parties on the flatter parts of the coasts and estuaries" holds good at the present time. Peak numbers usually occur in the period October/November with a pronounced, but lesser, spring passage during May.

The Sanderling's distribution in Wales emphasises its dependence on long sandy shores and is markedly concentrated on three distinct areas, one in the north and two in south Wales. In north Wales the favoured area stretches from the Dee estuary to the Clwyd estuary and the coastline from the Clwyd towards Llandulas (Denbighshire) with average winter peaks in the years 1986–90 of 574, 172 and 255 birds respectively. In south Wales a very important wintering area has been identified on the wide sandy beach at Cefn Sidan (Carmarthenshire), now recognised as the second most important estuarine wintering site in Britain, after the Ribble. A substantial wintering population is resident at Cefn Sidan throughout the winter with peak numbers of 1000 or more on occasion (Prys-Jones & Davis 1990). Some of the Cefn Sidan birds may well move at times to nearby sandy beaches at Pendine Sands to the west and Whiteford to the south east. The third major wintering area is Swansea Bay, especially near Blackpill and the Ogmore estuary where respective average winter peaks (1986–90) were 262 and 107. Other than in the three areas referred to above Sanderlings are very scarce in winter.

The Sanderlings wintering in Britain are believed to consist primarily of birds that have bred in Siberia but an unknown (though probably small) proportion are from the population breeding in Greenland. Although there are many recoveries of birds ringed in Wales, they throw little light on natal areas in relation to wintering, relating mainly to movements between passage sites in Wales and winter sites in Africa (ten birds ringed on passage in Wales recovered in the period August/March from the African coast between Morocco and Ghana). One bird ringed at the Point of Air (Flintshire) on 16 May 1972 was recovered in the USSR at Murmansk on 16 July 1974 proving that there is at least a connection with the Siberian population. A bird ringed in Iceland on 14 May 1972 and recovered at the Point of Air on 11 May 1975 suggests a connection with the Greenland breeding population.

Semipalmated Sandpiper *Calidris pusilla*
Pibydd Llwyd

A vagrant.

There are two Welsh records of this small wader, which breeds in northern North America and winters in Central America south to Ecuador and Brazil.

20–21 July 1964 – An adult moulting into winter plumage, Skokholm. It was on a drying up pond, with a Dunlin and was trapped on the 21st

6–17 September 1990 – Juvenile on Ogmore estuary (Glamorgan)

This is a very difficult species to identify but the number of records from Britain and Ireland, totalling 64 up to 1991, has risen considerably in the 1980s thanks to the advances in identification skills and improved knowledge of the characteristics that separate this from similar species.

Little Stint *Calidris minuta*
Pibydd Bach

An uncommon passage migrant in small numbers to coastal areas of Wales, principally in autumn, and has occasionally wintered. It is rare inland.

The status and distribution of Little Stints does not appear to have changed substantially in the past ninety years except that there has been an increase in records from the 1960s.

In 1960 there was an unprecedented invasion of Little Stints in Britain including Wales, just as there had been of Curlew Sandpipers the year before (Ferguson-Lees & Williamson 1960). The total number in Britain reported from peak days was about 3000 and the main invasion took place from about the middle of September, when exceptional numbers appeared all along the east coast of Scotland and England. At Shotton Pools (Flintshire) there was a massive influx of about 350 on 22 September associated with about 80 Curlew Sandpipers (Raines 1961). There were still 20 present on 24 September and big flocks up to the second week of October on freshwater pools. Subsequently up to five birds wintered at Shotton, for the third year in succession.

At this time, too, and in early October less spectacular but still unusual numbers were noted in the rest of Wales from Anglesey, Caernarfonshire, Cardiganshire, Pembrokeshire and Glamorgan. On Skokholm, for example, where the species had rarely been recorded there were four on 17 September, two on the 18th and one on the 19th and 20th, with five at the end of the month.

An indication of the southward movement of Little Stints that pass through Wales in autumn is provided by the recoveries of two birds ringed at Shotton in the 1960 influx; the first recovered on 16 October the same year in south-west France and the second in February 1967 in Morocco. There have been no recoveries on the breeding grounds, which range from northern Scandinavia to Siberia.

Little Stints are very rare inland with only ten records from Breconshire (all from Llangorse Lake since 1966), two from Radnorshire and one from Montgomeryshire.

Temminck's Stint *Calidris temminckii*
Pibydd Temminck

A rare visitor to Wales with a total of 20 records involving 25 individuals up to 1991. It breeds from northern Scandinavia eastwards across northern Siberia and winters in west and central Africa and south Asia.

Records cover the months of March (1), April (1), May (6), July (1), August (8) and September (8) and are from the northern counties of Flintshire, Denbighshire and Anglesey and the southern counties of Pembrokeshire, Glamorgan, and Monmouthshire. The most favoured localities have been Malltraeth Pool in Anglesey and the Connah's Quay reserve and Shotton Pools in Flintshire although even at these sites occurrences have, of course, been very irregular and spasmodic.

Least Sandpiper *Calidris minutilla*
Pibydd Lleiaf

A vagrant.

There is just one record in Wales:

2 September 1972 – One, Aberthaw (Glamorgan)

This diminutive wader breeds in northern North America and winters from the USA south to Peru and Brazil.

White-rumped Sandpiper *Calidris fuscicollis*
Pibydd Tinwen

A vagrant to Wales.

There are six records, dating from 1957:

25 September 1957 – One, Blackpill (Glamorgan)
19 March 1970 – One, Swansea Bay (Glamorgan)
12–26 October 1974 – One, Pembrey/Kidwelly (Carmarthenshire)
25–27 September 1975 – One, Pembrey/Kidwelly (Carmarthenshire)
20 September 1977 – One, Gann estuary (Pembrokeshire)
30 July–8 August 1984 – One, Shotton Pools (Flintshire)

It is perhaps slightly surprising that there are not more Welsh records of this trans-Atlantic visitor in view of more than 300 seen in Britain and Ireland.

Baird's Sandpiper *Calidris bairdii*
Pibydd Baird

A vagrant.

There are nine records from Wales:

2–10 October 1967 – One, Gann estuary (Pembrokeshire)
22 October–6 November 1971 – One, Point of Air
9 September 1972 – One, Aberthaw (Glamorgan)
31 August 1975 – One, Whiteford Point (Glamorgan)
17 October 1976 – One, Tywyn Point (Carmarthenshire)
7 August 1979 – One, Connah's Quay (Flintshire)
10 August 1984 – One, Nevern estuary (Pembrokeshire)
15 September–1 October 1984 – One, Dale airfield (Pembrokeshire)
17 October 1987 – One, Pembrey Harbour (Carmarthenshire)

Most of the Welsh records are outside the September peak of British and Irish occurrences, which total 160 up to 1991.

The species breeds from north-east Siberia eastwards across northern North America to Greenland; it winters in South America north to Ecuador.

Pectoral Sandpiper *Calidris melanotos*
Pibydd Cain

A scarce visitor to Wales.

This species breeds in north-east Siberia and northern North America and winters in southern South America. On the basis that most of the visitors to Wales probably originate in North America this is by far the most frequent and widespread of the trans-Atlantic waders in Wales with 67 records (totalling 71 individuals) since the first record on 23 September to 12 October 1958 (Skokholm).

Records cover all months from April to November inclusive with a marked peak in mid-September (see histogram). Two-thirds of the records refer to birds that stayed at the same location from one to three days but on eight occasions the stay has been for two weeks or more in suitable freshwater habitats.

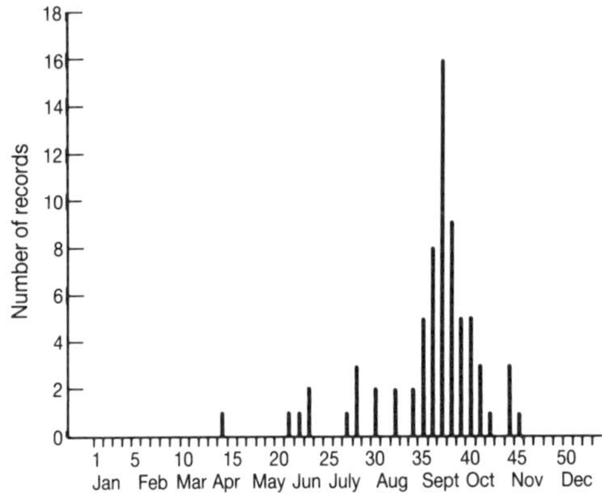

Weekly records of Pectoral Sandpiper to 1991

The distribution of the occurrences is quite wide, involving 23 sites with the majority from Pembrokeshire (28 records), Anglesey (11) and Glamorgan (12); other counties are Flintshire (5), Caernarfonshire (4), Cardiganshire (2), Carmarthenshire (1), Breconshire (1) and Monmouthshire (3).

Sharp-tailed Sandpiper *Calidris acuminata*
Pibydd Gynffonfain

A vagrant to Wales.

There are two Welsh records of this species, which breeds in north-east Siberia and winters in Australasia, both for the same first date:

14–25 October 1973 – One, Shotwick Fields (Flintshire)
14–15 October 1973 – One, Morfa Harlech (Merioneth)

There are only 20 records in all for Britain and Ireland.

Curlew Sandpiper *Calidris ferruginea*
Pibydd Cambig

An uncommon passage migrant to coastal areas of Wales, recorded in every month but principally in autumn, relatively numerous in some years (notably in 1959–60 and 1969). It is rare inland.

Forrest (1907) described the status of the Curlew Sandpiper in North Wales as an uncommon passage migrant and occasional winter visitor, highlighting especially the estuaries of the Dee, where it was not uncommon in autumn and the Dyfi where it was said to occur in varying numbers from year to year, usually in September. This description holds good today. Elsewhere in Wales, by contrast, it was a very scarce visitor to the coast up to the 1950s but the number of records have substantially increased since that time, partly perhaps as a result of increased coverage by observers.

Most records refer to single birds or small parties up to ten but there were exceptional numbers recorded at the head of the Dee estuary adjacent to Shotton Iron Works (Flintshire) in 1959. On 3 September a total of about 60 birds came in to the freshwater pools at high tide and there were large numbers amongst the small waders massed on the marshes during the tide. It was impossible to judge numbers but at least 200 birds must have been present (Raines 1960). Numbers at Shotton were also very large in 1960 with a peak of about 80 on 22 September.

Wales also shared, albeit to a relatively minor extent, in the quite unprecedented migration of Curlew Sandpipers through the British Isles in autumn 1969. Stanley and Minton (1972) have analysed the records relating to that period and showed that at least 3500 Curlew Sandpipers appear to have been present in Britain and Ireland on 31 August. Peak numbers in Wales were somewhat later than on the North Sea coast, as one would expect from a migrant from the east, reaching a maximum of about 200 during the period 9–15 September. Nine sites were involved in north Wales, from the Dee to the Dyfi estuary and five in south Wales in Glamorgan and Monmouthshire. Maxima were 20 at Blackpill (Glamorgan) on 13–14 September, 24 on the Dyfi on 5 September and 70 at Malltraeth on Anglesey (always a good locality for this species in autumn) on 10 September.

Curlew Sandpipers breed in northern Siberia and their main wintering grounds are the coasts of Africa south of the Sahara, India, Ceylon, Malaysia and Australasia. Stanley and Minton considered that the probable cause of the unprecedented passage of 1969 was a particularly persistent and complex low-pressure weather system over northern Europe, which coincided with the main migration of juveniles, possibly following a good breeding season. In overcast conditions large numbers were deflected west over the Baltic and northern Scandinavia and then south-west across the North Sea into Britain and Ireland.

Another good year was 1988, with records from most coastal counties and peaks of 66 at Malltraeth on 18 September and 21 at Broadwater (Merioneth) on 9 September.

Curlew Sandpipers are rare inland in Wales and the inland counties can only muster a handful of records between them. Of eight records from Breconshire, five are in September, which is clearly the peak month for occurrences in Wales.

Purple Sandpiper *Calidris maritima*
Pibydd Du

A fairly common winter visitor in small numbers to coastal areas with rocky shorelines, especially the North Wales coast from Rhos Point (Denbighshire) to Bardsey and south Pembrokeshire to the Gower peninsula.

It has always been very local in distribution owing to its feeding preference for the tidal edge of rocky shores but it shows a strong tenacity for favoured wintering sites. The map shows sites where maxima of ten birds were recorded during the period 1970–90. Several of these refer to island sites (Puffin Is., Bardsey, Skokholm, the Smalls and Grassholm) and others to long-established mainland sites (e.g. Rhos Point in Denbighshire, Aberystwyth Harbour in Cardiganshire, Trearddur Bay in Anglesey, Rhossili and Sker Point in Glamorgan).

For the period 1968–74 Atkinson *et al.* (1978) calculated average winter maxima of 229 birds in Wales although recog-

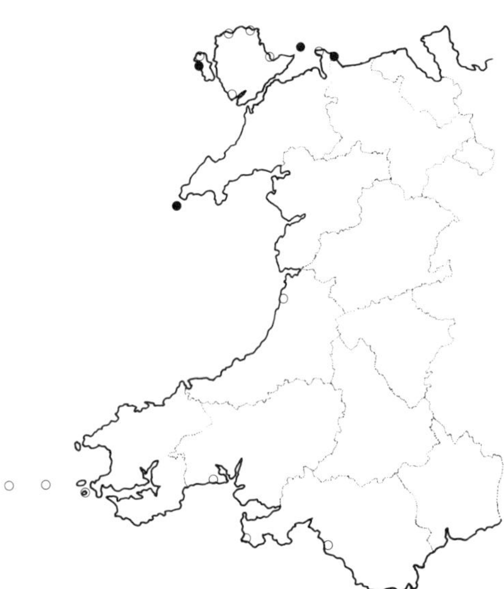

Purple Sandpiper – sites with maxima of 10+ birds 1970–90, ○ =10–49 pairs, ● =50+ pairs

nising that observer coverage may have been far from complete. The total was made up as follows: Anglesey (50), Caernarfonshire (70), Cardiganshire (10), Denbighshire (20) Flintshire (2), Glamorgan (67) and Pembrokeshire (10).

The evidence suggests that the contemporary figure is similar to the above calculation, increasing substantially in Anglesey, but declining in Glamorgan. The Winter Shorebird Count of December 1984 to January 1985 gave a figure of 162, in Wales, of which only five were in Glamorgan (Moser & Summers 1987).

The recent history of Purple Sandpipers on Bardsey is well documented (Roberts 1985). It was uncommon there in the 1950s, with spring maxima of just 10 birds in some years and often just a few singles in autumn. This increased to more substantial flocks of 20–30 in spring and autumn in the 1960s and continued to the present high levels since the mid-1970s, i.e. flocks of 20–40 are frequent and influxes increase the population up to 100 at times in both spring and autumn. Winter numbers are generally more stable at 30–50 birds. Similarly, likewise the population at Rhos Point substantially increased to a maximum of 73 on 10 November 1979. In Pembrokeshire the Winter Shorebirds Count in January 1985 recorded 25 birds on the mainland and at the same period there were up to 38 on the Smalls. An enthusiastic observer at the Smalls lighthouse from 1983 to 1985 discovered fluctuating numbers during the period August to May each year with maxima of 47 on 26 March 1982 and 1 April 1984.

Purple Sandpipers are recorded from all coastal counties but are scarce in Flintshire and Monmouthshire, where the shorelines are generally unsuitable for this species. There have been no ringing recoveries from Wales to prove that those birds which winter in Wales come from Greenland and/or Canada but there is one recovery of a bird ringed on Hilbre Island (Cheshire) in Greenland which tends to support this suggestion; the Icelandic breeding population has been shown to be mainly resident (Purple Sandpipers winter farther north than any other Wader).

Dunlin *Calidris alpina* **Pibydd y Mawn**

The Dunlin is a scarce and local breeder in Wales, restricted to a few poorly drained upland moors with scattered pools. It is also an abundant passage migrant and winter visitor, principally to coastal areas.

Records from the past one hundred years or so show that the Dunlin has probably never been a common breeding species in any part of Wales and it appears not to have bred in Glamorgan, Carmarthenshire, Pembrokeshire and Anglesey. There has, however, been a marked contraction of range, notably in Merioneth where at the turn of the century it was breeding at several widely scattered sites (e.g. near Trawsfynydd, Arenig, Llanderfel, Bala and Dinas Mawddwy) but also in parts of Monmouthshire, in Cardiganshire (e.g. Cors Caron) and in Radnorshire (e.g. Aberedw Hill). Coastal breeding was proved in Merioneth in 1953 and in Flintshire in 1961 (it nested on the Dee Marshes in the 19th century).

In view of the contraction of range, the overall breeding numbers must have changed quite significantly this century. The "stronghold", as with the Golden Plover, is the Elan/Claerwen catchment and adjoining areas of Cardiganshire. Surveys of this area found 15 pairs in 1976–77 (Elan/Claerwen catchment only), 41 pairs in 1982, 28 pairs in 1990 and 37–40 pairs in 1991. Dunlin numbers in this study area fluctuate considerably, a fact which may be linked to the level of rainfall and hence the wetness of the breeding grounds (Hack 1991). Breeding is confined to areas where the substrate is always wet and there are large puddles or pools of standing water.

Away from the "stronghold" there are small breeding populations on Mynydd Hiraethog, Migneint (Caernarfonshire/Merioneth), Berwyn, Mynydd Llangynidir/Mynydd Llangattock (Breconshire) and, until recently at least, on Llanbrynmair Moors

(Montgomeryshire). It appears that the present breeding population numbers from 50 to 70 pairs.

The Dunlin vies with the Oystercatcher as the most abundant wader wintering on the Welsh estuaries, the January average of the years 1987–91 amounting to more than 54 000. This is more than 4% of the East Atlantic Flyway Population in winter.

The mean winter peaks show wide fluctuations:

1969–71	41 870
1972–76	72 982
1977–81	45 269
1982–86	41 577
1987–91	55 305

As well as being one of the two most abundant species, the Dunlin is also the most widespread of the wintering waders in Wales and is regularly recorded on all estuaries and on non-estuarine coasts. The Winter Shorebird Count found more than 2000 birds on non-estuarine coasts in Wales in December 1984/January 1985 (Moser & Summers 1987). In several areas they occur throughout the year but they are relatively scarce in the period between the end of May and the end of June. Average winter peaks of more than 1000 in the years 1986–90 occurred at the following sites:

Severn (Monmouthshire)	26 458
Dee (Cheshire/Flintshire	17 588
Burry Inlet	7731
Taff/Ely (Glamorgan)	4800
Traeth Lafan (Caernarfonshire)	4718
Cleddau	3743
Swansea Bay (Glamorgan)	2131
Carmarthen Bay	1382
Inland Sea (Anglesey)	1346
Conwy (Caernarfonshire)	1129

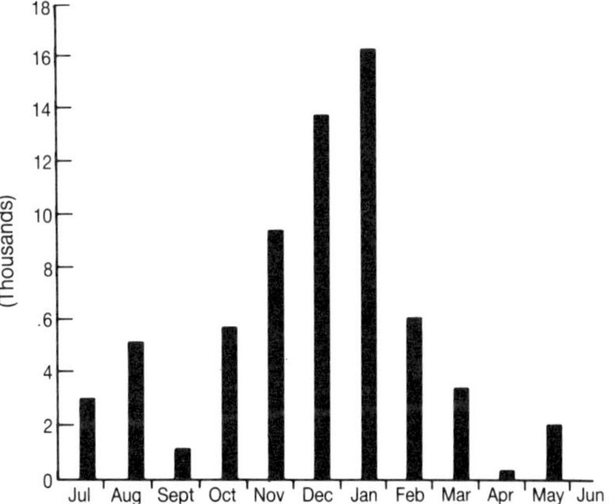

Average monthly counts of Dunlin on Dee, 1986–90

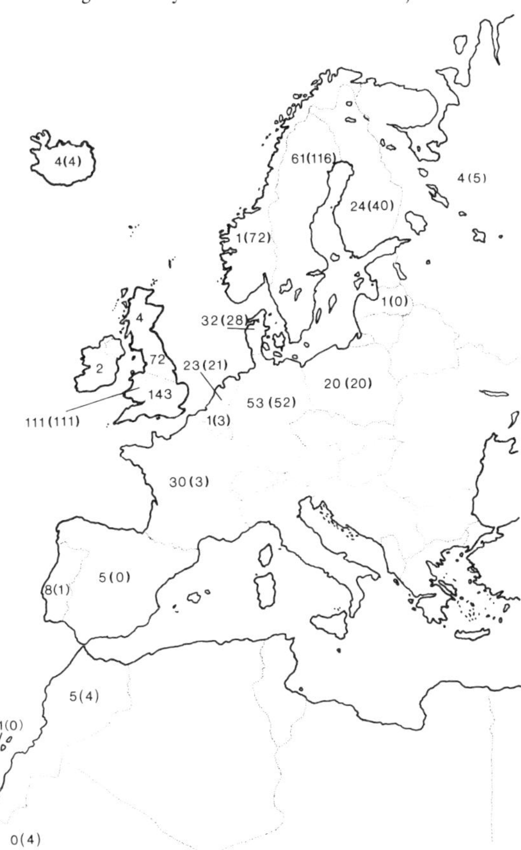

Dunlin – analysis of ringing recoveries. Open figure refers to the birds ringed in Wales, figures in parentheses to birds ringed abroad and recovered in Wales

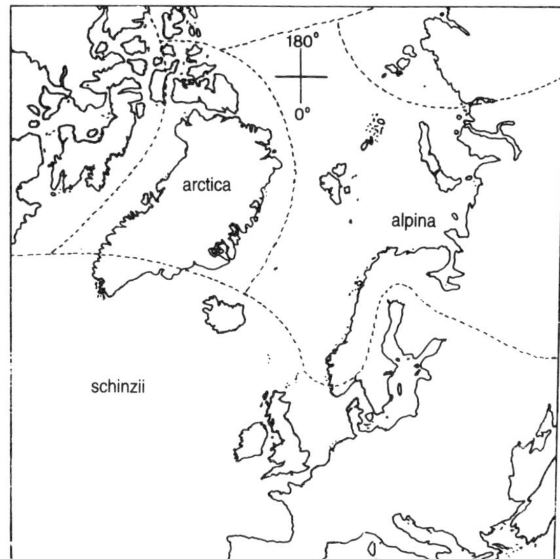

Breeding range of sub-species of Dunlin

It is only since the BoEE started in 1969–70 that it has been possible to quantify the numbers of waders using the Welsh shoreline. All that can be said with any certainty about the status of the Dunlin is that it has been a common visitor to all coastal areas throughout the present century. Inland it is a regular spring and autumn passage migrant to all counties in small numbers, also occurring occasionally in winter.

Ringing studies have revealed the Welsh wintering

population to be birds of the northern race *alpina*. Their breeding areas are in northern Scandinavia and north-west Russia, and a Welsh ringed bird has been recovered as far east as the Yamal peninsula in the former USSR at 71°E. This race starts to arrive in Wales in late September having moulted first on sites such as the Wash (England) or Waddenzee (Germany/Netherlands), and begins to depart from February onwards.

The race *schinzii* breeds in south-east Greenland, Iceland and southern Norway southwards and includes the Welsh breeding birds. Birds arriving in Wales in late June to September are mostly of this race, many coming from the Icelandic breeding population. The bulk of this population moults and winters in Morocco and Mauritania, moving south through France, Portugal and Spain.

The race *arctica* breeds in north-eastern Greenland and is believed to occur in Wales only as a passage migrant. The race is smaller and paler than *schinzii*.

During spring all three races pass through Wales in breeding plumage, as shown by ringing data from the Severn and the Dee (Ferns & Green 1979; Eades & Okill 1977). Retrap data of ringed birds showed that individuals of *alpina* retrapped on the Dee in May were mainly birds that had wintered there, especially juvenile and second-year birds. The presence of *arctica* in the May passage was confirmed by recoveries of Dee birds in north-east Greenland; *schinzii en route* to Iceland are closely associated with *arctica* in this movement. The figure gives the average monthly counts of Dunlin on the Dee for the years 1986–90. This clearly shows the August peak of *schinzii*, followed by a January peak of *alpina* and a May passage principally of *arctica* and *schinzii* from the Icelandic breeding population. This pattern is largely replicated on other Welsh estuaries, the spring peak being particularly notable in some years on the Severn (Ferns & Green 1979). On some estuaries the peak numbers occur in February or December rather than January but this can vary from year to year.

The maps show the breeding range of the three races of Dunlin that occur in Wales, and analyses of recoveries of birds ringed abroad and recovered in Wales and of birds ringed in Wales.

Broad-billed Sandpiper *Limicola falcinellus*
Pibydd Llydanbig

A vagrant.

Four confirmed records give this species a place on the Welsh list:

22 September 1960	– Two, Shotton Pools (Flintshire)
7 May 1979	– One, Peterstone Wentlooge (Monmouthshire)
4–6 June 1984	– One, Malltraeth (Anglesey)
15 May 1988	– One, Sluice Farm (Monmouthshire)

Broad-billed Sandpipers breed in Norway, Sweden, northern Finland and northern Siberia, wintering mainly in south Asia and Australia. The pattern of British records reveals a distinct peak on spring passage in late May.

Buff-breasted Sandpiper *Tryngites subruficollis*
Pibydd Bronllwyd

A rare visitor but the second commonest trans-Atlantic wader in Wales, after Pectoral Sandpiper.

It breeds in Alaska and north-west Canada and winters in northern Argentina and Uruguay. It has a particular preference for a dry grassland habitat.

The first record from Wales was one on Bardsey on 21 September 1968. Since then there have been 25 further records involving 30 individual birds. These have all been in the last two days of August or in September, with the exception of one at Tywyn Point (Carmarthenshire) on 26–29 April 1975, one on Skokholm on 3 June 1985, one which joined the Sharp-tailed Sandpiper at Shotwick Fields in 1973 and was present from 18 to 28 October and one at Mathern Pill (Monmouthshire) from 23 to 29 May 1990, which attached itself to a pair of Lapwings with their three chicks.

The records all refer to single individuals except for five at Dale airfield (Pembrokeshire) on 20–22 September 1984 (then three on the 23rd, two from 24th to 30th) and two at the same favoured locality on 9 September 1985. Most stays are of short duration but one individual was at Cemlyn (Anglesey) intermittently from 21 September to 17 November 1977.

Only two coastal counties (Merioneth and Denbighshire) are without a record; Pembrokeshire with eight records (13 individuals) is the most prominent.

Ruff *Philomachus pugnax*
Pibydd Torchog

A fairly common passage migrant in small numbers, principally in autumn, to coastal and inland areas throughout Wales; it occasionally winters.

Ruffs were very scarce in the early years of the century, chiefly recorded from the estuaries of the Dee and Dyfi on autumn passage. Numbers began to increase slightly in the 1930s and particularly from the 1950s onwards, reflecting a widespread increase over Britain as a whole. In

Carmarthenshire, for example, where the first record was not until 1931, there were two further records in the 1930s, two in the 1950s and it has been recorded almost annually since 1966 including wintering flocks. In 1987 there was a notable spring passage at the most favoured locality, Morfa Coedbach, Cydweli (Carmarthenshire), peaking at 70 birds on 18 April (the highest number recorded in Wales).

Although there was an isolated winter record from Glamorgan as early as 1937 it was not until the period 1960–65 that very small numbers began to winter in Wales, there being one to three birds recorded in those six years (Prater 1973). This was a period when unprecedented numbers began to winter in Britain, principally in southern and eastern England. By 1966–71 there was an average winter population of 26 in Wales out of a total of more than 1200 in Britain and Ireland; in Wales the most favoured locality was Shotton Marshes in Flintshire, with very small numbers in Anglesey, Caernarfonshire, Pembrokeshire and Glamorgan.

Wintering has continued in at least some years since 1971. In 1972 up to 22 wintered in Flintshire and 11 in Glamorgan but numbers in those counties and other wintering areas (Cardiganshire, Pembrokeshire, Carmarthenshire, Merioneth and Anglesey) have been very small in recent years.

Most records are in the period August to September but the spring passage can be very marked: 1987 was a particularly good spring, with considerable numbers noted at Cydweli. There was also a substantial movement through Pembrokeshire at this time, with peaks of six at Dale Airfield and Skomer, 12 at Skokholm, 32 at the Gann and 46 at Marloes Mere from 15 to 27 April. There were also seven on the Dyfi on 17 April and four on the Rhymney estuary on the same date confirming the widespread nature of the movement. Autumn passage involves maxima of up to 30–35 birds in good years (e.g. 35 at Shotton, Flintshire, in 1974 and 30 at Llyn Alaw, Anglesey, in 1977).

In 1970 a male and two females may have bred at a site in south-west Anglesey, where a few birds were wintering regularly. This was a period when there was a relatively strong breeding population on the Ouse Washes in eastern England.

A favourite habitat of Ruffs is the muddy margins of freshwater lakes and pools; sites such as Llyn Alaw (Anglesey) and Cors Caron are typical in this respect. It is not, therefore, surprising that comparatively few are recorded on the BoEE counts although there is a handful of records from 15 count sites in the period 1981–90, chiefly from August to October.

Few Ruffs are ringed in Wales but the recoveries of two individuals ringed at Shotton (Flintshire) give some clues as to the breeding and wintering areas of Welsh birds. One ringed on 17 August 1962 was recovered on 16 May 1966 some 1424 km to the north-east in Sweden, which is within the species' breeding range and one ringed on 31 August 1962 was recovered on 23 September 1962 in Morocco some 2242 km to the south, possibly *en route* to the major wintering area in West Africa.

Jack Snipe *Lymnocryptes minimus*

Gïach Fach

A fairly common passage migrant and winter visitor to all counties.

At the turn of the century records indicate that it was a common winter visitor in most districts from its breeding grounds in north-east Europe. For example in Pembrokeshire the Jack Snipe was regarded by Mathew (1894) as a "fairly numerous winter visitor" with occasional flocks of a dozen or more being encountered. Lockley *et al.* (1949) also considered it "fairly numerous" and added that cold spells brought "great numbers" to Pembrokeshire. Likewise in Breconshire where Cambridge Phillips (1882) called it very common and recorded that a shooting party had put up 30–40 in one day in a large snipe-bog near Trecastle in December 1881. In the same county Ingram and Salmon (1957) reported that one shooter had analysed his records, kept for about 60 years, which showed a proportion of one Jack Snipe shot for every three Snipe.

Owing to its crepuscular and nocturnal habits and the fact that it is reluctant to fly until nearly trodden on, it is probably substantially under-recorded at present. Most knowledge of the species numbers and distribution in the 19th century was derived from shooting records. Though it is difficult therefore to be definite, published information suggests that overall numbers are now appreciably less than in the 19th century. In Lack (1986) a sample of birds shot in the early 1980s throughout Britain showed a ratio of one Jack Snipe for every eight Snipe shot, a marked contrast to the Breconshire figure earlier in the century, assuming that numbers in Breconshire were not exceptional. Its favoured habitat is shallow freshwater pools with exposed mud and dense grass cover, a habitat which has been considerably reduced in Wales since the 1940s.

Most contemporary records refer to single birds or small numbers and are in the period October to February; the earliest record was 18 August and the latest 12 May. The distribution is largely coastal, favoured localities being freshwater marshes near the Dee estuary, on Anglesey and in Carmarthenshire. Exceptional influxes have been recorded in hard weather, notably 20 on Bardsey in January 1963, 50+ by the Tywi near Gelli Aur (Carmarthenshire) on 12 January 1969 and c40 at Gweunydd Cochion, Tumble (Carmarthenshire) on 19 December 1977.

Snipe *Gallinago gallinago*

Gïach Gyffredin

The Snipe is a widespread breeding bird, nesting in all counties but in small and declining numbers; it ranges from coastal marshes and lowland bogs to upland wetlands. Substantial numbers from Europe winter in Wales, particularly in lowland wetlands near the coast.

From being described as a common breeding species in several parts of Wales in the late 19th and early 20th centuries, the Snipe is now greatly reduced in numbers and range. Even as early as the first decade of the 20th century Forrest (1907) commented that "several observers record it as a decreasing species" in north Wales. Although quantitative information on breeding numbers is somewhat sparse for much of Wales, the picture that emerges from what is available is that the current breeding distribution is very patchy and confined chiefly to a relatively few favoured hill bogs and some lower marshy ground. It is clear that former breeding areas have been lost to drainage schemes and afforestation.

In the northern counties the principal concentrations are on Mynydd Hiraethog with 14 to 21 pairs (Bain 1987) and on a bog at Trawsfynydd (Merioneth) where an estimated 18 pairs were recorded (RSPB 1988). On upland areas such as Berwyn and Ruabon Moor (Denbighshire) fewer than 10 pairs were recorded in surveys in the 1970s and 1980s. RSPB surveys in 1986–87 reported 30–60 pairs on the Llŷn peninsula and 20–40 pairs on Anglesey.

In mid-Wales the Elan/Claerwen uplands (Radnorshire/Cardiganshire) formed an important site for Snipe with 34–66 breeding pairs in recent years (but declining) as did Mynydd Eppynt (Breconshire) with 12–15 pairs in 1990 (Peers & Shrubb 1991). Elsewhere in Breconshire and Radnorshire Snipe have become uncommon away from a few additional favoured locations such as Maelienydd, Penybont Common and Rhosgoch (Peers 1985).

In south Wales there has been a drastic decline in the Pembrokeshire breeding population; the Pembrokeshire breeding bird survey of 1984–88 (Donovan & Rees, in press.) found only about 10 pairs in bogs of the Preseli Mountains and possible breeding in the meadows around the confluence of the rivers Syfynwy and eastern Cleddau. In Carmarthenshire breeding records have only been received patchily in recent years from coastal marshes at Pendine, Cydweli and Witchett Pool, in mid-county from near Carmarthen and Tumble; in the north at Llanfihangel ar Arth and Cwmann; in uplands at Caeo, Esgairdawe, Rhandirmwyn and Llyn y Gwaith. In West Glamorgan there are small concentrations at Crymlyn Bog and in the lower Neath and Tawe Valleys and it is estimated that the breeding population is probably only in the order of 15–20 pairs although it is likely that the species is under-recorded in the more remote upland areas in the north and east (Thomas *et al.* 1992). In Monmouthshire the breeding population may not exceed 50 pairs (Tyler *et al.* 1987); the boggy areas in the hills in the north-west probably represent the breeding stronghold of Snipe in the county and the few remaining damp cattle-grazed pastures on the low-lying levels give an additional scatter of isolated records.

A reduction of breeding range in Wales is confirmed by a comparison of the *Breeding Atlas* and the *New Atlas*. The comparison shows an overall reduction of nearly 30% in the number of 10-km squares in which breeding was confirmed, in spite of an increase in Anglesey.

It is impossible to give an accurate figure for the breeding population at the present time but it is probably in the order of 300–500 pairs in an average year.

The Snipe is a numerous winter visitor to Wales, principally to low-lying bogs, rushy fields and marshes near the coast. There is a considerable influx in October and small groups scatter widely across coastal areas during the winter, concentrating at places where ground conditions become temporarily favourable. Groups of 50–200 are not unusual but harsh weather conditions cause major concentrations (e.g. 500 on Bardsey Island in January 1963, 1082 at Cydweli Flats, Carmarthenshire on 16 February 1974). When the ground becomes frozen, Snipe move to soft areas like the foreshore and springs, and numbers diminish steadily if the hard weather is prolonged, many presumably leaving the country altogether.

Ringing recoveries in Wales indicate that the winter influx comes from as far away as eastern and central Europe with records from Poland (1), East Germany (3), Scandinavia (7), Denmark (5), the Netherlands (6) and Belgium (2). Of the ten recoveries of birds ringed in Wales, five were local (Wales), two were from France and three from England.

Great Snipe *Gallinago media*

Gïach Fawr

A vagrant in the 19th and early 20th centuries; there are no recent records.

It is impossible now to verify the several records claimed from one hundred years or so ago. Forrest (1907) in listing records for North Wales provides the caveat that "large examples of the Common Snipe are so frequently mistaken for the present much rarer species that it is necessary to receive with caution any records not substantiated by actual specimens . . . they are presented as being probably correct but in most instances it has been impracticable either to verify or disprove them".

It is thus very difficult to determine which records are acceptable by modern identification standards and most recent county avifaunas conclude that claimed records from that county are not completely convincing and cannot be regarded as fully authenticated. Records which can, however, be given serious consideration on the grounds that they refer to shot birds which were examined personally, and it must be presumed, critically by contemporary authorities, include the following:

About 1876	– One shot, Mynydd Eppynt (Breconshire)
About 1876–77	– One shot near Cray (Breconshire)
About 1888	– One shot Plas ucha (Denbighshire)

About 1894	– Two shot locally in collections at Nanteos and Gogerddan (Cardiganshire)
Prior to 1907	– One shot near Betws y Coed (Caernarfonshire)
23 September 1911	– One shot near Holt (Denbighshire)
September 1947	– One shot, Ystrad Meurig (Cardiganshire)

Long-billed Dowitcher *Limnodromus scolopaceus*
Gïach Gylfin-hir

A vagrant.

There are five accepted records as follows:

12 October 1963	– One shot near Nefyn (Caernarfonshire)
12–14 September 1978	– One, Llyn Alaw (Anglesey)
14 September 1985	– One, Peterstone Wentlooge (Monmouthshire)
12 December 1987–3 January 1988	– One, Gann estuary (Pembrokeshire)
9 March–18 April 1989	– One, Sluice Farm (Monmouthshire) on 9 March, Rhymney estuary (Glamorgan) from 10 March

In addition, there are six records which refer either to this species or the very similar Short-billed Dowitcher, most if not all of which are probably the former. The additional Dowitcher sp. records are:

Pre-1899	– One shot, Penllergaer Common (Glamorgan)
19 November 1961	– One, Shotton Pools (Flintshire)
21 June–at least 12 November 1964	– One, Shotton Pools (Flintshire)
16 September–30 October 1967	– One between Rogiet and Magor Pill (Monmouthshire). (At least two birds were involved according to the *Gwent Bird Report*)
29 August 1976	– One, Collister Pill (Monmouthshire)
30 July 1978	– One, Gwendraeth estuary (Carmarthenshire).

Woodcock *Scolopax rusticola*
Cyffylog

A regular breeder in small numbers in all counties except Anglesey and Pembrokeshire; it is commonest in eastern counties. It is widespread and common/abundant in winter.

The breeding history of the Woodcock in the British Isles has been well documented by Alexander (1945–47) through a national inquiry in 1934–35 organised by the BTO and shows that numbers began to increase rapidly in the first half of the 19th century. By the middle of that century the Woodcock was well established as a breeder in Monmouthshire, by about 1880 in Merioneth and by 1885 in Breconshire. During the period 1885–1935 Woodcock nests were reported from all Welsh counties except Anglesey as part of a general increase in breeding range in the British Isles, albeit in quite small numbers. Alexander (1945–47) commented that the Woodcock "appears to breed in all suitable localities in the British Isles, except in Wales and the south-western peninsula (Somerset, Devon and Cornwall). The reason for its absence or scarcity in the breeding season in those two areas remains obscure". The Welsh breeding distribution of Woodcock was shown by Alexander as being confined on a regular basis to parts of Glamorgan, eastern Monmouthshire, north-eastern Breconshire and eastern Radnorshire.

Since the 1934–35 survey there has been a further extension of breeding range. For example, in Denbighshire, Harrop (1961) found that the Woodcock had become a regular breeder in the county and that the increasingly large area of young fir plantations provided undisturbed cover for it to breed in. Breeding has been recorded very infrequently from Anglesey and not in recent years from Pembrokeshire. Although there has clearly been a recent expansion of range, the stronghold continues to be in the border counties from Monmouthshire to Denbighshire, and there is evidence of a decline in coastal counties such as Cardiganshire.

Unfortunately, it is not possible to give a meaningful estimate of the size of the breeding population. The species is not only crepuscular but possesses an unusual polygynous non-territorial breeding behaviour. A secretive species, it is usually noted in spring only during its "roding" display flight at dawn and dusk, and nests are rarely discovered. It favours a variety of habitats, ranging from mature, damp broadleaved woodland to extensive conifer plantations with open rides.

Considerable numbers of winter migrants arrive in Wales at the end of October and the beginning of November from breeding areas in England and the continent. Owen (1603) referred to the Woodcock as an abundant winter visitor to Pembrokeshire at that time, in "almost incredible" numbers, outnumbering all other gamebirds together. They were taken by nets between trees on "Cockroades" in woods at dawn and dusk, sometimes two to four at a netfall. Owen himself often took six at a fall and 18 an evening and it was not unusual to take 100–120 from one wood in a day. Based on the above figures, it is not surprising that Owen was able to claim that it arrived earlier, was present in greater numbers in winter and stayed longer than in any other place in England and Wales!

By the time of Alexander's report (1945–47) there had evidently been a substantial decline in numbers on passage and in winter as shown by the average total season's bag for the period up to 1935 for 20 Welsh sporting estates. For nine estates in south Wales the average bag was 11 Woodcock and for 11 estates in north Wales the average bag was 21 Woodcock. A comparison of shooting records in Denbighshire for the period 1955–60 showed no change from the data up to 1935, suggesting that the wintering population was stable over the intervening years (Harrop 1961).

In Wales, winters are normally wet and mild, conditions that keep the ground suitable for Woodcocks but when hard weather occurs the birds concentrate at springs that remain ice-free, flighting to their feeding grounds at dusk. In extreme conditions they resort to roadsides and gardens and are forced to forage by day. In the severe frost of January to February 1963, 100 were found dead on Bardsey and over 200 on Skokholm. Judging by the records from the offshore islands of Pembrokeshire and Anglesey, many birds pass through to Ireland. Return passage on the islands takes place in March and, to a lesser extent, in April but numbers are less than in the autumn passage.

The natal area of birds recorded in Wales in winter is revealed by eight recoveries of Woodcock ringed as nestlings and found in the Principality. These originated from Norway (2), Sweden (2), France (1) and England (3). Other recoveries of birds found in Wales and ringed abroad come from West Germany, Latvia, Netherlands (2), Sweden, Finland and Norway.

Black-tailed Godwit *Limosa limosa*
Rhostog Gynffonddu

A common spring and autumn passage migrant to estuaries; it is common to abundant and increasing in winter on the Dee but it is generally uncommon elsewhere at that time of year. In several recent years there has been a non-breeding flock summering adjacent to the Dee estuary.

In the early years of this century, the Black-tailed Godwit was a somewhat rare autumn visitor, even scarcer in winter and spring. Since 1930, however, it began to increase in Carmarthenshire, becoming much more regular as an autumn passage migrant and winter visitor to the estuaries of the Taf, Tywi and Gwendraeth, often in some numbers, up to just under a 100 (Ingram & Salmon 1954). By the late 1960s large early autumn flocks were also noted in the Burry Inlet, e.g. 500 on 7 September 1968 (Salmon 1974). At about this time flocks of up to 100 or more were also being recorded on the Dee.

The birds wintering in the British Isles are from the Icelandic breeding population (four specimens from Wales are all of the Icelandic race, Vernon 1963). Increasing numbers of Icelandic birds have been wintering in the British Isles in the past forty years, an increase which is reflected in the numbers in Wales. On the Dee a large wintering flock averaging 670 in 1971–74 built up from the late 1960s and has maintained good, if fluctuating, numbers ever since. The average winter peaks on the Dee were 775 for the years 1981–85 and 911 for the years 1986–90. In 1989–90 the flock has numbered up to 1600 birds (33% of the British wintering population) and is concentrated on the mud-flats near Flint, roosting at high tide on Oakenholt Marsh. Elsewhere in Wales very small numbers are recorded in winter from the Monmouthshire Severn, Burry Inlet, Carmarthen Bay, Cleddau, Swansea Bay and Traeth Lafan.

Breeding of Black-tailed Godwit has not been proved in Wales but there are three possible occurrences. In 1938 two were seen near the Morfa Pools, Margam (Glamorgan) early in April; they later appeared to be a pair and under circumstances that subsequently suggested the possibility of breeding. The pair remained there until at least mid-May and by 10–15 May were behaving exactly like an anxious pair of Redshanks with young. The area was a difficult one either to search or even to observe adequately and proof of a nest; or young was impracticable. After 15 May it was not possible to visit the area again until June and by that time the pair had left (Heathcote *et al.* 1967). The second possible occurrence was at Afonwen (Caernarfonshire) in the late 1940s when an egg collector claimed to have found a nest; unfortunately confirmation is lacking. The third instance was in south-west Anglesey where probable breeding was recorded during the BTO Atlas Survey, 1968–72. The presence of a summering flock from 1988 onwards on pools near the Dee gives hope that a breeding population may be established in the future. Numbers reached more than 250 in 1992, making it the largest such flock in Britain.

Black-tailed Godwits have been recorded as scarce passage migrants in all three inland counties, particularly in spring.

Bar-tailed Godwit *Limosa lapponica*
Rhostog Gynffonfrith

A common passage migrant and winter visitor to flat parts of the coast, more numerous in autumn than in spring. It is rare away from the coast but has been recorded in all inland counties.

Historical data from the early part of the century show that the status of this species has not altered greatly up to the present although there is evidence of an increase from the 1960s. Numbers do, however, fluctuate considerably from year to year, partly perhaps in response to fluctuations in breeding success on its Arctic breeding grounds and partly to the severity of the winter conditions in the Waddenzee (Germany/Netherlands).

It is much more widely distributed than the Black-tailed Godwit because it is found both on sandy and muddy shores whereas the Black-tailed is very much more a bird of the muddy estuaries. The average January count for 1987–91 was 709 (only 1% of the British population), a marked decline from the figure of 1660 for the average January counts in 1971–74. The average winter peaks for the counting years 1986–90 do, however, show a much higher figure of 1899, still a very small percentage of the British population.

The main wintering areas (average winter peaks, 1986–90) are:

Dee (Cheshire/Flintshire)	725
Burry Inlet	455
Inland Sea (Anglesey)	274
Carmarthen Bay	151
Swansea Bay (Glamorgan)	111

On the Dee the number of Bar-tailed Godwits has declined drastically between the years 1976–77 to 1984–85. Mid-winter peak numbers were 11 149 in 1976–77 and 25 in 1984–85. Even though there has been some recovery, to an average winter peak of 725 in 1986–90, numbers are still at a relatively low level. The decline has been explained by human disturbance to the high-tide roost sites in the outer part of the Dee estuary, causing birds that formerly roosted on the Dee to switch to the Alt estuary, 20 km to the north-east.

Little Whimbrel *Numenius minutus*
Coegylfinir Bach

A vagrant.

This species, which breeds in central northern Siberia and winters in New Guinea and Australia, owes its place in the British list to a record from Glamorgan:

30 August–6 September 1982 – Adult, Sker (Glamorgan)

There has been one subsequent British record, in Norfolk in 1985.

Whimbrel *Numenius phaeopus*
Coegylfinir

The Whimbrel is locally common in coastal counties on spring and autumn passage, scarce inland; small numbers occasionally overwinter. The spring passage predominates.

The status of Whimbrel in Wales does not appear to have changed in coastal areas since the last century, although there has been an increase in inland Wales, e.g. Breconshire where it was first recorded in 1963, since when it has been found to be a scarce but fairly regular (perhaps annual) passage migrant (Peers & Shrubb 1990).

Occasionally small numbers have spent the winter in Wales (e.g. on Bardsey) but spring passage of Whimbrel moving north from the African wintering grounds normally begins in mid-April and peaks in early May on the Welsh coast. A particularly important area at this time is the Caldicot and Wentlooge Levels (Monmouthshire), which together with the Somerset Levels held about 2000 Whimbrels in the period 1973–77 (Ferns *et al.* 1979); during the month of peak spring passage (May) between 1972 and 1975 74% of the Whimbrels counted in Britain occurred at sites round the Severn estuary. The work carried out by Ferns *et al.* showed that during the day Whimbrels spread out to feed in small groups on grass fields, showing a preference for damp, tussocky pastures and at night gathered at two main roosts either side of the Severn. The roost on the Welsh shoreline at Collister Pill (Monmouthshire) peaked at 1000 on 30 April 1976. Ferns *et al.* suggested that the Levels on either side of the Severn constitute the last major feeding area for many Icelandic breeding birds.

A survey of the Monmouthshire Levels on 5 May 1991 confirmed the area's continuing importance for spring migrating Whimbrel, arriving at a total of 739 birds scattered throughout the area.

Apart from this unique concentration on either side of the Severn estuary, spring passage consists of groups of 1–50 birds, sometimes up to 200 stopping off on the coast or passing northwards, principally in the period late April to early May. Reverse migration is typically between mid-July and mid-September but carries on to a lesser degree in October and up to early November.

There have only been two recoveries of birds ringed in Wales. One ringed at Collister Pill (Monmouthshire) on 30 April 1976 was recovered in Senegal on 8 December 1979, and one ringed on Skokholm on 13 August 1967 was recovered in Finland in June 1969, an indication that not all the Whimbrels occurring on autumn passage are from the Icelandic breeding population. There is only one recovery in Wales of a foreign ringed Whimbrel: one ringed in Belgium in April 1975 was recovered at Llandewi (Radnorshire) on 15 June 1977.

Curlew *Numenius arquata*

Gylfinir

A fairly common and widespread but declining breeding species throughout most of Wales but local in Pembrokeshire and the coastal fringe of south Wales. It is an abundant winter visitor to coastal areas where small numbers are present year-round. Very small numbers winter inland.

Since the 19th century there have been considerable changes in breeding numbers and distribution throughout Wales. In north Wales Forrest (1907) describes the Curlew as a common breeder on all the moorlands with a few nesting on lowland bogs (e.g. occasionally in east and west Anglesey and the south shore of the Dyfi) and frequently in low-lying wet fields in central Montgomeryshire. Farther south in Wales the breeding distribution of Curlews in the late 19th and early years of the 20th century was decidedly patchy. In Pembrokeshire Mathew (1894) had "little doubt" that it nested on Preseli Mountains and occasionally on Skomer. In Carmarthenshire, Barker (1905) recorded it as fairly common and as early as 1880 Cambridge Phillips noted the spread from the hills to the adjacent enclosed fields for breeding. In Glamorgan, Curlews were not known to have bred in the first two decades of the 20th century although they may have done so in the 19th century in the north of the county (Heathcote *et al.* 1967). In Breconshire, Cambridge Phillips (1899) noted that Curlews were breeding on the hills and in swampy places in increasing numbers.

From the 1920s onwards there was a considerable increase in breeding range in south Wales coinciding at least to some extent, to an altitudinal re-distribution. In Pembrokeshire, Lockley *et al.* (1949) stated that Curlews were not common as breeding birds before the early 1930s, but greatly increased afterwards to breed throughout the county, including Ramsey and Skomer but not the Castlemartin peninsula, being most numerous in the north. Bertram Lloyd's Diaries indicate that they were a widespread breeding species in Pembrokeshire by 1927. The breeding range was extended to the Castlemartin peninsula about 1950. In Carmarthenshire, Ingram and Salmon (1954) recorded that as a quite numerous breeding species it was formerly mainly confined to the northern and eastern hills but that during the westward spread of the 1920s it began nesting in the wider river valleys which became the main breeding areas, the higher elevations in the hills being largely deserted; a few pairs were also found breeding in the coastal marshes and sand dunes. Glamorgan was colonised in the mid-1920s and numbers increased to the extent that by the end of the Second World War Curlews were breeding right across the county from the Rhymney valley to the Gower, chiefly in southern agricultural areas rather than amongst the hills of the coalfield. In Breconshire, numbers also increased greatly in the 1920s (Ingram & Salmon 1957) but they mainly occupied the lower ground.

The overall trend in breeding numbers in the past three decades is for a decline coupled with a diminution of range. The position can be summarised as follows:

Flintshire

Well distributed as a breeding species over most of the county (Birch *et al.* 1968).

Denbighshire

Fairly common breeder on rough grassland and moorland throughout the county and in meadows in the southern Vale of Clwyd and near Ruabon (Hope Jones and Roberts 1982). Bain (1987) found a minimum of 49–61 pairs on part of Mynydd Hiraethog in 1984 and 1985, birds breeding in both upland grassland and in the valley bottom where on the most favoured damp pastures there was a maximum density of 16 pairs per 1000 ha.

Anglesey

In 1986, 250–350 breeding pairs were recorded (RSPB 1988) making the island a very important area for Curlews.

Caernarfonshire

Breeding throughout the county except on the Orme peninsula, frequenting riverine meadows, lowland heaths but especially upland bogs, rough grasslands and rushy pastures up to 457 m (Hope Jones & Dare 1976). On the Llŷn the estimated breeding population was 200–300 pairs in 1986 (RSPB 1986 and 1987).

Merioneth

Fairly common and widespread breeder in lowland flats and, more abundantly, in rough upland marsh and grassland (Hope Jones 1974). Bog and rough pasture at Trawsfynydd supported an estimated 28 pairs in 1987 (RSPB 1988), a density of 50 pairs per 1000 ha on bog and 80 pairs per 1000 ha on the pasture. On the Migneint and Moelwyn (Merioneth, Caernarfonshire and Denbighshire) a survey in 1976 located 54 breeding pairs at an average density of 2.6 pairs per 1000 ha.

Montgomeryshire

A survey of the eastern half of the county in 1986 found a total of 176 breeding pairs, mostly on low-lying pastures and hay/silage fields (McFadzean & Tyler 1987).

Radnorshire

The above survey also included most of Radnorshire and 71 breeding pairs of Curlew were located. The overall density of the combined areas was only 1.3 pairs per 1000 ha. In 1981, 60 pairs were located in the county (Peers 1985).

Cardiganshire

Breeding comparatively sparingly near sea level (e.g. at Cors Fochno) and most numerously between 152 m and 305 m (Ingram *et al.* 1966). In 1988 a survey of the Cambrian Mountains ESA, largely in Cardiganshire, found only ten breeding pairs of Curlew on the upland semi-natural rough grazings, a very low density of only 0.7 pairs per 1000 ha (McFadzean 1988).

Pembrokeshire

Breeding numbers had declined dramatically by the mid-1970s. The Pembrokeshire Breeding Bird Survey of 1984–88 found only c20 pairs (13 on Skomer, the rest on bogland on the Preseli Mountains).

Carmarthenshire

A comparison of the results of the *Breeding Atlas* and the *New Atlas* shows a marked reduction in confirmed breeding records along the coastal fringe. Otherwise, there appears to have been little change since Ingram and Salmon's (1954) account.

Glamorgan

The *West Glamorgan Breeding Atlas* indicates that numbers in the late 1980s on the Gower appear not to have declined significantly since the 1950s but elsewhere in west Glamorgan (i.e. the rough pastures in upland areas) there has been a decrease as a result of large-scale afforestation. In the remainder of Glamorgan, as in Carmarthenshire, there is evidence of a decline in the lowland breeding population.

Monmouthshire

Ferns *et al.* (1977) noted that breeding was recorded widely over the upland areas in the north and west of the county but ten years later most pairs were found in the central valleys (Tyler *et al.* 1987). The county population in 1987 was estimated as being in the order of from 150 to perhaps as high as 300 pairs.

Breconshire

Peers and Shrubb (1990) indicate that Curlews were still quite widespread on Mynydd Eppynt but breeding birds were mainly found in damp pastures and rough grass in the lowlands. In addition, there had been a shift in distribution so that more occupied sites and pairs per site were north of Mynydd Eppynt than to the south. In the 1988–90 *New Atlas* work breeding Curlews were reported in 19% of the tetrads visited, giving a breeding population of about 200 pairs. It was not clear whether any change of status was occurring at this time but counts in the Llanwrtyd Wells area from 1986 suggested a possible decline.

The current Welsh breeding population is difficult to gauge with accuracy but taking the survey results outlined above, a reasonable estimate would be of the order of 2000 pairs, which equates to an overall density of about one pair per 1000 ha. The past three decades have been marked by a reduction in range, a general move to some extent away from the higher moorland and, probably, a reduction in overall numbers. This last point is, however, difficult to establish firmly in the relative absence of quantitative data; unfortunately the surveys of breeding waders on lowland wet grasslands do not help much in this respect because Curlews make only limited use of this habitat in Wales. On the same sample of 22 wet meadow sites surveyed in 1982 and 1989 numbers of breeding Curlew remained the same at 18 pairs.

In the Montgomeryshire study area (McFadzean & Tyler 1989) it was found that the distribution of Curlews was influenced by the presence of rank rushy vegetation or a hay/silage/cereal crop. Where pairs nested in cereals, five times as many nested within the taller cover of autumn-sown rather than spring-sown cereals.

Curlews face pressure in upland areas through drainage, loss of moorland vegetation to forestry and grassland reclamation, and in the lowlands through the drainage of wetland areas and intensive management of grassland for silage production where breeding performance can be limited by thc first early cut taking place whilst the young are still vulnerable.

At the end of the breeding season most Curlews resort to the coast until the following spring but small numbers do winter inland at a few favoured localities (e.g. the Wye valley near Glasbury, Breconshire and the Camlad/Severn confluence in Montgomeryshire). In Wales, the Curlew's preferred winter habitat is the large estuarine mud-flats and, to a lesser extent, neighbouring wet pastures. The average January count for the years 1986–90 for Welsh estuaries was 12 368 with concentrations of more than 1000 on the Dee (Cheshire/Flintshire), Cleddau, Burry Inlet, Traeth Lafan and Severn (Monmouthshire). The non-estuarine coasts counts in winter 1984–85 gave a figure of 3430 for Wales, principally in the north, so the total mid-winter population allowing for non-coastal birds is of the order of 15 000 to 20 000 birds. The estuaries hold more than 3% of the East Atlantic Flyway wintering population.

Winter peak counts for Welsh estuaries from the BoEE counts show a marked increase over the past twenty years:

1969–71	4415
1972–76	7549
1977–81	6273
1982–86	10 375
1987–91	12 227

Ringing recoveries have shown that most Curlews remain in one area throughout the winter and return to that area in successive years (Bainbridge & Minton 1978), confirmed in north Wales by 28 retraps and ten recoveries in years subsequent to ringing (Dodd 1987). However, a proportion of birds appear to have changed their moulting areas from north Wales to Jutland (Denmark), judging by three recoveries of moulting Curlew ringed in north Wales and recovered in Jutland on dates when moult would have been well advanced. Large autumn flocks on the coast comprise mainly moulting adults and there is only limited evidence of post-moulting migration. The proportion of local breeding birds that moult and winter locally is not known. Ringing recoveries show that birds wintering on the Welsh coast come predominantly from local and northern England populations and from Scandinavia (six recoveries of birds ringed as chicks), Germany (six recoveries of birds ringed as chicks) and Netherlands (two recoveries of birds ringed as chicks). This confirms the summary by Cramp *et al.* (1983) suggesting that Curlews generally move south-westwards from their breeding grounds, both within Britain and from Scandinavia. Most recoveries of 103 birds ringed in Wales are from Britain (73) but include some from Scandinavia (10), Denmark (7), Germany (7), Netherlands (2) and France (4); two of the last include nestlings ringed in Monmouthshire and Cardiganshire in June and recovered in Morbihan in the same autumn, which shows that at least some dispersal to the south takes place.

Peak numbers of Curlew occur in the autumn from late July to October, mid-winter numbers on the estuaries tend to be smaller as birds disperse locally. Returning emigration commences in late February and continues up to early April. A few non-breeding birds summer on the coast.

Upland Sandpiper *Bartramia longicauda*
Pibydd Cynffonir

A trans-Atlantic vagrant.

There are three Welsh records, all from the same corner of Pembrokeshire:

18 October 1960 – One, Skokholm
19–20 October 1961 – One, Skomer
1 September 1975 – One, Dale airfield

The Upland Sandpiper breeds in North America and winters from south Brazil to northern Argentina. There are only 43 British and Irish records up to 1991.

Spotted Redshank *Tringa erythropus*
Pibydd Coesgoch Mannog

An uncommon but regular passage migrant to coastal counties chiefly in autumn. Small numbers overwinter on the Dee estuary, Burry Inlet and the Cleddau estuary, occasionally elsewhere. It is rare inland.

Historical information shows that Spotted Redshanks were extremely rare even in coastal counties until the 1940s/1950s, as the following examples illustrate:

Caernarfonshire	– First record in 1912, second record 1948
Merioneth	– First record 1899, second record 1912, third record 1953
Cardiganshire	– First record in 1951
Pembrokeshire	– First record in 1896, second record 1944
Glamorgan	– Three records in 19th century, next in 1959

Since the 1960s, in all coastal areas with muddy estuaries, the species has occurred regularly, in small numbers, usually less than ten, principally in the period July to October. The most favoured site is on the Dee estuary at Oakenholt Marsh where there was an exceptionally large flock in the period August/September 1977 peaking at 180 but regularly since then between 60 and 100 birds. Numbers have steadied at a lower level but Oakenholt continues to be the prime site for this species in Wales.

The BoEE gave an average winter peak for Wales for the years 1986–90 totalling 26, principally for the Dee (11) and Cleddau (10) with Burry Inlet (2), Carmarthen

Numbers of Spotted Redshank at Oakenholt Marsh

Year	*Jan*	*Feb*	*Mar*	*Apr*	*May*	*Jun*	*Jul*	*Aug*	*Sept*	*Oct*	*Nov*	*Dec*
1988	11	10	2	–	–	–	10	22	37	42	13	10
1989	20	7	3	1	–	4	11	21	28	19	12	10
1990	10	10	3	–	–	2	6	15	25	17	12	12

Bay (1), Foryd Bay (1) and Traeth Lafan (1). The average January count for the years 1987—91 was 14, which compares very favourably with the 1971–74 January average of four (Prater 1976).

There are no inland winter records in Wales but in Breconshire following the first record in 1963 a total of 22 was seen in the period 1969–89; only three of these were in spring, the remainder in the period between 24 July and 3 October. Most were seen singly but two parties of five were recorded. In Radnorshire, the first record was on 6 August 1968 and there have been three subsequent records in August.

Scandinavia is the nearest point of the breeding range and is the presumed origin of most of the passage and wintering birds in Wales but there are as yet no ringing recoveries to link Welsh birds with any part of the breeding range. The Welsh contribution is reasonably substantial to the total British and Irish wintering population of 80 to 200.

Redshank *Tringa totanus*

Pibydd Coesgoch

A scarce and declining breeding species in Wales, confined chiefly to saltmarshes but with a few pairs inland. There is an abundant wintering population on the major estuaries, particularly on the Dee, where birds feed mainly on the bare mud, but also on damp pastures close to the estuary.

It is not possible to give an accurate picture of the breeding status of Redshank in the 19th century but it is known that the species was nesting as early as 1662 near Aberavon (Glamorgan) (Ray in his Third Itinerary)! At the beginning of the 20th century, Redshanks were nesting in several scattered localities on both the north and west coasts of north Wales as well as, much more rarely, inland (Forrest 1907). In 1919 Forrest noted that it had greatly increased as a breeding bird in recent years in Anglesey and Caernarfonshire, an increase which was part of a west and southwards spread in southern Britain during the 60 years 1865–1925 (Parslow 1973).

Along the south Wales coast breeding was proved at Kenfig (Glamorgan) in 1911 and by 1925 there were probably a dozen pairs nesting at Kenfig Morfa. At this time Redshanks were spreading westwards and from 1922 onwards pairs were found breeding as well in the lower Thaw valley in 1922, at Llanmadoc Marsh in 1924 and at Oxwich Marsh (all Glamorgan). In 1929 there was the first definite proof of breeding in Carmarthenshire and a few pairs started to breed at one coastal locality. This spread, however, began to slow down and the species only marginally colonised Pembrokeshire where breeding was suspected, but not proved, near St. David's in 1932 and 1948. Two pairs did, however definitely breed near St. David's in 1955, but not in the county subsequently.

Farther north in Cardiganshire the species, although not numerous, showed an increase in breeding numbers especially since the mid-1920s, with a few pairs nesting annually on the Dyfi estuary, Cors Fochno, Cors Caron and occasionally at higher altitudes e.g. Llyn Eiddwen and once at Bray's Pool in 1938, both at c305 m above sea level (Ingram *et al.* 1966).

Farther inland breeding was first recorded from Breconshire in 1913 and Radnorshire in 1914. In Breconshire it has been a rare and irregular breeder with a maximum of eight to ten pairs possibly present in the 1970s but no more than three to five pairs likely even in a good year in the late 1980s (Peers & Shrubb 1990). In Radnorshire its status (Peers 1985) is given as a regular, if local and uncommon breeder, with possibly fewer than ten pairs, even in the best years.

Since at least as far back as the 1960s a decrease has occurred in Wales both in range and numbers, partly at least perhaps triggered by the mortality caused by the severe winter of 1963. A study of wader habitats in a 7080 km^2 area of upland north Wales (Bain 1987) showed that Redshanks in that area were confined to damp pasture/meadow in the valley bottoms (e.g. with soft rush and sharp rush) and damp coarse grassland (e.g. with sheep's-fescue, mat-grass, heath rush and deergrass). Many suitable habitats have been lost through drainage and pasture improvement with higher stocking rates in the past thirty years. In such circumstances shallow pools and flooded ditches, so important as feeding areas for adults and young, together with the tussocks which provide nest sites, are no longer available. The BTO/RSPB survey of waders of lowland wet grassland in 1982–83 found only 53 pairs of Redshank on those, rather few, sites surveyed in Wales. Thirty-one such sites were in Monmouthshire (mainly on the Levels), ten in Montgomeryshire, Breconshire and Radnorshire combined, six in Anglesey, Merioneth and Caernarfonshire combined, four in Glamorgan and two in Cardiganshire, Pembrokeshire and Carmarthenshire combined (K. W. Smith 1983).

The coastal area has always held by far the greater part of the breeding population in Wales, where pairs are chiefly confined to saltmarshes and lowland wet grassland close to the sea (e.g. the Gwent Levels). As with inland areas, lowland wet grasslands close to the coast have been greatly modified by drainage in recent years, while overgrazing by stock, and human disturbance have been recognised as detrimental factors.

In 1985 a survey of breeding Redshank on British saltmarshes was carried out by the RSPB for the Nature Conservancy Council (Allport *et al.* 1986). This covered 13 sites in south and west Wales where a total area of 678 ha held 99–102 breeding pairs of Redshank. A more widespread survey of Wales in 1991 (Williams *et al.* 1992) included the same 13 sites and it was found that the population had decreased substantially.

In Britain as a whole the breeding population is estimated at about 30 000 pairs, principally on saltmarshes. The current Welsh population at about 180 pairs is only a little over a half per cent of the total and is, sadly, in a state of substantial decline. The 1991 survey located only 20 pairs inland, two in Radnorshire, four in Montgomeryshire, two in Merioneth, six in Cardiganshire and six in Monmouthshire. The coastal sites were made up as follows:

Dee estuary (Flintshire)	55 pairs
Anglesey coast	10–12 pairs
Foryd Bay (Caernarfonshire)	4 pairs
Glaslyn Marshes/Traeth Bach (Merioneth)	5 pairs
Morfa Harlech/Shell Island (Merioneth)	9–10 pairs
Mawddach/Dysynni (Merioneth)	8 pairs
Dyfi estuary	20 pairs
Taf/Tywi/Gwendraeth estuaries (Carmarthenshire)	2+ pairs
Pembrey Burrows (Carmarthenshire)	3 pairs
Loughor estuary (Carmarthenshire/Glamorgan)	16–18 pairs
Gower marshes (Glamorgan)	6 pairs
Afon Neath/Crymlyn (Glamorgan)	3 pairs
Kenfig Burrows (Glamorgan)	3 pairs
Rhumney to Chepstow (Glamorgan/Monmouthshire)	12–14 pairs
Total	156–163 pairs

Apart from the relatively small breeding population there are very few Redshanks around the Welsh coast in May and June although small parties of non-breeders do occasionally summer in favoured localities. Birds returning from the breeding areas begin to arrive at the end of June; these birds presumably stay to moult. BoEE counts show that the Welsh population increases to a peak in the period September to November and then decreases slightly in the period from December to March.

During the spring Redshank attain their summer plumage before departing to the breeding areas in April. There is no evidence of a spring passage. There is, however, an indication that some of the birds that arrive in autumn migrate farther south as is shown by recoveries in the period from August to April of birds ringed in Wales in southern England (5), France (2) and Netherlands (1). Ringing information from the SCAN Ringing Group indicates that birds are site faithful throughout the winter.

Recoveries show that the Redshank which winter in Wales either belong to a population that breeds in Britain (north England and Scotland) or in Iceland; the Icelandic data refer to six individuals ringed in Wales and recovered in Iceland and four individuals ringed in Iceland and recovered in Wales. The Icelandic birds are on average larger than British birds and whilst there is a considerable degree of overlap between the two populations there is a method of separating the two by plotting the moult score against the wing length. Two catches on Lavan Sands by SCAN at the crucial time of the year had a British: Icelandic ratio of 56%: 44% (Johnson 1985). Catches on the Severn show a similar result with two-thirds of the birds probably being British breeders and one-third Icelandic (Ferns, in litt.). There is a paucity of juvenile Redshank in west coast estuaries. Recent catches from the Welsh Severn found only 18 out of 211 (8.5%) birds were juveniles. The simplest interpretation of this is that British breeders are having very poor breeding success at the present time. An alternative explanation is that competition for space forces juveniles away from the better (milder) west coast wintering areas, to the harsher conditions in the east (where there is certainly a higher proportion of juveniles, but there are also more Icelandic breeders) where they suffer a higher mortality during hard weather (Ferns, in litt.).

An interesting comparison has been made between the weights of Redshank caught on the Wash and those caught in north Wales (Johnson 1981-82). This showed that the Wash birds are considerably heavier in all the winter months. This lends support to the theory of a compromise between the need for insurance against future severe weather and the need to minimise flight load during non-migratory periods: in the much milder climate of north Wales compared with the Wash (2°C different on average in January) the balance shifts in favour of lighter birds.

An analysis of Redshank retraps at Traeth Lafan (Moss 1985) where a high percentage of the population has been ringed and recaptured gave an estimated mortality of 18% per annum which is low compared with other published estimates. Most of these estimates are, however, from east coast sites where Redshank are especially vulnerable to the colder winters and the difference is therefore probably a realistic one.

The British wintering population numbers about 75,000 which is about 60% of the number wintering in Europe (Moser 1987). There has been an overall 25% decline in British wintering numbers since 1975–76 but this is not reflected in the Welsh figures. The average winter peaks for Welsh estuaries from the Dee (Cheshire/Flintshire) to the Monmouthshire Severn were as follows:

1969–71	7562
1972–76	7616
1977–81	5580
1982–86	8334
1987–91	12 667

The Welsh wintering numbers now amount to more than 8% of the East Atlantic Flyway population.

The Dee estuary has shown a particularly remarkable increase in numbers; taking the English and Welsh side as one unit the increase has gone from a five-year average peak of 3598 in 1981–85 to a five-year average peak of 8441 in 1986–90. This makes it the most important estuary in Britain for this species with 11.25% of the British and 5.63% of the European populations.

The estuaries which held an average winter peak of more than 500 birds in the five-year period 1986–90 were as follows:

Dee (Flintshire and Cheshire)	8441
Severn (Monmouthshire)	1686
Cleddau	1442
Burry	1063
Taff/Ely (Glamorgan)	638
Traeth Lafan	589

In addition there is a sizeable population on the non-estuarine coastal areas of Wales, amounting, in the 1984–85 Winter Shorebird Count, to 1035 birds.

Marsh Sandpiper *Tringa stagnatilis*

Pibydd y Gors

A vagrant.

It breeds in Bulgaria and Romania discontinuously eastwards through Kazakhstan and eastern Asia and winters in Africa south of the Sahara, south Asia and Australia. It owes its place on the Welsh list to three records:

30 June–2 July 1977	– One, Malltraeth (Anglesey)
7 May 1990	– One, Oakenholt Marsh (Flintshire)
19 May 1990	– One, Penclacwydd (Carmarthenshire)

Greenshank *Tringa nebularia*

Pibydd Coeswerdd

Locally common in coastal areas on passage, chiefly in autumn. Small numbers regularly overwinter in Anglesey, Caernarfonshire, Glamorgan and Pembrokeshire, occasionally elsewhere. It is uncommon inland.

From at least as far back as the beginning of the century the Greenshank was considered to be not uncommon on the coasts and estuaries of north Wales in autumn and spring, fewer in winter (Forrest 1907). Farther south the indications are that it was not so regular in occurrence in the early years of the century but by the 1950s it was considered to be a regular autumn visitor to all coastal counties.

At the present time, autumn passage is well marked between July and October in all the major estuaries. Peak autumn numbers of up to 80 are recorded on the Gwendraeth (Carmarthenshire), up to 76 on the Dee and up to 50 at Traeth Lafan. Numbers on spring passage are much less. Inland it is a scarce but regular passage migrant. In Breconshire, for example, about 164 were seen between 1969 and 1989, all but three records were for the autumn between 12 July and 7 November whilst the three spring records fell between 16 April and 5 May.

The main wintering areas are in the Cleddau/Milford Haven complex (average winter peak, 1986–90 of 22) and on Traeth Lafan (average winter peak 1986–90 of 14). Smaller numbers winter on the Dee, Foryd Bay, Glaslyn, Braint and Inland Sea in north Wales and Burry Inlet, Carmarthen Bay and Swansea Bay in south Wales. Regular wintering on the Cleddau/Milford Haven complex has been recorded since 1945–46 with up to 15 recorded by the winter of 1953–54 (Barrett & Davis 1958). Wintering numbers are stable or slightly increasing at present, the average January count of 27 for the years 1971–74 comparing with 31 for the years 1987–91. The highest winter count was of 51 birds on the Cleddau in February 1985, presumably reflecting a cold weather influx. There were 24 at Aber Ogwen (Caernarfonshire) in January 1991.

There are no ringing recoveries to elucidate the breeding areas of Greenshanks that migrate through, or winter in, Wales. However, the timing of spring migration provides a strong indication that Welsh wintering birds are drawn from the Scottish breeders. Scottish Greenshanks are back on their territories in early April, about a month earlier than Scandinavian breeders and in approximately the same period that birds leave the Welsh, English and Irish estuaries (Lack 1986). Autumn passage involves considerably more birds than the total which remains in winter and is likely to include Scandinavian and Russian birds wintering farther south. The small spring passage in April and May presumably involves Scandinavian birds returning north. For example, there was a marked spring passage through Monmouthshire in 1990 between 1 May and 11 May, amounting to at least 24 birds along the coast.

Greater Yellowlegs *Tringa melanoleuca*

Melyngoes Mawr

A vagrant.

Two records from the same part of Flintshire add this species to the Welsh list.

23 July 1961	– One, Shotton Pools (Flintshire)
24 November 1983	– One, Shotwick Lake (Flintshire)

A third record from the same area, from Shotton Pools on 3 September 1962 was not accepted by the British Birds Rarities Committee.

Greater Yellowlegs breed in northern North America and winter in the USA south to southern South America. There have been 29 records in Britain and Ireland.

Lesser Yellowlegs *Tringa flavipes*

Melyngoes Bach

A vagrant.

This elegant wader has been seen nine times in Wales:

11–13 September 1953	– One, Oxwich
9–10 October 1961	– One, Skokholm
7 August 1969	– One, Skomer
11–21 November 1969	– One, Ynys-hir
9–10 September 1975	– One, Pwllheli Harbour (Caernarfonshire)
11 October 1977	– One, Bardsey
25 September 1981	– One, Peterstone Wentlooge (Monmouthshire)
7–22 October 1984	– One, Bosherston Pool (Pembrokeshire)
27 September–18 November 1990	– One, Cydweli Marsh (Carmarthenshire)

The Lesser Yellowlegs breeds in northern North America and winters in the USA south to southern South America. There have been a total of 215 records in Britain and Ireland up to 1991.

Green Sandpiper *Tringa ochropus*

Pibydd Gwyrdd

The Green Sandpiper is a regular passage migrant and winter visitor in small numbers to all counties.

Described as "not very common" and "rare" in different parts of Wales at the turn of the century, it appears that there has been an increase in numbers since that time. Records span all months of the year and two were present at Llanishen Reservoir (Glamorgan) from October 1931 to September 1932. An analysis of the Breconshire records from 1969 to 1989 inclusive (Peers & Shrubb 1990) showed that 64% of the total records (the total was 264) were in the period July to September with the largest number (36%) in August.

Most records are of one or two birds but parties of five to ten are not infrequent, particularly on passage in August on the coast. Wintering numbers are relatively small and usually involve one or two birds in scattered inland and coastal sites ranging from the Dee estuary, the Cleddau estuary and the Taf/Tywi/Gwendraeth estuary (Carmarthenshire) to inland sites at Llandegfedd Reservoir (Monmouthshire), the river Wye at Glasbury (Breconshire) and the river Severn near Welshpool (Montgomeryshire).

There are as yet no ringing recoveries to indicate the origins of the passage and wintering birds found in Wales; the sole recovery is of a bird ringed in August 1964 at Llanigon (Breconshire) and recovered 3 km to the north-east in September the following year.

Wood Sandpiper *Tringa glareola*

Pibydd y Graean

The Wood Sandpiper is a scarce passage migrant recorded in every county, principally occurring in autumn but with a scattering of spring records; most records are of single birds.

At the turn of the century Wood Sandpipers were very rare vagrants and it is only in comparatively recent years that the first record was noted in several Welsh counties (e.g. Carmarthenshire 1976, Merioneth 1976, Breconshire 1964, Glamorgan 1962). Whether this reflects a genuine recent increase or general improvement in field identification skills is uncertain but the prevalence of specimen shooting in the early years of the century suggests that the former is probably the case.

The bulk of the records come from coastal areas, especially freshwater or brackish pools near estuaries, and August is by far the most favoured month. The data from Pembrokeshire (Donovan & Rees, in press) are typical of a coastal county. In Pembrokeshire the Wood Sandpiper had only been recorded once prior to 1955; since then it has been recorded in 21 autumns and in 10 springs, one or two occurring between 17 April and 5 June and 2 July and 2 October. Most of these records were from the coast but inland occurrences were noted at five sites.

In the inland county of Breconshire the distribution of records is April (1), May (1), July (1), August (12), September (2) (Peers & Shrubb 1990). It is very rare in the other inland counties of Radnorshire and Montgomeryshire.

There are no ringing recoveries relating to Wales. Migrant Wood Sandpipers from their northern Europe breeding grounds largely favour a more easterly route to the African wintering grounds.

Common Sandpiper *Actitis hypoleucos*
Pibydd y Dorlan

The Common Sandpiper is a summer visitor which breeds in all counties except Pembrokeshire, albeit sporadically in Anglesey and Flintshire. Very small numbers over-winter in Wales.

A few pairs of Common Sandpipers nest close to sea level near river mouths (e.g. in Caernarfonshire and Merioneth) but the vast majority breed along the banks of upland rivers, streams and lakes in the hills up to a height of 762 m. Since the turn of the century there has been a marked shift away from lowland sites near the coast; e.g. it is a far cry from the time when in 1892 three nests were found within 90 m along the river Conwy at Glan Conwy (Denbighshire). In Pembrokeshire, Mathew (1894) stated that Common Sandpipers bred near Maenclochog but there have been no breeding records since.

Historical information on breeding numbers is very sparse but it is clear that there has been a general decline in the 20th century in addition to the slight contraction of range into the favoured upland areas. In common with some areas of England there was a marked decrease noted in Wales about the 1950s in Radnorshire, Breconshire and Carmarthenshire (Ingram & Salmon 1954 & 1957). In Caernarfonshire Hope Jones and Dare (1976) felt that the breeding population "could well be decreasing steadily due to increasing disturbance of nesting haunts by tourists and anglers".

The present distribution of Common Sandpipers is largely determined by their need for fast flowing streams and rivers with shingle or gravel shores, bare stony lake/reservoir edges, which are not too disturbed by human activities. Survey work carried out by the RSPB in 1991 suggested that densities were higher on rivers in south and mid-Wales (Usk, Wye, Tywi, Severn and Vyrnwy) than on the Dee in the north. This was thought to reflect availability and suitability of shoals rather than any geographical preferences (Tyler 1992a).

Some data on mean breeding densities exist for different Welsh rivers, notably the Tywi in 1991 (Friese & Stewart, pers. comm.) and the Severn, Vyrnwy and Wye in 1977 and 1978 (Round & Moss 1984). These produced densities of:

Tywi	10–35 pairs per 10 km	(58 territories)
Wye	4.86 pairs per 10 km	(107 territories)
Vyrnwy	3.0 pairs per 10 km	(16 territories)
Severn	2.77 pairs per 10 km	(36 territories)

In addition, many upland lakes hold concentrations of breeding pairs; for example lakes on Mynydd Hiraethog and at Trawsfynydd (Merioneth) which held 19 pairs around 14 km of shoreline in the 1980s and Llyn Conwy (Caernarfonshire) where eight pairs were recorded in 4 km around the lake.

Few estimates exist for breeding populations of counties but recent figures include fewer than 100 pairs in Monmouthshire (Tyler *et al.* 1987), 100–120 pairs in Breconshire (Peers & Shrubb 1990) and 30 pairs in west Glamorgan (Thomas *et al.* 1992). It is possible that the total Welsh population is in the order of 1000 pairs.

Spring passage is less obvious than that of autumn and is concentrated in the period from late March to May, principally from mid-April to mid-May. Return passage takes place between mid-June and mid-October, predominantly in July and August when parties of 10–20 birds are not unusual on the estuaries and coastal pools. A handful of birds overwinter on favoured inland and estuarine sites in most years, notably on the Wye at Glasbury (Breconshire) and on the Cleddau estuary, and Taf/Tywi/Gwendraeth estuaries (Carmarthenshire); the total probably does not amount to more than five to ten birds in most years.

The normal winter range of Common Sandpipers is south of the Sahara and there are only two foreign recoveries of Welsh-ringed birds to give some indication of migration routes: one ringed on Bardsey in July 1957 was recovered near Nantes in France in July 1959 and an adult ringed at Shotton (Flintshire) in July 1981 was recovered at Rabat in Morocco in May 1989.

Spotted Sandpiper *Actitis macularia*
Pibydd Brych

A vagrant.

Since the first in 1960 there have been a further eight records of this trans-Atlantic visitor.

15–18 May 1960	– One, Whitland (Carmarthenshire)
27 August–4 September 1973	– One, Oxwich
24–25 August 1974	– One, Aberthaw (Glamorgan)
9 October–20 November 1975	– One, Ynys-hir
16 September 1977	– One, Bardsey
7–12 August 1979	– One, Ynys-hir
5 October–2 November 1980	– One, Bosherston Pools (Pembrokeshire)
26 October 1980–25 April 1981	– One, Peterstone Wentlooge (Monmouthshire)
29 September 1981	– One, Porth Colmon Caernarfonshire)

Spotted Sandpipers breed in North America and winter from the USA south to Uruguay. There has been a total of 102 records from Britain and Ireland up to 1991.

Grey-tailed Tattler *Heteroscelus brevipes*
Pibydd Gynffonlwyd

A vagrant.

One on the Dyfi estuary (Cardiganshire/Merioneth) from 13 October to 17 November 1981 was the first, and so far only, record for Britain and Ireland and new to the Western Palearctic.

This medium-sized wader breeds in north-east Siberia and winters in south-east Asia to Australasia. Acceptance of the record was delayed for a while because of the identification problems associated with separating this species from the closely allied Wandering Tattler, which has not yet been recorded from the Western Palearctic.

Turnstone *Arenaria interpres*

Cwtiad y Traeth

An abundant wader on the Welsh coast, a passage migrant in spring and autumn but most numerous and widespread in winter. A few non-breeding birds are present throughout the summer. It is rare inland.

The evidence suggests that the Turnstone is now more numerous in several parts of Wales than it was at the turn of the century. In Anglesey it was recorded as common on the coast as early as the first half of the 19th century (Eyton 1838) but, by contrast, farther south in Pembrokeshire, Mathew (1894) considered it a rather rare autumn visitor to the coast and knew of no spring record. In Carmarthenshire it was not recorded by Barker (1905) whereas Ingram and Salmon (1954) were able to comment on its status '"formerly less frequent, it is now to be met with in most years".

Turnstones are now almost ubiquitous in coastal areas, small parties forage along most rocky shores as well as, to a lesser extent, along the tide wrack on sandy and shingle beaches. The main concentrations are along the Flintshire/Denbighshire/Anglesey and Llŷn coasts in north Wales and the Gower in south Wales. Wintering birds show considerable site tenacity and occur in discrete flocks at favourite sites such as mussel beds where they forage for shrimps, winkles and barnacles. Flocks at most wintering sites usually number 50–100 but much higher numbers are sometimes recorded at the most favoured locations (e.g. 710 at Rhos Point, Denbighshire on 10 November 1979 and 700 at Whiteford, Glamorgan on 1 September 1963).

Site fidelity to the same wintering site was clearly demonstrated by Dodd and Moss (1983). At Rhos on Sea there was an exceptional catch of 660 Turnstones by the SCAN Ringing Group on 10 November 1979. In four subsequent catches in 1980–82 out of 243 birds trapped no less than 187 (77%) were controls/retraps. Likewise on Bardsey retraps of ringed birds have clearly shown that the island is a regular migration stop-off and wintering area for the species in successive years (Roberts 1985).

The average winter peak for Welsh estuaries in 1987–91 was 1789 birds but approximately the same number is found on non-estuarine coasts. The Winter Shorebird Count in 1985–86 located 1989, principally in Anglesey (473) and Glamorgan (593). The wintering population of Turnstones in Wales amounts to approximately 9% of the British population. Average winter peaks in 1986–90 of more than 100 occurred at the following estuaries:

Dee (Cheshire/Flintshire)	925
Burry Inlet	435
Clwyd Coast (Denbighshire/ Flintshire)	300
Swansea Bay (Glamorgan)	287
Cleddau	133
Traeth Lafan	107
Severn (Monmouthshire)	106

Birds start to arrive in Wales from breeding grounds at the end of July and at least some of these are on passage farther south, as shown by recoveries of birds ringed in Wales and recovered in France (2), Morocco and Gambia. Those wintering in Wales are likely to be from the breeding population in Greenland and north-east Canada as proved by recoveries in Wales of birds ringed in Iceland (1) and Greenland (1). The wintering population also includes Scandinavian birds, as shown by recoveries in Wales of birds ringed in Sweden (1) and Finland (1). Emigration of adult birds wintering in Wales commences in early March, ringing recoveries indicating that there is a northward movement to the estuaries of the Dee and Mersey and Morecambe Bay from March to May. There is a marked spring passage in May at which time most of the very few inland records occur. It is not known whether the May passage birds belong to the Greenland/Canada breeding population or Fenno-Scandia and Russian breeding population.

Wilson's Phalarope *Phalaropus tricolor*

Llydandroed Wilson

A vagrant.

Since the first record in 1958 there have been seven further sightings of Wilson's Phalarope, a species which breeds in North America and winters in Peru south to Argentina and Chile.

The Welsh records are:

15–16 June 1958	– Female, Malltraeth (Anglesey)

30 August–4 September 1959	– One, Shotton Pools (Flintshire)
11 October–1 November 1964	– One, Bettisfield Pools (Flintshire)
17 November 1974	– One, Pembrey (Carmarthenshire)
6 September 1975	– One, Llyn Heilyn (Radnorshire)
5 September 1982	– One, Connah's Quay (Flintshire)
22 September–8 December 1989	– One, Glan Conwy (Denbighshire)
29 September 1991	– One, Penclacwydd (Carmarthenshire)

The same observer (J. P. Wilkinson) had the good fortune to find the Malltraeth and Shotton Pools individuals; the Malltraeth record was the third for Britain. The finding of the Llyn Heilyn bird must have greatly delighted a field meeting of the Herefordshire Ornithological Club.

Red-necked Phalarope *Phalaropus lobatus*
Llydandroed Gyddfgoch

A rare and sporadic visitor to Wales in the period from March to November.

There have been at least 31 dated records involving a total of 35 individuals in the past one hundred years, from all coastal counties except Denbighshire and Caernarfonshire, with two inland records from Monmouthshire and Breconshire. Most of the records come from Glamorgan (12) and Anglesey (5).

The span of records covers every month from March to November with the exception of August, with peaks in September (10) and October (8).

It is quite possible that these few records relate to birds from the substantial Icelandic breeding population whose wintering area is still unknown. It is notable that this species does not produce the substantial west coast wrecks that occur with Grey Phalaropes.

Grey Phalarope *Phalaropus fulicarius*
Llydandroed Llwyd

An uncommon and irregular passage migrant, predominantly to coastal areas and in the period September to October.

Most of the records refer to singletons but occasionally there are marked influxes when birds from their Greenland breeding grounds drift into Welsh waters in autumn following extended spells of strong north-westerly winds. Autumn 1891 was one such period when no less than 17 were shot on the Dyfi estuary during rough weather between 15 and 24 October and several were shot near Bala (Merioneth) in December.

A particularly notable influx occurred in September to October 1960, part of a widespread incursion into south-west Britain. The movement was first noted on 15 September when there were six on the Glamorgan coast and two at Bardsey. Numbers at the Smalls Lighthouse (Pembrokeshire) were vastly higher than have been recorded in Wales before or since:

22 September	14
23 September	53
25 September	18
27 September	21
29 September	227
3–5 October	Small numbers, maximum 30 on the 3rd

Even these unprecedented numbers were dwarfed by the records from St. Agnes (Scillies) where there was a minimum of 1000 birds, St. Ives (Cornwall) where there were between 500 and 1000 and Torquay (Devon) where there were at least 700.

Grey Phalaropes have been recorded in all Welsh counties except Montgomeryshire, principally from the coast. There has only been one record for the period May to July inclusive, the vast majority being in September and October.

Pomarine Skua *Stercorarius pomarinus*
Sgiwen Frech

A fairly common and increasing passage migrant to coastal areas, principally in autumn. It is a vagrant inland.

This species breeds in the Arctic and winters at sea in the Tropics/sub-Tropics. It is recorded in Wales chiefly from headlands in Anglesey and Pembrokeshire where systematic sea-watching from the mid to late 1970s to date has given a new perspective on its occurrence. For the years 1967–75 inclusive there were records of fewer than three birds on average per annum whereas the years 1989–91 averaged 159 birds per annum.

The autumn totals from Strumble Head dominate the records, averaging 22 for the years 1980–83, 60 for the years 1984–87 and 143 for the years 1988–91. A particularly productive year was 1991, with no less than 97 birds passing Strumble Head on 17 October and 130 on 18 October. Donovan and Rees (in press) suggest that the path of Pomarine Skuas flying out of the Irish Sea is blocked by the north coast of Pembrokeshire and they follow this until they can head out to sea again when clear of the Bishops, a few individuals thereafter wandering inshore.

Point Lynas on Anglesey is another location where Pomarine Skuas are of regular occurrence in autumn, albeit in much smaller numbers and less frequently than at Strumble Head. In contrast to other areas records from South Stack are markedly concentrated in spring from mid-April to early June. In 1983 a total of 125 passed South Stack during the period 8–20 May, spring figures unprecedented and not approached since then.

Arctic Skua *Stercorarius parasiticus*

Sgiwen y Gogledd

Passage bird off coastal counties; scarce in spring, more numerous in autumn, numbers recorded probably being closely correlated with the intensity of effort in sea-watching.

Although the Arctic Skua has long been recognised as an offshore bird of passage and the most frequent of the four northern skua species, it was poorly recorded until recent decades of the present century, with the upsurge in sea-watching off coastal headlands. Forrest (1907) mentions only a handful of individual records for north Wales counties and in south Wales only five records had been made in Glamorgan up to 1929. Mathew (1894) could cite only one for Pembrokeshire as could Barker for Carmarthenshire. As there has been no major change in the north European population of this species (BWP vol. 3) it seems unlikely that the offshore passage of birds in autumn is much different now from what it has been in the past but that it has remained essentially undetected. Southward passage in autumn, as the birds make for wintering areas on both sides of the south Atlantic, is much more strongly marked than the return migration in spring; even so it is believed that only a relatively small percentage of the passage is visible from land as is demonstrated in periods of strong onshore gales when the number seen increases considerably.

Autumn passage is strongest off the north Pembrokeshire coast. Here, as on other parts of the coast, the first birds appear in late July and early August, with numbers building up rapidly in the latter part of that month and reaching a peak in September before falling off again through October; very few are seen by the end of that month and they are rare in November.

Passage off the north Wales coast is well marked off Bardsey and the Anglesey headlands, especially Point Lynas and South Stack and is regularly observed off the Orme headlands (an exceptional maximum of 44 on 19 September 1919) and Point of Air. Strumble Head, which is the most intensively watched of Irish Sea headlands, has produced as many as 500 individual sightings in an autumn (July–October) with a day-maximum of 103 on 3 September 1983 during onshore gales. Off the south Wales coast a scattering of late summer and autumn records is made each year, with individuals occasionally occurring well up the Severn estuary off the Monmouthshire coast; a total of 96 birds has been recorded in Glamorgan between 1980 and 1989. The species is also regularly recorded in small numbers off the coasts of Cardiganshire and Merioneth, again especially after onshore gales in autumn. Winter records are extremely rare.

In spring a few birds are seen in April and May off the Glamorgan coast in most years and the regularity with which such records occur has led to suggestions that occasional non-breeding birds may summer in the Bristol Channel. Spring records in south-west Wales, Cardigan Bay and north Wales are very sporadic, almost always in April or (mostly) May. There have been a number of June records, particularly on the south Wales coast, including several enigmatic inland ones e.g. Llandegfedd Reservoir (Monmouthshire) on 10 June 1990 and two birds in the Doethie Valley (Carmarthenshire) on 13 June 1991. Inland records are rare but, in addition to those mentioned above, have occurred in Montgomeryshire (several) and Breconshire.

Of 292 individuals seen well enough to determine details off Strumble Head in the autumn 1983, 35% were dark phase birds, 32% light phase, 5.5% intermediate and 26.5% were juvenile birds.

Long-tailed Skua *Stercorarius longicaudus*

Sgiwen Lostfain

An uncommon visitor recorded chiefly on autumn passage along the western seaboard during westerly gales, spring records are very infrequent. It breeds in the circumpolar Arctic south to southern Norway and winters at sea south of the Equator.

Up to 1991 there are 91 dated records involving 180 individuals. In the 19th century there were records from Denbighshire/Flintshire in 1869, Pembrokeshire in autumn 1889 or 1890 and Monmouthshire in January 1892. From 1900 until 1979 there were only ten further sightings whereas from 1980 to 1991 there has been a remarkable increase with 81 records.

This apparent sharp increase owes a lot to the intensive sea-watching carried out by dedicated enthusiasts since 1980 at a strategically located watch point on Strumble Head, which accounts for 57 of the records. Observer awareness, particularly of the identification features of immature birds, may also have played a part. The majority of the records are of one or two birds but at Strumble Head there were four on 5 October 1985, three on 13 September 1988, seven on 7 October 1988 and four on 19 September 1990. A total of 20 records involving 89 individuals made 1991 an unprecedented year; maxima were 18 at Strumble Head on 15 September and seven at Red Wharf Bay (Anglesey) on 7 September. The histogram of the timing of occurrences shows a wide scatter of autumn

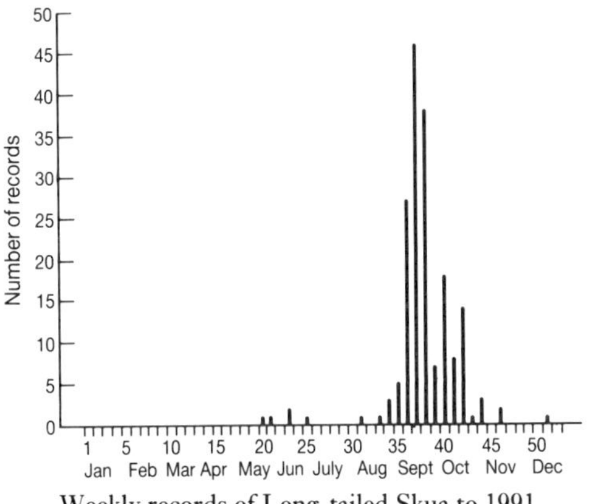

Weekly records of Long-tailed Skua to 1991

dates with peaks in mid-September, and to a lesser extent, early October. There are dated records from all maritime counties except Flintshire, and Merioneth. Perhaps the greatest surprise is that there have only been three records from Point Lynas where autumn seabird movements have been systematically recorded since 1976. Presumably the species passes south farther out to sea than the other three skua species, which are all regularly seen at Point Lynas, but in view of the large number of autumn records from the Cheshire coast it is still puzzling that there are not more from this north-east extremity of Anglesey.

Great Skua *Stercorarius skua*

Sgiwen Fawr

Regularly recorded on autumn passage along the west coast where small numbers are also seen in winter and spring. Numbers vary annually, the highest counts dependent on the strength and frequency of onshore gales.

Until the last 30 years, the Great Skua was regarded as a scarce passage migrant in Wales. Around the turn of the century, it was recorded from several coastal counties in north and south Wales, but there were few subsequent sightings until the 1960s. This increase in the number recorded in Welsh coastal waters is undoubtedly due to an increase in the population in Scotland and helped by a greater interest in birdwatching, particularly in certain coastal areas. A largely pelagic migrant, the Great Skua is rarely recorded inland.

During the 19th century, it was recorded from the counties of Glamorgan and Pembrokeshire but Forrest (1907) could quote no confirmed sighting from the north Wales coast. Mathew (1894) quoted Sir Hugh Owen as saying that it was always to be found in Goodwick Bay (Pembrokeshire) in a good herring season and in 1894 one was shot at Solva Harbour in this county. In Glamorgan it was said to have been seen off Swansea, Cardiff and the Gower coasts.

On 20 July 1903, one was seen off Holyhead, and on 25 August 1912, Richard Johns watched one in the Conwy estuary (Caernarfonshire) for over an hour (*British Birds* vol. 6, p. 163). There were several records from Anglesey during the 1920s and 1930s but elsewhere, it had become very scarce and was not recorded in counties such as Pembrokeshire and Caernarfonshire for over 30 years. It is perhaps surprising that very few birds were recorded around the Welsh coasts at a time when the small Scottish breeding population was slowly increasing (Dickens 1964).

Great Skuas were not recorded from Caernarfonshire between 1915 and 1947 when one was seen off the Great Orme and the first confirmed records for Carmarthenshire, Monmouthshire, Cardiganshire and Merioneth came in 1939, 1935, 1940 and 1951 respectively. Indeed sightings were so scarce during this period that the first four Carmarthenshire records and four of the first six Merioneth records involved birds picked up dead on beaches. An autopsy on one of the Carmarthenshire birds, found on Cefn Sidan beach in March 1975, revealed that death was probably due to the toxic effects of oil in the gizzard. In the adjacent county of Glamorgan there were only 16 records between 1926 and 1966, the majority seen between August and November and only two records from Monmouthshire up to 1976.

Frequent sea-watching at favoured sites such as Point Lynas and Strumble Head during the 1980s has revealed that the Great Skua occurs annually off the Welsh coast and that it is often as numerous as the Arctic Skua. At the former site, 33 were recorded flying west between 8 August and 23 November 1987, but at Strumble Head between 2 and 25 pass offshore most days during the autumn, with larger numbers following south-westerly gales. The maximum recorded here was 503 during autumn 1983, with 198 on 3 September alone. Smaller numbers are recorded in most autumns from other coastal headlands such as Rhos Point in Denbighshire (11 on 30 September 1978), South Stack, Great Orme, Aberdysynni and Aberdyfi in Merioneth and Port Eynon and Worm's Head in Glamorgan. Although the vast majority of records come from the autumn months, and in particular between August and October, smaller numbers are recorded in spring and in winter (e.g. three off Strumble Head on 26 January 1990).

The number of records has also increased markedly from the offshore islands of Wales in recent years as birds fly to and from wintering grounds mainly offshore or coastally in the Atlantic, particularly off Iberia. On Bardsey, a total of 17 birds recorded in autumn 1967 was thought at the time to be exceptional, but recent totals have been much higher, with 114 recorded in 1983, 36 of these on 3 September. Until 1985, there had been only four spring records including three birds seen on 4 May 1981. The Great Skua is recorded as being an uncommon passage migrant to Skokholm and Skomer, although it is now an almost annual visitor offshore to both islands in the autumn. An exceptional gathering of 14 was seen attending the Kittiwake feeding flock at Broad Sound, Skokholm on 28 September 1978. There has only been one record from the inland counties, a bird picked up near Bettws Cedewain (Montgomeryshire) on 23 May 1973; it had been ringed in south-east Iceland on 30 July 1971.

There are six records of nestlings ringed on Shetland and recovered in Wales and one of a bird ringed on Handa (Sutherland). Additionally, there are two records of pulli ringed in Iceland, which were recovered in this country, the farthest travelled involving a bird ringed at Breidanerkursandur on 30 July 1969 and recovered in north-west Wales on 18 November 1980, a distance of 2211 km south-south-east.

Mediterranean Gull *Larus melanocephalus*
Gwylan Môr y Canoldir

An uncommon but increasing visitor to coastal counties at all times of the year.

Although its main breeding area is around the Aegean and Black Seas the Mediterranean Gull has bred in the Netherlands, and in southern England since 1976. In Wales the first county records were in 1964 (Glamorgan and Caernarfonshire), 1968 (Pembrokeshire), 1970 (Anglesey), 1971 (Denbighshire), 1975 (Monmouthshire and Flintshire), 1976 (Carmarthenshire), 1980 (Cardiganshire) and 1981 (Merioneth). In view of the relative recency of these first county records, it is surprising how many records there have been in the past few years. From 1978 to 1987 there was a minimum of 483 records in Wales, principally from Glamorgan (46%) but with good numbers also in Monmouthshire, Cardiganshire and Pembrokeshire. The distribution is markedly coastal, usually not more than five miles inland, with no records yet from Montgomeryshire, Breconshire and Radnorshire.

Hume (1976) analysed the pattern of Mediterranean Gull records at Blackpill for the period 1970–75. He found that adult Mediterranean Gulls appeared briefly in spring, presumably on their way to breeding grounds. Wintering birds also left by early April. Then in April and May there was an influx of first-year birds, which coincided with a marked spring peak of first-year Common Gulls. These Mediterranean Gulls may have moved through or remained among a scattering of other gulls until a fresh arrival of adults and second-year birds began in late June. This pattern was confirmed by an analysis of Glamorgan records for the 15-year period 1970–84 (Hurford 1985) when there were no records of adult birds between 20 April and 20 June.

Most records are of single birds but there were 12 at Blackpill on 11 July 1984 (when it was noted that there were more Mediterranean Gulls present than Black-headed Gulls!) and ten at Taff/Ely estuary (Glamorgan) on 13 May 1989.

There is some indication of the origin of birds seen in Wales through sightings of birds ringed as nestlings: one ringed in East Germany in 1985 was seen at Roath Park, Cardiff on 26 December 1985 and 23 February 1986, and one ringed in the Netherlands in 1990 was seen at Blackpill on 15 August 1991.

Laughing Gull *Larus atricilla*
Gwylan Chwerthinog

A trans-Atlantic vagrant to Wales, breeding in North America and the Caribbean.

There are three Welsh records:

19 May 1978	– One, adult Nant y Moch Reservoir and same bird at Llyn Syfydrin Reservoir on 20th May (Cardiganshire)
27–28 December 1981	– One, Colwyn Bay (Denbighshire)
12 September 1988	– One, Newton Point (Glamorgan)

Franklin's Gull *Larus pipixcan*
Gwylan Franklin

A vagrant.

This trans-Atlantic species owes its place on the Welsh list to a single record in 1986. This was a second summer or adult bird seen at Aber Dysynni (Merioneth) on 22 March.

Little Gull *Larus minutus*
Gwylan Fechan

A fairly common, sometimes common passage migrant and winter visitor to the coast, scarce inland.

Formerly rare there has been a considerable increase in numbers seen recently. There were less than ten records in the 19th century, associated with stormy weather in winter. Records continued to be sporadic until the mid-1960s when there was a dramatic upturn in numbers in common with other parts of Britain and Ireland (Hutchinson & Neath 1978). Smith (1987) supports the suggestion that the increase is probably due to an expansion of the breeding population in Europe with notable increases in Finland and the Baltic states as well as a westwards spread to south-west Norway and increasing numbers in the Netherlands.

In Wales the period 1973–75 was particularly notable for a marked spring passage with maxima in May 1973 of 25 on the Glaslyn Marshes (Merioneth/Caernarfonshire), nine at Ynyslas (Cardiganshire) and 24 in the Kenfig area (Glamorgan); in April 1974, 73 at Eglwys Nunydd

(Glamorgan); in May 1974, 12 at Llandegfedd Reservoir (Monmouthshire); and in April 1975, 28 on the Glaslyn Marshes.

Since that period, numbers have continued at a relatively high level although the spring passage is not nearly so marked as in the early 1970s. There are very few records in June and July (e.g. Hurford 1985, Fox 1986b). Fox has shown that there is a distinct habitat preference according to the season in his analysis of Cardiganshire records, 1968–1983. Spring and autumn passage records came predominantly from brackish water and lagoons near the coast (as in other parts of Wales) whilst winter records came mainly from steeply-shelving storm beaches. Birds appeared after severe winds, frequently associated with Kittiwakes, and were dip-feeding amongst the white water of waves crashing onto the storm-beaches; when conditions ameliorated they immediately disappeared.

Sea-watches at Strumble Head throughout the 1980s showed, similarly, that whilst small numbers of Little Gulls passed southwards out of the Irish Sea each autumn, larger numbers occurred during severe gales late in the season, e.g. 69 on 24 December 1984, 75 on 10 November 1985, 76 on 13 November 1987.

Numbers recorded off the north Wales coast are comparatively small. For example the sea-watches at Point Lynas (Anglesey) averaged only four records per annum between 1987 and 1991. Similarly Eades (1982), during transects from Eastham/Garston on the Mersey to Point Lynas found only seven of a total of 1879 birds in the sections from the Bar Lightship in Liverpool Bay to Point Lynas. Smith (1987) gave further information on the important gatherings of Little Gulls on a 9 km stretch of sandy coast in the outer part of the Mersey estuary. Spring passage, mainly of adults in breeding plumage, increased substantially from 1978 at a time when numbers in Wales were not reaching the figures of the early years of the decade. Liverpool Bay now appears to be an assembly point of birds that have wintered in the Irish Sea before departure, probably overland across northern England, for European breeding grounds.

Inland, the Little Gull is a scarce visitor, with records from Breconshire and Radnorshire. In Breconshire it was first recorded in 1968 and there have been 14 records in March/May, none in June and July, 37 in August/September and seven in October/November.

Sabine's Gull *Larus sabini*

Gwylan Sabine

An uncommon visitor to the coast in autumn, numbers vary according to the frequency of westerly gales.

This species breeds in Greenland, Arctic North America and north-east Siberia. The Atlantic population winters at sea off Namibia and western South Africa. In Wales it occurs close inshore as a result of Atlantic storms as it moves south in autumn and the peak numbers are in September and the first half of October.

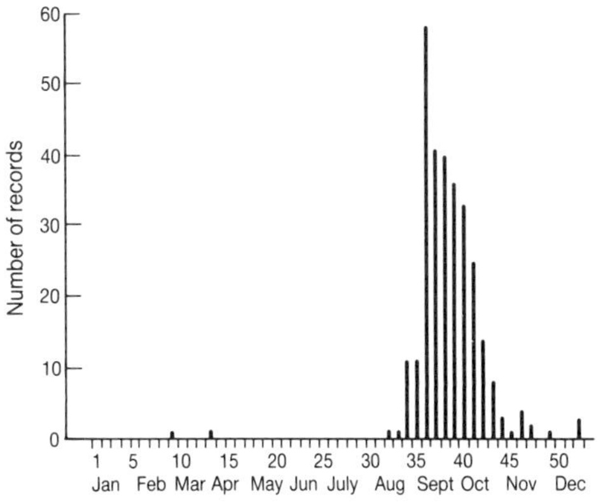

Weekly records of Sabine's Gull to 1991

There are 19th century records from Pembrokeshire (1839 and 1892), Flintshire (1884) and Cardiganshire (1891 and 1896). The last includes six which were shot and two or three others seen during the gale of 24–26 September 1896. There are few records during the period 1900–79 inclusive (29 in all) but a dramatic increase thereafter such that the Welsh total now stands, up to the end of 1991, at 183 records totalling 295 individuals. This increase is probably related, in part at least, to the recent interest in sea-watching.

There was a pronounced peak in numbers in 1983 as a result of an exceptionally severe Atlantic storm on 3 September which brought many individuals, predominantly adults, close inshore in south Wales and south-west England. During the period 3–6 September there were 37 birds recorded in Wales (one in Caernarfonshire, three in Cardiganshire, four in Pembrokeshire, five in Carmarthenshire, up to 20 in Glamorgan and four, the first county records, in Monmouthshire). Over 100, almost all adults, were seen at St. Ives, Cornwall, on 3 September.

There are records from all Welsh coastal counties, notably Anglesey and Caernarfonshire in the north (19 records of 25 individuals and 32 records of 44 individuals respectively), and Pembrokeshire (73 records of 138 individuals) in the south, reflecting the distribution of coastal headlands jutting out into the Irish Sea. Strumble Head provides the vast majority of the Pembrokeshire sightings. There are, not unsurprisingly, no records from inland counties.

Bonaparte's Gull *Larus philadelphia*

Gwylan Bonaparte

A vagrant

There have only been four records of this rare trans-Atlantic vagrant which breeds in northern North America and winters from the USA to northern Mexico and the West Indies.

The first was one shot at Solva (Pembrokeshire) in spring 1888 and the second did not occur until nearly one hundred years later when an adult in winter plumage was recorded feeding off the West coast of Bardsey Island on 6 and 7 November 1984.

In 1986 a first-summer bird was seen in the North Cornelly/Kenfig Pool/Sker Point area from 14 to 17 March and in the Tremorfa Park and Rhymney River area (Glamorgan) on 21 March and from 8 April to 29 May.

In 1990 an adult was seen at Inner Marsh Farm (Flintshire/Cheshire) on 2 December.

Black-headed Gull *Larus ridibundus*

Gwylan Penddu

A locally common breeding resident, scarce only in south and south-west Wales in summer months. A common and widespread passage migrant and winter visitor.

Unlike other British breeding gulls, the Black-headed Gull is a scarce breeder on the rocky coasts and offshore islands of Wales, preferring instead to nest in colonies on the edges of marshes and pools and on small islands in the middle of remote lakes. In some situations, the nests are vulnerable to changes in water level and ground predation, and accordingly colonies are often very mobile, with birds vacating traditional breeding sites for several years before returning. Heavy persecution by shooting and the harvesting of eggs almost exterminated this species during the 19th century, but numbers have increased gradually since that time.

Much of our knowledge of Black-headed Gull breeding colonies in Wales is derived from three national surveys undertaken in 1938 (Hollom 1940), 1958 (Gribble 1962) and 1973 (Gribble 1976). Gurney (1919) also undertook a survey throughout Britain in the early years of the current century although, as was so often the case, coverage for several Welsh counties was poor. An accurate census of the number of breeding pairs was not always possible, but these four surveys demonstrated that the number of occupied colonies showed a gradual increase from 15 colonies in 1919 to 34 in 1938, 33 in 1958 and 64 in 1973. Unfortunately, the locations and sizes of individual colonies are not always given in the 1958 and 1973 surveys and therefore comparisons can only be made for the number of colonies and total number of breeding pairs for each county.

Anglesey

Two colonies were occupied in 1938, Llyn Llywenan and Newborough Warren. The former was colonised sometime around 1895 and in 1904 T. A. Coward (Diaries) found at least 50 pairs nesting. This must have been a low estimate however as, strangely, Coward counted at least 500 pairs here the following year, although by 1936 the population had decreased again to c300 pairs. C. Oldham found a colony of more than 50 pairs on marshy ground between Llyn Dinam and Llyn Penrhyn, but Hollom (1940) thought it doubtful that this site was occupied in 1938. The second active colony located during the 1938 survey was a small one consisting of some four pairs in Newborough Warren.

In 1958, two colonies are known to have been occupied, Llyn Llywenan and Mynydd Bodafon. The former was the larger of the two and contained about 300 nests whereas only about six pairs were recorded at the latter. By 1973, the maximum number of pairs had increased to 849 at five colonies, the largest being at Beaumaris and Bodedern with 225–245 and 360–445 nests respectively.

Caernarfonshire

Pennant recorded that Black-headed Gulls were nesting at Llyn Conwy and Llyn Llydaw during the 18th century but both of these colonies had been deserted by 1900. In 1938, between 90 and 330 pairs were recorded at four sites, the largest colony at Llyn Ty'n-y-Mynydd numbering a maximum of 150 pairs. A small colony at Llyn Ystumllyn was destroyed by gamekeepers each year and a few pairs had returned to Llyn Conwy by the late 1930s. By 1958, 230 pairs were found at four sites but the 1973 survey showed that the population had increased to between 481 and 698 pairs at three sites. The largest colony was located at Llyn y Dywarchen, where 400–600 nests were located. Several other sites had been used intermittently, including Llyn Elsi which held 100 pairs in 1969.

Denbighshire

In 1904 a colony of about 400 pairs was recorded at Llyn-y-Foel Frech where six years earlier only two or three pairs had been present. A decade later, a colony at Cyfynwy was destroyed by gamekeepers who believed that the gulls were depleting fish stocks in the lake. In 1938 between 590–840 pairs were recorded at three or four sites, the main colony at Llyn-y-Foel Frech numbering several hundred pairs. The 1958 survey revealed a total of only 50–55 pairs at two colonies, several of the formerly occupied sites having been lost due to drainage and human persecution. By 1973, however, the number of Black-headed Gulls breeding in the county had increased dramatically to 1181–1442 pairs at eight colonies, the largest being Fawnog Fawr with 560–675 nests.

Flintshire

Although almost exclusively an inland breeder in Wales, a single Black-headed Gull nest was discovered by Dobie at the Point of Air around 1870. It was near here, at Mostyn Marsh, that the only colony was discovered in the 1938 survey and although up to 20 pairs were present, all the eggs were taken. By 1958, Llyn Helyg was the only colony in the county, numbering some 400–500 pairs but in 1973 three colonies numbering a total of 305 pairs were recorded. Llyn Helyg was still the most important site, with 283 nests.

Merioneth

Forrest (1907) recorded how the colony at Llyn Mynyllod grew from only two pairs in 1888 to "quite two thousand birds" in 1904 when pairs were nesting on the lake shore as well as the small islets. This colony was still occupied in 1938 when 315–365 pairs were located at five or six sites in the county. Three coastal sites at Harlech, Mochras and near Llanaber were occupied during the survey although a colony of about 50 pairs on Cefndu Moor was shot out and no nests were located at Llyn Trawsfynydd where the birds had been "thoroughly discouraged" over a number of years. In 1958, between 300 and 355 pairs were recorded at four sites but by 1973, the county total had increased to 1434–1514 pairs at six sites. Several of these were small colonies on upland lakes and bogs but the one at Llyn Trawsfynydd then contained 1116 nests, making it one of the largest in Wales.

Montgomeryshire

By 1910 small colonies had become established at Llanllugan and near Llandinam (Forrest 1919) and on 5 May 1939, C. Oldham reported "about 1000 on Llyn Mawr and 3000–4000 on Llyn Tarw where the majority at any rate nest on two boggy islets". The 1938 survey showed that Montgomeryshire held the bulk of the Welsh breeding Black-headed Gull population at that time with 3050 pairs located at four sites in the county; two of these (Llyn Tarw and Llyn Hir) supported over 1000 pairs each. The population had declined to around 1690–1750 pairs by 1958 with two sites holding over 700 pairs, and in 1973 1710–1883 pairs were located at 12 sites. The biggest colony located by the survey was Llyn Tarw which supported some 600 pairs, although large numbers were also located at Llyn Hir (400 pairs), Llyn Coethlyn (250–280) and Llyn Coch Hwyad (200–220).

Cardiganshire

In 1905, Salter believed that Black-headed Gulls were nesting on Cors Fochno and certainly breeding had been recorded at Cors Caron prior to 1925. However, by 1924 the only colony in the county was at Llyn Du where 200 pairs were nesting although three pairs had been found at Llyn Eiddwen in the previous year. By 1932, over one hundred pairs were nesting at Cors Caron and a colony of about 200 pairs was present at Rhos Cilcennin in 1933. F. C. R. Jourdain noted in his diaries that an enormous colony numbering several thousand pairs was found at Cors Goch Glan Teifi on 10 May 1934, a record which refers to the colony at Cors Caron.

In the 1938 survey, between 615 and 715 pairs were located at seven to ten sites, the largest being Rhos Cilcennin and Llyn Du with 150 nests each. The latter site held 200–300 pairs in 1927, was deserted in 1932, but recolonised in 1935 or 1936. In 1958, it was the most important colony in the county, supporting about 600 pairs out of the Cardiganshire total of 1000–1050 pairs at five sites. In 1973, 13 colonies held a total of 1739–2054 pairs, by far the largest colony being Cors Caron with 1000–2000 nests.

Radnorshire

The first recorded breeding for this county occurred around 1900 at Llanwefr Pool and by 1934 some 750 pairs were present here. In 1904, a few pairs nested on mawn pools near Rhulen which by 1930 had spread to adjacent boggy areas and numbered about 600 pairs. F. C. R. Jourdain recorded a colony of 5000 birds at Rhosgoch Common on 10 May 1932, although a figure of only 300 pairs is given for this site in 1934. In 1938, 342 pairs were recorded at four Radnorshire sites, the mawn pools at Rhulen having been deserted after all the boggy areas had dried out. The most important site was Llanwefr Pool, with about 300 pairs, although 2000 birds were often seen in the evenings. This site suffered heavy persecution during the Second World War when great quantities of eggs were taken.

In 1958, the county total had decreased to 250–300 pairs at six sites, Llanwefr Pool again the most important with 150 pairs. In 1973, 251–265 pairs were located at nine sites, Force

Hill Pool being the major colony with 113 nests. Numbers on Rhosgoch Common had increased to over 2000 pairs in 1954, but the population declined to 40 nests in 1958 and by 1973 the colony had been deserted as it dried out. Recently, birds have returned to this site and several colonies in the county now number over 50 pairs.

Breconshire

Black-headed Gulls first colonised Breconshire in 1908 when a few pairs nested on Mynydd Eppynt. By 1923 about 190 pairs were present at two separate sites and although one of these was deserted in 1926, pairs had already colonised a third site nearby. The 1933 survey located 80 pairs at these three sites on Mynydd Eppynt and in 1958, 70–80 pairs were found at two new sites, Cefn Lechid (c50 pairs) and Mynydd Illtyd (c12 pairs). By 1973, the county population had doubled to 152–162 pairs with three sites supporting over 35 pairs. One of these (Llyn Login) was one of the original sites colonised on Mynydd Eppynt. In 1989 and 1990, the most important site (supporting 150 pairs) was at Llyn Traeth Bach and the county total in any one year is not thought to exceed 200 pairs.

Carmarthenshire

A coastal colony was known to have been occupied near Witchett Pool on the Laugharne Burrows since 1926, when two nests were found. Although the actual site changed more than once, birds nested in the same general area until the 1960s with 52–60 pairs in 1938 and a maximum of 70 nests in 1953. Owing to Ministry of Defence restrictions, nothing further is known about the colony until 1967–69 when 20–25 pairs were suspected to be breeding at nearby Ginst Point, but no nests have been found subsequently. The colony may have moved to the Gwendraeth estuary as birds were seen on eggs here during the early 1970s. A colony of c22 birds was discovered at Llyn y Gwaith on 17 May 1971 and although no breeding occurred in 1972, two pairs nested in 1975. Although other inland sites have been used intermittently (including 60–70 birds at Pontfen Isaf in 1951), by the late 1970s summering birds were reduced to a small number of non-breeders near the coast.

Pembrokeshire

There is said to have been a large Black-headed Gull colony breeding on Caldey Is. in 1662 but there are no other records for the county until three or four pairs were seen near St. David's in 1947. The following year, Dowry Pond and Trefeiddan were colonised, the former building up to 80 pairs by 1958 but dying out shortly after 1966. Trefeiddan was also deserted around the same time after supporting 18 pairs in 1958. Between three to four pairs bred on Skomer between 1965 and 1970 and similar numbers bred on Ramsey in 1962, with 16 pairs in 1964. There has been no subsequent breeding in the county.

Glamorgan

Five pairs nested among the dunes at Kenfig in 1899 and although the colony was relocated in 1911, by 1913 the pairs had moved to nearby small islets as a result of the unusually high water level. Birds last bred here in 1936 or 1937, the decline almost certainly caused by increasing disturbance. In the summer of 1942, a few pairs were found nesting on a reed-fringed pool at Eglwys Nunydd; this whole area was later flooded with the creation of the steel works' reservoir. Two colonies were located in 1946, one at Oxwich and another at Broad Pool on the Gower. The latter never numbered more than nine nests, but the former reached a peak of 200 pairs in 1950. Numbers were much reduced by the 1960s (c40 pairs in 1964) and it is believed that its eventual extinction was caused by fox predation. In 1956, a small colony was located at Crymlyn Bog NNR, but breeding was not recorded again until 1962. That year, c50 pairs bred but all failed because of predation by rats and none returned to breed in 1963. However, the site was recolonised and 18 pairs bred in 1984 (25 in 1985) but it had become deserted once more by 1990. A few pairs bred on a small upland lake at Seven Sisters during the 1970s and by 1991 an estimated 60 pairs were present.

Since the last national survey in 1973, few data have been collated other than for the breeding atlases of Pembrokeshire, Breconshire and Monmouthshire. In the last county, breeding was first recorded on a reservoir near Brynmawr in 1982 when six nests with eggs were found. There has been sporadic breeding at this site more recently although the number of pairs involved is always small. In 1985, one nest was located at the nearby Garnlydan Reservoir but it failed because of flooding.

Over the past twenty years there has been a marked decline in the population of breeding Black-headed Gulls in Wales. In Montgomeryshire, two sites, which between them held over 1000 pairs in 1973, Llyn Hir and Llyn Tarw, now hold fewer than 100 pairs, and at least three other sites have been deserted due to afforestation. In Cardiganshire, the colony at Cors Caron had no birds breeding in 1984 although over 300 pairs had been recorded in 1981. In 1991, however, a new colony of some 80 nests had become established at a newly created pool. At least eight colonies were abandoned in the county during the 1970s and 80s, some owing to the growth of vegetation, others because of agricultural improvement, but most because of unknown factors. In Anglesey, the number of breeding Black-headed Gulls decreased during the 1970s as a result of the spread of foxes onto the island, and many former inland haunts are now deserted. However, a flock of several hundred breeding pairs is present at Cemlyn and sporadic breeding has occurred in recent years on the Skerries.

The declines in Montgomeryshire and Cardiganshire, as far as is known, are reflected in most other Welsh counties and it is likely that the current breeding population numbers fewer than 4000 pairs.

In winter the Black-headed Gull frequents a wide range of habitats, both coastal and inland. In Wales, it is absent from most of the central and north Wales uplands

Location and size of Black-Headed Gull colonies in Wales in 1973, 1958 and 1938

	1973 No. of colonies	1973 Numbers Min.	1973 Numbers Max.	1958 No. of colonies	1958 Numbers Min.	1958 Numbers Max.	1938 No. of colonies	1938 Numbers Min.	1938 Numbers Max.
Anglesey	5	703	849	2	306	506	2(1)	305	305
Caernarfonshire	3	481	698	4	230	230	4(1)	90	330
Denbighshire	8	1181	1442	2	50	55	3(1)	590	840
Flintshire	3	305	305	1	400	500	1	10	20
Merioneth	6	1434	1514	4	300	355	5(1)	315	365
Montgomeryshire	12	1710	1883	4	1690	1750	4	3050	3050
Cardiganshire	13	1739	2054	5	1000	1050	7(3)	615	715
Breconshire	5	152	162	2	70	80	3	80	80
Radnorshire	9	251	265	6	250	300	4	342	342
Carmarthenshire	–	–	–	–	–	–	1	52	60
Pembrokeshire	–	–	–	2	98	98	–	–	–
Glamorgan	–	–	–	1	50	50	–	–	–
Total	64	7956	9172	33	4444	4974	34(7)	5449	6107

although it is numerous in many coastal areas. In England the species is more frequently encountered on inland waters during the winter months and Bowes *et al.* (1984) estimated that out of a total population of 1 877 000 Black-headed Gulls wintering in Britain in January 1983, over 1 000 000 were found inland. Ringing recoveries indicate that most of the breeding population remains in Britain throughout the winter (Flegg & Cox 1972) and that large numbers of immigrants from as far afield as Poland, USSR and Scandinavia augment the resident population (Horton *et al.* 1984).

Forrest (1907) mentions only one large flock of Black-headed Gulls in north Wales on 10 July 1903 when "the whole estuary of the Dovey was dotted with them", an estimated total of over 1000 birds. At that time, it was said to resort almost exclusively to the coast in winter. The increase in the breeding population led to a parallel increase in wintering numbers and, although national censuses organised by the BTO in 1953 and 1963 did not provide coverage for Wales, some coastal counts of roosting gulls were undertaken in the counties of Merioneth, Cardiganshire, Pembrokeshire and Carmarthenshire in 1955–56 (Hickling 1960). The largest numbers of Black-headed Gulls were found at Aberystwyth (Cardiganshire) and Black Scar (Carmarthenshire), with totals of up to 1750 at the former and 1500 at the latter.

A further BTO survey of winter gull roosts at inland sites in 1973 produced a total of 10 635 Black-headed Gulls in Wales. Most roosts were on the larger lakes although, again, coverage for the Principality was not comprehensive. A more complete census in 1983 gave a total of 15 559 birds at coastal and inland sites (Bowes *et al.* 1984). The true population in Wales had probably changed little during the intervening ten years, although larger roosts were found inland during the 1980s. On 19 March 1982 the roost at Lake Vyrnwy (Montgomeryshire) contained over 6000 Black-headed Gulls. In February 1986, 4750 were present at Blackpill (Glamorgan), whereas during the early 1960s, the largest flocks generally had numbered no more than a few hundred (e.g. c400 on Hensol Castle Lake on 26 November 1964). An estimate of the north Seven coast population in the winter of 1974–75 revealed a total of 32 650 birds.

Recoveries of Black-headed Gulls ringed in Wales

Currently, sizeable winter roosts are found on several inland and coastal waters, the largest being at Blackpill (Glamorgan), where 6320 were recorded in November 1991, and Llangorse Lake where 5600-10 000 are sometimes recorded in late winter.

Non-breeding birds and immatures are seen in every

Welsh county throughout the year (e.g. c150 first-year birds summer on the Cleddau estuary in Pembrokeshire) often feeding in large flocks in the main river valleys. Passage birds are also encountered throughout Wales as they make their way to and from the breeding grounds and when continental birds arrive in October/November. Passage flocks have been recorded at South Stack in Anglesey (55 flew north on 15 July 1987) and all the major offshore islands. Small numbers are recorded every month on Bardsey, but a huge influx is experienced in October and November, with a maximum of 4000 on 1 October 1961. On Skokholm, Black-headed Gulls are usually recorded between August and April, but here again the highest numbers are in late autumn.

During the 1980s, several observers recorded small flocks of pink-plumaged Black-headed Gulls in Wales. Four were present on Malltraeth (Anglesey) from August until November 1986 and a small flock was also present at Beddmanarch Bay in the same county in late August 1982. Pink gulls were also seen in other parts of Britain during this period and were thought to be part of a ringing programme.

Relatively few foreign-ringed birds have been recovered in Wales compared with the numbers in eastern England (Horton *et al.* 1984), but have come from as far afield as Norway (5), the Baltic States (12), Poland (6), Denmark (1), the Netherlands (5) and Belgium (2). A total of 49 Welsh-ringed birds have been recovered abroad, 35 of these in Ireland.

Ring-billed Gull *Larus delawarensis*

Gwylan Fodrwybig

An uncommon but regular visitor to coastal areas of Wales, chiefly in the period from mid-February to the end of May.

This species has a special place in the annals of Welsh ornithology. On 14 March 1973 R. A. Hume noticed a pale gull amongst a large flock of Common Gulls resting on the beach at Blackpill, Swansea Bay. From detailed notes he identified it as Britain's first Ring-billed Gull, an American species whose occurrence had long been expected but never proved (Hume 1973). During the subsequent three years a small group of Swansea students and the local enthusiasts found more Ring-billed Gulls at Blackpill and by the end of 1975 a total of 11 individuals had been recorded (Vinicombe 1973, 1975). In 1976 the Blackpill monopoly was finally broken and by the end of 1980 the British and Irish total had risen to 44 individuals from 11 counties, exactly half of these having been at Blackpill (Vinicombe 1985).

In following years numbers in Britain and Ireland increased dramatically with 55 in 1981, 75 in 1982 and 84 in 1983. There can be little doubt that Ring-billed Gulls were occurring here well before 1973 and the initial records at Blackpill resulted in a wider understanding of the subtle field characters of the species, a process assisted by the coincidental appearance in the March 1973 issue of *British Birds* of a paper on its identification. The 1981 influx was set against a background of a large increase in the breeding population in the United States and Canada coupled with a widespread dispersal caused by prolonged severe weather in 1980–81 in the main wintering areas in the south-eastern United States.

The early Blackpill records soon established a regular pattern of occurrence:

i) wintering adults from about late November to late March;
ii) additional "passage" adults in late March and April;
iii) "passage" second-year birds in April and May; and
iv) summering first-year birds in June and July.

At Blackpill in the mid-1970s wintering adult Ring-billed Gulls were occurring in Common Gull flocks at a rate of one or two in about 2000 to 3000.

The Ring-billed Gull was removed from the list of species regarded as British rarities in December 1987 since which time records have been assessed by the Welsh Records Advisory Group. It is clear, however, that the recent picture is far from complete since not all observers now submit records and/or field descriptions. There therefore has to be a caveat that the published figures to date give an under-estimate of the true status. Between 1973 and 1991 the total figure amounts to 168 of which no less than 132 (79%) come from Glamorgan. Other coastal counties from which accepted records originate are Cardiganshire, Pembrokeshire, Carmarthenshire, Monmouthshire, Merioneth and Flintshire. There are only four records from counties north of the Dyfi, confirming that the Ring-billed Gull is very much a south Wales speciality.

This species has rarely been recorded away from coastal localities and there is only one record from an inland county (Breconshire in 1988). Records are most numerous in the three-month period between 10 February to 10 May though the species has occurred in all months of the year. Adult birds are recorded far more frequently than any other age group and records of first-year birds are few.

Common Gull *Larus canus*
Gwylan y Gweunydd

Small numbers formerly bred for a short period on Anglesey. It is now a numerous and widespread winter visitor with small numbers recorded on passage.

The Common Gull is a common breeding bird on inland hills and coasts in the north and west of Britain but sporadic breeding outside the main Scottish and Irish nesting grounds is thought to stem either from Continental winter visitors which remain to breed, or an expansion from the British stronghold. However, nothing is known about the origin of the birds that formed the small Welsh colony in Anglesey during the 1960s and 1970s. Elsewhere in Wales, there have been sporadic claims of Common Gulls breeding, but these records have almost certainly involved birds that were summering here but not attempting to breed.

Forrest (1907) knew of no breeding records in north Wales and in 1919, he believed that adults and juveniles seen together around Criccieth in June 1908 and at Conwy Sands (both Caernarfonshire) on 28 June 1913 were not Welsh-bred birds. It was not until around 1960 that breeding was first proved at the Flagstaff quarry near Penmon on Anglesey although confusion surrounds the exact year of the establishment and subsequent extinction of the colony.

Between 1960 and 1963, less than half a dozen pairs of Common Gulls nested at this site, scattered among a larger colony of Herring Gulls. Breeding continued, probably annually, until by the late 1960s only three pairs were present. During the early 1970s, the numbers of ground-nesting gulls on Anglesey decreased markedly with the reappearance of foxes on the island (E. I. S. Rees, pers. comm.), and the number of Herring Gulls and Common Gulls at Flagstaff quarry also declined during this period. The exact year of extinction of the Common Gulls is not known, but pairs are unlikely to have nested here after the late 1970s. The *Breeding Atlas* also reported nesting Common Gulls at three other sites in Anglesey and north Caernarfonshire, but these records almost certainly involved immature birds from the Penmon area.

It is only in winter that the Common Gull becomes numerous and widespread in southern Britain owing to an influx of Continental birds. In Wales, its winter distribution is mainly coastal although sizeable flocks roost on some inland reservoirs. Immigrants arrive from July until late October with return movements occurring mainly in April.

Forrest (1907) found Common Gulls widespread along the north Wales coast in winter but rare inland except following gales. In Pembrokeshire at the turn of the century, however, Common Gulls were often encountered inland, especially in association with Rooks. During this period, it was considered to be an uncommon winter visitor to Glamorgan and the inland county of Radnorshire, but it was described as common following storms in Breconshire. Most inland records were of single birds, although four were seen together near Builth (Breconshire) on 28 December 1881.

The number of wintering birds increased gradually over the first two decades of this century and by 1925, it was said to be common along the north coast of the Bristol Channel. Flocks were often seen feeding on estuaries and coastal fields throughout much of south Wales at this time, although there is little subsequent information on numbers from any Welsh county until a survey of coastal gull roosts in 1955–56 (Hickling 1960) in which, sadly, coverage for Wales was poor, with records collated from only four coastal counties. The largest number of Common Gulls recorded in Wales was 2000 at three roosts on the estuaries of the Taf and Tywi (Carmarthenshire). Smaller roosts of up to 400 birds were recorded on the Dyfi estuary and at Broadwater (Merioneth). Throughout the 1950s, small numbers were also recorded inland with four seen together in Montgomeryshire on 4 December 1957 (*Montgomeryshire Field Society Report* 1957).

During the 1960s, up to 1000 were regularly recorded at roost in Aberystwyth (Cardiganshire) and there was an exceptional count of c500 birds at Oxwich Marsh NNR on 19 January 1963. In the winter of 1963–64, large numbers were observed flying from the Hereford area over Abergavenny (Monmouthshire) along the River Usk and across the Bristol Channel to roost at Avonmouth. Towards the end of the decade, and throughout the 1970s, a small roost of 10–60 birds was occasionally recorded either at Llangorse Lake or Talybont Reservoir and although this disappeared during the 1980s, a small roost was observed here again in 1990.

A survey of inland roosts in England and Wales in 1973 revealed a total of 2960 Common Gulls at five roosts in the Principality (Hickling 1977). Three of the roosts contained fewer than 50 Common Gulls although 200 were counted among a mixed flock of 1450 gulls on Ty Mawr Reservoir in Denbighshire. By far the largest roost in Wales however was one of 2700 among a mixed flock of 3108 gulls on Llyn Tegid (Merioneth). No common gulls were recorded in the counties of Caernarfonshire, Glamorgan, Montgomeryshire or Pembrokeshire.

In 1983, another midwinter gull roost survey organised by the BTO revealed a total of only 87 Common Gulls inland and 25 447 at coastal sites in Wales (Bowes *et al.* 1984) and although coverage in the Principality was not complete, it was better than for previous surveys. By far the largest numbers (11 787) were recorded on the Dee estuary, 10 177 of which were in Wales. These consisted of three roosts, the two largest located at Flint (5530) and Gronant (4300). Elsewhere, sizeable roosts were found along the south Wales coast, where a large roost of several thousand Common Gulls has been present at Blackpill on the Gower (Glamorgan) for over 20 years, with 3100 recorded in February 1986. Numbers here have increased from a maximum of only a few hundred birds during the 1960s.

The *Wintering Atlas* showed that the largest concentrations in Wales were found along the north-east coast with smaller numbers around the north-west and Cardiganshire coasts and on the north shore of the Bristol Channel. In the winter of 1974–75, the total population of the north shore of the Severn estuary was estimated at 1020 birds, but the largest flocks in Monmouthshire are usually found inland (e.g. 4000 at Llandegfedd Reservoir in February 1990). In Cardiganshire, a roost of several hundred is present on the Dyfi estuary most years, generally near Borth, but an exceptional roost of 1349 was recorded here on 25 November 1988 and a huge influx of Common Gulls was reported from Borth Bay the following day although no total was given.

In Pembrokeshire, the winter total is thought to number c4000 birds, with the principal roosts at Fishguard Harbour, the Cleddau estuary and the Amroth Saundersfoot area. Elsewhere, roosts are generally much smaller, particularly inland where it is usually seen in flocks of less than 30, often in association with other gull species.

Common Gulls are regularly recorded on passage from all counties. Vernon (1969) described a northern and westward movement of birds through Wales in spring, and large flocks recorded in Radnorshire during February are thought to be the result of pre-migratory flights from the Severn estuary roosts. These birds are believed to follow a route eastward across Oxfordshire and presumably to Holland, Germany and Denmark, or north-west, reaching the Continent via Cheshire and Lancashire (Vernon 1969).

Most passage birds are recorded along the coast and on occasion, large flocks are recorded immediately prior to the main eastward movement in late February and March (e.g. 1593 at Cydweli Marsh in Carmarthenshire on 13 February 1977). In the autumn, some adult birds arrive in August, sometimes accompanied by immatures, although the main westward movement occurs in October. Several thousand birds are regularly recorded at the Point of Air in August, the maximum recently being 3000 on 20 August 1989. During the autumn, small flocks are also recorded inland and it is during this period that Common Gulls are most numerous on the offshore islands. On Bardsey and Skokholm, numbers peak in October and November with up to 30 recorded daily at the former island.

Ringing records confirm a westward movement of Continental birds into Wales over the winter with recoveries involving birds ringed in Norway, Finland and Denmark. The records also indicate that Scottish birds show a southerly movement in winter, six birds from Scotland having been recovered in Wales.

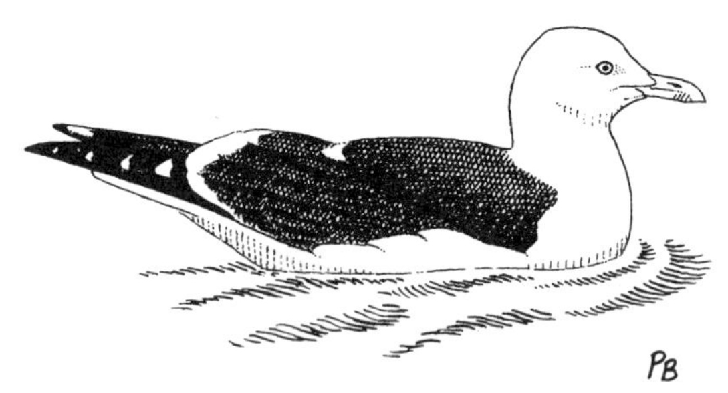

Lesser Black-backed Gull *Larus fuscus*
Gwylan Gefnddu Leiaf

Breeding summer visitor, passage migrant and winter visitor.

The Lesser Black-backed Gull is a numerous breeding species around the Welsh coast albeit somewhat restricted in its distribution, with its major colonies on the Pembrokeshire islands and lesser ones in Anglesey; that on Skomer, peaking at c17 500 pairs in 1988, is the largest in Britain. Breeding birds return to the colonies during March and depart again in August and September. Forrest (1907) recognised the spring and autumn passage of birds in north Wales, both inland and along the coasts, but found it very restricted as a breeder, limited to colonies on Little Orme, Puffin Is. and to the north coast of Anglesey; only that on Puffin Is. appears to have been of substantial size (no figure given). In south-west Wales, Mathew (1894) described it as a common resident in colonies of 20–30 "on the various islands". In Glamorgan, the only colony in the early years of the century was a small one of 12–14 pairs on Worm's Head, Gower, which had dwindled to three or four pairs by 1936 and has since ceased to exist. An interesting inland colony of up to 50 pairs existed on Cors Caron up to 1933–34 (when they were driven away by game shooting interests); this colony was already well established when first recorded by Salter in 1893. It can be speculated that the total of Welsh breeding pairs at the beginning of the 20th century may well have been no more than 500.

The expansion of this species, poorly documented in Welsh literature, seems to have started in a limited way around 1900 and continuing to 1940 or thereabouts. After that the population sank again to a low point in 1955 (probably due to persistent heavy egg collecting for human consumption) before going into a period of very rapid increase up to the late 1970s. Since then numbers, at least in the large Pembrokeshire colonies, have fluctuated year by year but have been more or less stable overall. Coastal breeding was first noted in Cardiganshire in the 1930s (possibly with birds displaced from Cors Caron) and by 1951 there were several small colonies there. In Glamorgan, while the Worms Head colony declined, occasional pairs appeared elsewhere on the southern coast in the 1940s and 1950s; the Flatholm colony off Penarth was unknown before 1956 but had risen hugely to c1000 pairs a decade later. Lockley (1949) recorded 1000 pairs on Skomer and 800 pairs on Skokholm compared with the groups of 20–30 Mathew described fifty years earlier. In north Wales, they first bred on Bardsey in 1930 and by the same date the modest colony on Great Orme, established some years earlier, had increased further. The Puffin Is. birds are recorded as increasing by 1953 although it was noted as having been "considerably reduced" again by 1956; other Anglesey colonies – Bodorgan Head, Newborough Warren – continued to increase across the same time.

It is relevant to stress that large numbers of Lesser Black-backed Gull (and Herring Gull) eggs were collected for human consumption in the first half of the present century. Annual outings to the Pembrokeshire islands by Marloes villagers and fishermen to gather eggs were a feature of local life until the 1950s; in the late 1930s and up to 1946 R. M. Lockley alone took in the region of 3000 eggs annually, latterly to feed his visitors on Skokholm.

Concurrent with the increase in coastal colonies in the 1950s, the habit – shared with Herring Gulls – of nesting on the flat roofs of buildings developed at one or two sites in Glamorgan and Monmouthshire. First discovered at Merthyr Tydfil in 1958 – where birds had apparently bred for 13–14 years unrecorded – the roof-nesting habit spread by 1969 to Cardiff, Barry, Newport and even Hirwaun (29 km inland). By 1975, there was a total of 204 roof-top pairs, rising to 260 pairs by 1980 and spreading to Penarth. In more recent years, roof-nesting has also occurred sparingly in Pembrokeshire and the Swansea area.

The Seafarer Survey of 1969–70 produced an overall total of 11 500 pairs in Wales and in the years following this there were impressive increases in the size of breeding colonies, exemplified especially on the islands of Skokholm and Skomer and Flatholm (table below) (Ferns & Mudge 1981). This great increase in numbers since the 1950s is seen (Lloyd *et al.* 1991) to be in parallel to the rapid development of a *Nephrops* (scampi) fishery on the Smalls Grounds 65 km west of Pembrokeshire, which resulted in a large number of unwanted fish being discarded overboard and being made readily available to the gulls, certainly those from the Pembrokeshire islands, if not from farther afield.

On Skokholm and Skomer, Lesser Black-backed Gulls suffered disastrous breeding seasons in 1989, 1990, 1991 (about 50% of normal) and 1992 with remarkably few chicks reared in any season. It is subsequently considered that 1987 may also have been very poor and 1988 only slightly less so. Breeding failure is largely attributed to changes in the availability of fish in the offshore sea areas. The amount of discarded fish (including a Whiting *Gadus merlangus* and Poor Cod *Trisoperus minutus*) declined substantially with the increase in legal net size used for the Irish Sea *Nephrops* fishery after 1986. Whiting and Poor Cod (obtained in the sea area around the Smalls Grounds) are known to have formed substantial percentages of the food supplied to nestling Lesser Black-backed Gulls up until that time (Todd 1987).

The Seabird Colony Register 1985–87 showed an increase of some 75% in Wales since 1969, producing a total of 19 000 pairs of which 16 560 were in

Numbers of breeding pairs of Lesser Black-backed Gulls on Skokholm, Skomer and Flatholm 1969–90

Year	*Skokholm*	*Skomer*	*Flatholm*
1969	c2000	5559	1100
1975	c3000	7000	4055
1979	4600	9600	NC
1980	NC	13 030	2379
1985	4186	13 200	1740
1988	2706	17 500	1400
1990	2605	12 800	1400
1992	3017	16 000–18 500	1900

NC = no count

Pembrokeshire (compared with 20 300 for the county in the peak year of 1983). Other than the Pembrokeshire colonies and the adjacent one on Cardigan Is. (c1500+ pairs in 1990 rising to a huge total of 3517 nests in 1993), the concentration of breeding Lesser Black-backed Gulls is to be found mainly around the Anglesey coasts (see table). In the early 1990s small numbers of breeding pairs are to be found in Monmouthshire (c50 pairs on Denny Island), Merioneth (c20 pairs on Llyn Trawsfynydd), somewhat larger numbers in Glamorgan (c500 pairs), Cardiganshire (1500+ pairs) and Caernarfonshire (c240 pairs).

Lesser Black-headed Gull colonies of 100 pairs or more 1969–92

Colony	*Maximum count*		*Minimum count*	
Flatholm	1975	4055+	1969	1,100
Cardiff rooftops	1987	200+ (still increasing)		
Skomer	1988	17 500	1969	5559
Skokholm	1979	4600	1969	2000
Caldey Island	1992	198	1987	71
Cardigan Island	1990	1500	1969	8
Bardsey Island	1989	260	1969	4
Skerries	1986	505	1969	116
Carmel Head (Anglesey)	1969	800	–	–
Bodorgan Head (Anglesey)	1986	2000	1969	65
Middle Mouse (Anglesey)	1987	110	pre 1986	0
Porth Llanlleiana (Anglesey)	1969	119	post 1970s	0
Newborough Warren (Anglesey)	1969	c400	NK	0
Amlwch (Anglesey)	1986	156	NK	0
Freshwater Bay (Anglesey)	1969	281	1986	6
Penmon Point (Anglesey)	1969	200	1984	3
Llyn Trawsfynydd	1986	115	1992	30–40

NK=date not known

Most of the breeding birds leave the colonies once the young have fledged and depart southwards to winter in waters off Iberia with some Welsh birds (substantiated by ringing recoveries) reaching as far as Senegal, the Canary Is. and the Gulf of Guinea. Other individuals ringed on Skomer or Skokholm have been found subsequently as adult birds in Iceland (1), Faeroe Is. (3) and Denmark (2). Up until the 1960s, few Lesser Black-backed Gulls were recorded as overwintering in Britain (although Barnes (1961) pointed out that in ornithological literature up to 1910 the species was usually referred to as "resident") and they were recognised as genuine breeding-season visitors.

From the 1960s, increasing numbers have wintered in Britain, visibly taking advantage of the abundance of inland roosting sites such as reservoirs and increased supplies of food from artificial sources (Lloyd *et al.* 1991). Griffiths (1971) has shown how passage numbers in north Glamorgan and south Breconshire burgeoned around 1970, with autumn totals rising to 1300 at Taf Fechan reservoirs from earlier maxima of no more than 50–100). By 1979–80 Lesser Black-backed Gulls were recognised as regular overwintering birds in many parts of Wales with principal roosting concentrations in Pembrokeshire (Llysfran Reservoir 700, Rosebush Reservoir 130) and Merioneth (Trawsfynydd 50). In the national survey of inland roosting gulls in January 1983, the number recorded in Wales had reached 1582. Numbers have continued to increase in the 1980s particularly in Pembrokeshire, where wintering maxima rose to 4500 in 1984, 700 in 1987 and 10 000 by 1988.

Within the relatively small global breeding range of the Lesser Black-backed Gull (Iceland/northern Russia/Portugal) there are five well-defined subspecies. All the birds breeding in Britain and Ireland are of the race *Larus fuscus graellsii*. The race *L.f. intermedius* breeds in western continental Europe and *L.f. fuscus* in northern Scandinavia. Although occasional individuals of the darker-mantled *L.f. fuscus* have been recognised as long ago as 1928 (Glamorgan) and 1942 (Carmarthenshire) in winter, the number recorded in more recent times has increased and individuals of that race and *L.f. intermedius* are now recorded annually in very small numbers. Two birds ringed in Wales have later been recovered in the Faeroe Islands (July and August in subsequent years) and one in Iceland (August).

Herring Gull *Larus argentatus*

Gwylan y Penwaig

An abundant resident in coastal areas. Frequent in many inland areas, breeding on several freshwater sites.

From the earliest records it is clear that the Herring Gull has always been recognised as the most numerous and widely distributed of gulls around the Welsh coast; as early as the third or fourth century, Herring Gulls were prominent among the remains of a range of bird species subsequently excavated from middens at Roman Caerleon (Zienkiewicz 1986).

Forrest (1907) recognised the ubiquitous occurrence of the species in north Wales, but identified Bardsey, the Gwylan Is. ("Gull Islands"), Puffin Is. and the Orme headlands as the principal nesting colonies. He pointed out that at the turn of the century there was no breeding on the coasts of Merioneth, Denbighshire or Flintshire. In south Wales, at the same time, breeding was unknown in Monmouthshire and Carmarthenshire and in Glamorgan occurred only on the Gower cliffs. However, Mathew (1894) testified to the numbers breeding on the Pembrokeshire islands and mainland cliffs and this country evidently held the bulk of the south Wales population. Herring Gull eggs have long been collected for human consumption and the introduction of the Sea Birds Preservation Act (1869) and subsequent legislation undoubtedly signalled an upturn in the species' fortunes and provided the springboard for 20th century expansion although up to the 1950s the annual harvest was still heavy in some areas, e.g. Pembrokeshire.

In common with other *Larus* gulls, the Herring Gull has enjoyed a rapid expansion of population during the second half of the century. Between the 1940s and 1970s, the rate of increase was at a steady 10%–13% per annum with traditional colonies expanding, new ones being formed and even inland breeding becoming a regular habit. This great increase has been strongly linked to the all-year-round availability of food from two principal sources: the post-war increase in the fishing industry and the proliferation of open refuse tips. However, as the fishing industry at Milford Haven actually decreased in the latter part of this period it may be that the reduction of human persecution (e.g. egg collecting after the Second World War) was also a contributory factor. Up to 1975, as the population increased, these same tips then became the source of summer outbreaks of botulism poisoning and a steep decline in numbers in the 1980s is principally attributed to this cause, the catalyst for which may well have been the increasing use of black polythene refuse bags (generally introduced in 1974) which provide ideal incubators in summer heat for the bacterium *Clostridium botulinum*. Other factors obviously had a part to play, e.g. the introduction of foxes onto Anglesey and substantial vegetation changes on Puffin Is. following myxomatosis, which affected the huge colony there. The net result of this reversal of fortune was a decrease in the breeding population of 78% in Wales between the two complete censuses of 1969–70 and 1985–87.

Data annually recorded on Skokholm since the 1940s are the most accurate for any of the Welsh colonies. In summary they show a marked increase up to around 1950 followed by a decrease over the next ten years or so (possibly owing to the effects of heavy egg collecting?) and then another rapid increase in the 1960s, levelling off in the 1970s (S. J. Sutcliffe, in litt.).

Breeding numbers were at their highest in Wales in the early 1970s. The Seafarer Survey (1969–70) produced a total of some 48 500 pairs concentrated in the counties of

Pembrokeshire, Caernarfonshire and Anglesey. By the time of the Seabird Colony Register (1985–87) the population had fallen to fewer than 11 000 pairs. The fluctuations of the population this century are exemplified by figures for a selection of individual colonies.

Numbers of breeding pairs of Herring Gulls at selected sites in Wales 1934–91

Year	*Skokholm*	*Skomer*	*Puffin Is.*	*Grassholm*	*Caldey*	*Ramsey*
1934	270	–	–	104	–	–
1945	–	660	–	–	–	–
1949	572	–	–	–	–	–
1963	–	–	–	203	–	–
1969–70	1350	1985	15 500	–	3250	204
1973	1400	–	–	500	–	1003
1975	–	–	–	–	3857	–
1979	–	2350	–	–	–	–
1981	600	1409	–	40	–	–
1987	335	8000	1000	–	–	–
1991	–	465	–	–	802	–
1992	350	525	–	–	876	c400

The habit of rooftop-nesting dates back to the 1940s when it was first discovered at Merthyr Tydfil (Glamorgan), thereafter spreading into other urban areas in south Wales up to the the mid-1970s (Hurford 1984) when the Cardiff colonies were the largest rooftop ones in Britain. Elsewhere in Wales the habit also spread to other coastal towns until Herring Gulls became a common feature – frequently with high nuisance value – in many coastal towns from Barry (Glamorgan) to Milford Haven (Pembrokeshire), Barmouth (Merioneth) and Llandudno (Caernarfonshire).

During the 1960s and 1970s, when numbers were at their highest, inland nesting on upland lakes in north Wales became a feature, at least temporarily in most cases although at some sites the habit was perpetuated and sporadic breeding still occurs. The principal site was on Llyn Trawsfynydd where up to 100 pairs nested in 1973. On most other lakes,e.g. Llyn Arenig Fach, Llynnau Gamallt (Merioneth), Llyn Elsi, Llyn Edno (Caernarfonshire), only single pairs were normally involved although at Llyn yr Adar in the mountains above Blaenau Ffestiniog up to 30 birds were nesting in 1973 (Hope Jones 1976). The habit faded as numbers nationally decreased and such breeding is now very sporadic.

Although, like most of the southern British population, Welsh Herring Gulls are principally regarded as being sedentary, there is an element of dispersal, particularly of young birds, with ringing results producing recoveries from western France and Iberia (36), Poland (1), Norway (1) and East Germany in years subsequent to ringing. Winter concentrations, often best focused at roost sites, are common in many coastal areas. The habit of inland winter roosting was noted as early as 1953 (Hickling 1954). Although coverage was not complete in Wales, the 1973 winter survey (Hickling 1977) showed the extent to which the habit had increased, with some 8500 Herring Gulls counted in six counties. By the 1983 winter survey (Bowes *et al.* 1984) this had fallen to 2267, again reflecting the drop in numbers after the mid-1970s.

The Herring Gull has a wide Holarctic range and some ten subspecies are recognised. The subspecies breeding in Britain (and Iceland, Faeroes and southern North Sea countries) is the subspecies *L. a. argenteus.* Individuals of the paler nominate race *L. a. argentatus* (breeds Scandinavia) are occasionally identified in winter months usually between December and February and have been recorded in Pembrokeshire, Carmarthenshire and Glamorgan. Of more regular occurrence – records perhaps aided by the distinctive leg colour – are small numbers of *L. a. michahellis*, the Yellow-legged Herring Gull from western France, Iberia and the western Mediterranean. In the summary of occurrences of this subspecies, published in 1983 (Grant), no records were forthcoming from Wales up to that time. Grant suggested that the north-westward post-breeding dispersal of this population may be a phenomenon that has increased only in recent years. Since then, however, birdwatchers in most coastal counties, notably those in south Wales, have produced a handful of records in most years, mainly in the months August–March; it is clear that the numbers of individuals of this race that are identified around Welsh coasts at present are correlated with the amount of effort put into examining flocks of Herring Gulls outside the breeding season.

Iceland Gull *Larus glaucoides*

Gwylan yr Arctig

An uncommon and irregular winter visitor to coastal areas, formerly rare.

Prior to 1983 this species, which breeds in Greenland and winters in Iceland, was rare in Wales and recorded far less regularly than the Glaucous Gull. However, substantial influxes took place in the early months of 1983, and in particular, 1984, when there were 40+ individuals in Wales (21 of these in Glamorgan), part of an exceptional arrival of Iceland Gulls throughout western Britain. It has been suggested that severe weather on the east coast of North America or the decline of the Icelandic fishing industry may be reasons why these birds have arrived in Britain in such good numbers. Numbers in Ireland and Scotland are substantially higher than in Wales.

Iceland Gulls are now recorded annually in small numbers (11 records in 1990, at least 13 in 1991), principally in Glamorgan and other south Wales counties, very infrequently in north Wales. It has therefore reverted to the pattern of being less numerous than the Glaucous Gull (15 in 1990, 45 in 1991).

All records are from localities within close proximity of the coast and there are no sightings from the inland counties of Montgomeryshire, Radnorshire and Breconshire. An analysis of the pattern of monthly occurrences in Glamorgan, 1970 to 1984 (Hurford 1985) showed that 34 of the records (79%) were during the period 20 December to 20 March with only two records in the six-month period from June to November. The same pattern is apparent for Wales as a whole but there have been records in every month of the year.

Glaucous Gull *Larus hyperboreus*
Gwylan y Gogledd

An uncommon but regular visitor to coastal areas of Wales, chiefly in winter but occasionally in summer. It breeds all around the Arctic and as close as Iceland.

Even in coastal counties Glaucous Gulls were rare until the 1970s, e.g. in Carmarthenshire the first county record was in 1974 and in Monmouthshire the only record up to 1974 was one shot in 1893. In Glamorgan there were only eight records prior to 1970 but a total of 41 during the period 1970–84, illustrating the marked increase in occurrences at about this time (Hurford 1985).

There was a substantial influx in the early months of 1984 with a total of 26 birds reported from all around the Welsh coast (14 of these in Glamorgan). This coincided with a spell of strong north-westerly winds and severe cold weather, there was also an exceptional number of Iceland Gull records at this time.

Most records are from the coast or estuaries but there are occasional sightings of birds feeding with other gulls up to a few kilometres inland. There are, however, no records from the inland counties of Breconshire, Radnorshire and Montgomeryshire. The majority of the records fall in the period from November to March but small numbers have been seen in every month.

In general this species is more likely to be encountered in Wales than the Iceland Gull with a more even spread of records in recent years.

Great Black-backed Gull *Larus marinus*
Gwylan Gefnddu Fwyaf

Common and widespread in coastal areas in winter; locally common breeding bird.

Thomas Pennant, visiting Snowdonia in 1773, found Great Black-backed Gulls nesting on both Llyn Llydaw and Llyn Conwy (deserted before 1900). (This brief statement conceals Pennant's colourful account of the beating which a man took from the birds in trying to swim to their island). More (1865) listed it as breeding in north Wales and Forrest (1907) names some 16 sites in Caernarfonshire, Merioneth and Anglesey where pairs nested around the turn of the century or had done so in earlier times; he knew of no nesting in Denbighshire or Flintshire, a circumstance that persists to this day. It is evident that one hundred years ago, even accepting gaps in Forrest's knowledge, the Great Black-backed Gull was a very scarce breeding species in north Wales, albeit well recognised in coastal areas. In south Wales one or two pairs bred on Worm's Head (Gower) and possibly the nearby Burry Holms at the end of the 19th century and Mathew (1894) knew of only a few pairs on the Pembrokeshire islands where Montagu had not made mention of them in 1802. Harrison and Hurrell (1933) record that the species was very rare in Pembrokeshire at the end of the 19th century "but never became wholly extinct, though it has come very near to it". They catalogued two or three pairs on Skomer in 1875 rising then to c100 pairs by 1926; on Skokholm around the same time there were no more than three pairs in 1925 but up to 28–30 four years later, with increasing numbers on many of the other offshore stacks and islands. In Cardiganshire, Salter found it breeding for the first time in 1902.

In common with the other breeding *Larus* species in Britain, the Great Black-backed Gull has enjoyed a great increase in its numbers in Wales during the 20th century, followed after the mid-1970s by a sudden decline probably closely linked to botulism poisoning, as in the Herring Gull (q.v.).

T. A. W. Davis (1958) conducted the BTO survey into the breeding distribution of Great Black-backed Gulls in England and Wales in 1956. In this survey he indicated population estimates of 510–540 pairs in south Wales (c95% in Pembrokeshire) and 205 pairs in north Wales – Anglesey (75%) and Caernarfonshire (25%). He catalogued the increase in numbers since the 1930s which he suggested had stemmed from the "few pairs on Steepholm and the Isle of Man".

On the Welsh side of the Bristol Channel, the species first bred on Flatholm in 1962, doubtless as an overspill from the adjacent colony on Steepholm; the number of pairs here still does not exceed three to five annually. On Denny Island, off the Monmouthshire coast, a single pair first nested in 1954 and the colony there increased through the 1960s and 1970s and has since stabilised at 30–40 pairs, having reached a peak of 60 pairs in 1982. One or two pairs have traditionally nested at the western end of Gower but during the past decade of 1980s–early 90s none has been proved. In Pembrokeshire the breeding colonies have been well recorded since the late 1920s although the trends in breeding at the two main colonies on Skokholm and Skomer are disguised by heavy control measures that took place from 1958 and in particular on Skomer from 1960 in successful attempts to stem the increasing populations and thereby reduce predation on other breeding seabird species. Control measures were ended in the late 1970s, which coincided with the outbreak of botulism thus stimulating a major decline in breeding numbers.

Numbers of breeding pairs of Great Black-backed Gulls on Skokholm, Skomer and Grassholm, 1926–92

Year	*Skokholm*	*Skomer*	*Grassholm*
1926	–	100+	–
1928	31	–	–
1929	–	50	–
1934	45	–	67
1940	61	–	–
1945	–	40	–
1949	72	–	60 (1948)
1962	5	235	37
1969	12	160	50
1973	12	117	55
1985–87	10–15	25–33	10–15
1990	c20	41	–
1992	25	40	<10

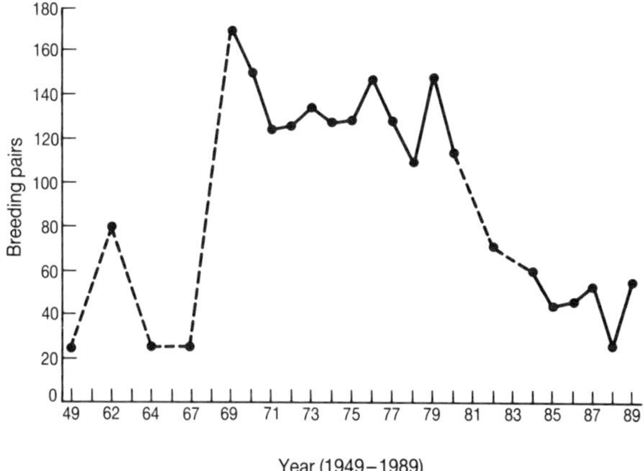

Numbers of breeding Great Black-backed Gulls, St. Margaret's Island, Pembrokeshire

In the Seafarer counts (1969–70), 542 breeding pairs were found in Pembrokeshire compared with 198 pairs during the Seabird Colony Register (1985–87). Evidence of this reduction is clear from some of the other Pembrokeshire colonies (see table below). By 1985–87 several colonies which had materialised in the 1960s had gone completely – St. David's Head, Strumble Head, Stack Rocks.

Reduction of breeding pairs of Great Black-backed Gulls at selected sites in Pembrokeshire 1969 and 1986–87 (Seabird Colony Register)

Site	*1969 pairs*	*1985–87 pairs*
Carreg Rhoson	32	2
North Bishop	40	7
Ramsey & satellites	27	4
Middleholm	109	23
Sheep Is.	16	1
St. Margaret's Is.	169	58

On other parts of the Cardigan Bay coastline, small colonies exist on Cardigan Is., St. Tudwal's Is. and Ynys Gwylan Fawr, with scattered pairs on mainland cliffs elsewhere (the small Bardsey population – one to three pairs – was essentially defunct by the late 1980s). On the mainland north Wales coast the only breeding site is at Little Orme.

The main breeding area in north Wales is around the coasts of Anglesey where up to 15 sites are sporadically used, none of them nowadays supporting more than 20–30 pairs (Puffin Is.) and most only having pairs in single figures. The breeding population on Anglesey is less than 100 pairs. In 1977–78, D. North (in litt.) found a population of 220 pairs there with a maximum of 85 pairs on Puffin Is. in 1977. As in Pennant's day, occasional pairs will nest inland in north Wales; the most recent site being Llyn Trawsfynydd (Merioneth) where one or two pairs have bred sporadically for several decades.

Outside the breeding season large flocks of this gull are recorded at many sites around the coasts where it remains principally dependent on natural marine feeding. The largest winter numbers are to be found in the vicinities of the Burry Inlet (up to 400 individuals), and north-east Wales/Dee estuary (individual flocks up to 250 individuals). S. L. White (in litt.) found 500 roosting regularly each day on Grassholm between 5 and 14 October 1982, with a maximum of c1000 on 13 October; a flock of c800 – many oiled – was on Newgale Beach, Pembrokeshire, in the days after the *Christos Bitas* oil tanker incident in autumn 1978. During winter, Great Black-backed Gulls will feed inland although, being confined to lowland areas, it is less prevalent over much of Wales than it is in England. These inland feeders make frequent use of refuse tips and other urban feeding opportunities. The Welsh birds are predominantly sedentary and 275 ringing recoveries in Wales confirm the fact that most winter birds are of local origin or from proximal parts of England; one bird ringed on Fair Is. (Shetland) in November 1959 turned up on Skomer ten months later. Eight birds (six of them less than one year after hatching) have travelled as far as northern France and one as far as northern Spain. Many juvenile birds from St. Margaret's Is. are shown, by ringing, to spend their first winter off the Cornish coast.

Ross's Gull *Rhodostethia rosea*

Gwylan Ross

A vagrant.

There is only a single Welsh record of this vagrant to British waters which breeds in north-east Siberia and winters in the Arctic Ocean. Most of the British records show, not surprisingly, a distinct northerly bias, especially Shetland. The one Welsh record rewarded a field meeting of the Dyfed Wildlife Trust to Fishguard Harbour (Pembrokeshire) on 15 February 1981, it was also present on the 16th.

Kittiwake *Rissa tridactyla*

Gwylan Goesddu

A breeding species in coastal counties from Glamorgan to Caernarfonshire, except Merioneth. Large numbers occur off-shore on passage.

Kittiwakes are the most oceanic of our gulls, spending much time outside the breeding season far out to sea and occupying a circumpolar breeding range reaching northwards to the limits of land in the Arctic and having southern Britain as the southerly limit of this range. It is another gull that has enjoyed a rapid and fairly continuous increase in breeding numbers throughout the greater part of the present century. Prior to the enactment of bird protection legislation, mainly in the years following the founding of the RSPB in 1889 (but the first Seabirds Preservation Act was passed as early as 1869), huge numbers of Kittiwakes were killed around the British coasts both for sport and, notably, for the millinery trade. Bristol Channel colonies were among the most heavily plundered (but 9000 allegedly shot in a fortnight on Lundy is probably apocryphal) but once the succession of Protection Acts was in place, the Kittiwake's fortunes rapidly improved. Although More (1865) said that they were not known to breed in Wales, this was clearly incorrect; the colony on the Orme headlands was certainly known as far back as 1835 and other long-established colonies are those on Grassholm and (probably) Skomer and Cilan Head (Caernarfonshire). Even by the middle of the present century there were no more than a dozen colonies round the Welsh coast and in the census of 1959, Coulson (1963) found some 3745 pairs on ten sites (two sites not counted would probably have added another 250 pairs). Coulson has shown that the annual increase in the 20th century, up to 1959, was between 3% and 4% and that this continued up to the next decadal survey in 1969 (Coulson 1983); this produced a Welsh figure of approximately 5250, pairs in 14 colonies.

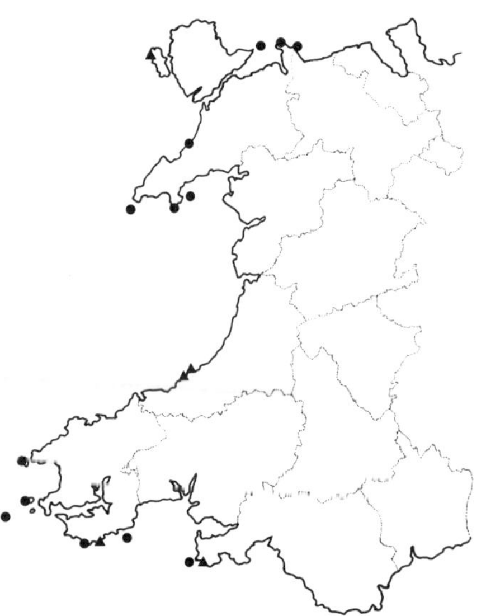

Kittiwake colonies in 1959 (●), additional colonies by 1991 (▲)

During this decade, however, three of the Welsh colonies showed reductions (out of a total of only five decreasing colonies around the whole of the British coasts): St. Tudwal's –28%, Bardsey –97%, Grassholm –21%. In their way these decreases were to prove prophetic because in the ten years to the next survey (1979) – and probably centred on the years 1973–76 – most of the Irish Sea colonies went into decline, counter to the continuing upward trend elsewhere around the western-seaboard and, especially, in the North Sea colonies. Coulson (1983) does not isolate the Welsh colonies in the 1979 survey but concluded as follows: South Wales and Lundy 1959–69 increase +59%; 1969–79 decrease –3%; North Wales, north-west England and Isle of Man 1959–69 increase +65%; 1969–79 decrease –20%.

Counts of Kittiwake pairs at selected colonies

Colony	*Year*	*Number of apparent breeding pairs*
Worm's Head (Gower)	1943	first proved breeding
	1959	75
	1969	560
	1991	114
Grassholm (Pembs)	1925	50
	1948	75
	1959	200
	1969	173
	1991	125
Skomer (Pembs)	1945	1500
	1959	1754
	1979	2296
	1991	2466
Flimston/Stack Rocks (Pembs)	1959	87
	1969	48
	1990	486
St. Tudwal's (Caern)	First recorded 1935	
	1959	130–140
	1969	97
	1986	395
Bardsey (Caern)	First bred 1920s	
	1959	100
	1969	3
	1979	100
	1989	318
Great Orme (Caern)	1959	750
	1969	1037
	1979	884
	1989	1399

Between 1979 and the Seabird Register counts (1986) no continuing decline was evident overall in Wales although some individual colonies have shown further reductions (see table). The Seabird Colony Register gives a total ot 9120 pairs at 20 colonies in Wales, an increase of some 29% throughout Wales since 1979.

Outside the breeding season the Kittiwake is an extremely numerous offshore passage species. In late autumn birds passing southwards are a feature of almost any sea-watch off western headlands: up to 3500 per hour have been seen off Bardsey; feeding flocks of up to 800 birds are recorded off Skokholm and up to a maximum of 30 000 per day have been counted at Strumble Head in

November. These birds evidently originate from colonies farther north and, on the evidence of a bird ringed at Penally in Pembrokeshire (4 March 1982) and shot at Inerssvat in Greenland three summers later, may well include a proportion of birds from colonies deep in the Arctic. Kittiwakes are long distance trans-oceanic travellers and recoveries of Welsh birds from Heligoland, southern Spain and western Canada (the latter one less than three months after fledging) give confirmation of this. Considerable numbers of first-summer birds assemble on offshore rocks and even join tern colonies (E. I. S. Rees in litt.) in late summer and early autumn.

Usually strictly marine, the Kittiwake is unlikely to be found in estuaries or inshore bays except in severe weather and it is very rare inland with only a handful of accidental records from inland counties. The most inexplicable inland record was a flock of 14 adults in the Elan Valley (Radnorshire) in May 1976 (M. Davies, pers. comm.). In common with other pelagic species, Kittiwakes are occasionally prone to serious "wrecks" when protracted winter storms cause birds to be driven ashore in an emaciated condition, often resulting in heavy casualties. Thomas Pennant records "many thousands" being picked up dead or dying in February 1776 in the Criccieth area. In the present century, the most notable wreck was in 1957 which resulted in over 100 dead birds in Glamorgan, 40 in Cardiganshire and 30 in Caernarfonshire. At the same time, eight dead or dying birds were found in inland counties (McCarten 1958).

Ivory Gull *Pagophila eburnea*

Gwylan Ifori

A vagrant.

It is difficult to assess the number of occurrences of this Arctic breeding species in Wales since, apart from one recent record, published details are very sparse.

In Glamorgan Heathcote *et al.* (1967) quote three possible occurrences: two of them, both sight records, dating a long way back into the last century, the latest one in 1883, in Gower and Swansea Harbour respectively. The third was a bird seen, with other gulls for comparison, in Roath Park, Cardiff on 3 April 1921; possibly the same bird was seen again over Llanishen Reservoir on 8 May and at Sully Island on 13 June 1921.

In Carmarthenshire Ingram and Salmon (1954) refer to two records from the Gwendraeth estuary: an adult on 17 December 1947 and a similar bird at exactly the same place on 21 October 1948 until at least 28 October. The observer claimed that there was "no possibility of misidentification, species is unmistakeable" but it is a pity that some details of the sightings were not available for future reference.

In Cardiganshire Ingram *et al.* (1966) refer, in square brackets, to the fact that an Aberystwyth taxidermist had had one certain specimen, at least, but with no date.

In Merioneth Hope Jones (1974) refers to a bird found dead at Broadwater in August 1954 which was identified as an Ivory Gull. He concluded that the record was not sufficiently well documented for acceptance, the specimen was not preserved and no detailed notes were kept.

In Pembrokeshire there is, fortunately, a good description of an adult near Giltar Point on 24 August 1950 (Cusa 1951).

There is, also, a recent and well-substantiated record supported by excellent photographs. This was a first-winter bird at Burry Port (Carmarthenshire) from 10 to 12 October 1988. It was found with Black-headed Gulls feeding on a drift net full of fish that had been towed in from Carmarthen Bay and left anchored to a pontoon in the little harbour.

Gull-billed Tern *Gelochelidon nilotica*

Morwennol Ylfinbraff

A vagrant.

This species breeds from Denmark, southern France and Iberia discontinuously eastwards to eastern Kazakhstan and south to north-west Africa, Pakistan, south-east China and Australia; also North, Central and South America.

There are eight records in Wales:

1 August 1960	– One, Shotton (Flintshire)
21 September 1968	– One, Dale Roads (Pembrokeshire)
16 September 1974	– One, Cydweli Marsh (Carmarthenshire)
16 July 1978	– One, Point Lynas
5 July 1979	– One, found shot, St. Brides Wentlooge (Monmouthshire)
16 October 1983	– One, Strumble Head
30 July 1985	– One, Strumble Head
20 April 1987	– One, Llanelli (Carmarthenshire)

Caspian Tern *Sterna caspia*

Morwennol Fwyaf

A vagrant.

This large tern, about the size of a Herring Gull, breeds on Baltic and Black Sea coasts, Tunisia and very discontinuously eastwards to Manchuria and south to southern Africa, Australia and New Zealand; also North America.

There are five records from Wales.

20 July 1962	– One, Llangorse Lake
19–21 August 1973	– One, Blackpill
8–28 May 1974	– One, Dyfi estuary (Cardiganshire)
26 May 1980	– One, Cemlyn Bay (Anglesey)
5 August 1988	– One, Cemlyn Bay (Anglesey)

There is an insufficiently documented record of a specimen believed to have been obtained at the mouth of the River Conwy (Caernarfonshire) some time before 1907.

Royal Tern *Sterna maxima*

Morwennol Fawr

A vagrant.

There have been only two records of this species, which breeds in North America, Caribbean and north-west Africa:

24 November 1979 – One, immature, Kenfig Pool
21 December 1987 – One, first winter, Mumbles, Swansea (Glamorgan)

The first record was at first thought to be a Lesser Crested Tern but much subsequent discussion and research produced some new identification criteria for large terns and revealed that the ring number, which had been partially read, came from one of six breeding colonies on the eastern seaboard of the USA (*British Birds* vol 76, pp 335–339).

Lesser Crested Tern *Sterna bengalensis*

Morwennol Gribog Leiaf

A vagrant.

The first British record of this species, which breeds on the coasts of Africa, south-east Asia and Australia, was from Anglesey:

13 July 1982 – Adult in summer plummage, Cymyran Bay (Anglesey)

Sandwich Tern *Sterna sandvicensis*

Morwennol Bigddu

A summer visitor and passage migrant to coastal areas in spring and autumn, scarce inland; it is a regular breeder on Anglesey in variable numbers up to 1080 pairs.

In the early years of the 20th century Sandwich Terns were very rare visitors to Wales. Forrest (1907), for example, quotes only one record for north Wales, relating to a pair which appeared on the Skerries (Anglesey) but did not stay to breed. The first county record for Denbighshire was not until 1914, for Pembrokeshire 1928, Carmarthenshire 1929, Glamorgan 1933 and Monmouthshire 1967, a reflection of its rarity in Welsh waters until the second half of the century.

The initial, and eventual, breakthrough that established the Sandwich Tern as a regular breeding bird was on Anglesey, consistent with the island's position as the main centre for breeding terns in Wales. The first breeding records were from Llanddwyn Is., where on 15 June 1915 C. Oldham and S. G. Cummings found between 30 and 40 nests on top of the offshore stack, in a loose colony but closely associated with nests of Common and Roseate Terns; the Llanddwyn pilots reported that a few pairs of Sandwich Terns had bred for the first time in 1914. At Llanddwyn there were still a few pairs in the mixed tern colony in 1920, 1921 and 1923 but only one pair in 1924 and 1925.

For the next 45 years there were occasional records of a few pairs (usually of one or two pairs) breeding at different sites on the north and west coast of Anglesey, including Newborough, Rhoscolyn Beacon, the Inland Sea, Cemlyn Bay and the Skerries. Colonisation proper came in 1970 when 20 pairs bred at Ynys Feurig, where there had been a single pair in 1969; numbers at this site then increased rapidly, to 100 pairs in 1971 and 200 pairs in 1972. Although there was a decline in 1973 and 1974, numbers built up again in subsequent years up to 1979; the sort of fluctuation that has occurred since then is characteristic of this species, which is notoriously fickle in moving its breeding grounds.

The total Welsh breeding population for the years 1969–92 was as follows:

1969	3 pairs	(two colonies)
1970–74	average 98 pairs	(one to three colonies)
1975–79	average 321 pairs	(two to three colonies)
1980–84	average 201 pairs	(one to two colonies)
1985–89	average 653 pairs	(one to two colonies)
1990–92	average 539 pairs	(one to two colonies)

The three colonies since 1970 are the same three sites in the north-west corner of Anglesey, predominantly at the North Wales Wildlife Trust reserve at Cemlyn Bay where the large counts of the late 1980s occurred.

The Welsh population is only a small percentage of the British breeding population, which amounted to 18 400 pairs in 1985–87.

Ringing recoveries provide some interesting indication as to the breeding areas from which the influx of Sandwich Terns originated. Up to 1992 there had been 31 recoveries of birds ringed in Britain and Ireland and recovered in Wales, all of which had been ringed in the months of June and July. Of the 31, seven had been ringed in Wales, one in Scotland, seven in northern England and 16 in Ireland.

This is usually the first Tern species to arrive on spring passage, earliest sightings usually being noted in the last ten days of March. Records of small parties are frequent along the coastline from Pembrokeshire northwards on passage and feeding groups remain for long periods in summer especially at estuary mouths. Records are progressively less numerous eastwards along the south coast but even in Monmouthshire there are a good scattering of sightings (e.g. two records of 14 individuals in 1988, five records of 13 individuals in 1989). On the west coast the main passage concentrations are from the Dyfi to the Dee, where substantial gatherings occur at favoured roosting sites. The summer influx takes place from mid-July onwards, peaking in late July to early September. Main localities are the mouths of the estuaries of the Dyfi, Dysynni and Mawddach in Cardigan Bay and the Point of Air at the mouth of the Dee. In these areas shallow warm waters with gently shelving sandy shores attract good numbers of fish which in turn attract terns. These parts of Wales are clearly of considerable importance as a nursery area for Sandwich Terns, adults resorting to the outer estuaries and inshore waters to feed young before moving south. Maxima include 1100 at Ynys las (Cardiganshire) on 14 August 1981, 570 at Fairbourne (Merioneth) on 30 July 1983, 400 at Aberdysynni (Merioneth) on 8 September 1983 and 800 at Point of Air on 30 July 1982 and 29 July 1984.

Autumn passage on the Welsh coast continues into October with occasional records from November and overwintering individuals have been noted, e.g. in Anglesey in 1988–89. On sea-watch stations Sandwich Terns feature in the autumn passage movements, particularly at Strumble Head where, for example, an average annual total of 1596 birds passed westwards on 43 dates between the years 1981 and 1984; for the same period at Point Lynas (Anglesey) the comparative figures were 142 birds on 35 dates. The difference is explained, at least in part, by the fact that most Sandwich Terns probably migrate south via the Menai Straits rather than circumnavigating Anglesey.

At Aber (Caernarfonshire) a remarkable movement of Sandwich Terns, totalling 4000 birds, was recorded on 20 August 1988, in a north-west gale. Another exceptional observation was of 120 birds flying inland at Rhos (Denbighshire) during a gale on 9 August 1979. There are no records from the inland counties of Radnorshire or Montgomeryshire but there is a series of records from Llangorse Lake where it was first noted in the 1940s, recorded almost annually from 1962 to 1978 but with only eight records from the 1980s. Spring records fell between 6 April and 8 June and autumn records between 22 July and 13 October, parties of up to 12 have been seen (Peers & Shrubb 1990).

Recoveries of young ringed in Wales show that in their first winter they migrate southwards along the west coast of Africa. Recoveries in the period from November to February inclusive showed that most were concentrated in the northern tropics from 4°N to 14°N (ten recoveries) and a lesser number had reached points farther south in the belt 4°S to 14°S (five recoveries). This is a similar pattern to the much larger sample covering all British ringed birds analysed by Langham (1971).

Roseate Tern *Sterna dougallii*

Morwennol Wridog

The Roseate Tern is confined to Anglesey for breeding and even there it is now down to fewer than ten pairs; elsewhere it is occasionally recorded on spring and autumn passage along the coasts.

The history of the Roseate Tern in Wales is difficult to piece together because published statements have been deliberately vague on account of the threats from egg collectors who greatly prized the eggs of the species. By collating information from unpublished sources with what has appeared in print, it is possible, however, to give an indication of changes that have taken place since the late 19th century. These confirm that Roseate Terns have moved their main breeding sites several times in the past one hundred years (no less than ten different sites, mostly offshore islets, have been used in that period on Anglesey alone) and that throughout much of the time the breeding population was of the order of 150–300 pairs until a sharp decline from 1987 onwards.

The Roseate Tern has always been a local breeding species in the British Isles, but the Irish Sea area has consistently been a stronghold for the species. At the end of the 19th century the Roseate Tern had been reduced almost to the point of extinction in Britain (Parslow 1973). At that time there was, however, a colony on the Skerries, described by the lighthouse keeper as "a very large colony" in 1892, the keeper having shot four adults, two of which were preserved in the Grosvenor Museum, Chester (Curators' Journal, Grosvenor Museum). Nine clutches of eggs of this species taken on the Skerries on 9 June 1896 are in the Oxford University Museum. Forrest (1907) confirmed that "it has been known for a good many years back that this beautiful species breeds on the Skerries. In the interests of the birds it was thought best not to publish the locality, but latterly so many egg collectors have obtained knowledge of it that further secrecy seems useless". In Pembrokeshire this species also used to breed in the 19th century but it is not known in what numbers; Lockley *et al.* (1949) state that there was a small colony of Roseate Terns on Grassholm in 1885 and that it formerly bred on Skokholm Stack. By 1908 it was recorded that "the colony on the Skerries has been considerably reduced but steps have been taken to preserve the birds. There is another colony, the locality of which is not divulged and the numbers here are well maintained" (F. C. R. Jourdain, in litt, *British Birds* vol. 2, 1909). The presence of rats on the Skerries, recorded in 1905 may have had something to do with its desertion by the Roseates.

The new colony was in fact on a rocky stack off Llanddwyn Island where on 24 June 1902, T. A. Coward recorded "a colony of several hundred pairs of Terns, majority were Common Terns but there were many Roseates". By 1910 Charles Oldham noted that at Llanddwyn "the Roseates now far outnumber the Common Terns, perhaps in the proportion of 6 to 1, and at a low estimate now number five hundred birds". In that year William Jones, one of the Llanddwyn pilots acting as watcher for the RSPB, in describing the birds of the island wrote that "none of these disturbed the terns only one that we call in Welsh barent, its bigger than a hawk, he kills a good many". This bird is presumed to be a Peregrine and this early reference, endorsed by the observation of the remains of two Arctic and one Common Tern near a Peregrine eyrie on the north coast of Anglesey (Coward & Oldham 1904), was a precursor to the substantial predation by this species on Roseate Terns in 1980 and following years.

The Llanddwyn colony continued to thrive for the next few years for in 1915 there were "several hundreds" (Forrest, 1919, stated that it "now numbers about three hundred breeding pairs"), 105 pairs were noted in 1920 and 100 pairs in 1924. That year marked the beginning of a move away from Llanddwyn since the main breeding stack (Ynys yr Adar) was occupied by Herring Gulls and the Terns were breeding on a small stack, which was accessible at low tide. In 1925 there were very few terns left and only one pair of Roseates and by 1928 there were no terns at all, even the small stack having been appropriated by Cormorants. Other tern species did, however, continue to breed on promontories on the island through the 1930s.

Roseate Terns have a preference for sheltered nesting sites, more so than other *Sterna* species, and it is interesting to note therefore the quotation relating to Llanddwyn by T. A. Coward (1922): "On a headland is a tower, used as a landmark before the lighthouse was erected, and beyond the lighthouse on a couple of stacks hundreds of terns lay their eggs on the bare, jagged rock or amongst the dense tangle of tree-mallow and sea beet". The plants referred to would have provided ideal cover for the Roseates. In 1929 Mrs Jones, the RSPB Watcher, made reference to the fact that "the Cormorants have made away with their ferns (!) and the island is bare"; this process of the vegetation being denuded, if started earlier than this report, may have made conditions less suitable for the Roseates and been one possible factor in the movement of the colony. In the meantime Roseates also bred at Rhoscolyn Beacon ("a good many" in 1916 and three nests in 1925) and in 1922 they were nesting "in some numbers" on a reef in Cymyran Bay in West Anglesey and at Dulas Island off the north-east Anglesey coast. At the latter site 20 breeding pairs had been noted in 1912.

By 1925 Roseates had re-colonised the Skerries. In 1928 a visit on 16 June notes that "we estimated on a very conservative basis that we actually examined or verified some 75 Roseate Tern nests and we could undoubtedly have seen at least 3 or 4 times that number had we wished or given the time" (A. Whitaker, Diaries, Edward Grey Institute). This would give a population of 225–300 pairs. Characteristically for the Skerries the nest sites were mainly on small rocky outcrops and knolls, often in shallow cracks or crevices as well as the entrances to rabbit burrows. The Roseate nests were in discrete groups amongst the much more numerous Arctic colony. In 1933 there were at least 150 pairs of Roseates on the Skerries and "several hundred" in 1935. In 1942 it was noted by the lighthouse keepers that "the terns were continuously worried by the gulls this season, many of the eggs were eaten also dead chicks seen about. A pair of Buzzard Hawks were also very active during the Terns mating season" (Letter from the Assistant Keeper, RSPB file). The "Buzzard Hawks" were probably Peregrines from an eyrie on the mainland opposite since Buzzards have always been rare on Anglesey. On his 1933 visit Charles Oldham had noted a Peregrine passing to and fro along the Skerries cliffs sometimes going through a cloud of Terns "which obviously resented its presence", later seen flying across the channel to the mainland with its quarry, believed to be a Tern (Oldham, Diaries at Edward Grey Institute). On the Skerries numbers diminished to 75 pairs in 1947 and 30–50 pairs in 1957; in 1952 P. E. S. Whalley and M. J. Wolton recorded that "although eggs had been laid all of the birds had deserted and the clutches had been cleared by gulls".

In the period from 1952 colonies were recorded at several localities in different areas of Anglesey, particularly East Mouse island off Amlwch (60 pairs in 1957), and islands in the straits between Holy Island and the main island of Anglesey as well as small numbers at Abermenai, Ynys Gorad Goch in the Menai Straits and a small stack off the south-east side of Llanddwyn Island and, briefly, the Skerries again (30 pairs 1961, unsuccessful because of rats and gulls).

From 1959 onwards the main and sometimes only colony has been at one site in west Anglesey, Ynys Feurig, a small reef luxuriantly vegetated with Marram, Thrift, and Red Fescue which provide good cover for Roseate nests. The total breeding population from 1969 to 1992 has been:

1969	202 pairs	(2 colonies)
1971	205 pairs	(2 colonies)
1972	155 pairs	(2 colonies)
1974	251 pairs	(3 colonies)
1975	173 pairs	(2 colonies)
1976	197 pairs	(3 colonies)
1977	176 pairs	(3 colonies)
1978	158 pairs	(3 colonies)
1979	189 pairs	(3 colonies)
1980	180 pairs	(1 colony)
1981	130 pairs	(2 colonies)
1982	170 pairs	(1 colony)
1983	174 pairs	(1 colony)
1984	150 pairs	(1 colony)
1985	200 pairs	(1 colony)
1986	208 pairs	(3 colonies)
1987	77 pairs	(4 colonies)
1988	64 pairs	(3 colonies)
1989	96 pairs	(3 colonies)
1990	45 pairs	(3 colonies)
1991	5 pairs	(2 colonies)
1992	7 pairs	(1 colony)

From the above it can be seen that numbers in Wales held up well until 1987. It is notable that the main colony on the east coast of Ireland increased by 70 pairs in that year. The peak of perhaps 3500 pairs in Britain and Ireland was reached in the late 1950s and early 1960s but numbers had declined to an average of 1075 pairs in 1970–74 and 919 pairs in 1975–79, falling further to an average of 488 pairs in 1988–92. Part of this serious decline is probably a result of the trapping of birds on their wintering grounds; the trapping of terns for sport or food in West Africa has been identified as the major recorded cause of death of terns outside the breeding season.

At Ynys Feurig from 1959 onwards there have been notable problems in productivity through loss of eggs and young to a variety of predators, rats, foxes, gulls and Peregrines. In 1978 rats invaded the colony and killed 17 adult terns, including 14 Roseates, as well as taking virtually all the eggs and young that year. In 1987 a fox got onto the breeding site in late May and took 52 roosting adult terns, including 12 Roseates; this was the year that numbers slumped at that site, from 200 pairs in 1986 to 40 pairs in 1987. Predation by a small number of gulls resulted in a complete failure of the colony in 1990. The biggest menace, however, has been Peregrines, starting in 1980 when an immature bird killed at least 36 Roseates and took at least five young birds prior to fledging. Predation by Peregrines, chiefly but not by any means exclusively from the nearest breeding site, continued in subsequent years with varying degrees of severity; quite apart from the taking of adults and young birds the disturbance to the colony as a whole during raids by Peregrines was substantial.

Amongst the gloom of this recent picture of substantial decline there is one encouraging item to be noted. The only substantial colony of Roseate Terns remaining in Britain and Ireland is a small island off the east coast of Ireland, numbering 378 breeding pairs in 1992 (i.e. 73% of the total).

In 1990 BTO ring numbers were read on 241 birds in the colony using telescopes; of these no less than 65 had been ringed at Ynys Feurig including a long-surviving bird, 19-years-old. A good number of birds from Anglesey were still therefore breeding successfully albeit in another part of the Irish Sea. Hopefully the continued success of the Irish colony will provide a reservoir of birds for the recovery of the Welsh population, reduced now to seven pairs at one site.

Considering that there has been a reasonably large breeding population on Anglesey for at least one hundred years, there are surprisingly few records away from the breeding area, suggesting that the arrival and departure routes are either well out from the coast or along the western shoreline of the Irish Sea. For example, even on Bardsey which is well watched, and relatively close to Anglesey, there were only eight records, involving 21 individuals, between 1967 and 1984. Similarly, there were only 15 records, involving 34 individuals in Pembrokeshire in years of intensive sea-watching at Strumble Head between 1981 and 1989 (11 of the records were from Strumble). It is quite probable that many of the Welsh Roseates gather with the substantial tern roosts that congregate at Sandymount Strand, Dublin and Broad Lough, Wicklow, where birds apparently move in September into the area off the Dublin and north Wicklow coast to feed on the rich fishing banks offshore before going south on autumn migration. In addition to the Irish connection established in 1990, there are three instances of birds ringed as young in Anglesey controlled at a breeding colony in Wexford in July 1967 and July 1974 (2).

The most favoured locality away from Anglesey is the Dee estuary where there have been a series of records from Point of Air and Gronant, and even up river to Shotton in company with other Tern species principally in July.

Indeed the only hint of possible breeding away from Anglesey this century was on 15 July 1916 at the Point of Air when Charles Oldham recorded that "there were several Roseates, about six pairs in all. We were unable to mark any Roseates down to their nests and the birds were less clamorous than the Commons but sometimes singly and sometimes in pairs they mingled in the screaming crowd above the nesting ground" (Oldham, Diaries).

Records of passage birds come sparsely from all coastal counties and there are inland records of an immature bird shot near Llanymynech (Montgomeryshire) on 21 September 1914, one at Llangorse Lake on 22 August 1985 and two there on 21 May 1989, one remaining the following day.

Up to 1992 there were 170 recoveries of birds ringed as young in Anglesey, including 92 recoveries in Britain and Ireland. Away from British waters there were two September records, of first-year birds, one from Cadiz (Spain), the second already off Sierra Leone at 7°57′S a movement of 5097 km south. By October three first-year birds were already at 5° North in Ghana and Ivory Coast but five others were still relatively far north from Huelva (Spain) to Morocco. In November all four records of first-year birds were between 4°N and 5°N and for the period December to April inclusive between 4°N and 6°N (37 records). In May and June there were six records between Ghana and Senegal (4°N and 16°N) suggesting that in their second summer birds remained in the tropics although there is evidence of a slight movement northwards. The remainder of the records indicate that the winter quarters were the same for all age groups, i.e. just north of the equator but not penetrating farther south, very much centred on Ghana although it is possible that there is a bias through the concentration of Tern trapping in that country. One bizarre recovery was of a bird ringed on 5 July 1963 and recovered at Budbrooke in Warwickshire on 28 September 1965. That so much ringing information exists for Roseate Terns is due to the dedicated efforts every year since 1959 of Arthur Jones and his colleagues, Chris Ellis for early years and latterly Bill Ashby.

Common Tern *Sterna hirundo*

Morwennol Gyffredin

A summer visitor which breeds in Anglesey and Flintshire. It is a spring and autumn passage migrant to all coastal areas and is the commonest of the Sterna *species inland.*

Anglesey has always been the main breeding area for Common Tern, as for other tern species, in Wales. The breeding range in Wales has contracted since the 19th century. In south Wales there is some doubt about its early status in Glamorgan. Ingram and Salmon (1967) say that "in 1848 Dillwyn stated that it bred regularly but it has never been possible to confirm where or when or even if it occurred at all". However E. Doddridge-Knight recorded it as common in summer in the Sker/Kenfig area in the 1830s and with suitable shingle habitat (there was a colony of Little Terns at Sker Point in the 19th century), breeding must have been at least a likelihood.

In Pembrokeshire there was a small colony on Skokholm Stack at the end of the 19th century, believed to have died out in 1916. In Cardiganshire the only breeding record was in July 1914 when several pairs were found nesting on the south side of the Dyfi estuary near Ynyslas. In Merioneth in the period 1911–14, two or three pairs nested near Tywyn and another in 1927, otherwise there were no records until 1968. On the Caernarfonshire side of the Menai Straits at Morfa Dinlle there were 40 nests in 1902 and the colony was still active until at least 1932.

On Anglesey it is difficult to deduce accurate information about the early years but the general picture that emerges for the first three decades of the present century is that numbers and breeding sites varied quite considerably. On the Skerries none was seen on some visits but in other years it was noted that there were small groups scattered amongst the much more numerous Arctic terns; in 1933 it was estimated that there were 50 pairs. There were also a few scattered pairs nesting in 1928 along the strait separating Holy Island from the mainland. On Rhoscolyn Beacon there were a few nesting amongst the more numerous Arctics in 1916, but 56 pairs in 1925 outnumbered the latter. Llanddwyn Island was a stronghold at least as far back as 1903 "when hundreds of eggs" were noted on an offshore islet and "so thickly was the surface covered with nests that it was impossible to avoid treading on some of them" (Forrest 1907); the main island and offshore islet continued to be a breeding site, albeit in lesser numbers until at least 1958. At Llanddwyn, Common Terns consistently outnumbered Arctics, likewise at the Newborough Warren sand dune complex (including the shingle point of Abermenai) where there was a colony of up to 120 or 130 pairs from at least 1902 to 1974. On the east coast of Anglesey the main colony in the early years of the century was at Ynys Dulas where there were "large numbers" in 1903 and at least sporadic breeding until 1952; from 1923 to at least 1939 there were terns "in abundance" on the shingle ridge at Traeth Dulas and the colony was still in existence up to 1953 at least. Forrest (1907) stated that Common Terns bred at Ynys Moelfre as well as Ynys Dulas. In the early 1950s there was a colony of up to 55 pairs nesting in the low heath at Llugwy Bay on the east coast and from 1953 for a few years there was a small mixed colony of Common and Arctic Terns on top of a limestone spoil dump by the Menai Straits at Penmon.

In Flintshire there were two colonies recorded in the early part of the century. The Point of Air colony was in existence at least as far back as 1898 and reached a peak of about 150 pairs in 1916; it continued, perhaps sporadically, at the Point of Air and/or Gronant until the early 1950s with an isolated nesting occurrence in 1965. In the inner estuary of the Dee a colony was recorded near Connah's Quay in about 1915 and was still in existence in the 1960s with up to 75 pairs on the saltmarshes near Shotton Steel Works.

Recent figures for the Welsh breeding population are as follows:

1969	302 pairs	(seven colonies)
1978–82	average 472 pairs	(five to six colonies)
1983–87	average 476 pairs	(five colonies)
1988–92	average 471 pairs	(five to seven colonies)

The Welsh breeding population is only a small percentage of the total British figure, which amounted to about 14 700 pairs in 1985–87. All but two of the above colonies were in Anglesey. One of the exceptions is in Flintshire, which is now the largest colony in Wales. Since 1970 the Merseyside Ringing Group has operated a highly successful project by providing firstly platforms, and subsequently islands, in a reservoir at the British Steel works at Shotton. This attracted birds from the nearby saltmarsh where eggs and young were frequently lost through tidal inundation. The number of birds nesting at this site has gradually increased with a minimum of 150 pairs each year from 1976 onwards and 312 pairs by 1992. Substantial numbers of young have been reared from this secure site, freedom from predators has proved to be a major benefit in this respect (Norman 1987).

Common Terns are less maritime than other *Sterna* species and this is reflected in the number of inland colonies in different areas of the British Isles, particularly in Scotland, Ireland and eastern counties of England. In Wales there was an inland colony with up to ten pairs in Merioneth, on the lake at Trawsfynydd from 1968 to the early 1970s. Less transient but with greatly fluctuating numbers, an inland colony was established at about the same time on islets in the newly flooded reservoir at Llyn Alaw on Anglesey; maxima were 141 pairs in 1969 and 100 pairs in 1983 stabilising at between 40 and 50 pairs in the last five years of the 1980s but declining to nil in 1991 and only two pairs in 1992. Both these inland sites are less than 10 km from the nearest coastline, enabling birds to catch fish at sea. At the two inland sites no other species of breeding terns were associated with the Common Tern colony and at Shotton, at the head of the Dee estuary, only the occasional Arctic Tern mixes with the large Common Tern colony. Elsewhere, although usually in discrete groups, the Common Terns are frequently associated with Arctic Terns and in some locations with Roseate and/or Sandwich Terns as well.

The Common Tern occurs more frequently in inland areas on migration than the other *Sterna* species. Even allowing for the large number of instances where it is not possible to separate it from the Arctic Tern, it is clear that it is a regular passage migrant to inland areas of Wales. Recoveries of Common Terns ringed as young in Wales show a gradual move southwards in August, September and October, although even as early as September two had reached as far south as 4°N. By November and through to May all of the 35 recoveries of first-winter birds up to 1992 were off the West Africa coast between 15°N and 4°N.

In subsequent winters, recoveries occur between the same latitudes suggesting that adult Common Terns return to the same general region, as shown by Langham (1971). An adult bird ringed at the Point of Air on 6 September 1975 and recovered on 9 December 1980 as far south as Cape Province, South Africa (33°S), may have been from the Scandinavian population which appears to winter in a zone farther south than British, including Welsh, birds. However, a young bird ringed at Shotton in June 1986 was recovered in Cape Province (34°S) in February 1990 so at least some of the Welsh population reach that far south.

Arctic Tern *Sterna paradisaea*

Morwennol y Gogledd

The commonest of the tern species breeding in Wales. It is confined to the Anglesey coast for breeding. Away from Anglesey the Arctic Tern is a regular passage migrant to coastal areas, generally scarcer than the similar Common Tern, although the position regarding numbers is difficult to establish precisely as they can be difficult to separate in the field. It is scarce inland.

The Skerries (Anglesey) has been the predominant breeding site for most of the period since the 19th century and although precise figures of breeding pairs are not known until recent years, descriptions by visiting naturalists give a reasonable picture of the importance of the colony.

Numbers of Arctic Terns on the Skerries

1893	"Thick" colony
1905	Several thousand pairs
1923	About 700 pairs nesting
1927	"Vast numbers" breeding
1928	"Enormous colony"
1933	At least a thousand pairs
1935	"Thousands of nests"
1947	About 750 pairs
1951	300 to 600 pairs
1952	Although eggs have been laid all the Terns deserted
1961	A colony of 150 pairs of Common/Arctic with 30 pairs of Roseates. Rats and gulls were causing problems. No subsequent records until 1979
1979	Re-colonisation; a small number bred
1980	60 to 70 pairs
1981	About 200 pairs
1982	About 100 pairs
1983	About 150 pairs (RSPB reserve established)
1984	168 pairs
1985	155 pairs
1986	190 pairs
1987	260 pairs
1988	250+ pairs
1989	380 pairs
1990	464 pairs
1991	700 pairs
1992	670 pairs

The recent build-up of the colony is encouraging even though there is clearly a long way to go to emulate the numbers of the period up to the 1930s. Breeding success has been high in recent years in contrast to the contemporary situation in Shetland, where a marked food shortage has been a cause for concern.

Away from the Skerries the only substantial colony in the early years of the 20th century was at Rhoscolyn Beacon (Anglesey), where there were of the order of 200–300 pairs in the years 1884–1910, declining to "a few" by 1925. Small numbers up to a maximum of 75–100 pairs were also nesting between 1909 until at least 1924 with other tern species at Llanddwyn Island. Other sites occupied by small numbers at least sporadically during this period were at Ynys Dulas and a reef in Cymyran Bay. In Caernarfonshire there was a small colony of up to 12 pairs nesting, at least in some years with Common Terns at Morfa Dinlle from 1915 to 1950.

The total number of breeding pairs in recent years, all from Anglesey except for the occasional one or two pairs that have attempted to nest, with unknown success, in the Shotton Common Tern colony, is as follows:

1969	436 pairs	(five colonies)
1978–82	Average 733 pairs	(four to six colonies)
1983–87	Average 656 pairs	(three to four colonies)
1988–92	Average 889 pairs	(three to five colonies

These figures are a very small percentage of the total British breeding population, which amounted to 80 200 pairs in 1985–87.

It is difficult to evaluate the numbers that occur away from breeding areas because the preponderance of records do not differentiate between Arctic and Common Terns. Assessing Common/Arctic records as a whole they are much more numerous in autumn than on spring passage and from the cases where the species is identified it appears that Common are much more numerous than Arctic Terns but this generalised statement has to be treated with some caution of course. Within the last few years autumn sea-watches at Strumble Head and Point Lynas have had annual passages totalling maxima of 1221 Common/Arctic Terns passing west at the former and 827 passing west at the latter.

There are few records away from the coast apart from in Breconshire where its status is given as probably a regular passage migrant (Peers & Shrubb 1990) with 11 spring records between 24 April and 24 May and 61 autumn records between 1 July and 24 October; most of the records were from Llangorse Lake.

Annual migration for this species spans a greater distance than for any other bird and although the ringing recoveries of Welsh birds total only 17 outside Wales up to 1992, they confirm an autumn movement along the coast of western Europe and West Africa as far as South Africa and beyond. The three most interesting recoveries were:

18 April 1961 – One ringed in Anglesey on 9 July 1960, recovered in Cape Province, South Africa (a move ment of 10 022 km south-south-east).
1 November 1969 – One ringed in Anglesey on 7 July 1968, recovered in Cape Province, South Africa (a move ment of 10 099 km south-south-east).
31 December 1966 – One ringed in Anglesey on 28 June 1966, recovered in New South Wales, Australia (a spectacular movement of 18 056 km south-east – the first recovery for this species in Australia).

Ringing recoveries within Wales showed that there were several cases of movements away from natal colonies to breed elsewhere in Anglesey and there were six cases where adults breeding in Anglesey had been ringed elsewhere (two on the Farne Islands, Northumberland and four in Eire).

In 1979 a number of adult and young Arctic Terns at a colony in Anglesey were killed by an outbreak of pseudotuberculosis, the first time that this had been reported in any species of tern.

Forster's Tern *Sterna forsteri*

Morwennol Forster

A vagrant.

It breeds in North America and winters from the United States of America to Mexico.

The first British record was in 1980 and there has been a series of sightings of wintering birds in Wales from 1984 onwards, many of which probably refer to the same individual as the same favoured sites, the Holyhead area, Penmon (Anglesey) and the mouth of the Dee estuary feature in the records:

3 July to at least 6 August 1984 – One, Point of Air
30 September–20 October 1984 – One, Penmon (Anglesey)
19 January–6 February 1986 One, Holyhead (Anglesey)
27–28 September 1986 – One, Point of Air
7 October 1986 – One, Abergele (Denbighshire)
22 October–23 November 1986 – One, Penmon area (Anglesey)
20–30 August 1987 – One, Gronant (Flintshire); two, 31 August–20 September
23 August 1987 – One, Great Orme (Caernarfonshire)
17 October 1987 – One, Penmon
10 January–26 February 1988 – One, Holyhead area
15–16 October 1988 – One, Penmon
21 January–1 March 1989 – One, Holyhead area

Bridled Tern *Sterna anaethetus*

Morwennol Ffrwynog

A vagrant.

This species breeds in the Caribbean, west Africa, Red Sea, Persian Gulf, Indian and Pacific Oceans. There are 15 British records, up to 1991, of which two are from Wales:

11 September 1954 – One found, freshly dead, in Three Cliff Bay, Glamorgan. Unfortunately the specimen was mutilated before it could be critically examined so that the race, and consequently, its origin, could not be established.
1–23 July 1988 – One frequented the tern breeding colony at Cemlyn Bay, Anglesey.

Sooty Tern *Sterna fuscata*

Morwennol Fraith

A vagrant.

It breeds on tropical and subtropical islands in all oceans and the Red Sea.

The only definite record in Wales is a male that was knocked over by a boy with a stick on the Barmouth Golf

Links (Merioneth) on 17 August 1909. The specimen was preserved (*Zoologist* 1909, p. 438).

In *A Hand List of the Birds of Carmarthenshire* by G. C. S. Ingram and H. Morrey Salmon there is a bald statement that one was picked up alive at Llandovery, August or September 1923, and released later. Unfortunately no details are given and the record must be regarded as unsubstantiated.

Little Tern *Sterna albifrons*

Morwennol Fechan

A summer resident which now breeds only at one colony in Flintshire, otherwise a spring and autumn passage migrant to other coastal areas, rare inland.

At the beginning of the 20th century the Little Tern was widely distributed as a breeding species in Wales, nesting on coastal sand and shingle bars in two areas of Glamorgan and at many sites from Cardiganshire northwards to the Dee estuary. Since that time, there has been a substantial contraction of range as well as a reduction in the total number of breeding pairs. Unlike other tern species, which have tended to move from site to site, Little Terns have shown a pronounced site fidelity for a few favoured breeding locations, but with one exception even long-established colonies have now been abandoned.

In Glamorgan there were two breeding sites during the 19th century. The first was on the Leys shingle bank in the Thaw estuary, with between two and eight pairs; the colony died out about 1911, was re-occupied in 1925, but finally deserted after 1929. The site has now been destroyed by the construction of a Power Station. The second was at Sker Point, where there were 40–50 nesting pairs in the 1880s; numbers of breeding pairs fluctuated considerably thereafter with a maximum of at least 30 pairs in 1910 at the Point and Kenfig Sands. In 1911 the birds were subject to much disturbance from holidaymakers and left the area until 1922 (six pairs) and 1923 (five pairs); apart from 1936 when one pair nested, the site has since been deserted by breeding birds.

Northwards from the two isolated Glamorgan colonies the nearest breeding location was on the sand dune/shingle area of Ynyslas in north Cardiganshire, first established about 1896 or 1897. Numbers fluctuated widely, between one and 15 pairs. By 1938 the colony was down to a single pair and during the Second World War the area became a firing range and the birds totally deserted the site.

In the early part of the 20th century there were small breeding colonies scattered all along the coast from the Dyfi estuary northwards, wherever low-lying and flat sand and shingle shores predominated. In Merioneth there were breeding colonies at Aberdyfi, Aber Dysynni, Fairbourne, and in the vicinity of Mochras as well as sporadically at other locations between Barmouth and Porthmadog; most of the colonies were usually of fewer than ten pairs. The Aber Dysynni colony was formed in 1894 or earlier and remained active until 1988 at which time it was the only remaining site in the county; it reached a maximum of 36 pairs in 1977 after which, in spite of a protection scheme organised by the RSPB, breeding success declined because of losses of eggs and young to a variety of predators, which included Oystercatcher, Kestrel and Stoat.

In Caernarfonshire there were colonies on several of the flatter shingle shores between Conwy and Porthmadog. One of these at Aber Dwyfor numbered about 50 pairs (Forrest 1907) and this colony was in existence until 1982. The other large colony in the county, at Morfa Dinlle, dating back to at least 1915, was occupied sporadically until 1988. The first recorded breeding in the county was at Traeth Lafan in 1865 (Ecroyd Smith 1866).

Small colonies, usually between 10 and 20 pairs, were a feature of several coastal areas of Anglesey, one at the Crigyll estuary, Rhosneigr, dating back at least as far as 1886. Regular sites also included Red Wharf Bay and Abermenai/Newborough. The last successful breeding on the island was at Abermenai Point in 1978. Unfortunately, the Anglesey colonies were particularly susceptible to disturbance by increasing numbers of holidaymakers in the second half of the century, a factor which no doubt played a substantial part in their desertion.

In Denbighshire Little Terns have not bred since the late 1950s. Up until then it is possible that several sites used to be occupied (e.g. 20 pairs at Abergele in 1923).

Flintshire is now the only county where breeding still takes place and has long been a stronghold for Little Terns. The Point of Air colony was first described as far back as 1866 and the shingle spit there and at nearby Gronant have featured substantially in the history of this species in Wales. In some years both sites have been occupied, in other years only one; this depends largely, it appears, on the current stage of development of the respective shingle ridges, for the coast in this part of Wales is an accreting one where the shingle bars are covered by sand dunes and are replaced by new ridges on the seaward side. At early stages of this cycle the developing shingle bars are usually too low not to be inundated by high waters of spring tides.

From 1989 onwards Gronant has been the sole surviving breeding colony of Little Terns in Wales and the retention of the species as a Welsh breeding bird appears to depend on the success or otherwise of this site. The Gronant colony has been wardened since 1975 by the RSPB, with help from the Clwyd Ornithological Society and, initially, the North Wales Naturalists Trust, and fortunately it has been successful during this period with more than 680 young fledged in the 18-year period up to 1992. The Gronant colony has increased from 15 pairs in 1975 to 52 pairs in 1992; the main problems encountered in this period have been predation by Carrion Crows and foxes, and adverse physical factors such as tidal inundation and sandstorms during prolonged gales.

As to numbers, historical information indicates that most colonies were of the order of 5–20 pairs; by far the largest was the Point of Air where on 17 July 1916 Charles Oldham and S. G. Cummings estimated that the colony between the lighthouse and the colliery numbered 200 pairs (Oldham, Diaries).

Recent totals for the Welsh breeding population are:

1967	35 pairs in eight colonies
1969	25 pairs in six colonies
1975–79	average 71 pairs in four to six colonies
1980–84	average 56 pairs in three to six colonies
1985–89	average 50 pairs in one to two colonies
1990–92	average 51 pairs in one colony

The Welsh population is therefore only a very small percentage of the British breeding population, which amounted to 2800 pairs in 1985–87.

Outside the breeding area Little Terns are scarce but regular passage migrants in small numbers, chiefly in the period from the last week of April to the end of May in spring and from late July to the end of September on the return passage. There are fewer records from the southern counties from Monmouthshire to Pembrokeshire than from the more northerly counties. Numbers seen on passage are usually small but there is a notable exception in the regular post-breeding build-up from mid-July at Gronant (Flintshire), where each year there is a gathering of between 300 and 500 birds in what must be a favourable feeding area for Little Terns nesting farther north.

There have been very few inland occurrences of this markedly coastal species but there are a scattering of inland records from May to October from Monmouthshire, Breconshire, Cardiganshire, Radnorshire and Merioneth, chiefly in September. Llangorse Lake has been the most favoured inland locality, with Little Terns recorded there fairly regularly during the period July to October.

There are only two recoveries of birds ringed in Wales, both of young ringed at Gronant: one ringed on 24 July 1975 was recovered at Lymington (Hampshire) on 13 October 1979 and one ringed on 24 July 1983 was recovered at Millom (Cumbria) on 20 June 1988.

Whiskered Tern *Chlidonias hybridus*

Corswennol Farfog

A vagrant.

This species breeds from south-west Europe discontinuously eastwards to Manchuria, and south to southern Africa, Australia and New Zealand; the European population winters in Africa south of the Sahara.

There are five records from Wales:

21–22 April 1956	– One, Llan Bwlch-llyn (Radnorshire)
11 July 1965	– One, Glandyfi (Cardiganshire)
13 May 1974	– Two, Blackpill
7 September 1974	– One, Eglwys Nunydd Reservoir (Glamorgan)
5 July 1976	– One, Shotton (Flintshire)

Black Tern *Chlidonias niger*

Corswennol Ddu

A fairly common passage migrant, chiefly to coastal counties in autumn but also recorded on inland waters; spring records are usually fewer in number but there are substantial influxes in some years.

At the beginning of the century Black Terns were seen much less frequently in Wales than now. Forrest (1907) described it as "somewhat rare" in north Wales; the first record in Caernarfonshire was in 1909, the next was not until 1956. Similarly in Pembrokeshire Lockley *et al.* (1949) referred to the only recent record in 1904, in Carmarthenshire the first record was not until 1937 and in Radnorshire until 1949.

Heathcote *et al.* (1967) indicated that the Black Tern was seen much more frequently in Glamorgan than it had been some 50 years earlier; there were only four occurrences in the last quarter of the 19th century; some were seen and two shot in May 1901, but the next was not recorded until 1921 and 1925. Then there was another gap until 1933 but after that the species became almost annual until the Second World War. Some were seen in 1943, 1948 and 1950 and from 1954 onwards Black Terns have been recorded in the county every year.

Now of regular occurrence in singles and small groups, especially in southern counties, there are widespread movements in some years involving birds moving to or from their breeding sites in continental Europe to wintering areas in Africa. Notable concentrations have occurred in recent years as follows:

1969	Influx in mid-August when there were 75 at Llangorse Lake (11 August) and in Glamorgan 22 off Whiteford and 27 off Port Talbot Breakwater
1970	23 at Whiteford and 36 at Llandegfedd Reservoir (Monmouthshire) on 17 August and 24 at Llangorse on 21 August
1974	In early September 77 at Eglwys Nunydd Reservoir (Glamorgan) and a peak of 43 at the warm water outfall at Carmarthen Bay Power Station, Burry Port (Carmarthenshire)
1981	100 off Strumble Head on 12 September
1985	98 at Strumble Head on 14 August
1990	An exceptional concentration in the Severn estuary: 286+ off Black Rock (Monmouthshire) on 2 May, 270+ on 3 May and 70+ on 4 May.

> There were also 56 at Ynyslas (Cardiganshire) on 28 April and 27 at Llangorse, 30–35 at Llyn Trawsfynydd (Merioneth) on 2 May. This was part of a massive movement of Black Terns in southern Britain.

Spring passage is most marked in May and autumn passage between mid-August and mid-September but there are records from all months between April and December with an exceptionally late bird at Ynys-hir on 17 December 1989. Black Terns have been recorded in all counties with many inland records but the bulk of records currently come from the coast; the annual pattern of records suggests there has been a genuine decline in the numbers occurring in Breconshire, one of the most favoured inland areas in the 1980s (Peers & Shrubb 1990).

White-winged Black Tern *Chlidonias leucopterus*
Corswennol Adeinwen

A rare passage migrant to Wales.

It breeds from Hungary and Bulgaria eastwards through Asia; the Western Palearctic population winters in Africa south of the Sahara. There have been 22 dated records in Wales plus two seen in Cardiff (one shot) in March 1891.

Fifteen of the Welsh records are in the period July to October but there are seven spring occurrences including one which pre-dates all contemporary British records by five weeks: this was one in almost full summer plumage on the Gwendraeth near Llandyfaelog (Carmarthenshire) on 8 March 1958.

Glamorgan is the most favoured county with eight dated occurrences followed by Anglesey (4), Flintshire (3), Pembrokeshire (3), Carmarthenshire (2), Breconshire (1) and Cardiganshire (1).

Guillemot *Uria aalge*
Gwylog

Resident breeding species in south-west and north-west counties, moving out to sea to moult after breeding and returning to the vicinity of colonies as early as December.

Two races of the Guillemot *Uria aalge* occur in Welsh waters. The race *U. a. aalge* breeds northwards from southern Scotland and is a scarce but possibly regular winter visitor, as shown by tideline corpses from time to time. It is the paler southern race *U. a. albionis* of south-west Europe that breeds extensively on the headlands and offshore islands of west Wales from the Great and Little Ormes in the north, southwards to Worm's Head on Gower. Parslow (1973) has suggested that numbers in Wales (and southern Britain overall) fell during the first half of the present century, a trend which continued up to

the 1960s. Lockley's counts on Skomer between 1939 and 1958 support this contention. Another blow to these populations was dealt in the Irish Sea Seabird disaster of autumn 1969 in which large numbers of Irish Sea Guillemots perished (Holdgate 1971). About 15 000 dead or dying birds were accounted for but Saunders (1976) believed that the true number may have been nearer 35 000. Recovery followed, however, and the Seabird Register Survey, 1985–87, produced a total of some 34 300 birds* in Welsh colonies compared with the total of approximately 15 000 recorded in Operation Seafarer, 1969–70. This represents an overall increase of about 130% in that period, which bears out the increasing numbers which were widely recorded at Welsh colonies, as elsewhere around British coasts, during the 1970s and early 1980s. However there is some evidence (E. I. S. Rees in litt.) that fewer birds bred in 1969–70 because they were in poor condition, so that the Seafarer counts may have been under-estimates of the actual population at that time. All the same, the Welsh population comprises no more than some 3% of the British and Irish total (cl 200 000 birds in 1985–87).

The Operation Seafarer report (Cramp *et al.* 1974) emphasised that there were decreases in colony size in the years prior to the survey (1969–70). Birkhead (1974) has shown how the number of recorded tideline corpses increased enormously in the Irish Sea from 1920 to 1970, by which time oiling accounted for some 50% of all tideline deaths. The *Christos Bitas* (1978) and *Bridgness* (1988) oil spills off Pembrokeshire caused temporary setbacks to the general increases that took place in the 1970s and 1980s. Earlier, however, the species almost certainly benefited from the enforcement of bird protection laws at the beginning of the century, as Guillemot eggs had been widely collected at Welsh colonies; for example, a local man is reported to have collected as many as 500 eggs from Ramsey in 1908, and such collecting was doubtless commonplace.

* Counts of Guillemots have been done by varying methods across the years and care should therefore be taken in differentiating between *pairs* and *birds* (i.e. individuals).

Because of the birds' nesting requirements, Guillemot colonies are predictably traditional in their location over the years, despite the small size of some of them. Thus almost all the colonies now present have been in existence since records were first kept and are not subject to the sudden changes characteristic of those of some other marine species such as Puffin, terns or *Larus* gulls.

The most impressive and most easily viewed colony in Wales is undoubtedly that on the Elegug (Guillemot) Stacks (now usually referred to as Stack Rocks) below the limestone cliffs on the south coast of Pembrokeshire, near Flimston Bay. Here in 1793, Sir Richard Colt Hoare (M. W. Thompson, Diaries) saw "an astonishing quantity of Eligugs . . . the number of birds so great that the two rocks are so completely covered with them that in parts no rock whatever is to be seen". This colony is still extant and remains an impressive sight from the mainland cliffs above. The main Pembrokeshire colonies are on the offshore islands. Grassholm supports about 400 birds and Ramsey and its satellites around 1500-1700 individuals, which had risen to a maximum of 2169 individuals in 1992 (T. Sutton, unpublished report to RSPB). Skokholm has only small numbers, generally suggested as no more than 300 birds, having risen by some 100% in the 1980s. Middleholm has a small colony of c60–100 birds, St. Margaret's Is. supports around 600 birds, having increased from 118 in 1969, and Caldey Is. a mere 20 or so. The Pembrokeshire stronghold is on Skomer where, again, the number of birds counted has gradually increased across the 1970s, 1980s and early 1990s, albeit not to the extent that has been recorded at some other colonies. Buxton and Lockley (1950) estimated 5000 *pairs* in 1946 and Lockley and Saunders (1967) confirmed a similar figure for 1966. The Seafarer counts (1969–70) produced 3920 *individual birds* - an apparent discrepancy which can probably be accounted for by the *Torrey Canyon* oiling incident in 1967 and the Irish Sea seabird disaster in autumn 1969 (see above). Since then full counts in the 1980s and early 1990s have indicated the rate of gradual increase (Poole & Sutcliffe 1992):

1981	4526 individuals
1982	4711 individuals
1983	5112 individuals
1986	>5% reduction after *Bridgness* oiling incident in 1985
1988	6532 individuals
1989	5556 individuals = 15.1% decrease
1990	6051 individuals = 9.1% increase
1991	7516 individuals = 24% increase
1992	8032 individuals = 6.9% increase

Small numbers breed on the North Bishop, two or three kilometres north of Ramsey. On the south Pembrokeshire coast there are several small colonies scattered on the mainland cliffs between Stackpole Head and Linney Head. The famous colony on Elegug Stacks has reduced considerably since the earliest records, for unknown reasons. Lockley estimated 1750 pairs in 1960 which had reduced to no more than 520 apparently occupied sites in 1969–70. The Seabird Register illustrates the subsequent resurgence and put the colony size at about 2700 birds in 1990 (3553 in 1989) and 5802 in 1992.

The only breeding site in Glamorgan is the small long-standing colony (still c150 birds) on the Worms Head.

The largest Welsh colony is that on Carreg-y-Llam on the north coast of Llŷn: 2750 birds in 1969 and 7586 counted in 1989. Other Caernarfonshire colonies are on Little Orme (251 birds in 1969; 840 in 1986), Great Orme (736 in 1969; 1120 in 1986), Bardsey (34; 176), Aberdaron Head (17; 60), Gwylan Is. (60; 132), Cilan Head (947; 2403) and St. Tudwals Is. (137; 377). Coward (Diaries) obtained eggs from Cilan Head in 1887 "from a dealer named Kirby. He told us they worked the cliffs with a gang of men. Their last draw was on 25 May when they got over 900 eggs."

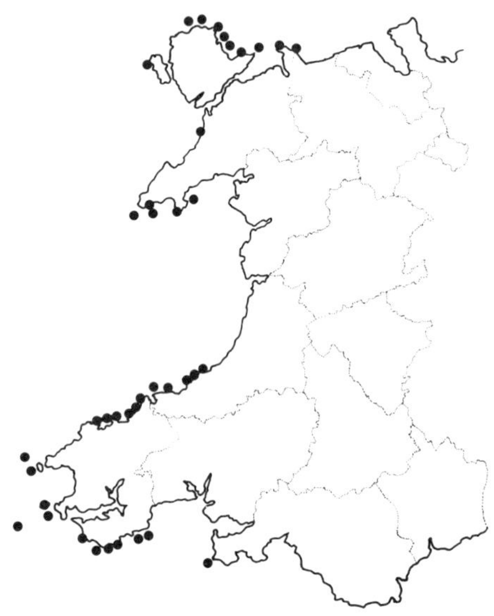
Guillemot colonies on the Welsh coast

The Bardsey colony has never been large; Aplin recorded about 20 "pairs" in 1901 and Coward no more than 50–60 in 1922. Since the Bird Observatory opened in 1953 the numbers have generally been between 50 and 100 pairs annually, increasing – in common with the overall trend elsewhere – to maxima of 321 birds in 1987 and 400 in 1989. Most of these Caernarfonshire colonies had increased further still between 1986 and 1992, some, e.g. Little Orme, by as much as 40% (although the unreliability of single year counts as measures of overall trends should be borne in mind).

The only substantial colony on Anglesey is that on the cliffs of South Stack, where counts of individual birds have risen 125% from 1363 in 1976 to 3071 in 1992. Other colonies around the Anglesey coast exist at Carmel Head, West Mouse and Middle Mouse (300–400 birds), Point Lynas, Ynys Dulas and Ynys Moelfre, the vicinity of Penmon and Puffin Is. The total involved in these colonies is no more than 1000 individuals. On Puffin Is.

there were 269 birds on the Seafarer counts (1969–70) which had risen to 552 by 1982 and 872 in 1990.

The low coasts of Merioneth do not support breeding Guillemots but the cliffs in south Cardiganshire and north Pembrokeshire hold a scattering of small colonies, some five or six sites in the former county and about the same on the Pembrokeshire coast westwards to Fishguard. In total these Cardigan Bay colonies support no more than 3700 birds, the largest concentration being on New Quay Head. A reflection of the increasing fortunes of Guillemots may be found in the fact that a pair bred on Cardigan Is. for the first time in 1989, and had increased to five pairs by 1990 and a maximum of 17 birds counted in 1992. Other small colonies along this coast which have been monitored regularly show strong increases through the 1980s, roughly in line with known trends elsewhere.

At end of August, after breeding is completed and the colonies deserted, adult Guillemots move out into the Irish Sea and Western Approaches with the young birds for the period of moult during which they are flightless. Welsh birds mainly go to the inshore waters off the eastern coast of Ireland although at least some of the Great Orme birds go to the mouth of the Mersey estuary (E. I. S. Rees, in litt.). After the moult (October–November) birds move back into inshore waters and some revisit the nesting cliffs as early as December: Marloes (Pembrokeshire) fishermen traditionally regard "the birds" as being back by Christmas each year. Passage through the Irish Sea is well marked in autumn, and as many as 26 000 and 35 000 per day have been estimated as passing Strumble Head on occasion in October and November.

When tideline corpses of Guillemots are found in winter it is often possible to determine the race. Predictably, most are of the southern race, i.e. the breeding stock, but a small percentage can be identified as *Uria a. aalge*, indicating that some individuals from the northern colonies – presumably from western Scotland – penetrate as far south as the southern Irish Sea. Two recoveries on the beaches of Cardigan Bay of young birds from western Scotland bear out this trend.

Individuals of the "bridled" morph are very scarce in the Welsh colonies, accounting for no more than 1% of all birds (S. J. Sutcliffe, in litt.).

Although many birds, particularly adults, remain for the winter in the sea areas relatively close to the colonies, others – mainly young birds – disperse more widely and ringing recoveries show movements into the North Sea (as far north as Norway), the Bay of Biscay and western coasts of Iberia.

Razorbill *Alca torda*

Llurs

A breeding resident on the cliffs and islands of north-west and south-west Wales. Some disperse south to the Portuguese coasts in winter with a proportion entering the Mediterranean; part of the population remains in Welsh waters.

The Razorbill is a fairly numerous breeding bird on the western coasts of Wales and there is no evidence to suggest that there have historically been any major long-term changes in numbers or distribution. The Seabird Register survey (1985–87) produced a total of c180 000 birds* in British and Irish colonies of which slightly more than 5% (c9600 birds) breed in Wales. There is little evidence of major population change since the previous national seabird survey (1969–70), although most colonies have shown a modest increase from the low point of 1970 (see below). Pairs are scattered around the rocky cliffs and headlands of north-west and south-west Wales, paralleling very closely the breeding distribution of Guillemot (q.v.). There are few sizeable colonies of note, the largest being on Skomer, which supports about half the entire Welsh population.

Razorbill populations showed a marked decline from the early 1940s for the next three decades or so (Parslow 1973) similar to the decrease which affected southern populations of Guillemots over the same period. This decrease has been linked to the increasing frequency of oil spills at sea (Lockley 1949; Saunders 1962) but it is likely other factors such as changes in food abundance and drowning in fishing nets (e.g. Portuguese waters in winter, are also implicated (Lloyd *et al* 1991)). Between the Seafarer Survey (1969–70) and the Seabird Register Surveys (1985–87) the Welsh population stabilised and then increased fairly dramatically at many colonies during the 1980s. The Irish Sea seabird disaster which occurred immediately after the Operation Seafarer surveys in 1969 killed numbers of Razorbills which were one of the two species (second only to Guillemot) most affected, although the effect on numbers in breeding colonies was temporary and a modest increase in the population was detected through the 1970s.

Razorbills breed on most of the Pembrokeshire islands – albeit in varying numbers – and at a handful of sites on the mainland cliffs of the county. On Skomer, which is the largest Welsh colony, no estimate of numbers was made until 1963 (c2100 pairs), after which there is evidence that the population was depleted by the Irish Sea seabird disaster in 1969 (certainly numbers were at their lowest in 1970). Subsequently there was a sharp rise to an estimate of 2350 pairs in 1981 and 2348 apparently occupied sites by 1983. Regular counts in more recent years have given totals of c2600–2700 *individuals*. The Seabird Register count in 1985–87 was 3000–3500 individuals and Sutcliffe and Poole (1992) recorded a maximum count of 3135 birds in 1992.

* Counts of Razorbills have historically been done by varying methods and care should be taken in differentiating between *pairs* and *birds* (i.e. individuals).

Other Pembrokeshire colonies supporting more than 100 individuals are Ramsey (c600–1000), Skokholm (c700), Middleholm (c170), St. Margaret's Is. (c200), Stackpole Head (c100), and Elegug Stacks (c300) and Needle Rock (c250). The Skokholm estimates and those on St. Margaret's Is. showed wide variations throughout the 1980s, although this may be more a function of vary ing methodology and observer bias than actual fluctuations: e.g. Skokholm 109 pairs in 1981; 300 pairs 1982; 442 sites 1983; 380 *birds* 1990. Other colonies, including ones on Carreg Rhoson and North Bishop, north of Ramsey and some mainland cliffs are very small. The total for Pembrokeshire is now (1993) in the region of 6000 birds.

The small colony on the Worm's Head in west Glamorgan – the only one on the south coast outside Pembrokeshire – has been in existence for many years; it fluctuates around 50 birds.

There are seven small colonies along the cliffs of south Cardiganshire, with the largest (c250 birds) on New Quay Head. Birds bred on Cardigan Is. for the first time in 1983 and had increased to c19 pairs by 1990. The total count for Cardiganshire gives around 700 birds.

In Caernarfonshire approximately 1300 individual birds were recorded in the Seabird Colony Register surveys (1985–87) (600 pairs in 1969) at the same nine or ten sites as Guillemot. Bardsey numbers have apparently increased considerably since pre-1950 when estimates suggested only a score or more pairs; the numbers in the early 1990s indicate about 250 pairs. Similarly at Carreg y Llam (Llŷn) there have been significant increases of around 100% between 1969 and the early 1990s. On Anglesey the largest numbers are on Holyhead Is. where around 400 birds are present on South Stack with a further 300 on adjacent cliffs. Other modest colonies occur at Middle Mouse (c100 birds), the Penmon area (c60 birds) and Puffin Is. (c200 birds). Otherwise there are small numbers of breeding pairs scattered around the north coast at sites such as Porth Llanlleiana, Carmel Head and Trwyn Cilan to give a total for the whole of Anglesey of probably no more than 1100 birds.

After fledging in July, young Razorbills remain mainly in the Irish Sea for the first month or so before moving southwards through the autumn to the Bay of Biscay and the coasts of Iberia. An unknown proportion enter the Mediterranean as evidenced by ringing recoveries; six young birds from Wales have been recovered on the Italian shores of the Gulf of Genoa and others on the Algerian and Tunisian coasts. In their second and subsequent years it appears that many of these birds move again to northern waters, distributed along the coasts of the English Channel, North Sea and as far north as Scandinavia. Between breeding and the completion of their subsequent moult, few adult Welsh Razorbills occur outside the Irish Sea. Ringing recoveries show that some adults then also wander as far as the southern North Sea for the winter.

Passage off the headlands of west Wales in autumn is sometimes heavy, as exemplified by an estimated 10,000 birds passing south off Strumble Head (Pembrokeshire) on 29 October 1990; such movement is evidently from northern colonies as numbers such as seen on this one day dwarf the entire Welsh population.

Black Guillemot *Cepphus grylle*

Gwylog Ddu

A very scarce breeding resident in Anglesey and rarely elsewhere in north Wales, representing the south-eastern outliers of the larger populations of the Isle of Man, south-west Scotland and Ireland. Occurs sparingly off the coasts of west Wales outside the breeding season.

There is a very small but well-known breeding population of Black Guillemots in south-east Anglesey, which has been established since at least the early 1950s. However, the history of breeding in this corner of Wales goes back a very long time. Pennant (1778) recorded "a few" on the Great Orme over 200 years ago and there is general acceptance that this was once a regular breeding site. RSPB Watchers' Reports (1920) record "previous breeding but they are now gone". Similarly Dobie (1893), quoting Price's *Guide to Llandudno* adds that they formerly bred at the western end of the Little Orme. In the early years of the 19th century breeding was even claimed in Pembrokeshire (Montagu 1802) where there were allegedly a few pairs near Tenby each year (summer birds have again been reported in south Pembrokeshire in 1990 and 1993 (R. Haycock, pers. comm.). At present, and during the remainder of the present century, however, it is clear that breeding Black Guillemots have been restricted to a few sites in Caernarfonshire and Anglesey, principally around Conwy Bay. The Great Orme birds ceased to breed there sometime prior to the Second World War, although they are still seen in the vicinity from time to time in summer; in 1954 nesting was suggested and Whalley (in litt.) watched birds in 1953, even carrying material, but asserted that they did not breed. It seems reasonable to suppose that the pairs which were subsequently found on Puffin Is. (probably breeding 1953) and Fedw Fawr – first reported in 1948 – were the same population abandoning the mainland sites for unknown reasons and re-establishing across the Menai Straits on south-east Anglesey. However, the picture is incompletely understood for, as early as 1912, Oldham (Diaries) found birds behaving as though they were breeding on Moelfre Is. a little farther up the east coast of Anglesey. The small colony of four or five pairs is now well established in the Fedw Fawr cliffs: the maximum number of adults recorded there being 16 in April 1987. Elsewhere around the Anglesey coast breeding has once been proved (1970s) in Beddmanarch Bay on the Inland Sea near Holyhead, which reflects the regularity with which birds have been recorded both there and in Holyhead Harbour over the years, both outside the breeding season and, less frequently, during the summer months. Winter observations of Black Guillemots, although not numerous, are made at many locations around the Anglesey coast. At Llanddwyn Is. a single bird was seen in June 1901, offshore from Ynys yr Adar, and there have been a number of other breeding season records from the same site in intervening years; breeding has not been proved, however, although it is interesting that Pennant alluded to breeding here as long ago as the second half of the 18th century.

On the coast of south Caernarfonshire there have been

some 30 records off Bardsey Is. since 1954, mainly in September and October. Writing in the *Merseyside Naturalists' Journal* in 1957 J. Williams reported being unsurprised at seeing a pair at St. Tudwal's Is. because he had seen them there regularly since 1954. Further sporadic summer records are published for the St. Tudwal's/Abersoch/Cilan Head area and in the early 1980s D. Mayes (in litt.) found at least two pairs there entering crevices, displaying agitation and showing every sign of breeding.

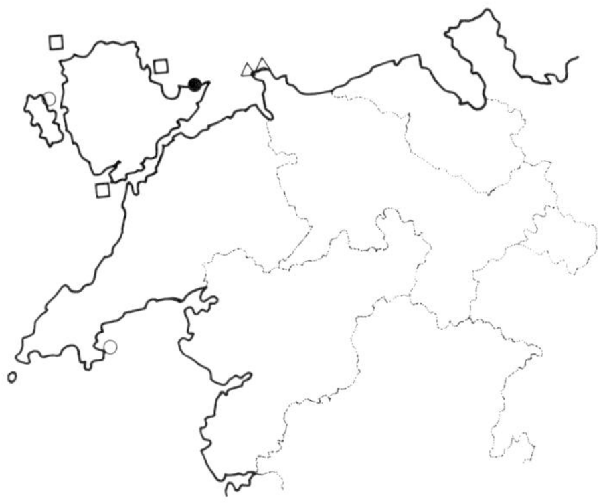

Black Guillemot – ● regular breeding, ○ sporadic breeding, □ possible sporadic breeding, Δ former site now deserted

Outside the breeding season, Black Guillemots remain fairly sedentary, which presumably accounts for the predominance of winter records around Anglesey and not Caernarfonshire (the record of c30 off Nefyn in March 1972 is exceptional). Nonetheless, autumn and winter records elsewhere around the coasts of Wales have become increasingly regular in the 1970s and 1980s, doubtless because of the increase in the number of active birdwatchers and, particularly, systematic sea-watching. Merioneth has produced only a handful of winter records, mainly as tideline corpses. In Cardiganshire only three records were known up to 1974 but since then there have been some 19 sightings, in some cases probably involving the same individual birds. Similarly in Pembrokeshire, despite the alleged breeding in the 19th century, Lockley (1949) could list only five records, whereas one or two sightings are now made in most years, usually between September and March, although there was one aberrant record from Skokholm in June 1968. Black Guillemots are vagrant to the coasts of Carmarthenshire and most of Glamorgan, with each county producing only two or three records. The exception to this is Gower, where records have become a little more regular in recent years, e.g. four in 1990 in March, August, September and November. It is presumed that these birds seen round the coasts of south Wales are more likely to be wandering individuals from the Irish breeding population than from north Wales.

Little Auk *Alle alle*

Carfil Bach

Scarce winter visitor in irregular numbers offshore on the western seaboard. Most sightings in late autumn and early winter, often associated with strong westerly gales.

Records of Little Auk around the Welsh coast have increased considerably in recent years. Without doubt this is almost wholly due to the increased effort that is put into regular sea-watches, particularly in Anglesey and Pembrokeshire, rather than any shift in behaviour or spatial distribution of the birds themselves. Thus the *Pembrokeshire Bird Report*, and to a lesser extent the *Gwynedd Report* (covering Anglesey and north-west Wales), produce records of Little Auks in most years. A few of these records are for September but most occur between October and March; exceptionally, vagrants may occur in other months, e.g. one in full summer plumage off St. Govan's Head (Pembrokeshire) in May 1983. In the six north Wales counties Forrest (1907) could list only 17 records across all the years up to 1907, together with a reference to "many" washed ashore in Merioneth in 1877; nowadays, as many as that are recorded in north Wales in an unexceptional season. Similarly in south Wales Mathew (1894) referred to it as infrequent, whereas in 1990 alone 16 were either found ashore or seen offshore from headlands in Pembrokeshire – a number which is now regarded as unexceptional annually. As many as 21 were seen passing Strumble Head on 20 November 1983. Nor are Little Auks particularly rare on the south Wales coast. The Carmarthenshire and Glamorgan Bird Reports show increased numbers of reports in recent years. In the latter county, where only five records were known in the last century, nine birds were found dead or dying along one mile of coast in the "wreck" of 1950 and a number of years since then have provided individual sightings.

The Little Auk is particularly vulnerable to protracted severe weather at sea during winter and such conditions can produce "wrecks" of the species on windward shores. Strangely such "wrecks" appear to occur only at the very southern limit of their winter range. Evidently one such year was 1877; another was was 1912 (February) – when many appeared on the north Wales coastline of Denbighshire and Flintshire and subsequently 1932. A particularly notable "wreck" occurred between 8 and 11 February 1950 mainly in south-west Britain after a period of powerful south-westerly winds. It was clear that large numbers of Little Auks were wintering in the Western Approaches and that many were driven onshore in south-west Ireland, the south-west peninsula and Wales. Dead and dying birds were found in all the western coastal counties, totalling over 40 individuals, presumably many others were never found, even on shorelines, at this mid-winter time.

The massive seabird "wreck" in the North Sea in February 1983 was associated with unusual weather conditions and north-easterly gales when 1200 Little Auk corpses were picked up, but, unsurprisingly, no effect was seen on western coasts. Although there was no "wreck" as such in 1988, repeated south-westerly gales in January and February produced an unusually high number of records. Exhausted Little Auks, if found alive, seldom survive and occasional storm-driven birds may be picked up as far inland as the English border; all Welsh counties can produce one or two vagrant records of this nature. With a world population of perhaps as many as 30 million, the Little Auk may well be the world's most numerous seabird. The numbers that winter in Welsh waters are not known but Britain is at the southern edge of the normal wintering areas and in any case the species is a specialist feeder on large planktonic crustaceans, notably *Calarium hyperboreus*, which are poorly represented in Welsh waters, where virtually all the copepod species are rather too small for Little Auks (E. I. S. Rees, pers. comm.).

Puffin *Fratercula arctica*

Pâl

A breeding bird of offshore islands and a few western headlands. Formerly bred in huge numbers.

The British and Irish population of this popular seabird was put at approximately 512 000 pairs in 1985–87 (Lloyd *et al.* 1991) of which fewer than 3% – some 14 000 pairs – breed in Wales. This was not always the case, however, because up to the end of the 19th century and early years of the present century, Wales supported several enormous colonies on islands off Pembrokeshire, Caernarfonshire and Anglesey; in the decades leading up to the end of last century there were probably as many as 400 000 pairs of Puffins breeding round the Welsh coast and Grassholm was one of the largest Puffin colonies in British waters. The chequered history of these colonies is consistent with the pattern which has occurred elsewhere around the coasts of southern Britain. Although most, but not all, of the recent history of individual colonies is a story of decline and sometimes demise, the causes are not always understood. Harris (1984) attempted to review likely reasons for colony decline and concluded that whereas factors such as rats and exploitation by man can be attributed in some instances, these causes and underlying factors such as sea pollution, e.g. oiling, are less consequential than changes in the marine environment – and thus food supplies – due to climatic shifts, over which man has no influence. In specific instances, e.g. Grassholm (see below), a physical reason for abandonment is clear. E. I. S. Rees (in litt.) suggests that a direct parallel can be drawn between the declines of Puffin numbers around the Welsh coasts and the changes in the herring fisheries; winter spawning stocks declined markedly in the 1920s and 1930s probably due to oceanographic events. There were large fisheries in Cardigan Bay, off the north coast of Llŷn and off Moelfre (Anglesey) and the larvae/post larvae would have provided rich feeding for Puffins.

The Puffin is restricted in its breeding to offshore islands and a few mainland cliff coasts; its breeding distribution in Wales, both historically and currently, extends from Worms Head on the Gower Peninsula northwards to the Great Orme (and formerly Little Orme) headland. The map shows the distribution of Welsh colonies.

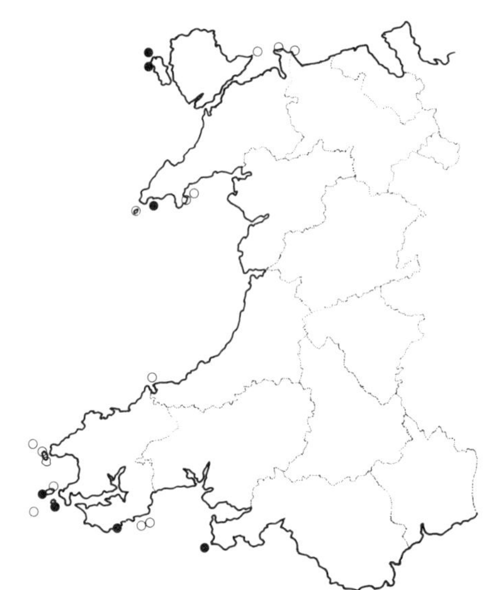

Puffin colonies on the Welsh coast – ● current, ○ former

Glamorgan

Worm's Head. Dilwyn (1848) found Puffins breeding here and since then they have been confirmed periodically, although not regularly, albeit in very small numbers. It is not certain whether breeding has been continuous since it was first recorded although it seems probable that this is the case up to the present time (D. K. Thomas, pers. comm.) although numbers remain extremely small, probably no more than one to five pairs. These pairs are now restricted to rock crevices on the outer head, although Thomas believes that in Dilwyn's day, when they were recorded as breeding in burrows, this may well indicate that they were on the inner head where soil is deeper and burrowing easier. This remarkable small colony seems to have maintained its toehold at much the same level for at least 150 years.

Pembrokeshire

Mathew (1894) suggested that "we do not believe that we should exaggerate were we to say that the Puffins in number are equal to all the other birds in the county added together". The populations are now greatly reduced from Mathew's day, although the main Welsh breeding stations are still in the county, some 7400 pairs on Skomer and 5000 individuals on Skokholm (Lloyd *et al.* 1991). The rise and fall of numbers on the Pembrokeshire colonies is shown as follows:

Caldey

1804	"Many"	E. Donovan (1805)
1860	"Numerous"	T. Dix, in Mathew (1894) (Dix relates how the men of Tenby indulged in a ritual Puffin slaughter each Whit Monday)
1902	"A small colony"	J. Walpole-Bond (1904)
1923	"Still breeding"	W. J. Wintle (1925)
1927	A few pairs	B. Sage (1956)
1949	Present: no breeding proved	B. Sage (1956)

Breeding appears to have ceased finally on Caldey around the time of Sage's visit.

St Margaret's Island

1903	"A small colony"	J. Walpole-Bond (1904). His photograph shows 28 birds on land
1949	Birds present: no nests found	B. Sage (1956)
1952–55	c10 pairs	B. Sage (1956)
1962	17 birds on the sea	S. J. Sutcliffe (1963)
1983	4–10 pairs	*Pembrokeshire Bird Report*
1984	Breeding confirmed	*Pembrokeshire Bird Report*
1986	2–3 pairs	Seabird Colony Register
1989	2 birds only	Seabird Colony Register
1992	2–3 pairs still present	S. J. Sutcliffe (in litt.)

St. Margaret's Island has long been infested by rats, which have presumably been responsible for restricting the population to a few pairs deep in crevices.

Grassholm

Up to the turn of the 19th century there was a vast puffinry on Grassholm, at the time claimed as one of the largest in Britain. In the 1890s, there were several wild claims of the numbers breeding on this remote 9-ha island. J. J. Neal estimated over half a million Puffins and R. Drane (1893) believed there were as many as 700 000 birds. It was left to Williams (1978) to demonstrate that such totals were not possible and that the maximum number that could be accommodated on the island was probably around 200 000 birds. The collapse of the colony, which took place in the early decades of the present century, was literal in so far as the whole island catacomb caved in where the birds had burrowed in the shallow *Festuca* peat. The Puffins then moved elsewhere and the period of the 1920s and 1930s saw a correlated build-up on Skokholm and, although unrecorded, probably on Skomer at the same time.

1928	200 pairs	
1934	130 pairs	
1940	20 pairs	
1946	max 50 pairs	R. M. Lockley
1948	75 pairs	
early 1960s	2–3 pairs	
1973	1 or 2 adults with fish	G. A. Williams
1990	1 bird in burrow; possibly 3 occupied burrows	S. J. Sutcliffe

Since 1973 a few Puffins have been seen on or close to the island during the breeding season leading to suggestions that occasional breeding may still occur.

Skokholm

Numbers rose steadily in the 1930s as the Grassholm colony burrowed itself out of existence, and birds moved to Skokholm where the population rose to unexpected levels, only to fall sharply again by the 1950s. The population in recent years appears to have stabilised at 2000–2500 pairs.

1928–40	40 000 pairs	R. M. Lockley (1953b)
1953	5000–10 000 pairs	P. J. Conder
1955	6000 pairs	H. Dickinson
1969	2500 pairs	Seafarer Survey
1971	2500 pairs	Bird Obs. Report
1985–87	5000 birds	Seabird Register
1990	2500 pairs	M. Betts (1992)

Skomer

The most important Welsh colony since the demise of Grassholm, Skomer has supported a large Puffin colony since the first records were kept. Recent numbers have been substantially unchanged since 1970s.

In the early 1990s July counts, including many nonbreeders rose to 15 000–17 000 birds (S. J. Sutcliffe, in litt.).

1894	"Scarcely a yard of ground free from them" Mathew (1894)	
1946	50,000 pairs	R. M. Lockley
1963	c6000 pairs	D. R. Saunders
1971	c6000 pairs	P. Corkhill
1975	stable at c5200–6500 pairs	Birkhead and Ashcroft
1985–87	c5000–6000 pairs	Seabird Register
1989	c5700 pairs	*Pembrokeshire Bird Report*
1990	5000–6000 pairs	*Pembrokeshire Bird Report*

Middleholm

1933	Probably 1000 pairs	Lockley and Saunders (1967) A very large colony was rumoured "in former times"
1954	c300 birds	Lockley and Saunders
1964	c150 birds	Lockley and Saunders
1966	estimated 80 pairs	Lockley and Saunders
1969	c100 pairs	Operation Seafarer
1983	c200 pairs	Seabird Colony Register

Ramsey

Lockley (1947) recounts that vast numbers allegedly bred on Ramsey during Elizabethan times in the second half of the 16th century. In *A Survey of the Cathedral Church of St. David's* (1715) it is recorded that Ramsey supported

> ". . . vast numbers of Puffins, which living upon fish . . . were allowed to be eaten in Lent. The vast number of eggs laid in these rocks are, in a time when they are in season, the great subsistence of the poorer inhabitants of St. David's. The people hazard their lives to get them; for whilst one stands upon a steep and high rock, another is let down on a rope which is held by him that stands above. Sometimes they both fall and are dashed to pieces among the rocks."

(Doubt is cast as to whether this account truly relates to Puffins or whether it is a case of mistaken recording and the species involved was actually Guillemot. Nonetheless (see below) Puffins certainly were involved on Ramsey.)

By Mathew's time (1894), they were confined to the northern headlands and they evidently disappeared some time after 1927 when Bertram Lloyd (Diaries) found c200 pairs still breeding. The probable site of this last colony is still clearly discernible. The date of arrival of rats on Ramsey is not known but it is generally assumed that they came ashore after a shipwreck; Bennett (1982) lists 28 shipwreck incidents between 1831 and 1903. Whatever the date of their arrival, they presumably hastened the demise of the Puffin colony if not being directly responsible for its extirpation.

Ynys Bery (Ramsey)

1810	". . . stocked with Puffins . . . and other sea-fowl"	R. Fenton

This is the only reference to Puffins on Ynys Bery. The rats which so devastated seabirds on Ramsey did not apparently reach this islet off the south end; the reason for its desertion is not known.

Bishops

Lockley (1953b) reported "a few pairs" but also quoted Owen (1603) who referred to them as "breeding in numbers". A few pairs of Puffins still inhabit the North Bishop, put at 40–50 pairs in 1969 and 1972 and perhaps as many as 50 pairs by 1990 (S. L. White, pers. comm.). There were c30 pairs recorded there in 1992 (T. Sutton, pers. comm.).

Mainland cliffs of Pembrokeshire

Small numbers of Puffins breed on stacks and islets around the south Pembrokeshire coast. These are mainly in the area between Stackpole Head (ten pairs in 1983) and Linney Head in crevices in the carboniferous limestone cliffs. Lockley mentions breeding at one site on the north Pembrokeshire coast between Fishguard and Cardigan but there is no recent evidence of occurrence.

Cardiganshire – Cardigan Island

There was formerly a small colony here, until the wrecking of the merchant ship *Herefordshire* in 1934 from which rats that landed on the island soon accounted for several of the breeding species, including Puffin, which had gone by 1930. Attempts were made to attract the species after the island was cleared of rats in 1969 but to date this has been without success.

Mainland coast of Cardiganshire

Lockley (1953b) and others have referred to the presence of a few Puffins around the seabird colony on New Quay Head but breeding has never been proved.

Caernarfonshire

Historically there were several important colonies on the south coast of the Llŷn peninsula and a few pairs on the

Orme headlands. As has been the case elsewhere, rats were responsible for the demise of some colonies, although in other cases the disappearance of the breeding Puffins is once more difficult to attribute.

Bardsey

1662: John Ray recorded the presence of breeding Puffins on Bardsey and breeding was suspected on the north-eastern headland until the 1930s, based on the reports of experienced visitors such as Bertram Lloyd, Ticehurst, Aplin and Coward. No proof of breeding was forthcoming, however, and the Puffin remains only a regular visitor to surrounding waters.

St. Tudwal's Islands

These two islands formerly held enormous puffinries. Thomas Pennant was the first to record them here in 1773 and William Bingley also found them breeding when he landed in 1798. Forrest's estimate of "hundreds of thousands" on both islands in 1903 was probably an exaggeration, although there was clearly a very large colony. Aplin (1910) found enormous numbers on the east island in 1905 and also considerable numbers on the west island. By 1935 there were still large numbers breeding but after rats reached the east island by the 1940s the Puffins' days were numbered and they had gone by 1951.

It should be noted that the colony on the west island although not subjected to depredations by rats, still declined in parallel with the colony on the east island. Today the islands remain deserted of puffins.

Gwylan Islands

These two small islands have long supported modest colonies of Puffins, reaching as many as 1000 pairs on the larger island in the early years of the present century (Aplin 1910) with small numbers on the adjacent Ynys Gwylan Fach.

Ynys Gwylan Fawr

1915	"A large colony"	
1958	450–500 pairs	J. M. Harrop
1966–69	500 birds	Operation Seafarer
1985–87	600 pairs	Seabird Register

Ynys Gwylan Fach

1905	"A fairly large number"	Aplin (1910)
1961	15–20 pairs	Bardsey Bird Observatory Report
1968	None breeding	Bardsey Bird Observatory Report

In the early years of the century there was a small overflow colony on the mainland between Trwyn-y-Penrhyn and Ogof Lwyd but this was not recorded thereafter.

Great and Little Orme headlands

At the end of the 18th century Pennant recorded large numbers breeding on the Little Orme headland and a few on the Great Orme. The former colony was soon extinguished but that on the Great Orme has clung on, however, in much the same way as that on the Worm's Head (Glamorgan), until c1980.

1920s	A few breeding	RSPB Watchers' Reports
1950s	One or two pairs	P. E. S. Whalley (1954)
1976	A few pairs still breeding	Hope Jones and Dare (1976)

No breeding has occurred on the Great Orme in recent years (T. Gravett, pers. comm.).

Anglesey

Three sites on Anglesey have held Puffin colonies for as long as records have existed. That on the Skerries is the smallest and may possibly not have been continuous throughout the current century. The colony on South Stack cliffs on Holyhead Island is also fairly small whereas that on Priestholm (Puffin Is.) was formerly very large and was heavily exploited by local inhabitants.

Forrest (1907) claimed that Puffins were not breeding on the Skerries at the time he was writing, although he accepted the reports of Eyton (1839) and, before him Thomas Pennant, that they "formerly bred", Pennant explaining that the islets "afford food for a few sheep, rabbits and Puffins". They were certainly breeding in 1911 when G. R. Humphreys collected eggs on Ynys Awr where Charles Oldham (Diaries) found several pairs still nesting in 1933. RSPB correspondence from lighthouse keepers indicates that in 1941 Puffins were regarded as "more numerous than ever" whereas in 1945 there were allegedly no more than a dozen or so nests, a figure echoed by Whalley (1954) for 1951. In the late 1980s and early 1990s, since the islands have been managed as a reserve by the RSPB, annual monitoring has shown a regular population of some 80 pairs (but almost double that number in 1989, which indicates that fluctuations in these small north Wales populations still occur).

Forrest records breeding on South Stack cliffs in 1902 and an earlier record for 1898 (A. Whittaker, Diaries) suggests several hundreds of birds, although Forrest gives only two pairs for the same year! T. A. Coward found "large numbers on the cliff face opposite the (lighthouse) steps" in 1902 and confirmed this in several subsequent years noting that they were more numerous than ever in 1905. Later this century the colony has stabilised reasonably at an average of around 100 individuals after having fallen to no more than 10-15 pairs in the 1950s.

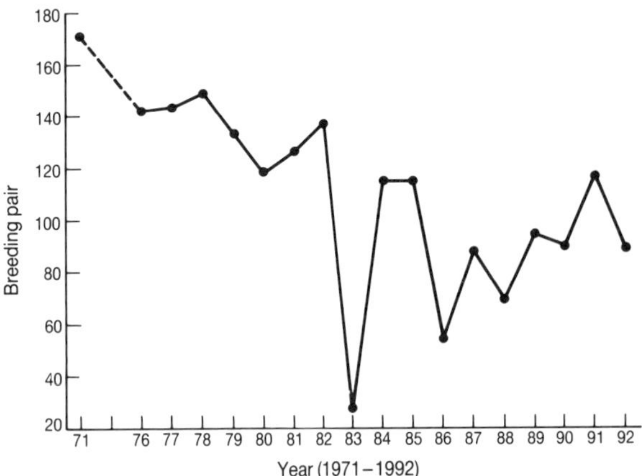

Numbers of breeding Puffins, South Stack Cliffs (RSPB reserve)

The major colony on the north Wales coast was always that on Ynys Seiriol (Puffin Island). As long ago as 1662 Ray found huge numbers on the island; a hundred years later Pennant likened them to swarms of bees and in 1804 they allegedly covered some 50 acres of the island with an estimated population of 50 000 pairs (Bingley 1800). Around this time there was heavy harvesting of the birds by local people each July with up to 15 dozen birds being taken in a single morning for pickling and selling in England for three shillings per barrel. Around the same time, 1817, a Prussian vessel was wrecked onshore and rats escaped onto the island. The fortunes of the Puffin colony fluctuated thereafter, apparently being exterminated by 1835, returning around 1880 after, it is reported, the rats were eliminated, although numbers still seemingly fluctuated unaccountably. (Rats have certainly been numerous on the island in the 20th century.) By 1907 there were again some 2000 pairs (Forrest). By 1926, however, Coward found that the expanding colonies of Lesser Black-backed and Herring Gulls had encouraged vegetation changes on the Puffin slopes which resulted in Puffins being restricted to a much smaller area of the northern slopes. Great Black-backed Gulls and Ravens were also shown to be slaughtering considerable numbers. Through the 1950s and 1960s, there were generally several hundred pairs annually and this still applied up to 1972 and 1973 when R. W. Arnold counted maxima of 450 each year. Numbers fell gradually thereafter, with a maximum of 225 in 1975 and minima of 120 in 1976, 120 in 1977 and 100 in 1979. Since then a total approaching 100 birds has been counted on only one occasion (1983). Numbers have fluctuated between maxima of only four (1981) and c90 (all figures R. W. Arnold). In the late 1980s and early 1990s, small numbers still remain each year (SCR).

Outside the breeding season the Puffin is a pelagic species, wintering well out to sea with some of the young birds moving even farther west or north-west than most of the adults (BWP vol 4). Although large numbers have been ringed in the colonies over the years, notably in Pembrokeshire, recoveries are less numerous than with other auks, because of the more pelagic winter habit of the Puffin. Many move out into the North Atlantic where they are very unlikely ever to produce recoveries. A spread of recoveries ranging from Norway (3) and Greenland (1) in the north to Iberia (10) and the Canary Is. (1) in the south, gives an indication of their wide distribution. There is one recovery of a Skomer bird from the Mediterranean (Balearic Is.).

Onshore or inshore records of stranded birds in autumn and winter are infrequent and the Puffin is the least frequently affected of the breeding auks by oiling incidents at sea at these times of year, mainly again because of its pelagic habits. Inland records or storm-driven individuals are rare but have occurred in most counties some time during the present century as far east as the English border.

Pallas's Sandgrouse *Syrrhaptes paradoxus*

Iâr y Diffeithwch

A vagrant to Wales which occurred several times in the 19th century. It breeds from the Kirgiz and Aral-Caspian steppes eastwards to Manchuria and is only partially migratory.

The first record for the British Isles was one shot from a party of three near Tremadoc (Merioneth) on 9 July 1859.

Large invasions occurred in the British Isles in May 1863 and May 1888 after heavy snowfall on a hard snow crust had prevented feeding in the breeding area. There was a scattering of records in Wales in the years of the two invasions, from Pembrokeshire, Merioneth and Glamorgan. The largest flock seen was 16 birds at Llanrhidian (Glamorgan) in 1888 out of which a pair was shot.

Rock Dove and Feral Pigeon *Columba livia*

Colomen Wyllt
Colomen Ddôf

Now extinct as a truly wild species. Individuals showing characteristics of Columba livia *are invariably derived from feral stock.*

The frequency of Welsh place-name elements such as Colomendy (see Welsh name above) give testimony to the former familiarity of this species and its derivatives. The demise of the Rock Dove in Wales is not the familiar story of gradual extinction of a species at the hand of man or his activities; on the contrary, it is a somewhat ironic outcome in view of man's interest in promoting and encouraging the species for his own benefit. The gradual disappearance of genuinely wild Rock Doves has occurred through interbreeding with feral strains which themselves had their origins in the wild species, *C. livia*. Birds originating from wild Rock Dove stock – with its convenient attraction of breeding throughout the year – have been exploited as a source of food for man for many hundreds of years, notably through such devices as the provision of dove cotes and breeding platforms in natural caves. The renowned artificial walling of Culver Hole on the cliffs of south Gower (Culver is an old name for pigeon or dove), dating from mediaeval times, is an outstanding example of the extent to which men would go to provide optimum conditions in which this strongly colonial species could breed in numbers so that the young "squabs" could be harvested with ease.

Culver Hole, Gower

The progressive disappearance of the original strain of Rock Dove from Wales is difficult to chart both because of the paucity of written accounts of Welsh ornithology for 18th and 19th centuries and also the fact that, even to the present day, the numerous feral populations inhabiting cliff nesting areas regularly contain individuals that exhibit the double black wing-bars, white rump and underwing and unspotted wing coverts of the true Rock Dove. Forrest, writing in 1907, was the first to allude to these difficulties in trying to determine the position with the species *vis-à-vis* its feral derivatives in north Wales and concluded that it was "practically impossible" to do so. He summarised the situation by listing those areas – mainly coastal cliffs in Anglesey and Llŷn – where apparently pure Rock Doves still existed (although Oldham and Coward were unable to find any themselves in the 1900s). These areas included Holy Is., Bull Bay, Valley and Rhosneigr on Anglesey and several sites on Llŷn (Aplin claimed two flocks of wild birds around Pen Cilan in 1903) together with Great and Little Orme's Head (where Pennant had found them abundant in the 18th century) and Conway Mountain in Caernarfonshire. Forrest further refers to records of earlier occurrences inland in Denbighshire and Montgomeryshire. In Merioneth Forrest knew of former records in the Barmouth area and Pennant mentioned them as breeding not far away at Bird Rock, Tywyn; A. W. Boyd saw two or three at Llwyngwril in 1906 which he adjudged to be genuinely wild birds. Other than the two or three inland instances mentioned above together with one or two hearsay accounts in county avifaunas, there are no records to be traced or other inland occurrences in Wales, although presumably the Rock Dove inhabited inland sites widely in the distant past as elsewhere in its range (BWP vol. 4).

In Glamorgan it was apparently still breeding on the sea-cliffs in the south of the county at the end of the 19th century and is recorded as being fairly numerous on Gower at the same time, with records continuing in the latter area until 1925. Elsewhere in the county the final claim to breeding was an alleged pair on Sully Is. as far back as 1903. Carmarthenshire too had Rock Doves breeding on the coastal cliffs around Llanstephan but they had vanished long since to be replaced by feral pigeons by the time of Salmon and Ingram's *Handlist of Birds of Carmarthenshire* (1954); no date for their demise is known.

Rock Doves were formerly to be found around most of the rocky coasts of Pembrokeshire. However, Mathew, writing in 1894, had never seen one himself although he quoted another observer reporting Rock Doves in the Tenby area (1887) and he believed that one or two pairs were possibly still breeding on Ramsey Island. Lockley *et al.* (1949) include the Rock Dove on the list of species breeding on Ramsey as late as 1948, adding that they believed the numbers were probably much the same as in Mathew's time fifty years earlier. At the same time Lockley *et al.* conceded that "it is doubtful if any truly wild descendants of the original Rock Dove are left" and that it was by then "perhaps extinct (in Pembrokeshire) as a pure race". In the years immediately after the Second World War other observers were reporting apparent Rock Doves on other Pembroke headlands – Linney, Stackpole, Strumble, Newport; at St. David's Head there was reputedly a small colony of six pairs as late as 1947.

The *Dyfed Bird Reports* and *Ceredigion Bird Reports* (1967–91) record that "a considerable proportion" of the feral birds on the coasts of Pembrokeshire, Carmarthenshire and Cardiganshire still show the characteristics of wild Rock Doves.

From the scant evidence available, it seems likely that the last truly wild Rock Doves in Wales were probably to be found on the more remote headlands of Pembrokeshire and offshore, on Ramsey, during the 1930s and 1940s. Birds conforming to the plumage of the wild race can still be found on these coasts but there is no possibility that any of them are of the original pure wild stock.

The feral replacements of the original wild stock are widely spread and numerous throughout Wales. Few towns of any size do not have resident populations but they are, unsurprisingly, least numerous in areas of the central mountain spine. Feral pigeons are abundant in the major cities and towns – Cardiff, Swansea, Newport, Wrexham – and in addition they are particularly numerous around the cliffs of Pembrokeshire and Anglesey. In the industrialised areas of south Wales and north-east Wales they are hardly less common. Their populations are constantly supplemented by the large number of "lost" homing pigeons which join the feral ranks each year.

Stock Dove *Columba oenas*

Colomen Wyllt

Breeding resident occurring widely but not numerously up to approximately 305 m; most common in lowland areas. Formerly scarcer than at present.

The Stock Dove is a resident species, strongly sedentary, which breeds in all Welsh counties although it is nowhere particularly numerous and its distribution is somewhat patchy. Although occurring sparingly in suburban areas and town parks, it generally tends to shun built-up areas and avoids open moorland (but will nest in crags on the flanks of the hills). It is absent from much of the coalfield area of south Wales, parts of Snowdonia, Llŷn, inland Anglesey and the north Wales coastal belt, although in the last area it is not infrequent on the arable land, with flocks of up to 30 in the lower Conwy Valley.

Stock Doves in Wales are particularly well distributed in some coastal areas notably Pembrokeshire and Anglesey, on both low coasts and cliffs, for they nest in a wide variety of sites ranging from cliff crevices, barns and derelict buildings to rabbit burrows and other ground-nesting sites as well as the more traditional use of holes in hollow trees. In many areas (e.g. Anglesey, Glamorgan) ground-nesting is apparently much less frequent than it was in the earlier years of the present century.

Coincident with the expansion of arable farming in the 19th century, Stock Doves increased and spread northwards and westwards from their former strongholds in south and east England. Although arable farming has never predominated over the traditional grassland system in Wales, the westward expansion of the species was clearly reflected here and in the last quarter of the 19th century up to 25% of farmed land was tillage and a further 13% ley (ploughed annually in rotation) which clearly suited Stock Doves. In Glamorgan in 1900 "it was certainly on the increase in recent years" (Paterson *et al.*) and in Breconshire Cambridge Phillips found it greatly increasing in 1899 where it had been "a most rare bird, hardly ever observed" 20 years before. By the turn of the century he had observed as many as 80 in one flock in a turnip field at Talybont-on-Usk.

In north Wales too there were indications of increase about the same time. Whereas T. C. Eyton could only write in 1838 that "perhaps it bred in N Wales" (he knew the Holyhead and Capel Curig areas best), by 1907 Forrest was able to quote it as "quite common along the coasts but less so inland". O. V. Aplin recorded it on Bardsey in 1903 and by the same time S. G. Cummings had encountered a winter flock of some 200 at Sealand (Flintshire) and in 1905 a further flock of 400.

The sudden decline, attributed to the use of organo-chlorine seed dressings, which occurred in eastern England between 1951 and the early 1960s, was not evidenced in Wales for the simple reason that it remained predominantly a grassland country, essentially free from the use of those particular chemicals as seed dressings. In fact the population almost certainly continued to increase in Wales between this period and the 1970s (O'Connor & Mead 1984). In Monmouthshire little or no change in status was detected between 1939 and 1977 and the species is still common over most of the county (62% of tetrads, 1981–85). The *Pembrokeshire Breeding Atlas* records Stock Doves in 33% of tetrads in that county with a suggested population of 300–500 pairs and in west Glamorgan, parts of which are heavily built up, it occurs in 23% of tetrads, most numerously on the more agricultural areas of Gower. RSPB surveys on Anglesey found Stock Doves present in 30 one-kilometre squares in 1986 and in a minimum of 32 one-kilometre squares on Llŷn, in 1986–87.

In most counties the Stock Dove is best described as a widely spread but somewhat localised species, more numerous on lower agricultural land than in the hills; coastal breeding, mainly in cliff sites, is habitual in most counties although not necessarily moreso than inland breeding, e.g. Pembrokeshire. Numbers are highest in some of the fertile agricultural areas of the lowlands, e.g. Vale of Glamorgan, Severn Valley, Tywi Valley. Large winter flocks are unusual nowadays, however, and 400 at Pontiets Carmarthenshire) in December 1980 is exceptional. Winter maxima of feeding flocks are seldom more than 100 birds: 200 were seen at Llys y Fran (Pembrokeshire) in March 1986; 100+ near Kenfig (Glamorgan) in several winters; 250 at Uskmouth (Monmouthshire) January 1979; 140 in Radnorshire, winter 1987 and 160–200 in linseed stubbles at Talgarth (Breconshire) in January 1993. In the north Wales counties there are fairly regular records of flocks of up to 90 but no recent published counts in excess of 100.

In west Wales a remarkable set of changing fortunes has occurred on Skokholm. One pair bred for the first time in 1967 and by 1971 the number had increased to 20 pairs; seven years later the population had increased to 60 pairs on an island of only 100 ha, only to fall again to ten pairs in 1980, one pair in 1983 and none by 1984. This extraordinary history is difficult to interpret although it might be noted that the final rapid decline coincided with the return of Peregrines and the original colonisation occurred at around the time of the falcon's earlier demise. However, the pattern had a reasonable parallel on Skomer, where nesting pairs rose to c50 pairs by 1975, the numbers then declining to two or three pairs in the 1980s and breeding finally ceased in 1989.

On Ramsey a ground-nesting pair bred in a rabbit burrow in 1927 and by the 1970s three or four pairs were regular in most years. Whalley (1954); recorded burrow-nesting as regular on Anglesey. Breeding has never been recorded on Bardsey where Stock Doves are of scarce occurrence, only 35 records of passage birds having been obtained up to 1984 since the observatory opened (1954); most records have been in October although a corresponding concentration, albeit with smaller numbers, in April gives an indication of passage. Of the handful of ringing recoveries of Welsh birds, none has been recovered further than 27 km from the place of ringing, emphasising the sedentary nature of the population. The fact that a small amount of passage occurs (the origin of such birds is not known) is evidenced by the offshore occurrences with occasional individuals turning up as far out to sea as the Smalls lighthouse (Pembrokeshire) 30 km offshore, e.g. 9 March 1984.

Woodpigeon *Columba palumbus*

Ysguthan

Abundant breeding species: passage migrant and winter visitor.

The Woodpigeon is an abundant and familiar species, occurring numerously throughout Wales. It breeds in at least 97% of tetrads, urban or otherwise, in Monmouthshire and in a minimum of 88.7% of tetrads in Pembrokeshire. It is typically a bird of open farmland but it is also common in urban areas and – as its name implies – in woodlands. Its nests can be found at almost any time of the year, although the majority of pairs in rural areas breed between July and September, with urban birds nesting somewhat earlier. The existence of abundant small woods and copses, amongst the mosaic of hedgerows and fields which characterise so much of the Welsh countryside does not in fact provide the most ideal conditions for a species which requires woodland cover in which to roost, nest and find some of its food but a *mixed* agricultural pattern for the bulk of its feeding. In view of the demonstrated importance of access to cereals in order to breed successfully (Murton 1965), it is quite likely that many of the Woodpigeons in Wales do not breed regularly and may well include numbers of birds from farther east at all seasons; this hypothesis has not been tested, however.

The Woodpigeon probably rates second to the Carrion Crow as the most disliked of avian species amongst Welsh farmers and is a considerable pest of agriculture. The new generation of conifer plantations in the uplands has extended the opportunities for the species, particularly where such plantings are adjacent to "reclaimed" moorland sown with grass/clover leys on which the pigeons feed. Some of the largest concentrations of Woodpigeons regularly found in Wales now occur in association with these new forests and improved upland pastures. A roost of some 8000 birds was recorded in Carmarthenshire in such circumstances in 1971.

The Woodpigeon is a species that has adapted successfully from its origins as a forest species to take advantage of the agricultural patterns that developed in the 18th and 19th centuries, profiting initially from the growth of roots as winter forage crops, and also from the increasing use of clover. The original, smaller woodland populations subsisted largely on acorns and beechmast with ivy berries and other fruits as secondary foods but becoming particularly important in late winter and early spring. The new agricultural practices of the 19th century enabled far greater numbers to survive the winters and the population increased accordingly. The habit of urban living has developed only over the past 150 years or so and many of these town birds – often associating with feral pigeons – are substantially dependent on food provided by man. Woodpigeons are now plentiful in almost all urban areas which can offer parkland, allotments, large gardens or similar suitable conditions.

The breeding population is densest in the more fertile areas of east and south-east Wales particularly the great river valleys of the Dee, Severn, Wye and Usk. The species is, for example, much more numerous in the Vale of Glamorgan than in the adjacent areas of the coalfield valleys. Proof that Woodpigeons are not wholly restricted to areas offering woodland for roosting and nesting is forthcoming from several areas of Wales. Nesting occurs in low thickets on coastal cliffs in both north and south Wales and nests are sometimes found in rank heather on open moorland, some nesting on the ground among bracken and brambles (e.g. Ruabon Moor in Denbighshire). Nesting also occurs regularly on Skomer (c11 pairs in 1992), Bardsey (since 1966 and now 8–12 pairs annually), Caldey Is. and occasionally other offshore islands (e.g. Puffin Is., Ramsey).

Passage movements are regularly observed, most readily in coastal areas and the offshore islands, although large influxes from the east are sometimes visible in late autumn. Up to 2000 birds have been seen (November) passing south from Lavernock Point and westwards at Merthyr Mawr, e.g. 1000 on 29 October 1988 and 5800 heading south-west over Cardiff in 30 minutes on 6 November 1989. Similarly large movements have been seen in Monmouthshire, usually involving birds from the east; 4600 were counted moving west in 1½ hours, November 1987 at Goldcliff and other similarly sizeable counts exist. On the west coast, passage is small on Skokholm in autumn but a little better defined in spring both there and on Skomer. Individual birds have even been found as far out to sea as Grassholm (October, May) and the Smalls lighthouse (May, July) 30 km offshore. On Bardsey passage is variable from one year to another, sometimes undetectable and in other years producing autumn flocks of 150–300. Evidence of the emigration of some Welsh birds is provided by the winter recovery of ringed birds as far south as Cornwall and France (2); the most distant recovery to date (830 km south-south-east) was of a nestling ringed in Caernarfonshire and recovered in Charante-Maritime (France) the following winter.

Winter numbers in Wales are clearly swollen by immigrants from the east although there is not yet evidence from ringing recoveries to prove the point. Large-scale cold-weather movements are a regular feature of hard winters: 4000 birds moved north in Denbighshire on 14 December 1984; 3500 were recorded at Marloes (Pembrokeshire) on 7 December 1986; on 14 January 1988, 2500 were watched moving east in Monmouthshire and on 17 December the same year a flock of 1750 flew west in the same county. At any time of year a flock of 1000+ is a large number to see at one time in Wales; in the 1970s and 1980s flocks of such size have been recorded from Ffairfach, December, Llanarthne, February, Rhandirmwyn, December (all Carmarthenshire), Nelson, November (Glamorgan), Penrhyn (Caernarfonshire), Corris (Montgomeryshire); doubtless many other such flocks occur unrecorded.

Collared Dove *Streptopelia decaocto*

Turtur Dorchog

Now a widespread and common resident, having first appeared in Wales in 1959.

The explosive expansion of range of the Collared Dove across Europe has been well documented (e.g. Fisher 1953, Stresemann and Nowak 1958, Hudson 1965). At the end of the 19th century it was restricted to southern Asia with western outposts in the Middle East and as far west only as the Balkans. After 1930 it began a rapid spread north-westwards across Europe and by 1947 had reached the North Sea coast. First confirmed arrival in England was 1955 (but it was possibly there three years earlier) and within four years the first pioneer colonists had reached as far west as Bardsey; one, and then two birds were present there between 6 and 16 May that year (1959) and a few months later (November) a third bird was seen at Llandudno (also Caernarfonshire). These were the first records of the species in Wales, but colonisation thereafter was extremely rapid (see table). The main colonisation of Wales took place in the early 1960s.

The somewhat later recorded arrivals in Breconshire and Radnorshire are possibly a result of poor observer coverage in these two sparsely populated counties although the species is still rather scarce in both and the upland nature of the majority of both counties is not conducive to supporting high numbers of Collared Doves. The absence of proved breeding in the coastal county of Carmarthenshire until 1968 is more surprising.

The *Breeding Atlas*, covering the years 1968–72, con-

Arrival and spread of Collared Dove in Wales (mainly from Hudson 1965 and 1972 and county bird recorders)

County	*First recorded*	*First proved breeding*	*Status by 1970*	*Current status 1993*
Anglesey	1960	1962	Common resident	Widespread and common
Caernarfonshire	1959	1966	Spreading west and south from coastal strip	Widespread in coastal lowlands near habitations; very local in hill areas
Denbighshire	1962	1962	Well established in six centres	Widespread throughout lowlands
Flintshire	1963	1963	Breeding around several towns/villages	Widespread and evenly distributed in the lowlands, particularly in urban and surburan areas
Merioneth	1961	1963	Regular Tywyn. Present six other centres	
Montgomeryshire	1962	1964	Breeding six towns/ villages in the east	Widespread and fairly common below 180 m notably in the east
Cardiganshire	1960	1961	Present around eight villages on coastal plain or inland	Widespread in towns, villages and farmsteads
Radnorshire	1965	1966	Small numbers Presteigne and Knighton only	Rather local. Mainly in the east and south
Breconshire	1965	1969	Only one or two known pairs	Fairly widespread. Found in 12% tetrads 1988–90
Monmouthshire	1962	1964	Widespread except in Wye Valley	Well established south central areas. Found in 66% tetrads 1981–85
Glamorgan	1962	1963	Well established around large towns south of coalfield and on Gower	Now locally common in many valleys and in the Vale. Common on Gower
Carmarthenshire	1965	1968	Suprisingly scarce. Carmarthen, Llanelli, Cwmann, Pembrey only	Present in at least 22 out of 25 10-km squares
Pembrokeshire	1961	1961	Well distributed	Widespread but patchy. Found in 56% tetrads c1600–2000 pairs

firmed the lowland distribution of Collared Doves breeding in Wales at that time, with almost continuous distribution around the coasts and adjacent lowland areas and the whole of Anglesey and the Marches; only throughout the central upland spine of Wales, on land over approximately 200 m are Collared Doves not to be found. The *New Atlas* shows no major alteration in this distribution. One was seen at 640 m on Rhinog Mountains (Merioneth), June 1970 (Hope Jones 1974) but this is an exceptional height at which to see them. Although the distribution pattern at the present time is little changed from 1968–72, the *New Atlas* shows that some in inland areas, particularly in south central Wales have been further infilled since then. However, distribution locally is nowhere uniform, but invariably patchy reflecting the bird's characteristic association with town suburbs, villages, rural farmsteads and other human habitations. There is some evidence that numbers not only stabilised after the expansion of the 1960s and 1970s but may even have settled at a lower level in some parts, e.g. Anglesey, Montgomeryshire. Population estimates are not easy to establish as few regular counts are available from those habitats which are the main stronghold of the species. On the basis of the three or four CBC plots in Monmouthshire, Tyler *et al.* (1987) estimate average densities of up to eight pairs in tetrads where the species is common and postulate a population of perhaps 2000 pairs. Monmouthshire probably has the highest number of Collared Doves in Wales, whereas in the adjoining county (Breconshire) – principally an upland county – only 12% of tetrads visited in 1988–90 were found to contain Collared Doves (no estimate of numbers). In west Glamorgan 47% of tetrads were occupied (1984–89). The only other population estimates are for Pembrokeshire (1985–88) where Collared Doves were found in 265 tetrads (56.4%) with an estimated number of 1600–2000 pairs. Twelve Collared Doves ringed in England have so far been recovered in Wales, all in the 1970s except one in 1965 and another in 1982. The distance and direction of travel of these birds from ringing site to recovery location in Wales help confirm the belief that immigration in the 1970s added to the rate of population increase attributable to successful breeding in Wales. The fact that Collared Doves will breed almost continuously round the calendar is itself a major contribution to population expansion.

Heavy westerly passage is recorded regularly at South Stack, Anglesey, notably in Spring (April–June) with flocks, often up to 40–60 strong, heading out to sea towards Ireland; a particularly large flock of 151 was seen here on 8 May 1985. There is a similar, but smaller, passage seen in August in some years and on at least one occasion a reverse, eastwards migration was observed.

The Collared Dove has prospered as a near neighbour of man and has exhibited the resourcefulness and boldness to feed at garden bird tables, amongst poultry and at grain stores and similar sites. Although rather surprisingly the species does not readily exploit ripening corn or residual grain in stubbles – and there is little corn now grown in Wales – its habit of concentrating around bulk grain supplies soon brought it into conflict. After it first bred in Britain, the Collared Dove was placed under special protection on Schedule 1 of the Protection of Birds Act 1954. By 1966, however, it was removed from special protection and under the 1981 Wildlife and Countryside Act it is listed as a pest species. Little direct persecution has occurred in Wales, however, and large concentrations are rare. Flocks of around 200 are recorded from Valley (Anglesey) 1967 and Aberdaron (Llŷn) 1978 and Porth Lisqi (Pembrokeshire) 1973. A flock of c400 was present at a poultry farm near Cydweli (Carmarthenshire) in September 1974. Flocks of this size have become increasingly unusual since these dates.

Turtle Dove *Streptopelia turtur*

Turtur

Very scarce summer visitor in eastern Wales with numbers declining further; passage migrant, mainly spring, in small numbers.

As a breeding species the Turtle Dove now has only a tenuous foothold in the eastern parts of the border counties of Wales. Farther west Turtle Doves breed very sporadically but occur, especially in coastal areas, as a spring migrant, principally in May and June. In Wales, as elsewhere, the species is primarily associated with open farmland, especially arable cultivation, nesting in shrubbery, woodland edge or other trees.

Marchant *et al.* (1990) have summarised the former status of the species in southern Britain and have shown that the general increase in numbers and range which took place in the mid-19th century was reflected in an expansion into Wales. Although this increase was not necessarily continuous into the next century – marked fluctuations were noted in Glamorgan for example – up to around 1900 it produced breeding as far west as Bangor, in the north, the Aberystwyth area (annually 1895–1912) and south Pembrokeshire. Following a subsequent withdrawal in the first two decades, there was a further expansion in the 1920s, probably linked to the recession in agriculture which produced a more abundant supply of field weeds. Once again there was sporadic breeding in

Caernarfonshire, Merioneth, west Montgomeryshire and probably Carmarthenshire but this was relatively short-lived. By 1950 breeding records, seven in eastern Wales, were notably fewer (although there was one late-summer flock of 40 in Montgomeryshire in 1952). Marchant *et al.* (1990) indicate that a modest resurgence then took place from the mid-1960s until around 1978; once again this was reflected in increased reports of birds in Montgomeryshire and other border counties and produced breeding as far west as Tywyn (Merioneth) in 1969, (probably) Carmarthenshire (1968–77), Gower 1967, Caernarfonshire 1974 (pers. obs.). Turtle Doves were recorded breeding on Anglesey at the time of the *Breeding Atlas* 1968–72 and there was one singing bird there in June 1986.

The CBC index shows a marked decline in the population nationally since the late 1970s and this has been mirrored in a clear withdrawal from the border counties in which it had maintained a tenuous hold. Such fluctuations as outlined above are no less than to be expected in a species which is here at the extreme north-western edge of an enormous Eurasian breeding range.

At the present day the Turtle Dove remains a regular breeding bird only in Monmouthshire. The *New Atlas* shows breeding in Denbighshire but it is now very scarce and probably sporadic there and in Flintshire, where it last bred in 1988. In Montgomeryshire it was still present in small numbers up to the early 1980s in the lower Severn Valley but there have been no records of suspected breeding since then. Up to ten pairs were breeding in one locality in Breconshire in 1964 but breeding in the county has not been recorded since 1981 although Griffiths (1971) believed there was a population of 10–20 pairs in the early 1970s. Regular breeding seems to have stopped in Glamorgan since 1972 with only one proved pair since then (1985). Breeding was suggested but not proved at several sites in the lowlands of Carmarthenshire in the late 1960s and early 1970s and the last case of possible breeding was at Pembrey in 1981.

Spring records of birds on passage occur mainly through the second half of May and first half of June with early records from late April. Such passage is clearly marked on the Pembrokeshire islands and Bardsey; even as far offshore as Grassholm Turtle Doves are not uncommon in spring. Coastal movement is evident on the south Wales coast – up to 41 birds per day have been counted in spring – although in Monmouthshire it is claimed to be more regular in autumn than in spring. In the north too, spring passage is often detectable on the north Wales coast on headlands such as Great Orme and South Stack. Autumn movement around the coasts is less well marked but there are regular records for July and August, with occasional autumn birds far offshore (Smalls lighthouse, September 1982) and one exceptionally late bird at Stackpole (Pembrokeshire) on 4 November 1983. The last autumn birds normally have passed through by the end of September. There was one unexpected record at Pilton (Glamorgan) on 13 and 20 February 1977 which may have been an overwintering bird.

The overseas recoveries of Turtle Doves ringed in Wales predictably indicate movement southwards to France (2) and Spain (1) in autumn. A bird ringed in Pembrokeshire in April 1969 was found dead one month later in Galway (Eire).

Ring-necked Parakeet *Psittacula krameri*
Paracit torchog

Feral species. Free-flying birds originate from escapes from captivity. Proved to breed twice in one location.

The Ring-necked Parakeet is an indigenous species across a wide band of north tropical Africa and the Indian sub-continent (with small isolated outposts elsewhere), reaching pest proportions in some areas, e.g. West Africa (BWP vol. 4). Because of its attractiveness and adaptability it is frequently kept as an aviary bird. Despite its tropical origins, however, escaped birds have shown themselves to be well able to survive in the wild in temperate zones, to breed successfully and even to endure severe winters. In western Europe free-flying populations have been established and proved self-sustaining since the 1960s or 1970s in Britain, Holland and Germany. By 1985 the feral British population was estimated at 1000+ individuals with core breeding areas around London and Merseyside.

The existence of a genuine feral British population, although not then known to be self-sustaining, prompted the British Ornithologists' Union Records Committee to recognise the fact formally and accord the species category D status in 1978 – effectively a "holding" category for admission to the British list. This was later (1983) upgraded to category C – species originally introduced by man, now having established a regular breeding population in the wild and proved itself to be self-sustaining.

The number and regularity of records of the species in Wales has predictably increased over the years despite the absence of any proof of breeding (except two instances in Denbighshire) and notwithstanding the distance from regular breeding concentrations. It has been pointed out (Lansdown 1984) that despite the startling appearance of this species, identification requires care as several similar species are regularly kept in captivity and known to escape on occasion. By way of illustrating the point, only 55% of parakeet records in Wales between 1978 and 1984 were positively attributed to *P. krameri*: 9% specifically referred to other similar species and 36% were not specifically identified.

The list of published records of parakeets is reproduced below, including – for completeness – those (indicated by *) which may not have referred to *P. krameri*. (Three records of other confirmed parakeet species from Glamorgan 1983 and 1984 are omitted.)

Caernarfonshire	1 Bardsey	15 October 1976
	1 Llanfairfechan	6 July–end August 1986
Anglesey	no records	

Denbighshire	Gresford Flash	Bred successfully 1979. Failed breeding 1980
(male bird		shot)
Flintshire	1 Shotton	20 September 1989
Merioneth	1 Aberdyfi	3–20 February 1988
Montgomeryshire	1 Brooks	September–November 1985
Breconshire	no records	
Cardiganshire	1 Ynys-hir	1–12 March 1988
	1 Aberystwyth	17 March 1988
	2 Aberaeron	January–March 1989
Radnorshire	1 (no location given)	28 June 1986
Carmarthenshire	1 Felinfoel	18 February 1984
	1 Felinfoel	24 August 1986
Pembrokeshire	1 Newport	30 January 1988
	1 Crundale	20 March 1988
Monmouthshire	1 Peterstone Wentlooge	1975
	3 Peterstone Wentlooge	30–31 October 1976
	1 no location	13 August 1981
	1 no location	25 November–12 December 1985
	1 no location	14 December 1985
Glamorgan	1 Penarth	summer 1974
	1 Rumney (Cardiff)	6 December 1975
	*1 Cardiff	30 April 1976
	*1 Cardiff	August 1976
	1 Penarth	19 September 1977
	*1 Hengoed	October 1978
	1 Glamorgan Canal	15 January 1980
	1 Cosmeston	26 August 1980
	1 Llanishen (Cardiff)	14 October 1981
	1 Roath (Cardiff)	21 October & 14 November 1981
	1 Cosmeston	25 July 1982
	2 Llanrhidian	23 & 27 January 1983
	*2 Llanishen (Cardiff)	4 August 1983
	1 Bryncoch	25 August 1983
	*1 Llangennith	September 1983
	1 Blackpill	26 February 1984
	1 Derwen Fawr	28 February 1984
	*2 Blackpill	May 1984
	2 Dinas Powis	15–20 July 1984
	2 Blackpill	25 August 1984
	1 Llanishen (Cardiff)	4 April 1986
	1 Penclawdd	1 January 1987
	*1 Dinas Powis	7 February 1987
	*1 Roath (Cardiff)	2 May & 21 August 1988
	*1 Llanishen (Cardiff)	4 May 1988
	1 Pontardulais	10–12 May 1988
	*1 Tongwynlais	11 May 1988
	*1 Birchgrove (Cardiff)	19 May 1988
	*1 Aberthaw	9 July 1988
	*1 Pontcanna (Cardiff)	9 September 1988
	1 Penarth	13 March 1989
	1 Llanedeyrn (Cardiff)	7–21 May 1989
	1 Roath (Cardiff)	27 August 1989
	1 Bute Park (Cardiff)	12 April 1990
	1 Penarth	2 May 1990

The concentration of records from the Cardiff area over a period of years – and latterly Swansea on a smaller scale – indicates that undiscovered breeding may possibly occur in either area. The dependence of the species on supplementary suburban food supplies is implied in many of these records.

Great Spotted Cuckoo *Clamator glandarius*

Cog Frech

A vagrant. It breeds from Iberia and southern France discontinuously east to south-west Iran and south to South Africa.

There are two Welsh records:

1 April 1956 – Male, dead, Plas Penhelig, Aberdyfi (Merioneth)

3–15 April 1960 – One, Newborough (Anglesey)

Cuckoo *Cuculus canorus*

Cog

A locally common breeding summer visitor to all counties; also recorded on passage.

The Cuckoo generally arrives from Africa during April, with females returning to the same areas in consecutive years. It is found in many habitats from the upland plateaux to coastal dunes, although numbers of Cuckoos in Britain and Ireland have declined since the 1950s (Brooke & Davies 1987), probably due to changing agricultural practices affecting both the Cuckoo and its host species, although it is still relatively common in the Welsh uplands (Seel, pers. comm.).

Forrest (1907) found the Cuckoo abundant on the moors of north Wales where it was generally seen "going about in parties of three – two cocks and one hen". It was numerous in the sandhills at Abersoch (Caernarfonshire) and west Anglesey, and Mathew (1894) found it particularly common along the banks of the Cleddau (Pembrokeshire). All county avifaunas at the turn of the century described the Cuckoo as a common summer visitor, and thus it remained until at least the 1940s.

Although a decrease was recorded in some south Wales counties during the 1940s (Parslow 1973), it was not until the 1950s that a general decline was noted throughout most of the Principality. The severity of the decrease is hard to assess, however, due to the lack of data specific to Wales. Where declines have been recorded, this is generally as a result of anecdotal information; in Glamorgan, for example, Heathcote *et al.* (1967) stated that by the 1960s, the number of Cuckoos were "certainly far less than they were 30–40 years ago".

Although the extent of the decline is impossible to assess, numbers have clearly decreased in several areas due to agricultural intensification leading to the loss of semi-natural habitats (Seel, in litt.). Despite this, Cuckoos were still regarded as fairly common in most Welsh counties during the 1970s, the *Breeding Atlas* indicating that the species was widely distributed throughout the Principality. The *Gwent Breeding Atlas* recorded Cuckoos in over 80% of the tetrads surveyed, although breeding was proved in only 7% of these. The county population was estimated at 300–350 breeding females, with most concentrated in the uplands and on coastal flats.

The *New Atlas* confirms that there may have been a decline in some of the upland areas of central and north-West Wales although Marchant *et al.* (1990) found the results between 1963 and 1988, from both woodland and farmland CBCs, to be at variance with this trend. In Breconshire, Cuckoos were found to be very scarce in the valley bottoms (Peers & Shrubb 1990) but they were still recorded in 50% of all tetrads visited, with high numbers encountered on hillsides and in young forestry plantations. High densities have also been noted in young conifer plantations on the uplands of north and mid-Wales (pers. obs.), although these areas are suitable only during the pre-thicket stage. The Pembrokeshire Breeding Birds Survey of 1984–88 showed that Cuckoos were most frequent in the south of the county and in the Preseli Mountains with an estimated population of about 210 breeding females (Donovan & Rees, in press.).

At Oxwich Marsh NNR Cuckoos are common, but extensive surveys of Reed Warblers occupying *Phragmites* reedbeds found no evidence of Cuckoos using this host although over 400 pairs frequent the marsh (Thomas, in litt.). Here, it almost certainly parasitises Meadow Pipits and Skylarks in the adjacent dunes. The latter two species are also thought to be the principal hosts in the Welsh uplands although the range of species is very wide.

Cuckoos generally depart in July although stragglers are recorded in September. On the offshore islands, numbers on spring passage are always greater than in the autumn although they rarely exceed two birds per day. Breeding has been recorded from several of the islands, including Bardsey, Ramsey (20 breeding females in the early 1970s), Skokholm and Caldey Is. Two May records have been noted from Grassholm and on Bardsey, extreme dates are 14 April (1980) and 5 October (1981). The only winter record is of a bird seen by several observers at Little Milford (Pembrokeshire) on 21 and 22 December 1954.

There are several ringing records to indicate a migration route through France and Italy to Africa, the longest involving a bird ringed on Bardsey on 21 May 1990 and recovered at Malo in Italy on 20 July 1990, a movement of 1415 km south-east.

Yellow-billed Cuckoo *Coccyzus americanus*

Cog Bigfelen

The Yellow-billed Cuckoo is a vagrant which has been recorded three times in Wales, all in the 19th century. It breeds in North, Central and South America.

The three records were:

Autumn 1832	– One killed, Stackpole (Pembrokeshire)
26 October 1870	– One found dead near Aberystwyth (Cardiganshire)
10 November 1899	– Female found dead, Craig y Don (Anglesey)

The dates are very typical of other British occurrences, many of which have also concerned birds found dead or dying.

Barn Owl *Tyto alba*

Tylluan Wen

A breeding resident in all counties, but showing evidence of a continued decline in many areas.

The Barn Owl is found in a wide variety of mainly agricultural land, favouring vole-rich damp pasture and unimproved meadows. Formerly a familiar sight throughout lowland Wales, it has shown a gradual decline over most of its range this century, mainly because of changing agricultural practices. At the turn of the century, Barn Owls were also heavily persecuted by gamekeepers and were particularly susceptible to pole-trapping on account of their habit of hunting from fence posts and similar perches. Short-term fluctuations caused by hard winters or variations in the density of small rodents sometimes mask long-term trends.

Much of Wales during the 19th century was ideal habitat for this species. The extensive tree cover had already been cleared several centuries before and agriculture was still non-intensive. Vole-rich hay meadows, rough grazing and hedgerows abounded and the main limiting factor on Barn Owls was probably persecution by game-rearing estates. Persecution appears to have been heaviest in the counties of Breconshire, Glamorgan and Monmouthshire, although it also had a local effect on the population in Cardiganshire. In north Wales, gamekeeping was concentrated mainly on the grouse moors and therefore did not seriously affect a predominantly lowland species such as the Barn Owl.

Even while persecution was at its height, farmers recognised the value of this species as a means of controlling small rodents and most stone barns built at the time often contained small windows to allow access for the owls. Forrest (1907), however, noted that in north Wales, Barn Owls preyed mainly on shrews, but that one pair fed almost exclusively on House Sparrows and that the skull of a Dunlin was found in the pellet of another bird. Thus, at the turn of the century, the Barn Owl was common throughout most of lowland Wales, probably surpassing the Tawny Owl in numbers in the less wooded areas. Mathew (1894), however, describes it as a far from common resident in Pembrokeshire, although it is likely that it was more numerous than he imagined, and that its true status was obscured by its secretive nature during the day.

The first nationwide census of this species was undertaken by George Blaker for the RSPB in 1932, although this looked mainly at breeding in English counties (Blaker 1933). The results showed that the counties of Anglesey and Pembrokeshire held some of the highest breeding densities in Britain, Anglesey in particular supporting up to 20 breeding pairs per 100 km^2. High densities were also recorded in parts of Denbighshire, Caernarfonshire (Llŷn), coastal Merioneth, Montgomeryshire, Cardiganshire, Carmarthenshire, Glamorgan and Monmouthshire, but Barn Owls were breeding at very low densities in Snowdonia and most of the Cambrian Mountains.

At the time, the species was showing a gradual decline throughout most of the country, mainly as a result of climatic factors, probably combined with the decrease in the number of stack yards associated with less cereal and arable farming (Shawyer 1987). Blaker believed that the population had decreased by as much as 33% since 1922, and attributed the decline to several additional factors including severe winters and extremes of weather during the summers of 1921 and 1927. The population had in fact increased during the First World War when most gamekeepers joined the forces, but the downward trend continued throughout the 1920s, 1930s and 1940s.

By 1950, Barn Owls were scarce breeders in many Welsh counties, and with agriculture becoming increasingly intensive resulting in the loss of favoured feeding areas and breeding sites, the decline accelerated. In Glamorgan, a large influx in 1949–50 resulted in a partial recovery with birds seen in areas where they had been absent for over a decade. The origin of these birds remains a mystery as no such influx was recorded in any of the other Welsh counties. Despite the loss of habitat, Barn Owls were said to be increasing in less intensively farmed counties such as Cardiganshire and Breconshire during the 1950s, although set-backs were noted nationally following each hard winter.

Over the past 25 years or so, the population appears to have recovered in some areas, possibly connected to milder winters and a more enlightened attitude by landowners towards the species. Although no comprehensive surveys were undertaken during the 1970s, some monitoring was being undertaken either studying the Barn Owls' diet (Plant 1976; Bowman 1980; Brown 1981) or population studies within defined areas (Seel *et al.* 1986). The former work was valuable as it often gave an insight into population densities in different areas, but Seel's work on Anglesey is the only attempt to repeat a population census in a Welsh county since 1932.

However, G. Cundale (pers. comm.) organised a complete survey of the Brecon Beacons National Park in 1991 and found 29 occupied sites with breeding proved in 14 of these.

Seel *et al.* (1986) estimated the Anglesey Barn Owl population between 1978 and 1982 to be in the region of 30 pairs, most of which were located in the east and south of the island. They concluded that the population was not limited by nest-site availability but probably by food, the most favoured hunting areas being rough grassland, heathland and woodland edge.

Between 1982 and 1985, the Hawk Trust (or Hawk and Owl Trust as it now is) organised the first complete survey of breeding Barn Owls in Britain since Blaker in 1932. The census revealed that the total Welsh population had declined from c1436 pairs in 1932 to a mere 462 in the early 1980s (Shawyer 1987). The highest numbers were found in Cardiganshire and Anglesey with the highest densities in the latter county. The population had declined by 40% or more in every county, the most marked decrease in numbers being found in Carmarthenshire. The strongholds in Wales were found to be west Anglesey, the Llŷn peninsula (Caernarfonshire) and lowlands associated with many of Wales' larger rivers. In Breconshire, the association with river valleys is particularly marked, especially along the Usk.

Breeding Barn Owls in Wales in 1932 and 1982–85 (from Shawyer 1987)

County	*1932*	*Population estimates 1982–85*	*% change*
Anglesey	126	68	–46
Breconshire	70	17	–76
Caernarfonshire	90	47	–48
Cardiganshire	138	78	–43
Carmarthenshire	187	34	–82
Denbighshire	109	24	–78
Flintshire	60	16	–73
Glamorgan	105	36	–66
Merioneth	60	30	–50
Monmouthshire	120	25	–79
Montgomeryshire	93	20	–78
Pembrokeshire	210	55	–74
Radnorshire	48	12	–75
Total	1416	462	

The majority of Welsh nest sites, where the location was specified in Shawyer's study, were found in nest-boxes, although obviously the sample is thus biased. In the past, several Pembrokeshire nests were found in cliffs, with a pair breeding at one time in a cave amongst seabirds on Skomer and another breeding in a crevice on St. Margaret's Is. (Whittle 1924). Elsewhere in Wales, nest-site locations vary depending on the availability of old buildings, barns and hollow trees, the easiest and most frequently located being in buildings.

There has been much debate as to the accuracy of Shawyer's population estimates, but recent work for the *New Atlas* and surveys by the newly formed Wales Raptor Study Group (1991) has shown Shawyer's figures to be reasonably accurate for many counties. The *Pembrokeshire Breeding Atlas* found more Barn Owls than was expected at the outset, the total county population being over 100 pairs. In west Glamorgan, the population is thought to be fewer than ten pairs. It is extremely difficult to give a more accurate population estimate for Wales than that of Shawyer's, and despite the fact that his figures for several counties are considered by many to be too high, Cayford (pers. comm.) has shown that in Suffolk even intensive fieldwork over a number of years does not locate every breeding pair within a given area. Shawyer's figure of 400–500 breeding pairs was based on the results of a census which followed a severe winter in 1981–82 and is probably too low. The true figure may be as high as 600–700 pairs.

Barn Owls are fairly sedentary birds and therefore their winter distribution varies little from that of the breeding season. There may be small movements, however, during hard winters when several birds sometimes congregate at lowland bogs or coastal marshes. Evidence that some birds move farther afield outside the breeding season comes from the offshore islands, where most of the 14 Bardsey records involved single birds seen during September and October. Up to ten birds are recorded regularly during the winter months from Cors Caron (Cardiganshire), and Barn Owls are known to visit the Pembrokeshire islands at this time of year. One exceptional record is of a bird seen on the Skerries (Anglesey) in the autumn of 1941.

Mathew cites two instances of communal roosting in Pembrokeshire, the most remarkable containing between 40 and 50 Barn Owls in the connected roof space of a row of cottages. The only modern equivalent record is a roost of 12 in the ruins of Butterhill Mansions (Pembrokeshire) in 1987.

Ringing recoveries show that young Barn Owls rarely disperse far from their natal areas, although there are a few records of birds having moved longer distances, one of a bird ringed near Devils' Bridge (Cardiganshire) on 9 August 1987 and recovered dead on the Alresford bypass in Hampshire on 14 February 1988, 233 km south-east. Another nestling ringed at Deeping St. Nicholas in Lincolnshire on 29 June 1988 was recovered 274 km west south-west on the M4 near Margam Park (Glamorgan) on 28 April 1989.

A specimen of the dark-breasted race of the Barn Owl *T. a. guttata*, which breeds in northern Europe, was shot at Blaenavon (Monmouthshire) in 1903.

Scops Owl *Otus scops*

Tylluan Scops

A vagrant.

It is included on the Welsh list on the strength of two records from Pembrokeshire.

Spring 1868 – One was caught near Pembroke
25 April 1955 – One was trapped, Skokholm

In Britain and Ireland as a whole there were 64 records prior to 1958 compared with only 19 since which suggests a marked decrease. The species breeds from France, Iberia and north-west Africa east to Japan and Indonesia; most of the European population winters in Africa south of the Sahara.

Snowy Owl *Nyctea scandiaca*

Tylluan yr Eira

A vagrant.

There are only eight records in Wales:

Winter 1902–03	– One in the neighbourhood of Cwmffrwd (Carmarthenshire)
Winter 1915–16	– One near Cross Ash (Monmouthshire)
7 January 1947	– One near Bwlch (Breconshire)
23 December 1953	– One, Bishton (Monmouthshire)
27 March 1959	– One, Valley (Anglesey)
28 March 1972	– Immature, Penarth Moors (Glamorgan)
3 May 1972	– Adult female found dead, Mynachdy (Anglesey)
28 January 1976	– One, Abergavenny (Monmouthshire)

Snowy Owls are Arctic breeders from Iceland and Scandinavia east to Siberia, Alaska, Canada and Greenland and are partially migratory. A pair bred on Shetland between 1967 and 1975.

Little Owl *Athene noctua*

Tylluan Fach

Locally common breeding resident in most eastern counties, scarcer but increasing in the west.

The Little Owl is not indigenous to Britain but became established following a series of introductions, mainly in southern England, during the 19th century. A bird of well-wooded agricultural countryside, parklands and orchards, the population soon increased in range and numbers until, by the 1930s, it had colonised every county in England and Wales. Since then, however, populations have shown marked local fluctuations, this species being particularly susceptible to severe winters.

Little Owls were first introduced into Yorkshire in 1842, but became established only after further introductions in Norfolk, Suffolk and Northamptonshire. The first Welsh record is of a bird killed at Merthyr Mawr (Glamorgan) around 1860–70. However, it only became established as a breeding species in Wales in the 1920s, despite the fact that at the turn of the century it had been recorded from Breconshire (first at Ffrwdgrech in 1890), Carmarthenshire (one heard at Cilycwm on 18 April 1903), Monmouthshire (one at Chepstow on 5 December 1901) and, remarkably, Anglesey, where one was shot by a pheasant-shooting party in the winter of 1899–1900. Forrest (1917) suspected that these early Anglesey records originated from birds released on the island, although there are no definite records of anyone having imported owls.

An influx was reported from many counties in south and south-east Wales during the First World War, probably due to many gamekeepers leaving the large estates to join the forces. By the early 1920s, it had become established as a breeding species in all counties except Caernarfonshire, although a pair bred on the Great Orme in 1930. By the late 1920s, it was certainly common on the farmland and orchards of south and south-east Wales, although it was still absent from the coalfield areas. At the same time, it was described as the most common owl in Pembrokeshire.

First sightings and first recorded breeding of Little Owls in Welsh counties

County	*First sighting*	*First recorded breeding*
Glamorgan	Merthyr Mawr c1860–70	?
Breconshire	Ffrwdgrech 1890	?
Anglesey	Winter 1899–1900	Penmon(?) 1908
Monmouthshire	Chepstow 5/12/03	Chepstow 1914
Carmarthenshire	Cilycwm 18/04/03	?
Radnorshire	Knighton 04/14	Knighton 1918
Cardiganshire	Llanon 1918	1918?
Montgomeryshire	Llanllugan 25/10/19	?
Flintshire	Holywell Jan/Feb 1920	?
Pembrokeshire	Solva 1920	1920
Caernarfonshire	Gt. Orme 10/10/20	Gt. Orme 1930
Merioneth	Aberdyfi 29/06/22	Aberdyfi 1922
Denbighshire	Denbigh 1924	?

The population of Little Owls in Wales reached its peak during the late 1930s, for during the next decade a decline was reported from many counties, possibly due to a succession of hard winters. The particularly severe winter of 1946–47 appears to have caused local extinctions in many areas, especially in south Wales. The population was almost wiped out in Radnorshire and was severely depleted in Pembrokeshire, Carmarthenshire, Glamorgan and Breconshire. A steady recovery followed in most counties during the 1950s and 1960s, but a decline was reported in Cardiganshire from 1954 and 1955 onwards, becoming more marked during the early 1960s, and no birds were seen here from 1961 until 18 December 1964, when one was sighted at Cnwch Coch.

Fieldwork for the first *Breeding Atlas* between 1968 and 1973 revealed that the Little Owl was confined mainly to coastal areas in north and south-west Wales and to lowland farmland on the Marches and that it was common only in the counties of Monmouthshire and Glamorgan. It was absent from most of Pembrokeshire, Cardiganshire and Merioneth and in Caernarfonshire was confined almost exclusively to the Llŷn peninsula.

The *New Atlas* shows that there has been an increase on Anglesey, along the coastal strip from Pembrokeshire north to Harlech (Merioneth) and in south Carmarthenshire, although the population appears to have declined in its former stronghold of Glamorgan and Monmouthshire. In Glamorgan, intensive fieldwork in one 10-km square revealed a minimum of 13 territories in 1982. The population of Monmouthshire was estimated to be between 500 and 1500 pairs in the mid-1980s, with the greatest densities in the orchards of the central and eastern parts where nests were less than half a kilometre apart. The *West Glamorgan Atlas* estimated a population of 100 pairs, the majority on the coastal lowlands of the Gower and the Breconshire population has been estimated at 25 pairs, confined almost exclusively to the mixed farmland of the Usk and Wye Valleys. The Pembrokeshire population has been estimated at approximately 50 pairs (J. Donovan, pers. comm.) and a recent study by Julian Moulton (pers. comm.) on Anglesey revealed a minimum of 54 territories in the county.

Little Owls bred on Skokholm between 1937 and 1954, where they fed almost exclusively on Storm Petrels, and about half a dozen pairs still breed on Skomer and on Bardsey, where they nest almost exclusively in rabbit burrows.

Little Owls are widespread but uncommon over most of Wales, confined to coastal areas, river valleys and the eastern lowlands. The total Welsh population is likely to be in the region of 2000 pairs.

The winter distribution of Little Owls is very similar to its breeding range as this species is largely sedentary, but the *Winter Atlas* shows that it was apparently absent from almost the whole of central and western Wales. Much of this is probably due to coverage as little work was undertaken in central Wales and on the Llŷn peninsula compared with other more populous areas. However, coverage for Anglesey was good and its apparent absence from much of the island during the winter is puzzling.

Their sedentary nature means that most recoveries of Little Owls involve movements of less than 10 km but one bird ringed at Rhydycroesau in Denbighshire on 31 December 1981 was recovered at Pantydwr (Radnorshire) 60 km south-south-west on 4 August 1982.

Tawny Owl *Strix aluco*

Tylluan Frech

A common and widespread breeding resident in areas of woodland in all counties.

Tawny Owls are common throughout most of Britain, but are absent from Ireland, the Isle of Man and some of the Scottish islands. It is a bird of deciduous and mixed woodland, although it may also occur in mature coniferous woodland, particularly where nestboxes are provided. It is also found on farmland and in urban and suburban areas, with pairs nesting successfully in gardens and parkland in the largest towns and cities.

Owing to its territorial nature, the Tawny Owl's winter distribution is almost identical to that of the breeding season, with birds relying on a good memory of their territory's structural components for winter survival (Southern 1970). All ornithological literature on this species' distribution in Wales has dealt with the breeding population, and this species account will do likewise.

Early Welsh naturalists recorded the Tawny Owl as being common and widespread throughout most of Wales, surpassed only by the Barn Owl in low-lying counties such as Anglesey. Forrest (1907) stated that it was most common in Montgomeryshire and west Merioneth, but rather scarce on the Llŷn and on Anglesey. The grey form of this species was apparently encountered more frequently than the commoner brown form around Tywyn (Merioneth) at the turn of the century.

As with all other owls and diurnal birds of prey, Tawny Owls were very heavily persecuted during the 19th and the first half of the 20th century, and in Glamorgan it appears that the population was much reduced by persecution at the end of the last century. However, it was able to survive, and numbers increased considerably in most Welsh counties from the 1920s or 1930s onwards. In June 1930 R. P. Sanderson reported that while examining Jackdaw nests near Crickhowell (Breconshire), he found four occupied Tawny Owl nests within about 256 m of each other (F. C. R. Jourdain Notebooks held at the EGI). In many counties, this population increase continued until the 1950s, although local declines were reported following the hard winters of 1947 and 1963.

In 1955, the myxomatosis epidemic greatly reduced the Welsh rabbit population and may have caused local declines in the number of breeding Tawny Owls, but this setback was minor and shortlived. Today, the Tawny Owl is one of our commonest raptors, surpassing even the Common Buzzard in numbers in counties such as Radnorshire. High densities are recorded from some areas, particularly on nature reserves such as Lake Vyrnwy (Montgomeryshire), where nestboxes have been erected. It is undoubtedly more common in many upland areas than at any time over the past 200 years as mature coniferous woodlands have provided nesting and feeding sites. Its adaptability has allowed it to colonise all wooded areas, and where trees are largely absent, unusual nest sites are chosen on occasion. In 1968 a pair nested on the ground beneath a zinc sheet at Llanfachreth in Merioneth.

Comparison between the two national Breeding Atlases show that the breeding population in Wales may have declined in Merioneth, Flintshire, Radnorshire, coastal Pembrokeshire, northern Glamorgan and on the Llŷn.

However, it is not known how much this constitutes a true decline in the population or whether it is due to less intensive coverage. Marchant *et al.* (1990), have shown that on the whole, the Tawny Owl population has remained fairly constant for the past 25 years, although there are fluctuations on occasion.

Population estimates have been given for Breconshire (320 breeding pairs), Monmouthshire (600–1200) and Pembrokeshire (800–1000), and the total Welsh population is likely to be between 5000 and 6000 pairs.

The Tawny Owl is occasionally recorded on the offshore islands, although the sea provides an efficient barrier to dispersal. There are two records from Bardsey, both involving birds hunting migrants attracted to the lighthouse on 19 and 21 October 1960, and at least one bird was present on Caldey Is. (Pembrokeshire) each breeding season between 1984 and 1988.

Ringing recoveries throughout Britain have shown the Tawny Owl to be the most sedentary of British owls. However, one Welsh record of a bird ringed at Bayston Hill in Shropshire on 28 April 1983 and recovered at Llanwern (Monmouthshire) on 15 August 1983 represents a movement of 121 km.

Long-eared Owl *Asio otus*

Tylluan Gorniog

A scarce resident in some counties. Also a scarce winter visitor and passage migrant.

Long-eared Owls are the most nocturnal of all the resident British owls and this, combined with their preferernce for coniferous woods and commercial plantations, means that they are difficult to locate. In many parts of the country they are largely confined to small, isolated clumps of trees, probably due to their inability to out-compete the larger Tawny Owl, which is associated mainly with broadleaved woodlands.

In Wales, the population has declined since the 1900s, and it is now a scarce breeding bird, confined mainly to the denser coniferous woodlands of the Welsh uplands. Here, it lays its eggs in the old nests of other birds, notably corvids, with both the incubating female and the roosting male sitting tightly throughout the day. Although they are certainly under-recorded in the Principality, nonetheless they are still uncommon, even in areas with an abundance of mature coniferous woodland.

During the second half of the 19th century Long-eared Owls were considered by many authors to be widespread and locally common. Salter (1900) described it as a fairly numerous resident in parts of Cardiganshire, particularly around Nanteos, and Dobie (1893) recorded that it bred near Corwen (Merioneth) and used to be widespread around Colwyn Bay (Denbighshire). Barker (1905) noted that the species was not uncommon in Carmarthenshire, and that a pair nested regularly in the grounds of his house at Cwmffrwd. Although not widespread in the neighbouring county of Glamorgan at that time, Dillwyn wrote that they were occasionally shot in the thick woods on the Gower peninsula.

Forrest (1907) recorded pairs nesting in all six counties of north Wales including Anglesey, where breeding was mainly confined to the woods along the north shore of the Menai Straits. Bolam (1913) thought the species to be less numerous around Llanuwchllyn than in some of the neighbouring valleys where there was more coniferous woodland, and it appears that it was scarce at the turn of the century in Pembrokeshire, Breconshire and Monmouthshire.

Although its subsequent history is vague for much of north Wales, in south Wales it is known that the population declined between 1900 and 1950, coinciding with an increase in the number of Tawny Owls in Wales. Unfortunately, Long-eared Owls also suffered heavy persecution over much of the country, often falling foul of illegal pole-trapping. Even when the population had declined drastically however, egg collectors were still able to locate nests in many counties. J. H. Howells took three clutches from Pembrokeshire between 1907 and 1911, and W. P. Dodd, in association with W. Maitland Congreve, is known to have collected at least three clutches from Denbighshire between 1910 and 1924. At one nest near Llangollen (Denbighshire), the female was found poisoned on her eggs, but the nest was used again the following year.

Evidence that this species was much more common earlier this century is the fact that eggs were taken from several other localities during the same period, including a clutch taken from Aberdaron (Caernarfonshire) on 17 April 1947. More recently, the late G. Ireson (pers. comm.) located nine nests at various sites in Radnorshire between 1952 and 1965, but by this time, it was a scarce and sporadic breeder throughout most of Wales.

One particularly interesting record from this period comes from Carmarthenshire where, in 1956, a farmer, eager to collect a bounty for having a successful Red Kite nest on his land, was sorely disappointed when told that the nest in a birch tree contained two young Long-eared Owlets!

It is surprising, considering the proliferation of coniferous plantations throughout much of upland Wales, that breeding numbers of Long-eared Owls have not increased. In Breconshire, one or two pairs were attracted to the Irfon Valley by the new plantations, but these birds moved on as much of the open ground disappeared. During the past 15 years, breeding has been proved in only seven counties, two of the most regular sites being at Lake Vyrnwy (Montgomeryshire) and near Llangollen (Denbighshire) where two pairs were present in 1991. Only four confirmed breeding pairs were located during the fieldwork for the *New Atlas*, half the number found by the previous Atlas. Throughout the 1980s, one pair bred on the ground amongst rank heather near Abergynolwyn (Merioneth), although this site has since been abandoned following the construction of a log cabin nearby.

It is not known to what extent Long-eared Owls use forest restock areas in Wales, but a more uneven age structure in coniferous woodlands should benefit a species which requires open ground. The total Welsh population is difficult to assess, but is probably less than 30 pairs.

On occasions, Britain experiences a large influx of birds from the Continent, particularly Scandinavia, when the resident population is augmented by several hundred individuals. Large numbers rarely reach Wales although influxes were recorded on Bardsey in 1975 and 1989, with a maximum of six birds seen on 6 December 1989. In winter this species is even more elusive than during the breeding season, making any attempt to census the population very difficult. The majority of Welsh records are of single birds, often recorded on passage, although there were several records of more than one bird from the Aberystwyth (Cardiganshire) area at the turn of the century. Two were shot out of five birds seen at Llantrithyd (Glamorgan) in January 1913, and many birds were shot and pole-trapped during an influx into the county in 1934.

Today, the only substantial winter roost known in Wales is on the Wentlooge Levels (Glamorgan), where up to 12 birds were present from 12 November until 31 December 1990. In 1991 and 1992, however, this roost had apparently moved to a site near the Rhymney estuary where 18 birds were present in February 1991, and ten in November of that year. The roost peaked at 18 birds in February 1991.

Adult Long-eared Owls are sedentary, whereas youngsters often disperse widely during their first year. There is only one Welsh ringing recovery for this species, reflecting the difficulty in locating nests. This involves a bird ringed at Abbeycwmhir (Radnorshire) on 5 June 1956 and recovered near Knighton in the same county on 6 August 1957, a distance of only 16 km.

Short-eared Owl *Asio flammeus*

Tylluan Glustiog

Scarce breeding resident. Winter visitor and passage migrant in small numbers to all counties.

In Britain, Short-eared Owls are confined to the remoter tracts of moorland, bogs, sand dunes and young plantations, where there is an ample supply of small rodents. Nest sites are not easy to locate as incubating females sit very tight, although adults are often forced to hunt during daylight hours when they have fully grown young in the nest. The diet usually consists mainly of field voles, although one study in north Wales showed that on one particular heather moor, pigmy shrews were the most frequently taken prey item (Roberts & Bowman 1986). Numbers of breeding Short-eared Owls are known to increase dramatically when there is a vole plague (Lockie 1955b), and in Wales it is only on Skomer that the resident population is generally unaffected as the owls here prey mainly on a stable population of Skomer voles. In 1993, however, the population on the island had increased from the usual three to five pairs to 12 pairs, with fledged birds seen in early May (S. J. Sutcliffe, pers. comm.).

During the 19th century this species was a scarce breeding bird in Wales, confined mainly to the Welsh uplands and the coastal marshes and dunes of Anglesey. It is possible that many breeding attempts went unrecorded in the uplands of north Wales as there are surprisingly few records of pairs from seemingly suitable heather moorland such as the Berwyns, Migneint and Mynydd Hiraethog (Montgomeryshire, Merioneth, Caernarfonshire and Denbighshire).

Salter, writing in the *Zoologist* in 1895, stated that several breeding pairs had been located on Sir Pryse Pryse's land (probably Nanteos, Cardiganshire) around 1874. Forrest (1907) recorded several pairs breeding on Pumlumon (Montgomeryshire/Cardiganshire) in 1886, and during the 1900s breeding was recorded at several lowland bogs on Anglesey. At that time Short-eared Owl were also known to nest on Skomer and on mainland Pembrokeshire, with two nests found near Pembroke Dock in May 1908. During the 1920s, at least one pair may have nested in sand dunes on the Gower Peninsula (Glamorgan) and in 1926 a pair bred for the first time in Carmarthenshire. In Cardiganshire, sporadic breeding records were recorded from the uplands in the north of the county, Cors Caron and Cors Fochno, and in 1927 a pair bred at Gwbert Burrows.

During the 1940s and 1950s, breeding pairs were reported principally from the counties of Anglesey and Pembrokeshire, but during the next two decades this species became increasingly common elsewhere, mainly due to the spread of afforestation on the Welsh uplands. In 1954, a pair nested in a young plantation at Waun Marteg (Radnorshire), and by 1957 the population had increased to five pairs. As the trees matured, however, the site became less suitable, although breeding was still recorded here in 1966 when a nest containing three young was located (the late G. Ireson, pers. comm.). Since about 1970 at least one pair is suspected to have nested in the uplands of north Carmarthenshire, and following the first confirmed breeding in Breconshire in 1963 up to seven pairs were present in any one year during the 1970s.

In north Wales several pairs were found nesting on open moorland and in young plantations in Merioneth, Montgomeryshire and Denbighshire, and up to three pairs nested during the 1960s in young conifers on Newborough Warren on Anglesey (Hope Jones 1965). Since the mid-1970s there have been declines in many counties and, apart from the Pembrokeshire islands, where up to 12 pairs breed on Skomer and one pair bred on Ramsey until 1982, the species no longer nests in coastal areas in Wales. The current strongholds are the uplands of Merioneth and Denbighshire, the Berwyns, Skomer and, more recently, the Llanbrynmair Moors (Montgomeryshire) which were afforested during the early 1980s and where a minimum of three pairs bred in 1990.

Williams (1988) estimated the Welsh breeding population in that year to be approximately 20 pairs, with similar numbers again the following year (Williams 1989). A desk survey undertaken by the RSPB in 1992 revealed a minimum population of 23 pairs, but it is impossible to give an accurate population estimate as numbers fluctuate with the numbers of field voles. However, the Welsh breeding population is likely to be between 15 and 40 pairs in different years.

In winter, most records of Short-eared Owls in Wales come from coastal dunes, lowland bogs and estuaries, with few birds remaining on the uplands throughout the year. Adult birds travel far and wide in search of food, with youngsters dispersing widely within Britain and, on occasion, even farther afield. From late August to November, a variable number of owls arrive from the Continent, although most of these wander no farther than the east coast of England. Larger influxes are recorded during hard winters, as experienced in north Wales in January and February 1979 when more than 20 birds were present on Anglesey.

The winter distribution of the Short-eared Owl in Wales appears to have changed little since the turn of the century. Inevitably, some favoured haunts have been lost to drainage (e.g. Llyn Ystumllyn in Caernarfonshire) and afforestation (e.g. Malltraeth dunes on Anglesey), but a general increase in the breeding population has resulted in a corresponding increase in the number of winter records.

Even during the turn of the century, Anglesey was an important wintering area for this species, and it remains so today, with a maximum of ten birds at a roost near Beaumaris in January and February 1979. The Dee estuary (Flintshire) is also an important site, and up to six birds can sometimes be seen quartering the saltings at high tide searching for small rodents and small passerines. On the west coast, the estuaries of the Glaslyn/Dwyryd (Caernarfonshire/Merioneth) and the Dyfi support two or three birds most winters, and in the latter county the lowland bogs of Cors Fochno and Cors Caron have held good numbers on occasion. In the winter of 1956–57 an influx of birds into Cardiganshire coincided with a peak vole year.

The Skomer birds regularly visit the mainland during the winter months, and the coastal dunes and estuaries of Carmarthenshire are a favoured winter haunt of this species. In Glamorgan, Kenfig dunes and the north Gower coast support good numbers each year, with eight birds at the former site in December 1970 and seven at Llanrhidian on the Gower in October 1982. High vole numbers on the Levels in Monmouthshire every winter between 1972 and 1974 attracted between four to eight birds, and four birds were seen together at Llyn Heilyn in Radnorshire on 18 December 1978. In this county, as in Denbighshire, Short-eared Owls will often remain on the moors throughout the winter period as long as the weather stays mild.

By far the largest number recorded together in Wales, however, is 40 in the air at the same time on Ramsey in autumn 1979 (T. Sutton, pers. comm.), and it has also been recorded on more than one occasion from Grassholm. Most Bardsey records are of single birds recorded mainly in October and November, some of which will be passage birds. On 21 October 1978 six were seen together on the island, and on several occasions owls have been observed hunting migrant birds attracted to the lighthouse at night. Small numbers have been recorded coming in off the sea at Strumble Head and on the Smalls (Pembrokeshire) in October and November, and records from the Smalls and South Bishop (Pembrokeshire) in March and April are probably birds on return passage.

Recoveries involving birds ringed or recovered in Wales are few and mainly involve birds having moved fairly short distances. There are exceptions, however, the longest distance recorded being a bird ringed at Newburgh in the Highland Region of Scotland on 10 September 1988 and recovered 648 km south on 27 October 1988 near Cardiff (Glamorgan).

Nightjar *Caprimulgus europaeus*

Troellwr

Widespread but scarce summer visitor. The breeding population has increased over the past ten years and is now mainly dependent on clearings in forestry plantations.

In Wales, the Nightjar was typically a bird of bracken-covered slopes, open woodland, sand dunes and, to a lesser extent, heathland. It was common and widespread up until the 1920s and 30s in most Welsh counties, but climatic changes, loss of habitat and increased recreational pressure in coastal areas resulted in a prolonged decline until the 1970s. From about 1980 onwards, the population has increased once more as Nightjars colonised areas of newly planted and/or clearfelled conifer plantations.

Nightjars were common throughout Wales during the latter half of the 19th century, their more common name of "fern owl" referring to one of their preferred breeding haunts on bracken-covered slopes. At the turn of the century, they were particularly common on Anglesey, where they were said to be "abundant throughout the whole district where ever we went we heard them calling" (T. A. Coward, Diaries). During this period, they were nesting in coastal dune systems all along the Welsh coast, and several were said to be breeding on Cors Fochno and Cors Caron. In Glamorgan, up to six churring males were recorded at Merthyr Mawr, on Kenfig dunes and on the Gower dunes. Matthew (1894) recorded them breeding in many parts of Pembrokeshire, and even believed them to be "not uncommon in September turnip fields". Bolam (1913) found them most common on the hazel-covered hillsides near Dolgellau (Merioneth). Several observers had seen Nightjars caught in pole traps, their habit of perching on prominent perches making them vulnerable to this barbaric method of controlling "vermin".

Despite their widespread distribution, Walpole-Bond (1903) remarked that they were becoming less plentiful in mid-Wales at the turn of the century, and numbers certainly declined markedly in several counties from the 1920s and 1930s onwards. Nightjars were scarce in Breconshire by 1940, and were irregular visitors to Carmarthenshire by 1953. Declines were recorded in every Welsh county as the population began to retreat eastwards. Climatic changes, particularly cold wet springs, and habitat loss are thought to have been the two major factors responsible for the decline (Gribble 1983), but recreational pressure, particularly in coastal areas, was also important. The first attempt to quantify the extent of the decline came in 1957 when Stafford (1962) collated data on the status of the Nightjar in the British Isles and compared his results with those of Bannerman (1955) and Norris (1960). However, coverage achieved in Wales by all three authors was poor and often relied on the knowledge of only one naturalist.

Stafford (1962) found that Nightjars were decreasing in seven Welsh counties, with regular breeding still occurring only in Glamorgan, Pembrokeshire, Cardiganshire, Montgomeryshire, Merioneth and Anglesey. It was recorded as not uncommon only in Merioneth where it was numerous on bracken-covered slopes towards the coast. Lockley *et al.* (1961) however, recorded Nightjars as fairly common and widely distributed in Pembrokeshire, where a sharp decline was not noted until the 1960s. In 1967, churring males were heard at five localities in Glamorgan. The *Breeding Atlas* revealed that breeding was confined almost exclusively to the lower-lying valley sides and heaths of coastal counties. Shrubb (pers. comm.) believes the Nightjar probably never disappeared completely as a breeding species in Breconshire and that during the late 1960s and early 1970s, it was still found at seven sites in the county. In Merioneth, the main contraction in numbers occurred between 1925 and 1940, although they were still regarded as widespread. They were still fairly common in Caernarfonshire during the 1960s where they were known as "adar naw o'r gloch" (nine o'clock birds) in Welsh due to their habit of churring at dusk (H. J. Hughes, pers. comm.).

Parslow (1973) knew of no recent breeding records from Carmarthenshire, Breconshire and Radnorshire and asserted that pairs nested irregularly in Denbighshire and Monmouthshire. In the last county, however, it was discovered that Nightjars had colonised open, scrubby areas within conifer plantations and up to eight pairs were located at Wentwood in 1970. A churring male at Cors Goch on Anglesey in May 1976 is the last record from the island and the alarm at the continued decline of this species throughout its range in Britain resulted in a full survey being undertaken by the BTO in 1981 (Gribble 1983).

The results of the survey, which included the first detailed census to be undertaken in Wales, revealed that the population had diminished to a mere 57 singing males, 26 of which were in Monmouthshire (not 33 as stated in Gribble's paper). Elsewhere, numbers were shown to be at their lowest ebb. However, during the 1980s, new pairs were discovered in conifer plantations throughout Wales. It became apparent that clearfell areas and young plantations were providing ideal breeding habitat for Nightjars, especially as so much of the Welsh forests were then being felled and replanted. A sample survey organised by the Forestry Commission in 1990 located a total of 36 territorial males in seven counties (Westlake 1991) and better monitoring resulting from this census revealed a minimum of 67 churring males in 1991.

In 1992, a further national census of Nightjars was undertaken, the results of which confirmed that the recovery was continuing (T. Morris, pers. comm.). In Wales, 193 churring males were recorded at 107 sites, Merioneth being the most important county with 41 males at 19 sites. Anglesey was the only county where no birds were found, although suitable sites still exist there. It is likely that the total population in 1992 may have been more than 200 males as coverage in some counties was not complete. The Welsh Nightjar population will probably continue to increase, particularly as the cycle of felling and replanting in Welsh forests is now permanently established; this will ensure that clear-fell areas are available to the species at all times.

Total number of calling male Nightjar recorded in Welsh counties in 1982 and 1992

	1982		*1992*	
County	*Males*	*Sites*	*Males*	*Sites*
Anglesey	0	0	0	0
Breconshire	4	1	14	7
Caernarfonshire	2	2	20	8
Carmarthenshire	2	2	5	4
Cardiganshire	3	2	5	4
Denbighshire	1	1	26	15
Flintshire	0	0	2	2
Glamorgan	7	6	33	22
Merioneth	9	9	41	19
Monmouthshire	26	10	28	16
Montgomeryshire	0	0	7	5
Pembrokeshire	3	3	1	1
Radnorshire	0	0	11	4
Total	57	36	193	107

Nightjars are more often encountered on the offshore islands on spring rather than autumn passage, although numbers throughout both periods have declined over the past thirty years. It formerly bred on Skomer, Ramsey and Caldey Is., but it is now no more than a rare passage migrant. Until recent years, birds were recorded almost annually on spring passage from Skokholm, but there were only five records between 1960 and 1991. On Bardsey, a total of 22 records involved 16 in spring, the last one in 1977. The maximum number recorded here was three birds together on 3 June 1910.

Swift *Apus apus*

Gwennol Ddu

A widespread and plentiful summer visitor.

Swifts are one of the most familiar and conspicuous summer visitors, noisy and self-advertising near the breeding colonies as they wheel in fast, screaming parties. They occur in towns and villages throughout the length of Wales. For all their conspicuousness however their summer stay here is shortlived and they spend less time with us than any other breeding visitor – except the Cuckoo. For only 16 weeks or so – from the early days of May until mid-August – do the Swifts enliven the summer evenings. Swifts are more exclusively urban in their choice of nest sites than any other British breeding species and this fact alone virtually determines their local distribution in Wales. Most towns and villages of any size support nesting populations and although they may tend to favour larger towns with an abundance of nest sites – Swansea and Cardiff both have high numbers – they also occur in many small villages as well as occasionally in isolated buildings.

Cliff nesting colonies of swifts still occur at one or two sites round the Welsh coast. They formerly nested in the Lias cliffs on the Glamorgan coast but have not now done so for many years; they still nest in small numbers in the limestone cliffs at Falls Bay and Mewslade on the south coast of Gower. In south Pembrokeshire too they nest in small numbers in the Stackpole/Linney Head/St. Govan's area where the carboniferous limestone provides a suitable structure of crevices and recesses. In the early part of the present century there were claims of nesting on several coastal cliffs and mountain crags in Snowdonia (e.g. North *et al.* 1949, who saw swifts entering cliffs near Betws-y-Coed) but this has never been proved and Hope Jones and Dare (1976) could find no modern evidence to support the claim or indicate that inland cliff nesting took place anywhere in that part of Wales. Williams however (in Forrest 1907) testified that swifts nested in the cliffs of the Great Orme (no longer suspected in modern times) and T. A. Coward and C. Oldham recorded breeding in cliffs at Red Wharf Bay in 1902 as well as "other cliffs on the east coast of Anglesey in the following year". Again, there is no recent evidence of this still taking place. On Llŷn, however, it is possible (RSPB surveys 1986–87 unpublished) that nesting on cliff sites still occurs in the far west.

Swifts are seldom censused accurately on any scale, and evidence of long-term changes in status is particularly difficult to establish. Earlier accounts certainly suggest that Swifts have always been plentiful in Wales. The spread of industrialisation and urbanisation in the 19th century presumably brought widespread opportunities for expansion in, for example, areas such as the coalfield valleys, which had been sparsely populated agricultural cul-de-sacs until the discovery of coal brought linear building booms. Subsequently the increase in air pollution probably produced a check. Additionally the replacement of old buildings by those of modern design also tends to operate against the interests of Swifts and there are claims of substantial local reduction in some of the major towns and cities by the middle of this century. Newtown (Montgomeryshire) is a case in point where the Development Corporation has considerably rebuilt the town centre in the 1970s and 1980s and caused a reduction of probably 70% in the Swift numbers (pers. obs.).

Although the distribution of Swifts in the breeding season is usually considered on the basis of their nesting colonies, it must not be forgotten that they are fast, aerial and highly mobile birds and frequently feed a long way from the colonies, depending on weather conditions and the availability of their insect food. For this reason they can more accurately be regarded as virtually ubiquitous throughout Wales during the summer. They often feed even over the highest hills and only in some western areas, e.g. Llŷn and west Anglesey, can they be claimed to be anything other than numerous. In all counties they breed in villages high up in the hills, certainly to 313 m or so (e.g. Llangurig, Montgomeryshire).

Large concentrations of Swifts sometimes occur, particularly in early August as they are preparing to leave, and although they are likely to be encountered anywhere (especially in the vicinity of lowland waters or water meadows) they are noteworthy for regularity at Kenfig NNR, Shotton Pools (Flintshire), the Valley Lakes (Anglesey) and Llangorse Lake (Breconshire) where, at the last site, up to 500 have been regularly recorded in recent springs. Spring arrivals in south Wales consistently begin during the last days of April and the increase in numbers is thereafter rapid.

In Monmouthshire in spring Swifts can be seen travelling up the river valleys to disperse and similar movements are clearly visible in the upper Severn Valley in Montgomeryshire. Most birds have emigrated by mid-August although September sightings are by no means rare and a scattering of late records in October also exists: 5 October (five birds) Montgomeryshire, 8 and 15 October Pembrokeshire, 12 October Flintshire, 23 and 24 October Caernarfonshire and the latest record to date on 28 October on Skokholm in 1976. Overseas recoveries of Swifts ringed in Wales have been reported from Spain (1) and Morocco (2) while on post-breeding passage farther south.

Pallid Swift *Apus pallidus*

Gwennol Welw-ddu

A vagrant.

This is a difficult species, which requires extreme care in identification; it has been recorded once in Wales:

12–13 November 1984 – One, Strumble Head

There were two other records in November 1984, in Kent and Dorset (two birds) from a total of only five in Britain and Ireland. The species breeds in north-west Africa and eastwards from Iberia through the Mediterranean basin to south Iran, wintering mainly in north tropical Africa.

Alpine Swift *Apus melba*

Gwennol Ddu'r Alpau

A rare visitor to Wales.

There have been 25 records, involving 27 individuals, of this species which breeds from Iberia and north-west Africa, through southern Europe eastwards to India and eastern and southern Africa.

The records fall chiefly in the periods April to May and late June to September but there are two exceptional dates: a very early record from Borth (Cardiganshire) on 1 March 1987 and an extremely late record, the first for Wales, of two birds (one shot) at Angle Bay (Pembroke) on 20 November 1908.

Ten of the 25 records came from Pembrokeshire which is as one would expect from the geographical position of that county in relation to Mediterranean migrants overshooting their breeding areas from African wintering grounds. Other records come from Caernarfonshire (4), Anglesey (2), Breconshire (1), Glamorgan (5), Cardiganshire (1) and Monmouthshire (2).

Little Swift *Apus affinis*

Gwennol Ddu Fach

A vagrant.

Two of a total of ten British records of this small swift, which breeds in Africa and southern Asia, come from Wales:

6 November 1973	– One picked up (and released next day), Llanrwst (Denbighshire)
31 May–1 June 1981	– One, Skokholm

Kingfisher *Alcedo atthis*

Glas y Dorlan

Resident in lowland areas.

Lowland waters that offer soft earth banks for excavating nesting tunnels and an abundant supply of small fish are the main requirements for this popular and colourful species. Its distribution in the breeding season is therefore strongly associated with slow-moving lowland rivers, canals and lowland lakes and reservoirs. Close to the western edge of a vast Eurasian range, the Kingfisher is generally a well-distributed breeding species in suitable habitats below about 243 m. (Highest breeding altitude known in Wales is c305 m in north Montgomeryshire.) In the southern half of Wales the principal exceptions are much of Pembrokeshire, Gower, industrial Glamorgan and the coastal fringe from the Gwent Levels to Cardiganshire and the Cambrian Mountains. Thomas *et al.* (1992) suggest a maximum of no more than ten pairs annually in west Glamorgan. In north Wales it is absent from Anglesey, scarce in the mountainous counties of Merioneth and Caernarfonshire, plentiful in the catchments of the Severn, Vyrnwy and other tributaries (Montgomeryshire) and the Dee (Denbighshire and Flintshire).

Parslow (1973) suggests no evidence of long-term major changes in status or distribution on a UK basis. Although this probably holds true for Wales in general terms, there have indisputably been local changes in the population this century, which, at least to date, seem irreversible. In Glamorgan numbers declined on many of the rivers from the middle of the present century due to urban intensification, increased waterside disturbance, gross pollution of waterways and loss of breeding sites through river engineering works; until about 1940 even within the Cardiff City boundary pairs were regular on the stream between Llanishen and Roath Park Lake, where none now exists. The *New Atlas* indicates a marked reduction in the same county since the earlier Atlas (1976) and also a noticeable withdrawal from north-west Wales, which was always marginal ground for Kingfishers. The strongholds for the Kingfisher in Wales are to be found on the slow, eastward-flowing rivers – Dee, Vyrnwy, Severn, Wye, Usk and their tributaries. On similar ones in south-west Wales – Taf, Tywi, Teifi, the species is not nearly so numerous. RSPB surveys of some of these rivers in the 1970s and 1980s produced the following figures:

1976 Severn (Montgomeryshire)	– 97 km, 29 pairs = 1 pair per 3.4 km
1976 Vyrnwy (Montgomeryshire)	– 67 km, 16 pairs = 1 pair per 4.1 km
1977 Wye (Radnorshire/ Breconshire) downstream to Redbrook (Monmouthshire)	– 187 km, 49 pairs = 1 pair per 3.8 km
1981 Teifi (Cardiganshire)	– 4 pairs, all concentrated in one 30-km length

In Breconshire 46 different Kingfisher sites were identified between 1969 and 1988, although it is not suggested that all were occupied at the same time, or each year (Peers & Shrubb 1990). The Usk and Wye were predictably the most important rivers (14 and 18–19 sites respectively) and a population of 40–50 pairs was estimated for 1988–90 – similar totals to that suggested for Monmouthshire (Tyler *et al.* 1987) and Pembrokeshire (Donovan & Rees, in press). Kingfishers are very scarce in Cardiganshire outside one section of the R. Teifi (see above) where breeding is probably annual.

Similarly in Merioneth it is known to breed only sparingly on the rivers Dee, Wnion, Dwyryd, Dysynni and Mawddach. It is a scarce but regular breeding bird in Flintshire but more numerous on the principal rivers and lowland streams in Denbighshire with perhaps 80 pairs in both counties combined. A breeding total of perhaps 400 pairs is probably the maximum in a good year in Wales.

In late summer there is a dispersal of young birds as the adults begin to defend individual winter territories on the favoured river sections. Young birds are pushed into secondary areas and some travel across watersheds to other river systems: near Newtown (Montgomeryshire) as many as 35 juvenile birds have been caught in one month (July 1965) on one small stream, all travelling southwards up to the head-waters. One bird from this site travelled as far as northern France within two months (all other recoveries of those birds have been less than 10 km from the place of ringing). By contrast, a juvenile bird ringed in northern France in July was recovered at Abersoch (Caernarfonshire) almost two years later, and there have been three long-distance movements (92–183 km) westward into Wales from ringing locations in central England; all involved birds of the year. The late summer tendency to dispersal is emphasised by 14 casual records from Bardsey, all except one of which were between mid-July and mid-September. There is one August record from Grassholm, sporadic records from Skokholm, Skomer and Ramsey, and one from the Skerries off north Anglesey (March 1936). This dispersal is supplemented by a withdrawal of some birds from the upper reaches of breeding rivers to spend the winter on lower ground, frequently around lakes and reservoirs but also on estuaries and other low coastlines where individual birds are regular and many locations are used habitually.

Smith (1969) lists three main coastal wintering areas in north Wales – Dee estuary and the Point of Air, Conway estuary and area and the coast near Portmadoc. They are also regular in Merioneth in winter – Dysynni and Mawddach estuaries (three or four birds), Traeth Bach and Dyfi all hold birds under normal weather conditions (R. Thorpe, pers. comm.). Similarly, on Anglesey, Kingfishers are regular in winter at Malltraeth and along the Menai Straits (P. Bellamy, pers. comm.), although they do not regularly breed on the island. In south Wales favoured wintering places include the Burry Inlet, Carmarthen Bay, Milford Haven and Bosherston Pool (Pembrokeshire).

The most serious threat to Kingfishers is severe winter weather and they are seriously at risk when harsh conditions are protracted and shallow inland waters freeze over. A series of hard winters in this century severely affected Kingfisher numbers: 1916–17, 1939–40 (up to 75% died nationally), 1946–47 (Welsh populations probably less affected than in other parts of Britain), 1961–62, 1962–63, 1978–79 and 1981–82. Kingfishers were the most affected species in the bitter and protracted winter of 1962–63, being completely eliminated in some areas and reduced by up to 90% in others. The winter temperature in south Wales averaged 0.4°C for the three winter months (normal average 5.0°C) with air frost on 71 days (average 29 days) (Smith 1969). Recovery of numbers can be surprisingly quick as the surviving pairs are habitually at least double brooded. However, in Caernarfonshire, Hope Jones and Dare (1976) suggested that numbers were still recovering ten years after the 1962–63 crash, but in other counties the build-up in numbers was thought to be quicker than this, aided by a run of mild winters between 1964 and 1975. Smith attempted to catalogue the recovery in numbers: he could establish only about 16 possible or confirmed breeding pairs throughout Wales in 1963, mainly in the south and west, although he does not claim this total as complete.

Bee-eater *Merops apiaster*

Gwybedog y Gwenyn

A rare visitor to Wales.

It breeds from Iberia, southern France and north-west Africa east to Kashmir and eastern Kazakhstan, also South Africa.

The first record was a single bird killed near Johnston (Pembrokeshire) in about 1854 and since then there have been a further 15 records involving 28 individuals. Most of the sightings are of one or two birds but there was a party of four at South Stack on 11 July 1987 and a party of seven near Mumbles (Glamorgan) on 21 June 1973, later seen in Cornwall. There is indication of an increase in sightings, seven of the 16 records were in the 1980s.

The records are fairly evenly divided between coastal counties with four from Pembrokeshire, three from Caernarfonshire and Anglesey, two from Monmouthshire and Glamorgan, one from Cardiganshire; Radnorshire is the only inland county with a record.

Roller *Coracias garrulus*

Rholydd

A vagrant to Wales.

It breeds in Iberia, southern France and north-west Africa, and from Germany and Italy north to Estonia and east to Kashmir and south-west Siberia.

There are 19th century records from Denbighshire (one shot at Abergele on 19 October 1874 and one seen near Colwyn Bay on 7 October 1897), and Flintshire (one shot near Holywell on August 1855 and one shot near

Holywell in September 1857); there is some doubt as to whether these are actually two different birds, the note on the 1855 record was not published until 1860 and there could be some confusion over dates.

The only 20th century records of this brilliantly coloured bird are:

12 July 1962	– One, Pen y Garreg Reservoir (Radnorshire)
4–6 July 1965	– One, Pentre-Bach (Carmarthenshire)
7–8 June 1970	– One, Talybont (Cardiganshire)
30 September 1987 (probably since early September)	– One, Castleton (Monmouthshire)
2nd to at least 11 August 1991	– One, near St. Nicholas (Pembrokeshire)

Hoopoe *Upupa epops*

Copog

An uncommon but regular passage migrant to Wales.

It breeds from France and Iberia eastwards across Asia to China and southwards to south-east Asia, Sri Lanka and South Africa. A striking and conspicuous bird, it has been recorded throughout much of Wales since the early years of the 19th century, unfortunately all too often to end up as a mounted specimen in a collection.

The earliest records are from Pembrokeshire and Flintshire. In Richard Fenton's *A Historical Tour through Pembrokeshire* (1811) there is a description of one "lately shot" at Penyrhiw near Fishguard, with sufficient detail to satisfy the most stringent modern records committee, the more so since it is accompanied by an excellent engraving made by the author. Fenton comments that "the coloured prints in the best works on ornithology which I have consulted, are not at all happy in conveying the likeness, and give you very little of the character of the bird" – a comment very much ahead of its time in suggesting the need to illustrate the "jizz" of a bird! A good description was also provided for one shot at Nannerch (Flintshire) (T. Pennant, *British Zoology*, 1812, vol. 1, no. 90).

There are very occasional breeding records from the British Isles, mainly from southern counties of England as would be expected of a migrant with a principally south European origin. There is no evidence of recent breeding attempts from Wales but at the end of the 19th century there was a claim that a bird was taken, with a nest and eight eggs in Pontrhydyrhun Wood, Monmouth. This, together with a number of other contemporary records from the neighbourhood of Pontypool are, however, now deleted from the county records (Ferns *et al.* 1977) on the grounds of doubts about the origin of the record.

There have been sightings in Wales in every month of the year but an analysis of the 156 records (involving 162 individuals) during the period 1967–86 inclusive shows

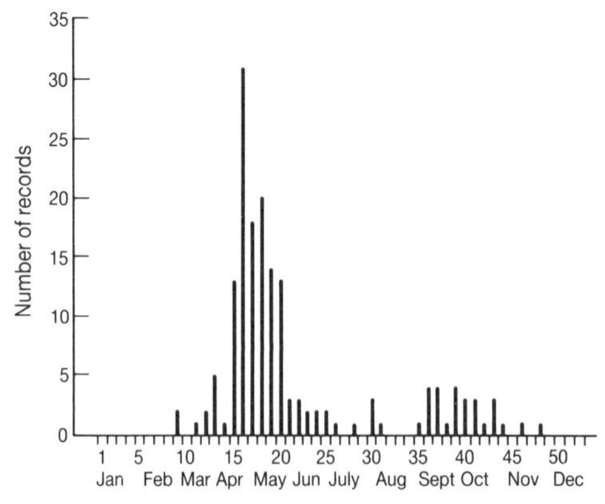

Weekly records of Hoopoe, 1967–86

that there is a pronounced spring passage, peaking in the third week of April.

Most of the records are from the south coast, particularly Pembrokeshire, but during the twenty years of the analysis there are plenty of sightings from other coastal counties on the western seaboard and a handful from Breconshire, Montgomeryshire, Flintshire and Denbighshire. There are no records from the period for Radnorshire, although there are 11 occurrences noted before 1967.

Wryneck *Jynx torquilla*

Pengam

An uncommon but regular passage migrant in Wales, chiefly to coastal counties in autumn.

Wrynecks bred regularly in Wales, albeit in very small numbers, in the mid- to late 19th century but in common with other parts of southern Britain it showed a steady decline and in Wales it probably ceased to breed in the early part of the 20th century. Breeding records came from a wide scattering of counties:

Pembrokeshire	Two 19th century breeding records

Monmouthshire	Ingram and Salmon (1939) quote four breeding records around 1900, e.g. Abergavenny in 1891 and 1898
Glamorgan	The only fully authenticated case of breeding was in 1890 when a clutch of nine eggs was taken from a nest at Lower Penarth
Carmarthenshire	A nest was recorded at Ammanford in 1904
Breconshire	A nest was reported to have been taken in the county some time before 1882 and a pair nested at Llansantffraed in 1903 (a pair or two were seen annually at that time in the Wye Valley)
Montgomeryshire	A pair nested near Llandinam about 1860
Denbighshire	Frequently at Plas Heaton between 1864 and 1869, eggs were found about 1866

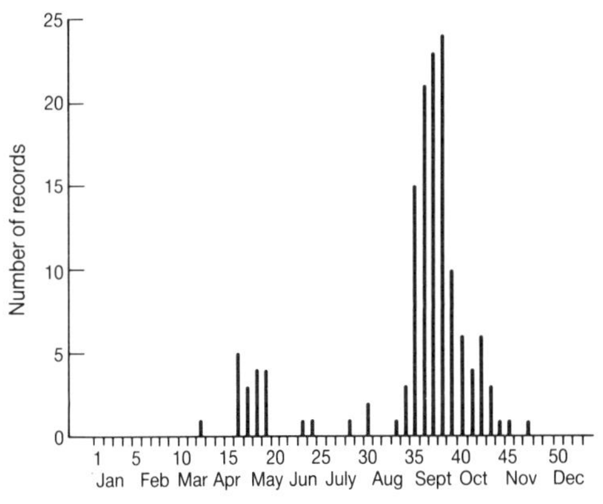

Weekly records of Wryneck, 1969–91

In the early part of the 20th century county avifaunas show that it was an uncommon but probably regular spring passage migrant at that time. From the 1920s onwards there were exceedingly few records but from the 1950s the number began gradually to increase, predominantly in the autumn, and more markedly from the late 1960s. The BTO's Wryneck survey covering the years 1954–58 inclusive (Monk 1963) had no breeding season records for Wales and concluded that it was unlikely that Wrynecks bred anywhere in Wales at that time. It is, therefore, perhaps rather surprising that the BTO's Wryneck Census 1964–65 revealed two possible breedings in South Wales (no locations given) and one observation, given as just possible breeding, in North Wales. The record of a bird holding territory in Merioneth in June 1988 does, however, renew hope for the possibility of a breeding record in the future.

An analysis of dated records 1969–91, 136 in all involving 141 individuals, shows a marked autumn passage peaking in September and a less obvious spring passage (see histogram). Most of these records are from the coastal areas of South Wales, with a scattering from elsewhere, particularly from Bardsey. Outside the analysis period, one was found dead at St. David's (Pembrokeshire) on 31 January 1965 and one wintered in the Manorbier/Penally area of Pembrokeshire from 23 December 1965 to 1 February 1966.

Green Woodpecker *Picus viridis*

Cnocell Werdd

A common breeding resident throughout lowland Wales excepting Anglesey where it is scarce. It favours open parkland or areas of deciduous woodland adjacent to old grasslands with an abundant supply of meadow ants.

The overall distribution in Wales has not varied very greatly over the past hundred years although there have been some local changes. For example, Forrest (1907) noted that Green Woodpeckers were rarely seen in the Holyhead (Anglesey) area up to 1890 but were then increasing; latterly they were present there until the 1970s but have since disappeared.

More significantly, the species has been seriously affected by severe winters such as those in 1947, 1962–63 and 1984–85. In Glamorgan it was noted that it almost vanished from the county following the severe winter of 1962–63 (Heathcote *et al.* 1967) and in Pembrokeshire Lockley *et al.* (1949) noted a considerable recent decrease attributed to the effects of the severe winter of 1947. Recovery to previous numbers takes several years after particularly hard winters.

A comparison of the *Breeding Atlas* and the *New Atlas* shows a roughly similar distribution, but one feature that is apparent is the reduction of confirmed breeding records per 10-km square in parts of Pembrokeshire, and part of the west coast of Caernarfonshire as well as Anglesey. This suggests that there is possibly a slight retraction of range taking place and it will be of interest to see if this is confirmed in future years.

In tetrad surveys of breeding birds in the 1980s the following percentages were occupied in different counties:

Monmouthshire	84%	(= 324 tetrads)
West Glamorgan	78%	(= 182 tetrads)
Pembrokeshire	30%	(= 139 tetrads)
Breconshire	29%	(= 130 tetrads)

With an estimated average of one or two pairs per tetrad, the Pembrokeshire population was considered to be about 140–280 pairs at that time whilst in Monmouthshire a density of four or five pairs per tetrad, as suggested by Common Birds Census data, gave a population of 1200 to 1500 pairs. The Monmouthshire figures are, perhaps, untypical of Wales as a whole and an overall density of one to two pairs per tetrad would give a current breeding population in Wales of the order of 2500 to 5000 pairs.

Outside the breeding season Green Woodpeckers forage widely into more open country and some may then even explore heathlands up to 396 m above sea level. There have been 15 sightings on Bardsey up to 1985 and occasional records on Skomer and Caldey Is. There have only been three ringing recoveries of Welsh birds, which confirm the sedentary nature of the species, involving movements of 2 km, 3 km and 17 km.

Great Spotted Woodpecker *Dendrocopos major*
Cnocell Fraith Fwyaf

A common breeding resident in wooded parts of Wales but scarce on Anglesey and the Llŷn peninsula.

In north Wales, Forrest (1907) noted that it was fairly common in wooded parts, occurring in all counties except Anglesey (Anglesey was added in 1912). At the same period it was also far from common in south Wales. In Glamorgan, for example, it was regarded as not very common but was considered to be increasing at the end of the 19th century (Heathcote *et al.* 1967). Similarly, in Breconshire in the 1882 edition of Cambridge Phillips' book it was classed as rare but in that of 1899 it was increasing.

Increases were noted in many parts of Wales during the first half of the 20th century (e.g. in Merioneth (Hope Jones 1974), Cardiganshire (Ingram *et al.* 1966), Pembrokeshire (Lockley *et al.* 1949) and Carmarthenshire (Ingram & Salmon 1954). Heathcote *et al.* recorded that in Glamorgan the increase at the turn of the century continued and during the next fifty years it became very widespread and in good numbers in the eastern half of the county.

Other than in Monmouthshire and, probably, west Glamorgan the Great Spotted Woodpecker is the commonest Woodpecker, outnumbering the Green Woodpecker. In Cardiganshire (Ingram *et al.* 1966) Salter first noted it as "more common than the Green Woodpecker now" in 1932. The Great Spotted Woodpecker is aided in this advance by being more catholic in its habitat requirements: it nests more frequently in mature conifer plantations and more often in birch and oakwoods at higher altitudes than does its larger relative. The Great Spotted Woodpecker is also less susceptible to prolonged hard weather than the Green Woodpecker, partly because of its habit of feeding on garden bird tables; another important factor is the extent to which Green Woodpeckers feed on the ground. Even so, severe winters can take their toll as shown in Pembrokeshire in 1979 (e.g. five pairs bred in Hylton Woods in 1978, reduced to a single bird in summer 1979 and even by 1983 there had only been a 50% recovery).

At the present time Great Spotted Woodpeckers are common breeders throughout wooded areas of Wales, absent only from north Anglesey and western parts of the Llŷn and mountainous areas above about 366 m. In tetrad surveys of breeding birds in the 1980s the following percentages were occupied in different counties:

Monmouthshire	74%	(= 289 tetrads)
Glamorgan (W)	52%	(= 172 tetrads)
Pembrokeshire	37%	(= 122 tetrads)
Breconshire	28%	(= 125 tetrads)

Taking these figures as representative of Wales as a whole and a density of three to four pairs per tetrad, this gives a breeding population in the order of 7000–9000 pairs. The Monmouthshire population is estimated as around 1000 pairs (Tyler *et al.* 1987) and in Pembrokeshire is about 500–700 pairs (Donovan & Rees, in press).

Great Spotted Woodpeckers sometimes wander onto the cliffs of the seacoast in autumn and there have been eight records from Bardsey up to 1985 (one in April, one in July, one in August, four in September and one in November). The November bird was thought to be of the Scandinavian race. Individuals have also reached both Skomer and Skokholm.

Within Wales this is very much a sedentary species. Of five ringing recoveries three were in the same location after one, two and four years respectively, one had moved 6 km and one 41 km.

Lesser Spotted Woodpecker *Dendrocopos minor*
Cnocell Fraith Leiaf

A scarce and local breeding resident in all counties, principally distributed in the lowlands and along the main river valleys.

In the course of the 20th century the Lesser Spotted Woodpecker has gradually extended its range to cover all lowland areas although it remains very scarce in Anglesey and the Llŷn peninsula. Its preferred habitat is mature deciduous woodland, old orchards and parkland and although it has been recorded at up to 274 m above sea level it is not regularly recorded above 152–182 m.

Forrest (1907) recorded that it was not uncommon in the eastern half of north Wales but unknown in the west. He did, however, note that there were indications that it was gradually extending its range westwards, e.g. none were seen at Llandderfel (Merioneth) during the 25 years preceeding 1898 but several since then. In Caernarfonshire the first record was at Llandudno in 1926, subsequently it was recorded at Penmaenmawr in 1928 and Deganwy in 1929 (Hope Jones & Dare 1976). The first record for Anglesey was also in 1926 when one was shot at Bodorgan Hall in October; the second record was not until June 1935 when one was seen feeding young at Penmon.

In south Wales at the beginning of the century it was scarce in Pembrokeshire but had been seen at Goodwick (Mathew 1894). In Carmarthenshire Barker (1905) recorded that he saw one near Golden Grove in October 1902 and that a Carmarthen taxidermist considered it occasional which probably fairly reflected its status at that time. In Glamorgan it bred regularly at Merthyr Mawr and Margam (Cardiff Naturalists Society 1900) but had not been recorded farther west. In the border counties it was, at about the turn of the century, far from plentiful but quite widely distributed in suitable areas.

In the past twenty years or so there has been little evidence of any change in status and both the *Breeding Atlas* and the *New Atlas* show a generally similar pattern of distribution and a similar number of occupied 10-km squares. Estimates of breeding population are difficult to carry out since the species is notoriously shy and retiring and therefore difficult to census. Recent estimates are:

Monmouthshire

Probably not fewer than 100 pairs "and may well be considerably higher than this". Recorded in 113 tetrads (29%) 1981–85.

Glamorgan

In west Glamorgan the Breeding Birds Survey recorded the species in 18 tetrads (7%) and suggested the population "could be in excess of 30 pairs".

Pembrokeshire

The Breeding Bird Survey of 1984–88 recorded the species in 20 tetrads (4%).

Breconshire

In 1988–90 recorded in 7% of the tetrads visited suggesting that at least 31 were occupied. There "may be no more than 30 to 50 pairs" (Peers & Shrubb 1990).

Based on these data and the results of the BTO Breeding Atlas Survey of 1988–91 it is felt reasonable to suggest that the breeding population is in the order of about 400–600 pairs.

Short-toed Lark *Calandrella brachydactyla*
Ehedydd Llwyd

A rare visitor to Wales. It breeds from Iberia, southern France and north-west Africa east to Manchuria.

There are 11 Welsh records distributed in April (3), May (3), June (2), July (1) and October (2). The first six records commencing on 9–13 April 1952 and up to 1 June 1970 are all from Skokholm whereas the subsequent five from 17 to 28 October 1975 are all from Caernarfonshire (four on Bardsey and the only mainland record, from Llanfairfechan).

In only two cases was the bird seen only on a single day, the average stay was five days.

Crested Lark *Galerida cristata*
Ehedydd Copog

A vagrant.

There has only been one record of this large lark, which breeds throughout continental Europe and southernmost Sweden, east to Korea, Arabia and northern Africa:

5–6 June 1982 – One, Bardsey

Woodlark *Lullula arborea*
Ehedydd y Coed

Formerly resident but now extinct as a breeding species.

Few passerine species attracted more attention from ornithologists in Wales one hundred years ago, or indeed in the first half of the 20th century, than the Woodlark and many written records testify to its former occurrence in most lowland areas. It is a species that has been subject to wide fluctuations in both numbers and distribution over the past two hundred years. In the early 19th century it is believed to have been universal in most counties of England and Wales (Parslow 1973) but then declined thereafter in the second half of the century: Dobie (1893) quoted a Liverpool taxidermist who regarded Woodlarks as "plentiful twenty years ago (in Denbighshire and Flintshire) but now (1854) never seen". However, at the same time it appears still to have been well represented in some parts of Wales, for example in Pembrokeshire where Thomas Dix, writing in 1866, found it "very generally distributed and a constant breeder. . . I have heard it singing in every month of the year." He lamented the fact that Woodlarks were much persecuted by bird catchers in the county who could command "the princely sum of 36 pence per dozen for freshly caught birds".

A further resurgence of fortunes began from the 1920s and continued up to the 1940s. J. H. Salter (1900) lived in the Aberystwyth area where he found Woodlarks local in small numbers around the turn of the century but noted that they had increased and extended their range by 1924; certainly his records for the late 1920s and early 1930s indicate a steady increase across that period despite occasional serious setbacks, as after the severe frosts of the 1928 winter. W. M. Condry (in litt.) confirms it as having been a familiar bird in that part of Wales – and including south Merioneth – in the 1940s and 1950s being especially typical of the near-coastal hillsides and valleys but also extending well inland, nesting up to at least 305 m above sea level. He regarded the Woodlark as being as numerous in north Cardiganshire and north Pembrokeshire as anywhere in Britain. By 1949 Lockley found them widely distributed in Pembrokeshire, most numerous in the east and north-east of the county and equally as common as Skylarks around Crymych on the Preseli Mountains! In Cardiganshire there was an abrupt decline after the 1961–62 and 1962–63 winters with very few records up to 1969, none at all in the 1970s, two records in 1980–81 but none since (P. E. Davis, in litt.). The story in Glamorgan across the same period is somewhat conflicting with that for other counties, although Dilwyn had found it abundant in the Swansea area around a hundred years earlier. Heathcote *et al.* (1967) knew the Woodlark at only four sites in the county in the first quarter of the century, after which it slowly disappeared up to the 1940s, although odd pairs reappeared briefly in the Merthyr Tydfil and Swansea areas in the late 1950s.

Breeding in other counties was widespread, particularly throughout the 1950s. In Radnorshire the Woodlark bred in a variety of sites, some of them well known – Rhayader (lost after the heavy snows of 1947), Llanelwedd, Llowes (lost after severe winter 1962–63) – and others less so: the late G. Ireson (pers. comm.) knew the species as a breeding bird in eight other sites, usually regularly, between 1951 and 1961. In earlier years O. R. Owen, the well-known egg collector from Knighton, recorded Woodlarks' first appearance in the Knighton area in 1920 and found them there in subsequent seasons. The species was evidently quite common in the east of Montgomeryshire until about 1960 but also nested in a number of other localities as far west as the Dyfi Valley and as high up as Llangurig village (313 m) – as late as 1960 – and on the Kerry hills south of Newtown. The last one recorded in the county in this era was a male singing at Aberhafesp in June 1965. Breconshire supported a reasonable population between 1920 and the time of the Second World War after which they were more difficult to find (Ingram & Salmon 1957). A. T. L. Wilson remarked on their increasing scarcity in the county in 1934, a situation no doubt assisted by the fact that he had helped himself to at least 15 clutches in previous years from the area around Garth, near Brecon. Congreve, another inveterate collector, also had several clutches from the same area in the 1920s. Griffiths (1971) wrote of the species' fortunes in the county since the Second World War in those areas for which he had information:

"The 1947 winter apparently wiped out the Woodlark in this county but slowly it appears to be returning to former haunts with reports from various sites: Irfon valley 1964–1970 during the breeding season (sometimes several), three pairs near Hay in 1966, nests at Ystradgynlais, Dyffynog, Aber Car and Glan Wye 1969–70."

In Carmarthenshire it is rumoured to have occurred (breeding was suspected) in Crychan Forest in 1991 where it is said to have been present regularly in early 1980s (H. Ostroznik, pers. comm.). There are occasional winter records from coastal areas in the same county, usually in association with Skylarks. Carmarthenshire had been a relative stronghold of the species in the second quarter of the present century with regular nesting on the coastal lowlands but also at many inland sites, including areas well up into the hills in the Tywi Valley in the far north of the county. However, Barker (1905) had found it scarce and local at the beginning of the century. D. H. V. Roberts (in litt.) has recorded the decline since the mid-1950s, accelerated by the 1962–63 winter, with the last proven breeding at Cynwyl Elfed in 1972, although singing birds were present until 1975 at Tumble and Cilycwm: a single male was singing again at the Tumble site in 1981 and there have been occasional records at some former sites throughout the 1980s. Breeding in Pembrokeshire was widespread up to the early 1960s after which the 1962–63 winter once again apparently accounted for most pairs. One pair bred in 1965 on the south coast of Pembrokeshire, but breeding has not been proved since then. In Monmouthshire, breeding was still occurring as late as 1963 (after two severe winters) and continued sporadically up to 1974.

In north Wales, Forrest (1907) knew of no records in Anglesey or Caernarfonshire but Hope Jones and Dare (1976) mention records in the east of the latter county between the 1930s and 1950s and believe that breeding was possibly implied; Woodlarks certainly bred just over the county border at Penrhyndeudraeth (Merioneth) as many other breeding pairs were to be found on the Merioneth coastal plain, lingering on at Dyffryn Ardudwy until about 1968. The Dyfi and Mawddach estuaries were important strongholds until the 1960s but populations then dwindled rapidly and were extinguished by the end of the decade.

Although Forrest did not know the species on Anglesey it had clearly reached there by the 1950s when Whalley (in litt.) regarded it as an occasional visitor to the island; a pair was present at Newborough in 1951 but breeding was not then proved.

In Denbighshire and Flintshire it was fairly common in the 19th century but has been no more than a rare visitor since the middle of the present century. Sitters (1986) who summarised the status of the Woodlark in Britain between 1968 and 1983 suggested three reasons for the serious decline which had clearly taken place: severe winter weather, climatic change, reduction of suitable habitat. The last of these three cannot truly be seen to apply in Wales as abundant habitat, apparently suitable, still exists. Sitters infers a significant parallel between the rise and fall of Woodlark fortunes and long-term climatic changes which have occurred, particularly relating to spring weather conditions. Severe winters, too, are known to have a serious effect on Woodlark numbers, e.g. 1928,

1947, 1962–63 and it is clear from many published accounts that the final demise in many counties occurred after the severe weather in winter 1962–63. Few birds survived after that (e.g. Merioneth, Breconshire, Monmouthshire) and the last nesting in Wales was probably at the Radnorshire site at Llanelwedd near Builth Wells in 1980 or 1981; no birds have been seen at the site since the hard weather early in 1981. Although harsh winters have indisputably had a part to play, they do not satisfactorily account for the final demise of the species in Wales, certainly as small populations clearly survived after the 1962–63 winter but then were extinguished before the next severe one. It seems likely that the real cause of extinction – like that of Red-backed Shrike and Cirl Bunting – was long-term climatic change accelerated by severe winters.

By the early 1990s the Woodlark in Wales has been reduced to the status of an occasional visitor with a scattering of records, invariably of individual birds, from both inland and coastal counties mainly in spring and autumn months. Although it may seem a pious hope, there is suggestion that – as in East Anglia – careful management of suitable areas of forest holding could entice the species to recolonise: the recent unconfirmed report from Crychan Forest lends some encouragement to such a possibility.

Skylark *Alauda arvensis*

Ehedydd

A widespread but declining breeding species; abundant passage migrant in autumn and winter.

Over the long term, at least throughout the 19th and 20th centuries, there has been no major change recorded in the status or distribution of the Skylark until the past two decades of the present century. Parslow (1973) reflected this view up to his time of writing but points to local losses due to land-use change, notably urbanisation. In Wales there has been a slow, progressive loss of Skylark habitat since the Second World War not simply to urban development but especially to rural land-use changes – agricultural intensification and afforestation.

Afforestation in the uplands of Wales now occupies some 25% of the land, thus rendering it unusable to Skylarks on ground which was previously part of the species' stronghold. Only slightly more subtle is the direct effect caused by the rapid post-war expansion of sheep numbers in Wales as shown in the graph on page 27. The intensity of grazing throughout the country, particularly in improved (i.e. "reclaimed") fields in both uplands and lowlands produces a sward so closely cropped that it prevents adequate cover for nesting. A clear example of this effect is Bardsey, where up to seven pairs bred regularly until the mid-1960s after which they steadily declined and last bred in 1978. This demise has been attributed directly to the intensification of sheep grazing over the island leaving no rough, long grass for nesting. This is the situation on the many thousands of hectares planted with rye-grass mixes and habitually grazed to ground level by sheep. At the same time it is well known that pastoral agricultural systems support lower populations of Skylarks than arable or mixed systems where the supply of spilt grain and weed seeds is much greater. O'Connor and Shrubb (1986), showed that the mean density of Skylarks on stock-rearing farmland in Wales, at only 7.9 pairs per km^2, is lower than anywhere else in Britain (c20 pairs per km^2 in north England; 24 pairs per km^2 in East Anglia). It has been suggested (Schlapfer 1988) that crop diversity, even within single-farm units is a major determinant in producing high numbers of Skylarks, notably as it provides more secure nest sites. Wales certainly has small farm units but they are almost totally devoted to grass crops and do not have diversity. The amount of tillage in Wales does not now exceed 7% of the total farmed area. The net result of these factors has been a decrease in population to about 50% of the 1970s population (Marchant *et al.* 1990).

However, the Skylark is still one of the most widespread breeding birds in Wales as shown by the *New Atlas*. It is an open country species, shunning urbanised areas, wooded countryside, mature conifer plantings, narrow valleys (which abound in Wales) and farmland which is characterised by a pattern of very small fields separated by tall overgrown hedgerows-with-trees. It prefers flat or gently undulating countryside of rough grassland with wide vistas. Thus it is able to exploit a wide variety of sites including open farmland, coastal marshes, dune systems, ffridd, grasslands, rolling moorland and mountain grasslands. It breeds on most ungrazed offshore islands, e.g. Ramsey, Skomer, Skokholm, Worms Head. In upland sites Skylarks breed to hill-top levels in south and mid-Wales, e.g. 800 m on Brecon Beacons, 770 m on Pumlumon and even higher in Snowdonia and the other mountains of north Wales where they reach a height of at least 762 m.

Forrest (1907) first commented on Anglesey as an area with particularly high numbers of breeding Skylarks. Being an area which has suffered relatively less intensive agricultural change than most other counties in Wales, the same still holds on Anglesey although elsewhere other fairly high population estimates are also given: 10 000 pairs and c8000 pairs have been given for Monmouthshire (Tyler *et al.* 1987) and Pembrokeshire (Donovan & Rees in press) respectively. Most county reports acknowledge the decline that has clearly taken place; this is particularly noted in Montgomeryshire, Radnorshire and Breconshire. In the last county Peers and Shrubb (1990) recorded it in 59% of tetrads in 1988–90 with highest numbers found on Mynydd Eppynt where the minimum mean density was 3.5 pairs per one-kilometre square . On the Pumlumon massif, RSPB surveys in 1988 (RSPB 1988) found maximum densities of 20 pairs per one-kilometre square.

Skylarks move away from the upland breeding areas and many other inland locations at the end of the summer and are then often distinctly uncommon away from coastal sites; Montgomeryshire, Breconshire and Radnorshire, for example, support very few Skylarks in most winters. However, large numbers occur both as passage migrants and winter visitors, especially associated with cold weather movements. Such influxes and movements are recorded from all counties and can be very large on occasions. Along the south Wales coast autumn flocks of several hundred are not uncommon and in Pembrokeshire flocks of 1000+ are regular with strong visible passage to Ireland frequently seen from September to November. Numbers moving south on Bardsey in October regularly reach 250–500 per day with occasional counts as high as 1000 (6 October 1962). In the winter months hard weather movements are characteristic of the species as large numbers move westwards or southwards ahead of heavy snow or severe frost. There are many reports of movements involving 1000 birds or so and one exceptional count in excess of 4000 in the Camlad Valley (Montgomeryshire) in February 1969. In normal winters the population of Skylarks in Wales is strongly focused in coastal areas as exemplified by the *Winter Atlas*. For those Skylarks that remain in winter severe weather takes its toll although recovery of numbers is reasonably rapid in subsequent seasons.

Shorelark *Eremophila alpestris*

Ehedydd y Traeth

Regular annually at one north Wales site in very small numbers, otherwise a scarce and irregular winter visitor and scarce passage migrant.

The vernacular name of this species refers only to its winter habitat for, in northern Europe, the Shorelark is a breeding bird of high altitude and arctic tundras in Scandinavia, moving south in winter to coastal areas around the shores of the southern North Sea and western Baltic. In Britain most wintering takes place on eastern coasts, from Kent to Northumbria and in the west the only regular wintering is around the Dee/Mersey estuaries where records since around the mid–1960s have indicated an extension to the former winter range (*Winter Atlas*). The numbers are small, usually singles or groups of two or three but outside this one site on the Dee estuary the species is little more than accidental and irregular. The numbers visiting Britain appear to have increased in the late 1960s and early 1970s, which coincided with the extension into north Wales and a similar spread into coastal sites on the south coast of England. The numbers have subsequently contracted again and the species has withdrawn from the English Channel coasts, although the small north Wales population has remained, certainly up to the early 1990s. In their winter habitat Shorelarks are normally restricted to coastal areas with extensive saltmarsh and associated sand dune systems – such as those around the Dee – where they are found in company with larger flocks of Skylarks, pipits and finches.

The history of occurrence on the Dee estuary goes back as early as 1905, in which year Forrest recorded one being watched at close quarters on Hilbre Is. (Wirral). This was then the only record known to him in that area and since the tiny wintering population became regular in the late 1960s, Shorelarks have almost invariably been found on the marshes and dunes at Point of Air and Gronant, across the water from Hilbre Is. In the 1966–67 winter – which coincided with the time of short-lived expansion in southern England and even as far as Ireland – as many as 70 were estimated to be in the Gronant area and on the Dee Training Wall at Shotton but that total has never been approached again and in the most recent years no records have been made. The only other areas where Shorelarks are seen with even a degree of regularity are Anglesey, where there has been a scattering of records in recent years and a small group overwintered at Cemlyn in 1973–74, and Bardsey, where autumn passage birds are reasonably regular although again in very small numbers; a group of ten or so birds present between October and December 1966 was exceptional. On mainland Caernarfonshire there have been some four other records, all in the 1980s, with birds at Llanfairfechan in two successive years.

In Merioneth, single Shorelarks have been recorded on five occasions in the 1970s and 1980s usually on the Mawddach estuary or Morfa Harlech. Cardiganshire provides only three records: 1905, 1967, 1972, all of single birds. There are seven records from Pembrokeshire, two on Skokholm in April 1957 and April 1967 and one there in June 1990; the other records – all from the mainland coasts – are at times of autumn passage in September and October. Shorelarks have been seen on the Glamorgan coast only three times, in October 1976, October 1982 and November 1984. Two Shorelarks were reported in Monmouthshire, on the Rhymney estuary in November 1972 and there is one Carmarthenshire record of a single bird on Llangennech Marsh in November 1964.

Sand Martin *Riparia riparia*

Gwennol y Glennydd

Locally common breeding visitor, most numerous in eastern counties. Passage migrant in spring and autumn.

Although Sand Martins breed in all of the counties of Wales, their distribution is not even and in most areas of the west the colonies are small and frequently scattered; they are normally absent from upland areas above c365 m. The Sand Martin's strongholds in Wales are on the lowland sections of the meandering eastward-flowing rivers of the border counties – Dee, Severn, Vyrnwy, Wye, Usk and Monnow – and the River Tywi in Carmarthenshire. Most of the main colonies in the Principality are in "natural" sites (as opposed to the predominance of sand pits and similar man-created habitats in southern England), notably in the soft vertical banks of rivers. In western counties sea-cliff nesting colonies have long been recorded and although the practice is less frequent than formerly, presumably due to higher levels of human disturbance, it still occurs in several coastal areas. Some well-known cliff colonies such as those on the sand cliffs at Porth Dinllaen and the low clay cliffs at Porth Neigwl (both Caernarfonshire) have certainly existed for a century or more (Forrest 1907) and remain of good size today. Sand Martin colonies are frequently ephemeral, dependent on the availability of freshly available sandy faces, particularly on river banks where they quickly take advantage of newly exposed faces created by shifting channels.

Thus the pattern of nesting may shift considerably over a period of time although actual numbers may remain reasonably static. Transitory breeding of this sort is sometimes characteristic of sand heaps, dune faces, roadside cuttings, and breeding in such sites occasionally occurs at altitudes of up to 305–365 m. In Glamorgan and Monmouthshire Sand Martins have bred in drainage pipes in retaining walls along riversides.

There is little recorded information of the historical status of Sand Martins in Wales although it is suggested – Glamorgan – that numbers rose towards the end of the last century. In the same county, they had apparently declined markedly 25 years later due to increasingly frequent riverbank works and human disturbance, leaving no more than four colonies as large as 100 pairs or so.

A well-documented and dramatic crash in numbers occurred after the 1968 breeding season (c.f. Whitethroat and Redstart). In 1969 it was estimated that the Sand Martin population in Britain was only c40% of that in previous years, a crash of numbers which was subsequently attributed to the prolonged drought in the Sahel region on the southern flanks of the Sahara desert and neighbouring savannahs, where the Sand Martins winter. Numbers remained at low levels through much of the 1970s only to be worsened again by another drought-related crash in 1984 after which Mead (1984) estimated that the overall British population had fallen to no more than 10% of that in 1968. Improved rainfalls in the Sahel in the late 1980s and early 1990s have gone some way to building up the population again: Mead (pers. comm.) estimated that by 1991 numbers were perhaps still only some 20% of the 1968 level throughout Britain overall. However, he also speculates that the Welsh population, nesting a little later and probably producing more young *pro rata*, may have fared better. The major population collapses were well reflected in Wales although in some areas, e.g. Pembrokeshire, it was noted that whereas the numbers of spring passage birds was much lower than usual, the modest breeding population was probably unaffected.

In Montgomeryshire, the entire lengths of the River Vyrnwy (67 km) and the River Severn downstream to the English border (137 km) were censused in 1969, the year following the winter crash in numbers, producing counts of 585 apparently-occupied burrows on the Vyrnwy and 1271 on the Severn. A similar count on the River Usk in Monmouthshire and Breconshire in 1965 (before the crash) found 1850 occupied holes in 116 km, compared with only 500 counted in 1985 on the three rivers in the county which support colonies – Usk, Monnow and Wye. Surveys in Breconshire (1988–90) suggest a total of around 400 pairs. In Carmarthenshire the River Tywi supports the main population, where Friese (in litt.) found 517 pairs in 1987 and 1092 pairs in 1988 (including 100 pairs at sites not counted in 1987), providing evidence of the rising population at that time. He estimated a population in the early 1990s of around 2000 pairs on the main river and its tributaries. The River Llwchwr, in the same county, also supports good numbers.

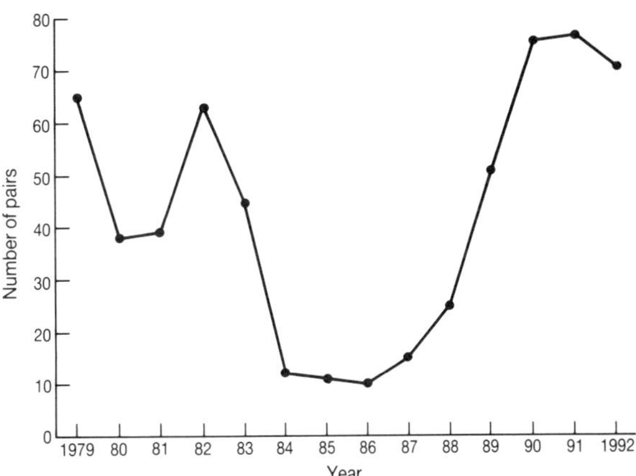

Numbers of Sand Martin pairs, colony at Rhandirmwyn (R. Tywi) (data RSPB and J. Friese 1988)

In the counties of the western seaboard the Sand Martin is thinly distributed and colonies are generally small. Pembrokeshire supported no more than 70 pairs or so annually between 1984 and 1988, in a maximum of ten tetrads; the species "breeds locally" in Cardiganshire, is "widespread and locally common" in Merioneth with probably not more than 100 pairs in the western half of the county (Thorpe, pers. comm.) but higher numbers on the eastward flowing Dee – one sand quarry colony near Bala has been known since 1895. On Anglesey the Sand Martin is a scarce breeding bird with no more than eight colonies located at the time of the *Breeding Atlas*

(1968–72), reduced to only three, in the east of the island at the time of RSPB surveys in 1986. Similar surveys on Llŷn in 1986–87 located only seven breeding colonies, with possible breeding at a further six; proved breeding occurred as far west as Porth Neigwl.

The Sand Martin is one of the earliest migrant species to appear in spring, the first arrivals reaching the south coast in late March (earliest date 23 February in Pembrokeshire). The main passage – strongly characterised by coastal movements – continues through April and May with late stragglers occurring into June. Up to 400 birds per day are recorded on Bardsey and Skokholm. Autumn passage, which starts as early as July, is usually considerably smaller although occasional counts (Skokholm) reach 500 per day. Later birds have been seen into the first week of November. Autumn roosts in reed beds are a regular feature of the return passage; the roost at Llangorse Lake (Breconshire) varies from season to season between maxima of 500 and 3000; the estimate of 10 000 in August 1979 was exceptional. Ringing recoveries from France (29), Spain (11), Portugal (1), Morocco (1) and Senegal (5) indicate the route to the African wintering areas. A bird ringed at Llangorse in August and recovered on Malta the following spring indicates that some birds may move northwards by a more easterly route.

Crag Martin *Ptyonoprogne rupestris*

Gwennol y Clogwyn

A vagrant.

This chunky hirundine breeds in north-west Africa and Iberia through southern Europe and south Central Asia to the Far East. The northernmost populations winter farther south in Europe and North Africa but most Western Palearctic birds are resident.

The first two records for Britain were in 1988, closely followed by the only Welsh record, from Llanfairfechan (Caernarfonshire) on 3 September 1989.

Swallow *Hirundo rustica*

Gwennol

A widely distributed and common summer visitor and passage migrant.

The Swallow is the traditional harbinger of spring and its welcome arrival on Welsh shores each year occurs from the late days of March to a peak in the second half of April. It shares with only a handful of other species the distinction of being present and presumed to breed in every 10-km square in Wales. This broad distribution reflects the fact that the Swallow is more dependent on man and his buildings than any other breeding species except Swift. Nesting occurs on the great majority of farms, from sea level to well over 430 m; although in fine weather birds may forage up to the summits of the high hills. Although nesting sites are traditionally in open-fronted farm buildings and the like, Swallows will also resort fairly readily to porches, garages, garden sheds and other outhouses. One can only postulate what the size of the population of Swallows may have been before the convenience of man-made buildings, for it is only rarely now that a pair is found breeding in a "natural" site; such nesting is still recorded occasionally on sea-cliffs in Pembrokeshire (where, interestingly, both Sand Martins and House Martins use some sea-cliff sites also). Forrest (1907) reported a nest on a ledge several yards inside a cave at Tremeirchion (Flintshire) in 1903 and a nest with a late brood of young was found 4 m down a derelict mine adit at 430 m on Pumlumon in August 1974 (pers. obs.). Nesting is not restricted to mainland sites and most islands with buildings on support a pair or more of breeding birds. Urban centres are normally avoided for nesting – although not shunned for feeding – and the record of a pair nesting successfully off Queen's Street in Cardiff City centre in 1965 is noteworthy.

Changes in population levels of Swallows are difficult to assess because of the absence of regular monitoring of any of the Welsh population. Nationwide, the CBC reveals a broadly stable position until the beginning of a steady decline in the 1980s. Most references in county bird reports imply reductions in numbers and this is the subjective view of many other observers. Although reliable breeding data are generally absent, some indication of population levels may be deduced from reedbed roost sites in late summer and autumn: in the 1960s, for example, the roost at Llangorse Lake regularly reached as many as 5000 individuals but was down to 1000 in 1981 and since 1984 the number has not exceeded 250. One or two other reedbed sites confirm the same trend, e.g.

Cosmeston Country Park (Glamorgan): 8000 in 1980; 2000 by 1984
Aberthaw reedbeds (Glamorgan): 1200 in 1981; 600 by 1988

In Breconshire, Peers and Shrubb suggest a population of around 2500 breeding pairs on the basis of the surveys carried out for the *New Atlas* (1988–90). In Monmouthshire the estimate in 1987 was around 4000 pairs and the Pembrokeshire population was suggested at a minimum of 8500 pairs (Donovan & Rees, in press). Swallows have been dependent on an ample supply of nesting sites in traditional farm buildings, and the replacement of these by new generation cattle sheds and barns does not favour the Swallows. At the same time, as has been suggested by several writers, the improved storage and handling of farm slurry has brought about reductions in the numbers of available flying insects – the Swallow's staple diet. Although data are not available to support the contention, the widespread opinion is that numbers in Wales in the early 1990s are noticeably lower than previously.

The spring and autumn passage of Swallows is well marked, particularly on coastal headlands and offshore islands. In spring the earliest arrivals are in the second half of March and numbers are at a peak in April and early May. On the Pembrokeshire islands daily totals of 75–100 are unusual and normally the northward passage is no more than half that number. Coastal movement is clear on headlands such as Strumble, where 167 passed by on 29 April 1988. Farther north, on Bardsey, day totals of up to 250 are quoted as commonplace in May, with as many as 400–500 occasionally and 1000 on 9 May 1961 being exceptional. In autumn the return passage is smaller on Bardsey – 50–150 the usual daily maximum – although 2000–3000 were present on one day in October 1982. Coasting movements in south Wales are well demonstrated at Lavernock Point, where an estimated 40 000 Swallows passed along the coast during 11 hours watching on four consecutive afternoons in September 1965; what the total count would have been for that period is conjecture. On the north Pembrokeshire coast a massive landfall of Swallows took place from the north-west on 22 September 1983, when a figure of 90 000 birds was calculated for the coastline between Strumble Head and St. David's.

The winter destination of British Swallows is well known from ringing recoveries and borne out by the results of birds ringed in Wales. Twenty-eight Welsh-ringed birds have been reported from South Africa in the months December to March, most from Cape Province, but six from Transvaal and one or two from Natal and the Orange Free State. Birds ringed in Wales on autumn passage (all October or November) have been recovered from Nigeria (3), Cameroon (1), Zaire (2), Algeria (2), Morocco (1). Two spring passage birds (ringed in April) have been reported from Morocco and one from Tunisia subsequently.

Late birds are not infrequent into November and even December (4 December 1967, Breconshire). Early December birds are difficult to interpret but the assumption of attempted overwintering must be made in respect of a bird at Bala on 24th of the month in 1891, as with one in Monmouthshire on 23 January 1990. Another overwintering bird was seen on several dates at Haverfordwest in the early weeks of 1989.

Red-rumped Swallow *Hirundo daurica*
Gwennol Dingoch

A vagrant to Wales.

It breeds in Iberia, southern France and north-west Africa, Balkans east to Japan and south to Sri Lanka and central Africa.

The first Welsh record was as recent as 1973 but there have been seven further records since then, this increase coinciding with the spread of breeding areas northwards from extreme southern Spain into France.

The Welsh records are:

15 August 1973 – One, Eglwys Nunydd Reservoir (Glamorgan)
5 March 1977 – One, Dale (Pembrokeshire)
30 June 1977 – One, Bosherston Pools (Pembrokeshire)
3 May 1980 – One, Bardsey
6 & 20 September 1985 – One, Gronant (Flintshire)
26 October 1987 – At least five, possibly seven, point of Air; four there on 28 October
30 April 1990 – One, Skokholm
29 August 1990 – One, Mynydd Rhiw (Caernarfonshire)

The unprecedented records from Flintshire in 1987 coincided with a remarkable influx into 15 counties of England and Scotland in late October and early November, extending from the Scillies as far north as Shetland.

House Martin *Delichon urbica*
Gwennol y Bondo

Summer visitor, well distributed and locally common.

The House Martin breeds throughout Wales from sea-cliff sites up to altitudes of at least 400 m in some hill areas. Nothing is known about the numbers of House Martins nesting in the distant past and one can only conjecture that on account of the species' predilection for nesting on human habitations, its numbers must be considerably greater now than in former centuries. Colonies are usually small – fewer than a dozen nests and often only three or four, or single nests – and are to be found in urban, suburban and rural communities. Interestingly there is a clear, albeit irregular, gradation of increasing colony size from urban situations through to those on isolated rural properties. Although most colonies are small, there are occasional larger ones, the biggest known in Wales being at Aber (Caernarfonshire): 100 pairs in 1967, 180 in 1980 (no recent published count).

The return to urban areas in south Wales in centres such as Cardiff and Swansea has been marked since the introduction of the Clean Air Act in 1956. Although the great majority of colonies are on buildings or under bridges, "natural" nest sites on sea-cliffs and quarry faces are still used, albeit the habit has declined throughout the present century. Cliff-nesting House Martins are habitual in several coastal counties. In Glamorgan colonies were known on the Southerndown cliffs and near Porthskerry at the beginning of the present century and the habit extended over the years with numbers of nests along the coast rising from 75 in 1962 to 330 in 1966. By the late 1980s the colonies had reduced again to a handful of no more than a dozen or so nests. In Pembrokeshire there are many regular colonies both on the limestone cliffs in the south and rather less numerously on the northern coastline, recorded as long ago as 1894. There are several small colonies on the Cardiganshire cliffs between Cardigan and Llangranog, which again have been known since the last century. In Caernarfonshire cliff-nesting was referred to by Forrest (1907) in the Penmaenbach area and on the cliffs of Great Orme. Similarly on Anglesey, cliff pairs are well known from a number of localities dating back to the last century: Lligwy Bay, Point Lynas, Bull Bay, Moelfre, Castell Mawr near Amlwch and Red Wharf Bay. Cliff-nesting is not currently known on Llŷn although the species is well distributed throughout the peninsula, being known to occur – breeding or possibly breeding – in 56 one-kilometre squares (RSPB surveys 1986–87).

There are conflicting opinions as to whether long-term changes in House Martin numbers have taken place. On one hand there is very little data in Wales across the years on which to base comparisons, and on the other hand it is recognised that the species is a very mobile one, shifting breeding areas at will and easily leading to confusion in assessing numbers. Tatner (1978) suggested that, because of this high degree of mobility, very large survey areas are needed to pick up overall, rather than local, trends. The concensus opinion leans towards there being a stable population at present but one which is prone to both seasonal variations (as shown in the CBC index) and regional mobility (Marchant *et al.* 1990). Bearing these factors in mind, the population estimates which have been arrived at in one or two south Wales counties can only be taken as tentative. In Monmouthshire 4000 pairs has been suggested (Tyler *et al.* 1987), in Pembrokeshire c4000 pairs and in Breconshire around 1600 pairs (1988–90). M. Doe (1976) surveyed nesting House Martins in Cardiff. His survey area included 181 streets with 7721 houses (904 of which were unsuitable for Martins) in a predominantly suburban area of approximately two square miles. He found a total of 68 nests being used by House Martins out of 205 located: one nest per 18 acres or one nest per 1.7 mile of street length. Tallack (1982) found 522 occupied sites on Gower (42 tetrads) in 1981. RSPB surveys on Anglesey in 1986 found breeding, probable or confirmed, in 24 one-kilometre squares.

House Martins are rather late arrivals, the main waves of immigrants not appearing until late April and early May. Isolated individuals are recorded in March in some years (earliest dates 4 March Radnorshire, 12 March Glamorgan.) Rare occurrences are not unknown in February. In Pembrokeshire, for example, individual birds have been seen on 10 February 1982, 17 February 1958 and 23 February 1984. At the other end of the season House Martins – often producing up to three broods – are among the latest of the summer visitors to depart, lingering regularly into October and not infrequently November. Several December records exist: 8 December (Glamorgan), 7 December (Caernarfonshire). Visible emigration is seen on the offshore islands and coastal headlands; 520 were watched heading out to sea from Mumbles (Glamorgan) in a two-hour period at the end of September 1974 and 750 passed Lavernock Point on 25 September 1965. Similar passage is recorded from other coastal sites.

Richard's Pipit *Anthus novaeseelandiae*

Corhedydd Richard

A scarce autumn passage migrant.

This large pipit breeds from western Siberia east to Mongolia and south-east to New Zealand, also Africa.

There are 67 records involving 89 individuals spanning the period from 7 September to 25 November with three extreme records, on 9 April, 22 August and 26 December. There is a marked concentration from late September to October.

As with many of the rare migrants the bulk of the Welsh records are from Skokholm and Bardsey; there are some difficulties in determining actual numbers when a series of sightings occurs over relatively short periods.

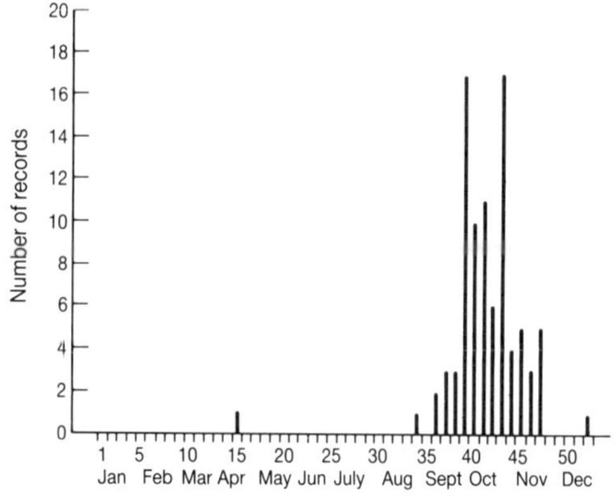

Weekly records of Richard's Pipit to 1991

Parties of four or more birds have been noted as follows:

30 September 1966 – Six or seven, Bardsey
25 November 1967 – Five, Wooltack Point, Marloes (Pembrokeshire)
26 October 1968 – Four, Skokholm
29 September 1970 – Four, Skokholm

It is notable that the records in the 1980s refer to single birds or, in one case only, two birds, so that a small party would now be considered exceptional.

Most of the records are from Pembrokeshire (27 records, 37 individuals) and Caernarfonshire (20 records, 31 individuals) but there are also records from Flintshire, Anglesey, Monmouthshire, Cardiganshire and Glamorgan.

Tawny Pipit *Anthus campestris*

Corhedydd Melyn

A rare visitor.

Up to 1991 there are 15 Welsh records of this species, which breeds from north-west Africa, Iberia, southern and eastern France and southern Sweden eastwards to Mongolia. The first was on Skokholm on 19 September 1961.

September is the peak month with seven records, the other occurrences being in April, May, August, October and November. The November record (Aber, Caernarfonshire 26–29 November 1977) was very late for this species. All the records are of single birds apart from two on Skomer on 10 October 1975.

Most of the records (eight) are from Pembrokeshire with three from Glamorgan, one from Monmouthshire and three from Caernarfonshire.

Olive-backed Pipit *Anthus hodgsoni*

Corhedydd Gwyrddgefn

A vagrant.

An unidentified pipit trapped on Skokholm on 14 April 1948, and staying on the island until at least 18 April was subsequently confirmed and officially recognised in 1977 as the first record of Olive-backed Pipit in Britain and Ireland. With a background of detailed notes on nine British records between 1964 and 1976 and advice from an observer with experience of the species in Mongolia assessment of the record in 1977 was much easier than in 1948 and the identification was clinched by a photograph of the bird in the hand – a just reward for some good detective work!

There have been no further records of this species in Wales although a spate of records since 1973 has now brought the Britain and Ireland total to 130 by 1991. Olive-backed Pipits breed in north-east USSR to central and east Asia and Japan, wintering in India, south-east Asia and Philippines. The spring record from Skokholm remains highly untypical with virtually all the other British occurrences in October and November although there are two May records.

Readers are referred to *British Birds* vol. 72, pages 2–4 for a full account of the 1948 record by Peter Conder.

Tree Pipit *Anthus trivialis*

Corhedydd y Coed

A widely distributed and locally numerous summer visitor. Passage migrant in fairly small numbers, best marked on the south coast in autumn.

The parachuting song flight of the Tree Pipit is one of the most familiar and characteristic features of the Welsh countryside in spring and early summer. The first arrivals reach Wales in the earliest days of April but the main influx does not take place until the middle of that month. Tree Pipits breed throughout Wales, with the exception of Anglesey, where they are at best limited to one small area annually. They are particularly characteristic of the ffridd lands on the slopes of the upland core of Wales, where bracken and other rough vegetation is studded with trees, bushes or other suitable song posts. A range of other habitats is also extensively used, notably mature open woodland, young forestry plantations, clear-felled areas, railway lines and similar sites which offer the requirements of open ground, rank vegetation and song posts. In such habitats Tree Pipits will readily breed up to altitudes of 454–516 m in the hills, for example in young forestry plantations. In fact, with the increasing amount of harvesting of conifer forests in Wales there is possibly a shift of population (or an expansion?) occurring into these clear-felled areas where Tree Pipits are often the first pioneer colonists.

During the past 100 years, Tree Pipits have gradually expanded northwards, extending their range in Scotland and Scandinavia. In Wales there is no evidence of any substantial change in status or distribution and it is therefore interesting that a northward extension has not resulted in the wider colonisation of Anglesey, where suitable habitat is available. In fact its status on the island may even have weakened in that T. A. Coward (Diaries) recorded it at several localities in the early years of this century where it is certainly not found at present; Forrest (1907) stated that it was confined to south-east Anglesey in the vicinity of the Menai Straits, which is still applicable today. Tree Pipits clearly shun the western seaboard.

In addition to their virtual absence from Anglesey, they are scarce on Llŷn – found in only 49 one-kilometre squares in RSPB surveys in 1986–87 and absent from the western end, west of Nefyn/Abersoch. They are also absent from most of Gower and much of Pembrokeshire, where most breeding is found in the eastern – inland – part of the county. A total of c800 pairs is suggested for Pembrokeshire (Donovan & Rees, in press.)

Elsewhere, where suitable habitat exists Tree Pipits can be expected, although, as is pointed out for Carmarthenshire (D. H. V. Roberts, in litt.) and south Cardiganshire, there are sometimes unexpected and unexplained gaps in the distribution.

The *Gwent Atlas* found 42% of tetrads occupied in Monmouthshire although the suggested breeding population of up to 1500 pairs (Tyler *et al.* 1987) is almost certainly too high. Similar work in Breconshire found 62% occupied tetrads and a suggested figure of c900 breeding pairs. On the basis of the above figures the Welsh breeding population is probably 8000–10 000 pairs.

Despite the spread of urbanisation, industrialisation and the tipping of waste in the past 100 years or more, the Tree Pipit remains well distributed in Glamorgan and much of the open countryside on the hills and the slopes of the industrial valleys offers ideal habitat. It is throughout the core of central Wales that the Tree Pipit is at its most numerous. Here it is locally abundant in Cardiganshire, Breconshire, Radnorshire and Montgomeryshire. In Caernarfonshire, other than on Llŷn, it is widely but thinly distributed, being most numerous in Snowdonia. It is well distributed in Denbighshire and Flintshire. where there is much suitable habitat: on 861 ha of Moel Famau Country Park 14 pairs were found breeding in 1986 (RSPB 1986).

The spring arrival of breeding birds from wintering areas in tropical Africa merges with the passage of others moving farther north. This spring passage is difficult to detect and even on the offshore bird observatories it is poorly marked. A daily maximum of 20 birds (only once achieved) is the most recorded on Bardsey, where spring passage is described as light. On the Pembrokeshire islands very few birds are seen each spring. Passage on the south Wales coast has been detected up to the end of May, for example the presence of resting birds on Flatholm. Passage is better marked after the breeding season than it is in spring but still fairly modest: double the spring numbers occur on Bardsey (120 on one August day was exceptional) and the Pembrokeshire islands produce only a handful of August and September sightings. Coasting movements on the north shore of the Severn estuary can be quite impressive in some years, with up to 70 birds a day passing Lavernock Point and 50 at Ogmore (22 August 1988). Late records occur up to October: a very late inland record was of a bird at Abergwesyn (Breconshire) on 24 October 1987.

Meadow Pipit *Anthus pratensis*

Corhedydd y Waun

A widespread and numerous breeding species, especially on moorland. Winter visitor and passage migrant. Most breeding birds are summer visitors.

The Meadow Pipit is one of the most widespread breeding species in Wales. The *New Atlas* shows an almost ubiquitous distribution, the only exceptions being in much of Anglesey, parts of the Severn Valley, south Cardiganshire and much of central Pembrokeshire. The species is certainly abundant in many hill areas but a distribution map by 10 km^2 disguises the fact that there are large tracts where Meadow Pipits are now thin on the ground. Upland afforestation (currently 25% of upland Wales) holds little attraction for Meadow Pipits once the trees grow and exclude the rough grassland below; neither do the improved and reseeded pastures in the uplands – usually reclaimed from moorland – provide habitat for them when these are tightly grazed and provide no cover for breeding. Such an effect is graphically demonstrated on the recently produced tetrad maps of Pembrokeshire and Monmouthshire where breeding is virtually absent from the intensive dairy and sheep-rearing areas and restricted to those parts of the counties which still have rough grazings. The graph on page 27 illustrates very clearly the decline in rough grazings on which the Meadow Pipit is dependent. For these reasons the Meadow Pipit, for all its localised abundance, is a much less numerous species in Wales than it was in the earlier decades of the century. Its population may well be less than 50% of its earlier level. It is essentially an open-country species, often sharing these habitats with the Skylark, but usually occurring on its own at the highest altitudes; Meadow Pipits prefer denser and longer vegetation whereas Skylarks usually show preference for more open ground with shorter vegetation. Above 400 m or so it is invariably the most numerous species on open ground; in such habitats it often forms a principal prey item of Merlin, Hen Harrier and sometimes Kestrel. Its nests are extensively parasitised by Cuckoos, for which it is the most important host species, not only on the open moors but also on ffridd slopes bordering the moorland where it is frequently an abundant constituent of the bird community and overlaps with the Tree Pipit. Meadow Pipits nest from sea level to altitudes as high as 915 m on montane grasslands in Snowdonia and to the summits of other north Wales mountain ranges. Their main breeding habitat requirements are wide open spaces with rough grassland, heath or marram. Rose (1982) showed that the main breeding habitats are moorland (35%), rough grazing (25%), lowland heath (13%) and saltmarsh/dunes (13%). In Wales they also take advantage of new forestry plantations at their early stages of development when rank grasses predominate but they are decidedly rare on the same sites when these are clear-felled and restocked (Bibby *et al.* 1985). They are much less common on enclosed field systems and show a marked reluctance to exploit leys, improved pastures or hay/silage fields.

The highest densities of breeding Meadow Pipits are to

be found on montane grasslands and heather moor. Seel and Walton (1979), in a four-year study, found densities which varied annually between 33 and 74 pairs per km^2 over four years (average 48 pairs per km^2) on upland sheepwalk in Snowdonia. On maritime grassland on Skokholm (96 ha) populations have been measured at as many as 58 pairs per km^2 in the 1950s but more recently falling to fewer than 20 pairs per km^2.

In a mixture of lowland habitats and moorland in Monmouthshire it is estimated that there is an average of c20 pairs per tetrad (= 5 pairs per km^2) to give a total for the county of some 4000 pairs. The only other comparable figure is a similar total suggested for Pembrokeshire (Donovan & Rees, in press).

At the end of the summer Meadow Pipits abandon the hills and do not return until the following March or April. Lack (1986) has suggested that up to 80% of British Meadow Pipits emigrate in autumn and that the residue is supplemented in winter by an enormous immigration from the north. Ringing recoveries of Welsh birds confirm an exodus to France and Iberia (38 recoveries). In autumn there is a very well marked passage of birds southwards through inland and, particularly, coastal areas. Flocks such as the 1000 seen near Claerwen Reservoir in October or 4000 over Llangynidr (both Breconshire) in September are unusual but not rare. At coastal stations this autumn emigration, sometimes confused by the arrival of passage birds, is equally well defined. Peak counts on Bardsey occasionally reach as many as 1500+ per day and flocks of several hundreds in September and early November are regular. On Skokholm and Skomer daily peaks of over 100 are regular, with maxima of 800–1000 reached occasionally. Along the Severn estuary coast the visible autumn movement is also sizeable. It is stated that up 300 birds per hour are not unusual on many days in September and October on the Monmouthshire shore, while 600 per day at Lavernock Point are frequent and a maximum of 800 has been recorded in October on Flatholm. The return spring passage is smaller; maxima on the Pembrokeshire islands reach 100–150 per day on occasion.

In winter Meadow Pipits are widely distributed through the lowland areas of Wales, usually in gatherings of fewer than 100, foraging on coastal marshes and dunes, ploughed fields, root crops, pasture and wetland margins. The largest concentrations are in the lowlands of Glamorgan and Monmouthshire, and on Anglesey and Llŷn.

Red-throated Pipit *Anthus cervinus*

Corhedydd Gyddfgoch

A vagrant to Wales.

It breeds in extreme northern Eurasia, from Norway to north-eastern Siberia.

There are three records from Wales:

13 October 1970 – One, Skokholm
24–26 October 1988 – One, Bardsey
19–23 September 1989 – One, Skokholm

Rock Pipit *Anthus petrosus*

Corhedydd y Graig

A breeding resident of rocky coastlines in all coastal counties except Flintshire and Monmouthshire. Winter visitor in small numbers.

The Rock Pipit/Water Pipit complex is a difficult one and it was only as recently as 1986 that the two were separated into distinct species. Added to this, the Scandinavian race of Rock Pipit (*A. p. littoralis*) occurs as a winter visitor to the British coast, albeit only rarely in Wales so far as is currently known. Accordingly there is a degree of confusion about some of the earlier records of wintering birds. However, there is no confusion about the situation in the breeding season as the Water Pipit (q.v.) breeds far away from British shores. Despite this, the Rock Pipit is a species which has been largely neglected and precise information about numbers is lacking in all but one or two locations, notably the island bird observatories.

The Rock Pipit is the most exclusively littoral of all our birds, spending virtually its entire existence in the salt-spray zone almost within touching distance of the high tide line. Its breeding distribution around the Welsh coast is strongly correlated with the presence of rocky shores, and consequently it is absent as a breeding species only from Monmouthshire and Flintshire; even in the latter area it breeds sporadically a mile or so offshore where a rocky environment is offered by Hilbre Island (Wirral). Otherwise in these two counties the Rock Pipit is a regular winter visitor in small numbers. In other Welsh counties the species is a common inhabitant of the coastline, occupying the rocky stretches and spurning such low, sandy coastlines as Morfa Harlech, Morfa Dyffryn (Merioneth) and Pendine and Pembrey Saltings (Carmarthenshire). Artificial "rocky" sites such as that provided by the causeway at Porthmadog (Caernarfonshire) will also induce the species to colonise otherwise unsuitable low coasts. It is an inhabitant of virtually all offshore islands including Grassholm, some 14 km distant, where six to eight pairs breed on the 9-ha site.

So far as is known Rock Pipit numbers have always remained much as they are at present; no long-term changes in distribution or status are recorded. In Wales the breeding population appears to be strongly sedentary. More than 2000 Rock Pipits have been ringed on Skokholm over the years but none of these has produced a long-distance migrant (Evans 1966). In fact, no Welsh Rock Pipit from other sites is known to have moved as far as 10 km from its place of ringing (total 15 recoveries). On Skokholm up to 41 pairs breed (1990) at a density of just over 10 pairs per mile of coast (see table), a figure which is slightly higher than that quoted by Gibb (1956) for the same island and is at the highest end of other density figures so far produced. Skomer, with its slightly longer coastline, supports up to 45 pairs, and Ramsey 30+ pairs. On Bardsey, a steady population of 40–50 pairs up to 1970 reportedly fell to 20–40 pairs thereafter (only 13 pairs were present in 1963 after the severe winter), for reasons which are not indicated; by 1989 a figure of 45–55 pairs was given and a precise count in 1990 produced 56 pairs. The only mainland density figures produced are those for Gower where Thomas (in litt.) estimates about 30 pairs on the rocky sections of the south Gower coast from Mumbles to Rhossili.

Density of breeding Rock Pipits on rocky coastlines in Wales

Coastline	*Rocky coast length (km)*	*Pairs*	*Density (pairs per km)*
Skokholm (1992)	5.9	53	11.1
Skomer (1990)	9.9	c45 (estimate)	4.5
Grassholm (1978)	1.8	6–8	3.3–4.4
Bardsey (1990)	8.2	c55	c6.7
South Gower (1992) (cliff sections)	23.0	30+ (estimate)	1+
Flatholm (1991)	2.0	11	c5.5
Ramsey (1992)	9.8	30+	3.1

The record of a pair of Rock Pipits nesting in Radnorshire in 1902 is extremely difficult to accredit, albeit the observer, Walpole-Bond, was a well-respected ornithologist. Although the record has been reprinted on many occasions, Forrest (1907) dismissed it and referred to the voluminous correspondence which ensued in *The Field* (1902) on the correct identification and interpretation of occasional pipits of dusky plumage found on the Welsh hills.

In winter Rock Pipits are found in some of the nearby estuarine and low coasts which they shun at other times of year. Many of these are juvenile birds from the Welsh breeding population, displaced by the territorial adults, but they are also supplemented by immigrants from farther north. A census on the Dyfi estuary in November 1984 revealed 107 birds, with a further 51 on the sandy foreshore at Ynyslas nearby. Observed autumn passage is very light on Bardsey – although there was a count of 70 birds on one beach in October 1988 – whereas on Skokholm sizeable influxes of up to 200 birds have been recorded in mid-October. Specific identification of birds of the Scandinavian race have been confirmed occasionally in autumn and winter, e.g. Pembrokeshire, December 1981 and November 1990 but otherwise these northern birds are slightly more numerous on spring passage. There is one ringing recovery of a young bird from Kalmar (Sweden), found at Cable Bay in Anglesey the same autumn.

Inland records of Rock Pipits are rare, although Peers and Shrubb (1990) list 18 records in Breconshire since 1969, mostly at Llangorse Lake between October and March.

Water Pipit *Anthus spinoletta*

Corhedydd y Dŵr

A winter visitor in small numbers, mainly recorded in coastal areas but also, less frequently in Wales, inland. Passage migrant in small numbers.

Unlike the closely related shore-dwelling Rock Pipit, the Water Pipit breeds far inland in the mountains of central Europe and migrates to Britain and the countries bordering the western seaboard of Europe in autumn. In winter plumage there is similarity between this species and the Scandinavian race of the Rock Pipit (and more so in spring), with the result that some records in the past have doubtless been confused. The Water Pipit has long been recognised as a winter visitor to parts of Wales; Caton Haig shot three on the marshes at Porthmadog at the end of the last century, a place far removed, as Forrest commented, from other known locations. Had he known then what we now know about the species' winter distribution, he would have been less surprised; there seems every reason to suppose that the numbers and distribution of this bird in winter have changed little over the years. It is simply that, like several other difficult species, it is better recognised now by a far greater number of competent birdwatchers but is still doubtless under-recorded. However, Water Pipits seldom occur in the same winter habitats as Rock Pipits, preferring freshwater margins, estuaries and river banks. The only habitat in which the two regularly overlap is *Spartina* marshes.

County bird reports include a thin scattering of records through the 1960s and 1970s, principal among which are regular occurrences on the *Spartina* beds at Peterstone Wentlooge (Glamorgan). The number of sightings has increased dramatically in Wales in the 1980s and early 1990s as more competent birdwatchers have specifically looked for the birds. It was only in 1986 that Rock Pipit and Water Pipit were afforded independent specific status. The confirmed records of Water Pipit sightings in Wales since then (1986–91) are summarised in the tables below.

Total of published records of Water Pipit 1986–90 (from county bird reports)

County	*Number of records*
Anglesey	c15
Caernarfonshire	7
Denbighshire	0
Flintshire (1980–90 only)	3
Merioneth	0
Montgomeryshire	0
Cardiganshire	2
Radnorshire	0
Breconshire	0
Monmouthshire	c40
Glamorgan	20
Carmarthenshire	3
Pembrokeshire	12

Monthly distribution of published Water Pipit records in Wales 1986–90 inclusive

Jan	*Feb*	*Mar*	*Apr*	*May*	*Oct*	*Nov*	*Dec*
10	11	35	21	4	6	10	5

Yellow Wagtail *Motacilla flava*

Siglen Felen

A summer visitor, now breeding regularly in only six counties.

Belonging to a complex group of some 14 geographically identifiable subspecies, *M. f. flavissima*, the Yellow Wagtail, breeds only in southern Britain and the immediately adjacent coastal areas on the mainland side of the English Channel and southern North Sea. The Blue-headed Wagtail, *M. f. flavissima* breeds throughout western and central Europe and occasional pairs or individuals are found in the British breeding population. Other races are recorded only as vagrants:

M. f. thunbergi, the Grey-headed Wagtail from Scandinavia and northern Russia;
M. f. cinerocapilla, the Ashy-headed Wagtail from Italy, the Adriatic and Tunisia;
M. f. feldegg, the Black-headed Wagtail from Balkans, Turkey and the Middle East.

Almost all records of these vagrants which occur in Wales are on spring passage in May. (Individuals of these races are much more easily overlooked in autumn plumages.)

The Yellow Wagtail is a summer bird of water meadows, coastal saltmarshes, marshy fields and occasionally other lowland waterside sites such as reservoir edges (Llandegfedd). Some pairs in south-east Wales nest in arable fields. In Britain it is principally confined to England – but absent from the south-west peninsula and the chalk hill areas – with relatively tenuous outposts in Scotland and Wales. As it shuns most uplands (it has occasionally occupied flat wet meadows up to c360 m, e.g. Elan Valley (1990) and Llyn Tegid) and also avoids areas bordering the western seaboard, its opportunities in Wales have always been limited. At present its regular breeding distribution in the Principality is restricted to the counties of Denbighshire, Flintshire, Montgomeryshire, Breconshire, Radnorshire and Monmouthshire. The total estimated populations in these six counties is given in the following table.

Probable breeding totals of Yellow Wagtails in Wales, 1989–92

County	*Pairs*	*Reference*
Denbighshire	5–10	J. M. Harrop *et al.* (pers. comm.)
Flintshire	c40	C. Wells (pers. comm.)
Montgomeryshire	max. 40	Pers. obs.
Breconshire	35–40	Peers & Shrubb (1990)
Radnorshire	c10	M. Peers (pers. comm.)
Monmouthshire	max. 50	S. J. Tyler (pers. comm.)
Total	180–190	

It is not believed that there were any major changes in status in Britain during the 19th and the majority of the 20th century, although it is clear that locally the species did decrease to some extent in Wales in the first half of the present century; for example, some of the traditional nesting areas in Glamorgan – Margam and Aberavon moors, Llanishen Marsh – were taken for building development. From the middle years of the century the decrease in range in Wales has accelerated as can be seen on the accompanying map and although the full reasons are not clear, change of land use and especially the draining and ploughing of meadowlands and the reduction of cattle and increase of intensive sheep grazing are doubtless implicated. This decrease has occurred gradually and not necessarily evenly as populations tended to hold up in the 1970s when national levels were higher than usual. The main downturn was in the 1980s, which was a period of further agricultural intensification in Wales, particularly for sheep, although the losses of some breeding sites was clearly not for that cause (Cefn Mably (Glamorgan), Penclacwydd (Carmarthenshire)).

The Yellow Wagtail's main stronghold in Wales has always been in the south-east, in Monmouthshire and the coastal belt of Glamorgan as far west as Swansea – where Dilwyn listed it in 1849. The withdrawal from most of the Glamorgan sites took place slowly but inexorably from the early decades of the century. The most recent breeding has been very sporadic, even at traditional sites and none is now regular. In Monmouthshire its main population is on the Gwent Levels and in the Usk Valley and its tributaries. At the time of the *Gwent Atlas of Breeding Birds* (1987) the population was estimated at upwards of 200 pairs, although S. J. Tyler (pers. comm.) subsequently believed that the total was probably no more than 100 pairs. Since then the estimate has been revised downwards to a maximum of 50 pairs. Peers and Shrubb suggest a maximum of 40 pairs in Breconshire. The Radnorshire population, never more than about 12–20 pairs, is now put at the bottom end of this estimate, with a few pairs near Boughrood and in the Teme Valley near Knighton (M. Peers, in litt.).

In Montgomeryshire the population historically is not known (strangely Forrest did not even record it from the county) but in the 1960s and 1970s there were impressive numbers in the Severn Valley and its tributaries. RSPB surveys in 1978 suggested 126 occupied territories in the Severn system and 43 in the Vyrnwy Valley. In those two decades pairs were found well into the foothills as far as Caersws and Llandinam in the Severn Valley and Dolanog in the Vyrnwy Valley. One pair was even found on the flanks of the Berwyn Mountains at Llanrhaeadr-ym-Mochnant. Today, although no recent survey has been undertaken, the population may be as low as 40 pairs. Farther west in Cardiganshire it appears that there was a thriving breeding nucleus on Cors Caron and other pairs on Cors Fochno and in wet meadows on the south-side of the Dyfi estuary in the vicinity of what is now the RSPB's Ynys-hir Reserve (Salter, Diaries); these colonies had gone by 1977 at the latest, but one pair nested again in the county in 1981. In Carmarthenshire, where breeding was once fairly widespread on the coastal lowlands, the residual breeding pairs were on the east side of the Tywi Valley; previously breeding had been recorded as far north as Llandovery and up the Gwili Valley as far as Cynwyl Elfed. The last breeding was at Penclacwydd,

Machynys in 1987. In north-east Wales the main concentration of Yellow Wagtails has long been on the low-lying agricultural land around the Deeside Industrial Estate at the head of the Dee estuary. A small population still exists on these meadows, gravely threatened by continuing development. In the early 1990s the population was no more than 40 pairs and declining. Outlying pairs have been found in the past in the Gronant area (Flintshire), St. Asaph, the Clwyd Valley and near Nant y Ffridd (Denbighshire). At present no more than 50 pairs nest in the two counties.

Old records refer to occasional breeding in the lower Conwy Valley (1892) in Caernarfonshire. A single pair bred at Cemlyn Lagoon on Anglesey in 1986 and 1987, the first proved breeding record for the island. Sporadic breeding has also occurred in Pembrokeshire, e.g. 1977 and possibly 1983.

Decline of breeding Yellow Wagtails in Wales – ○ prior to 1968, Δ 1968–87, ● current breeding, early 1990s

The Yellow Wagtail, which is a long-distance migrant, is the only one of the family regularly found breeding in Britain. It winters south of the Sahara on the savannahs of tropical and equatorial Africa. Passage birds begin to arrive in the southern counties of Wales in the first week of April but usually not until the middle of the month in north Wales; the main arrivals are in late April and early May. This land-fall is well marked in Glamorgan and Monmouthshire but, farther west, is represented by relatively few birds on the well-watched offshore islands. A few records have been made as far offshore as Grassholm. The return movement begins as early as July and peaks at the end of August. Again it is a light passage in north Wales and the counties bordering Cardigan Bay but notable for the build-up of numbers in the coastal stretches of Monmouthshire and Glamorgan. The size of night-time roosts gives some indication of the numbers involved: 150 (maximum 190) are not uncommon on the Levels at Peterstone Wentlooge with other counts during the 1980s of 60 at Kenfig, 53 at St. Athan airfield, 60 at Aberthaw and 50 at Lavernock (all Glamorgan). By contrast the record of 22 sightings at 11 locations in Anglesey in autumn 1986 was regarded as "exceptional". It is interesting that the size of the south Wales gatherings has not noticeably diminished during the 1980s, at a time when breeding numbers have fallen considerably. In Monmouthshire, dawn counts of passage birds on 14 days in August/September 1981 gave a total of 711 birds, moving westwards on Peterstone Wentlooge; when the count was repeated in 1989, 906 birds were counted in 11 days. On the Glamorgan coast recent maximum daily counts at the end of August have reached 170 at Kenfig, 150 at Ogmore (both 1990) and 166 in 1½ hours at the Rhymney estuary (1989). In mid-Wales roost sizes have certainly declined since the 1960s when c60 birds were regular at sites such as Newtown sewage farm, where they are now infrequent. The long-standing Llangorse roost site now totals about 50 birds at maximum and similar numbers are to be found at Glasbury in the Wye Valley (maximum count 120 in early 1970s). Several recoveries of Welsh-ringed birds illustrate the southward autumn movement. Individuals from the breeding population in Flintshire have been recovered in France, Portugal and Spain and birds from the Llangorse Lake roost site have reappeared in Portugal and Morocco (and one in north Yorkshire three years later).

The last autumn birds have usually gone by the end of September but stragglers are occasionally recorded up to the first week or so of October. Winter records in Britain are very rare and two birds seen in Anglesey on 4 November 1982 must be presumed to be extremely late migrants.

The nominate continental form, the Blue-headed Wagtail, *M. f. flava*, has been recorded on many occasions in Wales – 30 records involving at least 36 individuals between 1978 and 1987. This is mainly at times of peak spring passage in early May and records have occurred in all counties. Occasional pairs breed, sometimes in a succession of years, e.g. Montgomeryshire 1968–71, probably involving the same individuals. Records of interbreeding with the British race also exist.

The race *M. f. cinerocapilla*, Ashy-headed Wagtail, is a rare wanderer from the Adriatic. Three autumn birds were recorded by N. F. Ticehurst on Bardsey in autumn 1913, the first Welsh records; another was identified there in September 1956 and yet another in April 1984 and one was present between 17 and 20 May 1992 on Skomer. One in Glamorgan on 3 May 1981 is the only record so far from mainland Wales.

The Grey-headed northern European race, *M. f. thunbergi* has also been recorded on a handful of occasions, invariably at well-watched coastal locations such as Bardsey (2), Little Orme (1), Skokholm (5), Strumble Head (1), always in May, when off course on northward migration.

One record of the Spanish race, *M. f. ibericae*, has been made, on Skokholm on 19 April 1989.

The Black-headed Wagtail, *M. f. feldegg* from the

Balkans and Middle East is a vagrant in western Europe which has been recorded twice in Wales:

8 May 1976 – Adult male on Bardsey
7 May 1986 – Adult male on Skomer

Grey Wagtail *Motacilla cinerea*

Siglen Lwyd

Locally common breeding resident, mainly along fast-flowing streams in all mainland counties of Wales, but very scarce on Anglesey. Most move away from hills in the autumn and either emigrate or frequent sites on lower ground.

Although the Grey Wagtail is to be found throughout all the counties of mainland Wales, its local distribution is governed by the availability of the fast-flowing, rocky streams which form the Wagtail's principal breeding-season habitat. This predetermines that its main stronghold is in the hill areas where it is frequently found alongside the Dipper, although its habitat preference is less restricted than that species and it is also found in a wider range of waterside and other sites, e.g. stream outflows from lakes, disused quarries. Neither is it restricted exclusively to upland sites for breeding but will fairly readily exploit lowland watercourses if these provide such features as weirs, mill races or canal locks and culverts. In west Glamorgan, for example, which boasts less upland than most other areas of Wales, Grey Wagtails are found in 117 tetrads (50%) out of a total of 234 during the breeding season, including 12 tetrads on Gower with breeding proved in three (Thomas *et al.* 1992). They breed regularly on lowland sections of rivers such as the Conwy, Teifi, Wye, Tywi and Severn (downstream of Newtown). One of their principal requirements is the presence of plenty of deciduous bankside trees which guarantees an ample supply of insect food, to supplement that of aquatic origin (Ormerod & Tyler 1987a). Accordingly, at a local level, pairs are invariably absent from those stream sections bare of trees.

No long-term changes in status are known to have occurred since records were first made. Forrest (1907) recognised it as common in all the north Wales counties except Anglesey (where it has always been rare) and Llŷn (where RSPB surveys in 1986–87 located it in only nine one-kilometre squares). Writers of the same era at the beginning of the century reflected a similar status for the species in Glamorgan, Breconshire, Pembrokeshire, Carmarthenshire and Monmouthshire. In Pembrokeshire, however, Mathew's (1894) claim that it was abundant may have been misleading on the evidence of its more recent distribution, where it is shown to be very patchy. In common with a number of other passerine species the population of Grey Wagtails appears to have been at high level following the succession of mild winters in the 1970s. Since the 1978–79 winter and, later, those of 1981–82 and 1984–85, numbers have fallen somewhat, emphasising the effect which hard winters have on the species. Ormerod and Tyler (1987b) have shown that Grey Wagtails, with a less specialised diet than Dippers, are unaffected by acidified streams even when these are grossly polluted.

Densities of Grey Wagtails on some Welsh streams are sometimes impressive especially in those years following a succession of mild winters. The *Breeding Atlas* gives densities of 114 pairs per 100 km of watercourse in north Wales – a higher figure than for elsewhere in England or Wales. RSPB surveys on the rivers Wye (1976), Severn and Vyrnwy (1978) found densities of 74 pairs in 223 km, 31 pairs in 107 km and 28 pairs in 67 km respectively.

On the river Irfon (Breconshire) – a severely acidified stream – 19 pairs were found (1990) on a 20 km stretch. On the river Ogwen the population has been measured (1970s) at between one and five pairs per km. Donovan and Rees (in press) suggest a total of c450 pairs in Pembrokeshire, Peers and Shrubb c490 for Breconshire and Tyler *et al.* up to 800 pairs for Monmouthshire, which last figure we believe may be a generous estimate.

Breeding distribution throughout the Welsh counties is confirmed by the *New Atlas*, although its scale tends to give a false impression of the evenness of distribution. For example, only 24% of tetrads (mainly upland and inland areas) are occupied in Pembrokeshire, 55% in Monmouthshire, 44% in Breconshire and 50% in west Glamorgan. The pattern of occurrence is strongly related to river systems rather than the apparently even distribution suggested by the Atlas. In the hills, Grey Wagtails occur up to altitudes as high as 490 m. It is only in western Llŷn and Anglesey that it is a scarce species, breeding regularly in only three or four sites on the island. Although it can be argued that Anglesey is a lowland area, this is too simplified a reason; a number of other passerine species, numerous on the mainland opposite, e.g. Redstart, Pied Flycatcher, Whinchat, are virtually absent from the island also. In the case of the Grey Wagtail, however, it is clearly a lack of reasonable gradient on Anglesey streams which fails to provide the type of water course necessary during the breeding season.

Most Grey Wagtails move away from the uplands in autumn and either emigrate (perhaps more so when numbers are high) or resort to lower ground, where they can be found around farms, slurry pits, sewage works, lowland water courses and coastal marshes. At this season the distribution becomes almost the obverse of the breeding season distribution, with maximum numbers in the south Wales lowlands, Anglesey, lowland Caernarfonshire and Flintshire, and the broad river valleys of the Marches. The winter population is almost certainly supplemented by immigrants from the continent.

It is known that some British Grey Wagtails emigrate to France or Iberia in autumn but to date only one recovery of a Welsh bird has occurred across the English Channel: a nestling ringed at Tintern in Monmouthshire was recovered in northern France the following winter. All other recoveries of Welsh-ringed birds (all nestlings or juveniles) outside Wales have come from southern England indicating the positive southward movement of many young birds in autumn. One bird ringed in the Channel Is. in October was found dead in Monmouthshire nine days later. Birds from Scotland are known to move south and south-west in autumn to winter in places as distant as Devon and Ireland; doubtless many of these birds pass through Wales.

Pied Wagtail/White Wagtail *Motacilla alba*
Siglen Fraith

A common and widespread breeding resident in many habitats. Autumn passage migrant and winter visitor. The continental race, White Wagtail, M. a. alba, *is a regular passage migrant, occasionally breeding.*

The British race, *M. a. yarrelli*, the Pied Wagtail, is a widespread and common resident throughout Wales. The "grey-backed" continental and Icelandic race, *M. a. alba*, the White Wagtail, is a regular spring and autumn passage migrant occurring widely in Wales in late spring and autumn, notably on coasts. Pied Wagtails exploit a very wide range of habitats, not necessarily associated with watersides, despite popular belief; nonetheless it is likely that nesting territories in the proximity of water are the preferred sites. Otherwise pairs will breed in cavities and recesses in urban centres, working (or disused) quarries, sand dunes, sea shore, roadsides, farms, industrial plants and even derelict tanks used as targets on firing ranges in Breconshire! In upland areas stone walls, either as field boundaries or roadside features, offer further opportunities for nest sites. It is only areas of woodland and wide open countryside such as moorland, saltmarsh and the smaller offshore islands which are unattractive to Pied Wagtails. Altitude is not a particular bar to the species where suitable nesting sites and feeding opportunities exist and pairs can be found in several counties up to at least 500 m.

There is no evidence that the status of Pied Wagtails has changed substantially over the past two centuries or so. The first reliable written records of birds in Wales did not start appearing until the middle of the 19th century and then became a modest flood around the turn of the century; all these records testify to the common status and widespread distribution of the species. In recent times it has become well appreciated that the Pied Wagtail, as a resident insectivorous bird, is vulnerable to severe and protracted winter weather. It is thought that the population may have been reduced by as much as two-thirds in the harsh winter of 1962–63. Further drops in population attributable to the same reason can be seen in the CBC indices after the 1978–79, 1981–82 and 1985– 86 winters (Marchant *et al.* 1990). Winter conditions are evidently the main controlling factor on population levels. In between these harsh winters it is clear that, like other resident insectivores, the Pied Wagtail population rose to artificially high levels in the 1970s after a longish succession of mild winters; a result of this was an expansion of breeding pairs into drier, sub-optimal sites which are the first to be abandoned once the population recedes again. Breeding occurs regularly on the larger offshore islands: Ramsey (1–2 pairs), Skokholm (1), Skomer (1–2), Caldey (1), Bardsey (2–3). Numbers appear to have become fairly stable at a slightly lower level in the late 1980s and early 1990s.

The *Breeding Atlas* suggests an overall figure of 1.4 pairs per km^2 in the 1970s but it is not known how accurately this relates to Wales. Peers and Shrubb (1990) have suggested a population of c1250 pairs for Breconshire where 62% of tetrads are occupied; the *Gwent Atlas* estimates c2000 pairs (89% occupancy) for Monmouthshire and the population in Pembrokeshire is put at 1400–1700 pairs (Donovan & Rees, in press). In surveys of three major Welsh rivers in 1976 and 1978 the RSPB found 243 pairs of Pied Wagtails on 223 km of the Wye, 114 on 137 km of the Severn and 45 pairs on 67 km of the Vyrnwy.

After the breeding season there is a defined movement of many Pied Wagtails away from the breeding areas, especially those at higher altitudes. This shift in the population becomes confused by the passage of other birds from breeding areas further north in Britain (*Winter Atlas*) and the arrival of White Wagtails also from the north (Iceland and northern Europe). The passage of these Wagtails occurs on a broad front and can be clearly identified as numbers vary from day to day in September and October on favoured sites, especially mown grassland such as parks, sports grounds, school playing fields as well as coastal headlands and offshore islands. These shifting movements sometimes produce large concentrations, e.g. 160 at Builth Wells (Radnorshire) on 10 October 1983, 150 Newtown High School (Montgomeryshire) on 21 September 1968 and 182 at Newtown on 11 October 1991. Communal roosts established in autumn in reedbeds, willow shrub, on buildings or elsewhere often carry over through the winter and not infrequently number several hundred birds. Sites at Oxwich NNR, Llangorse Lake, Carmarthen County Hall (c500 in November 1979) and many others are regular. One at Newport (Monmouthshire) in 1990 contained about 400 birds but the largest counted roost in Wales has been on a factory roof at the Treforest Industrial Estate in the Taff Valley (Glamorgan). Here a maximum of 857 birds was counted going to roost in February 1970 which had increased to 1040 by December 1972, reducing to 500 nine years later and then disappearing by 1982.

The number of Pied Wagtails in Wales in winter is lower than the breeding season total and fewer than those remaining in the proximal lowland areas of west and south-west England (*Winter Atlas*). Some birds of Welsh origin clearly emigrate, especially when weather is particularly hard; most emigrants are juvenile birds although the proportion of adults moving out doubtless increases when conditions are notably severe. This exodus is evidenced by several ringing recoveries of Welsh breeding birds and nestlings from France, Spain and Portugal. In addition, other autumn-ringed birds which are not specified as being either of the race *yarelli* or *alba* have been recovered as far south as Morocco.

It has been clearly recognised since the late 19th century that the distinctive race of White Wagtail *M. a. alba* passes through Wales on spring and autumn migration, most individuals presumably being of Icelandic stock but doubtless including some from Scandinavia. Spring passage centres on late April and early May and the return movement peaking in September when 50–100 per day (exceptionally up to 250–300) are not uncommon on Bardsey (not necessarily all *M. a. alba*). The passage of these White Wagtails is readily visible at certain waterside sites in all counties and is not confined to a coastal movement, strong though this is. Numbers are sometimes quite large, e.g. 100 at Aberdyfi on 24 April 1932; 150 at Trawsfynydd on 26 April 1968 (both Merioneth). Occasionally birds showing the characteristics of this race have been recorded as breeding: a pair apparently did so successfully at Kenfig Dunes early this century and a recently fledged family with adults still feeding young was seen near Criccieth (Caernarfonshire) in 1903; it is reputed to have nested at Little Orme in 1926; a female of this race bred successfully with a "normal" male Pied Wagtail at Penrhyndeudraeth (Merioneth) in 1970.

Waxwing *Bombycilla garrulus*

Cynffon Sidan

An irruptive winter visitor from northern Scandinavia and north-east Europe; occurs erratically in Wales in small numbers, most usually in years of major irruption in eastern England.

The Waxwing is a breeding bird of the great boreal forests of Scandinavia and northern Russia. It forsakes the most northerly parts of its breeding range in autumn to feed on a winter diet of berries, especially rowan, mainly in the countries bordering the Baltic Sea and in eastern Europe. Britain, and especially Wales, thus lies far to the west of the normal winter range. However, the Waxwing is a celebrated "irruptive" species which is forced to move far from its preferred wintering areas in years when the continental rowan crop has failed or when breeding numbers have been particularly high; in such years Britain can sometimes enjoy large-scale invasions of this beautiful and colourful bird although, being on the west side of the country, Wales usually receives relatively fewer of the visitors than areas farther east.

"Waxwing years" have been well documented for a long time (quoted as far back as the 17th century – Fisher 1955). The hard winter of 1788–89 was one such and in this winter the first documented record of Waxwing in Wales occurred, a bird killed at Garthmeilio in Denbighshire, recorded by Thomas Pennant and which was still in existence as a mounted specimen in 1917. Nineteenth-century records are sparse but it can be assumed that this represents the lack of observers rather than a necessary absence of birds. Parties were seen at Roath (Cardiff) in 1859 and Waxwings are mentioned by Mathew in Cardiganshire pre-1865 and by Forrest near Corwen in 1898. Cambridge Phillips (1899) recorded one undated record from Llanwrtyd Wells (Breconshire) earlier in the century.

Cornwallis and Townsend (1961 and 1968) listed the principal Waxwing years this century, up to 1965–66: 1903–04, 1913–14, 1921–22, 1931–32, 1932–33, 1937, 1941–42, 1943–44, 1946–47, 1949–50, 1957–58, 1959–60, 1965–66. Waxwings were recorded in Wales in the majority of these years (11 out of 15) although numbers were never high and the birds occurred in very small parties or as individuals. In the major influx of 1946–47, when thousands of Waxwings moved out of northern Europe ahead of bitter late-winter weather, fewer than 30 individuals were recorded in Wales. Similarly, the next really large immigration, in 1965–66, produced widespread sightings across the Welsh counties but only small numbers of individuals. The largest group to be recorded in Wales was 52 at Llysfaen in Denbighshire during that winter. Since Cornwallis and Townsend's accounts, other good Waxwing years have occurred in eastern England in 1988–89 and 1990–91. The very large influx of birds which occurred in late autumn 1988 resulted in at least 100 birds being recorded in Wales in the succeeding winter, in all counties except Carmarthenshire, Merioneth and Pembrokeshire. This contrasts with a total of only 11 records in the previous decade (1978–87). The winter influx of 1990–91 produced only six Welsh records.

Outside the irruption years Waxwings have been recorded in twenty other winters, usually on the basis of single records although 1955–56 and 1961–62 each produced sightings from several counties.

The principal winter food of the Waxwing is rowan but, by the time the birds reach this country on their sporadic invasions, most of that crop has already been stripped by winter thrushes and other birds, as has much of the hedgerow hawthorn crop. Waxwings are often obliged to feed in gardens and other suburban areas where their bright colours make them readily noticeable. They are then often pleasantly confiding and approachable on bushes of cotoneaster, berberis or other cultivated berry shrubs. There is little geographical pattern to their occurrence in Wales, records from lowland areas and the eastern half of the country being only slightly more frequent than those from the western counties. Upland areas are avoided.

Dipper *Cinclus cinclus*

Bronwen-y-Dŵr

Common resident of fast flowing streams, particularly in hill areas. In winter some individuals move into the lowlands and also make use of estuaries, rocky coasts and lake edges.

The Dipper is the most characteristic bird of the innumerable, boulder-strewn streams which drain the uplands of Wales. It is evenly distributed throughout the hill areas wherever such streams are found although its density varies dependent on such factors as gradient, availability of food and – more recently – the acidification of some watercourses. The Dipper's heartland in Wales, as is well shown in the *Breeding Atlas*, and the *New Atlas*, extends from the coalfield valleys of south Wales to the northern foothills of Snowdonia and from the steep streams running westwards into Cardigan Bay to the hills of the Welsh Marches. It shows a preference for tree-lined (but not conifer) watercourses where food supplies are likely to be richest, but will also occur in the hills on open streams as narrow as 1 m.

On the larger rivers Dippers occur downstream so far as there are shallows, weirs, gravel runs or rocky outcrops to produce a combination of riffles and deep pools with the clear, shallow water they require for their unique underwater feeding. Thus on the Dee they are to be found as far downstream as Chirk, on the Severn as far as Welshpool and on the Wye to tidal reaches at Tintern. In the west they are irregular below Llandeilo on the Tywi and on the Teifi usually absent below Cenarth. Even so occasional pairs may be found lower down these larger rivers, especially where fast-flowing side streams join the main river or where weirs, old mill walls or other structures produce broken water and provide potential nest sites.

Because of the nature of the waters which Dippers require there is a strong correlation with slope and altitude, most breeding pairs being found between 3 m and 650 m above sea level. Only a few pairs occur as high as 650 m with 75% of pairs occurring below 250 m, where the feeding is particularly suitable in streams on base-rich rocks, e.g. Heol Senni (Brecon Beacons), Black Mountains (Monmouthshire), although Forrest (1907) reported Dipper as having been found as high as 670 m on the Carneddau (Caernarfonshire).

There is little reason to suppose that, until recently, the status of the Dipper in Wales had changed substantially in historical times, although clearly local factors such as river works, impoundments and urbanisation have had the effect of making some river sections unsuitable. Dippers are fairly tolerant of human proximity, however, and many of the Welsh towns regularly have Dippers nesting on their riversides, e.g. Newtown and Llanidloes (Montgomeryshire), Tregaron (Cardiganshire), Monmouth (Monmouthshire). Earlier writers testify to the ubiquity and abundance of Dippers but there are two respects in which the more recent interference of man has evidently contributed to declines and, in some cases, local extinctions: pollution of watercourses in the industrialised valleys and, more recently, the acidification of streams draining on some of the heavily afforested catchments on underlying base-poor rocks in mid and north Wales.

The Dipper was regarded as common in south Glamorgan, even on streams within three or four miles of the centre of Cardiff (including Penylan stream through Roath) until the early years of this century. However, increasing industrial pollution of rivers such as the Taff, Tawe, Ebbw and Neath drove the species away from many areas from the 1920s onwards. Some stretches have been recolonised since the 1960s as water quality improved again.

Tyler and Ormerod in their extensive work on Dippers in mid-Wales in the 1980s established a clear chain of reactions between atmospheric pollutants, the exacerbating effect of upland conifer forests, increasing stream acidification with attendant aluminium concentration and finally the resultant paucity of aquatic invertebrates and a decline in Dippers. They showed that on some acidified rivers, e.g. Irfon (Breconshire), the population had declined by c80%.

The highest densities of Dippers in Wales are on the streams running off the old red sandstone rocks of the Black Mountains in Monmouthshire. Here Tyler has recorded densities up to one pair per 300–400 m on streams such as the Grwyne Fawr; in 1989 on the Grwyne Fawr and Grwyne Fechan systems (c24 km) there were 37 pairs. In north Wales Schofield (pers. comm.) found an average of 1.7 pairs per kilometre on an 11 km length of the river Ogwen. RSPB river surveys between 1976 and 1982 found the following mean densities (m.d.) of Dipper pairs per 10 km.

R. Severn	60 km	15 pairs	m.d. 2.5
R. Wye	80 km	42 pairs	m.d. 5.25
R. Vyrnwy	40 km	16 pairs	m.d. 4.0
R. Teifi	60 km	12 pairs	m.d. 2.0

Tyler and Ormerod quote an average of 9–10 pairs per 10 km on smaller rivers which have greater proportions of suitable Dipper stretches. On tributaries of the river Teifi, Dippers were found to be almost twice as numerous as on the main river. However, it is relevant to note that the Teifi tributaries survey was carried out in 1982, immediately after bitter weather in January, which is known to have had a serious effect on Dippers among other small resident species in south-west Wales. On this basis the "normal" population on these tributary streams may well be higher than indicated by the figure above. Tyler *et al.* (1987) have estimated 200–300 breeding pairs in Monmouthshire and Peers and Shrubb estimated the same sort of levels in Breconshire while in Pembrokeshire Donovan and Rees (in press) calculated between 70 and 140 pairs (1984–88). On these bases it is likely that the Welsh population is in the region of 2000–2500 pairs. Breeding is scarce on most of Anglesey (probably no more than one or two pairs regularly) and the species is absent from the Llŷn west of Pwllheli and Trevor (and only 15–18 sites on the "inland" part of the peninsula – RSPB surveys 1986–87). Elsewhere in Wales it is scarce or absent only in the lowlands of south Cardiganshire (e.g. lower Teifi catchment, south-west Pembrokeshire, most of Gower, the south of Glamorgan and the broad lowland valleys of the Welsh Marches. In some places where the streams run steeply down to the coast, e.g. Old Colwyn (Denbighshire), Talybont (Merioneth) Pwll (Carmarthen) and Blackpool Mill (Pembrokeshire) pairs breed almost to sea level.

Most Dippers are sedentary and remain faithful to their breeding streams through the year. The principal exceptions to this are the birds from the higher streams; they are most readily forced to lower ground in the winter, abandoning these upland areas, the extent and timing partly depending on the severity of the weather. Although they are essentially sedentary, and given that there is a degree of altitudinal migration along watercourses, some young Dippers clearly disperse from their natal areas in their first autumn. Of 113 recoveries of ringed birds 69 have moved less than 10 km and only one of the remaining 44 recoveries involved a movement of more than 50 km; the most distant recovery of a Welsh bird to date is 64 km from Glasbury (Radnorshire) to Pitching Green (Gloucestershire). Recoveries over 20 km have been mainly of juvenile females (Tyler *et al.* 1987).

Dippers are frequently found on lake and reservoir edges, rocky coasts and estuaries in winter, such individuals most usually being young birds. Only once has a Dipper been recorded on one of the offshore islands – an individual which was seen on Skomer in May 1946.

Severe and prolonged winter weather can have a serious effect on a sedentary species that relies entirely on the sub-aquatic stream bed for its feeding. Although Dippers are well known to be able to feed under ice so long as open stretches permit entry, they can be badly affected if the conditions are protracted. P. E. Davis (in litt.) records that Dippers were particularly hard hit in Cardiganshire in the winter of 1981–82, a Dipper sighting being "quite an event" in 1982 and some regular streams not having birds again until 1984. Other winters in which Dippers were severely hit by hard weather were 1947 and 1962–63.

Ormerod and Tyler (1990) have shown the widespread communal roosting habit of Dippers outside the breeding season, mainly under bridges, and birds are sometimes recorded roosting in porches, under eaves and similar unexpected sites by streamside dwellings.

Wren *Troglodytes troglodytes*

Dryw

Abundant resident.

The Wren is one of the most successful and numerous of Welsh birds, breeding in a wider variety of habitats than any other species. Although its preferred areas are woodland and watersides (Williamson 1969) it also occupies a wide range of other urban, rural and coastal habitats, including farmland, mountainside, ffridd, gardens and urban parkland, crags, quarries, conifer forest and offshore islands. Wrens breed to a higher altitudinal level than most other passerines in Wales, to at least 660 m in north Wales and on Pumlumon but will forage even higher and have been recorded from the summit of Cadair Idris at 893 m (Forrest 1907). Elsewhere on the lower mountains of central and south Wales they occur regularly up to 500 m in Breconshire, Radnorshire, Carmarthenshire and Cardiganshire. The *Breeding Atlas* and the *New Atlas* confirm Wrens as being ubiquitous throughout Wales, which pattern is reflected similarly in the *Winter Atlas*. Apart from its stronghold on the mainland, the Wren also breeds on many of the offshore islands including Puffin Is., Bardsey, Ramsey, Skomer, Skokholm, Caldey and has even been shown to breed on North Bishop and Carreg Rhoson – small islets off St. David's Head (Pembrokeshire). On Flatholm, in the Bristol Channel, 10–12 pairs nest.

No historical evidence exists to suggest that the status of this numerous small bird has ever been different from the present picture. What is well documented, however, is the extent to which the size of the breeding population is affected by the severity or otherwise of winter weather conditions. The Wren, as its scientific name implies, is a successful forager in places not exploited by other species and it will not only enter deep recesses in its hunt for spiders and hibernating small insects but can feed in the depths of bushes and vegetation under a thick mantle of snow. Despite this, its small size means that it suffers high rates of heat loss and a low capacity for storing winter fat so that high numbers of Wrens die when conditions are particularly severe or protracted. For example, the numbers of Wrens fell dramatically throughout Wales after the 1962–63 winter; even on Bardsey the number of breeding pairs dropped from 20 to 25 in 1962 to four in 1963; in Glamorgan no measurable recovery from a position of severe depletion was detected until 1965 and in other areas of Wales populations did not recover fully until 1967 or 1968. Other severe reductions took place after the hard winters of 1947 (Wrens "almost wiped out in Cardiganshire" (Ingram *et al.* 1966)), 1961–62, 1978–79, 1981–82 (numbers fell by 79% at Ynys-hir, Cardiganshire) and 1985–86. Recovery from these low points can be rapid and within 2–3 years populations have often returned to average levels, or above, although this is not always the case as was seen after the 1962–63 crash (see above). 1982 population levels in Monmouthshire were reduced by 60–70% on the previous year and were not back to normal before 1985. Conversely, after successive mild winters populations may reach abnormally high levels; for example, on one Monmouthshire CBC farmland plot with a mean population of 3.8 pairs per 10 ha over 12 years, the index had fluctuated between 0.6 pairs and 9.3 pairs per 10 ha in individual years. Tyler (1990) has shown that the mean of 3.8 is higher than that recorded on CBC plots elsewhere on mainland Britain and second only to Ireland (5.9 pairs per 10 ha). RSPB surveys on two unimproved farmland plots in Carmarthenshire in 1987 produced densities of 3.65 pairs per 10 ha and 1.5 pairs per 10 ha respectively. In forests in north Breconshire Shrubb (in litt.) found Wrens the third most numerous species, representing 12.5% of the total birds counted.

Apart from the extremes of population caused by winter weather, the Wren is normally one of the most numerous birds in Wales. In Breconshire it was found in 90% of tetrads sampled between 1988 and 1990 and in west Glamorgan 98.5% of tetrads (230). In Pembrokeshire it was recorded in 95.3% of tetrads (448) and in Monmouthshire 98% (379 tetrads) where it was estimated that an average of 160 pairs occurred in each tetrad after a mild winter to give an estimated total for the county of some 20 000–60 000 pairs. An estimate for Pembrokeshire is given as around 40 000 pairs.

Densities of breeding Wrens in selected woodland sites in Wales (pairs/10 ha)

Site	*Density*
1. Deciduous woodland (Cardiganshire)	21.8 pairs
2. 326 point counts, restocked conifer (N. Wales)	28.3 individuals
3. Sessile oakwood (Caernarfonshire)	10.7 pairs
4. Four pre-thicket conifer plantations (N. Wales)	1.0–14.0 pairs
5. Six 100-year-old conifer stands (N. Wales)	5.0–12.5 pairs
6. Mature redwood plantation (Montgomeryshire)	4.8 pairs

(Adapted from CBC Ynys-hir reserve (1); Bibby *et al.* 1985 (2); Gibbs and Wiggington (3); Currie and Bamford 1981 (4); Currie and Bamford 1982 (5); Williamson 1971 (6).)

O'Connor and Shrubb have shown that there is an east-west increase in numbers in Britain with the highest populations in Wales and the south-west, which echoes the results of the Monmouthshire CBC work. This pattern doubtless reflects both agricultural practices, small field sizes and the slightly milder weather experienced in the west.

Marchant (in *Winter Atlas*) estimates the British and Irish population to be in the order four to five million pairs in a "normal" year; the Welsh proportion of this total is probably in the region of 10% to give a population of around half a million pairs. Extrapolating from the median figure given in the *Gwent Atlas*, approximately the same total is arrived at.

Although the Wren is rightly regarded as principally a sedentary species, a limited amount of movement does take place after the breeding season. On the one hand there is an apparently random dispersal of young from areas in which they were reared, once territorial activity intensifies again in October. Of 112 recoveries of Wrens ringed in Wales, only four have moved more than 10 km. However, the occurrence of increased numbers of Wrens on the offshore islands in late summer and autumn gives weight to the evidence of some degree of dispersal or more defined southward movement (c.f. Hawthorn and Mead 1973). On Grassholm, Wrens appear regularly between late August and November with at least 15 present on 24 October 1974 and 25 on 27 September 1972. On Skokholm, where three–five pairs breed it is regarded mainly as an autumn and winter visitor, and ringing showed that some individual birds returned to the island in successive winters. Up to 80 individuals have been recorded in October. Two birds have been recovered in Wales in autumn/winter from ringing locations in England, one having travelled 147 km south-west from Stafford and the other 227 km south-west from Yorkshire. Peers and Shrubb (1990) and Hope Jones (1974) record movements from upland breeding areas to lower ground in winter, a trend which has also been noted in Carmarthenshire and several parts of north Wales.

Dunnock *Prunella modularis*

Llwyd y Gwrych

Abundant breeding resident. Passage migrant and possible winter visitor in small numbers.

The Dunnock is one of the most widespread and abundant birds in Wales, breeding throughout the country in a wide variety of habitats. Being a somewhat unspectacular and skulking species it is not as readily remarked upon as some of our other common residents although it is in fact one of the most numerous birds in Wales. The Dunnock exploits an extremely wide range of habitats including farmland, suburban and urban gardens, town parks, cemeteries, scrubland, woodland fringes, dune systems, rough cliff-top vegetation, ffridd and, in north Wales, even rank heather on open moorland – in which respect comparison with the Hebridean race *P. m. hebridium* is noteworthy. In several Welsh counties it is regularly found up to about 500 m where suitable habitat occurs and has been found in rank heather up to 530 m in the Rhinog Mountains (Merioneth).

The Dunnock has almost certainly been numerous in Wales since historical times and has probably benefited from the long-term change from forest cover to agriculture. It is not a numerous bird in deep woodland, especially where this is devoid of any understorey as most Welsh woodland is, and it prefers coppice or the more open scrub on the woodland edge. It is very well suited to the hedgerow system that developed with the enclosures and the agricultural revolution.

It is a successful pioneer species and has been quick to move into areas of new conifer afforestation in the uplands, which have replaced large areas of open hill previously unavailable to it. Such plantations are occupied for 12–15 years until the canopy closes and Dunnocks are then absent until clear-felling and replanting take place and they are able to recolonise. By this means there is essentially a permanent and sizeable forest population. It is likely that the Dunnock is more numerous in Wales at present than at any time in the past.

This is one of the passerine species that has readily colonised offshore islands and it breeds on all the larger ones. Skokholm has 8–10 pairs in most years (although only one or two recently), Skomer some 30+ pairs, Bardsey 15–25 in most years and Flatholm c8 pairs.

The CBC index indicates a gradual decline in Dunnock numbers nationwide since the mid-1970s for reasons that are poorly understood (Marchant *et al.* 1990). As indicated above, this may well not apply to Wales and the relatively few CBC plots that have been worked for 10 years or more do not indicate a decline.

In Pembrokeshire, Donovan and Rees (in press) recorded Dunnocks in 434 tetrads (92.3%) in the breeding survey 1984–88, estimating an average of 50 pairs per occupied tetrad to give a county population of some 22 000 pairs. Thomas *et al.* found an occupancy of 89.5% tetrads in west Glamorgan but give no estimate of numbers. In Breconshire Peers and Shrubb found Dunnocks in only 15% of tetrads and much more common in the lower land of the south and east of the county. Tyler *et al.* found 95% occupancy in Monmouthshire estimating the same density as in Pembrokeshire to give a likely total in excess of 38 000 pairs.

Welsh Dunnocks are essentially sedentary and of 90 or so recoveries of birds ringed in Wales only two have subsequently been recovered more than 10 km from the place of ringing. Autumn and spring movements on the mainland are exceptionally difficult to identify but some evidence of a small-scale southerly movement down the west coast can be detected in autumn. Such a movement is suspected in some years on Bardsey, Skokholm and Skomer. Occasional individuals even turn up from time to time on Grassholm, e.g. two present on 21 August 1965. The actual scale of any autumn movement – or a return in spring – either coastally or inland is unknown but clearly is not large; whether such movement is intentional or is random dispersal is similarly not known. The populations in northern continental Europe are migratory and some of these birds are evident in eastern Britain in autumn but have not yet been identified in the west. The origin of the few birds detected in coastal movements in Wales is not known.

Alpine Accentor *Prunella collaris*

Llwyd y Mynydd

A vagrant.

This is a bird of the high mountains of central and southern Europe. There has only been one record from Wales:

20 August 1870 – One on the Llanberis side of Snowdon (Caernarfonshire)

Robin *Erithacus rubecula*

Robin Goch

A common and widespread resident in all counties and a passage migrant.

The Robin is one of the most abundant and widely distributed birds in Wales and both the *Breeding Atlas* and the *New Atlas* show it as breeding in every 10-km square throughout the country. Within this generalised picture, however, the numbers vary considerably from one habitat to another and with altitude. As its widespread distribution implies, the Robin occurs in a wide range of habitats from coastal cliffs, farmland and woodland to urban, suburban and industrialised areas, forestry plantations, watersides, offshore islands (occasionally) and even uplands. In this last habitat they will occur to altitudes of 457 m from the south Wales coalfields northwards through mid-Wales to the mountains of Snowdonia, wherever residual woodland or wooded dingles persist and they can even be found occasionally on the flanks of more open areas of moorland.

There is no evidence of long-term changes in status and most early writers testify to the Robin's abundance. Originally a forest species (as it still is predominantly on the Continent), one can postulate a similar distribution in Wales in times before extensive woodland clearances; however, Robins have adapted very successfully to the more open countryside that succeeded the woodland clearance. Evidence of their woodland origins is still to be found after severe winters such as 1962–63 when the surviving population tends to occupy woodlands and gardens first and only subsequently recolonises farmland and other more open sub-optimal habitats.

Short-term fluctuations in numbers occur, notably after harsh winters. The species is susceptible to protracted hard weather in winters such as those experienced in 1880, 1947 and 1962–63 and its population can then be drastically reduced. Recovery is usually fairly rapid, however; for example, it had regained its numbers in Monmouthshire within 12 months of a 50% reduction in the winter of 1962–63. In periods between severe winters, once the population has stabilised, the numbers appear to remain remarkably stable as is shown by the CBC index nationally.

The Robin is not the wholly sedentary bird that many assume it to be. Marked autumn passage can often be identified on coastal headlands and, particularly, the offshore islands. On Bardsey, for example, daily passage of up to 75 birds has been recorded in October and is similarly marked on the Pembrokeshire islands from late August to October. Evidence makes it clear that this passage is not restricted to the west coast but occurs on a wider front as demonstrated by observed concentrations, e.g. Lavernock Point (Glamorgan), and by ringing recoveries. Recoveries of Welsh-ringed birds from Spain and south-west France give evidence of the destination of some of these autumn Robins. Some clue to the origins of other individuals comes from one April recovery from central Scotland of a bird ringed on Bardsey the previous autumn. An adult bird ringed in Shropshire in March 1986 had moved across the Welsh mountains to reach Aberystwyth (Cardiganshire) a month later. Return migration in spring is most clear in April and May but invariably less well marked than in autumn.

The supposition is that some of the Robins from farther north remain to winter in Wales but there is only meagre evidence so far to support this; 20–30 birds move onto Skokholm each winter and remain there until March, Lockley (1947) remarking that in some cases the same individuals were involved in successive years.

Breeding occurs on some of the islands, albeit spasmodically. Forrest (1907) recorded breeding on Puffin Is. in 1903 and Lockley quotes 1939 and 1940 for Skokholm where a pair also bred in 1980. Breeding is regular on the more wooded island of Caldey.

Thrush Nightingale *Luscinia luscinia*

Eos Fronfraith

A vagrant.

This species, which breeds from Denmark and southern Sweden east to south and central Siberia and winters in east Africa, owes its place on the Welsh list to a single record:

20 September 1976 – A first-year bird was killed at the Bardsey lighthouse

Most of the British records are from the North Sea counties of England and Scotland which is what would be expected from its breeding range. A dramatic increase has occurred since 1970 in line with an increase in breeding numbers in north-west Europe, and it seems likely that there will be further records in Wales in the future.

Nightingale *Luscinia megarhynchos*

Eos

Scarce passage migrant; formerly bred locally in the east. No breeding since 1981.

Sadly the Nightingale has recently been lost to Wales as a regular breeding bird. Formerly it occurred annually as a scarce breeding species in Glamorgan and in the counties of the Welsh Marches but its range and numbers gradually declined throughout the present century and it is now no more than a scarce and irregular passage migrant, mainly in coastal counties. In the past, breeding has been recorded with varying degrees of regularity in all the counties along the English border and on at least one occasion as far west as Carmarthenshire. Regular breeding finally ceased in Monmouthshire in the early 1980s; one pair nested successfully in Glamorgan in 1981 after an interval of 50 years but it seems unlikely that the long-term contraction of range, which has accounted for its decline in Wales, will be reversed. The species has also withdrawn this century from other peripheral parts in the north and east of its European breeding range – northern England, France, eastern Germany, Poland and USSR.

The Nightingale is at the north-west edge of its range in Britain and has seen a slight but progressive overall contraction in its range this century despite the fact that there have been intermittent periods when the population has temporarily prospered again. The late 1920s and 1930s was one such period although by the 1950s the population in Wales was declining again until breeding ceased by the early 1980s. Even in periods of maximum numbers the Nightingale had only a tenuous foothold in Wales. It is recorded as having increased in the Vale of Glamorgan in the last decade of the 19th century with 10 localities listed and up to five pairs noted within a 3 km radius of Llanmaes near Cowbridge. The only record from the north of the county was in 1951 when one was heard in the upper Taff Valley – perhaps significantly within 2 km of a location called Llwyn-yr-Eos (Nightingale grove)! Its situation was relatively strong in Monmouthshire, where pairs frequented the valleys of Wye, Monnow and Usk, occasionally as far north as Abergavenny (nest with five eggs in 1898) and in some numbers around Newport. The Nightingale's stronghold in Wales – if the modest population merits the term – was always in this south-east corner of the country. Very occasionally odd birds or pairs strayed farther west as far as Merthyr Mawr (1903) or Gower (one singing in May 1924) and into adjoining Carmarthenshire as at Llandovery (report of nest and eggs in 1898), Alltwalis (1934) and Nantgaredig (1925); there is also an interesting sequence of records from Brechfa in the same county. The egg collection of the late A. F. Griffiths, in the National Museum includes a Nightingale's egg labelled "Brechfa 7 June 1845". A second record is listed by Ingram and Salmon (1954) of a bird singing at Brechfa in May 1927 and A. Lindsay (*British Birds* 1945 p. 318) reported that, "I examined a Nightingale's nest on a laneside at Brechfa which contained two typical eggs and later produced young." His account is then made more intriguing and somewhat less convincing when he adds that "odd males have been heard singing in previous years but this season several pairs were in the district with males singing lustily." He further erroneously claims the record as the first for the county. An egg in the National Museum collection is, however, attributed to this nest.

As an extension of the corner of south-east Wales where Nightingales were regular annual breeders, Breconshire supported a meagre scattering of pairs in its southern areas, mainly the Usk Valley as far north as Brecon. It was doubtless always rare in the county and in the present century resident ornithologists as competent as Walpole-Bond and H. A. Gilbert never met with it, the latter over a very long period of years. Over the border in Radnorshire there are only four old records: nesting was reported at Glasbury (1903, 1926), a pair claimed at Llandrindod in 1912 and a male singing at Llanbadarn Fawr in 1926; there have been no records since. There are four old records for Cardiganshire, all of which are considered reasonably credible: a Nightingale was reported singing for a few nights near Cardigan town in 1886 and another in 1895 and there is a similar record from the Tregaron area in 1910. In 1929 Salter (Diaries) records a secondhand but seemingly acceptable note of a bird singing at Falcondale near Lampeter. It was to be another 56 years before the Nightingale was heard in the county – see below.

Forrest documents 19th century records for one or two sites in the Severn valley of Montgomeryshire and there were sporadic records from places such as Abermule, Berriew, Guilsfield, Middletown, Leighton and Castle Caereinion until the 1950s. Farther north, in Denbighshire and Flintshire, Nightingales probably had a slightly firmer toehold where the lowlands occupy wider areas than in the hilly counties of mid-Wales. Forrest records them as being heard "many times" in the Wrexham area and elsewhere near Mold, Overton-on-Dee and sporadically at other sites near the English border, while other writers add locations near Chirk (Glyn Ceiriog) and reputedly on one occasion along the coast as far west as Rhyl. Such records almost certainly represent the zenith of the Nightingale's fortunes in Wales for by the late 1920s the species was already noticeably withdrawing southwards and the pairs in north-east Wales were no longer to be found.

The first national survey of the distribution of Nightingales was undertaken in 1910 (Ticehurst and Jourdain 1911), the findings of which accord with the above summary so far as Wales is concerned. Further national surveys organised by BTO in 1976 and 1980 produced only one pair (Monmouthshire) in 1976. Since that date the only known nesting to have occurred in Wales was the pair that bred successfully at Llanishen (Cardiff) in 1981. (A second bird sang at Llanedeyrn for a month the same year.) A singing bird was present at Llanishen in 1972 and males have been heard at other sites in the county in 1969, 1971, 1979, 1985 (three birds) and 1986 (two birds).

At the present time the Nightingale is no more than a scarce passage migrant in Wales, not necessarily recorded in every year. It is rarely reported on the well-watched offshore islands (only four records throughout the 1980s on the Pembrokeshire islands) and on the mainland of Wales records are equally few and far between. One sang for a month at Bodfari (Denbighshire) in 1981, two were seen on passage at South Stack (Anglesey) in 1987 and one on the Great Orme in 1988 with another on Bardsey in late summer the same year. The one and only record of Nightingale in Merioneth was of a bird caught and ringed at Arthog in July 1982. Pembrokeshire produced a crop of four spring and eight autumn passage records in 1988, one in 1987 and three in 1982 and a male was singing at Ynyslas (Cardiganshire) in May 1985. In Montgomeryshire a male was singing at Middletown (a former site) for two weeks in 1983 but there have been no recent records from Breconshire (since singletons in 1968 and 1969) or Radnorshire.

Wales is marginal ground for Nightingales and with a climate which tends to mild and wet springs. It is a position summed up by Giraldus Cambrensis who, travelling through Wales with the Archbishop of Canterbury and writing in 1188, left us the earliest written reference to Nightingales; he alluded to the Archbishop's powerful dislike of the Welsh weather and commented that the absence of Nightingales in Wales clearly indicated that the bird was wiser than he.

Bluethroat *Luscinia svecica*

Bronlas

A rare and irregular passage migrant in spring and autumn with a total of only 30 records, all of single birds, since the first Welsh record on 1 May 1946 (a male of the White-spotted form on Skomer).

Where separated, the majority of records refer to the Red-spotted form, which breeds in Scandinavia and Russia, whilst the White-spotted form breeds in central and southern Europe.

Fifteen of the records fall in the period 26 April to 29 May and the remaining 16 in the period 10 September to 20 October with a marked preponderance (nine records) in the period 10–16 September.

Pembrokeshire has been the most favoured county with 15 records (all from the islands), five from Anglesey, three from Caernarfonshire, two from Flintshire and singles from Breconshire, Cardiganshire, Radnorshire, Montgomeryshire and Glamorgan.

White-throated Robin *Irania gutturalis*
Robin Gyddfwyn

A vagrant.

The only record for Wales, and the second for Britain and Ireland, was a bird seen on Skokholm in 1990:

27–30 May 1990 – Female, Skokholm

This species breeds from Turkey eastward to Iran. The previous British record was a male on the Calf of Man in June 1983.

Black Redstart *Phoenicurus ochruros*
Tingoch Du

Spring and autumn passage migrant. Winter visitor in small numbers. Has bred at least twice.

This species was once regarded as being sufficiently rare for individual records to be published in journals such as the *Zoologist* and *British Birds* (e.g. *Zoologist* 1850 p. 2641; *British Birds* 1910 p. 308). The earliest published records of Black Redstart in Wales date from the middle of the 19th century: in Pembrokeshire it was described as "very rare" with two records quoted for autumn 1847 in the south of the county, one killed on the roof of the Coburg Hotel in Tenby and the other shot by Mr James Tracey "with an air-cane loaded with small shot, on the water trough of my neighbour's house in Pembroke". However, only twenty years later a correspondent to the *Zoologist* considered it a regular visitor to the county while by the end of the century Mathew recognised it as "a winter visitor, although not common". The first record in Carmarthenshire was of a female shot at Kidwelly in 1864 (*The Field* 1864 p. 190) and in Glamorgan none was recorded until 1884. In north Wales Forrest mentions five records from Merioneth (4) and Flintshire (1) between 1886 and 1889. The first record for Caernarfonshire was on 27 December 1910, and from Anglesey, one in October 1912.

Nowadays the Black Redstart is recognised as a spring and late-autumn passage migrant in Wales. Small numbers overwinter most years, almost exclusively in coastal areas. It is from these same coastal areas that most of the records of passage birds derive and whereas this is probably a genuine reflection of the pattern of occurrence, it is also likely to reflect the emphasis of birdwatchers in coastal districts through the autumn and winter months. First arrivals in spring are consistently seen around the third week of March, although the picture is slightly obscured by the small numbers that are present throughout the winter. The main spring passage continues for a month or so and only a few stragglers remain by the end of the month or the first few days of May and these spring birds may well be overshooting birds from a northerly movement in western France (Langslow 1977). In autumn, Black Redstarts are fairly late migrants with passage not usually apparent until October; records are then more numerous than at any other time of year and continue until well into November before they again become confused with overwintering birds. In 1982 as many as 111 autumn individuals were reported from 38 different localities in Pembrokeshire. Between the months of May and September records are few and in June, July and August they are rare. Although birds may appear at almost any suitable coastal site in winter, there are numerous locations where overwintering individuals are almost regular, e.g. Cardiff docks, Pengam Moors, Porthcawl, Gower cliffs (all Glamorgan), Tenby, Fishguard (Pembrokeshire), Aberporth, Aberystwyth (Cardiganshire), Holyhead Is. (Anglesey), Bangor, Great Orme (Caernarfonshire), Gronant, Connah's Quay (Flintshire). Usually it is single birds that are encountered in winter although occasionally there are two or more, as shown by a group of five on Little Orme (Caernarfonshire) on 11 January 1981 and six at the Rover Works, Pengam Moors (Glamorgan) in 1987. The numbers wintering in any particular year appear to have some correlation with the strength of the autumn passage, more winter individuals being present when, as in 1977, passage was heavy.

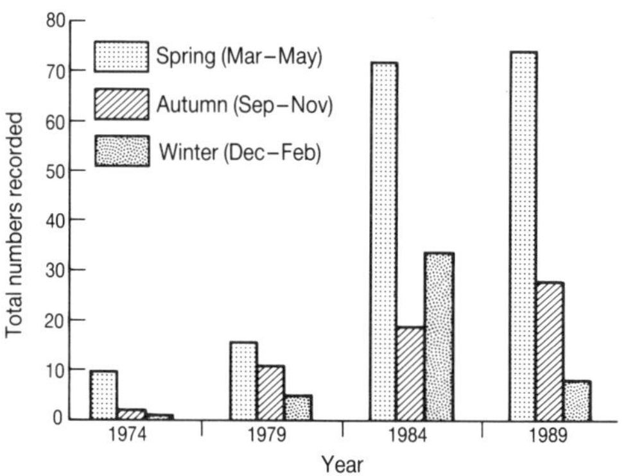

Black Redstarts recorded in coastal counties at five-year intervals 1974–89

Although the number of birdwatchers in the field and their proficiency in identification, has increased greatly over the years there is little doubt that the increase in records of Black Redstart in recent decades is principally a genuine reflection of the fact that the species is a more numerous passage migrant than it was in earlier years, especially in spring. This is consistent with the spread of the species in north-west Europe which eventually resulted in breeding in Britain (annually from 1920s) and Scandinavia (Sweden from 1910) (BWP vol. 5). Since the 1970s the population has continued to expand very gradually (c100 pairs breeding annually in England, none regularly elsewhere in Britain) and the general increase is reflected by the growing frequency of records in Wales. The 1977 national survey of breeding Black Redstarts carried out by Morgan and Glue (1981) not unexpectedly produced no pairs in Wales. However, a pair did breed successfully soon afterwards in Cardiff in 1981, followed by another pair at Point of Air colliery in 1984; this pair had overwintered on site and eventually successfully reared two broods. Occasional singing males have appeared during May or June in scattered locations (e.g. Llanymynech (Montgomeryshire) 1977, Porthcawl (Glamorgan) 1985, Tretower (Breconshire) 1978, Abergele (Denbighshire) 1924, Worm's Head (Glamorgan) 1974), but to date no further proof of breeding has been forthcoming.

Redstart *Phoenicurus phoenicurus*

Tingoch

A numerous breeding species in deciduous woodland and ffridd in many parts of upland Wales. Locally common elsewhere; scarce in Anglesey and Pembrokeshire.

The ancient sessile oakwoods of upland Wales have a rich community of breeding birds, among which the Redstart, in company with Pied Flycatcher and Wood Warbler, is one of the most attractive constituents. The Redstart is a numerous bird in most parts of upland Wales and, consequently, the Principality is one of the main British strongholds for the species. The upland element to its distribution is demonstrated by its avoidance of most coastal areas and the pattern is again strongly emphasized by the tetrad breeding surveys of Monmouthshire, west Glamorgan and Pembrokeshire. Although Redstarts are essentially summer birds of the upland oak woods, they are by no means confined to these woods but are sometimes almost as common in several other habitats. On ffridd for example they can be very numerous on bracken-, gorse- or heather-dominated slopes where there are scattered trees, derelict buildings, stone walls or similar nesting sites. Almost any site in upland valleys which boasts a handful of mature trees is likely to support a pair of Redstarts.

Throughout the hill areas of Wales they will readily breed up to at least 450 m; in surveys in the Brecon Beacons National Park in 1978 M. Davies (RSPB 1978) found Redstarts common above the top wall line of the mountains, i.e. on the open hill, locating 77 pairs in such habitats on the Black Mountains alone despite that habitat being presumably sub-optimal, while lower down the slopes they were everywhere abundant as soon as hawthorns or other bushes/trees were present. They are regularly to be found in the vicinity of farms and cottages, not infrequently nesting in holes in the buildings themselves. Redstarts are also found – albeit more sparingly – in upland conifer forests where nesting sites in the form of derelict buildings, isolated old deciduous trees or redundant stone walls are present. Bibby *et al.* (1989) found Redstarts present at 47 out of 253 survey points in upland conifer forests when assessing the effects of isolated broadleaved trees or small stands of them in plantation conifers. Bibby *et al.* (1985) only found Redstarts in a minority of point-count sites in conifer restock areas, where the reduced occurrence is doubtless a result of the paucity of nest sites. In lowland agricultural areas Redstarts are not uncommon where hedgerow trees, waterside willows or old alders provide a choice of nest sites.

Nationally, Redstart populations have fluctuated over the years although the pattern has sometimes been confused and subject to varying interpretation. In southern England there was certainly a steady decline of Redstarts in the early decades of the century but there is no account of this being parallelled in Wales; in fact this period was one when numbers were shown to be rising in south Wales and the breeding range apparently expanding, at least in Glamorgan. The overall rise documented nationally (Mason & Hussey 1984) in the 1950s was certainly reflected in Wales, building up towards a peak in the mid-1960s. From 1969, however, a marked decline in numbers followed, continuing to a low point in 1973 which was subsequently linked to the long-term failure of rains in the Redstart's wintering area in the Sahel region south of the Sahara (Marchant *et al.* 1990). These reductions were specifically noted in Snowdonia and Breconshire. In most areas recovery in numbers was fairly rapid and it is apparent that the effects of the decline in the strongholds of Wales were neither as severe nor long-lasting as in the lowlands of southern England.

The density of Redstarts in some breeding areas in Wales is impressive. In the Rhayader area up to as many as 40 singing males have been recorded per km^2 (RSPB 1979). Garnett (in prep.) found Redstarts present on 45 out of 119 ffridd sites (38%) throughout Wales where it made up 2.2% of the total songbird population (where Meadow Pipit formed 37% and Tree Pipit, Wheatear, Whinchat approximately 10% each). In Monmouthshire the tetrad breeding survey (1981–85) found Redstarts present in 196 tetrads (51%) and estimated a population of c1500 pairs. Peers and Shrubb (1990) estimated an average of four pairs per tetrad and give a calculated total of c1500 pairs in 352 tetrads.

In deciduous woodland at Ynys-hir (Cardiganshire) Squires (in litt.) has shown a mean population density of 5.5 pairs per 10 ha over seven years. Stowe (1987) has demonstrated that the provision of nestboxes can sometimes increase populations by as much as 40% in individual woods. In Radnorshire, where the recovery from the 1969–73 decline was rather slow (Peers 1985), the Redstart is considered to be the second most numerous summer visitor (after Willow Warbler). All other upland counties of mid and north Wales boast similarly high populations and it is only in north-west Wales (Anglesey and Llŷn) and south-west Wales (Pembrokeshire, Gower and the lowlands of Cardiganshire and Carmarthenshire) that the Redstart is either absent or at best scarce. On Anglesey the first proven breeding for many years was recorded in 1969 and the following year three pairs were found in woodland on the Menai Straits. The Redstart remains, however, a rare and irregular breeding bird on the island; RSPB surveys in 1986 located only three singing males. Similarly, on Llŷn there are very few regular sites and in Pembrokeshire breeding is virtually restricted to the uplands of the north, on and around the Preseli Mountains. Redstarts nest only very sporadically on Gower (one record during the 1984–89 breeding bird survey).

Arrival in spring is concentrated around the third week in April, with a scattering of early records in some years from the last week in March. The rather modest peak of spring passage on the offshore island observatories – Skokholm and Bardsey – is likewise concentrated in April (Hope Jones 1975). Autumn passage peaks in September and by the end of that month only occasional stragglers are to be found on mainland Wales. Evidence from ringing recoveries – (Morocco (6), Iberia (5), southern France (3) – confirms the known west-coast migration route in autumn on the way to wintering grounds south of the Sahara. Although most Redstarts, including passage birds from farther north have passed through by October, very late individuals do from time to time occur, e.g. 11 November; 8 December; 11 December (all Glamorgan); 17 December (Anglesey).

Several recoveries of breeding adults in areas where they had been ringed in previous years (in two cases as nestlings) indicates a tendency to return to the same breeding area in successive years. One individual ringed in the nest at Cilfor (Caernarfonshire) in June 1979 was found breeding 15 km away six years later.

Moussier's Redstart *Phoenicurus moussieri*

Tingoch Moussier

A vagrant to Wales.

Along with the Grey-tailed Tattler the most spectacular discovery of a rare bird in Wales must surely be awarded to the sighting of a male of this species on Dinas Head (Pembrokeshire) on 24 April 1988 (Barrett 1992).

This strikingly handsome bird was a completely unexpected addition to the British and Irish avifauna because it breeds in north-west Africa where it is normally migratory only within its very limited breeding range. Vagrants have, however, recently reached Italy and Malta and the first record in Greece was on 30 March 1988.

Whinchat *Saxicola rubetra*

Crec yr Eithin

Summer breeding visitor; now predominantly a species of upland areas.

The Whinchat is one of the most characteristic Welsh upland birds in summer. It is among the later summer visitors to arrive, not appearing in numbers until late April. Once here, it is a conspicuous species, perching in the open on fence lines, overhead wires or similar prominent song posts, on bushes or other tall vegetation. It breeds throughout Wales, although it is scarce in Anglesey and is absent from large parts of south-west Wales. In addition, it shows a marked inland bias in its distribution and in these respects provides almost a mirror image of the distribution of the closely related Stonechat. The Whinchat's stronghold in Wales is the extensive areas of bracken-dominated ffridd, which cloaks the sides of so many of the valleys and hillsides in rural Wales. Whinchats nest in tussocky grassland or similar rough ground vegetation, almost invariably where such vegetation is found in combination with bushes, trees or fence lines as available song posts. On ffridd, ideal conditions are provided where bracken areas are dotted with occasional hawthorns, rowans or other trees. Garnett (in prep.) has shown that on a sample of 483 Whinchat territories on ffridd throughout mainland Wales, there was a mean of five trees/bushes per territory. Bracken hillsides with few trees or those heavily tree covered had correspondingly lower numbers of Whinchats. Whinchats are one of the four most numerous songbird species in such areas (after Meadow Pipit, Wheatear and Tree Pipit), forming about 10% of the total songbird population. They are also often numerous on the margins of rough, boggy areas scattered among improved fields.

Both the English and Welsh names refer to the Whinchat's association with gorse which still holds good, particularly in hill areas, although gorse brakes – no longer valued in agricultural terms as they formerly were – are somewhat scarce and many are regularly burnt before they become over-mature and thereby most attractive to the species. The Whinchat is one species that has benefited from the planting of conifer forests which, in their early stages, produce the perfect combination of rank grass and song posts. In both habitats Whinchats can occur as high as 580 m in the hills, e.g. Black Mountains (Monmouthshire). Whinchats will also recolonise clear-felled/restocked sites where Bibby *et al.* (1985) showed densities of around 0.5 individuals per 10 ha, although distribution in such sites is clearly not even. Numerically, ffridd and new forestry plantations are the two most important habitats for Whinchats in Wales and although the total Welsh population is difficult to assess, it is likely to represent a considerable percentage of the overall British population (see below). Whinchats are not exclusive to these habitats but are also found in some lowland areas on residual commons and heaths, e.g. Cors Caron and Cors Fochno (Cardiganshire) as well as on a variety of derelict land sites, rough pastureland and similar uncultivated areas. Wide roadside verges and railway cuttings and embankments once much favoured by Whinchats are seldom used now, the former being much more disturbed and often heavily managed and the latter overgrown and scrub-covered.

This species has suffered an overall decline in Britain since the onset of agricultural intensification following the Second World War. How far this general statement is true of Wales is more difficult to determine. Certainly there have been changes in the distribution mainly resulting from a displacement of the population from the lowlands and an expansion of range into new habitats in the uplands. In north Wales Forrest (1907), regarded it as "rather a common bird" found it "numerous on the belt of flats from Pwllheli to Tywyn . . . (while) inland its favourite haunts seem to be railways and the skirts of moors". This certainly does not hold true today as the Whinchat is virtually absent in the southern half of Llŷn. In south Wales Heathcote *et al.* (1967) recorded it as common and widespread at the turn of the century, diminishing gradually over the next 25 years although still occurring in the lowlands and still breeding around the outskirts of Cardiff. The decline continued until the 1950s, when areas on the flanks of the coalfield began to be reoccupied, aided by the expansion of forestry plantations in the same areas, the development of rough uncultivated vegetation on the spoil tips and the abandonment of many smallholdings in the hills.

In ideal habitats Whinchats can occur in densities of up to 2.5 pairs/territorial males per 10 ha as indicated by R. Knight (in litt.) on Corngafallt Common in Breconshire. RSPB surveys on Moel Famau Country Park on the Clwydian Hills (Denbighshire) in 1986 found 118 pairs on the 861 ha of the Park, almost all of which were clustered on the sides of the westward-facing valleys running off the north–south spine of the hills. The densities reached a maximum of 3.2 pairs per 10 ha. In Breconshire, Peers and Shrubb suggest an average of three pairs in occupied tetrads (39%) implying a total of about 600 pairs for the county. Tyler *et al.* (1987) suggest an average of up to five pairs per occupied tetrad in Monmouthshire, which may give a slightly high figure of some 450 pairs.

Although it breeds in all counties, the Whinchat's heartland is on and around the uplands of central Wales, from the northern parts of the coalfield north to southern Snowdonia – the counties of Breconshire, Montgomeryshire, Cardiganshire, south Merioneth and south Denbighshire. In Pembrokeshire (present in only 8.9% of tetrads) and Carmarthenshire it is more or less confined to the hill areas in the northern parts of those counties, plus a few lowland heaths, and on the Gower it is a no more than a scarce passage bird. Breeding in Monmouthshire (23% tetrads) is almost entirely restricted to the hill lands, particularly in the north and west. In north Wales it is not numerous in Flintshire but occurs widely in the hills of Denbighshire, Caernarfonshire and Merioneth; on Llŷn, where hill habitats are scarce but many damp lowland hollows are still uncultivated, Whinchats occur around the margins of such sites but are less frequent on the limited areas of ffridd (A. J. E. Seddon, pers. comm.). On Anglesey the species is distinctly scarce (although frequent on passage); RSPB surveys there in 1986 found birds present in only 16 out of a total of 762 one-kilometre squares, although Hope Jones found up to five pairs in one site at Newborough in 1965, when forest plantations were newly established. The total breeding population in Wales is probably in the order of 5000–6000 pairs.

Spring and autumn passage, often evident in coastal areas, is well marked on the offshore islands, e.g. Bardsey, where there is a heavy concentration of birds in the first two weeks of May and again from late August to early October. Most birds have left by this time and only stragglers are recorded each year up to the end of the month. The winter record of a bird at Whiteford (Gower) on 25 December 1966 is unprecedented.

Whinchats winter in a wide band of tropical and subtropical Africa north of the equator and ringing data predictably confirm the migration routes through Iberia and north Africa to and from these wintering areas.

Stonechat *Saxicola torquata*

Clochdar y Cerrig

Breeding resident, locally common in coastal areas and thinly distributed inland. Numbers regularly reduced after hard winters.

The distinctive and colourful Stonechat is a familiar bird of undeveloped areas almost all round the coasts of Wales; its extrovert habits of perching conspicuously and calling incessantly when agitated, make it easy to locate and difficult to overlook. It is especially numerous in cliff-coastal areas where gorse, brambles, often mixed with heather or bracken, predominate. These western coasts are now one of the strongholds of the Stonechat in Britain and the breeding distribution in Wales is almost exactly the complement of that of the Whinchat (q.v.) which is predominantly an inland species and is very scarce in Anglesey and south-west Wales where Stonechats are numerous.

Stonechats, however, are notoriously susceptible to prolonged hard winter weather. Their numbers fluctuate considerably, falling dramatically in severe winters and then rising again and recolonising "lost" areas over the next few years. At such times the inland pairs are the first to succumb and the residual population effectively retreats to surviving pairs in a narrow coastal belt. A series of severe winters can be identified which had serious effects on Stonechat numbers: 1838, 1855, 1878–79, 1880–81, 1889 (when according to Forrest dead Stonechats were picked up in dozens at Barmouth (Merioneth), 1895, 1916–17 (almost 40 consecutive nights of frost), the three winters 1939–42 (in which 1939–40 caused most severe reductions), 1947, 1955–56, 1961–62, 1962–63, 1978–79 and 1981–82. In recent times the severe spell in March 1947 was probably the most damaging for this species in Wales, eliminating Stonechats completely from several counties including virtually the whole of south Wales; in Breconshire only one pair was found between 1947 and 1959 and even in a coastal county such as Glamorgan none was to be found in the 1947 breeding season where there had been perhaps 100+ pairs beforehand and by 1948 a maximum of only five pairs could be located. Stonechats are multi-brooded each year and the potential for recovery is therefore considerable. Records indicate that recovery is invariably in the favoured coastal areas first, subsequently spreading inland relatively slowly when a succession of mild winters permits numbers to be maintained. Although a small number of Welsh Stonechats move away after the breeding season it is believed (e.g. Heathcote *et al.* 1967 and Hope Jones 1974) that a considerably higher proportion of the population was formerly migratory – Forrest suggested up to 80% – which naturally enabled the species to withstand the effects of severe winters better than if the entire population were resident. The extent to which this change of habit may have occurred, or the causes of it are unclear and in any case it would appear to conflict with the levels of mortality claimed in the past in harsh winters.

Magee (1965) has demonstrated the overall reduction that has taken place in the Stonechat population in Britain this century, attributing it mainly to loss of habitat associated with agricultural intensification. Certainly the Stonechat is much reduced in inland areas from the levels it enjoyed up to the early decades of the current century.

The Stonechat's partiality for gorse is well known and the plant is usually, although not invariably, a dominant component of the vegetation in Stonechat territories. RSPB surveys on Moel Famau (Denbighshire/Flintshire) in 1986 demonstrated how the 12 pairs present on the mosaic of habitats on the 861 ha of the country park, occurred specifically only wherever there were small stands of gorse. Where gorse is not present Stonechats will resort to rough, overgrown areas of bramble and bracken such as are found on many of the rough headlands along the western seaboard. Mathew (1894) recorded them in Pembrokeshire "to be seen everywhere, roadside, coasts and farmland". Inland pairs are also found on heather areas including some of the mountains of north Wales, e.g. Berwyns, Rhinogs, Ruabon Moor. Short-cropped grassland is also an essential constituent of their habitats, as they feed Robin-like on open ground, so that areas where gorse, heather and bramble are extensive and dense are avoided. Stonechats have taken limited advantage of the establishment of young forestry plantations both in uplands and lowlands although Magee (1965) points to the fact that the spread of such afforestation onto rough areas previously frequented by Stonechats accounted for the total disappearance or drastic reduction of breeding pairs in parts of Cardiganshire, Merioneth, and west Montgomeryshire in the 1950s and 1960s.

A further hazard to Stonechats is the traditional burning of gorse and heather in spring. Even adherence to the legal burning dates (1 November–31 March in coastal areas) – which is regularly ignored – can cause serious problems for a species in which the first pairs commence nesting as early as the end of March/early April.

Number of pairs of breeding Stonechats in Wales (Magee 1965; Gibbs & Wood 1968)

County	*Pairs found in 1961*	*Pairs in 1968*	*Additional (non-breeding?) birds in other sites 1968*
Anglesey	28	143	3
Caernarfonshire	18	73	28
Denbighshire	12	1	1
Flintshire	0	7	2
Merioneth	1	32	2
Montgomeryshire	0	1	2
Cardiganshire	10	no data	no data
Radnorshire	0	no data	no data
Pembrokeshire	264	97	3
Carmarthenshire	0	2	15
Breconshire	0	5	2
Glamorgan	150	59	33
Total	483	420	91

(Monmouthshire not covered in either survey but Tyler *et al.* (1987) estimate a maximum of 20–30 pairs in the county between 1981 and 1985.)

Few data exist on which to estimate the numbers of Stonechats in Wales, especially at periods when the population is relatively high. Heathcote *et al.* quote 15 pairs on 12.8 km of Gower coast in 1951, four years after the population was obliterated by the 1947 winter. In the 1961 national census (Magee 1965) Pembrokeshire (264 pairs) had the highest total for any county in Britain, followed by Hampshire (178 pairs) and Glamorgan (150 pairs). The Glamorgan figure however was clearly an overestimate as 50 pairs were counted in the main part of the county (excluding Gower), and the population for that peninsula never approaches the 100 pairs implied by these figures. The Pembrokeshire figure was probably equally flawed (see table). Conversely, in north Wales the survey was considered seriously to under-estimate numbers and Gibbs and Wood (1974) accordingly re-surveyed the species over the whole of Wales in 1968, again six summers after a severe winter had depleted the species. Their survey yielded a total of 420 pairs with additional birds present at a further 91 sites. Maximum density recorded was 7.2 pairs per km^2 on Ramsey.

The *Pembrokeshire Breeding Atlas*, 1984–88 found Stonechats in 126 (26.8%) tetrads in the county but what is a noteworthy fact is that the species occurred in 82 out of a total (including built-up areas) of c102 coastal tetrads (80.4%) with a county population estimated at 200 pairs.

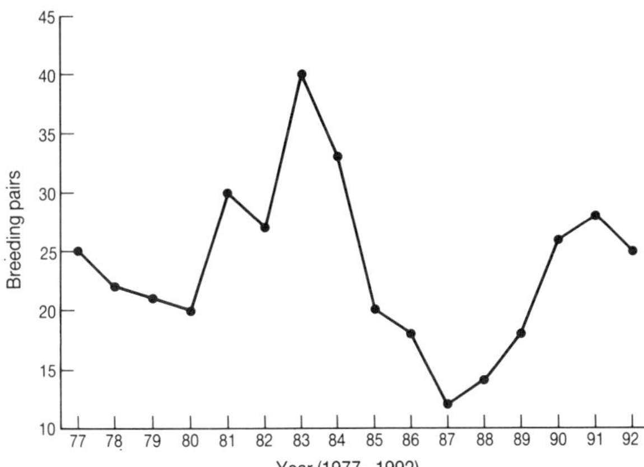

Numbers of breeding Stonechats, South Stack Cliffs (RSPB reserve)

A light passage of Stonechats is detectable on the offshore islands in spring and slightly more obviously in autumn (including Grassholm). A party of 15 near St. David's on 7 March 1989 may well have been immigrants. The origins and destinations of most of these birds are unknown although the ringing recoveries of five autumn passage birds have occurred in Spain (3) with two birds as far south as Alicante and Seville, south France (1) and Dorset (1). One young bird hatched in Denbighshire had moved 210 km north-east by autumn.

As long ago as 1913 it was recognised that individuals of the highly migratory eastern races of Stonechat occurred sporadically as far west as Britain. In recent years the increase in birdwatchers and the advances in identification skills have resulted in an increase in the number of confirmed records of these eastern birds belonging mainly to the western Siberian race *maura* but occasionally the eastern Siberian race *stejnegeri*. Individuals from one or other of these races have been recorded as follows:

25–27 October 1983	– One, Bardsey
6–7 November 1983	– One, Llanfairfechan (Caernarfonshire)
12 October 1986	– One, Strumble Head
10–18 October 1987	– One, South Stack
15–18 October 1987	– One, South Stack (separate bird)

Wheatear *Oenanthe oenanthe*

Tinwen y Garn

Summer visitor and passage migrant. Locally common, favouring upland areas with sheep-grazed turf.

The Wheatear is a numerous visitor to Wales between the months of March and October. Many of the birds which are recorded in March and April and later from August to October are on passage, while others remain to breed throughout all the counties of Wales. The principal requirement for these breeding birds is a combination of sheep- or rabbit-grazed grassland and rabbit holes or stony crevices in which to nest. Such features occur plentifully in Wales, particularly in the uplands but also on offshore islands, coastal headlands and less frequently other sites in the lowlands. Wheatears may thus be found nesting almost from sea level to the tops of some of the highest hills – 790 m on Cadair Idris and over 910 m on the Carneddau – where the harsher climate substitutes for stock grazing to produce a close sward. It was near here, at Ffynon-las on Snowdon in 1778, that Thomas Pennant first noted this species: "Here we observed the Wheatear, a small and seemingly tender bird; and which is almost the only small one, or indeed the only one, except for the Rock Ousel, that frequents such heights".

Other upland habitats, notably heather moors, are not attractive to Wheatears except where they are intermingled with rocky outcrop and patches of grassland and the afforestation of uplands has clearly had a substantial effect locally on both distribution and numbers by producing a land use which is incompatible with their needs. The intensification of hill land improvement for agriculture in recent decades has also had an impact on Wheatears which will use such ground more sparingly, principally then being restricted to the headlands and other margins.

The long-term decrease of Wheatears, which is known to have occurred in southern England for reasons of agricultural land-use change, is less clear in Wales. Nonetheless there is evidence from several counties of shifts in population from the lowlands and low coastal areas, such as dunes, where they formerly bred commonly but where (mainly) human disturbance has rendered such sites untenable. However, in view of the fact that some 25% of the Welsh uplands is now covered with conifer afforestation it is safe to assume that the overall population is smaller now than it was formerly, notwithstanding that much of the afforestation has taken place on ground which was not favourable to wheatears, e.g. heather moor. This likelihood is echoed in the lowlands by Heathcote *et al.* (1967) who state that the Wheatear was considerably more numerous in Glamorgan in the last century when there was a strong coastal population along the cliff tops and with many pairs in the rabbit-infested sand dunes between Ogmore and Swansea. Overall, myxomatosis, which struck Wales in 1954, played a part in diminishing Wheatear populations in the lowlands, through the removal of grazing in areas not otherwise covered by sheep and thus reducing the availability of both nest sites and short-cropped areas of grass. More recently a similar effect is evident along sections of the Pembrokeshire coastal path where reduced grazing regimes have rendered former Wheatear habitat less suitable.

The *Breeding Atlas* illustrates the scarcity of breeding Wheatears in Anglesey other than along some coastal areas, the dairy lands of Carmarthenshire and Pembrokeshire, the lowlands of Cardiganshire and the wide river valleys of the Marches. On Gower there are only a few pairs at the extreme western end. The core of the Wheatear's stronghold is thus in the uplands of the mountain spine of Wales, northwards from the coalfield valleys to the hill slopes bordering the north Wales coast. Even on Llŷn there are strong populations with particular concentrations in the north-eastern quarter (breeding in at least 70 one-kilometre squares) and the south-west corner. Populations are high in such areas as the Carmarthen Fans (the Black Mountain) in the western half of the Brecon Beacons National Park where RSPB surveys in 1978 found a minimum of 281 pairs in 87 one-kilometre squares (28 other squares with suitable habitat were not surveyed) with a maximum density of 13 pairs per km^2. In a repeat survey of the same area in 1992, the total of pairs was raised to 345 – probably a function of more intensive coverage than a genuine increase in numbers. On Ramsey Is. (Pembrokeshire) Wheatears nest at a density of c60 pairs on 268 ha (=23 pairs per 1 km^2). Garnett (in prep.) found Wheatears the second most numerous species (after Meadow Pipit) in his survey of 119 ffridd sites in mid-Wales (six counties), being present on 76% of all sites and yielding 10.3% of all songbird pairs (Meadow Pipit 37%). Peers and Shrubb (1990) suggest a breeding population of about 500 pairs for Breconshire where it was found in approximately 50% of tetrads sampled and was particularly numerous on Mynydd Eppynt. For 1981–85 Tyler *et al.* (1987) suggest a population of c200 pairs in the hills of west Monmouthshire. It is stated (Peers 1985) as having been "much commoner before the 1940s" in Radnorshire and is now absent from the sand dune areas of Cardiganshire where it formerly flourished but where there are now only scattered pairs, although it is still common in the hill areas of that county. The Pembrokeshire mainland breeding population is virtually confined to three discrete areas, the Preseli Mountains (where Mathew recorded it as very numerous in 1894), a few headlands on the north coast and the limestone cliffs of the south, particularly the Castlemartin tank training ranges.

Wheatears are much scarcer than formerly in Montgomeryshire where extensive tracts of the uplands have been surrendered to afforestation and agricultural improvement; there are probably no more than 400 pairs remaining in the county. Wheatears are relatively numerous on the hills of Merioneth and Caernarfonshire while on Llŷn they breed on many of the wild headlands as well as on the few inland hills, The Clwydian Hills are well populated (RSPB surveys in 1986 found 12 pairs on 861 ha on Moel Famau country park) but otherwise the species is scarce in neighbouring Flintshire. In Anglesey, RSPB surveys found Wheatears in only 48 one-kilometre squares in 1986, mainly on the north coast and in the south-west quarter of the island.

The offshore islands provide ideal breeding opportunities for Wheatears. On Bardsey breeding pairs normally average 12–15 pairs per year although as many as 25 pairs have been recorded on the 178 ha island. On Skokholm – where P. J. Conder did the bulk of his work on Wheatears over 25 years – there was an average of 32 pairs annually in the 1950s, which fell to an average of 24 in the 1960s and only 8–9 in the period 1973–76, rising again to an average of 19 pairs by the late 1980s. On Skomer the number of pairs increased gradually through the 1970s and 1980s to a peak of 48 pairs and Ramsey supports up to 60 pairs (268 ha), as mentioned. The total Welsh population is probably in the order 3500–4500.

Wheatears begin to arrive in mid–late March, among the earliest of summer arrivals. There are a number of records of exceptionally early birds across the years with the earliest being one seen at Aberystwyth on 23 February in 1921 while at the other end of the season late birds can still be encountered in some years in the first few days of November (one as late as 23 November in Glamorgan). Spring and autumn passage is evident throughout Wales as migrating Wheatears pass through on a broad front, often pausing to rest and feed in fields, pastures or golf courses on sites where they do not necessarily remain to breed. Spring passage on the islands is well marked. On Bardsey, for example, 50–100 birds per day is not unusual in April and the first half of May. Autumn passage is

Recovery locations, with months (e.g. 9 = September) of Wheatears ringed in Skokholm in July and August

equally well marked in August and September although numbers – normally 30–60 per day – are lower than in spring. The pattern on the Pembrokeshire islands is similar.

There are a few exceptional records of Wheatears apparently overwintering. For example one in Glamorgan on 12 December (year unrecorded) in which county they are also reputed to have wintered sparingly last century in sheltered coastal areas. Breconshire has produced several mid-winter records, December 1971, 20 January 1972 and a remarkable record of five near Trecastle in January 1966. There was a bird at Templeton airfield in Pembrokeshire on 7 January 1989.

A large number of Wheatears has been ringed in Wales, mainly at the observatories and ringing stations on Skokholm and Bardsey. In addition to 14 recoveries elsewhere in Wales and England, these birds have yielded recoveries from France (9), Spain (8), Portugal (2) and Morocco (10); the majority of these recoveries occurred in the months August to October. Birds of the race *leuchorhoa*, familiarly called the Greenland Wheatear, are regularly identified in late spring and autumn in many locations in Wales although the number specifically identified is usually not large. The first record was of one collected on the Helwick lightship (Glamorgan) on 2 October 1909. On Skomer in 1988 "Greenland" birds were identified on 36 dates with a total of 50 on 26 September and 164 birds in all. A bird ringed at Tasmussaq in west Greenland in August 1980 was found dying at Colwyn Bay (Caernarfonshire) on the return migration in May the following year. Another ringed on Skokholm in mid-April 1946 was found breeding in Iceland two years later.

Pied Wheatear *Oenanthe pieschanka*
Tinwen Fraith

A vagrant.

There is only one Welsh record of this small Wheatear:

27–29 October 1968 – Female, Skokholm

The species breeds in Bulgaria and Romania eastwards across south Central Asia and winters in East Africa.

Black-eared Wheatear *Oeananthe hispanica*
Tinwen Clustiog Du

A vagrant.

There is only one definite Welsh record of this species which breeds from Iberia and north-west Africa through southern Europe east to Iran:

18 April 1970 – One male, Bardsey

On 19 April 1951 a Wheatear seen near Rhandirmwyn (Carmarthenshire) was considered to be probably this species.

Desert Wheatear *Oeananthe deserti*
Tinwen y Diffaethwch

A vagrant.

The only Welsh record of this species was a first summer male at Penclawdd (Glamorgan) on 21–22 November 1989. This is a typical date for this species on its rare occurrences in Britain.

The species breeds in north Africa from northern Sahara, Arabia and southern Caucasus east to Mongolia.

Rock Thrush *Monticola saxatilis*
Brych y Graig

A vagrant.

This species breeds from Iberia and north-west Africa, through southern Europe (north to Switzerland) eastwards to Mongolia and winters in the Sahel zone of south Sahara to Kenya.

There have been two records in Wales:

21 June 1981 – Male at Ynyslas (Cardiganshire)
4–6 June 1986 – Female at Llyn Alaw (Anglesey)

The two Welsh sightings fall well within the general pattern of occurrences in Britain.

A record of a pair at Lavernock on 23 August 1868 published in *The Birds of Glamorgan* (1900) was reviewed in Heathcote *et al.* (1967) and assessed as "extremely doubtful and almost certainly of mistaken identity".

Swainson's Thrush *Catharus ustulatus*
Carfonfraith

A vagrant.

This species breeds in North America and winters from Central America south to Argentina. There is one Welsh record:

14–19 October 1967 – One, Skokholm

Grey-cheeked Thrush *Catharus minimus*
Bronfraith Fochlwyd

A vagrant.

There have been three records in Wales:

10 October 1961	– One trapped, Bardsey
31 October 1968	– One attracted to lighthouse, Bardsey
20 October 1971	– One found dead, Bardsey

Ring Ouzel *Turdus torquatus*
Mwyalchen y Mynydd

Summer visitor breeding in the uplands in moderate numbers; passage migrant.

The Ring Ouzel breeds in almost all Welsh counties although nowhere can it be considered numerous. It is selective in its breeding habitat, requiring gullies, cliffs or screes with a good mixture of heather and bilberry, and frequently with isolated hawthorns or rowan. Most pairs are found between 250 m and 350 m although some occur in Snowdonia as high as 760 m; at the other extreme, pairs nesting on Yr Eifl, Llŷn peninsula, are well below 250 m and have been seen foraging almost down to sea level during the breeding season. They do not necessarily shun coastal areas and where suitable high hills fall steeply to the sea, as at Penmaenmawr and Llanfairfechan in Caernarfonshire, Ring Ouzels can regularly be found in residence. Nesting areas are strongly traditional in all the breeding locations in Wales. The presence of rocky outcrops seems a prerequisite for breeding pairs and disused quarries and old mine workings, where vegetation has had time to redevelop, are not infrequently used. However, many other sites which seem to appear eminently suitable throughout the uplands, are not tenanted. There is a tendency for pairs to group together in the most favoured sites so that breeding pairs may not infrequently be encountered in close proximity to each other, then making identification of actual numbers difficult to ascertain because birds can be surprisingly elusive and secretive once territories are established. Other than quarries and mine workings, the only other man-made habitat occasionally exploited is forest edge, where a few pairs remain faithful to breeding gullies after the adjacent encroachment of afforestation and may nest in the forest edge. The Ring Ouzel's main strongholds in Wales are in the Brecon Beacons and Snowdonia National Parks. RSPB estimates of the Welsh population in 1972 (unpublished) gave a maximum of 600 pairs; subsequent work in 1988 revised this downwards to 450–500 pairs (Tyler & Green RSPB unpublished). Our belief is that even this figure is too high and the annual breeding population may be no more than 365–425 pairs, confirming that the species has suffered a gradual and continuing decline over the past twenty years.

An overall decline of the species in Britain has been demonstrated this century (e.g. Baxter & Rintoul 1953, Alsop 1975, respectively for Scotland and northern England) and it is evident that this is reflected in Wales, albeit the evidence is largely anecdotal. The historical perspective is slightly confused, although it is clear that the species is scarcer now than it was formerly. By the nature of its breeding-season requirements, the Ring Ouzel inhabits sites which are amongst the most difficult either to manipulate for agriculture or even to afforest. Notwithstanding this, Ring Ouzels have certainly been lost from a wide range of sites for either of these reasons, e.g. in Montgomeryshire (see below). On the other hand "artificial" sites such as quarries have helped to compensate by providing new opportunities. In the 17th century Willoughby and Ray knew the species well in north Wales, Ray regarding them as frequent in the high mountains of Snowdonia, where Pennant also (1778) recorded the Snowdonia mountains as being "much frequented by the Ring Ouzel, a mountain bird". Forrest regarded it as one of the characteristic birds "more abundant on the northern slopes of the Berwyn Mountains than elsewhere". It seems likely that, despite widescale afforestation and agricultural improvement in much of upland Wales the status of the Ring Ouzel in the four mainland counties of north Wales (Caernarfonshire, Denbighshire, Flintshire and Merioneth) although declined locally, has not changed substantially this century. Today Ring Ouzels are well distributed throughout the mountains of

Snowdonia, with a western outpost on Yr Eifl on Llŷn (maximum three to four pairs). In Merioneth the other strongholds are on the Rhinog Mountains (where Hope Jones (1980) found 36 territories over the years 1965–70), the Arans, the Arenigs and the quarry systems around Blaenau Ffestiniog. Several former sites – well known to egg collectors in the early decades of the century – have been lost in Denbighshire where pairs are limited to relatively few on Hiraethog, a few on the flanks of the Berwyn Mountains, two (maximum three) pairs on the Clwydian Hills and fewer than half a dozen on Eglwyseg and Llandisilio Mountains. One of the Clwydian Hills pairs is normally on the Flintshire side of the boundary.

In mid and south Wales the situation is different where the more shallow gradients of the hills have led to much more extensive habitat changes. Ring Ouzels have been lost from a number of former sites in Montgomeryshire (e.g. Mynydd Fynyddog, Llanbrynmair Moors, Hengwm, Esgair Geulan), Carmarthenshire (particularly upper Tywi Valley), Cardiganshire (upper Ystwyth and – with Carmarthenshire – upper Tywi Valley), and Radnorshire (e.g. Llanbwchllyn, Llyn Heilyn, Marteg and Black Yat). In Radnorshire there was a noticeable decline in the 1950s from which the species has never recovered its former status; H. McSweeney (pers. comm.) believes that Ring Ouzels may have declined by as much as 50% on the hills around Aberedw in the 1970s and 1980s. In Breconshire Cambridge Phillips (1899) regarded the species as "fairly common on the heathery areas", nesting regularly on the Eppynt (heather is now scarce on Eppynt and Ring Ouzels have not bred there since 1983); Walpole–Bond (1904) thought the Ring Ouzel to be more numerous in this county than others in mid-Wales, basing his statement principally on his knowledge of the hill around Abergwesyn. Massey (1976) thought it still reasonably common in the early 1970s. It is regarded now as being at best uncommon and declining (Peers & Shrubb 1990). In the first half of the present century Ring Ouzels were thinly scattered on the hills of Glamorgan between Port Talbot and the heads of the eastern valleys, such as the Rhondda; by the 1950s they were also found to be breeding at sites in the north of the county where they had not previously been known to occur. By 1965 no fewer than 14 pairs were known to be breeding and this loose scattering of pairs continues to nest in the same valleys each year although the annual total probably does not regularly exceed a dozen pairs or so. In Carmarthenshire, Ingram and Salmon stated that it had much declined by the 1950s and it is now confined to breeding on the Mynydd Du (Carmarthen Fans) where a reasonable population still remains. Mathew (1894) found the Ring Ouzel to be frequent on the Preseli Hills in Pembrokeshire during the breeding season (eggs collected there in 1867) and although the *Breeding Atlas* recorded the species in two 10-km squares between 1968 and 1972 it was not located during the tetrad breeding survey, 1984–88 and no longer breeds in the county – or at best only very sporadically; last proved breeding was in 1971.

RSPB surveys in the Brecon Beacons National Park in 1978 found 91 pairs, mainly on the western Black Mountain (Mynydd Du, see above) and the Black Mountains of the Monmouthshire/Breconshire border; it was further suggested that up to ten more pairs might have been present in areas not covered by the survey. As the *Gwent Atlas* and Tyler and Green have subsequently shown for the eastern areas, these pairs have subsequently declined. A breakdown of the current breeding population county by county is shown on the map below.

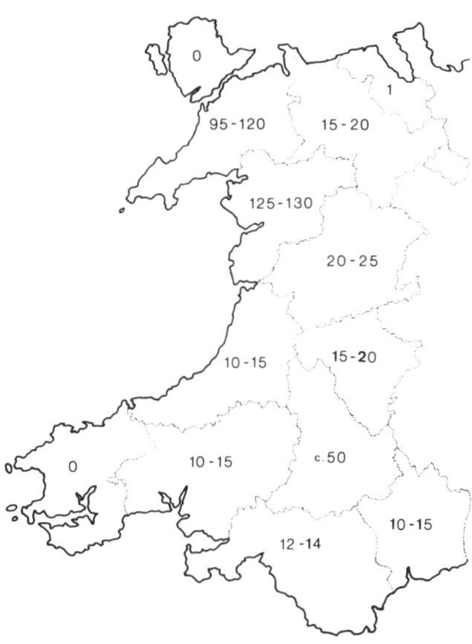

Estimate of breeding pairs of Ring Ouzels in Welsh counties, early 1990s

A variety of reasons has been suggested for the declines. In addition to the land-use changes referred to above, competition from the closely related Blackbird has been cited as a factor in Ireland and could equally well apply in Wales in those areas where populations of Blackbirds in the uplands have increased considerably with the development of afforestation. The effect of a burgeoning Peregrine population from the 1980s has also been suggested, as Ring Ouzel remains have not infrequently been found in Peregrine eyries. Merlins too will occasionally take them (Bibby (1986) quotes three examples from 6366 birds taken by Merlins at Welsh sites). Tyler and Green postulate the effects of degraded habitats both here and in the wintering areas, e.g. the loss of Juniper, a principal food source in winter, in the Mediterranean area. Hang gliding and other human recreational disturbance may also be implicated at some sites – a factor which is believed to be responsible for a decline on the Clwydian Hills.

Being relatively short-distance migrants, Ring Ouzels are among the first of the spring arrivals with the first birds arriving as early as the middle of March in some years. Birds continue to arrive throughout April and clearly include passage migrants *en route* for breeding areas farther north in Britain and Norway. Passage is quite well marked on the offshore islands ending the first week of May, with its peak (Pembrokeshire islands and

Bardsey) in mid-April. A remarkable gathering of 33 passage birds was seen at Amlwch (Anglesey) on 12 April 1988, but such concentrations are rare, especially in spring. Exceptionally, very early individuals may be encountered, as one on Skokholm on 29 February and an even stranger record of five birds which Bolam saw in a snowstorm on the Aran Mountain summits (Merioneth) on 19 February 1905. Autumn passage occurs throughout September and October (late dates up to 21 November (Skokholm)). There are one or two aberrant winter records: a bird was seen in December (year unknown) in Glamorgan and in January 1945 two individuals were watched in Merioneth and a third picked up dead at Aberdyfi on 26 January (*British Birds* vol. 38, p. 237). A January record also exists from Pembrokeshire, a bird having been watched at St. David's on 6 January 1985 and in Cardiganshire there have been two reasonably recent winter records, one of a bird at Llangrannog between 14 and 30 December 1985 and a previous bird at Mwnt on 3 February 1979.

Recoveries of Ring Ouzels ringed in Wales either as nestlings (3) or autumn passage birds on Bardsey have been reported from France (3), Morocco (1) or farther south in Britain (2). One of the nestlings from Trefil Quarries (Monmouthshire) was retrapped on the Sussex Downs in its first autumn, indicating a migratory route involving the shortest sea crossing.

Blackbird *Turdus merula*

Mwyalchen

Abundant breeding resident, passage migrant and winter visitor.

Few birds in Wales are more familiar than the ubiquitous Blackbird. It is an inhabitant of a wide spectrum of habitats from suburban gardens and parks to farmland, ffridd, woodland, watersides, sea-cliffs and the fringes of moorland. The *New Atlas* and the original *Breeding Atlas* unsurprisingly both record Blackbirds breeding in every 10-km square in Wales.

Originally outnumbered by the Song Thrush until the 1940s, the Blackbird has subsequently taken over as our most numerous thrush species and in overall national terms outnumbers the Song Thrush by a ratio of at least 4:1 (Marchant *et al.* 1990). It thrives particularly in suburban areas where nest sites abound in gardens and shrubberies, and where earthworms are abundant both on lawns and playing fields, and other supplies of food are artificially provided in winter. It is here that its population density is highest (Batten 1973), although the move into built-up areas was not really manifest until the second half of the 19th century. Farmland levels too may reach as high as 250 pairs per tetrad locally (e.g. parts of Monmouthshire) after mild winters.

The new generation of conifer forests now occupying about 25% of the uplands of Wales, as well as the many lowland areas, has produced an added opportunity for the Blackbird, which has been quick to colonise such habitats and thereby extend its range into many upland areas which were not otherwise available to it. Nesting regularly occurs in these forests up to at least 500 m in many parts of Wales and it has been suggested that its presence on the margins of such forests at the interface with heather/bilberry in some areas may have been to the disadvantage of the smaller and less dominant Ring Ouzel; there is no proof of this however. Neither is the Blackbird confined to the mainland, for breeding regularly occurs on most of the larger offshore islands e.g. Bardsey (up to 45 pairs in the 1960s; slightly fewer since), Skokholm (normally two to five pairs, with a maximum of nine pairs in 1990), Skomer (25–30 pairs), Flatholm (14 pairs, 1991).

The CBC indices for both woodland and farmland sites show a gradual decrease in numbers nationally after the relatively high levels which were reached in the early and mid-1970s although the number of regular CBC sites in either habitat in Wales is inadequate to judge whether this applies here; indeed it may not be the case in the grassland-dominated agricultural regimes which prevail in Wales.

Tyler (1990) found Blackbirds at a mean density of 5.9 pairs per 10 ha on farmland in Monmouthshire (1979–89) and Squires (CBC survey) found a mean of 11.9 pairs per 10 ha in mature, ungrazed woodland at Ynys-hir (Cardiganshire) (1985–91).

Spring and autumn passage is well marked in coastal

Overseas recoveries of Blackbirds ringed in Wales and (in parentheses) countries of origin of individuals subsequently recovered in Wales

areas and on the offshore islands with the highest numbers in October and November; on Bardsey, day-totals regularly reach as many as 400 at this season with occasional influxes as large as 2000 (22 October 1962) while spring maxima (March–April) rarely exceed 100 per day. Autumn maxima on the Pembrokeshire islands also reach four figures daily on occasion with up to 1000 present on Skokholm on 18 October 1964. Ringing recoveries show the complex composition of the wintering and passage populations, which includes individuals from farther north in Britain, several continental countries and a percentage of the local breeding populations which remain resident. Masked within this picture is the fact that many Welsh Blackbirds also move south in autumn.

Dusky Thrush *Turdus naumanni*

Brych Tywyll

A vagrant.

This species is a Song Thrush-sized bird, which breeds in Siberia and winters in Assam to south-east Asia and Japan. It has been recorded once in Wales:

3–5 December 1987 – One, Skomer

Fieldfare *Turdus pilaris*

Socan Eira

A winter visitor and passage migrant in variable numbers; usually abundant.

The numbers of Fieldfares visiting Wales in winter vary from one year to another, depending on weather conditions in other parts of their wintering area and the supply and availability of the fruit crops, notably rowan and hawthorn, both in the countries of origin and in Britain. Fieldfares leave their Scandinavian breeding areas in autumn once the crop of rowan berries has been exhausted and although small vanguards appear in September (rarely late August), the main influx into Britain occurs in October with flocks generally reaching Wales from the middle of the month onwards although dates can be very variable. For example, in years of heavy rowan crop and mild weather in Scandinavia the flocks may remain there until well into winter. In years of more normal winter weather Fieldfares tend to advance southwards in autumn as the weeks progress and northern food stocks are depleted. In severe weather, however, the pattern may well be different with larger numbers of birds present and a quicker dispersal southwards with many birds passing straight through Wales *en route* to Ireland or France and Iberia.

Fieldfares are strongly gregarious, usually moving around in single species flocks, often several hundred strong, occasionally with smaller numbers of Redwings or Starlings in company. The *Winter Atlas* shows that the larger flocks are almost invariably found on the lower ground of river valleys and lowland farmland. As winter progresses and the crops of hedgerow fruits become exhausted they feed in loosely scattered groups in open pastures when the ground is not frozen. They have a fondness for fallen apples and will readily exploit supplies when they can locate them. Despite their northern breeding range Fieldfares are not particularly hardy nor very versatile feeders and are therefore vulnerable to protracted spells of harsh weather which they usually escape by moving farther south ahead of the cold conditions; this was evident in the 1962–63 winter when great numbers appeared in Glamorgan in December but quickly moved on by the time the bitter weather arrived at Christmas.

Return migration takes place in March and April, with late birds occurring up to the early days of May. Extreme examples have been recorded as late as 1 June on Bardsey and 13 June on Skokholm. By the end of April there is occasional evidence of courtship and even territorial behaviour as instanced by a male holding territory in Breconshire on 17 May 1988. As a breeding species the Fieldfare has expanded its range slowly westwards across Europe over the past hundred years or so and now nests sparingly (or sporadically) in Holland, Switzerland, France, Denmark and northern Britain (since 1967). There has not been serious suspicion of breeding in Wales although one report from north Glamorgan in July 1972 was suggestive. There is also a puzzling record of two birds in moult in west Glamorgan in late summer 1983.

Song Thrush *Turdus philomelos*

Bronfraith

A widely distributed and fairly common resident, formerly abundant. Winter visitor and passage migrant.

In the early decades of the present century the Song Thrush comfortably outnumbered the equally familiar Blackbird, but since the 1920s and 1930s the position has changed as Blackbird numbers increased. The trend was further exacerbated as the Song Thrush suffered a decline in numbers in the 1940s which has continued erratically up to the present (Marchant *et al.* 1990).

Although resident, the Song Thrush is surprisingly vulnerable to harsh winter weather which is believed to have been a principal factor in the periodic reduction of populations. Marchant *et al.* (1990) cite the series of cold winters around 1940 as being particularly damaging and subsequently those of 1947, 1961–62, 1962–63 and 1981–82 having further depressing effects on numbers from which the species appears to recover less rapidly than many other resident passerines.

The *Breeding Atlas* (1968–72) demonstrated that Song Thrushes were then found in all counties and in fact in every 10-km square in Wales. The *New Atlas* paints much the same picture, although, perhaps significantly, there are one or two squares in which Song Thrushes were not found or were not proved to breed. They occur in a wide variety of habitats that have bushes, thick hedges or trees and they are therefore most evidently missing only from the substantial areas of open hills and mountains in the uplands.

They do not breed, except spasmodically, on any of the offshore islands with the exception of Caldey but have taken advantage of the new generation of forestry plantations to expand into the hills in many counties up to at least 450 m. Here they are usually more numerous than Blackbirds and can be surprisingly common in younger areas of forest and restocked areas, occurring in north Breconshire in 34% of sample sites and comprising some 5% of total bird populations.

In recent surveys Song Thrushes have been found in 83% of tetrads in Pembrokeshire, with a suggested population of some 6000 pairs. In west Glamorgan Thomas *et al.* (1992) found them in 87% of tetrads; comparable figures for Breconshire and Monmouthshire were 62% and 97% respectively. Densities on farmland can reach as many as 20–40 pairs per tetrad but are considerably lower than this in most areas.

It is considered that populations in the counties of Montgomeryshire, Breconshire and Cardiganshire and possibly elsewhere may still be declining and the population in Wales at present has probably never been lower. Many of the Song Thrushes move away from the breeding areas, particularly those in the hills and upland forests, at the end of the breeding season and an unknown percentage emigrates. Ringing recoveries of breeding birds or fledglings ringed in Wales have occurred from Italy (1) and western France (1). Autumn passage, which is clearly marked in some coastal areas (e.g. Lavernock Point, Great Orme, coastal plain of Cardigan Bay) as well as on the offshore islands, sees the arrival of immigrants from Europe and the northern and eastern parts of Britain. Individuals of both the British/proximal continental race *T. p. clarkei* and the nominate race, *T. p. philomelos*, typical of the central parts of the European continent, are recognised among these migrants and again ringing recoveries confirm the origins of some of these individuals. Some of these autumn migrants move considerable distances, e.g. Bardsey to Seville (Spain); Bardsey to Rome. Most recoveries of these autumn passage birds have come from Ireland (6), France (3), Spain (1) and Portugal (1). One Bardsey migrant was recovered in Denmark the following summer and another was found breeding in Finland the following May.

Redwing *Turdus iliacus*

Coch Dan-aden

An abundant winter visitor and passage migrant; particularly numerous in severe winters.

The Redwings that visit Wales in large numbers in winter are predominantly from the Scandinavian breeding populations although a number of the darker, more heavily streaked Icelandic birds are also regularly identified. Thus, a nestling ringed in southern Finland was found at Tondu (Glamorgan) the following winter and an individual ringed on Bardsey was subsequently recovered in the same area of Finland. Similarly, an adult ringed at Dale (Pembrokeshire) was reported at Hedmark in central Norway on southern migration at the end of the breeding season a year later. A series of other recoveries of Redwings ringed in Wales gives evidence of southerly movement of passage birds (and sometimes the onward progress of wintering ones) with numerous recoveries from France, Spain, Portugal and Ireland. However, the interesting recovery of a bird which had been ringed in Merioneth in January 1971, and subsequently found 4000 km east in the Ural Mountains in July the same year, is proof that at least some of the birds from much farther east in Europe occasionally reach Wales.

Redwings arrive in force from late September through to early November and the soft "seep" flight call is a common feature of autumn nights as the flocks pass unseen over rural areas or above the noise of busy conurbations. The smallest of the British thrushes, the Redwing is notoriously susceptible to hard weather. Flocks are very mobile, moving from one area to another as food supplies are exhausted and making long hard-weather movements if necessary. They are also highly nomadic from year to year, often wintering in wholly different parts of Europe from one winter to the next as shown by ringing recoveries: a bird wintering in Swansea in January 1965 was found in southern Italy in March 1966; another ringed at Gresford (Denbighshire) was also in Italy two years later; another bird caught in Flintshire in December 1976 was in southern France within three months; a first-winter bird caught on Bardsey in October 1959 was *en route* for Iberia, where it was found dying in southern Portugal in February 1960.

In the early part of the winter flocks of Redwings are a

familiar sight feeding in large parties on the fruits of hawthorn and other berry-bearing trees. Later in the winter, depending on weather as well as food supplies, many birds move elsewhere while those remaining spend more time foraging in pastures for earthworms in company with Starlings or Fieldfares. Those that stay and are caught in severe weather can suffer heavy mortality. Forrest records how birds were picked up dead in the woods around Little Orme in the bitter weather of January 1902. In the memorable winter of 1962–63 most Redwings moved on and escaped mortality whereas in 1947 they were caught by the sudden heavy snowfall in March and "died in their thousands under the hedges" in the Cardiff area (Heathcote *et al.* 1967), as elsewhere.

Feeding flocks in autumn frequently number as many as 2000 in any areas of Wales but as most move on the numbers are smaller as winter progresses and flocks of 50–100 are more usual. Autumn influxes are sometimes heavy on the offshore islands, e.g. 3000 on Bardsey on 15 October 1966 and 1500 on Skokholm on 28 October 1964.

Although Redwings have colonised Scotland slowly since 1925, there has never been any suggestion of breeding as far south as Wales. The report by Saxby (in More 1865) of "a nest in north Wales" where the bird was "repeatedly observed sitting and the eggs collected" must presumably be disregarded.

Mistle Thrush *Turdus viscivorus*

Brych y Coed

A fairly common resident; many young birds emigrate in their first autumn.

The Mistle Thrush is a relatively common bird throughout Wales, widely distributed in countryside and suburban areas but nowhere numerous. Although it is less plentiful than the other two familiar resident thrushes – Blackbird and Song Thrush – there are areas locally where it outnumbers the latter, e.g. parts of Montgomeryshire and Snowdonia, especially in the foothills. In this respect the nature of many of the woodlands and other tree-covered areas favours the Mistle Thrush whilst operating against the bush- or shrubbery-nesting Song Thrush. The absence of understorey, thick hedgerows or bushes, often through long-standing overgrazing by sheep, tends to exclude the Song Thrush while not affecting the tree-nesting Mistle Thrush. Garnett (in prep.) found nesting Mistle Thrushes on 12% of ffridd sites surveyed in 1985–87.

The Mistle Thrush was virtually unknown in much of north and west Britain two hundred years ago and although no contemporary accounts exist relating specifically to Wales, the inference has to be drawn that most of Wales was colonised by the species during the period of fairly spectacular expansion which occurred in the first half of the 19th century (Ireland was colonised for the first time during this period also).

Nationally, the Mistle Thrush has shown a marked decline in numbers since the severe winter of 1981–82 and the subsequent reasonably harsh ones in the mid-1980s. However, during this period Welsh numbers seem to have kept up fairly well and county bird reports do not identify any noticeable declines, except those to be expected temporarily after severe winters. Mistle Thrushes do particularly well in stock-rearing areas like Wales, benefiting from such factors as the double cuts of silage which make for easy feeding on the harvest of ground invertebrates thus made available (O'Connor & Shrubb 1986). The two *Breeding Atlases* (1968–72 and 1988–91) both indicate ubiquitous distribution throughout Wales although the rather coarse picture produced (breeding presence or absence in 10-km squares) disguises the more patchy distribution, particularly in the west where the Mistle Thrush is more thinly distributed than in other areas. Donovan and Rees (in press) suggest an average of two pairs per tetrad in the western half of Pembrokeshire contrasted to perhaps 10 pairs per tetrad in the east. Little work has been done in Wales to determine numbers or densities of breeding Mistle Thrushes. On the basis of figures given above a population of c1400 pairs is suggested for Pembrokeshire and Tyler *et al.* (1987) estimated a total of up to 3500 pairs in Monmouthshire (1981–85).

Mistle Thrushes will also take advantage of the new generation of conifer forests where suitable areas exist within them. At 236 count points in restocked conifer plantations in north Wales in 1983 Bibby *et al.* (1985) found Mistle Thrushes at a density of approximately 0.7 *individuals* per 10 ha; in mature plantations with occasional broadleaved species in association, Bibby *et al.* (1989) found them slightly more numerous at just under one bird per 10 ha.

The species is well known to be highly territorial during winter, adult pairs (or single birds) fiercely defending selected berry-bearing trees/bushes against congeners and other berry-eating species. However, many of the young birds are migratory, crossing the Channel to France and Iberia during their first autumn. The sudden arrival of autumn flocks on the offshore islands presumably involves migratory gathering of such birds. Occasional daily maxima of 50 have been recorded on Bardsey; 100 were seen on Skomer on 3 October 1983 and there was an astonishing flock of c500 on Ramsey on 12 October 1992. Prior to this emigration of young birds, combined family parties are frequent in hill areas after the breeding season where they forage in roving bands for bilberries and rowan berries. Such parties break up once the adults begin to defend winter territories and the young then move away. The roving parties, often as large as 50 strong, forage well up into the hills and have been recorded above 660 m in Merioneth and Caernarfonshire. Despite its habit of defending food supplies, the Mistle Thrush is somewhat vulnerable to hard weather; Forrest referred to substantial kills of these birds as long ago as 1895 and the CBC identified the severe fall in numbers after the 1962–63 winter. O'Connor and Shrubb (1986) have recognised the fact that Mistle Thrush populations in pasture-dominated agricultural areas such as Wales remain more constant, and have better breeding results, than in the more arable areas of England where they are prone to fluctuations and probably a steady continuing decline to the present.

Cetti's Warbler *Cettia cetti*
Telor Cetti

A scarce breeding resident in south Wales, occasionally recorded on passage.

Although in Europe this warbler was practically confined to the shores of the Mediterranean at the turn of the century there has been a gradual northwards extension which culminated in the confirmation of breeding in Kent in 1972.

The first record in Wales was one seen on Bardsey from 26 to 30 October 1973. There was then a gap until the second record at Rhosneigr (Anglesey) from 11 to 18 December 1976. In 1977 one was found in Glamorgan (Oxwich, NNR 9 July) followed by the first in Pembrokeshire at Bosherston Pools from 24 January to 28 March 1979.

By 1985 there were up to nine singing males at Oxwich NNR and breeding was confirmed in Wales for the first time. In 1987 breeding was also confirmed in Pembrokeshire where the species was present at seven sites south of the Cleddau estuary.

In 1988, in addition to records from Glamorgan and Pembrokeshire, there was one singing at Pwll in February and May and one singing at Llangennech from 15 October to 31 December (both sites in Carmarthenshire). In 1989 breeding was proved for the first time in Carmarthenshire at Ffrwd Fen Nature Reserve whilst in Monmouthshire there was one at Greenmoor Pool from 13 April to 20 June and one at Newport from 5 to 7 June.

In 1990 there was again one singing at Greenmoor Pool, from 18 March to 7 June and in addition to the usual site at Oxwich there was one at Square Pond Neath from 14 to 20 November. In Carmarthenshire breeding was noted again at Ffrwd Fen and one was singing at Witchett Pool from 10 May. An extension of range was noted in Pembrokeshire where in addition to the regular sites in the south of the county breeding was recorded at Pentwd Marshes. Birds were also present on the Cardiganshire side of the Teifi Marshes during March to July and in October to November.

It therefore appears that a further extension of breeding range northwards along the Welsh coast can be expected provided that there is no setback through severe winters. The preferred habitat of areas of low tangled vegetation in wet or damp situations is available at many sites in mid and north Wales.

Outside the breeding period there have been recent records of single birds at Lavernock Point, Skokholm, Skomer and Strumble Head.

Lanceolated Warbler *Locustella lanceolata*
Telor Rhesog

A vagrant.

This skulking species, which breeds in east Eurasia from central Russia to north Japan, owes its place in the Welsh avifauna to one record:

18 October 1990 – One caught at Bardsey lighthouse

Grasshopper Warbler *Locustella naevia*
Troellwr Bach

Summer visitor, widely but thinly distributed in all counties. Passage migrant.

E. A. Swainson writing in the *Zoologist* in 1899, summed up the requirements of this intriguing and elusive summer visitor. Speaking of Breconshire he wrote, "We have here most of the conditions in which this little summer migrant delights, such as rushy meadows with grass tussocks here and there, neglected fields with stunted clumps of blackthorn, bushes and brambles, dingles furnished with little alder bushes and dry wastes of low cover". To this concise summary we can now add the newly established conifer plantations, which have proved to be attractive to the Grasshopper Warbler in recent decades, enabling it temporarily to extend its range into many areas especially upwards into the hills to altitudes of 305 m or rarely 457 m.

The Grasshopper Warbler breeds throughout Wales, although its local distribution has necessarily shifted considerably from pre–war years as a result of agricultural intensification, the drainage of much wetland and the wholesale removal of many pockets of rank vegetation on farmland which formerly held pairs. The losses caused by these factors have to a small degree been offset by the expansion which has taken place in new areas of upland afforestation at both first-generation and restocking stage of conifer plantations. No definitive figures exist to suggest whether the Grasshopper Warbler is now scarcer or more numerous than formerly, particularly through the absence of data from the past. The suspicion is that the losses in lowland sites have not in any way been matched by the gains in upland forests, certainly since new plantings virtually ceased from the late 1980s. Distribution shown in the *New Atlas* demonstrates a markedly western bias to the breeding population, with strongest numbers in Carmarthenshire, Anglesey and all round the coast of Cardigan Bay from Pembrokeshire to Llŷn. Most accounts at the end of the 19th and beginning of the 20th centuries speak of the Grasshopper Warbler as being generally common throughout the country.

Forrest (1907) suggested that it was a more common bird in the uplands of north Wales than on the low ground but it is certainly to be found in a wide variety of lowland sites, particularly those which offer patches of rank vegetation on wetland margins.

Swainson, in Breconshire, plundered nests in several years following his arrival in the county in 1884 and regarded it as reasonably common, at least locally, although not all writers were consistent in this view; Cambridge Phillips, for example, actually thought it rare in Breconshire while a correspondent in the *Zoologist* (1850) thought it "very uncommon" in Pembrokeshire, a view which is not greatly at variance with Mathew (1894), who considered it "rather scarce and local" and not unlike the situation that apparently prevailed in Carmarthenshire. Similarly, another correspondent in the *Zoologist* in 1895 expressed the view that "there is some reason to think that the bird may only have reached this part of Wales in recent years", writing of Cardiganshire. In Merioneth around the same time it was stated to be "first heard six years ago (in 1888) and now annually increasing". In the other counties of north Wales different writers regarded it as widespread and fairly common in the lowlands; on Anglesey Coward found it abounding in the north-east of the island and only slightly less so in the west in the 1920s.

With a species that is prone to fluctuations of numbers from one year to another (as demonstrated by the CBC indices) it is more difficult to detect longer term trends. Nonetheless there was clearly a serious fall in numbers nationally around 1972–74 (Marchant *et al.* 1990) which may possibly be attributed to the failure of rains in the African wintering quarters. However, there is little evidence in Welsh reports to relate this event to the Welsh population. This overall drop in numbers followed a steady but measurable increase during the 1950s and 1960s. Throughout the 1980s the population appears to have strengthened again in some parts of Wales although the further intensification of agriculture in the 1980s has removed more areas of suitable habitat.

It is interesting to consider the present status of the species and compare this with some of the above statements made (mostly) about one hundred years ago:

Anglesey

RSPB surveys in 1986 located a minimum of 133 singing birds; widely spread throughout the island, which clearly remains one of the main strongholds for the species in Wales.

Caernarfonshire

Thinly but widely distributed (Hope Jones & Dare 1976). RSPB surveys found 87 singing birds on Llŷn in 1986–87.

Denbighshire / Flintshire

Thinly distributed as a breeding species, predominantly in coastal scrub.

Merioneth

Still fairly common in suitable habitats (Thorpe, in litt.).

Montgomeryshire

Now very scarce and local. Lost from many lowland sites and not now common in upland conifer plantations since new plantings have virtually ceased; doubtfully as many as 50 pairs in most years (pers. obs.).

Cardiganshire

Fairly common breeding visitor in coastal scrub, valley bogland and young conifer plantations (*Ceredigion Bird Report* 1982–83).

Breconshire

Very local and uncommon; probably around 50 pairs, Shrubb (in litt.).

Radnorshire

Very scarce. Probably fewer than 10–12 pairs (Peers, in litt.) now that young conifer plantations have matured and become unsuitable.

Pembrokeshire

Located in 89 tetrads (1984–88). Perhaps a population of c400 pairs annually.

Carmarthenshire

Frequent along the coast occupying rushy meadows, fens and other areas of rank growth. More local inland utilising areas of rough vegetation such as valley mires and young plantations.

Glamorgan

Fewer than 50 pairs in west Glamorgan (Thomas *et al.* 1992). Locally common in suitable habitat elsewhere in the county.

Monmouth

Very localised. Probably fewer than 50 pairs (Tyler *et al.* 1987).

The overall population in Wales at present (1993) is probably not much more than 1300 pairs. Several ringing recoveries give support to the suggestion that Grasshopper Warblers return to the same localities in succeeding seasons, although, as Parslow (1973) has pointed out, they are seemingly prone to local fluctuations and erratic in their occupation of sites. One recovery from Algeria (February) of a bird ringed as an adult in Cardiganshire the previous April gives a clue to the likely wintering area for some birds in the Mediterranean basin and north Africa. However, much still has to be learned about the main areas in which the population winters which are believed to be principally in west Africa south of the Sahara (BWP vol. 7). Grasshopper Warblers arrive from mid-April onwards, and leave in August–September, with stragglers occurring up to early October.

River Warbler *Locustella fluviatilis*

Telor yr Afon

A vagrant.

The only Welsh record was of one found dying at the Bardsey lighthouse on 17 September 1969. This was the third British record, the second having been on Fair Isle the previous day. This species breeds from Poland eastwards to western Siberia and winters in East Africa.

Savi's Warbler *Locustella luscinioides*

Telor Savi

A vagrant to Wales even though it breeds as close as northern France and south-east England.

There are three records from Wales:

31 October 1968 – One seen, Skokholm
18 June 1983 – One singing, Dowrog (Pembrokeshire)
13–20 May 1987 – One singing, Oxwich Marsh

Aquatic Warbler *Acrocephalus paludicola*

Telor y Dŵr

A scarce autumn passage migrant.

This is a species that breeds from Germany eastwards, also northern Italy and Hungary and winters in Africa south of the Sahara. The first Welsh record was at Skokholm on 5 September 1949, there was then a lengthy gap until the second, at the same location, on 20–21 September 1961. The total up to 1991 stands at 35 occurrences, all of single birds, except for records of two at Kenfig on 20 and 31 August 1989, and 20 August 1991, three there on 16 August 1991 and two at Skomer from 7 to 9 September 1990; there are records for 16 of the last 22 years from 1970 to 1991 inclusive. The apparent increase is probably a reflection of a developing interest in south Wales in ringing at reedbed sites, the preferred habitat for this skulking species.

August is the most favoured month but there are also some records from September and October. There is a strong bias towards south Wales, with 19 records of 24 individuals coming from Glamorgan, the remainder from Pembrokeshire (7 records of 8 individuals), Breconshire (4), Caernarfonshire (2), Monmouthshire (2) and Carmarthenshire (1).

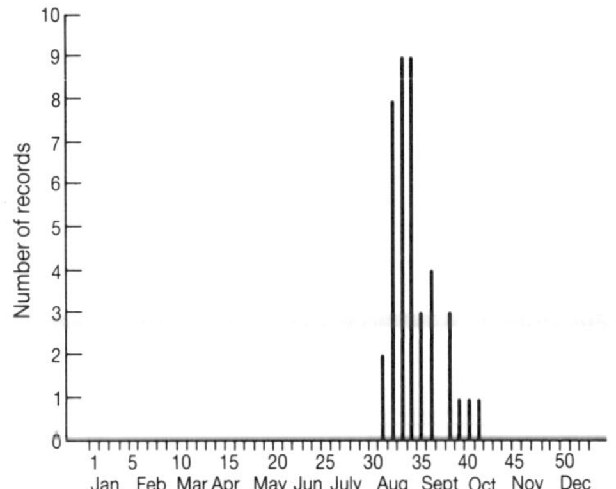

Weekly records of Aquatic Warbler to 1991

Sedge Warbler *Acrocephalus schoenobaenus*
Telor yr Hesg

A locally distributed summer visitor, numerous in some sites; plentiful passage migrant.

Although overlapping in some reedbed sites with the Reed Warbler (q.v.), the Sedge Warbler is much more catholic in its choice of habitats, preferring low dense vegetation on the margins of wetlands, and is therefore more widespread and generally far more numerous as Thomas (1984) has demonstrated in his detailed work at Oxwich NNR. It breeds in all counties although usually avoiding upland areas (above approximately 260 m) and is very scarce in the three mid-Wales counties of Radnorshire, Montgomeryshire and Breconshire. There are, however, occasional records of birds at altitudes as high as 420 m, e.g. Waun y Griafolen (Merioneth) in 1988 and 1989 (pers. obs.).

Many of the earlier writers regarded the Sedge Warbler as particularly common in lowland areas, as indeed it probably was in days before the extensive drainage of wetlands, which reached its peak in Wales in the second half of the present century; Forrest recorded it as being generally distributed (in the six north Wales counties) and particularly numerous in many parts of the lowlands, even if somewhat localised. Mathew believed it was the second most numerous warbler in Pembrokeshire, after the Chiffchaff and Lockley (1949) similarly found it ubiquitous and numerous in the same county.

There is no evidence of long-term changes in status of the Sedge Warbler in historical times although there have been peripheral withdrawals from some areas where habitats have been lost through urban development and agricultural land improvement in post-Second World War years. However, since the late 1960s there has been a major collapse in numbers, followed later by a sharp (partial) recovery since about 1986. The CBC clearly shows how the population nationally reached a peak in 1968 but thereafter fell progressively, albeit with some annual fluctuations, to a low point in 1974 and then even lower in 1985. As with the Whitethroat and Sand Martin, there is no doubt that the decline since 1969 has been as a result of the serious rainfall deficit in the species' wintering area, the Sahel region in the savannah zone of north tropical Africa. Conversely, the recovery of numbers in the later 1980s followed the more recent improvement in rainfall in the region. The effects of these major fluctuations are reflected in county bird reports over the past two decades or so, e.g. Glamorgan. In many of the marginal areas where populations were always rather sparse – Montgomeryshire, Radnorshire, Breconshire – the species has now been lost from various sites in which it was previously regular, although the habitats remain apparently suitable.

In the years prior to the decline in the 1970s, numbers were impressive at some sites, e.g. 100+ pairs at Kenfig NNR, 160+ at Crymlyn Bog. Even during the poorer years prime sites held relatively high numbers. Thomas (1984) located up to 87 pairs at Oxwich NNR in 1977 (which was a relatively "good" year among poorer ones both before and after); in the five seasons 1975–79 an average of 68 pairs was counted on the 62-ha site. In 1992 46 pairs were found at the same site.

In Carmarthenshire, where the post-1968 crash in numbers was also evident the main breeding sites are found in the wet habitats on the coastal lowlands, e.g. Pendine, Cydweli, Pembrey and Bynea/Machynys. It is also present, but more thinly distributed, in the main river valleys including the principal inland waters.

Current population estimates are available from three south Wales counties. The *Gwent Atlas of Breeding Birds* estimated 100–200 pairs in 1981–85 (present in 12% of tetrads in the county, almost exclusively on the Gwent Levels). In Pembrokeshire, Sedge Warblers occurred in 23% of tetrads and the population is considered to be around 650 pairs (1988). The present population in west Glamorgan is estimated at 150–200 pairs (D. K. Thomas, pers. comm.).

In the north Wales counties the Sedge Warbler is still generally well distributed in suitable areas even as far west as the extremities of Llŷn, including breeding on Bardsey up to the years of the population crash. Since then the numbers have clearly improved again in some of these north-western areas. RSPB surveys on Llŷn (1986–87) found a minimum of 370 pairs and 705 on Anglesey in 1986 where Forrest found it "extremely common in the marshy lands in the west of the island". He further recorded it as very common around western estuaries – Dyfi, Dysynni, Glaslyn, Dwyryd – and on the Dee marshes which still generally holds good. Far up the Dee Valley around Llyn Tegid and Bala (Merioneth) (c160 m) Sedge Warblers seem always to have been more numerous than they are at other sites well inland in Wales. Cors Fochno, Cors Caron and the Teifi and Dyfi estuaries are the principal sites in Cardiganshire but it is in the neighbouring counties to the east, along the Welsh Marches, that numbers are lowest. In Montgomeryshire the modest population which existed pre-1969, mainly along the Montgomeryshire canal, has diminished and there are now probably fewer than 25 pairs in any year. Radnorshire produced only 14 pairs at six sites in 1981 (34 pairs in the early 1960s) and even fewer in the worst year, 1984. The Breconshire population is put at fewer than 25 pairs. On the basis of the above figures the Welsh breeding population is probably in the region of around 3000 pairs.

Sedge Warblers are one of the most numerous summer visitors recorded on passage on the offshore islands. Atypically, they are considerably more numerous on spring passage, between late April and the end of May, than they are in autumn; "falls" of these migrants can sometimes be quite spectacular, for example 750 present on Bardsey on 14 May 1967, whereas autumn maxima on the same island have not exceeded 150 (30 August 1968 – in which autumn 319 were killed at the lighthouse). Autumn passage occurs from late July, with the last birds passing through to the end of September or the early days of October.

On the Pembrokeshire islands spring passage is well marked and until the 1950s daily maxima of up to 100 were not unusual with 250 present on Skokholm on one day in May 1953. In the 1960s and 1970s maximum daily counts of 40–50 were recorded, falling to 20 or so in the 1980s. Both on account of this strong passage through given locations and the ease of trapping Sedge Warblers on their breeding sites, large numbers have been ringed in Wales and recoveries of ringed birds are relatively numerous. The bulk of recoveries and retraps have been from the counties of southern England but there have also been 26 from western France and the Channel Islands and one from Morocco. More distant recoveries have traced birds as far as Upper Volta (1) in the same autumn, and Senegal (1) on return migration four years after being ringed at Llangorse. An adult bird ringed on Bardsey on spring migration was found in southern Norway the following spring. Evidence that some birds move north on a more easterly route than in autumn is provided by a bird ringed at Shotton (Flintshire) and retrapped on Malta the following spring.

Marsh Warbler *Acrocephalus palustris*

Telor y Gwerni

An extremely rare summer visitor and passage migrant. Has bred at least once.

The Marsh Warbler has only the most tenuous foothold in Britain, its present breeding distribution being restricted to four or five pairs in Worcestershire and perhaps four times that number in Kent. Other sporadic pairs occur, together with occasional individuals or singing males, as has been the case in south Wales.

The only proof of Marsh Warblers breeding in Wales is a pair that successfully reared young at Llanwern steelworks (Monmouthshire) in 1972. Eight years earlier two birds were present nearby at Magor but it is not known whether breeding was attempted. Singing males have subsequently been heard briefly at Wentlooge (1981), Caerleon (1983) and Magor (possibly two singing males) (July 1991) all in Monmouthshire. The *Zoologist* (1915) published the claim of a pair nesting near St. Fagans in 1914 but the record was not accepted by the *Birds Of Glamorgan* (1967) and cannot be substantiated. The only valid record from Glamorgan is a bird at Oxwich on 18 September 1963.

The only other records in Wales are of Marsh Warblers trapped and ringed on Bardsey on 30 June 1959, 2 November 1959 and 7 September 1988 and one singing for several days at a site in Anglesey in June 1986.

Reed Warbler *Acrocephalus scirpaceus*

Telor y Cyrs

Local summer visitor, breeding sparingly in reedbeds in many counties; scarce passage migrant.

The Reed Warbler has a highly specialised habitat requirement, being principally reliant for breeding purposes on beds of the reed *Phragmites*, although it will occasionally nest in other habitats (e.g. oil seed rape in Monmouthshire). Furthermore the species is at the north-western limits of its range in southern Britain and until a measurable expansion took place into Wales, Lancashire and south-west England from the 1960s onwards, it was a rare breeding bird in most of Wales and a very scarce passage migrant.

Historically, the Reed Warbler was confined to very restricted areas of south Wales and one or two sites in the border counties. Around the end of the last century it bred quite extensively on the Gwent Levels and at two or three sites – Grangetown (Cardiff), Crymlyn Bog (Swansea – where Dilwyn knew it as long ago as 1840) and Margam (Glamorgan). The reed beds at Llangorse Lake (Breconshire) were always a well-known site with a high population. Forrest (1907) testifies to "numerous breeding" in the meres area on the Flintshire/Shropshire border – Fenns and Whixall mosses – near Ellesmere, and refers to a nest which was allegedly found near Holywell (Flintshire) about 1870. Nesting may have occurred in the lower Conwy Valley and possibly elsewhere in Caernarfonshire from time to time (Forrest 1907). Elsewhere the Reed Warbler was unknown in Wales other than as an accidental migrant; even on Anglesey, which possessed extensive reedbeds, there were no records before 1950; in 1986 RSPB surveys found 19 singing birds on 12 sites on the island and on Llŷn in 1986–87 it was located only at one site, the reedbeds near Pwllheli.

Exactly when the modest expansion into western Wales began or what stimulated it, is unknown. Two of the three original Glamorgan sites – Margam and Grangetown - had been destroyed by 1946 although by 1950 Reed Warblers were breeding at two "new" sites – Cadoxton and Eglwys Nunydd (until the latter site was flooded for the new reservoir in 1962). On Gower it is now known that they had been breeding at Oxwich NNR since at least the 1920s, increasing to 100 pairs or so as the water levels rose with neglected drainage in the Second World War and the population there had reached at least 450 pairs by 1977. Crymlyn Bog NNR has become another important site in the county with a large but uncounted population and Reed Warblers were nesting at 12 other sites in Glamorgan by 1978.

The Reed Warbler's stronghold in Monmouthshire remains on the Gwent Levels and surveys in 1978–79 revealed a minimum of 169 singing males, which may well indicate a slightly higher number of breeding pairs as population counts on the basis of singing males are known to produce under-estimates of actual pairs (Thomas 1979; Bell *et al.* 1968).

Ingram and Salmon (1954) were unconvinced of its proved breeding in Carmarthenshire although I. K. Morgan (in litt.), disputes this and has produced evidence of long-established breeding in some numbers on the south coast, east from Cydweli to Bynea. Records increased in the 1960s showing breeding at sites such as Pendine, Bynea, Pinged and Cydweli. By 1971 breeding was also proved at the Witchett Pool and near Trostre. Since then breeding has been shown at virtually all reedbed sites and the population in the county is probably well in excess of 400 pairs, with some 200 pairs in one site alone, Llangennech reedbeds, near Pontardulais. In 1989 a pair bred successfully at Dryslwyn Castle, inhabiting reed grass, reed sweetgrass and reedmace (Friese 1992).

Mathew (1894) did not record the bird in Pembrokeshire and Lockley (1949) could only instance four spring migrant records. Bertram Lloyd (Diaries) records that a nest was found at Goodwick Marsh in 1934 (or 1935). This record apart, Reed Warblers were first proved breeding in 1975 (but suspected in 1974). Nowadays, Reed Warblers breed in at least a dozen sites in the county with a likely population of perhaps 60 pairs each year. Farther north, in Cardiganshire, breeding was first recorded in 1978 and by 1984 there were at least 17 pairs in three different sites in reedbeds near the Teifi and Dyfi estuaries (the only reedbeds in the county) although the population is currently probably twice that level. The Llangorse colony in Breconshire contains up to 90 or so pairs and it bred at one other site in the county in 1971. In Radnorshire it is a rare summer visitor, having bred at one site in at least two seasons (1942 and 1963) although it still breeds irregularly at Llanbwchllyn (e.g. 1991). In Montgomeryshire a few pairs bred in the small reedbeds at Montgomery in the 1960s until the reeds were destroyed; the species no longer occurs as a regular breeding species in the county.

In north-west Wales, some evidence of the increasing numbers is to be gained from the number of passage birds on Bardsey, the modest total of 42 individuals having been recorded in the 30 years between 1954 and 1984; between 1985 and 1990 a total of 37 individuals were recorded. Breeding in Caernarfonshire was claimed historically at Deganwy and near Dolwyddelan at the turn of the century but cannot be substantiated. In more recent times breeding has been known to occur at Pwllheli for 20 years and a colony of some 20 pairs exists in Dolgarrog; reedbeds in the lower Conwy Valley. In Merioneth old records exist of claimed breeding in the Mawddach Valley and near Talsarnau. Since the 1970s a small number of pairs, four to five, has bred at Penmaenpool near Dolgellau and in the Dysynni Valley five to six pairs.

Breeding was known at Abergele in Denbighshire in 1922 and some six to ten pairs still breed there. There is also a small number of pairs around the Afon Ganol on the Denbighshire (east) side of the Conwy estuary probably no more than ten or so. Similarly in Flintshire there is only one regular breeding site, at Shotton, where a colony of up to 50 pairs has been long established; occasional birds are found at one or two other sites, e.g. Point of Air colliery and near Wrexham.

Observed passage is very light in both spring and autumn. As on Bardsey, the number of records on the Pembrokeshire islands is small but has become more evident in recent years: between 1947 and 1967 only five records were made on Skokholm, since when records have been almost annual with a maximum of five birds in any one year. At all the island sites, spring records reach only about a third of late summer/autumn ones.

Birds have been ringed in considerable numbers at several main sites in Wales – Oxwich, Kenfig, Llangorse, Shotton – and have accordingly produced a wide range of recoveries and retrapped birds. In France and Iberia 23 birds have been retrapped on autumn passage and another five have been recovered in Morocco and one in Spanish West Africa *en route* for wintering areas. Individual birds ringed in Denmark (August) and Guinea Bissau (February) have later been recovered respectively on Bardsey and breeding at Kenfig Pool. British and west European Reed Warblers are believed to winter in central parts of Africa southwards from about 12°N (BWP vol. 7).

Great Reed Warbler *Acrocephalus arundinaceus*
Telor Mawr y Cyrs

A vagrant to Wales.

It breeds in north-west Africa and most of continental Europe, north to southern Sweden and east to Sinkiang – Vigur, wintering in Africa south of the Sahara.

Up to the end of 1991 there have been seven records from Wales:

13–14 May 1967	– One, trapped, Skokholm
11 May 1970	– One, trapped, Skokholm
28–29 May 1976	– One, trapped, Bardsey
30 July 1978	– Male trapped, Penmaenpool (Merioneth)
8 June 1979	– One seen, Bardsey
15 June 1988	– One seen, Llangynog (Montgomeryshire)
22 May 1991	– Male trapped, Bardsey

A less likely habitat for this denizen of reed swamps would be difficult to imagine than the June 1988 record where a bird was singing lustily from a garden on the edge of the Berwyn Mountains! An excellent tape recording of the bird's song was made by the lucky owners, complete with a background of sheep and dogs.

Olivaceous Warbler *Hippolais pallida*

Telor Llwyd

A vagrant.

A record of this species on Skokholm from 23 September to 3 October 1951 has the distinction of being the first in Britain and Ireland. There have been no further records in Wales. This warbler breeds in Iberia, North Africa and the Balkans east to Pakistan and Kazakhstan and winters in Africa south of the Sahara.

Icterine Warbler *Hippolais icterina*

Telor Aur

A scarce spring and autumn passage migrant to coastal areas of Wales, half as numerous as the very similar Melodious Warbler, due presumably to a more easterly breeding distribution (from north-east France to Norway and Sweden and east to western Siberia).

The first occurrence in Wales was a bird on Skokholm on 31 August 1955 since when the total of 70 records involving 72 individuals have been recorded up to 1991.

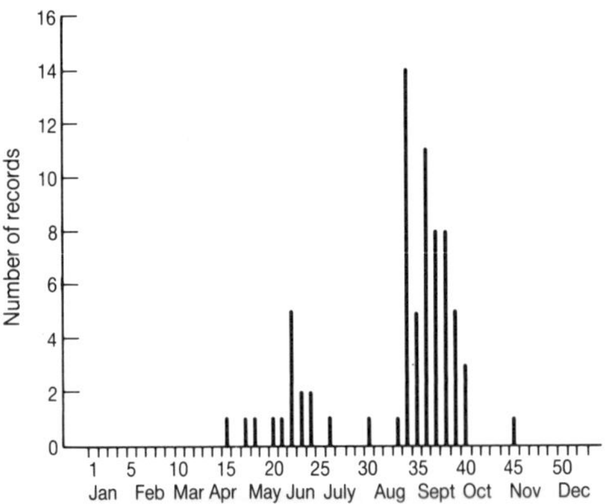

Weekly records of Icterine Warbler to 1991

There are more records than Melodious Warbler in spring and these have included individuals in full song but the majority are in the period 20 August to 23 September. There is an exceptionally late record from Cosmeston Lakes (Glamorgan) from 11 to 15 November 1983. Other than one from mainland Pembrokeshire, one from mainland Caernarfonshire, three from Glamorgan and one from Cardiganshire, all records are from Bardsey (principally), Skomer and Skokholm.

Melodious Warbler *Hippolais polyglotta*

Telor Pêr

A scarce but regular migrant to the Welsh coast, principally in the autumn but with a scattering of spring occurrences.

It breeds in north-west Africa, Iberia, France, Switzerland and Italy and winters in Africa south of the Sahara.

The first for Wales was appropriately at Bardsey, for this is a Bardsey species *par excellence*, on 27 August 1954, from which time there are records from every year except 1972. There is some difficulty in interpreting the Bardsey data in years of relative abundance, especially in 1962 and 1977 when there were lengthy series of sightings with up to five daily in 1962 and up to three daily in 1977. With that proviso, there is a minimum of 137 records up to and including 1991 involving at least 142 individuals.

The histogram for this species shows that most records

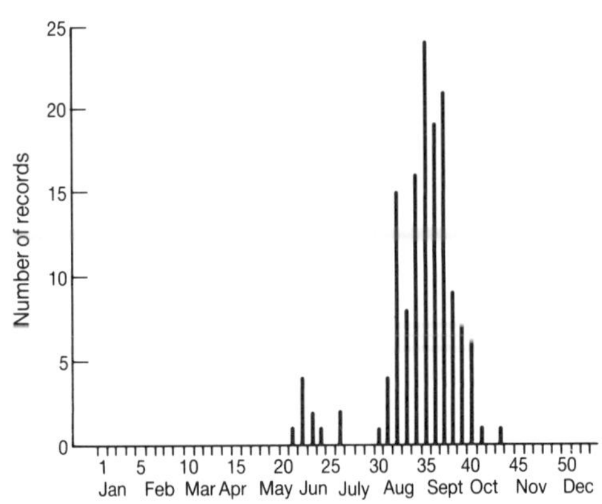

Weekly records of Melodious Warbler to 1991

are in the period 6 August to 16 September, presumably involving birds moving through the Irish Sea from breeding grounds on the continent. The bulk of records are from Bardsey and Skokholm and counties outside Pembrokeshire and Caernarfonshire have only a few records between them (Monmouthshire four, Glamorgan one, Anglesey three and Denbighshire one).

Dartford Warbler *Sylvia undata*

Telor Dartford

A vagrant to coastal areas of south Wales.

Although small numbers breed in the heathland of southern England as close as Devon and Dorset, the species is non-migratory and it is perhaps rather surprising therefore that there have been as many as six records in Wales:

25–27 December 1969	– One, Langland (Glamorgan)
14 September 1971	– One, Skomer
26 October 1975	– One, Eglwys Nunydd (Glamorgan)
14 April 1980	– One, Singleton Park (Glamorgan)
1–2 June 1981	– One, female, Skokholm
3 December 1984 to 1 January 1985	– One, Peterstone Wentlooge (Monmouthshire)

Subalpine Warbler *Sylvia cantillans*

Telor Brongoch

The Subalpine Warbler is a rare visitor to Wales.

It breeds in southern Europe, west Turkey and north-west Africa.

The first record was a first-winter female caught and ringed on Skokholm on 1 October 1953. There have been 17 subsequent records, 14 of which have been concentrated in the period mid-April to early June. The earliest record was one on Skomer on 26–27 March 1989.

Most of the occurrences were for a single day only but three records were of birds which remained on station for three weeks or more. The longest staying bird came to an unfortunate end: a male on Bardsey from 20 June to 16 August 1988 was seen to be taken by a Sparrowhawk.

This is very much a bird of the offshore islands: Skokholm (4), Skomer (6) and Bardsey (7); there is only one record from the mainland, a male at Porth Meudwy (Caernarfonshire) on 30 May 1985.

Sardinian Warbler *Sylvia melanocephala*

Telor Sardinia

A vagrant.

There is only one Welsh record of this species, which breeds in the Mediterranean basin eastwards to Afghanistan:

28 October 1968 – Male, Skokholm

This constituted the third record for Britain and Ireland.

Barred Warbler *Sylvia nisoria*

Telor Rhesog

The Barred Warbler is a rare autumn passage migrant in Wales.

It breeds from north Italy, Germany and south Sweden eastwards to Mongolia, wintering in north-east Africa south to Kenya and southern Arabia.

The first Welsh record was a bird killed at the Skerries Lighthouse (Anglesey) on 10 September 1910. There was then a very long gap until the second record, a bird ringed on Skokholm on 12 September 1956, following which there have been records in 21 years up to and including 1991. The Welsh total of 41 records have involved 43 individuals, all in autumn, from 21 August onwards with a marked peak in the period 10–23 September (see histogram). The latest record is of one seen at Penally (Pembrokeshire) on 3 and 10 November 1976, trapped on 21 November.

As would be expected with a species which occurs in the UK chiefly along the east coast, the Welsh records have a strongly coastal bias, principally from Bardsey, Skomer and Skokholm, possibly involving birds which are filtering south-west after arriving on the North Sea coast. Eighteen of the records are from Caernarfonshire (all Bardsey), 19 from Pembrokeshire, two from Anglesey and singles from Flintshire and Glamorgan.

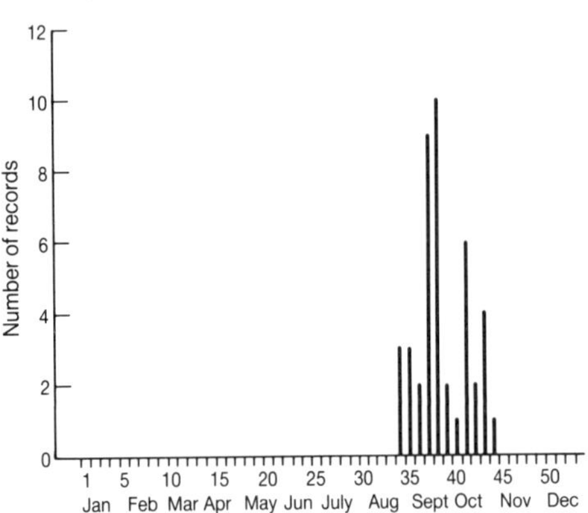

Weekly records of Barred Warbler to 1991

Lesser Whitethroat *Sylvia curruca*

Llwydfron Fach

A summer visitor, locally plentiful in the south-east but scarcer elsewhere although it is still gradually expanding westwards. Passage migrant in small numbers.

The Lesser Whitethroat has long been known as a breeding bird in Wales, although distribution in the Principality has been very localised. At the north-west edge of range here, the species has been primarily restricted to the lowlands and river valleys of the Welsh Marches although latterly there has been a clear extension of range during the present century, which trend has become more clearly marked in the past two decades. During this period, modest spreads in northern England and southern Scotland have also been recorded (Blackett & Ord 1962, Hutcheson 1986), as in south-west England (Marchant *et al.* 1990) and it is now clear that a parallel extension took place in Wales but has not been previously documented. Lesser Whitethroats occur in tangled thickets and hedgerows and as this is a habitat which is scarce in the hills, the species is strongly confined to lowland areas.

Nationally, the CBC indicates considerable annual fluctuations in the breeding population, with little evidence of any long-term changes until the past twenty years or so, since when there has been a gradual but clearly marked expansion westwards and into Anglesey. In the Welsh Marches there appears to have been very little change in the distribution of the species this century but evidence of a marginal increase in numbers in some counties. On the flat lands of Flintshire, along the Welsh side of the Dee estuary, Lesser Whitethroats have been known from such sites as Gresford and Chirk in Denbighshire for over 100 years. In Montgomeryshire they are not uncommon in the wide valley of the River Severn as far west as Newtown or in the low hilly country around Welshpool and Llanymynech on the Shropshire border. Lesser Whitethroats are found in a number of localities in the uplands (up to approximately 200 m). C. Oldham heard one singing near Kerry village (220 m) in 1936. Edgar Chance, O. R. Owen, and K. Wilson all collected clutches of Lesser Whitethroats from Radnorshire in the early years of the century, usually from the Knighton area in the valley of the River Teme, although Wilson also found nests around Rhayader; other known localities in the same county are mainly in the Wye Valley downstream from there – Glasbury, Boughrood, Llanelwedd. Across the Wye, Swainson (*Zoologist*, 1902) considered it a "not uncommon species" in the lowland parts of Breconshire although Cambridge Phillips clearly thought this a generous interpretation of its status. The Lesser Whitethroat has probably always been thinly scattered here and the BTO breeding survey, 1988–90, found it in only six localities, which suggested a population of no more than 25 pairs; in the hilly north of the county Shrubb (in litt.) found it only once in seven seasons, 1986–92.

In Monmouthshire it has one of its Welsh strongholds with a regular population identified by Campbell (1946) as long ago as the early 1940s. The tetrad survey (1981–85) located birds in 129 tetrads with main concentrations on the Gwent Levels (RSPB surveys located a minimum of 31 singing males there in 1984) and the Usk Valley where the first recorded Welsh nest was found in 1889 (Baker Gabb, Diaries). The other main strongholds for the species in Wales are in Anglesey and Carmarthenshire (*New Atlas*) (see below).

In the remaining Welsh counties the Lesser Whitethroat is – or has been – a much scarcer bird, unknown in many parts. The scattered Glamorgan pairs are restricted to the low agricultural land in the south of the county, with a maximum of 21 singing males having been heard in 1985. It was unrecorded as a breeding bird in the west of the county until 1984, although it was almost certainly breeding there from the late 1970s. The breeding survey which took place between 1984 and 1989 found confirmed breeding in five tetrads and presence in a further 23 (13.5%) tetrads.

Barker included Lesser Whitethroat on his list of birds of Carmarthenshire, where Ingram and Salmon considered it rare. In recent decades there has been increasing evidence of breeding at a number of sites, particularly in the south-east of the county and it is now to be found in many sites on the coastal strip and up the R. Twyi as far as Llandeilo. Mathew (1894) did not know the bird in Pembrokeshire in the last decade of the 19th century and Lockley *et al.* (1949) only noted breeding in 1932, around which time Bertram Lloyd (Diaries) also suspected breeding in the south of the county. There has been a very modest colonisation since about 1952, strengthening in the 1980s and it was known to be present in at least 38 locations during the five years of the tetrad survey of breeding birds (1984–88) with breeding established in at least a dozen cases and birds present in 69 tetrads (14.7%). In Cardiganshire the Lesser Whitethroat was essentially unknown until Salter recorded one in 1903 but by 1924 it was quoted as having become a regular summer visitor, albeit scarce. It is now well established in modest numbers in the lower Teifi Valley and locally along the coast, seldom inland. Merioneth produced only a handful of May records in the first seven decades of this century, mostly from the coastal strip, although nesting was proved around Llyn Tegid as long ago as 1867 and again in the years around the turn of the century; small numbers still breed in the Dee Valley in that part of the county. Small numbers are also now recorded each spring in the coastal strip with one or two singing males located each year and breeding proved at Arthog in 1991 (R. Thorpe, in litt.).

Breeding records have been reported from the lower Conwy Valley/Llandudno area since the early decades of this century (first nests found in 1922 and 1921 respectively) and it is widespread and reasonably common along the coastal strip. Eastwards in Denbighshire (e.g. Colwyn Bay, Abergele) it was well known in the 1920s, although many records may well refer to passing birds and not necessarily be indicative of breeding. It is suggested that the species is more widespread in recent years. On Llŷn the bird was unknown until Coward found a nest near Abersoch in 1892 and suggested that the species was extending its range westwards in Wales. RSPB surveys in 1986–87 found Lesser Whitethroats in 57 locations on Llŷn. In the early years of the century Lesser Whitethroats were known only as very scarce spring migrants on Anglesey although the RSPB watchers reports for 1933 recorded it as a "common victim at the Skerries light". One pair was proved breeding near Holland Arms in 1915 and there was a modest plethora of records from other parts of Anglesey around the same time (Forrest 1919). RSPB surveys in 1986 found territorial males at 75 sites, reinforcing the opinion that there has been a marked increase over the past two decades.

Lesser Whitethroats arrive in Wales from mid-April onwards and spring passage is most readily noticed at sites such as Lavernock Point and Great Orme on the mainland coasts, rather than on Bardsey or the Pembrokeshire islands where it is a very scarce migrant in both spring and autumn, and unrecorded in many years. Lesser Whitethroats are well known to make a long south-eastwards journey in autumn across mainland Europe to the Levant, to skirt round to northern Africa via the eastern side of the Mediterranean: as an illustration of this movement a bird ringed at Llangorse Lake at the end of August 1983 was shot on the Aegean island of Khios six weeks later. Winter records of Lesser Whitethroats in Wales are rare; there was one at Llandudno for several weeks in winter 1991–92, one was seen in a Cardiff garden on 24 December 1992 and an individual at Kenfig between 23 November 1982 and 9 January 1983 was ascribed to the Siberian race *S. c. blythii*, the first record of this race from Wales.

Whitethroat *Sylvia communis*

Llwydfron

A widespread summer visitor; formerly abundant. Common passage migrant.

The Whitethroat has long been known as an abundant and familiar summer visitor, possibly vying with the Willow Warbler as the most numerous breeding migrant in Wales. All the earlier accounts testify to its widespread distribution and numbers and there is no evidence of any major changes historically. Forrest (1907) in north Wales and Mathew (1894) in Pembrokeshire both testified to the abundance of the species around the turn of the century. This was the position up until 1968 in which year the CBC index showed the highest population level of Whitethroats which had been recorded up to then. As is well known, the Whitethroat then suffered a catastrophic collapse in numbers so that by the 1969 breeding season, the population was reduced by some 77% nationally. The causes for the collapse clearly lay in the wintering quarters and have subsequently been shown to be due to the onset of severe and continuing drought conditions which affected the Sahel region south of the Sahara where the Whitethroat population winters between approximately 12°N and 18°N. The effects of this crash were well reflected in various parts of Wales. In Gower, for example, one site holding 79 pairs in 1968 had only 22 the following year, falling still further to 13 in 1971 and six in 1972; by 1976 only 42 breeding localities were known in Glamorgan, 26 of them on Gower. Similar reductions were identified in other counties and recovery to pre-1969 population levels was still not achieved in many counties by the early 1990s.

Cambridge Phillips (1899) regarded the Whitethroat as "very common" in Breconshire, a view echoed by Salmon and Ingram fifty years later, although Peers and Shrubb could locate it in only 12% of tetrads visited in the 1988–90 BTO survey, twenty years after the collapse. They then estimated a population of no more than 100 pairs for the county. The complete tetrad survey in the neighbouring county of Monmouthshire (1981–85), however, found Whitethroats relatively more plentiful, occurring in 70% of tetrads with a population for the county estimated at as many as 3000 pairs. In Pembrokeshire the 1984–88 breeding bird census suggested a population of up to 8000 pairs for the county, with birds present in 86.2% of tetrads. Breeding densities in coastal areas of Glamorgan have been calculated at around 2.2 pairs per 10 ha. RSPB surveys on Llŷn (1986–87) found the Whitethroat a numerous breeding bird, especially on overgrown coastal cliffs, with a minimum of 390 pairs located. On Anglesey it is also very numerous, RSPB surveys there in 1986 located a minimum of 3198 singing males.

Contrary to the position with many other breeding passerines, O'Connor and Shrubb (1986) suggest that the density of breeding Whitethroats on the grassland-dominated farmland in Wales has probably always been lower than in most arable or mixed agricultural areas of England. However, they attribute this to geographical, rather than environmental (i.e. habitat) factors although the thin structure of the great majority of hedgerows in sheep-rearing areas of Wales may be a factor which limits numbers, as such hedgerows are unsuitable for Whitethroats. The Whitethroat is a scrub-loving warbler, inhabiting a wide variety of habitats including tangled hedgerows, scrubland, gorse and bracken hillsides (ffridd) – which are probably their main habitat in parts of mid-Wales – railway banks, rough cliff slopes, commons and overgrown wasteland, young conifer plantations and clearings; in fact any rough uncultivated areas with nettles and other rank vegetation up to altitudes of approximately 370 m is likely to support a pair. Roberts (1983) has shown how they can even successfully colonise open heather moor; he recorded up to 11 (possibly 17) pairs on moorland in Denbighshire between altitudes of 305 m and 400 m.

When population levels are high, pairs breed on several of the larger offshore islands. Single pairs (two in 1990) have bred sporadically on Skokholm over the years since 1931 (first recorded) while on Skomer the population has recovered from low points in 1981 and 1982 (ten pairs each year) and 1985 (nine pairs) (the last year reflected another low point nationally in the CBC index of numbers); 20-26 pairs bred on Skomer in 1990 and 1991. On Bardsey up to 15 pairs bred before the population crash; there was none in 1969 and breeding only recommenced in 1977 to climb to only three pairs by 1990.

The earliest Whitethroats do not normally arrive in Wales until mid-April and over the following four weeks numbers build up steadily, both with the influx of breeding birds and with the broad-front movement of migrants passing farther north, to produce a sharp peak in mid-May; some passage is still detectable into late June almost merging with the first southward-moving birds in July. The main southward movement takes place from early August onwards and the peak of passage on the offshore islands occurs at the end of August and in early September. Before the crash of 1968–69, the number of birds passing daily through these islands sometimes reached prodigious levels. On Bardsey, peaks of 150–200 birds were regular with big landfalls up to a thousand birds and a peak estimate of 2500 birds on 30 August 1968. Since then, numbers have peaked at around 20 birds daily (but there was an occurrence of 150 at the Bardsey light in autumn 1977). In the early 1990s numbers have picked up a little to give day totals of 60–70 on a few occasions. Baggott (1986) has demonstrated that autumn passage migrants passing through Bardsey have sufficient body-fat content to enable them to reach northern France in one direct flight in favourable conditions. British Whitethroats migrate on a well-defined route through south-west France and western Iberia (da Prato & da Prato 1983) and the recoveries of birds ringed in Wales reflect this, recoveries having come from south-west France (3), Spain (1), Portugal (6). It is also clear from ringing results that numbers of Irish breeding birds pass through Bardsey and the Pembrokeshire islands in spring.

Overwintering of Whitethroats is rare. A very late bird was seen at Kenfig NNR on 23 November 1983 and a male bird remained in Pembrokeshire throughout the winter of 1990–91.

A bird showing characteristics of the Eastern race *S. c. icterops* was trapped on Bardsey on 24 May 1989.

Garden Warbler *Sylvia borin*

Telor yr Ardd

A common and widely distributed summer visitor. Regular passage migrant.

The general distribution of the Garden Warbler in Wales closely mirrors that of the next species although the Blackcap is considerably more numerous and therefore more widely spread. Habitat preferences of the two species are very similar and there is some suggestion of interspecific competition for territories although in some woodlands the two species can certainly be found breeding in close proximity. The Garden Warbler, which is probably outnumbered by Blackcaps to a ratio of perhaps 3:1 overall in Wales, shares the preference for woodland and woodland edge where good understorey exists, but is also more likely to occur away from mature trees; garden shrubberies, rhododendron thickets, young forestry plantations and similar overgrown areas are readily exploited.

No long-term changes in Garden Warbler populations have been suggested historically, although the CBC clearly defines a steep drop nationally in the mid-1970s; in Wales, this trend, and its subsequent reversal in the late 1970s and 1980s, is reflected in the comments in various annual county bird reports. The steep drop in numbers has been linked with the extended drought conditions in the Sahel region of tropical Africa and although most Garden Warblers winter farther south than this, they presumably rely on the availability of good food supplies in that area to enable them to build up extra body fat before making the long Saharan crossing northwards in spring.

The results of tetrad breeding surveys in south Wales counties indicate the relative numbers of Blackcaps and Garden Warblers as shown in the table.

Percentage of tetrads found to contain Blackcaps and Garden Warblers in south Wales counties in recent (1980s) breeding surveys

County	*Blackcap*	*Garden Warbler*
Monmouthshire	85%	49%
West Glamorgan	78%	37%
Pembrokeshire	72%	41%
Breconshire	40%	63%

Mean densities of Blackcap and Garden Warbler in six woodland sites in Wales

Site	*No of pairs/(individuals) per 10 ha*	
	Blackcap	*Garden Warbler*
Pre-thicket conifer plantations (N. Wales)	0.5	1.5
Upland conifer with broadleaves (N. Wales)*	(0.8)	(5.3)
Restocked conifer plantations (N. Wales)*	(1.6)	(8.0)
Oak woodland (Breconshire)	0.7	1.6
Alder woodland (Breconshire)	1.0	0.3
3–5 year conifer (Caernarfonshire) 1968–71	1.3–2.6	0.3–0.6

* Figures in parentheses are for number of individuals recorded on point count surveys, not number of pairs.

The apparent anomaly in Breconshire is attributed to the greater proportion of upland in the county. Shrubb and Peers (pers. comm.) indicate that Garden Warblers almost invariably outnumber Blackcap in forest sites above 305 m. In Cardiganshire, Salter (Diaries) suggested that the species was less numerous than the Blackcap although Ingram *et al.* (1966) and the *Ceredigion Bird Report* 1982–83 both state the reverse view; P. E. Davis (pers. comm.) suggests that Garden Warblers are also more frequent than Blackcaps in upland areas of Cardiganshire where they seem to be more willing to breed in woodland with less dense undergrowth and more limited extent. Numbers are relatively high in Radnorshire and Montgomeryshire; in the former county the Garden Warbler nests at higher altitudes than the Blackcap and is more numerous in the north-western half of the county. In Merioneth and Caernarfonshire it is a widespread breeding bird although scarce on Llŷn and very scattered in many other parts of Caernarfonshire.

Several ringing recoveries (including France (3), Spain (1), Morocco (1)) confirm the southward route of Welsh Garden Warblers in autumn whilst also giving indication of passage birds from farther north (e.g. one bird from Orkney and another from Norway, both retrapped later the same autumns in north Wales). One autumn migrant ringed on Bardsey was found dead on spring passage in Denmark the following year. An east to west movement of some Garden Warblers in autumn is also instanced by three birds retrapped on Bardsey, Llŷn and the Skerries (off north Anglesey) which had been ringed earlier the same autumn in Belgium (2) and Lincolnshire. Overwintering birds are rare; one was seen at Pembroke on 28 February 1976, which must be assumed to fall into that category and there was another record (bird caught and ringed) in December 1991. The earliest spring arrivals are not usually seen until the end of the first week of April.

Blackcap *Sylvia atricapilla*

Telor Penddu

A plentiful summer visitor throughout Wales; passage migrant and small-scale winter visitor.

This fine songster is a relatively common summer visitor throughout all the Welsh counties (least so on Anglesey) although its local distribution is necessarily governed by the availability of a good shrub layer in its preferred woodland habitats. Relatively few Welsh woodlands provide such a feature due to continuous and excessive overgrazing and the Blackcap is therefore less characteristically a woodland bird than it would otherwise be, or than it is in much of England. It is more a bird of shrubberies, large gardens, overgrown wooded watersides, thickets and overgrown hedgerows-with-trees. Where woodland does provide a dense shrub layer – for example, where rhododendron has over-run the woodlands, as in much of Snowdonia, and on various neglected country estates elsewhere – Blackcaps are almost invariably present.

Although the Blackcap has always been referred to as plentiful in Wales (e.g. Forrest 1907, Mathew 1894) it is clear that it has been our most successful warbler in recent decades (certainly in terms of population increase) as shown by the steady rise in numbers on the annual CBC index. It is likely (Marchant *et al.* 1990) that the increase started as early as the 1950s and is now reflected in the occupation of new territories on farmland and also in forestry plantations as these pass through the early thicket stage or, later, reach clear-felling and restocking rotation. Blackcaps are certainly a feature of the new generation conifer forests once they reach the thicket stage, as demonstrated by Insley and Wood (1973) who found three or four pairs in a woodland plot of 14.2 ha on the Caernarfonshire shore of Menai Straits of which 12.2 ha was conifer planting between three and five years old. Bibby *et al.* (1985) found 0.8 *individuals* per 10 ha over a sample of 253 count points. In restocked forestry plots Blackcaps were at twice this density (Bibby *et al.* 1985). Williamson (1971) found a mean density of 2.4 pairs per 10 ha in mature redwoods in Montgomeryshire. In Breconshire Massey (1974) found densities of 1.0 pair per 10 ha in alder-dominated woodland and 0.7 pairs per 10 ha in oak-dominated wood.

Both Glamorgan and Gwent provide more woodlands suitable for Blackcaps – i.e. those in the lowlands which are less heavily grazed – than other parts of Wales and in these two counties Blackcaps are probably more plentiful than elsewhere in Wales. No figures exist for Glamorgan as a whole but in the west of the county Thomas *et al.* (1992) found tetrad occupancy at 78%. The *Gwent Atlas of Breeding Birds* (1981–85) found Blackcaps in 85% (330) of the tetrads in the county and suggested a breeding population of c9000 pairs. Farther west, in Pembrokeshire the breeding birds survey (1984–88) found Blackcaps in 72.1% (339) of the tetrads with the main concentration in the south of the county but with many other pairs scattered throughout the agricultural hinterland and a suggested breeding population of about 7000 pairs. Blackcaps are well distributed and fairly common in the lowland parts of the counties of mid and north Wales, usually concentrated in suitable habitats in the river valleys but more recently exploiting forestry plantations in the uplands up to at least 400 m in places. Peers and Shrubb estimated a population of perhaps 350–400 pairs in Breconshire following the BTO breeding bird survey in 1988–90 in which Blackcaps were found in 40% of those tetrads sampled. On Llŷn, Blackcaps are fairly well distributed, being found in a minimum of 136 one-kilometre squares during RSPB surveys in 1986–87, a total almost identical to that found for Garden Warbler. It is only on Anglesey that the species can be regarded as at all scarce as a breeding bird, for here they are only located in south and east Anglesey where suitable woodland and thicket exist.

It has long been recognised – certainly since early in the 19th century – that some Blackcaps overwinter in Britain and it is equally clear that the number doing so has increased considerably in the few last decades (average of 22 reported annually in Wales 1945–54; average 380 1970–77) as it has also in other western European countries (Simms 1985)). Winter records are well distributed from all the Welsh counties, a slight predominance from the south possibly being a reflection of the greater number of observers there as much as to the actual number of birds present. At least 15 different individuals were believed to be present in one small area of Mumbles (Glamorgan) in the winter of 1974–75. These wintering Blackcaps, which are mainly breeding birds from continental Europe, are well known for their willingness to use garden bird tables and they demonstrate an ability to survive even the most bitter weather. For example, there were still at least 13 on Gower after the severe 1981–82 winter and the species is relatively numerous in the Swansea/Gower area in all winters; elsewhere 18 were ringed at Stackpole (Pembrokeshire) in February 1991 and ringing at one site in Bangor (Caernarfonshire) in October–November 1992 produced at least 25 individuals. Although some exceptionally early migrants may arrive before the end of March, in general, Blackcaps arrive from mid-April onwards, with the main influx towards the end of the month and in the first days of May (earliest arrivals may be confused with overwintering immigrants) and most have departed again by the middle of September, well in advance of the late autumn arrival of passage birds from the continent, a small proportion of which remain to form the wintering population.

The main wintering area of British Blackcaps is in southern France, Iberia and north Africa, and ringing recoveries confirm the winter destination of Welsh birds. Recoveries between November and March have occurred in southern Spain (7), Portugal (1) and Morocco (3).

Greenish Warbler *Phylloscopus trochiloides*
Telor Gwyrdd

A vagrant to Wales.

It breeds from north-east Germany and Finland east to Sea of Japan and winters in India and south-east Asia. There are only eight records, all from Bardsey, Skomer and Skokholm:

16–17 June 1954 – One, Bardsey
11 September 1956 – One, Bardsey
31 August 1960 – One, Skokholm
30–31 August 1961 – One, Skokholm
31 August 1976 – One, Skokholm
7 June 1981 – Male in song, Bardsey
13 June 1988 – One, Bardsey
19 June 1990 – One, Skomer

Arctic Warbler *Phylloscopus borealis*
Telor yr Arctig

A vagrant.

There has only been one Welsh record of this warbler, which breeds from northern Scandinavia east through north Siberia to Alaska and winters in southern south-east Asia:

13 September 1968 – One was trapped on Bardsey

This is a typical date for this vagrant to arrive in Britain, on average there are three or four per annum, particularly at Fair Isle and on the North Sea coast.

Pallas's Warbler *Phylloscopus proregulus*

Telor Pallas

A vagrant.

This little gem of a bird, much prized by birdwatchers, has been seen nine times in Wales:

7 November 1975 – One, trapped Bardsey
30 October 1982 – One, Wooltack Point, Marloes (Pembrokeshire)
25–26 October 1983 – One, trapped Bardsey
31 October 1984 – One, Strumble Head
7–8 November 1987 – One, trapped Bardsey
15 November 1987 – One, South Stack
23 October 1988 – One, South Stack
27–28 October 1988 – One, Great Orme
2 November 1991 – One, Penmon Point (Anglesey)

It is noteworthy that all the records fall within a 22-day time span and all are recent, reflecting an increase in the prevalence of easterly winds during October. This species breeds from south central Siberia east to the Sea of Okhotsk and is, not surprisingly, much more numerous along the North Sea coastline of Scotland and England than it is in Wales. Nevertheless we can continue to look forward to a further small trickle of records, along with other Siberian rarities such as Dusky and Radde's Warblers.

Yellow-browed Warbler *Phylloscopus inornatus*

Telor Aelfelyn

A scarce to uncommon but rapidly increasing, autumn passage migrant.

It breeds from northern Siberia south to Afghanistan and north-west India and east to Sea of Japan, also eastern Tibet, and winters in India and south-east Asia.

The first Welsh record was not until 1959, a single bird on Skokholm on 2–3 October. There was then a sporadic pattern of occurrences throughout the 1960s and 1970s, with 28 individual records up to 1984, followed by unprecedented influxes in 1985 and 1986 (21 and 27 respectively), declining only slightly to 12 in 1987 and 16 in 1988 but with only three records in 1989, four in 1990 and two in 1991. The 1985 influx was mirrored along the east coast of England and Scotland, only with much higher numbers as would be expected from arrivals across the North Sea.

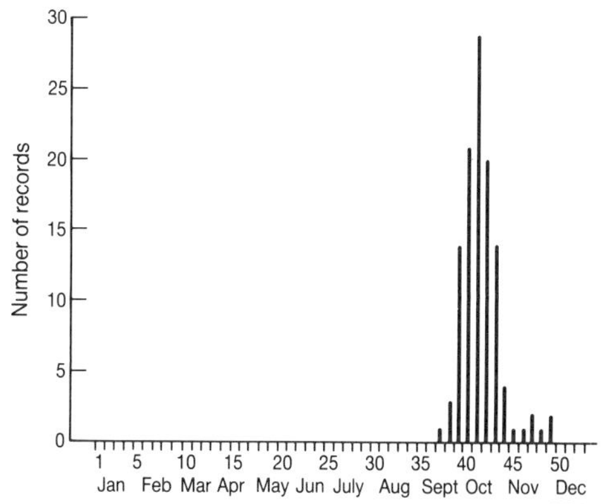

Weekly records of Yellow-browed Warbler to 1991

The timing of records is centred on the period 8–14 October, with extreme dates of 14 September and 4 December (see histogram).

The geographical distribution of records is dominated by Bardsey, which accounts for records involving 56 individuals. Other counties (including mainland Caernarfonshire) also appear on the distribution map and it can only be a matter of time before the counties which do not feature yet (Denbighshire, Merioneth, Montgomeryshire, Radnorshire and Carmarthenshire) have their first records.

Radde's Warbler *Phylloscopus schwarzi*

Telor Radde

A vagrant.

It breeds from south-central Siberia eastwards to Skokholm and winters in south-east Asia. It has been recorded three times from Wales:

22 October 1968 – One, Skokholm
29 October 1987 – One, Bardsey
18 October 1990 – One, Bardsey

Dusky Warbler *Phylloscopus fuscatus*

Telor Tywyll

A vagrant.

This species breeds from west-central Siberia eastwards to north-east Siberia and winters in north-east India and south-east Asia; there have been three records from Wales:

30 October 1982 – One trapped, Bardsey
7 November 1987 – First-winter male trapped, Bardsey
15 October 1988 – One, Strumble Head

These dates are very much in accord with records from the rest of Britain (chiefly from the English east coast counties and Scillies).

Bonelli's Warbler *Phylloscopus bonelli*

Telor Bonelli

A vagrant to Wales.

It breeds in Iberia and France east to Germany and Austria, also the Balkans, southern Turkey and Lebanon and north-west Africa.

The first record for Britain and Ireland was a female trapped on Skokholm on 31 August 1948. Since then it has occurred on a further seven occasions in autumn, all but two of these from Bardsey:

18 August–5 September 1959	– One, Bardsey
10 September 1959	– One, Bardsey
1–2 September 1962	– One, Bardsey
15–20 September 1962	– One, Bardsey
30 August 1963	– One, Lavernock Point
17 September 1968	– One, Llaniestyn (Caernarfonshire)
20 August 1984	– One, Bardsey

All the Bardsey birds were trapped and ringed.

Wood Warbler *Phylloscopus sibilatrix*

Telor y Coed

A summer visitor to oakwoods throughout most of upland Wales; passage migrant in small numbers at coastal sites.

The Wood Warbler is one of the characteristic summer birds of the hills and valleys of Wales. Its song is a feature of most areas of oak-dominated deciduous woodland in the hills, most notably throughout the counties of mid-Wales but also in large parts of the north and the upland areas of the south Wales counties. It is classically found in company with other characteristic members of the summer bird community in these woods – Redstart, Tree Pipit and Pied Flycatcher. As the *New Atlas* reveals, Wales and the borders are the most important breeding area in Britain for this species; only in Anglesey and the peninsulas of the west – Llŷn and Pembrokeshire – is the Wood Warbler very localised or wholly absent from extensive areas. It is typically a bird of well-developed, mature woodland, although it also occurs widely in scrub oak on hillsides and in relatively small copses and spinneys, often no larger than 0.5 ha. Wood Warblers require an open forest floor with little or no secondary vegetation and thus find the severely grazed oakwoods of Wales much to their liking. Primarily regarded as a bird of upland woods, the Wood Warbler is found up to the limit at which oak or beech occurs – usually about 305 m. It also occurs frequently in mature larch plantations (Shrubb, in litt., records it as forming 14% of songbirds counted in such plantations in Breconshire) and occasionally on the fringes and rides of other upland conifer woods – and some lowland areas, e.g. Wye Valley – particularly if some deciduous standards are present. In lowland Britain beechwoods are favoured but such woodlands are relatively scarce in Wales. The Wood Warbler is less numerous in many woodlands in lowland areas, e.g. Usk, Severn, Dee valleys, Vale of Glamorgan, while reappearing, often in abundance, in the woodlands on the valley sides. It is absent from much of Pembrokeshire and Llŷn and is represented on Anglesey only by a small handful of pairs in woodlands on the north side of the Menai Straits.

Parslow (1973) points to alleged decreases in parts of southern Britain between about 1940 and 1963 although he admits that no evidence to that effect referred to Wales. However, Heathcote *et al.* (1967) detail the apparent fluctuations in the breeding population in the southern parts of Glamorgan this century and identify some changes which are closely parallel with Parslow's conclusion (table below). The species is presumed to have been fairly numerous in other parts of the county – the north and north-west – all the time.

Apparent fluctuations in breeding Wood Warbler population, Vale of Glamorgan 1900–67

c1900–10	scarcely seen
1913–20	increasingly seen in spring
1920–27	breeding regularly 6–8 sites
1928–47	none suspected breeding
1947–50	pairs at 6–8 sites
1950–57	very few seen
1962–67	fairly widespread breeding pairs

In recent decades scattered pairs have bred in these lowland areas while the population in the uplands of the north (e.g. Aberdare, Merthyr Tydfil) has remained stable.

In Carmarthenshire, Barker (1905) recognised it as common in many parts, notably in the oakwoods of the upper Tywi Valley but, as now, it was apparently well spread on the lowlands although not so numerous as in the hills. Mathew (1894) doubted whether it occurred west of the Preseli Mountains in Pembrokeshire – a statement which is not so far removed from the picture today, which shows breeding in the north-west of the county, with a small concentration in woodlands at the head of the eastern Cleddau estuary and a total population in the county no greater than c250 pairs. Not dissimilarly, the population in Cardiganshire is concentrated in the north and west of the county, where it is numerous in both coastal and upland woods, leaving somewhat unexplained gaps in the south-west. In north Wales Wood Warblers abound in the woodlands of Snowdonia but, as remarked, they are absent from most of Llŷn (only 12–14 sites in 1986) and Anglesey. Denbighshire has a high population especially in the uplands but in Flintshire, although widely distributed, it is local, with the main population on the slopes of the Clwydian hills; it is the least numerous of the *Phylloscopus* warblers in the county. In the border counties of Montgomeryshire, Radnorshire, Breconshire and Monmouthshire, Wood Warblers are again numerous, especially in the hill areas, and least so in woodlands of the broad open river valleys. In Monmouthshire Wood Warblers occurred in 40% (155) of the tetrads in the 1981–85 breeding survey. Similar work in Breconshire (1988–90) indicated a county population of some 800 pairs. Bibby (1989) estimated a total British population of between c16 000 and 18 500 pairs. We estimate a total Welsh population of between 6500 and 8000 pairs.

Wood Warblers are notoriously difficult to assess and accurate figures are difficult to determine. Numbers fluctuate considerably from year to year and in addition an estimate of breeding pairs on the basis of singing males can be misleading as (i) a proportion of singing males may be polygynous and remain unmated and (ii) paired males may be polygynous and move outside their territories during incubation in an attempt to attract other females (Marchant *et al.* 1990). Bibby (1989) found median numbers of males in occupied 10-km squares in 1984–85 as follows: Wales (29), south-west England (22), Lake District (13), the Marches (12), Scotland (9), rest of England (3).

Wood Warblers are rather late arrivals in spring, early birds appearing from the second week of April but most not arriving in Wales until the end of the month or the first days of May. By early August they have normally departed again. Only two recoveries of Welsh-ringed birds have been reported to date; a nestling ringed at Glasbury (Breconshire) in June 1930 was found dying in Avellino in southern Italy in October the same year, clearly heading for the short sea crossing to north Africa via Sicily; a nestling ringed at Tintern (Monmouthshire) in June 1983 also headed south-east and was retrapped on the Dorset coast two months later.

Chiffchaff *Phylloscopus collybita*

Siff-saff

A locally common breeding visitor in most counties. Rare on Anglesey.

The Chiffchaff is predominantly a bird of mature woodlands where there is a well-developed ground cover. Despite the fact that deciduous woodland is one of the most widespread and characteristic semi-natural habitats in Wales, the Chiffchaff is not nearly as abundant as it might be for the single reason that overgrazing has deprived most woodlands of their ground cover, thereby removing the potential nest sites for this small leaf warbler.

Chiffchaff populations are highest in the south-east of the country, notably Monmouthshire where the woodlands are generally less degraded than elsewhere and often free from heavy grazing (e.g. Wye Valley). The *Breeding Atlas* shows the Chiffchaff present in almost all 10-km squares in Wales with the exception of the mountain spine of mid and north Wales. Unlike the distribution of Willow Warbler, however, this produces gaps when refined to tetrads: 87% occupation in Monmouthshire, 80% in Pembrokeshire, 76% in west Glamorgan. Although the completed tetrad surveys carried out to date refer to south Wales counties, the same pattern of broken distribution characterises counties in mid and north Wales. In some areas Chiffchaffs are not easy to find and are clearly locally scarce. Peers and Shrubb recorded them in no more than 37% of tetrads in Breconshire (1988–90). Similarly, in Cardiganshire they are mostly to be found on the coastal plain and in wooded valleys, sometimes locally numerous and infrequently above c200 m. Hope Jones (1974) considered it a rather uncommon species in Merioneth, even in the lowlands, which comment is probably fairly accurate for most other north Wales counties: Caernarfonshire with a higher proportion of mature woodland with understorey probably has better populations, especially on the coastal lowlands, than neighbouring counties. In this respect the infestation of rhododendron in many Snowdonia woodlands has one small benefit in producing an understorey attractive to Chiffchaffs. Population estimates have been given for three south Wales counties, Pembrokeshire c6000 pairs, Breconshire fewer than 350 pairs, Monmouthshire 7000–8000 pairs.

Not all Chiffchaffs leave Britain in winter and over the past few decades the number of birds recorded overwintering has increased. Not unexpectedly, south and south-west Wales are areas where winter records are most frequent. The map below summarises the number of published records for each county between 1962 and 1991, where available, although it must be appreciated that there are many more winter sightings of Chiffchaffs than ever find their way into the published annual county bird reports.

Chiffchaffs are one of the first spring migrants to arrive and one of the last to leave in autumn. The overseas wintering area of the British breeding birds is in the Mediterranean basin, west Africa as far south as Gambia

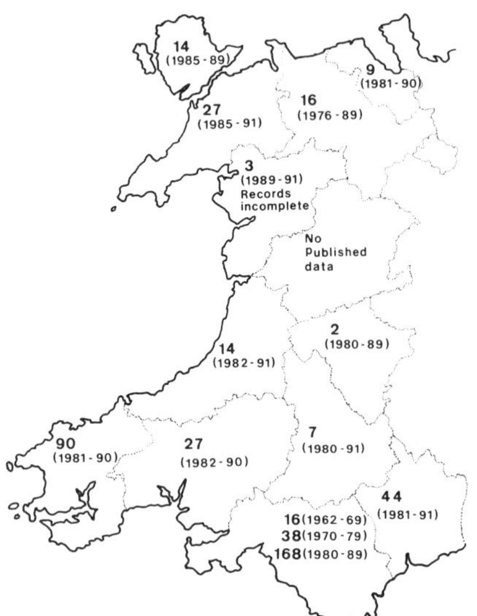

Winter records (December–January) of Chiffchaffs as shown by published accounts in annual county bird reports

and in a narrow belt south of the Sahara where its winter range almost overlaps with that of the Willow Warbler. The winter destination of birds ringed in Wales conforms to this pattern, with nine recoveries from France and Iberia and seven from north Africa (Morocco, Mauritania and Spanish West Africa); however, it should be noted that almost all of these individuals were ringed during the period of autumn passage and available data do not usually confirm whether they are birds of the race *collybita*, which breeds in Wales (as elsewhere in Britain) or the northern Scandinavian form *abietinus*. The latter population normally winters farther east so that the presumption is that these recoveries do indeed refer to British birds; however, Scandinavian birds certainly pass through Wales both in autumn and spring as is evidenced by two birds ringed on Bardsey on 14 April 1984 which were both subsequently recorded at the same place in Åland, Finland on 27 June and 21 September respectively, the same year. It is interesting to note that amongst 11 Chiffchaffs at Cydweli sewage works in December 1992 and January 1993 were at least two that showed characteristics of the eastern races. Although both were caught and ringed it remains unclear to which of the races – *tristis* or *abietinus* – they should be attributed. Switzerland features in two interesting recoveries: another bird ringed on Bardsey in June 1966 was wintering in Switzerland the following January and one ringed on spring passage in Ticino (16 April 1985) in the same country turned up at Bardsey one month later. Recent ringing expeditions to Senegal during winter have resulted in two subsequent recoveries of Chiffchaffs in Wales, one near Llanelli the following June and the other on passage on Bardsey the following April.

Willow Warbler *Phylloscopus trochilus*

Telor yr Helyg

The most abundant summer visitor. Numerous in suitable habitats up to around 670 m. Also numerous passage migrant.

The welcome sound of the Willow Warbler's song in April is one of the most familiar manifestations of spring and summer. The Willow Warbler is a very numerous species in Wales, widely distributed throughout a range of habitats and it can thus be heard between April and July in almost any corner of the countryside which affords rough vegetation with shrubs or trees. It breeds in all counties and intermittently on the larger offshore islands.

Although primarily a breeding species of young woodland, woodland edge and open scrubland, the Willow Warbler is more catholic in its choice of habitat than any other British warbler. In Wales, apart from the typical habitats outlined above, it is to be found in newly established forest plantations (and thereafter forest rides and clear-felled areas), moorland edge with isolated trees or scrub, farmland hedgerows, large gardens, wetland edge, railway banks and many similar areas offering the cover of rank ground vegetation in which to nest and trees/bushes for song posts and feeding sites. Where such features are present, even heavy urbanisation does not necessarily deter them and Willow Warblers breed in many urban parks and cemeteries in towns and cities such as Cardiff, Swansea, Wrexham and Llandudno. O'Connor and Shrubb (1986) have shown that on farmland the density of Willow Warblers increases westwards in the UK with the predominance of pastoral as opposed to arable agriculture. The mean density of 19.2 pairs per km^2 in Wales is higher than in any other part of the country.

The *Breeding Atlas* and the *New Atlas* show Willow Warblers breeding in every 10-km square in mainland Wales and this widespread distribution is further reinforced by the tetrad breeding surveys in Monmouthshire, west Glamorgan and Pembrokeshire which show occupation rates of 95%, 95.5%, 91% respectively. In Breconshire Peers and Shrubb (1991) estimated a total population in excess of 5000 pairs on the basis of the 1988–90 BTO sample tetrad survey.

In second-generation conifer plots, Willow Warblers can be extremely numerous. Bibby *et al.* (1985) found them the most numerous species at 326 count points in restocked conifer plantations 2–11 years old in north Wales with a mean density of 60.9 *individuals* per 10 ha. In mature conifer plantations with occasional broadleaved trees amongst the crop Bibby *et al.* (1989) found Willow Warblers the sixth most numerous species at a mean density of 23.7 *individuals* per 10 ha. Other selected densities (pairs) are given as follows:

	Mean number of pairs per 10 ha
1. Farmland in Monmouthshire (1 site, 12 years)	3.3
2. Unimproved farmland in Carmarthenshire (2 sites, 1 year)	6.1 & 5.3
3. Pre-thicket conifers in north Wales (2 sites, 2 years)	5.5–11. & 11–15.6

4. Upland sessile oak in Caernarfonshire (1 site, 1 year) 13.4
5. Mixed woodland in Breconshire (1 site, 1 year) 5.8
6. Ash wood in Breconshire (1 site, 1 year) 3.7

(References: 1 – S. J. Tyler (CBC); 2 – Tyler & Geach 1987; 3 – Currie & Bamford 1981; 4 – Gibbs & Wiggington 1973; 5 and 6 – Massey 1974).

There is no evidence of significant changes in population levels historically and annual indices for the UK as a whole have been remarkably constant over a period of the last quarter of a century or so (Marchant *et al.* 1990). However, in the 1980s and early 1990s, evidence from the BTO's Constant Effort ringing sites indicates a fairly steep and consistent decline; no obvious explanation for this can be offered to date.

In spring the first arrivals reach the south coast of Wales in the later days of March with the main influx occurring throughout the first two weeks of April. Male birds usually arrive a day or two ahead of the females. They quickly move on to occupy breeding territories and the pattern is temporarily confused as other Willow Warblers pass through on a broad front for breeding areas farther north. Some of these passage migrants in spring occasionally refer to the slightly more "brown-and-white" race, *P. t. acredula*, breeding in Scandinavia and eastern Europe.

Many pairs are double-brooded but once the breeding cycle is completed, southward movement is reasonably rapid and by mid-September most have departed. Only a trickle of passage birds occurs after this with the last individuals rarely occurring as late as 31 October (Skokholm and Bardsey).

Overwintering of Willow Warblers is very rare; a Willow Warbler was at Aberthaw (Glamorgan) between 12 December 1960 and 4 January 1961. Whether a bird seen in Glamorgan on 14 February 1983 was overwintering or an abnormally early migrant is not known; another bird (the same one?) was seen nearby on 9 March. An earlier record in February–March 1975 at Lisvane Reservoir was presumed to be of this species but not proved.

Willow Warblers winter in tropical west Africa and movement to and from Wales is by well-established west-coast routes through Iberia, as testified by an extensive series of ringing records. Welsh-ringed birds have later been recovered in France and the Channel Is. (8), Iberia (12), Morocco (6) with the most distant recovery to date being from Abidjan in the Ivory Coast, 5240 km to the south of its ringing site on Bardsey during spring passage earlier the same year. Ringing recoveries also confirm the northern destinations of many of the spring arrivals and clear coasting movements can be traced with some individuals, e.g. a bird on Bardsey on the last day of April retrapped in Merseyside two days later. However, the patterns are not always clear and there are several examples of ringed birds moving south again the same spring, e.g.:

Bardsey 12 April 1966 to Carmarthen 18 April 1966 (105 km south-south-east)
Bardsey 17 April 1964 to Skokholm 19 April 1964 (123 km south-south-west)
Isle of May 29 May 1948 to Caernarfon 10 June 1948 (338 km south-south-west)

Furthermore there is evidence of initial south-easterly movement of Welsh birds in late summer with retraps occurring in south-east England, mainly on the Channel coast. This conforms to a general pattern of emigration at the end of the summer, with birds mainly using the short sea crossing over the Channel (Norman & Norman 1985). A Willow Warbler ringed in Switzerland on 12 April 1985 was retrapped on Bardsey 12 days later, indicating that not all spring passage is through Iberia and western France. An individual ringed on Skokholm on 14 May (identified as a northern bird) was found dying on a boat off Esjberg (Denmark) ten days later.

The scale of some of the spring and (especially) autumn movements is sometimes very large. As many as 1000 birds a day have been estimated on spring passage on Skokholm and Bardsey although normal maxima in any year are 100–400. On Bardsey August numbers have sometimes run up to 1000 or more with an astonishing estimate of 5000 birds on 7 August 1982.

Goldcrest *Regulus regulus*

Dryw Eurben

An abundant resident, passage migrant and winter visitor.

This, our tiniest native bird species, is a remarkably hardy insectivore, which, unlike most other small insectivorous birds, is resident throughout the year. It is predominantly a bird of conifer woods wherein it can occur in almost prodigious densities, although it also breeds, more sparingly, in a variety of deciduous woods, possibly more so at times when populations are high and it overspills into such less favoured habitats. There are now some 171,000 ha of conifer forest in Wales and no species has benefited from this post-war change of land use more than the Goldcrest. From sea-level plantations such as those on the dune systems of Pembrey (Carmarthen), Newborough (Anglesey) or Harlech (Merioneth) to some of the highest upland plantations at almost 610 m the Goldcrest abounds as a breeding bird.

It is not confined to extensive tracts of conifer, however, and can readily be encountered almost anywhere that provides groups or stands of conifers, either exotic or native, and is a frequent inhabitant of churchyard yews, large gardens and urban parks. On account of its diminutive size, the Goldcrest is extremely susceptible to hard winter weather, when mortality can be very high. Low temperatures are tolerable but protracted periods, especially with severe frost and glazing which prevent access to the supply of minute insects in the axils of the conifer needles, are invariably fatal. Thus mortality in winters such as those experienced in 1947, 1962–63, 1978–79, 1981–82, 1985–86 produced sharp drops in the populations, particularly those of 1947, 1962–63 and 1985–86.

Other than these weather-related fluctuations, the only major long-term trend has been the increase in population coincident with the expansion of conifer forests in the past four decades.

Inadequate information exists on the actual numbers of Goldcrests in Welsh woodlands. Bibby *et al.* (1985) found Goldcrests at a mean density of 18.9 *individuals* per 10 ha at 326 count points in restocked conifer plantations in north Wales and (1989) at a mean of 97.7 *individuals* per 10 ha in mature plantations at 253 count points, also in north Wales. The BTO Atlas showed that populations can reach as high as 320 pairs per km^2 in Ireland and although they may well be of comparable numbers in Wales, no work has yet been done to establish this. In a mature plantation of redwoods at Leighton, Montgomeryshire, Williamson (1971) found Goldcrests the most numerous species with 138 pairs per km^2.

The only areas of Wales where breeding Goldcrests are normally absent are the urban and industrial centres, treeless areas such as the open uplands and parts of Anglesey and coastal stretches. Outside the breeding season individuals are liable to turn up in more unexpected places; one has been recorded at 816 m in Snowdonia and ten were present on Grassholm, 14.5 km offshore, in September 1972.

In autumn many Goldcrests from northern Europe arrive in Wales from the north. Passage is sometimes detectable inland but is very clearly marked on the offshore islands and coastal headlands. Bardsey is a key site in this respect with well-recorded observations over a long period of years which show the comparative peaks of autumn and spring movement in mid-October and mid-April respectively. Up to 200–300 birds per day are not exceptional on Bardsey in autumn with maxima reaching as high as 1000 on 19 September 1988 and 1300 on 3 October 1989. Spring numbers exceptionally reach as many as 500 daily but are not habitually more than 200 in "good" seasons. On the Pembrokeshire islands the passage is much more modest, with daily autumn totals over 50 on Skokholm or Skomer being rare; spring daily maxima are rarely in double figures. Many of these migrants are long-distance travellers and are passing on through the islands to distant destinations. Six Goldcrests ringed on autumn passage on Bardsey were subsequently recovered in Galloway, Dorset (2), Co. Cork, Cheshire and Merseyside.

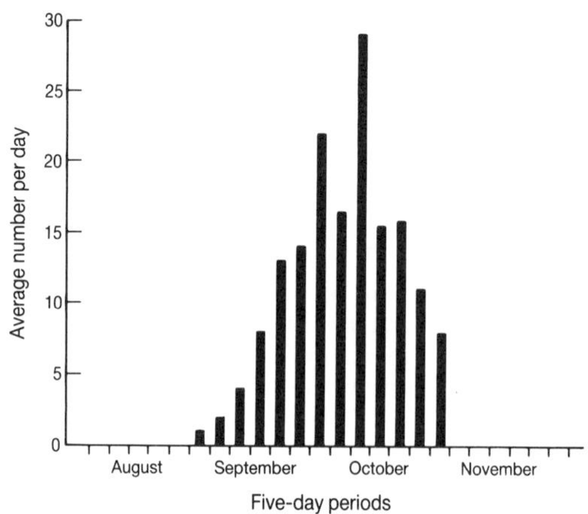

Autumn movements of Goldcrests on Skokholm Island

The Welsh resident population is supplemented by an influx of northern immigrants for the winter. Loose flocks of Goldcrests are usual in winter either unmixed or, frequently, in association with Coal Tits, Blue Tits, Great Tits, Long-tailed Tits and Treecreepers. What proportions of these are native birds and immigrants are not known. A bird found near Rhayader (Radnorshire) in December 1950 had been ringed on passage on the Isle of May (Fife) in October the same year.

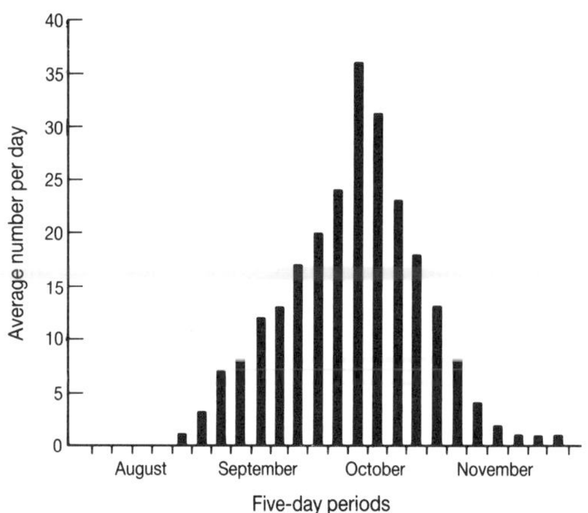

Autumn movements of Goldcrests on Bardsey Island

Firecrest *Regulus ignicapillus*

Dryw Penfflamgoch

A scarce breeding resident and uncommon passage migrant.

In the 19th century Firecrests were rare vagrants with only five records in Wales (two from Glamorgan and one each from Caernarfonshire, Carmarthenshire and Breconshire). This low level of occurrence continued until the 1960s and early 1970s during which time Firecrests began to be recorded with some regularity in small numbers from Bardsey and Skokholm, principally on autumn passage. The subsequent dramatic increase in numbers and range is documented in the Welsh Bird Reports from 1974 onwards:

1974

"A remarkable influx" with several records of overwintering in Caernarfonshire, Cardiganshire, Pembrokeshire, Glamorgan and Monmouthshire. The first county record from Merioneth and the second for Breconshire. One was singing at Llandegfedd Reservoir (Monmouthshire) in March, the first indication of what was to come.

1975

"Exceptional numbers" mainly in autumn and winter but several also in spring with up to three in the Wentwood area of Monmouthshire through May and June. The second county records for Merioneth and Carmarthenshire. It was "probable" that Firecrests bred at Wentwood in 1975 (Ferns *et al.* 1977).

1976

There were only four records in North Wales, all from Bardsey. In south Wales there was at least one bird singing in Wentwood on 25 June and there were two records from Pembrokeshire and four in Glamorgan.

1977

There were up to three at Wentwood, including a pair which built a nest, and one at Chepstow (Monmouthshire) on 31 May.

During the 1970s the Firecrest therefore established a foothold as a breeding species in Wales. This colonisation followed the first breeding in England in 1962 (in Hampshire), part of a northward extension of range in France and Germany. Monmouthshire has continued to be the main area for breeding Firecrests in Wales in proximity to a sizeable population in the Forest of Dean (Gloucestershire). It is characteristic of the species that breeding populations fluctuate markedly from year to year but by 1989 it was recorded that there were a minimum of 21 singing males at Wentwood and that 11 pairs reared at least 75 young. In 1990 new breeding sites continued to be located in Monmouthshire, with singing males at four locations in the Wye Valley in April and May and one or two near Cwmyoy; at least 12 singing males were recorded at Wentwood and successful breeding noted in three areas. In 1991 there was a complete collapse of the Monmouthshire breeding population and none was found at any site, possibly the result of cold weather in February.

Away from the main area in Monmouthshire it is difficult to ascertain the extent of the breeding distribution in Wales, not least because this tiny species, nesting high in conifers, is easily overlooked. In 1982 six territories were located at Lake Vyrnwy (Montgomeryshire) and at least four of these produced young; up to three pairs have been present in most, but not all, years since and several successful nests have been recorded. As in Monmouthshire there were no records in 1991.

In 1987 a pair bred at Nercwys (Flintshire).

In 1990 a singing male was noted at Mynydd Du (Breconshire) in the breeding season and a singing male was also present in the Elan Valley (Radnorshire) in April. It therefore appears that some further extension of range is taking place, although 1991 was clearly a setback in this. There is much suitable habitat (mature conifer plantations, often associated with mixed or even mainly broadleaved woodland) available for further colonisation in Wales, although the main breeding area continues to be in southern England.

Other than as a breeding species, the Firecrest is an uncommon passage migrant, principally in the period

October to November and at coastal sites but also, to a lesser degree, between March and May. In recent years there have been a good scattering of winter records, chiefly from coastal counties.

Spotted Flycatcher *Muscicapa striata*
Gwybedog Mannog

Common summer visitor; passage migrant, May and August/September.

The Spotted Flycatcher is one of the latest summer visitors to arrive each year, birds only rarely being recorded before mid-April and then only sparingly until the main influx in the first week of May. They remain with us only a short time and the exodus of breeding birds begins in August and is usually completed, even after second broods, by mid-September.

The Spotted Flycatcher is well distributed throughout Wales (47% occupation of tetrads in Pembrokeshire, 38% in Breconshire, 71% in Monmouthshire). The patchiest areas of distribution – excluding the hill ground above c266 m which is generally shunned (although Shrubb in litt. has found them breeding at 320 m in Breconshire) – are the western peninsulas of Pembrokeshire, Llŷn and Anglesey together with urbanised parts of Glamorgan. Spotted Flycatchers inhabit a fairly wide range of breeding habitats: essentially, they are birds of woodland edge and woodland glades but they are also common in large gardens, parkland, churchyards and cemeteries, farmland with copses and wooded hedgerows and particularly wooded watersides, which are rich in the large flying insects which form their exclusive food. Historically there has been no substantial change in status nationally, at least up to the 1960s. Since then a general decrease has been recognised over the southern half of Britain although evidence in Wales is too scant to show just how far this generalisation applies here. Peers (in litt.) suggests a decline of up to 25% in the breeding population in north Breconshire since the early 1980s. A minimum population of 1250 pairs has been estimated for Monmouthshire (Tyler *et al.* 1987) and Peers and Shrubb (1990) suggest c350 for Breconshire while Donovan and Rees (in press) estimate some 900 pairs for Pembrokeshire. Few other population estimates have been made. Massey (1974) found densities of 0.5 pairs per ha in oakwood in south Wales and 0.3 pairs per ha in alder. Currie and Bamford (1982) recorded between 0.5 and 1 pairs per ha in six mature conifer stands in north Wales.

The Spotted Flycatcher is a long-distance migrant, wintering well south of the Sahara, with some birds going as far as Cape Province. Ringing recoveries of birds caught in Wales illustrate these southern destinations; recoveries have been made in Eire (9), Channel Is. (1), France (1), Germany (1), Spain (6 + 1 at sea off Cape Finnisterre), Portugal (4), Morocco (1), Nigeria (2), Cape Province (1 – this bird had travelled a minimum of 10 015 km.

Passage is well marked on the offshore islands, particularly Bardsey (where breeding occurred regularly between the turn of the century and the early 1920s), notably in spring when up to 150 birds per day (exceptionally up to 200) can pass through. Passage continues up to early June. The autumn passage is similarly well marked in August and September, although the number of birds involved is considerably smaller with very late birds occasionally being recorded up to the third week of October. This passage is mirrored on the Pembrokeshire islands, although the numbers are much smaller, generally no more than daily maxima of 10–15 in spring or autumn, although an exceptional "fall" of 150 birds occurred on Ramsey in May 1967.

Red-breasted Flycatcher *Ficedula parva*
Gwybedog Brongoch

The Red-breasted Flycatcher is a scarce but regular autumn passage migrant to Wales, particularly to Bardsey where it can be considered something of an island speciality.

The first record for Wales was a female at Llanishen (Glamorgan) on 12 September 1943 and since that date there have been a total of 126 individuals up to 1991. Most of the records are of single birds, a record of 11 on Bardsey on 8 September 1988, following four on the previous day, is quite exceptional.

There is a sprinkling of spring records (see histogram) and two unusual dates well outside the normal pattern:

3–4 February 1983	–	One, Powys Castle (Montgomeryshire)
4 December 1983	–	Male, Whiteford (Glamorgan)

Passage is principally concentrated in the period late September to mid-October with 43% of the total between 24 September and 14 October.

No fewer than 58 of the records, involving 73 birds, come from Bardsey, where 36 had been ringed at the Observatory up to 1991. The vast majority of the remaining records are from coastal sites, including the offshore

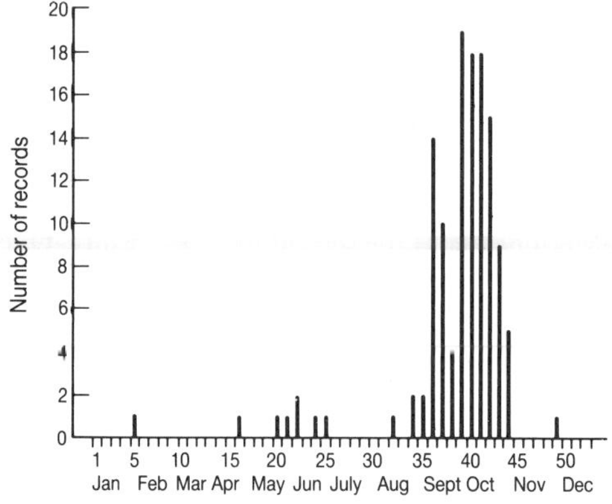

Weekly records of Red-breasted Flycatcher to 1991

islands/stacks of the Skerries (Anglesey), The Smalls, South Bishop Lighthouse and Grassholm (Pembrokeshire).

Collared Flycatcher *Ficedula albicollis*
Gwybedog Torchog

A vagrant.

This species breeds from eastern France and Italy through central Europe east to European Russia and winters in Africa south of the Sahara.

An adult male on Bardsey on 10 May 1957 is the only record from Wales. Full details are given in *British Birds* vol. 51, p. 36.

Pied Flycatcher *Ficedula hypoleuca*
Gwybedog Brith

A numerous summer visitor in most mainland counties, but a rare breeding bird on Anglesey.

No bird is more characteristic of the open sessile oakwoods of upland Wales than the delightful Pied Flycatcher. It is a summer visitor, which arrives in numbers from mid-April onwards, the males arriving a few days before the females and immediately proclaiming their presence with a short, pleasing and musical song. Numbers vary from one year to another and breeding success is dependent on the season's weather and its effect on the supply of caterpillars. The remnant oakwoods on the steep valley sides of mid and north Wales are the traditional stronghold of the Pied Flycatcher. Here, long-term neglect and overgrazing combine to produce perfect conditions with abundant insect food, nesting holes in decayed timber and a woodland floor sheep-grazed bare of understorey and field layer. Other wooded habitats are also used, with pairs frequently showing a predilection for waterside sites, particularly those with old alders and birches. Pairs take to nest boxes more easily than any other woodland species and can readily be induced into gardens, parkland and hedgerows or coppices with mature timber. Pied Flycatchers can be almost guaranteed in such woodlands in the uplands and foothills up to the tree limit but less so in similar sites in the broad valley bottoms.

The Pied Flycatcher was first recorded in Britain in 1676 breeding in northern England and eastern parts of north Wales and in 1766 Pennant, who lived part of his time in Flintshire, remarked that the bird was to be found "sometimes in Derbyshire, Shropshire and Flintshire". Dovaston (1832) was next to mention Welsh localities at Chirk and Valle Crucis, both in Denbighshire. In 1893 E. A. Swainson attempted a survey through correspondence in *The Field* magazine, which demonstrated the extent of regular breeding through the counties of Breconshire, Montgomeryshire, Radnorshire, Merioneth, Caernarfonshire and Denbighshire; there was a non-breeding record from Cardiganshire but none from Carmarthenshire or Flintshire.

As ornithological interest, often manifest in the upsurge in egg-collecting, developed in the later decades of the 19th century, knowledge of the Flycatcher's true numbers and distribution slowly grew. Forrest (1907) recognised it as common throughout the five mainland north Wales counties (but regarded it as absent on Anglesey) and Cambridge Phillips regarded Breconshire as a favoured area for the species, while Paterson *et al.* (1900) recorded it as "occasional" in Glamorgan with the first record in the county not made until 1888. Barker (on the evidence of J. H. Salter) listed it as abundant in the upper Tywi Valley while Mathew knew it only as a rare visitor in Pembrokeshire (where it is still only extremely local). It was not actually recorded in Monmouthshire until 1899, with only 19 records in the county between then and 1962, although it is evident that there was an expansion in range southwards and eastwards from the central Wales heartlands in the last two decades of the 19th century. This spread and an increase in numbers was revitalised again between 1940 and 1952 (Campbell 1954–55), which saw expansion into seven English counties, with reappearances after long intervals in ten others. The provision of nestboxes from about the 1960s has confused the true picture of expansion since then, as large numbers of pairs have been induced into nestbox schemes in Wales and the Marches.

In the counties of central and north Wales most suitable woodlands (i.e. in uplands and foothills) are occupied. However, in southern counties the situation is different. In Monmouthshire only 26% of tetrads are occupied and in Glamorgan the Pied Flycatcher shuns Gower completely and also avoids the coastal plain in the south of the county. Most of Pembrokeshire is beyond the normal breeding range, although breeding is now established with a population of at least 100 pairs by 1988, mainly occurring on the hill land in the north but also more sparsely in woodlands on the Cleddau estuary.

Pied Flycatchers on Anglesey were first recorded in woodlands along Menai Straits in 1910 and this is the only part of the island where a few pairs are habitual although it is numerous a stone's throw away across the Menai Straits in Caernarfonshire. On Llŷn, as in Pembrokeshire, it is very scarce, still being confined to those woodland areas at the eastern ("landward") side of the peninsula. Densities of breeding pairs can be impressive, especially where nestboxes are provided. On the RSPB's Gwenffrwd and Dinas Reserve the density was doubled to 120 pairs per km^2 after the provision of boxes.

Stowe (1987) gives mean densities of Pied Flycatchers in a series of deciduous woodland nestbox schemes:

	Pairs per hectare
Caernarfonshire	2.8
Merioneth	3.6
Denbighshire	3.1
	3.7
Radnorshire	4.0
	3.0
	3.2
Cardiganshire	3.4
	3.3
Breconshire	2.7
Carmarthenshire	3.0

In one isolated garden on farmland in Montgomeryshire three pairs regularly occupy nestboxes on the 0.1 ha site which possesses only ten mature trees within 100 m radius. Peers and Shrubb (1990) estimate c1000 pairs in Breconshire and it seems likely that the Welsh population may number as many as 13 000–15 000 pairs on the basis that the total of pairs in each of the eight counties within the "core" area is over 1000 pairs (pers. comm. and county recorders). Radnorshire is estimated to have between 1500 and 2500 pairs with at least 1000 in nestboxes (P. Jennings, in litt.) and a total of up to 2000 pairs is suggested for both Merioneth (R. Thorpe, in litt.) and Denbighshire (P. Rathbone, in litt.) Such a total would suggest that Wales supports well over 50% of the British population.

Once the breeding season is over and the young have left the nests (mid-June), Pied Flycatchers become surprisingly unobtrusive, although emigration does not begin in earnest until late July or early August. The return passage at coastal stations begins from the last days of July and continues until early October, evidently also reflecting the movement of birds from farther north. The peak occurs in late August although numbers are usually small on Bardsey and the Pembrokeshire islands as the species is predominantly an east coast migrant. The spring passage is very light and in some years negligible. Many Pied Flycatchers have been ringed in Wales, both pulli and adults, and numerous recoveries exist, which are tabulated below and demonstrate the southern destination of this trans-Saharan migrant. One bird, ringed on Orkney on 2 October 1982 was retrapped on Bardsey seven days later and another ringed in Sardinia in April on spring passage northwards was found breeding in Breconshire less than two months later.

Recoveries of Welsh-ringed Pied Flycatchers

Country	*Number of recoveries*
North England	10
South England	109
Netherlands	3
Belgium	1
France & Channel Is.	30
Spain	35
Portugal	15
Tunisia	1
Morocco	81
Algeria	31
Switzerland	1
Ghana	1

Bearded Tit *Panurus biarmicus*

Titw Barfog

An erratic breeding species in one or two locations in very small numbers from the mid-1960s to the late 1980s; otherwise a rare autumn and winter visitor to reedbed areas.

The Bearded Tit is an inhabitant of extensive areas of *Phragmites* reedbeds, its British population being mainly found in coastal reedbed areas of south and east England with occasional outposts farther west when numbers are especially high. There was marked decrease and contraction in range in the 19th century due to the drainage of wetland and by the turn of the century Bearded Tits were virtually confined to a limited area of Norfolk (Parslow 1973). The hard winters of 1917, 1940 and 1947 almost exterminated the residual population, reducing it to fewer than ten pairs. Since then numbers increased again in eastern England to reach several hundred pairs (Bibby & Lunn 1982). In Wales, it was only very rarely recorded until the 1960s during which decade there were several well-documented irruptions (Axell 1966, O'Sullivan 1976) in eastern England, resulting in a small crop of Welsh records through into the 1970s and 1980s and the establishment of a small breeding nucleus at Oxwich NNR. The link between the first Welsh records and erupting populations from the east is well confirmed by several significant ringing recoveries involving birds trapped at Shotton steelworks pools (Flintshire). 1965 witnessed a strong autumn influx of Bearded Tits from Holland and a party of at least 16 birds at Shotton in October contained two birds which had been ringed as youngsters earlier the same summer in Oost Flevoland reedbeds in Holland. Individual birds caught at Shotton in October 1971 and October 1972 were later recovered near Hanover (Germany) and Gelderland (Holland) respectively during 1973 indicating a positive return migration at least in some cases.

It is slightly strange that records from Wales were virtually non-existent before the 1960s even accepting the shortage of active field observers in the early decades of the present century. However, Axell argues that the paucity of records probably reflects genuinely low numbers on either side of the North Sea prior to the rapid expansion (initially in Holland) which took place soon after the devastating winter of 1946–47. Both he and O'Sullivan have demonstrated that the long-range eruptions which began in autumn 1959 were greater than at any time in the previous 100 years. They argue that the populations in England prior to this were insufficient to stimulate the emigration that would have produced anything more than occasional vagrants in the west.

Mathew (1894) mentions a record of birds in the reedbeds at Sealyham (Pembrokeshire) around 1860. It was a second-hand record and cannot be confirmed.

Similarly, two Bearded Tits alleged to have been present in the St. Clears area (Carmarthenshire) in November 1891 and during the following spring are unsubstantiated and should be discarded.

In 1883 and 1884, Cambridge Phillips (*Zoologist* 1884: p. 486) recorded the occurrence of birds in Breconshire in these years (a pair and a singleton were cited) but he did not include the reports in his own book (1899) and they too can be discounted.

After the eruptive activity which was a feature of the population on both sides of the North Sea in every autumn between 1959 and 1965, birds suddenly occurred in some numbers in Wales in 1965. The history of the species in Wales since that time is catalogued below.

1965	Oxwich, two birds present
	Anglesey, minimum 40 birds
	Flintshire, minimum 16 birds at Shotton
1966	Gwent, three birds, Newport Docks, January–March
1967	Anglesey, one pair bred Valley Lakes
1968	Glamorgan, present
	Anglesey, one pair bred Bodgylched
1969	Anglesey, one pair bred Malltraeth Marsh
1970–71	Glamorgan, eight Kenfig NNR
1971–72	Glamorgan, four present Oxwich NNR
	Monmouthshire, 14 present
	Flintshire, six Shotton
1972–73	Glamorgan, max. 10 at Kenfig NNR and Oxwich NNR
	Flintshire,five
	Carmarthenshire, small parties at two sites
1973	Anglesey, present in breeding season
	Glamorgan, present and presumed to breed Oxwich NNR
	Cardiganshire, two Cors Fochno
1973–74	Glamorgan, present
	Anglesey, present
	Flintshire, present
	Cardiganshire/Pembrokeshire, several Teifi estuary
1974	Glamorgan, bred Oxwich NNR
	Monmouthshire, two Uskmouth
1974–75	Flintshire, present Shotton
	Monmouthshire, three present Uskmouth
1975	Glamorgan, up to six present Oxwich NNR
1976	Glamorgan, two pairs bred Oxwich NNR
1977	Glamorgan, two pairs bred Oxwich NNR
	one pair Crymlyn Bog NNR
	max. four birds Kenfig NNR
	Breconshire, breeding suspected Llangorse Lake
1978	Glamorgan, three pairs bred Oxwich NNR
	Breconshire, 12 birds Llangorse Lake in January
1979–80	Cardiganshire, two Cors Fochno
	Monmouthshire, three present
	Glamorgan, three pairs bred Oxwich
1980	Glamorgan, two to three pairs bred Oxwich NNR
	up to four birds Kenfig NNR, November–December two, Comeston, October
	Breconshire, birds heard Llangorse Lake, September
	Monmouthshire, three Uskmouth, December
1980–81	Anglesey, two present
1981	Glamorgan, four birds Oxwich NNR – no breeding proved
	Carmarthenshire, one bird present
	Anglesey, two birds present
1982	Glamorgan, two pairs bred Oxwich NNR
	Carmarthenshire, one Witchett Pool, January
	Monmouthshire, five present, November
1983	Glamorgan, four pairs Oxwich NNR; at least two pairs bred
	two birds Kenfig NNR, October
1984	Glamorgan, presumed breeding Oxwich NNR, maximum 18 birds present
1985	Glamorgan, at least one pair bred Oxwich NNR
1986	Glamorgan, present Oxwich NNR
	one bird Kenfig NNR, April
	Monmouthshire, one Uskmouth, November
	Caernarfonshire, one bird Pwllheli, August
1987	Glamorgan, maximum nine birds Oxwich NNR
	Carmarthenshire, heard calling Burry Port
	Caernarfonshire, one Pwllheli, August
1988	Glamorgan, probably bred Oxwich NNR
	one bird Kenfig NNR
	Cardiganshire, two Ynys-hir
	Carmarthenshire, four Witchett Pool, November
	Monmouthshire, one Uskmouth, April
1989	Glamorgan, single birds only Oxwich NNR, April

	Cardiganshire, heard Ynys las, October
	Carmarthenshire, four Witchett Pool, November
1990	No records
1991	No records
1992	No records

Long-tailed Tit *Aegithalos caudatus*

Titw Gynffon-hir

A relatively common resident, widely distributed except on parts of Anglesey and Llŷn. Numbers fluctuate widely from year to year in response to the severity of winter weather.

The tiny and attractive Long-tailed Tit is a familiar woodland, garden and hedgerow species throughout most of Wales. Notwithstanding this it is not particularly numerous and it is extremely susceptible to hard winter weather. Thus its numbers tend to fluctuate considerably from year to year as they are suppressed by harsh spells and then build up again with a succession of mild winters. In the early 1990s, after a succession of mild winters, numbers were at a high level. The CBC index illustrates these fluctuations nationally very clearly with severe dips after the winters of 1978–79 and 1981–82.

The Long-tailed Tit is predominantly a bird of deciduous woodland – especially in association with understorey – scrub, thorn thickets, gorse and the like which the birds require both for the provision of nest sites and for foraging. Most Welsh woodlands are overgrazed and lacking understorey and the amount of lowland and hillside scrub remaining nowadays is also greatly diminished. The result is that Long-tailed Tits are probably somewhat less numerous than might be expected in such a rural country and almost certainly less plentiful than in the past, although this is not clear from the various historical accounts in county avifaunas or county bird reports. There are very few regular woodland CBC plots in Wales but those which have run for a period of ten years or more, thereby evening out any short-term fluctuations, show a mean density of approximately one pair per 10 ha. Like one or two of the *Parus* tit species, the Long-tailed Tit has certainly benefited, at least marginally, from the extensive conifer plantations which have been developed in recent decades. It is sometimes present, albeit sparingly, in areas of thicket-stage spruce and larch especially where pockets of gorse or deciduous scrub remain. However, it becomes very scarce over 335 m or so, although occurring sporadically in sheltered areas of moorland edge.

In most parts of Wales the Long-tailed Tit can be described as widespread and fairly common. In south-west Wales it was regarded by Lockley *et al.* (1949) as being common "even in the extreme west (of Pembrokeshire)" which is a slightly more optimistic picture than that revealed by the *Pembrokeshire Breeding Atlas* which shows it occurring in only 41% of tetrads (1984-88). It is equally scarce in the equivalent areas of north-west Wales and, as expected, it is distinctly uncommon in the barren parts of the western half of Anglesey and Llŷn. G. A. Humphreys, writing in 1903, commented that "in the absence of any breeding record from Anglesey, a nest at Llanfair P.G. was of particular interest". RSPB surveys in 1986 found it in only 36 one-kilometre squares on the island, with none in the northern parts.

The Long-tailed Tit is well known for being one of our most vulnerable species in the face of harsh winter weather. At times up to 80% of the population can be exterminated in such circumstances (*Breeding Atlas*) and may take several years to recover. A series of particularly cold winters is known to have had serious effects on the population in Wales. Forrest (1907) records the virtual elimination of local populations in the Conwy Valley and Corwen areas in 1895; the population took six years to recover. Other bad years for the species were 1939–40 and 1947. The severe spell in January–February 1963 was particularly severe, although not as bad as might have been expected due to the absence of freezing fog which produces rime and makes feeding impossible (Marchant *et al.* 1990). The winter of 1978–79 affected numbers markedly in parts of Dyfed, and 1981–82 was the most recent winter to affect numbers, although they were not seriously depressed and were already recovering well by 1983.

In milder winters, flocks made up of family parties establish a wide feeding territory and remain faithful to it, defending it against rival parties. Within these home ranges the flocks wander freely, sometimes associating with bands of other tits, Goldcrests or Treecreepers. Only occasionally do birds move any distance from their natal areas. There is one record only of a Welsh-ringed bird moving more than 10 km, being a young bird ringed near Cardigan in August and reported at Tenby eight months later (46 km south). There are only sporadic records from the offshore islands, most frequently in October or November (Bardsey 20 records, Skomer c5, Skokholm 2) and rarely in March–April (Bardsey 2, Skomer 1).

Marsh Tit *Parus palustris*

Titw'r Wern

A resident species, strongly sedentary, widespread but patchily distributed in the eastern half of Wales and in Pembrokeshire. Scarce or absent in other western and northern areas.

Marsh Tit is an unfortunate misnomer for a bird that has no particular association with wet areas. It is first and foremost a deciduous woodland species, insectivorous in summer but turning to seeds and especially woodland fruits – beechmast, spindle, rowan, yew – in winter. In common with the Willow Tit it is much more rural in its choice of habitat than the three commoner *Parus* species and although a reasonably frequent visitor to suburban bird tables it does not share the regular co-habitation with man so successfully exploited by Great Tit, Blue Tit and Coal Tit. In Wales Marsh Tits are most plentiful on the eastern side of the mountain spine although also reasonably numerous across substantial parts of south Wales and into the south-west peninsula. In these areas they are most frequent in the old deciduous woods dominated by sessile oak which, although now scarce on the flatter ground, still cloak many of the steeper valley sides. Such woods frequently have a mixture of birch, rowan, sycamore, wych elm and, particularly, ash but less frequently beech. In winter Marsh Tits are mainly ground feeders – although any other choice is effectively removed from them by the overgrazed nature of most Welsh woodlands – and they benefit from the easier foraging of an open woodland floor more than they do on one which has a dense ground layer or understorey.

The historical status of the Marsh Tit in Wales is difficult to establish, mainly because it was not separated from the very similar Willow Tit (q.v.) until almost the turn of the present century. Most early accounts of the status of "Marsh Tits" previous to then must become invalid as the two species overlap extensively and we can only surmise as to which species was actually involved in a given account. A few exceptions exist where egg collectors have left precise details of nest sites, which permit retrospective identification with good certainty. The two species are indeed extremely similar in appearance but they have important ecological separations (see under Willow Tit) which help throw light on their likely fortunes in the past. Globally, the Marsh Tit, primarily a deciduous woodland bird, has a slightly more southerly distribution than the Willow Tit. It has been shown (Perrins 1979) that in several areas of the Palearctic, e.g. Denmark, where the Willow Tit is absent, Marsh Tits will occupy the conifer habitat more usually occupied by the Willow Tit. In Wales, however, the Marsh Tit is rare in conifers; Bibby *et al.* (1985) failed to record it at 326 point counts in restocked conifer sites in north Wales or at 253 sites in mature conifer plantations with occasional broadleaved trees. Therefore as extensive natural conifer forest was negligible in Wales in the past it may well be that the historical separation here was more between the dry deciduous forest (Marsh Tit) and the wetter willow/alder carrs and alder/birch woodland in waterlogged valley bottoms (Willow Tit). This situation is evident in the modest population of the two species in southern Scotland (Thom 1986).

Marsh Tits are absent from most areas north and west of the mountain spine. They are scarce in the north-east half of Cardiganshire, and rare in west Montgomeryshire and Merioneth (valleys of Dyfi, Dysynni and Mawddach), the Llŷn peninsula and Anglesey. Anglesey birds refer mainly to single winter records in the wooded areas of the south-east of the island and presumably involve wandering juveniles from the adjacent mainland; even there they are scarcer than Willow Tits but are known to occur at Aber, Betws-y-Coed and a few other sites in dry woodland in the Conwy Valley, Llanfairfechan and Llandudno Junction areas.

The species' strongholds in Wales are found in the drier deciduous woodlands in a great sweeping arc from north-east Wales, through the counties of the Marches and across south Wales into south Pembrokeshire and southern Cardiganshire. Here they normally – but not inevitably – outnumber the Willow Tits; in Radnorshire, Peers (in litt.) recorded it in 42 locations (1980–87) compared with 38 for Willow Tit; A. R. Pickup, D. H. V. Roberts and P. E. Davis (all in litt.) consider it twice or four times as numerous as the Willow Tit in Carmarthenshire and Cardiganshire. In Pembrokeshire it was recorded in 36% of tetrads (118) compared with 22% (61 tetrads) containing Willow Tits and with an estimated breeding population of between 500 and 700 pairs (Donovan and Rees, in press.). In the southern half of Cardiganshire P. E. Davis (in litt.) regards the Marsh Tit as much commoner than the Willow Tit, being particularly numerous in the well-wooded valleys – Arth, Teifi, Wyre, Ystwyth and Aeron – running down to the coast of Cardigan Bay. The *Gwent Atlas* (1981–85) located birds in 45% of tetrads (175) in Monmouthshire and estimated a population of perhaps 500 pairs. In Glamorgan Heathcote *et al.* (1967) regarded it as the least numerous of all the *Parus* tits although in west Glamorgan it has been found in twice as many tetrads as the Willow Tit (Thomas *et al.* 1992) with a well-established population on Gower (28 tetrads out of 56).

Although virtually nothing is known about the territory size in Welsh woodlands, Marsh Tits are known to have fairly large territories elsewhere; in England (Southern and Morley) the average size was 2.5 ha while in Germany (Ludescher) they can be as large as 6 ha. The implication that pairs are fairly thinly distributed is borne out by work by Massey (1974) who found densities of one pair per 10 ha in oak woodland, 2.2 pairs in ash, 0.75 pairs in alder and 1.5 in mixed woodland. The apparent attractiveness of ash is mentioned from other parts of Wales also.

Marsh Tits are highly sedentary. Pairs remain together on their territories throughout the winter while the immature birds join the wandering flocks of other species of tits and woodland passerines. Ringing recoveries bear this out with only three birds having been recovered any measurable distance from the place of ringing – 7 km, 9 km and 11 km respectively.

Even during severe weather Marsh Tits are loath to leave their woodland territories when most other species move out to benefit from garden feeding stations, and the species does not take part in any of the irruptive movements that occasionally affect the three commoner *Parus* species. There have only been two offshore records of Marsh Tit, one on Skomer on 8 November 1961 and one on Skokholm on 12 October 1988.

Willow Tit *Parus montanus*

Titw'r Helyg

A thinly distributed resident, most frequent in south-east and east Wales and absent from most of north-west Wales.

The Willow Tit was not identified as a separate species from the Marsh Tit until 1897 and even thereafter for several decades there was not universal acceptance – and certainly not ready identification – of the fact that two distinct species exist. The result is that the Willow Tit is historically the most poorly recorded resident species in Wales; its past status is little more than conjecture. The two species are indeed so similar that both the *Breeding Atlas* (1976) and the *Winter Atlas* (1986) concede that some confusion of identification still exists. Today – and there seems no reason for doubting that the same applied in the past – the Willow Tit is a strongly sedentary resident. Although it overlaps both in distribution and occasionally habitat use with the previous species (q.v.) there are normally clear separations and, in particular, distinct preferences for the type of woodland in which they each occur.

The Willow Tit shuns the drier oak/ash woodlands preferred by the Marsh Tit and is characteristic of a wide variety of wetter woods, carr and wetland fringes. This choice is partly determined by the fact that it excavates its own nest hole and thus requires well-rotted alder, birch or elder stumps or decaying softwood fence posts for the purpose. In Wales it is found in a wide range of such habitats including alder and willow carr, mature alder woods, pre-thicket forestry plantings (where pockets of alder/willow or streamside trees remain, alder/willow choked dune slacks, thicket pine plantations on sand dunes, wet birchwoods, linear scrub with ditches, e.g. disused railway lines. Although such habitat preferences often determine a bias of numbers to the lower ground, birds do occur sparingly up to 457 m, e.g. on Radnor Forest, particularly in association with afforested areas. In such places they may be found in areas of windblown trees, clear-fell, ancient relic hedge lines, unplanted wetlands or young plantations but always where the territory has a core of willow/birch/alder scrub; in new plantations they are normally forced out by the time the trees are 2–3 m high. Only a small proportion of the Welsh population, however, is found in conifer woods of any form at present despite the fact that throughout its enormous Palearctic range the Willow Tit is predominantly a conifer forest species. With the increasing emphasis being put on sensitive management of the Welsh forests it may well be that the Willow Tit will be a primary beneficiary and will become a more common inhabitant of its natural forest habitat.

Its current status county by county in Wales can be summarised as follows.

Monmouthshire

First recorded 1942 but evidently overlooked. Now breeds in approximately 40% of tetrads with a population of perhaps 400 pairs.

Glamorgan

Salmon and Ingram "searched diligently for it for fifteen years" prior to finding two birds at Llanishen in 1938 (they surely overlooked it elsewhere?). They found it nesting there by 1946 and others were then located elsewhere in the county up to 1963 in which year, following a bitter winter, it was still reported over a wide area of south Glamorgan. It is now widely but thinly distributed, including the valleys of the coalfield area. In Gower and the Swansea Valley it is uncommon, found in no more than 14.5% of tetrads (Thomas *et al.* 1992), only five of them on Gower.

Carmarthenshire

First identified on the Dinas Hill, upper Tywi Valley in 1934. Still believed to be widely overlooked in the county and thus under-recorded. In north Carmarthenshire it is outnumbered by between 2:1 and 4:1 by the Marsh Tit but probably reaches parity on the lowland areas even possibly exceeding the Marsh Tit in numbers in some parts, e.g. Cydweli (D. H. V. Roberts in litt.) and Pembrey where it is particularly numerous (F. A. Currie, in litt.). On the RSPB Gwenffrwd/Dinas reserve neither the number of pairs (4) nor their location has varied since accurate annual counts began in 1966.

Pembrokeshire

First recorded (breeding) at Haverfordwest in 1925 which, together with a singing bird at Clarbeston in 1939, were the only records known to Lockley (1949). Today the Willow Tit is known to be widely but thinly distributed throughout the county and the tetrad breeding survey showed its presence in 22% of tetrads (Marsh Tit in 36%) and estimated a minimum of 200–300 breeding pairs.

Cardiganshire

First identified at Crosswood in 1926 (Salter). North-east of a line from Lampeter to Aberystwyth the Willow Tit is well distributed and in reasonable numbers (Marsh Tit virtually absent here) but south-west of this line it is scarcer, confined to some boggy valley woodlands and carr and is certainly well outnumbered by the Marsh Tit.

Breconshire

A specimen in Brecon museum was probably obtained around 1910. It is perhaps noteworthy that Walpole-Bond, who knew the species well in Sussex, did not find it in the county during his visits in the early years of the century. Now it is known to be thinly distributed in low-lying damp woodlands of the main river valleys but also, as in other counties, at higher altitudes than the Marsh Tit. Between 1980 and 1987 Peers (in litt.) found it in 38 localities (Marsh Tit in 42). Massey (1974) found six pairs per 40 ha in mixed woodland, three pairs in alder and two pairs in oak. Peers and Shrubb (1990) suggest a county population of perhaps 250 pairs.

Radnorshire

Known since 1934. Thinly distributed, mainly in river valleys of Wye, Edw, Elan and Ithon. Peers (in litt.) found it less numerous than the Marsh Tit in the eastern half of the county. In forest plantations on Radnor Forest it occurs up to 450 m.

Montgomeryshire

Reasonably numerous, mainly in wet woodlands in the main river valleys – Severn, Vyrnwy, Clywedog, Tanat, Dyfi and their tributaries. Quite numerous on Breidden Hills in larch, scattered with remnant hedge lines (F. A. Currie in litt). On the Lake Vyrnwy RSPB reserve in north Montgomeryshire Willow Tits outnumber Marsh Tits by a ratio of at least 2:1 (M. Walker, pers. comm.)

Merioneth

Widespread but uncommon. Recorded mainly from wet woodlands/scrub in the valleys of the rivers Dee, Arthro, Wnion, Mawddach, Dwyryd and Dysynnsi. Eleven individuals were caught at one site at Arthog in 1983 (F. A. Currie in litt.) and Thorpe (in litt.) knew it from only five sites in the Mawddach valley 1986–87.

Denbighshire and Flintshire

Very local and less numerous than Marsh Tit (Lawton Roberts in litt.). Probably extremely scarce in north and west Denbighshire, but found in the Dee Valley, Ruabon/Llangollen area and around Wrexham. Only one pair present on Moel Famau Country Park (861 ha) in 1986 (RSPB). There is little information from Flintshire.

Caernarfonshire

Recorded by H. G. Alexander at Capel Curig in 1919, but it is still virtually unknown throughout most of the county. It is found at only one or two sites in the Bangor–Caernarfon area (N. Brown *et al.* in litt.) but is more common in the Conwy Valley (e.g. Dolgarrog) and its tributary, R. Lledr, upstream of Dolwyddelan. Birds have been recorded from Llyn Cwellyn and the general impression is that it is scarce in most of the eastern half of the county. On Llŷn it is rare: in 1986 and 1987 RSPB surveys produced birds in only four sites, the most westerly being near Botwnnog.

Anglesey

Rare. Several pairs breed on Cors Erddreiniog (L. Colley in litt.). Whalley (1954) reported a breeding record as long ago as 1929. Two were seen in winter 1983 in willow scrub near Pen y Parc but other records have been very sporadic; the *New Atlas* records presence in four 10-km squares with proved breeding only at Cors Erddreiniog.

In winter adult Willow Tits remain on territory although they may extend their foraging area and more willingly leave the damp areas to feed farther afield. Occasionally individuals disperse as far as 5 km in winter (Sellers 1984) presumably in response to food shortage but evidence to support this in Wales is scant. They are less likely to be found in the winter flocks of tits, Treecreepers, Goldcrests than other *Parus* species even when these pass through their territories. Similarly, as they do not feed on nuts, e.g. beechmast as other *Parus* species do, they are less frequently found in high woodland, parkland or gardens and are uncommon attenders at garden feeding stations. Shrubb (pers. comm.) has found them often feeding in weedy root fields in Breconshire, making brief forays out from the hedgerows.

Coal Tit *Parus ater*

Titw Penddu

A widespread and fairly common resident. Most numerous in conifer woodlands. Has increased in the second half of this century with the spread of conifer plantations.

The Coal Tit is widely distributed throughout Wales although not of universal occurrence and it is certainly not nearly so numerous as either Blue Tit or Great Tit. Unlike all other members of the family in southern Britain, it is a conifer specialist, having a finer bill than its congeners which enables it to probe into the growing clusters of conifer needles more easily than the other *Parus* species are able to do. It is not surprising therefore that the Coal Tit is most abundant in the new generation of forests, which now occupy some 11% of the land surface of Wales, once the trees have advanced past the thicket stage. Despite this preference for conifers, the Coal Tit also occurs in other habitats, notably deciduous woodland; indeed in some of the higher altitude sessile oakwoods Coal Tits sometimes outnumber even Blue Tit and Great Tit locally. As a garden, parkland or suburban bird the Coal Tit is common and familiar but always more likely to be present if there are even a few coniferous trees in the vicinity.

All early references to the Coal Tit in Wales support the view that it has always been a fairly common and widely distributed bird. In 1894 Mathew was able to quote one of his sources in Pembrokeshire who claimed that it was a more common and widely distributed bird in the northern part of the county than he had ever met with in England. This general abundance, however, must not conceal the fact that it does not have a uniform distribution throughout all the counties, even where habitat appears to be suitable. In Glamorgan, Heathcote *et al.* (1967) describe it as somewhat uncommon in the southern part of the county (i.e. Vale of Glamorgan) which perhaps gives hint of an upland preference which is sometimes suspected; it is certainly plentiful in the extensive conifer plantations in the coalfield valleys throughout the county. In Monmouthshire it is well distributed but nowhere numerous and is virtually absent from the Levels. Ingram and Salmon (1957) suspected that it had increased in Breconshire "in the past twenty years" and noted pairs breeding well up in the hills. Partly because of the association with upland conifer forest, Coal Tits probably breed at higher altitudes – up to 550 m or so – than Blue Tits or Great Tits and even before the post-war plantings began to dominate the upland landscapes,there were various references to them breeding well up in the hills.

In north Wales, Forrest (1907) recognised that it was unevenly distributed. In Anglesey it is far from common, especially in the north and west and is similarly scarce in the western half of Llŷn beyond Pwllheli. In other north Wales counties and in Cardiganshire it is generally common, often noticeably more so on higher ground than near coasts or in major river valleys. At 253 count points in north Wales conifer plantations Bibby *et al.* (1989) found Coal Tits the fifth most numerous species (behind Goldcrest, Wren, Chaffinch, Robin) with a mean density of 25.4 *individuals* per 10 ha and in similar habitat in Monmouthshire Andrews and Bellamy (1988) recorded 27.6 *pairs* per 10 ha. On 326 point counts in restocked conifer plantations (Bibby *et al.* 1985) it was the tenth most numerous bird at a mean density of 4.1 individuals per 10 ha. Williamson (1971) found a density of 2.4 pairs per 10 ha in mature redwood plantations in Montgomeryshire. In oak woodland in Caernarfonshire, Gibbs and Wiggington (1973) found densities of 2.7 pairs per 10 ha and CBC surveys in woodland at Ynys-hir (Cardiganshire) show an average density of 4.3 pairs per 10 ha. Tyler *et al.* found Coal Tits in 75% (291) of tetrads in Monmouthshire and estimated a breeding population of about 3000 pairs. Donovan and Rees (in press) suggest 500–700 pairs in Pembrokeshire in 41% (194) of tetrads. Peers and Shrubb (1990) recorded it in 59% of tetrads in Breconshire.

Basically sedentary, the Coal Tit does, however, give some indication of local wandering, in some years more than in others. Coastal movements have been recorded at Lavernock Point and irregular passage has been recorded on offshore islands. On Bardsey some 34 occurrences are reported (1954–84) all between late September and mid-November, with the exception of four spring records (March–May). The pattern of occurrence on the Pembrokeshire islands is similar with almost all individuals appearing in October. One wintered on Skokholm in 1930–31 and during the large autumn irruption of tits in 1957 more than 100 Coal Tits passed through the island. The same phenomenon occurred on Bardsey where as many as 120 were counted in a single day; there was a smaller irruption in October/November 1975 with up to 40 per day for a short period. In the winter of 1972–73 Hope Jones and Dare (1976) recorded irruptive movement into western Llŷn. An adult ringed on Bardsey in October 1985 was retrapped on the Wirral (Cheshire) four months later (136 km).

Although a substantial part of the population remains in the conifer woodland in winter, where it may be joined by small numbers of other woodland species, Coal Tits also consort with wandering bands of other *Parus* species, Treecreepers and Goldcrests in deciduous woods, farmland hedgerows and coppices. The numbers of individual Coal Tits joining a group is usually small and it is unusual to find as many as ten together. A winter flock of 36 in Monmouthshire was certainly unusual. Hope Jones (1974) found that in the mature oakwoods in the Vale of Festiniog (Merioneth), Coal Tits were only slightly less abundant than Blue Tits in winter. It is slightly surprising that Coal Tits, seemingly cushioned from the worst effects of severe weather by their liking for the dense shelter of conifer plantations, should suffer such heavy mortality as they do. Numbers crashed in the hard winter of 1962–63 (Monmouthshire, Montgomeryshire, Caernarfonshire) and in 1981–82 (Cardiganshire, Radnorshire and other counties). However, populations quickly recover and are usually back up to the original level within two years. Winter mortalities are almost certainly higher in sub-optimal habitats, e.g. deciduous woodland, than in the preferred typical habitats.

Blue Tit *Parus caeruleus*

Titw Tomos Las

An abundant resident species, most numerous in deciduous woodland but not restricted to this preferred habitat, especially in winter when it is more ubiquitous.

As elsewhere in the British Isles, the Blue Tit is the most numerous and widespread of the titmice in Wales. The *Breeding Atlas* and the *New Atlas* confirm its breeding in every one-kilometre square in Wales and the *Winter Atlas* mirrors this universal distribution even though the winter numbers recorded in some areas, e.g. parts of Cardiganshire, are surprisingly low bearing in mind the apparent suitability of much of the countryside. Certainly all the county avifaunas and annual reports testify to the Blue Tit's abundance and ubiquity. By preference it is a species of deciduous woodland, especially oak, but it is extremely adaptable and versatile and occurs as a breeding species in a wide variety of habitats including farmland, parkland, alder and willow carr, riversides, urban centres, suburban gardens, conifer plantations (sparingly), scrubland and ffridd. In addition to nesting in natural holes in trees and buildings (as well as an almost infinite list of holes and crevices in other human structures), it is readily induced into nestboxes in gardens or woodlands, which can quickly result in considerable increases to the local breeding population.

There is no reason to suppose that the Blue Tit has ever been other than an abundant species in Wales. The gradual removal of deciduous woodland throughout the Welsh countryside has clearly not been in the species' interests but this has been compensated to a considerable degree both by expanding urbanisation, with its attendant gardens and parks, and also by the pattern of small fields, old hedgerows, coppices and hedgerow trees that characterises the Welsh countryside. Blue Tits are not particularly at home in conifers and Snow (1954) has shown that they are disadvantaged by their bill shape for feeding among conifer needles. As with Great Tit (q.v.) the presence of occasional stands of deciduous trees or even single mature individuals – especially oak – in conifer plantations is probably significant in determining the Blue Tit's presence. In conifer plantations, where there are deciduous trees present, Blue Tits may nest at a density of up to ten pairs per km^2 compared with anything up to 100 pairs per km^2 in oak-dominated deciduous woodland. In a study on the effects of broadleaved trees on birds in upland conifer plantations, Bibby *et al.* (1989) found Blue Tits only the 15th most numerous species, with a mean of 1.9 *individuals* per 10 ha across 253 count points. In restocked conifer plantations Bibby *et al.* (1985) showed Blue Tits to be even less frequent at a mean density of no more than 0.5 *individuals* per 10 ha.

Post-fledgling broods of young are often to be found feeding on dense bracken slopes away from woodlands in July and August. After the breeding season Blue Tits not infrequently wander in parties up onto the moorland fringes where they may be found foraging among the heather.

Although basically sedentary, Blue Tits have a tendency to wander more readily than the other *Parus* species. Modest autumn movements have long been recognised in some coastal areas, e.g. Lavernock Point, Goldcliffe (Monmouthshire) and Pembrokeshire headlands. On the offshore islands Blue Tits are more regular than Great Tits; on Skomer they have been recorded in all months in small numbers but predominantly in September/October. The pattern is similar on Skokholm but involving even fewer birds. They are reported as "regular visitors from the mainland" on Ramsey and on Bardsey they show even more erratic and unpredictable patterns of appearance than Great Tits. In some years there are no records at all but in most years up to ten or a dozen birds occur during the course of the autumn. The picture on the islands can be very different in years of irruption. In common with other *Parus* species, notably Great Tit, Blue Tits are irruptive in some autumns and at such times flocks may be found in many coastal areas and numbers on the islands can then be impressive: in 1957 50–75 birds were present on Bardsey in October (400 at the same time on Skokholm), 30–40 in October 1964 (50 on Skokholm), a similar number in late September 1974 and 120 on 6 October 1981. Ringing recoveries showed that many of the Bardsey immigrants were of local origin (north Wales) rather than continental birds. Birds of continental origin (race *P. c. caerulus*) do occur in Wales from time to time as shown by one taken at Erwood (Radnorshire) in 1939 (*British Birds* vol. 33, p. 159). Of 482 Blue Tits ringed in Wales and subsequently recovered, 397 have moved less than 10 km. Most of the longer distance recoveries involve birds that have been ringed in the winter half of the year (October–March) in Wales and have shown broadly easterly movements subsequently, implying some degree of immigration, with a return movement in spring.

In winter, Blue Tits exploit their versatility to the full and can be found in a very wide range of habitats. They forage in areas as diverse as farmland hedgerows, dune slacks, reedbeds and even beds of *Spartina* grass on the coastal marshes, where they feed on the standing seedheads. At this time of year they are often the main constituent members of the foraging parties of tits and other woodland passerines that hunt together in woodlands, hedgerows and gardens. The Blue Tit is probably the best known visitor to garden feeding tables and the incessant comings and goings of individuals often disguise the actual numbers of birds benefiting from a single site as Hope Jones and Dare (1976) showed at Criccieth, where over 100 were shown to be using one 0.04-ha garden.

Great Tit *Parus major*

Titw Mawr

A widespread and numerous species throughout Wales up to approximately 500 m.

The Great Tit occurs as a numerous breeding species in Wales not only in deciduous woodland but also on farmland, in urban and suburban environments and many other cultivated or developed areas where there are trees. Both *Breeding Atlases* and the *Winter Atlas* confirm its presence in virtually every 10-km square in Wales throughout the year. Its more precise distribution is strongly mirrored by the presence of deciduous trees for it occurs almost everywhere where there are small copses, gardens or heavily treed hedgerows as well as in larger woodlands, for it is first and foremost a deciduous woodland species. In these respects its status has probably changed little over the years although evidently it has contracted locally with the clear-felling of woodland for purposes such as agriculture and building development. Conversely, the post-war reafforestation of the uplands has contributed slightly in drawing Great Tits farther up into sub-optimal habitats in the hills. Although they make use of the new generation of conifer forests only sparingly, it has been shown (Bibby *et al.* 1985) that the presence of even a few mature deciduous trees within a conifer monoculture can then make the forest habitat acceptable for the species. Densities in these sub-optimal habitats are never as great as in deciduous woodland. Breeding occurs up to about 500 m above sea level.

Not surprisingly Great Tits are less numerous in the vicinity of the seaboard than they are in inland areas. This is emphasised on the Llŷn Peninsula and also on Anglesey, where they are plentiful along the Menai Straits and to a lesser extent on the eastern side of the island and throughout much of the hinterland. However, in the north and west of the island they are markedly less common.

In common with other resident passerines Great Tits can suffer badly in harsh winters although their winter woodland preference may protect them a little more than some other members of the family. In recent decades the winters of 1946–47 and particularly 1962–63 (Perrins 1979) and more recently 1981–82 took a heavy toll, but each time numbers soon recovered; in Monmouthshire, for example, they were back to the original levels by the autumn of 1965 after the severe winters in 1961–62 and 1962–63 had reduced numbers to as little as one-tenth of their spring 1961 population.

Great Tits are normally sedentary although some local movement may take place at the end of the breeding season, primarily involving juvenile birds. Lockley *et al.* (1949) accurately noted a marked westward passage in Pembrokeshire in late autumn with some birds crossing to Skomer and Ramsey. This certainly still occurs although Great Tits are rare on neighbouring Skokholm where, during the remarkable irruption in autumn 1957, over 200 were present in October. Coastal movement in autumn has been clearly documented at Lavernock Point and is evidenced by Hope Jones (1974) in Merioneth. On Bardsey its occurrence in autumn and spring is highly erratic: modestly numerous in some autumns and non-existent in others. A few individuals have been proved to overwinter but otherwise spring records (March–April) are usually few. In 1984 a pair bred on Bardsey for the first time, studiously ignoring nestboxes that had been put up in previous years to tempt them, but marking their success by, unusually, rearing two broods!

After the breeding season the males in particular are sedentary, normally remaining on territory through the winter while the juveniles and females forage communally throughout the local area with *Parus* species and others such as Treecreeper, Nuthatch and Goldcrest. Beechmast and hazelnuts are especially important at this season and Great Tits forage extensively amongst the woodland litter. Severe weather will, however, drive birds to urban and suburban gardens where they are one of the commonest and most familiar birds at garden feeding stations.

The essentially sedentary nature of Great Tits is demonstrated by ringing recoveries, which show that only 24 individuals ringed in Wales were subsequently recovered more than 10 km away from the place of ringing (162 were recovered less than 10 km away). One individual ringed in Monmouthshire in February was found dead 15 months later in Cheshire (178 km north-north-east), the most distant Welsh recovery so far.

Nuthatch *Sitta europaea*

Delor y Cnau

A widely distributed and fairly numerous resident of mature deciduous and mixed woodland throughout Wales.

In a century that has seen the loss of over 50% of deciduous woodland in Wales, the increase in numbers and expansion of range of the Nuthatch has been an interesting and little understood phenomenon. Formerly absent from most of England north of the Mersey and the Wash, and from the whole of Scotland and Ireland, the Nuthatch is a plentiful bird throughout most woodland parts of Wales, although it has by no means always been so. It is mainly a species inhabiting mature mixed or deciduous woodland, much of which is still to be found in Wales albeit mostly in relatively small remnants. Oak, beech and hazel are the three most important food sources in winter months and some combination of them, preferably with other tree species such as yew and hornbeam (scarce in Wales), is important in any Nuthatch territory.

The Nuthatch is well able to cope with even very small woodland areas, down to an acre or less provided that such areas are supplemented by belts of mature hedgerow trees, small copses or corridors of streamside trees, which is very much the pattern in lowland Wales. The Nuthatch can thus also exploit parkland, large gardens and well tree-covered suburban areas. However, its distribution is locally somewhat patchy, many apparently suitable woodlands not supporting pairs. It is strongly lowland in its distribution, being markedly uncommon above 245 m in most counties, a fact which of course also substantially mirrors the distribution of mature woodland.

In the 19th century the Nuthatch was unknown in many parts of Wales and distinctly scarce in others. Its strongholds appear to have been in the lowland areas of the border counties from Flintshire south through Montgomeryshire, Radnorshire and Breconshire to Monmouthshire and its westward distribution in these counties very probably showed a strong correlation with the low land of the main river valleys reaching into the Welsh hills. Elsewhere it was known only from the lower parts of Carmarthenshire (Davies 1858, Browne 1873 (in Barker 1905)) where Barker (1905) recognised it only from the Tywi Valley, nearby at Carreg Cennan and near Ammanford. Despite these few western outposts, the Nuthatch was unknown almost throughout Glamorgan until the late 1920s. The only known records were from Cwrt-yr-ala and Cowbridge in the 1890s and the Swansea and Merthyr Mawr areas in the first decade of this century. It first appeared on Gower in 1934 and was shown to be breeding there by 1949. In the counties of Pembrokeshire, Cardiganshire, Merioneth, Caernarfonshire, Denbighshire, and Anglesey it was essentially unrecorded other than as accidental occurrences up to the turn of the century (although Forrest documents a handful of records from the eastern parts of Denbighshire).

The westward spread of the Nuthatch in Wales, which was simultaneous with a northward expansion in England, can be detected from the early years of the 20th century, although no marked increase was clear until the 1920s. It then gathered momentum in the 1940s and 1950s. Some of the milestones can be listed in the newly colonised counties:

Glamorgan

Very rare in the early years of the century although occurring in all the surrounding counties. Nest found in 1905 but no other until 1925. Between 1930 and 1940 there was a fast westward spread (Gower by 1934). An enquiry in 1949 found only 12 pairs in the Cardiff–Merthyr Mawr area. It is now a widespread and fairly common species in the county; Thomas *et al.* (1992) record it in 62.5% (146) of tetrads in west Glamorgan.

Pembrokeshire

One shot 1893 otherwise unknown to Mathew (1894). Bertram Lloyd (Diaries) regarded it as being fairly widely distributed by 1930 and remarked on its increasing spread in the county. Lockley *et al.* recorded it as common in wooded areas by 1949, which is still an accurate summary of its status in the county where Donovan and Rees (in press) suggest a population of 600–800 pairs.

Cardiganshire

First recorded in 1899 by Salter who had also heard of other sightings. Also apparently seen occasionally on Gogerddan estate (near Aberystwyth) prior to 1894; began to increase noticeably in the 1920s. Now well distributed in wooded areas.

Merioneth

Probably present in the south of the county around the turn of the 20th century but absent in other areas although it possibly occurred in the upper Dee Valley in the vicinity of Bala; expansion of range and numbers was noted from the 1930s onwards. Now a widespread and fairly common species.

Caernarfonshire

First recorded at Caernarfon in 1902. Later recorded at Llandudno (1907, bred 1908), Bangor by 1912 and Llandwrog by 1914. The spread apparently occurred from the east rather than the south. The Nuthatch is now widespread and fairly common although still scarce in the western parts of Llŷn (e.g. Rhiw) with occasional vagrants such as one on Bardsey in October 1979.

Anglesey

First recorded in 1910 near Beaumaris and bred in one or two localities near the Menai Straits in subsequent years. One remained on the Skerries for a fortnight in May 1932. It is still a scarce bird on Anglesey, presumably mainly due to lack of suitable habitat and is therefore confined to the wooded parts of the south and south-east. Widespread RSPB surveys covering 762 one-kilometre squares in 1986 located Nuthatches in 28 squares with a maximum population of some 42 pairs, mostly in the south–east along the Menai Straits. Throughout the island, it has increased in both numbers and distribution in the past two decades. N. Brown (in litt.) estimates a population of 75 pairs.

Evidence from the border counties of eastern Wales and from Carmarthenshire supports the general trend of increasing numbers in the middle decades of the century. In Monmouthshire – one of the species' strongholds – it is estimated that the current population is 750–1250 pairs.

Nuthatches are sedentary birds, strongly territorial from autumn onwards and the eight recoveries of birds ringed in Wales merely emphasise this sedentary nature. The breeding and wintering distributions are virtually identical, as indicated by the *Breeding Atlas* and *Winter Atlas*.

Treecreeper *Certhia familiaris*

Dringwr bach

Resident. Widely distributed in woodlands throughout Wales.

This somewhat unobtrusive species occurs widely in woodlands throughout Wales. Because it is relatively easily overlooked, especially outside the months February–April when its thin reedy song is most frequent, it is more numerous in some areas than is readily assumed. Treecreepers are found in woodlands of most types as well as in wooded parkland and well-treed gardens and suburbs. The species is primarily a constituent of the deciduous woodland community but is in some circumstances at home in conifers (indeed on the continent it is predominantly a coniferous woodland bird). In Wales, a major limiting factor in the modern generation of conifer forests is not food supply but the availability of nest sites in forests where few old or broken trees exist, other than in areas of windblow. Even in six small stands (1.5–2.0 ha) of 100-year-old conifers in north Wales, Currie and Bamford (1982) located only one pair of Treecreepers in two years of survey (1979–80). Bibby *et al.* (1989) recorded Treecreepers on only four occasions during two visits to 253 count points in conifer plantations in north Wales, although in mature redwood plantations in Montgomeryshire, Williamson (1971) recorded a density of 1.2 pairs per 10 ha. (redwood is well known to be a particular favourite, especially for the Treecreeper's habit of excavating roost sites in the soft bark.) Flegg (1973) has indicated that nesting is a little earlier in conifers, which reinforces the point that food supply is usually abundant in that habitat. Throughout Britain as a whole, Treecreepers occur in about 60% of all CBC woodland plots at densities of between 50 and 100 pairs per 10-km square. In Wales the limited number of woodland CBCs which have been undertaken reveal a maximum density of c113 pairs per 10-km square.

On a 10-km square basis the *Breeding Atlas* indicates a fairly even distribution in the Principality. In those areas where tetrad surveys have taken place – Monmouthshire, Pembrokeshire, west Glamorgan - the patchiness of the distribution reflecting the presence of woodland is more truly revealed. Tyler *et al.* estimated a breeding total of some 100 pairs per 10-km square in Monmouthshire and a total population of some 1500 pairs for the county, which accords with the higher end of the estimate given in the *Breeding Atlas*. In some of the more western or mountainous counties (Pembrokeshire, Caernarfonshire) the distribution is more scattered, although in Pembrokeshire a total of 1000 pairs is estimated. On Anglesey the species is reasonably numerous in the wooded areas in the southern and eastern parts of the island but rather more thinly scattered in the more barren northern and western parts.

Although little exists by way of firm data, it is apparent that Treecreepers are reluctant colonists of woodland above c338 m, but it has to be remembered that such woodland in Wales is almost exclusively the highest zone of the modern generation of unpromising conifer forests.

There is no evidence to suggest any substantial change of status of the Treecreeper in Wales over the years and certainly most earlier accounts testify to its relative abundance. Salter (Cardiganshire) found it "rather rare" up to the 1920s, referring only to one site where he knew it as regular. Ingram *et al.* (1966) regarded it as fairly common throughout the same county and Salter's remarks are certainly unexpected. The Treecreeper's distribution has probably always mirrored the presence of mature woodland, contracting with the clearance of deciduous woods and now moderately re-establishing within areas of post-war replanting and, to a much lesser degree, 20th century afforestation.

The Treecreeper is one of our most sedentary species. Recoveries of Welsh-ringed birds are very few and none has been recorded moving more than 7 km.

Individual birds do occasionally turn up on offshore islands, indicating at least some degree of movement, albeit possibly accidental. On Bardsey there were some seven records in the 36 years between 1954 and 1990, mainly in September and October but with a few in April. Similarly, on Skomer there have been 11 records thinly scattered over the years between the months of April and September and on Skokholm it is recorded as very scarce and irregular with 16 records between June and early November, with most in July. Twice (August and September) single Treecreepers have been recorded on the Skerries off north Anglesey. Perhaps the most notable Welsh individual was one found climbing on the post of the reserve sign – the only wooden post available – on Grassholm, 18 km offshore in June 1957.

Penduline Tit *Remiz pendulinus*

Titw Pendil

A vagrant.

There is only one record from Wales. The species breeds discontinuously from Spain and southern France and from Holland and Germany to the Balkans eastwards to Manchuria; it is mainly resident but some dispersive and cold weather movements occur. The single record is:

9–13 May 1981 – Male, Bardsey

Golden Oriole *Oriolus oriolus*

Euryn

A scarce summer visitor, recorded in all counties except Radnorshire and particularly in Pembrokeshire. Most of the records fall in May and June as is to be expected with a species overshooting its continental breeding areas.

The first mention of Golden Orioles in Wales is the famous report of 1188 when Giraldus Cambrensis, while travelling with Baldwin, Archbishop of Canterbury, heard a bird calling in a wood near the Menai Straits (Caernarfonshire) which some of his party, endorsed by Giraldus, declared to be an "aureolus". Hope Jones and Dare (1976) however, concluded that this was not acceptable as a firm record.

Golden Orioles may have been regular breeding birds in England in the middle and late 19th century in very small numbers and there are two inconclusive references to possible breeding in Glamorgan at the same time when they "may have nested" at Penarth in 1863 and were "thought to have bred" at Coedarhydyglyn in 1883 and 1886. The possibility of breeding in Wales has been raised in more recent years especially as colonisation of England has taken place since about 1967. Golden Orioles in song have been recorded from several different parts of Wales in the past 25 years but there is, as yet, no definite proof of breeding and few sightings of females. Breeding is, however, notoriously difficult to prove with this species, as both sexes are remarkably well camouflaged in the leaf canopy which they frequent and the fluting song of the male is often the only clue to the bird's presence.

Pembrokeshire has by far the most Golden Oriole records of any Welsh county, recorded between 24 April and 18 June in 18 of the past 37 years and in every year since 1984 (Donovan & Rees, in press). Most of the Pembrokeshire records are from the coast (including Ramsey, Skomer and Skokholm). Elsewhere there have been 17 records from Bardsey, all but three in the period between 11 April and 13 June. Otherwise there is a scattering of records from all counties except Radnorshire, although the four Breconshire records are all from the 19th century! It is to be hoped that the Golden Oriole can be added to the list of Welsh breeding birds in the not too distant future.

Isabelline Shrike *Lanius isabellinus*

Cigydd Gwdw

A vagrant.

There is one Welsh record of this species, which breeds in Iran and Afghanistan, eastern Kazakhstan and Mongolia and winters in south and south-western Asia and north-east Africa:

25 October 1985 – One first-winter bird at Holyhead (Anglesey)

Red-backed Shrike *Lanius collurio*

Cigydd Cefngoch

A scarce passage migrant in spring and autumn chiefly to coastal areas, formerly a widespread breeding species in small numbers.

In the first half of the 19th century the Red-backed Shrike was described as very common in North Wales, particularly near Capel Curig (Caernarfonshire) and Barmouth (Merioneth). Forrest (1907) felt that it could hardly be described as common although he indicated that it was plentiful in certain districts (e.g. the west coast of Merioneth) and was still widespread; it was not as common on Anglesey although nests were recorded on different areas of the island (e.g. Aberffraw, Carmel Head, Red Wharf Bay).

In south Wales at the turn of the century it was common in several areas but with a rather irregular distribution. In Cardiganshire it was never numerous although it was regular around Aberystwyth at the end of the 19th century and was noted on the coastal areas and in the Teifi Valley. In Pembrokeshire it was apparently fairly common in the middle of the 19th century but had decreased considerably by the end (Mathew 1894). In Carmarthenshire Barker (1905) called it not uncommon. In Glamorgan in the Transactions of the Cardiff Naturalists for 1900 it was described as common, breeding regularly and to be increasing; for example at the end of the 19th century Red-backed Shrikes could be found at five localities within five miles (8 km) of the centre of Cardiff and in the early 1900s three pairs were nesting in the hedgerows of one 6-acre (2.5 ha) field near Radyr with others in nearby fields in the same neighbourhood (Heathcote *et al.* 1967).

In Monmouthshire it was widely distributed and was, for example, common at Abergavenny in the late 19th century.

In Breconshire it was recorded as common by Phillips

(1899) and said to be locally common by Walpole–Bond in the early years of this century (Ingram & Salmon 1957). In Radnorshire it was a regular although sparsely distributed breeding species; it was quite common on the Radnorshire/Herefordshire border in 1901–12 but quite rare by 1925. In Montgomeryshire it was plentiful on the borders of the county with Shropshire and Merioneth and in 1919 four males were noted in full song at Llandinam.

The demise of the Redbacked Shrike as a breeding bird in Wales (and England) is detailed in Peakall (1962) and is attributed, as with the similar but later situation in southern England, to long-term climatic changes rather than habitat destruction and/or predation. Other than isolated occurrences in the 1970s and 1980s, which are referred to later, Red-backed Shrikes declined most markedly from the 1920s in Wales and disappeared altogether from the late 1940s to early 1950s. Last recorded breeding was:

Caernarfonshire	Aber Valley in 1952
Merioneth	1951
Montgomeryshire	Llanymynech in 1950
Cardiganshire	Between Lovesgrove and Gogerddan in 1924
Pembrokeshire	1924
Glamorgan	2 pairs near Rhoose in 1948 and one pair near Llandough in 1949 and 1950
Monmouthshire	Nested Llanfoist 1944 to 1946 and probably 1948
Breconshire	Abergwesyn, nested certainly in 1940 to 1942 and "apparently as recently as the early 1950s" (Massey 1976)
Radnorshire	1944

After a gap of many years it appears that the Red-backed Shrike has intermittently made a slight comeback as a breeding species. The *Welsh Bird Report* 1978–87 records that it "has bred in at least two counties in the last ten years but no reports of breeding since 1985". This refers to successful breeding in Monmouthshire in 1981 and a pair present from about 1981 to 1986 at one site in Carmarthenshire but not actually proved to have bred. In addition, in 1990, a pair was found in suitable breeding habitat in Glamorgan in early June, so that the possibility of further breeding records still exists although the species now has only the most tenuous of holds as a British breeding bird, with one pair remaining (in England) in 1992.

Otherwise the Red-backed Shrike is recorded annually as a scarce migrant in May and the period July/October. Most of the recent records are from the coast, in particular Pembrokeshire.

Lesser Grey Shrike *Lanius minor*

Cigydd Glas

A vagrant.

There have been seven records in Wales as follows:

26 May 1961	– One, South Stack
21–22 September 1961	– One, Shotton Pools (Flintshire)
8 June 1967	– One near Pen y groes (Caernarfonshire)
18 September 1974	– One, Skomer
13 October 1975	– One, Ferryside (Carmarthenshire)
16 May 1982	– One, near Fan Pool (Montgomeryshire)
17 October to at least 15 November 1986	– Male near Abersoch (Caernarfonshire) 17–18 October and the same at Aberdaron (Caernarfonshire) 20 October to at least 15 November

These are characteristic spring and autumn dates for this species which breeds from France and Germany eastwards and winters in Africa south of the Sahara. The 1986 bird was unusual in its length of stay and its apparent tameness (it was even seen feeding on a bird table).

Great Grey Shrike *Lanius excubitor*

Cigydd Mawr

A scarce but annual winter visitor, occasionally recorded on passage.

Recent records refer to all counties. The *Winter Atlas* recorded Great Grey Shrikes in ten 10-km squares in Wales in the three winters 1981–82 to 1983–84. These were principally in a north/south sequence from eastern Montgomeryshire to Glamorgan/Carmarthenshire. Allowing for the fact that the mapping of records may have included some duplications of birds on the move at the beginning and end of winter the Welsh population may be no more than five to ten birds each year.

Characteristically, the first arrivals from their Scandinavian breeding grounds reach Wales about the middle of October and records span the winter period from then until early May. There is an unusual record which falls outside this timespan: one bird seen in Nant Ffrancon (Caernarfonshire) on 17 June 1949.

Woodchat Shrike *Lanius senator*

Cigydd Pengoch

The Woodchat Shrike is a scarce visitor to Wales.

It breeds from France and north-west Africa east to the Ukraine and southern Iran, wintering in Africa south of the Sahara. The first record was on 4 May 1923 at Tenby (Pembrokeshire) and there have been a total of 65 records involving 67 individuals up to 1991. Seven further records published in county bird reports for years after 1958 have not been submitted to the British Birds Rarities Committee for consideration and are therefore omitted from this analysis.

Records cover all months from April to November inclusive, with no marked seasonal peaks (see histogram). Many of the sightings from August onwards refer to birds of the year at least some of which stay off-passage for several days or weeks at Bardsey and Skokholm; an immature at Peterstone Wentlooge (Monmouthshire, the first county record) stayed for an unprecedented length of time, from 1 October to 8 November 1983. A bird ringed on Skokholm on 3 June 1976 was retrapped in Suffolk 17 days later.

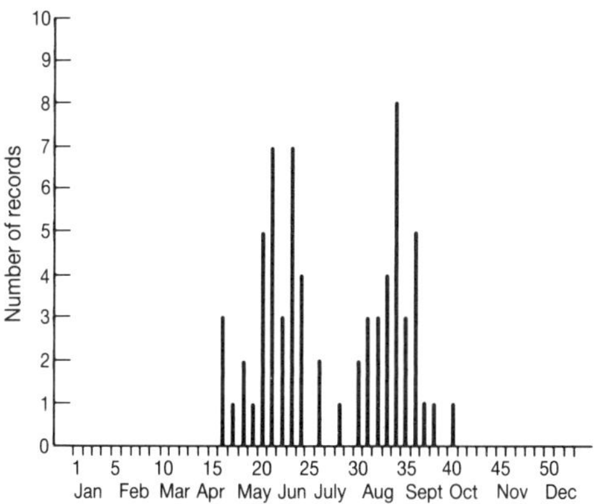

Weekly records of Woodchat Shrike to 1991

The preponderance of sightings come from Pembrokeshire (35 individuals) and Caernarfonshire (18 individuals), reflecting the importance of Skokholm and Bardsey as well-watched coastal migration points. Other records come from Flintshire (1), Anglesey (3), Cardiganshire (1), Glamorgan (5), Merioneth (2) and Monmouthshire (2).

Jay *Garrulus glandarius*

Ysgrech y Coed

Widespread resident, common in most deciduous woodland areas, least so in western coastal fringes; scarce in west Llŷn and absent from large parts of Anglesey. Autumn/winter influxes from the east occur irregularly.

The innumerable oakwoods of Wales provide ample first-preference habitat for the Jay and it is a widely distributed and fairly common bird throughout the country. Exceptions to this generalisation are the relatively treeless areas of west Llŷn and most of Anglesey. Although Jays are found commonly in other types of deciduous wood, the sessile oakwoods of Wales are ideal for a species whose staple autumn and winter food is acorns. There is an interdependence between the oakwoods and the bird rather than unilateral harvesting by the Jay, in so far as the Jays collect huge numbers of acorns in those autumns of heavy cropping and carry them away to bury as winter stores. In this way the otherwise heavy and immobile seeds of the oak are continually transported to new areas and the forgotten ones provide pioneer seedling stock outside the existing woods.

The Jay, the most arboreal member of the crow family, is a shy and wary species and therefore not as readily seen as the other, more conspicuous, crows; it is also the least numerous of the common crows. Heathcote *et al.* (1967) record how it occurred on the list of keepers' kills on one Glamorgan estate (1930–55) at a ratio of 1:4 with Magpies (but Jays are notoriously difficult to trap or shoot!) a ratio which, they suggest, still represented the relative abundances in 1967, but which, even if correct, must now be much too low for most other areas of Wales.

The Jay has benefited from the spread of afforestation in the uplands since the Second World War (late G. Ireson, pers. comm.) and has now colonised at least the fringes of the maturing forest up to an altitude of 400 m or so. Furthermore it has responded to increasing freedom from persecution since the early decades of the present century by showing a certain adaptability to urban areas – town parks, large gardens – and is increasingly seen at garden bird tables, especially in rural areas.

There is no doubt that the Jay (which is still on the list of "pest" species which can be legally killed at any time of year) was heavily persecuted, in recognition of its egg-stealing habits, during the principal game preservation era up to the period of the two World Wars. Not surprisingly it was therefore considerably less numerous in the 19th and early 20th centuries than it is now. Like the Magpie it benefited from the demise of gamekeeping that accompanied the onset of the First World War and which was never fully restored thereafter. Jay numbers responded to this respite by increasing and spreading, especially from 1940 onwards, which trend has probably continued up to the present, partly aided by the modern development of upland conifer forests. The number of Jays present in previous times in Wales can never be known but in so far as its population is largely dependent on the extent of woodland cover, the figures in the following table may mirror the fortunes of this woodland corvid, not forgetting the aberrations caused by the heavy persecution of the 19th and early 20th centuries.

Area of deciduous woodland coverage in Wales, 2000 BC to present

Date	*Area of deciduous woodland (ha)*
2000 BC	c2 000 000
16th century	c172 000
1871	62 044
1891	94 270
1913	86 598
1991	63 000

Figures from Agricultural Returns in Linnard (1982) and Forest Enterprise, 1992).

Monmouthshire and Montgomeryshire have long been recognised as the most heavily wooded (deciduous) counties in Wales and in both these the Jay is numerous in all woodland areas. In Monmouthshire the 1981–85 tetrad survey showed its presence in 84% of the tetrads with a suggested population of 1500–2000 pairs. In Glamorgan, Carmarthenshire, Cardiganshire (but less so the southern parts), Denbighshire, Flintshire and Merioneth, Jays are fairly common, although somewhat more local. The species is also fairly common although somewhat patchily distributed in Radnorshire and Breconshire where it is concentrated mainly in the woodlands of the main river valleys. The Breconshire population is put at a minimum of 300 pairs distributed in at least 31% of tetrads. Thomas *et al.* (1992) found Jays in 69% of tetrads in west Glamorgan. Woodlands are relatively few in Pembrokeshire and Jays correspondingly less numerous; Mathew (1894) and Lockley *et al.* (1949) found them uncommon and today they occur in only 42% of the tetrads in the county with an estimated maximum population of perhaps 600 pairs (Donovan & Rees, in press). Although plentiful in the woodlands of Snowdonia, Jays are much scarcer in the less wooded Llŷn Peninsula of Caernarfonshire, especially the most westerly parts, although they do breed regularly as far west as Rhiw. On Anglesey the species is found almost exclusively in the limited wooded areas of the south and south-east.

Jays have long been recognised for their irruptive behaviour in certain years, invariably linked to the annual abundance or scarcity of the acorn crop. In years of scarcity the Welsh population can be supplemented by considerable numbers of immigrants from farther east, although such increase is not necessarily always easy to detect. On occasion, however, small groups may even be found wandering on treeless areas of coastline, in dune systems or on cliff-tops. These movements are occasionally visible at coastal sites, such as the Gwent Levels and Lavernock Point or on the offshore islands. No year produced a more spectacular influx of birds than 1983, an autumn when unprecedented numbers were recorded moving westwards throughout southern Britain. Most counties in Wales witnessed the influx which was most marked in south Wales (John & Roskell 1985) and exemplified by the numbers on Skomer – a total of 77 between 6 and 26 October, where there had been only eight recorded in the previous 23 years. At Kenfig NNR 18 coasted westward on 6 October. In Pembrokeshire 200 passed over Martin's Haven on the same date, and on 19 September 127 flew westwards out to sea from Strumble Head. On Bardsey 19 occurred on 29 October, where only 16 records had been made in the previous 29 years.

Lesser influxes have been recorded – mainly in south Wales counties – in various years, including 1923, 1935, 1947, 1975, 1977 and 1981.

Few Jays have been ringed in Wales and only one of the recoveries deriving from these has been more than 10 km from the place of ringing. This helps to emphasise the basically sedentary nature of the Welsh breeding population.

Magpie *Pica pica*

Pioden

An abundant and widespread resident; numbers probably still increasing.

This familiar and distinctive bird is a widely distributed and successful species in Wales. It is not only plentiful across the countryside but, freed from organised persecution, has increasingly become a suburban and urban species as well. As with several other avian predators and scavengers (Buzzard, Carrion Crow, Raven) the traditional stock-rearing husbandry which predominates in Wales has produced a pattern of agriculture and a form of countryside that is ideal for the Magpie. The mosaic of mature hedgerows-with-trees, copses, thorn thickets, woodlands and upland shelterbelts provides an almost infinite range of nest sites. In addition, the rich feeding of the pasture land is supplemented by an endless supply of sheep carrion in the hills. Newton *et al.* (1982) found an average of about five ewe corpses and 45 dead lambs per km^2 in upland sheepwalks. Winter/spring stock feeding stations – from which food is freely stolen – are found on every farm (c30 000 farm holdings in Wales). The suburban Magpie is no less fortunate, with a wide choice of nest sites and ample food sources – parkland pastures, rubbish tips, garden feeding tables. There is no doubt too that the re-establishment of forest cover in the hills in the form of new conifer plantations has further widened the Magpie's opportunities by providing nesting and roosting cover in areas where food supply is plentiful. Magpies, although wary, are habitual roadside scavengers and a large percentage of pairs have road sections within their territories. It is common to see groups of up to 30 or more Magpies together nowadays in many parts of Wales.

All historical accounts testify to the Magpie's abundance in Wales. Such historical judgements are necessarily subjective and relative, however, and it is evident that the Magpie has increased in numbers in Wales this century as it has in other parts of Britain. The Magpie's habit of egg stealing made it a prime target for elimination on game-rearing estates and from the middle of the 19th century it was subject to heavy persecution by trapping, shooting and poisoning. As a consequence, it was probably at its lowest ebb in the early years of the present century. There is no doubt that it was then considerably aided in its recovery by three factors in particular; the Protection of Animals Act 1911 abolished the legal use of poisons in the open for the killing of wild birds; the two World Wars took many of the gamekeepers away from the shooting estates thus removing a substantial amount of persecution; the intensification of sheep numbers (see above) has doubtless also been a beneficial factor.

Accordingly, a marked resurgence has occurred since 1914–18 and has been accompanied by an increasing adaptability which has allowed the Magpie to expand its range, for example into urban/suburban areas, thorn scrub on cliffs of the western seaboard and across the tidal straits to offshore islands. On Skomer it first bred in 1967, reached ten pairs eight years later and has since settled at 9–13 pairs; similarly, it first colonised Bardsey in 1952 (now 11 pairs) and one or two pairs now nest on Ramsey. Although statistical evidence is lacking to date to support the claim, the everyday evidence is clear that Magpie numbers have very greatly increased throughout the 1970s and 1980s and are almost certainly continuing to do so. In some areas in the early 1990s increasing predation by Goshawk and Peregrine may be responsible for keeping numbers locally in check. Sadly the opportunity for monitoring the scale of this change appears to have slipped by. Gooch *et al.* (1991) have demonstrated that, overall, Magpies have been increasing in England at the rate of 4–5% per annum since at least 1966, with highest rates of increase in south-west England and in suburban habitats; there are no comparable data for Wales.

Magpies prosper especially well in areas where farmland is poorly managed or where marginal agriculture provides an intimate mosaic of small fields (especially of old pasture land), overgrown hedgerows and unimproved ffridd. Accordingly, there is still probably a greater density in such areas as Llŷn, the upland farms of mid and north Wales and the valleys of south Wales, than in the richer more intensively farmed land with larger fields in the larger river valleys on the Marches. In Glamorgan (areas unspecified) Heathcote *et al.* (1967) quote breeding densities of seven pairs on 80 acres (33 ha) and six pairs in 165 acres (68 ha). As mentioned, few estimates of Magpie numbers have been attempted which is particularly to be regretted in view of the dynamic changes that have evidently been taking place with this species. The principal ones are those produced in connection with the breeding birds surveys in Monmouthshire and Pembrokeshire. Tyler *et al.* (1987) estimated some 15 000 pairs in the former county (1981–85) and Donovan and Rees (in press) put the Pembrokeshire total at probably 22 000 pairs although the present authors believe that both estimates may be too high. These figures indicate densities of 11 and 14 pairs per km^2 respectively, which is considerably higher than the urban densities given by Marchant *et al.* (1990) who believe that urban populations (six to eight pairs per km^2) are generally higher than rural ones. Surveys by J. Moulton (in litt.) in the Rhyl area in 1992 accord exactly with Marchant *et al.* at six to eight pairs per km^2. Even if data on numbers are scarce, the density of occurrence is impressive with birds present in 95% of tetrads in Pembrokeshire, 99% in Monmouthshire, 98% in west Glamorgan (Thomas *et al.* 1992) and 84% in Breconshire (Peers & Shrubb 1990).

Magpies normally pair for life and are among the most sedentary of our resident species. The fact that few individuals wander far from their natal areas is shown by ringing recoveries, which indicate only nine movements farther than 10 km, and by the fact that very few passage spring and autumn reports come from those offshore islands where no breeding takes place (e.g. Skokholm: 7 records only, up to 1991).

Although pairs can readily be seen on their territories throughout the year, large numbers, presumably incorporating a majority of non-breeding birds, do congregate into communal roosts, sometimes involving well over 100 birds, e.g. Rhosgoch Bog (Radnorshire) up to 150; 177 in an osier bed at Abersoch, Llŷn, in 1977. In winter many pairs of Magpies tend to centre their territories close to buildings on individual farms and clearly use these as the primary source of their food supply. Magpies are by no means universally popular. Increasing concern is voiced publicly about the effects this burgeoning population may be having on songbirds and ground-nesting birds, in view of the species' habit of predating the eggs and young of others. To date there is no evidence to support the belief that Magpies are depressing songbird numbers, although there is gathering suspicion that, together with other predatory corvid species, magpies may be having a terminally detrimental effect on the declining local populations of some ground-nesting waders and gamebirds.

Nutcracker *Nucifraga caryocatactes*

Malwr Cnau

An irruptive vagrant recorded seven times in Wales.

The accepted records are as follows:

5 October 1753	– One shot in a garden at Mostyn (Flintshire). This was the first British record and was described in Pennants' *Tours in Wales* and his *British Zoology* (1812, vol. 1, p. 298).
Early 1800s	– One killed near Swansea (Glamorgan)
19 October 1954	– One in St. Julian's Wood, Newport (Monmouthshire)
Autumn 1957	– One, Llangattock (Breconshire)
26–31 August 1968	– Two near Henllan, Llandysul (Carmarthenshire)
30 August 1968	– One, Whiteford (Glamorgan)
12 November 1968	– One, Llanhilleth (Monmouthshire)

The last three records were part of an unprecedented irruption which resulted in more than 300 Nutcrackers reaching Britain, chiefly the east coast counties of England (Hollyer 1970). It was considered that all Nutcrackers reaching Britain were of the Slender-billed race of north-east Russia and Siberia. The cause of the irruption of 1968 was the failure of the cone crop of the Arolla Pine on which the bird depends and the arrival of birds in Britain was aided by the development over northern Europe in early autumn of an exceptional area of high pressure, providing calm conditions and clear skies.

Chough *Pyrrhocorax pyrrhocorax*

Brân Goesgoch

A resident in small numbers, mainly on cliff coasts of the western seaboard.

The Chough is a denizen of wild places in Wales, evocative particularly of the precipitous, uninhabited headlands of Pembrokeshire, Cardiganshire, Caernarfonshire and Anglesey – a bird truly of the Celtic fringe. It breeds in six Welsh counties and only irregularly now in a seventh (Denbighshire). At the present time it seems to be holding its numbers in Wales or even increasing slightly. It remains one of the most precarious breeding birds on the British list, and Wales is its most important stronghold (but there are also c750 pairs in Ireland).

It is tantalising to imagine what the historical status of this species might have been throughout the Principality. Sadly only fragments of the picture remain for us to build on, and most has to be conjecture. The clearest fact is that the Chough is a far rarer bird than it once was; it is possible that it was once well distributed throughout Welsh counties, where now only the slenderest inland remnants remain. Rolfe (1966) attempted to analyse the causes for the long-term decline that clearly took place. The confiding nature of the bird made it an easy target for trigger-happy gunmen (e.g. 30 were killed by one man in a morning on the Isle of Man). In the 16th and 17th centuries the Chough was listed in the Tudor Vermin Acts and bounties given for killing it. Egg collecting was rife in the last decades of the 19th century and continued in north Wales well into the present century: on Bardsey collectors were active annually up to the late 1940s. In fact, in the first three decades of the 20th century, on that island and the adjacent headland (where Wooley first collected a clutch as far back as 1846) there must almost have been an unseemly race each year between egg collectors, so numerous are the records that we have examined: E. W. Wade, Baldwin Young, O. V. Aplin, C. H. Gowland, K. Wilson, W. M. Congreve, F. C. R. Jourdain, Rees Jones, W. Hobson, G. Haines, F. C. Rawlings, Tomkinson. Other Chough areas, notably Pembrokeshire and Merioneth, as well as other parts of the Llŷn coast and Great Orme, were regularly relieved of their clutches. Rolfe (1966) also highlighted the fact that gin- and rabbit-traps were a regular cause of Chough deaths and there are several references to this from widely separated parts of Wales. Illegal gins were known to be responsible for Chough killing in Cardiganshire as late as the 1970s. Several birds now in the National Museum were found among a consignment of rabbits brought from the Pembrokeshire islands in the early decades of this century; they had been caught in the rabbit traps. Competition with Jackdaws, whose populations increased markedly between 1830 and 1890 has been cited as a cause of Chough decline for many years but it does not stand up to critical examination. Neither can it be taken seriously that Peregrines – which certainly prey on Choughs on occasion – even at their high levels of population pre-1939 or in the 1990s, have been responsible for holding back Chough numbers.

A lot more is understood now about the feeding ecology of Chough, particularly in its coastal strongholds, and it is appreciated that land use changes, which have reduced the areas of rabbit-grazed heathland on the cliff-tops, are probably the main factor directly responsible for suppressing the Chough population. As is ever the case, there is doubtless a complexity of factors which has caused the decline of Chough in Wales in the past two centuries or so; human persecution in the past may well have been a central cause. Agricultural intensification of traditional feeding areas is probably the main reason preventing a strong recovery at present. The Chough's distribution in Wales is markedly coastal and it should be remembered that the species is very susceptible to harsh weather in winter; a close association with the seaboard ensures that, even in periods of severe freezing, feeding areas on the coastal cliffs and close to the spray line remain available to the Choughs.

Evidence in the literature referring to the Chough's past status is fragmentary. In Monmouthshire there are reports of breeding in the north of the county around 1880, but these are disputed (Humphreys 1963) and seem somewhat improbable. A vagrant bird seen at Abergavenny in late September 1972 is the only other record. Cambridge Phillips could cite no records for Breconshire (1882) nor have there been any since his time. Neither are there published records for Radnorshire, although the Oxford University Museum has a skin labelled "Rhayader 1914, purchased from J. Betteridge . . . 1934". In Glamorgan Choughs bred "along the rocky cliffs of the . . . coast" in the 19th century according to Cambridge Phillips, and there is no reason to doubt that. Certainly they inhabited the area around Dunraven until the mid-1800s and the cliffs of Gower until almost the end of the century. Dodderidge-Knight, writing of the Newton Nottage area in 1853, regarded it as common, and Dilwyn (1848) lists it for the Swansea area. One story from the second half of the last century relates to "numbers exposed for sale in Swansea market", where they were known as Billy Cocks and much prized by French sailors as an article of food. The Gower population – and thus that of Glamorgan – became extinct around the turn of the century although one individual was shot on Aberavon Moors as late as 1945.

In Pembrokeshire Mathew (1894) even then lamented the demise of a bird which he claimed had been common "all the way from Tenby to Cardiganshire" 50 years earlier, and laid the blame firmly at the door of persistent egg collectors climbing even to the most inaccessible cavern nests. In the mid-19th century Mathew asserted that Choughs nested on the Bishop's Palace at St. Davids' but were robbed annually and soon disappeared. Another building where Chough nested around the same time was Manorbier Castle, allegedly "in great abundance" (*Zoologist* 1857). In 1948 Lockley considered the Pembrokeshire population to be c35 pairs. In neighbouring Carmarthenshire it is reputed to have nested in the Telpyn Point area up to the early 1920s. Since then occasional birds have been seen along the coast, most frequently near Ferryside with one (possibly two) reported from Llyn Brianne in April 1973 – the only inland record for the county. Salter (Diaries) was familiar with several well-established pairs on the Cardiganshire sea-cliffs in the first two decades of this century, especially in the Aberporth area but the numbers on this coastline were almost certainly higher than that. By 1966 Ingram *et al.* recorded an increased status, with two additional pairs on the cliffs and one or two possible pairs inland (not five as suggested by Rolfe). Surveying the whole coastline in 1976, Roderick (1978) located 15 pairs and also knew of two further sites inland.

In Montgomeryshire there is a small but well–established population in the west of the county but nothing of its history prior to 1950 is known. One bird (possibly two) lingered in the Llanymynech Rocks area, on the Shropshire border, in autumn 1974. One or two Merioneth sites around the Mawddach Valley were well known to egg collectors, notably nests at Bontddu and Arthog, which were regularly emptied, presumably helping cause their abandonment around the first two decades of the century (G. Haines, Diaries). Salter (Diaries) related that three to four pairs of Chough allegedly bred on Craig yr Aderyn up to 1887 but not subsequently (Choughs are certainly present there nowadays). Forrest (1907) gives accounts that indicated that they were tolerably common along the coastal belt up to about 1865, after which only a few pairs remained, as inferred above.

Caernarfonshire has long been the heartland of the Welsh Chough population, although there was a dramatic decline in the mid-19th century, disguising the size of the earlier population. It was apparently plentiful around Conwy in the 18th and early 19th centuries, and recorded there by Pennant (1778). On Anglesey, Forrest lists eight or nine coastal sites in the 19th century, including Puffin Is., where there was evidently a regular nest site in the ruined church tower and where Whalley (1954) believed Choughs nested in 1953. As elsewhere in its breeding range, young Choughs were regularly taken on Anglesey as pets. Pennant claimed breeding on Flint Castle, near his home, but Forrest believes this to be erroneous and to refer only to wandering birds. Up to 1810 one pair bred in the ruined Castell Dinas Bran ("Crow Castle") at Llangollen, the only known site in Denbighshire at that time.

The first systematic survey of Choughs was that organised by Rolfe (1966) in 1963, covering all the Chough regions of the British Isles. Further surveys, either specific to Wales, parts of Wales or embracing wider areas, have taken place in 1969 (Harrop 1970), 1971 (Donovan 1972), 1982 (Bullock 1983), 1990 (Williams *et al.* 1990), 1991 (Moralee, pers. comm.), 1992 (Green & Williams 1992). The results of these surveys are summarised in the accompanying table.

Outside the breeding season adult Choughs remain fairly faithful to their nesting areas, as has been demonstrated by recent winter surveys in Pembrokeshire and Anglesey. In the latter area, winter population counts of 32–36 in 1990 and 1991, rising to 58 in 1992 are believed to represent some two-thirds of the breeding numbers therefore roughly equating with the 14 pairs plus some additional 30% of juveniles and non-breeders (Moralee & Moralee, RSPB, unpublished). Communal gatherings and

Chough numbers in Wales as shown by surveys 1963–92

County	*1963*	*1969*	*1971*	*1976*	*1982***	*1990*	*1991*	*1992***
Anglesey	2	2	–	–	8	–	14[2]	13(3)
Caernarfonshire								
a) Llŷn	24	18	–	–	28	38	–	38(7)
b) Remainder	18	11	–	–	23	–	–	19(2)
Cardiganshire	9	–	–	17[1]	16	–	–	15(6)
Denbighshire	1	1	–	–	1	–	–	0(1)
Glamorgan	0	–	–	–	0	–	–	1
Merioneth	7	8	–	–	12	–	–	8(1)
Montgomeryshire	5–7	1	–	–	2	–	–	4(1)
Pembrokeshire	47–50*	–	c50*	–	49–54	–	–	53(5)
Total	113–118				139–144			151(26)

[1] H. Roderick 1978
[2] A. Moralee (pers. comm.)
* Adjusted (Bullock 1983) for areas thought to be missed in 1963 and 1971
** Confirmed and probable pairs combined in 1982 and 1992
(Figures in parentheses 1992 = possible additional pairs)

Note: Rolfe and Bullock both estimate an additional non-breeding population of c30% over and above the breeding pairs. Green and Williams give a figure of 99 non-breeding birds – equivalent to approximately 20% additional to the 1992 total of breeding birds.

roosts are also characteristic at this time of year. At Craig yr Aderyn (Merioneth) up to 65 have been found in recent winters, including individuals from breeding sites as far away as Newquay in Cardiganshire. In Pembrokeshire sizeable winter roosts have recently been located at West Angle (up to 25 birds) and St. Anne's Head although the latter was eventually deserted after several years of use. J. W. Donovan (pers. comm.) suggests that roost sites may well vary over the years dependent on the proximity of good daytime feeding conditions and thus the prevailing pattern of agriculture and other factors locally. Ramsey birds – up to 30 during daytime – depart for the mainland in late afternoon, presumably to join a roost there, at present not located.

The Choughs on Bardsey have long been ringed by observatory staff and have demonstrated the mobility of some of the young birds, three of which have been recorded within the following twelve months at Strumble Head (Pembrokeshire) 85 km south-south-west. Another young bird moved from Bardsey to south Lancashire (142 km east-north-east); other ringing recoveries confirm the essentially sedentary nature of the adults in Wales.

The breeding figures given in the above table indicate a gently rising population over the past three decades. This modest increase in Wales is almost certainly a genuine reflection of the well-being of the population, notwithstanding that the earliest estimates (1963 and 1969) were probably incomplete and that Bullock's survey in 1982 followed a particularly severe winter. The modest increase is mainly due to a slightly higher number of coastal pairs, whereas the inland population has fallen slightly across the same period. Wales now supports some 52% of the UK Chough population. Measures to improve the situation for Choughs, both in terms of artificial breeding sites in one or two areas and management of some coastal areas to provide more attractive feeding are now underway and in these respects the future prospects for the species are reasonably bright. Bullock *et al.* (1983) have instanced the serious effects which land use changes of the Chough's feeding areas can have, notably the change to arable or the removal of stock grazing which allows the feeding pastures to become overgrown.

At present, however, the Chough is faring moderately well in Wales, maintaining its numbers, increasing very slightly in coastal areas and even expanding its range modestly: birds reappeared on the Gower in 1990 and bred successfully in 1991 and 1992 while a pair also nested on Skokholm in 1992 for the first time since 1928.

Jackdaw *Corvus monedula*

Jac-y-do

An abundant and widespread resident.

The Jackdaw is an abundant and fairly cosmopolitan species throughout Wales except on open moorland, areas of conifer afforestation and the high mountain tops. It is the most urban of the crow family, living, feeding and breeding successfully in most cities, towns and villages from the metropolitan centres of Cardiff, Swansea and Newport, through the Valleys towns to the coastal resorts of the west and north and in innumerable small towns and villages of the hinterland. It is by nature a grassland and arable land feeder, regularly foraging in company with Rooks, and is the first of the crow family to exploit man-made sources of food, such as urban rubbish tips. It is also the boldest and most successful of the corvids at exploiting the high-quality food source available on stock-rearing farms. It steals continually from feeding troughs, is fond of concentrated feed blocks and readily enters lambing sheds to take food. This ease of feeding, together with its wide choice of nest sites, ensures its continuing success as a plentiful and widespread species in Wales.

As a nesting bird it frequents old woodland, parkland, mature hedgerow trees, inland and sea-cliffs, quarries and mines – both disused and working – old buildings, churches, disused nests of other species (e.g. Magpie), and it can cause considerable annoyance in its fondness for nest-building in household chimneys. In certain situations Jackdaws also nest underground in burrows, for example on the offshore islands. It was recorded nesting in burrows on Skomer as long ago as 1860 and continues to do so to this day (S. J. Sutcliffe in litt.). Jackdaws have nested in similar situations on Bardsey, and although the habit is less frequent on the mainland, it does occasionally occur, e.g. Cardiganshire (Ingram *et al.* 1966).

Alexander and Lack (1944) gave evidence of a great increase in Jackdaw numbers in the years up to 1940, accompanied by a northward and westward extension of range in Britain. It is likely that this increase continued through the 1950s and into the 1960s (Parslow 1973). Voous (1960) showed that this British increase was consistent with some northward (Scandinavia) and southward (Spain) expansion in the European range around the same time. However, as is often the case, the extent to which this expansion applied in Wales is by no means clear. On the one hand writers in the early part of this century already testified to its abundance and ubiquity; Forrest (1907) found it "common all over north Wales" and remarked on the "thousands inhabiting the rocky coasts (of north-west Wales)", as well as mentioning that the platforms of Dolgellau station were carpeted with the birds on Sundays! Equally in south Wales Jackdaws were then regarded as widespread and plentiful. On the other hand, while Bardsey has "always" had Jackdaws at a fairly constant population level of 30–50 pairs, on Skomer Saunders (1962) showed that there was a dramatic increase from 20 pairs in 1946 to 200–250 in 1961 (now stabilised at approximately 220 pairs). Apart from a single pair recorded as having bred in 1928, they first colonised Skokholm in 1965 and increased to 60 pairs by 1974 (Richford & Lawman 1978), although it has been shown that their productivity of young is almost certainly insufficient to maintain their numbers without annual recruitment from elsewhere (Richford, unpub. thesis) and they have latterly fallen back to 15–25 pairs: some 25 pairs breed on Ramsey. It is clear that between the 1940s and 1960s there was a genuine expansion onto the islands off Pembrokeshire, possibly caused by increasing densities in other west coast areas, although the inland population has generally remained stable and high throughout this century. Certainly such factors as the increasing area of pasture land, rising sheep numbers and the creation of new nesting sites (quarries and urban buildings) can only have operated in the Jackdaw's favour over the past one hundred years. The CBC index shows a steep upward trend in numbers nationally between 1970 and the mid-1980s.

Donovan and Rees (in press) estimated 9000–13 000 breeding pairs in Pembrokeshire while Tyler *et al.* (1987) give 7000–10 000 pairs for Monmouthshire which they suggest may be minimum. Peers and Shrubb (1990) give "a conservative population" of 5000 pairs for Breconshire; no estimates for other counties exist.

Although British Jackdaws are essentially sedentary, there is some movement westwards and southwards in autumn, as can be witnessed by such visible migration as that seen at Lavernock Point (up to 300 birds per day on occasion) and the arrival of numbers on the offshore islands. On Bardsey the breeding population often departs temporarily – presumably to the mainland – in July and August; late autumn is sometimes characterised by sudden immigrations of Jackdaws – 1850 individuals in October 1975 – which fly high out to sea south-westwards only to return later across the island to the mainland.

Ringing results for Welsh Jackdaws indicate some degree of movement although the number of recoveries (105) is small and only three have travelled over 100 km and only another 15 more than 10 km. A nestling ringed on Bardsey had moved 154 km west-north-west to Co. Meath (Eire) by the following February; an adult ringed at Deganwy in the breeding season was 105 km eastwards in the following January and another adult ringed at Rhos-on-sea in March was found dying 184 km away in the Rhondda Valley 12 months later. Further evidence of some emigration by Welsh Jackdaws is presented in that six wintering birds ringed on the North Slobs, Co. Wexford have been subsequently reported in Anglesey (1), Caernarfonshire (3) and Glamorgan (2) during the breeding season. At the end of the breeding season many birds move onto the improved pastures of the uplands and there is subsequently a considerable withdrawal of birds from the higher ground in winter to the more sheltered feeding of the coastal strips and the large eastward-flowing river valleys. Winter roosts, usually in the same woods as Rooks, can be large, involving several thousand birds from a wide foraging area, e.g. c10 000 in Pembrey Forest in November 1976; c4000 Leighton Woods, Welshpool, December 1981.

Rook *Corvus frugilegus*

Ydfran

A plentiful and widespread resident breeding mainly below 305 m.

Several full-scale surveys during the present century help to provide a fairly accurate picture of the Rook's past and present status in Wales. It is a widely spread resident breeding throughout all counties, usually concentrated on the lower land of the main rivers and their tributary valleys. Although Rooks rarely breed on land over 305 m they feed extensively in the summer on the ever-increasing areas of improved pasture reclaimed from the moorland. For this purpose they will regularly forage up to 610 m on the new pastures, frequently in company with Jackdaws and Carrion Crows. Arable farming occupies only about 7% of the farmed land in Wales and thus Rooks do not present the same degree of problem in relation to newly sown grain or standing corn that they do in the lowlands of England and elsewhere or as they did in

Wales in the past. Nonetheless they are not particularly popular with the farming community mainly because of their mass raiding of foodstuffs put down daily for the sheep flocks in winter and spring. As a result there is considerable local persecution of Rooks during the nesting season but, although satisfying human retribution, it has little effect on their numbers, mainly because it is the young birds which are shot and there is a high natural mortality of young in their first summer.

Rookeries, some of which may be of considerable antiquity (Chater & Chater 1974), are found in a wide spectrum of tree species. The 1975 National Rookeries Survey (Sage & Vernon 1978) recorded the range of trees used in Wales (see the following table). From this it is clear that species such as alder and silver birch, although very common, are avoided and even oak – by far the most numerous native tree species in Wales – is not preferentially selected. On the other hand Scots Pine – not particularly numerous in Wales – appears to be favoured, presumably because of its height, amenable form and the early-season protection it provides through its dense form and evergreen foliage.

Percentage of Rookeries by tree species in Wales

Tree species	*% used*
Ash	21.4
Beech	11.0
Elms	7.2
Oaks	22.1
Sycamore	18.1
Other broadleaves	3.9
Scots Pine	13.1
Other conifers	3.2

After Sage & Vernon (1978)

Although the highest known rookery in Wales is 347 m (in Radnorshire) the bulk of rookeries are well below 230 m; in Cardiganshire, for example, only six out of c110 rookeries are above 250 m.

The various surveys that have been carried out between 1930 and the present time give a better record of the variation of Rook numbers in Wales this century than exists for most other common passerine species. Although the Welsh figures which Fisher (1947) had to work on for his survey in 1944–46 were incomplete, he inclined to the view that numbers had declined since 1930, although in England he indicated a 20% increase. He estimated the Welsh population in 1946 at some 98 000 pairs. Subsequently there was certainly a marked reduction and by the time of the 1975 national survey only 39 000 nests were recorded. Evidence from one or two more local censuses which were carried out in the 1960s supports the belief that numbers did in fact increase up to the 1960s and the overall decline which was recorded by 1975 – and which, at 60.4%, was much greater than other parts of mainland Britain (England 44.6%, Scotland 36.4%) – actually occurred in the years immediately preceding that latter survey. One of the main reasons for the longer-term decline is not difficult to establish when the changing pattern of agriculture over the past one hundred years or so is considered. The graph on page 28 shows the reduction in the annual acreage of cereals grown in Wales since 1875. Cereals are an important food source for Rooks and a clear correlation can be drawn between the decline of both. Total acreage of cereals is not the only factor, as the method of harvesting is also important. The former binder-and-stook system guaranteed an abundance of available grain in the months of July and August – the crucial months of mortality for young Rooks; modern combine harvesting systems have removed that supply. The current Rook population in Wales is probably at the sort of density which is sustainable in a strictly pastoral environment.

Recent population estimates since the 1975 national survey have been made in only four counties of Wales. In Monmouthshire Tyler *et al.* (1987) showed a resurgence in numbers in the 1980s with the total in 1985 (4588 nests in 214 rookeries) being 30% greater than in 1980 (3233 nests in 155 rookeries). Donovan and Rees (in press) indicate an increase in the number of rookeries in Pembrokeshire since 1975 but do not estimate the numbers nesting during the breeding birds survey 1984–88. A third of the rookeries recorded during the 1975 survey in Montgomeryshire were revisited in 1991 (I. Williamson, RSPB unpub.); there was a mean rise in the size of this sample of rookeries of 52%. RSPB surveys in Anglesey in 1986 located 96 rookeries well distributed throughout the island but no count of nests was attempted.

The Rook's avoidance of uplands for breeding is well demonstrated by Sage and Vernon and by the Breeding Atlases, showing clear absence from such high open areas as the Brecon Beacons and Cambrian Mountains, and other mountain ranges. In addition it is absent from most of the highly developed industrial valleys of South Wales, e.g. Merthyr Tydfil borough where Griffiths and Griffiths found 100 pairs in 1953, diminishing to 30 by 1963. For this reason Merioneth, despite its large size (the sixth largest of the "old" Welsh counties) has the smallest Rook population of any county; here Rooks are not at all common west of the Dee Valley at Bala. Forrest, writing in 1907, considered them to be relatively scarce in Anglesey and also Llŷn although the latter comment does not apply today; indeed some of the larger rookeries in North Wales are near Tudweiliog (120 pairs) and Botwnnog (100 pairs). Few other rookeries in Wales now hold as many as 200 pairs.

British Rooks are basically sedentary whereas continental birds from northern and eastern parts of their breeding range are necessarily migratory. There is as yet no firm evidence that any of these birds reach as far west as Wales. Of the very few Welsh-ringed Rooks that have been recovered subsequently, only three have moved more than 10 km. One Rook ringed in winter on the North Slobs, Co. Wexford was found dead five years later near Corwen (Merioneth), a distance of 213 km. Offshore movements in autumn have been recorded on Bardsey and South Bishop (Pembrokeshire) from time to time, demonstrating some degree of dispersal, possibly of young birds. Confirmation of spring (March–April) and autumn (mainly September) movements is obtained by the scatter of records over the years from Skokholm,

Skomer and Bardsey; one bird has even been seen (October 1962) on Grassholm, 14 km offshore.

In winter Rooks usually forage within a few miles of the nesting rookeries and birds from a fairly wide catchment will gather together at dusk in favoured woodland roosts, often joining with Jackdaws and Carrion Crows to form gatherings of several thousand birds. From time to time severe weather in winter can cause heavy losses; quite large numbers perished in the heavy snow and bitter temperatures of January 1982 in Montgomeryshire and elsewhere and Mathew records "thousands dying" in Pembrokeshire in 1880. Normally, however, Rooks fare better in severe winters than many other resident passerines, no doubt partly helped by the easy access to sheep feeding stations throughout the countryside.

Carrion Crow/Hooded Crow *Corvus corone*

Brân Dyddyn

An abundant resident and partial migrant throughout Wales. Often regarded as the principal avian pest species in rural areas. The Hooded Crow (C. c. cornix) *is an irregular wanderer to Wales, mainly in winter and most frequently to coastal areas.*

The Carrion Crow is an extremely abundant species throughout Wales, and is a versatile and successful colonist in urban and coastal areas, in addition to the high numbers in its inland rural strongholds. Apart from its carrion-feeding habits (in which respect through sheer numbers it provides a genuine service on roads, coasts and the wider countryside), it feeds extensively on rubbish tips, grain, and grassland invertebrates. It also plunders nests of other bird species, and may eventually be proved to be a substantial factor in extinguishing the numbers of some ground-nesting species (e.g. waders). Together with the fox, it is the major predator challenging the future viability of the few remaining Welsh grouse moors, but nowhere is it more detested than among the sheep farming community. Its reputation for killing new-born lambs and maiming parturient ewes assures its status as the most disliked bird in the countryside, and it is still unremittingly shot, trapped and (illegally) poisoned, albeit to negligible effect on its annual population levels. The Carrion Crow is strongly dependent on the sheep industry, its very high numbers undoubtedly based on and sustained by the ever-increasing levels of sheep stocking in the hill areas. Furthermore, when winter conditions are harshest, the food which is provided for the sheep is stolen by crows (and other corvids) at a time when they most need it. The fact that its depredations on lambing flocks may well be overstated (see Houston (1977) for the situation in Scotland) does nothing to reduce the measures that are taken against it each spring, especially in those areas where stock husbandry is least efficient.

The Carrion Crow is a successful species also where other types of agriculture predominate (e.g. Pembrokeshire with its dairy farming), and it colonises urban areas freely, in some places nesting close to city centres (Cardiff) and even over busy streets. It is common on coastal areas and breeds freely on most offshore islands.

All historical accounts agree the ubiquity and abundance of the species throughout Wales and there seems no reason to suppose that it has ever been anything other than numerous. At the same time it was relentlessly eliminated in the 19th and 20th centuries on the numerous game preservation estates which then existed far more widely than they now do. Forrest (1907) found the Carrion Crow "astonishingly numerous" in the Penrhyndeudraeth–Llanbedr area of Merioneth at the turn of the century and reported "about a dozen nests in one small wood, each in a separate tree". Nesting at this implied density is impressive and certainly not usual. However, research in 1981 (RSPB 1981, unpublished) found a density of 117 nests over an area of 24 km^2 in Radnorshire on hill sheep farming land between 150 and 300 m (see map).

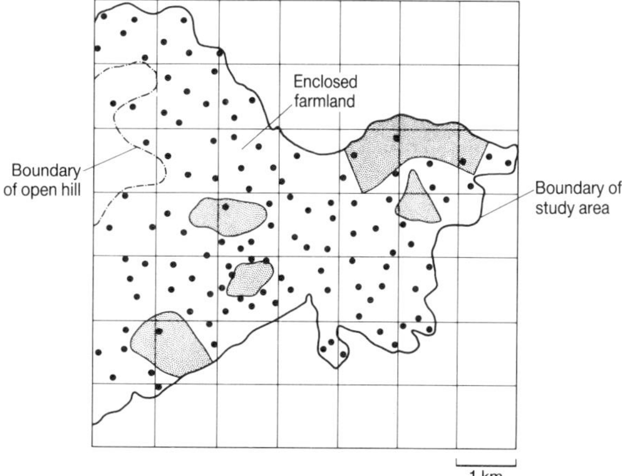

Density of Carrion Crow nests, by one-kilometre squares, in an RSPB study area, Newbridge-on-Wye, Radnorshire, 1982. Shaded areas show approximate boundaries of areas occupied by non-breeding flocks

The national CBC index illustrates how the population of Carrion Crows has risen almost continuously since the inception of the Census in 1961. It is known that a marked increase in numbers can be traced back to earlier origins in the reduction of gamekeeping activity in the two World Wars, the general lessening of persecution by farmers during agricultural recession (1920s) and the modernisation of farming practices since then. In Wales it is unlikely that there has been a further substantial increase in Carrion Crow numbers in the past two decades simply because they are almost certainly already at capacity throughout the Principality. Not only do both the original *Breeding Atlas* and the *New Atlas* record Carrion Crows in every 10-km square in Wales but it is the case that there are very few tetrads throughout Wales that do not support breeding Crows: 99% occupied in Monmouthshire, 98% Pembrokeshire, 94% West Glamorgan and 97% Breconshire. Population estimates exist only for the counties of Pembrokeshire (18 000–21 000 pairs) and Monmouthshire (15 000 pairs). On these bases the total breeding population in Wales must be in the region of 200 000 pairs.

In many areas breeding pairs are indeed at capacity, as a result of which, within the pattern of distribution of breeding pairs (birds maintain territories all year), there is a large non-breeding population. Frequently this totals as many as 50% of the breeding individuals (RSPB unpub.). These non-breeding birds gather into groups of variable size, often over 100 strong, comprising juveniles and adults – sometimes paired – which cannot find territory space. They tend to have their own foraging areas exclusive of the defended territories of breeding pairs and often centred on regular sheep-feeding points. (It is thus these non-breeding bands which may be responsible for much of the damage to lambing flocks and the common practice of shooting out nesting pairs may simply expand the area available to these marauding groups.)

Island breeding populations are well documented. On Skomer (120 ha) the population has reached as many as 26 pairs annually; Skokholm (96 ha) supported up to 12 pairs per annum until the 1960s since when they declined inexplicably until breeding ceased after 1984; Bardsey (178 ha) has five to eight pairs.

Ringing recoveries emphasise the sedentary nature of Carrion Crows in Wales, only 14 out of 63 recoveries indicating movement of more than 10 km; the most distant recovery is of a nestling from Bardsey later found in north Pembrokeshire, 91 km south-south-west.

In winter, Carrion Crows normally roost communally, frequently in favoured woods with other corvids. At some roosts the number of Carrion Crows may be considerable, 200 being not uncommon although up to 250 have been recorded at times. Winter feeding gatherings are often large, e.g. 350 on the tideline at Laugharne (Carmarthenshire) in February 1966. Winter numbers are supplemented by some immigration from farther east, although the extent of this is not known.

The Hooded Crow (*C. c. cornix*) has its nearest breeding area throughout Ireland and the Isle of Man. Inevitably occasional birds wander as far as Wales and there have been numerous scattered records from many counties over the past one hundred years or more. According to Dodderidge-Knight (1853) it was "common on the shores" (i.e. South Glamorgan) fifty years ago and Forrest (quoting Dobie) claimed that they were regular in the Dee estuary area in the first half of the 19th century. There have been several records of individual Hooded Crows hybridising with Carrion Crows (as they do in areas where the two races overlap), the most recent being in the Clarach area of north Cardiganshire from about 1985 onwards for several years.

Raven *Corvus corax*

Cigfran

Resident: distributed throughout the whole of Wales. Numbers are very high in many of the uplands and coastal-cliff areas.

Few birds in Wales have excited more interest or attention over the years than the Raven. It has been a focal species for egg collectors for over one hundred years and was formerly relentlessly persecuted by farmers and gamekeepers. Having been effectively eliminated from lowland England, Ravens in Wales have engaged the particular attention of both egg collectors and of the increasing public interest in birdwatching in the second half of this century. Its prominent place in folklore, legend and mythology still does much to enhance this public interest.

As Bolam (1913) mentioned, the Raven is very much the *genius loci* of the mountain areas of Wales. On the one hand it has always been loathed by sheep farmers and gamekeepers while on the other hand it has historically been regarded by other country people with awe and foreboding as being endowed with almost supernatural powers; Bolam gives many examples, from Merioneth, of those alleged powers and also traces the ancient lineage of Welsh Ravens back to the days of Hywel Dda in the 10th century. Many of the traditional nesting sites are of great antiquity and are reflected in Welsh place names such as Tap y Gigfran, Craig Nyth y Gigfran.

The Raven occurs throughout the length and breadth of Wales and in some areas, notably the mountain core and the cliff-coasts of the western seaboard, it is abundant and possibly more numerous than anywhere else throughout its Holarctic range. However, it is by no means confined to mountains and sea-cliffs for it is numerous in many of the lower, more productive agricultural areas as well as urban fringes and in some cases even nesting in city centres (see below). Much of the Raven's modern fortunes are closely linked to the stock-rearing pastoral pattern of agriculture that prevails in Wales, particularly the upland sheep industry to which the Raven is an inseparable camp-follower. The importance of sheep carrion for the inland Raven cannot be exaggerated and some of the effects of slipshod husbandry practices can be deduced from the work of Davis and Davis (1984). They found that in the mid-Wales uplands there were, on average, five dead ewes and 45 small lamb corpses in each one-kilometre square of hill sheepwalk per year. Coastal Ravens will forage well inland but they also find much of their food on or near the littoral.

The Raven has long been persecuted. As early as 1533, 1572 and 1597 Acts were passed to encourage the destruction of corvids and other "vermin". This period marked the beginning of three centuries of attrition for the Raven, during which it was eventually exterminated in lowland England, never yet to return to the vast majority of areas, and was considerably diminished locally elsewhere, including Wales. As an example of the levels of persecution Hope Jones (1974) details the numbers killed in Llanfor Parish (Merioneth) in the 18th century, between 1720 and 1757. With bounties being paid on annual totals varying between 15 and 120 Ravens each year (although it is not possible to confirm that all the birds claimed as "Ravens" were in fact of that species), it is hardly surprising that there was a steady decline from 1745. Killing was intense in the 19th century, not only by hill farmers and shepherds, but also by keepers on the great game estates. The total of 464 killed on Penrhyn Estate (Caernarfonshire) between 1874 and 1902 was doubtless matched on many other estates in the same period. Although the Raven is now fully protected by law, some persecution still continues, albeit at a greatly reduced rate compared with former years. Some accessible hill nests are destroyed (shooting, stones in nests) by farmers and shepherds (the old habit of lowering blazing gorse bundles into the cliff nests is no longer practised!). In addition, poisoned meat baits are still used by a minority of farmers, the Raven being particularly susceptible to such lures.

Ravens have always held a special fascination for egg collectors, and to this day they are more heavily robbed than any other species in Wales. The diaries of scores of erstwhile well-known collectors are studded with accounts of successive years' descents to traditional nests, and Welsh Raven clutches abound in virtually all notable collections, often by the score. The position is not much different today; the maps and diaries of modern egg collectors faithfully catalogue the same age-old nest sites which have been plundered year after year, in some cases since before the turn of the century. The principal areas for this egg collecting have always been mid-Wales, Llŷn, Anglesey, parts of Pembrokeshire and Snowdonia.

The level of persecution that the species suffered, culminating in an onslaught of firearms, traps and poisons in the late 19th and early 20th centuries, had a serious effect on the Raven's population. By about 1910 its numbers throughout the Principality were at their lowest ebb, with many writers lamenting the bird's local demise. Even in its mountain strongholds the numbers were reduced: R. W. Jones writing in the Llandudno & District Field Club Report in 1909 claimed that "so much and for so long have they suffered at the hands of unscrupulous egg collectors and others . . . that no-one knows when they last brought off their young with success". A. B. Priestley (1896) wrote of "a further decline in Caernarfonshire and K. J. P. Orton (1910) claimed that "it still manages to maintain a foothold (in north Wales) though in diminished numbers". Throughout Wales the story was consistently the same: Matthew (1894) believed there were as few as 20 or so pairs in Pembrokeshire one hundred years ago.

The reversal of the Raven's fortunes coincided with the period of the First World War and the decline of the game estates. From 1914 the species has enjoyed a slow but uninterrupted recovery, reclaiming some areas previously lost, such as the Marches, and increasing its numbers in its favoured upland heartlands. In some counties this recovery has apparently continued up to the 1970s and even the 1980s in parts of north Wales. With this recovery there has been a return to the former habit of tree nesting, which is now prevalent again in most of the lowlands and many of the upland areas.

In Breconshire there were no more than 20 pairs, all on crags, before the First World War, but by the 1920s the population had at least doubled (Salmon & Ingram 1957). In Radnorshire numbers were "small" at the turn of the century but had increased to 12 pairs by 1905 and to 30 or more by 1935. H. A. Gilbert (1946) noted that "the increase in the past 30 years in south Wales is well known but lately the increase has been startling".

In Pembrokeshire, Lockley *et al.* (1949) recorded that Ravens were numerous and increasing in 1949 – perhaps 80 pairs, of which 60 were coastal. Pairs have long nested on the Pembrokeshire islands, several on Skomer, usually one on Skokholm (two in 1966), two on Ramsey and even one pair traditionally on Grassholm where they scavenge on the ample wastage of the Gannet colony. At a mere 9 ha, Grassholm is probably the smallest breeding territory of any British pair of Ravens. In Carmarthenshire and Cardiganshire "considerable increases" were recorded by Salmon and Ingram by the middle of this century, and throughout the northern half of Wales the pattern has been much the same. Today it is scarcely possible to spend a day in the Welsh countryside without seeing or hearing Ravens. They occur in reasonable numbers in the fertile river valleys and low hill country of the Marches (and thereby indicate their willingness to spread readily into English counties were they to be spared the higher level of persecution there). A further manifestation of their success is an increasing willingness to nest on buildings and other structures. A pair reared young successfully on the tower of the Guildhall in Swansea City in 1973 and 1974, and in recent years other pairs have bred on pylons at Uskmouth and elsewhere in Monmouthshire as well as on the Britannia Bridge across the Menai Straits. In 1992 a pair nested on a roadside telegraph pole on Llŷn.

In parts of upland Wales and in some coastal areas it is clear that breeding pairs are at or near saturation density, resulting in a relatively low production rate of young (average 1.7 per pair per year, Newton *et al.* 1982) and in large non-breeding flocks throughout the year. Although these flocks comprise mostly one- to three year old birds, they also contain older birds of breeding age, often paired but (presumably) unable to find territories. Newton *et al.* have shown that in their 475 km^2 study area in the Cambrian Mountains between 1975 and 1979, there were 79 territories (not all occupied every year), or one pair per 6 km^2, which gives a greater density than has been recorded anywhere else in the Raven's enormous northern hemisphere range. If the upland part of the population in this study area is separated from the lower farmland pairs, the density is even greater at one pair per 4.9 km^2. Ratcliffe (1962) in the late 1950s and early 1960s occasionally found up to three occupied nests within a one mile (1.6 km) radius in 100 breeding pairs. In the Cambrian Mountains, however, Newton *et al.* believed the numbers of breeders and non-breeders to be almost equal and the impression from elsewhere in Wales tends to support this. The breeding population of Welsh Ravens is now fairly well known county by county from population studies and/or the current knowledge of county bird recorders and others: it is shown in the table below.

Estimated breeding population of Welsh Ravens

County	*Number of pairs*
Anglesey	c27
Breconshire	c200
Caernarfonshire	c140
Cardiganshire	165–190
Carmarthenshire	c100
Denbighshire	50–55
Flintshire	10–12
Glamorgan	60–80
Merioneth	c100
Monmouthshire	100–150
Montgomeryshire	120–140
Pembrokeshire	c140
Radnorshire	50–75

Accordingly the total number of breeding pairs in Wales in the early 1990s is probably in the order of 1250 to 1500, which indicates a total breeding season population (including non-breeders) of perhaps 4500 birds. Marquiss *et al.* (1978) have demonstrated that the Raven is unable to maintain its position when hill land is converted to extensive tracts of uninterrupted conifer forestry. Such is this species' reliance on sheep carrion that it is the least able of all upland scavengers and carrion feeders to compensate for loss of open foraging range. In Wales, however, although some 25% of the uplands are already afforested, the effect on the Raven is not yet seriously manifest, mainly because the Welsh forests, although fairly large in terms of overall percentage, are much fragmented and still interspersed with wide expanses of sheepwalk, improved pasture, enclosed farmland and ffridd. Newton *et al.* (1982) have suggested that mean clutch sizes may be smaller with increasing forest area around nest sites. In one case where the nest had 70% forest cover within 1-km radius and 50% within 3-km radius only two or three eggs were laid annually and no young were reared in three consecutive years.

Ravens are well known to congregate in large gatherings both for communal feeding and for roosting. Such gatherings are regular in Wales; feeding and roosting parties of 20–30 individuals are commonplace, and not infrequently up to 50 birds can be encountered together. Such groups, which may occur throughout the year, certainly contain non-breeding and unpaired individuals but also frequently a proportion of paired birds (Coombes 1978). Gatherings may sometimes be very large: the authors found well over 100 birds feeding on sheep carrion in the Tywi Valley in March 1975 and 150 on Mynydd Hiraethog in May 1992. W. A. Cadman knew an all-year roost in Merioneth between 1937 and 1939 which regularly had up to 70 individuals using it. In Pembrokeshire up to 80 have been noted on Skomer in autumn and as many as 250 on the mainland in August, with up to 170 – presumably non-breeders – still together in April. The largest roost we can trace involved some 250 birds near Roch (Pembrokeshire) in 1964 (J. W. Donovan pers. comm.).

Once they are adult, Ravens are extremely sedentary and site-faithful – although many pairs may not breed every year. Of 14 recoveries of ringed birds, 35 moved fewer than 10 km, and 71 between 10 and 100 km. Only eight were found more than 100 km away from the place of ringing. It is interesting to note that of the 20 birds which moved 50 km or more all but two were recovered in the arc between east-north-east to south-south-east. The two exceptions had moved north-west, one of them to Northern Ireland. The most distant recovery is of an Anglesey bird ringed in the nest and recovered two years later at North Berwick (317 km). It is tempting to postulate that some of the surplus of young birds each year head eastwards across the Marches to unoccupied ground, thereby abandoning the over-populated hill lands of Wales.

Starling *Sturnus vulgaris*

Drudwen

An abundant resident; passage migrant and winter visitor in large numbers.

The Starling is a ubiquitous and abundant bird throughout lowland areas of Wales. As a resident, it is most numerous in close proximity to man where buildings offer innumerable nesting opportunities, and surplus food, particularly in the form of refuse, provides ideal feeding. Its success in the modern world can be measured to some extent by the wide range of habitats it occupies. Apart from its urban and suburban concentrations, breeding pairs occupy woodland sites, inland and coastal cliffs, quarries, farmsteads and some pairs even resort to nesting sites in earth banks or stone walls (Bardsey) and occasionally the holes of such species as Rabbit or Sand Martin. Only in the extensive areas of open upland is the Starling not truly at home and such areas are largely spurned. Nonetheless the *Breeding Atlas* confirmed breeding in every 10-km square in Wales, although the *New Atlas* produces a few gaps in this distribution, notably in the upland areas. Even in the uplands, at the end of the breeding season flocks of Starlings are frequent on the invertebrate-rich improved grasslands and, less frequently, foraging on the moorland fringes for bilberries. However, there are still many upland farms and dwellings which do not have breeding pairs and where Starlings are not regular at any time of year.

The CBC index charts a sharp decline in numbers

nationally from around 1980, a decline that is still continuing. The root causes of this decline may lie in climatic factors, for parallel trends are well documented in northern European populations. The extent of any change in population in Wales is poorly recorded at present but there is a general view that in the early 1990s breeding numbers are certainly considerably smaller than they were 20 years ago, and in some counties (e.g. Breconshire) a decline is traced back distinctly to the early 1970s. In Pembrokeshire, where Saunders (1976) suggested that there were no more than 50 nesting pairs in 1948, the population is now put at some 2000 pairs (Donovan & Rees, in press). Tyler *et al.* (1987), with 95% occupation of tetrads in Monmouthshire (1981–85), suggested a population of over 18 000 pairs. Thomas *et al.* (1992) found a slightly lower occupation rate of tetrads (87%) in west Glamorgan, which was attributed to the lack of nest sites in large areas of mature conifer forest.

Among the few places where Starling breeding numbers have been accurately counted annually in Wales are one or two of the offshore islands. The trends of numbers breeding on Bardsey, Skomer and Skokholm provide contrasting pictures. The Pembrokeshire islands tend to mirror the general national trends, whereas the Bardsey population reflects the decline of the 1970s and 1980s, but then unexpectedly produced an all-time maximum in 1989. On Bardsey, Starlings ceased breeding in the early 1950s, not recolonising until 1962 and thereafter increasing rapidly to 40–60 pairs annually by the end of the decade, after which they declined to around 25 pairs in 1984 and then rose again to 40–50 pairs in the late 1980s, with an unexplained high count of 80–90 pairs in 1989. Breeding first occurred on Skokholm in 1946, and from the early 1960s numbers rose to a peak of over 50 pairs by the end of the decade declining rapidly thereafter to a level of around six to eight pairs in the early 1990s. Skomer numbers reached c50 pairs in the late 1960s and early 1970s but are currently around two to four pairs annually.

The Starling has not always been numerous throughout Wales. In the early 19th century it was virtually unknown as a breeding bird and occurred only as a numerous passage migrant and abundant winter visitor. Parslow (1973) has shown that nationally an expansion of range northwards and westwards took place between c1830 and 1880 and he suggested that increases were continuing in the counties bordering the Irish Sea: there is local anecdotal evidence to substantiate that this has been the case in west Wales.

Up until about 1880 Forrest regarded the Starling as unknown as a breeding bird in north-west Wales, although by 1907 he could cite it as being "now abundant and still increasing". Much the same pattern can be shown for the other western areas of Wales although colonisation in Pembrokeshire appears to have been underway somewhat earlier, presumably a reflection of that county's broad lowland connection with the remainder of south Wales where the species was strongly established well before the end of the 19th century. Although historical information by which the westward spread might be traced across Wales is far from complete, it is clear that the mountain core formed a genuine barrier to the rate of progress of colonisation and the species was well established in counties such as Montgomeryshire, Denbighshire and Flintshire before the colonisation of more western counties was achieved. For example, it was still a very scarce nesting species over much of west Wales in the late 1940s according to J. L. Davies (North 1949).

Very large numbers of migrant Starlings pass through Wales in the autumn, the first arrivals appearing in late September and then continuing into November. The offshore islands pick up this passage particularly strongly and it has been carefully documented over the years at the observatories on Bardsey and Skokholm and also on Skomer. Numbers attracted to the lighthouses, such as that on Bardsey, on nights of bad visibility can run into many thousands and the Starling is one of the principal casualties at such times. Southward passage on Bardsey has been counted at up to 17 000 per day and on Skokholm as many as 10 000 were moving over the island, south-westwards, on 6 November 1970. Autumn passage is prominent along the north Wales coast and to a slightly lesser degree along the Glamorgan coast also. Hard weather movements can produce spectacular numbers of birds in some years: 52 000+ birds passed over Milford Haven in 4½ hours on 14 January 1987 (Donovan & Rees, in press). At such times the size of flocks resting or feeding on the headland fields of Pembrokeshire, Llŷn and Anglesey may reach many thousands. These same headland fields are extensively used by huge wintering flocks. The return movement in spring (March) is equally well marked although the numbers are inevitably not as dramatic.

Post-breeding flocks of resident birds are common in many parts of the Principality from mid-June onwards and some can involve large numbers, e.g. at Penrhyndeudraeth in June 1886, seen by Caton Haigh – which example is slightly difficult to reconcile with the fact that the Starling was allegedly a very scarce breeder at the time.

The Starling is also an abundant winter visitor throughout the lowlands (one of its Welsh names, Adar-yr-Eira – "snow bird" – reflects this) except in severe winters when most of the immigrants move quickly out of Wales to milder areas farther west and south. Flocks break up from their night-time roosts and feed mainly in parties fewer than a hundred strong although bigger concentrations occur at sources of abundant food such as sewage treatment works, refuse dumps and similar; many parties raid garden feeding stations. Winter roosts – some of which may be traditional over many years, e.g. Llangorse Lake (reedbeds) – can be of immense size: that at Cwmogwr Forest (Glamorgan) has been estimated at c1 500 000 birds at times and other roosts estimated at over a million birds have been recorded at Brooks (Montgomeryshire) in the 1960s, Llyn Traffwll (Anglesey) in 1983, Spring Hill (Denbighshire) in the 1960s and Mynachlogddu (Pembrokeshire) in 1979 when over 2 000 000 birds were estimated to be present. Vernon (1963) has shown how summer and autumn roosts of resident Starlings are later deserted for the larger communal roosts dominated by immigrant birds and he details the distribution on known roosts in Glamorgan and Monmouthshire across the period 1948–63. Many of the feeding flocks make long evening flights back to the main roosts, in many cases well over 30 km; Gower birds are known to cross Carmarthen Bay to the big roost in Pembrey Forest, and birds have been watched in autumn (e.g. Lavernock Point) leaving the Glamorgan coast to cross the Bristol Channel for roosts in Somerset, returning the next morning.

The vast majority of these winter visitors originate from the Baltic states and north European Russia, as confirmed by many ringing recoveries, a summary of which is shown on the accompanying map. However, the numbers of these continental birds appear to be considerably smaller in recent years. Welsh breeding birds are much more sedentary, with little evidence of widespread dispersal.

Countries of origin of Starlings (ringed May–September) recovered in Wales (December–March) and (in parentheses) countries in which Starlings ringed in Wales (October–March) were subsequently recovered

---

Rose-coloured Starling *Sturnus roseus*

Drudwen Wridog

A rare visitor, most often recorded from Pembrokeshire and Caernarfonshire.

It breeds from Hungary and the Balkans eastwards and is a famous nomadic wanderer. There are seven records from Wales in the 19th century, all referring to shot specimens, from Flintshire, Anglesey, Carmarthenshire, Glamorgan and Monmouthshire.

The first record from this century was of two birds near Mostyn (Flintshire) on 18 August 1928. Otherwise all subsequent records are of single birds and refer to the period from June to December other than one record from January. One juvenile stayed at St. David's (Pembrokeshire) from late October to late December 1986. There are 21 20th-century records, eight coming from Pembrokeshire, eight from Caernarfonshire and the remainder from Flintshire, Monmouthshire, Glamorgan and Cardiganshire.

House Sparrow *Passer domesticus*

Aderyn y Tô

Resident, breeding throughout Wales in all counties. Not ubiquitous, however, and absent from many upland settlements.

The *New Atlas* gives the most up-to-date information on the distribution of the House Sparrow in Wales, showing that it is present as a breeding species in all but a tiny handful of 10-km squares (it was almost certainly missed in some of these). However, this distribution disguises the fact that it is not a species which is evenly distributed; principally it has such a strong association with man that it is scarce or absent in those areas, particularly in the uplands, where human populations are thin and scattered. In rural areas the association is even stronger where human presence is connected with stock-rearing or poultry as this guarantees a constant supply of spilled or stolen foods; examples exist where small populations of House Sparrows at isolated farms have disappeared with the change of husbandry away from livestock. In the upper Tywi Valley (170–276 m) A. R. Pickup (in litt.) indicates that House Sparrows are present at most of the larger farm complexes but absent from the smaller areas and from isolated habitations; this is a pattern which applies in many of the rural/upland areas. Its association with specific land uses – notably cereal growing in the lowlands is so strong that, although there have been no dramatic changes in status in Wales over the years, the ebb and flow of its population can be related to agricultural episodes. Marchant *et al.* (1990) have suggested that a plateau of numbers nationally in the 1970s was followed by a decline which could be related to regular crop spraying (reducing quantities of both invertebrates and weed seeds), "cleaner" cereal crops and the earlier clearing of stubbles. The species is certainly regarded as a pest on growing crops when large flocks visit ripening fields of corn in late summer. Such factors are, however, unusual in Wales, where less than 7% of farmed land is under tillage. Up until the 1960s House Sparrows presumably benefited from the amount of cereals then grown (c85 000 ha) especially oats (see graph on p. 28).

Although predominantly an urban and suburban species with its major strongholds in the largest towns and cities, nesting in buildings, the House Sparrow is also fairly adaptable. Not infrequently it produces colonies of straggling nests in "natural" sites such as overgrown hedgerows and has been recorded nesting in sea-cliff sites in several counties; Charles Oldham, in the early decades of the century, regarded such cliff-nesting on Anglesey as frequent. The habit of these "wild" colonies seems less regular now than it was formerly, possibly related again to the decline of cereal growing.

Nesting has occurred on several of the offshore islands in the past. It allegedly first took place on Bardsey in 1789 and there was a settled population of c20 pairs there until the demise of arable farming and poultry keeping which resulted in its disappearance by 1973. A similar situation applied on Ramsey, and breeding continued on Caldey Is. until 1985 since when it has only been sporadic.

The House Sparrow is a strongly sedentary species, little prone to movements, although occasional birds do occur on the offshore islands in spring and autumn and some autumn movement can sometimes be seen at mainland locations such as Lavernock Point.

Of 63 House Sparrows that have been ringed in Wales and later recovered away from the ringing site, no more than five had moved more than 10 km from the point of ringing.

There is no real knowledge of the size of the House Sparrow population in Wales. The *Gwent Atlas* speculated between 15 000 and 30 000 for Monmouthshire (1981–85), where the species was found in 92% (358) tetrads, whereas in Pembrokeshire (1984–88) House Sparrows occupied 79% (372) of tetrads and Donovan and Rees speculated some 3500 pairs.

Hybridisation between Tree Sparrow and House Sparrow has been recorded in Pembrokeshire (J. W. Donovan in litt.).

Tree Sparrow *Passer montanus*

Golfan y Mynydd

Breeding resident in small numbers with a patchy distribution. Absent from most western areas.

The Tree Sparrow has a very scattered distribution in Wales, a fact which has been recognised since the earliest records were made. As long ago as 1830 the Tree Sparrow was known in the Conwy Valley near Llanwrst, which colony was still present when Forrest visited in 1905. He remarked on the seemingly random occurrence of colonies all across the six north Wales counties. Mathew, writing in 1894, apparently was unfamiliar with the species which he did not include in his *Birds of Pembrokeshire* although later evidence suggests it was almost certainly present then, albeit only patchily. Barker (1905) included it in his list of birds in Carmarthenshire and it had been recorded in the Llanelli area in the first half of the 19th century. Although Ingram *et al.* (1966) knew of no records in Cardiganshire for over 60 years, Salter had seen birds near Clarach just after the turn of the century. Similarly, in Glamorgan it was regarded as uncommon and very local.

The Tree Sparrow has been prone to wide fluctuations in its populations over long periods, for reasons which are not well understood although Summers-Smith (1989) suggests that major increases are a consequence of irruptions from the Continent. In the early years of the present century its numbers nationally were probably at their highest. From the 1920s until the mid-1950s it then went into decline and disappeared from much of Wales. During the 1960s and 1970s it enjoyed a modest resurgence and began to expand again, regaining some of the ground it had lost earlier. This position was not held for long, however, and the CBC index shows that Tree Sparrows entered a period of strong decline again after 1976–77, a decline which continued until at least the late 1980s. The

New Atlas shows the random distribution throughout the Welsh counties, with only the most tenuous outposts remaining in the far west – Llŷn and Pembrokeshire – and substantial gaps, particularly in upland areas and Cardiganshire. It illustrates the losses along the south coast and on the western peninsulas which have occurred since the original *Breeding Atlas* (1968–72). Of the western areas, Anglesey remains the principal, albeit modest, stronghold, where numbers may actually have increased in the past two decades unless it is, once again, a case of better observer coverage. However, Walker (1960) regarded it as a rare visitor on Anglesey. Tree Sparrows were found in only three 10-km squares in 1968–72 but in seven in 1988–91. RSPB surveys on the island in 1986 found birds in 38 one-kilometre squares, well distributed across the island.

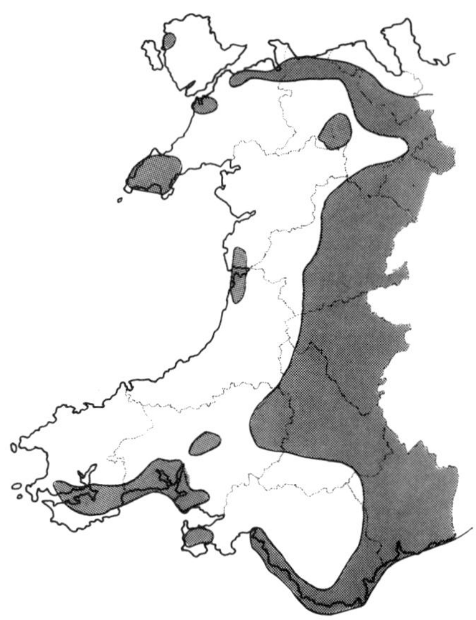

Principal areas of Tree Sparrow breeding in Wales 1968–72 (adapted from *Breeding Atlas*)

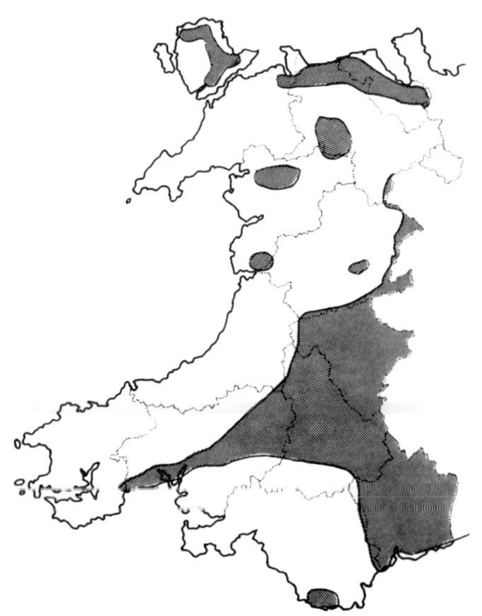

Principal areas of Tree Sparrow breeding in Wales 1988–91 (adapted from *New Atlas*)

Its stronghold in Wales has always been in the Marches, in those counties bordering the west Midlands. In Monmouthshire (Tyler *et al.* 1987) found 51% of tetrads occupied although we do not believe their estimate of the population (c4000 pairs) is valid. Peers and Shrubb (1990) suggest a maximum of only 250–300 pairs for Breconshire (12% occupation of tetrads) and it is similarly regarded as fairly uncommon, occurring mainly in the main river valleys, in Radnorshire, Montgomeryshire and Denbighshire/Flintshire. A two-year survey in the last two latter counties in 1990-91 found only two breeding sites, both in Denbighshire (which conflicts slightly with the wider findings of the *New Atlas*) involving only small numbers of birds. In these areas pollarded willows are favoured nest sites but old hawthorns, rowans and even Sand Martin holes are also habitually used. In Carmarthenshire the Tree Sparrow is relatively plentiful in the Tywi Valley, where there are several good colonies, often in association with old alders and willows alongside oxbows. There is a small but well-established colony in the exposed farmland of west Gower but the population for west Glamorgan (Thomas *et al.* 1992) is put at no more than 10–20 pairs. Tree Sparrows are notoriously fickle in their adherence to colony sites. A site may be used consistently for many years and then disappear totally (e.g. Penmon, Anglesey) to be re-formed elsewhere. The distribution maps demonstrate the species' avoidance of upland areas.

Tree Sparrows are highly sedentary birds and the winter flocks, often in company with finches and others, give a good indication of the size of local populations (e.g. c20 with finches in Conwy Valley, winter 1992–93; flock of 50 at Norton (Glamorgan) February 1993). Although, like House Sparrows, they feed readily on spilled grain, they feed more extensively on small weed seeds, notably in root fields, and thus may feel the effects of the "cleaner" arable crops of recent years in one or two areas of lowland Wales, e.g. Monmouthshire.

Only three ringing recoveries are recorded involving birds in Wales. All three refer to birds ringed in central England, later recovered between 27 km and 94 km westwards in Wales:

Ringed 1 December 1962, Wellington, Shropshire: recovered 17 August 1964, Montgomeryshire border
Ringed 11 October 1964, Eccles, Lancashire: recovered 13 March 1966, Prestatyn, Flintshire
Ringed 26 May 1980, Wilmcote, Warwickshire: recovered 20 December 1980, Abergavenny, Monmouthshire

---

Red-eyed Vireo *Vireo olivaceus*

Telor Llygatgoch

A vagrant.

There have been three records from Wales:

14 October 1967	– One, Skokholm
25–26 September 1975	– One, Aberdaron (Caernarfonshire)
15 October 1985	– One, Bardsey

The Bardsey record was one of a total of 12 in 1985, which was the largest influx ever of any American land-bird to Britain.

Chaffinch *Fringilla coelebs*

Ji-binc

An abundant resident, numerous passage migrant and winter visitor.

The Chaffinch is one of our most familiar and numerous birds. It is by preference a woodland species but its success is reflected in the fact that it also readily exploits a wide range of other habitats, including scrub, gardens, suburban parks, farmland, conifer forests and ffridd. In fact, wherever there are trees or bushes Chaffinches are likely to occur; they will breed in the hills up to the highest limit of isolated trees, up to 550 m in several counties. The intimate pattern of small fields with many coppices, larger areas of woodland and abundant hedgerow trees, which occupies much of the Welsh countryside produces an ideal countryside for Chaffinches.

CBC breeding densities for Britain overall, indicate an average of 19 pairs per km^2 on farmland and some 37 pairs per km^2 in woodland. Shrubb and O'Connor (1986) have shown that the populations of several of the most common farmland species are higher in the west of the country than those farther east. This is borne out by CBC data, which show even higher densities than those given by O'Connor and Shrubb. In Monmouthshire mean densities reach as high as 69 pairs per km^2 and RSPB surveys (Tyler & Geach 1987) on the western coalfields (Carmarthenshire/Glamorgan) approach similar levels in those areas which include a mosaic of old hay meadows, ancient pastures, well-developed hedgerows and coppices. Gibbs and Wigginton (1973) record slightly lower densities of 59.5 pairs per km^2 in sessile oakwoods in north Wales. The ubiquity of this species is demonstrated by its occurrence in 95.3% of tetrads in Pembrokeshire, 99% in Monmouthshire and 98% in west Glamorgan. It was one of the six most common species in forestry restock sites censused by Bibby *et al.* (1985) and can occur in densities of up to 60 pairs per km^2 on bracken-dominated ffridd with trees/bushes.

There is a mass emigration of Chaffinches from Scandinavia in September–October, many of these birds crossing the English Channel to winter in Britain. Although Wales lacks much of the arable land which principally attracts these large flocks, their presence is evident, not least in the form of visible migration, mainly in late autumn and less so in spring. Many Chaffinches, especially single-sex flocks of females, move onwards to Ireland and can be found on occasions in immense numbers in the headland fields of Pembrokeshire, Anglesey and Llŷn awaiting the right weather conditions to make the short sea crossings. Such emigration is clearly visible, usually in November and sometimes encouraged by the approach of hard weather, from headlands such as South Stack (Anglesey), Strumble and Marloes (Pembrokeshire) or Aberdaron (Caernarfonshire).

Coastal passage is particularly marked in autumn and has been well recorded from various sites around the Welsh coast, none more so than Lavernock Point. At this season movement at Lavernock is not necessarily westwards, in fact the majority of large-scale movements there have been to the east or south-east. Farther east in the Severn estuary westward passage has frequently been noted on the Rhymney estuary in autumn. Similar movements occur along the north Wales coast and the coastal areas of Cardigan Bay. Westward movement is very clear between Point of Air and Bangor along the north coast. Huge numbers pass through Bardsey in October, the

Overseas recoveries of Chaffinches ringed in Wales and (in parentheses) countries of origin of individuals subsequently recovered in Wales

largest daily count being c20 000 on 19 October 1966. On South Stack peak counts of up to 950 per hour (24 October) and then 1800 per hour (28 October) were recorded in 1987. In 1986 "it would have taken an army of observers . . . to count the visible birds but many more were heard . . . too high to see" as many thousands of Chaffinches passed north-west over Strumble Head in late autumn (*Pembrokeshire Bird Report*). There were similar mass movements observed in 1981.

The wintering flocks that remain in Wales resort to the lowland areas where they feed on open fields, or to coastal marshes such as those on Dee, Burry, Severn, Dyfi. Most flocks are usually no larger than a few hundred birds although flocks of 1000 or more are by no means exceptional in Wales. The origin of most of the autumn/winter immigrants is southern Norway and Sweden; however, three recoveries of Welsh-ringed birds have been from Finland, two of them from the far east of the country, close to the Russian border.

Welsh resident Chaffinches are remarkably sedentary, moving very little from their natal areas and probably remaining close to their breeding territories for the rest of their lives once they have bred. Family parties and small mixed-sex flocks foraging in hedgerows, field edges and roadsides characterise the catholic feeding strategy of these resident birds, in contrast to the large single-sex flocks of immigrants feeding more exclusively in open fields on a more specific diet of arable weed seeds.

Brambling *Fringilla montifringilla*

Pinc y Mynydd

Winter visitor in variable numbers; seldom numerous and usually locally distributed.

The Brambling, breeding from southern Scandinavia eastwards across Eurasia to Kamchatka, is one of the most migratory of finches, the European population leaving the scrub-birch woods and open conifer forests of the Arctic zone to winter farther south over large areas of Europe. Numbers in any area vary greatly from one winter to the next, depending mainly on the annual variability and availability of beechmast, the Brambling's principal food in winter. In most years numbers of this extremely numerous finch are relatively small in Wales and it is only in years when the beech crop is particularly poor on the continent that the birds occur here plentifully. In any year Bramblings are more likely to be encountered in the counties of eastern Wales than those of the west, as is borne out by the *Winter Atlas* results. In north Wales winter parties are relatively regular along the coastal strip and on Anglesey, but decidedly scarce elsewhere; similarly in south Wales Bramblings are more regular south of the coalfield valleys in Monmouthshire, the Vale of Glamorgan and lowland Carmarthenshire than they are farther north. They are scarce birds in most of west Wales from Pembrokeshire northwards to Llŷn.

Bramblings do not normally arrive in Wales before the second half of October and most have departed again by the early days of April; eight birds in Clocaenog Forest on 1 May 1978 were exceptionally late. Historically there is little evidence to suggest any change in status in Wales, although neither is there any reason why there should have been changes because its appearance here is entirely dependent on factors on the Continent. Heathcote *et al.* (1967) list the Brambling as recorded in only 17 out of the first 55 winters in the present century, but more regular and more numerous thereafter; this may well be a function of the great increase in birdwatchers in recent decades. The size of winter flocks in Wales does not often

Recoveries of Bramblings ringed in Wales. Figures in parentheses indicate subsequent month(s) of recovery

exceed 200 or so, the largest on record being 950 in Radnorshire in February 1986, 600 in Montgomeryshire in January 1980 and the same number at Rhosgoch (Radnorshire) also in January 1980, 550 in Glamorgan in December 1975, 400 in Denbighshire in January 1981, 350 in Anglesey in October 1985, 320 in Carmarthenshire in March 1976 and 300 at Lavernock Point in February 1979.

Bramblings are long-distance migrants and it is unsurprising that there have been several ringing recoveries from eastern and northern countries of Europe. A summary of these distant recoveries is given on the map. It is well known that individual birds may travel to widely different parts of their wintering range in successive years; the recovery of a bird from Modena (Italy) 12 months after it was ringed in Caernarfonshire in late October 1965 illustrates the point. Occasional large "falls" of migrating Bramblings occur, as evidenced by a flock of some 800 on Skokholm in October 1966. There is a published report of one summering at Conwy (Caernarfonshire) in 1953 (*Cambrian Bird Report* 1953–54).

Serin *Serinus serinus*

Llinos Frech

A rare passage migrant, recorded 14 times in Wales up to 1991.

The first record was on 21 October 1933, when there were two Serins on the west Pembrokeshire coast (Lockley 1933). There were three further records pre-1967 and a marked increase with 11 records post-1967; this was the year when the species first bred in England (in Dorset).

As might be expected with a southern species that is expanding its range northwards, the records in Wales show a markedly southern and western bias, with nine records from Pembrokeshire, two from Carmarthenshire and one each from Glamorgan, Cardiganshire and Flintshire.

Nine of the records are from the period March to June, the others are January (1), October (3) and November (1).

Greenfinch *Carduelis chloris*

Llinos Werdd

Common resident; partial post-breeding migrant.

The Greenfinch is a generally common species in Wales, occurring throughout lowland areas in all counties. It is particularly characteristic of suburban areas of towns and villages, city parks and those rural areas that afford old hedgerows, shrubberies or similar thick cover. Like other cardueline finches, the Greenfinch nests in loose groups, so that where one pair is found others may be expected nearby; conversely there are many apparently suitable areas locally without Greenfinches present, which represent the unoccupied areas between such nesting groups. There is a strong avoidance of uplands and although widely distributed in the lowlands, the Greenfinch becomes much scarcer above about 180 m. There is slight evidence that – through marginal deciduous plantings and the retention of some rough weedy areas – the expansion of forest plantations has given some scope for the species to move into the flanks of the hills, although it normally exhibits a strong avoidance of conifer plantations, despite its fondness for nesting in garden conifers in suburban areas.

Historical accounts give the strong impression that it was always more numerous in the eastern counties than the western ones. This pattern is still clear today, certainly so far as areas such as Llŷn and Anglesey are concerned. Even in Pembrokeshire it is regarded as breeding "in scattered pockets" (Donovan & Rees, in press), a description which is probably applicable to many other western parts. Certainly, in Merioneth it is uncommon all along the coastal fringe (Thorpe, pers. comm.). In south-east Wales a 12-year CBC survey showed an average 0.9 pairs per 10 ha on mixed agricultural land and the *Gwent Atlas* (1987) suggested a county population of up to 7000 pairs. Similar densities have been found in Glamorgan (0.87 pairs per 10 ha), although higher densities occur where habitat is ideal (e.g. 2.4 pairs per 10 ha, Lavernock Point, 1965 (Heathcote *et al.* 1967)). In Pembrokeshire the tetrad *Breeding Atlas* found Greenfinches in 55.3% of tetrads with an estimated population of some 4000–5000 pairs (this would equate to approximately 0.4–0.5 pairs per 10 ha overall). In west Glamorgan a tetrad occupation rate of 64.5% was recorded (Thomas *et al.* 1992). In north Wales the *Breeding Atlas* and the *New Atlas* both indicate a general distribution throughout the lowlands although, as noted, Greenfinches are least common in western areas and absent from the uplands.

Although Greenfinches are mainly resident and reasonably sedentary birds, a considerable amount of post-breeding dispersal of juveniles occurs. This is well shown by the results of birds ringed in Wales. Of 310 recoveries reported up to 1991, 131 were less than 10 km from the place of ringing, 114 between 10 and 99 km and 65 over 100 km, including four records of birds crossing the Irish Sea. Those birds that were recovered at a distance of 50 km or more show a marked tendency to disperse in a wide arc between north-east and south-east. Welsh Greenfinches, as elsewhere in Britain, are thus dispersive rather than truly migratory. Coastal movements have been noted as erratic at Lavernock Point in autumn but occur regularly on Bardsey and the Pembrokeshire islands, although numbers are usually modest, with even fewer in spring. Birds are occasionally seen as far offshore as the Smalls (three in November 1984).

In winter Greenfinches are widely distributed throughout Wales. The former habit of congregating at stockyards with other finches and buntings in hard weather has been replaced by a regular exploitation of garden feeding stations, where peanuts make up the staple winter diet for many individuals. Winter flocks can be large where spilled grain or other good sources of food are to be found. Five hundred in Radnorshire in December 1971 is exceptional in mid-Wales, but flocks up to 1000, and even 1500, have been recorded in Glamorgan (e.g. Sker January 1964).

Goldfinch *Carduelis carduelis*

Nico

Resident, summer visitor and passage migrant; possibly winter visitor in small numbers.

The Goldfinch is a moderately common species in Wales, breeding throughout the country, although infrequently occurring on land much above c305 m and avoiding heavily wooded areas, e.g. conifer plantations. The *Breeding Atlas* and the *New Atlas* confirm this widespread distribution, although it should be noted that Goldfinches are not evenly distributed but are locally restricted to suitable habitats on overgrown waste ground, farmland and woodland edge, gardens and parkland where open ground provides plentiful weeds, particularly thistles (*Carduus* and *Cirsium* spp.). Although firm data are scarce, the general impression is that Goldfinches are more numerous in the Welsh Marches than on the western side of the mountains. The distribution shown in the *New Atlas* is fairly uniform but this disguises the fact that numbers are higher in the eastern counties, e.g. 84% tetrads in Monmouthshire, 75% in west Glamorgan and 62% in Pembrokeshire.

Goldfinches were traditionally the most popular of British-caught cage birds, and large numbers were caught in parts of Wales in the 19th century. Forrest 1907 claimed that one bird catcher in the Llandudno area caught as many as 3000 Goldfinches in one year! As a result of even greater depredations in England, the species decreased severely there, only to pick up again under legal protection in the early part of the present century. In Wales a similar pattern was reflected (Forrest 1907; Heathcote *et. al.* 1967), although to nothing like the extent documented for England.

Witherby and Ticehurst (1907–8), reviewing the subject in 1907, also noted that "in the west the decrease does not seem to have been so marked and (it remained) common in parts of Wales in 1893". In Glamorgan, steadily increasing populations were recorded in the 1920s and 1930s, a trend that probably continued steadily in subsequent decades, not only there but in other Welsh counties also.

Many of the Welsh breeding Goldfinches emigrate at the end of the summer. Newton (1972) has suggested that as many as 80% of the breeding population in southern Britain comprises birds that are actually summer visitors. In this respect it is noteworthy that over half the ringing recoveries of Welsh Goldfinches so far recorded (seven out of 15) involve birds reported from Spain (5), France (1) or Portugal (1) between the months of October and February. Certainly Goldfinches, despite their extended breeding season, are much less numerous in Wales between October and March than in the other half of the year. Those that do remain are susceptible to severe weather conditions and considerable numbers may then die; winters such as 1895, 1962–63 and 1980–81 accounted for considerable tolls on Welsh Goldfinches.

The autumn passage of Goldfinches, as with many other species, is most clearly seen in coastal areas and on the offshore islands. Newton (1972) has identified the different strategies exercised by Goldfinches in autumn. Some British birds emigrate to Ireland and such passage can be seen on headlands in Pembrokeshire, Llŷn and Anglesey as well as on the islands; other birds migrate through France to Iberia as already shown, and part of this migration may involve initial south-east movement to enable birds to cross the Channel by the shortest route. Large-scale movements, visible in such places as the Rhymney estuary and Lavernock Point may be part of such passage: on 8 October 1976 over 1000 Goldfinches passed eastwards at Lavernock; in October 1973 a maximum 950 in four hours was counted, and similar impressive movements have occurred here in other years. Confusing the picture in some years is the fact that a reasonable westwards movement is also to be seen at times.

On Bardsey the main passage is also through October and early November, with a much smaller return migration in April and May; in only three years (1961, 1975, 1978) have more than 100 been seen here on a single day in autumn. Elsewhere on the north Wales coastal flocks of 100–200 are not uncommon in October and November and presumably represent part of a south/south-west movement of birds heading for Ireland. The return passage on these islands in April and early May is in much smaller numbers.

Siskin *Carduelis spinus*

Pila Gwyrdd

An increasing breeding species; numerous winter visitor and passage migrant.

The establishment of extensive areas of conifer forests in Wales in the middle decades of this century gradually opened up opportunities for colonisation by this species which it has been quick to exploit. Traditionally associated with remnant Caledonian pine forests in Scotland, the Siskin has diversified its habitat requirements to take advantage not only of new conifer plantations but also of parkland, gardens and similar areas where exotic conifers are grown and provide both food and nesting sites. It is an increasing and expanding species in Wales, now breeding in every county except perhaps Flintshire. Outside the breeding season the resident population is supplemented by fluctuating numbers of passage birds and winter visitors each year. By this time of year the birds have forsaken the conifer areas as the cone seeds become unavailable to them and they concentrate on birches, alder and larch, often consorting with Redpolls. Thus in winter Siskins are widely and ubiquitously distributed throughout Wales, most numerous in the south-eastern half, and are especially found in river valleys in the lowlands where alders and birches are most numerous. In recent years they have also learned to take advantage of garden feeders, notably for peanuts, which has obviously helped lead to improved winter survival. The habit is not constant however and in some years they may be scarce or even absent from "regular" feeding sites, perhaps indicating that these artificial food sources are used in inverse proportion to the supply of natural foods. Siskins are still expanding southwards in Britain and at present Wales holds the bulk of the southern British population (*New Atlas*).

Most of the old accounts of birds in Wales make reference to the Siskin but almost invariably as a winter visitor in small numbers, and often a target for the bird trappers. The only exception to this was Forrest (1907), who reported that it had long been suspected of breeding in north Wales. First proof that this was the case had come in 1872 (*Zoologist* 1873 p. 3410) when a nest was found in Denbighshire. There were further instances of suspected but unproved breeding in Caernarfonshire around the turn of the century and Forrest, referring to these north Wales counties, suggested that Siskins bred more frequently than was widely supposed. If this was the case, it is reasonable to postulate that breeding may indeed have been continuous since then, although unreported until much later this century certainly suitable habitat existed in north Wales across this period. Summer records occurred in areas such as Lledr Valley and Betws-y-Coed after 1945, where there were already extensive maturing conifer forests. Loose colonies of Siskins now nest annually throughout Wales, although in most counties breeding was only finally proved relatively recently (Monmouthshire – summering since 1960, proved 1984; Glamorgan 1973; Pembrokeshire (probably) by 1979, proved since 1986; Cardiganshire (probably) by 1977; Montgomeryshire c1960).

Although breeding numbers fluctuate annually, Siskins can be found during the breeding season in almost all the major Welsh forests as well as many of the smaller ones. They now nest regularly on Anglesey, on Church Is. and in Pentraeth woods and possibly at Newborough Warren NNR (H. Knott, pers. comm.). Early nests are invariably in conifers, especially spruce, but once the previous year's seed crop falls (spring), Siskins move elsewhere and rear young on seeds of pines or dandelion and other compositae; summer nests may be in birches or even oakwoods (e.g. Dinas RSPB reserve, Carmarthenshire, 1971 (J. Humphrey, pers. comm.)). Only very limited data on breeding densities yet exist for Wales. Bibby *et al.* (1985) recorded Siskins at a density of 4.7 *individuals* per 10 ha at 326 count points in conifer plantations in north Wales and subsequently at 10.3 *individuals* per 10 ha around restocked plantations. Andrews and Bellamy (1988) found similar densities in Mynydd Du Forest (Monmouthshire) in 1988 (11.2 per 10 ha). At one site in 100-year-old Douglas fir, Currie and Bamford (1982) recorded Siskins at the equivalent of 16.5 pairs per 10 ha.

Autumn passage varies considerably from year to year; in some seasons it is poorly detected and in others large numbers of birds may be seen in coastal areas from October. Such movements are visible on the north Wales coast and the north shore of the Severn estuary. Charles Oldham watched 30 passage birds feeding on knapweed at Holyhead Harbour in October 1913 and Shrubb (in litt.) has recorded flocks on sorrel. On Bardsey an exceptional 300 passed south between 12 and 14 October 1961. Similarly, on Skokholm and Skomer autumn numbers are normally small, although up to 100 have been recorded on a single day. In 1985 Siskins were unusually numerous in Pembrokeshire between September and December and in 1988 an unprecedented invasion took place between 17 September and 5 November, including 200 on Skomer on 27 October and an exceptional 1200 on Skokholm one day earlier. The return passage in March–April is much more poorly marked, usually involving small numbers and in many years none at all.

The recent habit which Siskins have acquired, of visiting garden feeding stations for suet and peanuts, began in the 1960s and is now commonplace throughout Wales. Winter parties are usually fewer than 50 strong; unusually large flocks have been recorded in Monmouthshire (250 in December 1975) and Breconshire (350 in larch in December 1990; 500 in February 1991 and 250 in December 1983). The origins of the passage and wintering populations in Wales can only be guessed at. In years of irruption presumably Continental birds are mainly involved but it is equally true that the British population itself is eruptive. The handful of ringing recoveries so far obtained suggests that Scottish birds are regular in Wales in winter.

Six birds ringed in Wales in the months February–April have subsequently been recovered in Scotland, the most distant being an individual recovered 490 km north at Alness (Highland Region) having been ringed at Rhos-on-Sea (Denbighshire).

Linnet *Carduelis cannabina*

Llinos

Breeding resident, summer visitor and passage migrant.

The Linnet is a widespread, common and familiar species throughout Wales. During the breeding season it occurs in all counties almost wherever the combination of plentiful weed seeds and suitable nesting habitat is found. Traditionally Linnets nest in loose colonies on rough commons, coastal heaths, overgrown verges, river banks, bramble thickets, and similar; inland, they are strongly associated with gorse in the early part of the breeding season and most gorse brakes of any size can be expected to support a small colony. Linnets are particularly common on coastal cliffs all round the Welsh coasts. They are also frequently found high in hill areas up to at least 457 m where gorse or rank heather is plentiful, and they have benefited from the expansion of young forestry plantations.

As with the Goldfinch, Linnet numbers fell markedly in the 19th century, due to a combination of agricultural intensification and the annual depredations of bird trappers, and although its decline in Wales has been noted by earlier writers (e.g. Heathcote *et al.* 1967), the extent to which numbers were reduced in Wales is not clear. Again, in parallel with the Goldfinch, numbers began to recover under increased protection in the early decades of the present century, probably helped by the decline in agriculture at the same time. The use of selective herbicides on arable crops in the past two decades has greatly reduced the volume of weed seeds available to Linnets (and others) and is held responsible for more recent declines, particularly the steep decline which has been noted nationally since the mid-1970s (Marchant *et al.* 1990). In Wales the herbicide factor applies to a lesser extent than in areas of south England as there is relatively little cereal production. Habitat loss through scrub clearance, hill "improvement" and the intensity of grazing by sheep (which has virtually the same effect as herbicides on field weed stocks!), is a more likely reason for keeping Linnet numbers in check, and the population in Wales at present is probably as low as it has ever been. However, the potential for this to be reversed may be presented through several developing agricultural changes, that is to say, the increase in set-aside land (albeit marginal in extent in Wales in the early 1990s), agri-environmental schemes and the increasing area of linseed, which provides a rich supply of spilled seed for finches and other seed-eating species.

In Monmouthshire an annual breeding population of around 6000 pairs was estimated for the years 1981–85 (Tyler *et al.* 1987), when Linnets were found in 78% of tetrads. In Pembrokeshire the tetrad breeding survey found Linnets in 73.2% of tetrads and a population of c8600 was estimated. Peers and Shrubb (1990) suggest a population of some 400 pairs in Breconshire and in west Glamorgan they were found in 75% of tetrads (no estimate of population). The Merioneth population is small and Thorpe (pers. comm.) regards it as scarce throughout the county with relatively few pairs scattered along the coastal strip. Elsewhere in north Wales the Linnet is widely distributed in suitable areas but no figures of breeding density exist.

Many Linnets move out of Wales at the end of the breeding season and emigrate south to France and Iberia via the short Channel crossing in south-east England. Coastal and offshore passage is well marked both in autumn and spring round the Welsh coast. On Bardsey, many Linnets pass through in September and October, occasionally reaching over 1000 per day. At Lavernock (Glamorgan) similar peaks have been recorded heading south-eastwards and a return passage in March–April is much smaller but still clearly evident. At Goldcliffe (Monmouthshire) a flock of 2000 was passing on 6 October 1968; on the Pembrokeshire islands numbers are usually fairly modest (mainly in autumn), only rarely as many as 200 a day on Skokholm.

Although many of the breeding Linnets depart in early autumn, the species is still well distributed throughout Wales in winter, although the numbers present depend on the severity of weather in any particular year. The majority of these winter birds are probably immigrants from farther north and east but evidence from ringing is, as yet, insufficient to confirm this belief. Resident Linnets certainly desert the hill areas, as shown by the *Winter Atlas*. If conditions are harsh numbers may increase temporarily in coastal areas such as saltmarshes, dunes and coastal fields but they are among the first birds to move out when temperatures drop sharply. At times flocks of 1000 or so can occasionally be found, e.g. Kenfig (Glamorgan), Llanrhidian (Glamorgan), Dee estuary (Flintshire) and Ginst Point (Carmarthenshire), although since the mid to late 1970s flock sizes have been noticeably smaller (mean size of 39 in Breconshire since 1976, with the exception of one flock of 200).

Twite *Carduelis flavirostris*

Llinos y Mynydd

Scarce breeding species, presumably resident, possibly increasing slightly; scarce winter visitor and passage migrant in very small numbers.

The breeding status of the Twite in north Wales has long been cloaked in mystery and even now, at the southern edge of its breeding range, the numbers and distribution of breeding pairs each year is not fully understood. It is clear, however, that a small population of Twite breeds regularly in the counties of Caernarfonshire, Merioneth and Denbighshire. At some places they are probably site-faithful while in other areas they may possibly be more mobile and irregular.

Writing in 1838 Eyton claimed that it was "common in North Wales, breeding there . . ." However there is no substantiation for this claim and Forrest (1907) concluded that it was "a winter visitor in fair numbers . . . but seldom if ever remains to breed". With the benefit of more recent experience, Forrest's considerable number of accumulated records might lead one to a more positive interpretation as some of his quoted records were by competent observers in areas where breeding has subsequently been established; for example T. A. Coward watched a pair of Twite on Mynydd Hiraethog in June 1905 and there were pairs on the hills above Llanuwchllyn (Merioneth) in May 1895; both are sites where at least intermittent breeding is known to occur. Other historical records of variable quality quoted by Forrest pertained to several other sites including moors above the Vale of Clwyd, several locations on the north Berwyns and even, more improbably, Corndon Hill on the Montgomery/Shropshire border. However, in this last respect it is interesting to note that More (1865) claimed intermittent breeding on the Hereford/Monmouthshire borders on the Black Mountains and it should be noted that breeding has been proved as far south as Devon in 1904; moreover a pair of Twite was seen at a site in north Monmouthshire in 1991. Intriguingly, Forrest (1907) also quotes summer records for several sites on Llŷn at the turn of the century, notably Yr Eifl and Carn Fadryn, but also one report on the coastal heaths of Pen Cilan; there have never been subsequent records of Twite from any of these Llŷn sites and, if breeding ever did take place there, it is reasonable to suppose that it ceased with the extensive agricultural changes which took place in the first half of the 20th century.

The first nest to be claimed in Wales was found by George Bolam near Llanuwchllyn on 20 May 1905. There was then a long interval until adults were watched carrying food on Tryfan (Snowdonia) in 1944. More regular breeding has been proved in Caernarfonshire since then, near Llanberis (first in 1953 but by 1988 birds were shown to be present at at least five sites with breeding proved at two and up to 20 birds present at one site in 1989), near Llanfairfechan (1965), on the Carneddau (from 1971) and in the Nant Ffrancon pass at least since 1973. In Merioneth the Arenigs have been known to support small colonies of Twite, at least sporadically, since the 1920s; nests have been found there in 1988 and 1989 and suspected the previous two years; birds have been seen there in small numbers throughout the 1970s and 1980s. Breeding may also occur on the Rhinog Mountains where eight birds were watched in July 1989 and a pair was present on Rhinog Fawr in May 1990 (both pers. obs.).

In Denbighshire and Flintshire, Forrest referred to possible breeding records in the Llangollen and Llandegla areas and there were later (1910 on) records from the Hiraethog (Denbigh Moors). In recent years it has been firmly established that breeding does indeed occur in these areas, again at least intermittently, and possibly regularly. Nests have been found on Hiraethog since 1973 (four pairs present at one site in 1978) and on moors near Llangollen for the first time in 1990 (J. L. Roberts, pers. comm.). There is one exceptional record of a nest on the saltmarshes at Shotton (Flint) in May 1967 in sea purslane (late J. R. Mullins, pers. comm.). A pair was present and presumed to breed near Llanarman D.C. in the early 1970s (pers. obs.).

In 1988 (RSPB unpublished report 1988) two pairs of Twite were found, presumed to be nesting, on the northern area of Pumlumon Mt. (Montgomeryshire).

At the end of the breeding season Twite may be encountered more widely on the north Wales hills and moderate-sized flocks are sometimes encountered: 30+ on Cadair Idris, October 1990; 30–50 in Nant Ffrancon, winter 1990–91 and 58 in September 1979; an exceptional flock of c150 was found at Llyn Ogwen in December 1990.

There is evidence of a small autumn passage from the offshore islands and, less regularly, returning birds in spring in the north: predictably these have been more numerous in the north: Bardsey has produced 19 records, all but three of them in autumn, and Skokholm four, all in autumn. However, small parties have been reported irregularly around Burry Port (Carmarthenshire) in early spring. Although some Twite remain on the hills in winter, their presence presumably depending on weather conditions, most winter records are from coastal areas (a flock of 19 in Pembrokeshire in December 1981 was unexpectedly large and probably due to the onset of a spell of severe winter weather). Isolated inland records from a wide range of counties (e.g. Monmouthshire, Radnorshire, Cardiganshire) are not unknown. Small numbers are recorded in most years from the coastal counties, in dunes, saltmarshes and other estuary areas. They are scarce but not rare in Glamorgan but more irregular in Carmarthenshire, Pembrokeshire and Cardiganshire. They are to be found almost annually at the Great Orme (Caernarfonshire) and nearby at Llanfairfechan (maximum recorded, 67). The most regular and reliable sites to find Twite in winter are on the Dee estuary. Here they can usually be found either at Point of Air, Oakenholt Marsh or, particularly, Connah's Quay. At this last site a maximum of 50+ was present in November 1988 and maxima of c20–40 birds have recently been recorded at the other locations and at Gronant dunes to the west of Point of Air.

Redpoll *Carduelis flammea*

Llinos Bengoch

Partial migrant, widely distributed as a breeding bird. Passage migrant and winter visitor in varying numbers.

The Redpoll is now a widely distributed bird in Wales, occurring as a breeding species in all counties, most numerously in those along the central mountain spine from the north Wales coast to the coalfields valleys, and least commonly in south-west Wales, Llŷn and the lower land of the Marches. In winter, when many of the breeding birds have emigrated, Redpolls – evidently supplemented by birds from farther north – are still widely distributed but then occur more regularly on lower ground of river valleys, especially where birch, and to a lesser extent alder, are common. At this time of year parties are not normally large; flocks of 300 in Cardiganshire (September 1988) and 200 in Denbighshire (February 1977) and several of the same size in Breconshire are unusually large. Redpolls are strongly dependent on birch throughout much of the year (but also feed in other places, such as root fields, in winter), and their distribution is thus closely linked to that of birch although they have recently started to exploit the expanding conifer forest habitats (see below). Despite the fact that the high birchwoods which once dominated the uplands above the oak/alder zone are now represented by only small degraded remnants, birch remains a common component of many tree-covered habitats, including deciduous woods, overgrown hedgerows, scrub, waterside sites and ffridd. Moreover, birch is an aggressive pioneer and an abundant seral species, and in particular gains an early foothold in unmanaged areas of forestry plantations and clear-fell/restock areas. All these sites are used by Redpolls.

Parslow (1973) has described how the Redpoll increased markedly in Britain around 1900–10, losing ground again after that and then enjoying a much more vigorous expansion from the 1950s onwards. Although first-hand evidence specific to Wales is sparse, it is fairly clear that this pattern was reflected here. Forrest (1907), in the north Wales counties, and Paterson *et al.* (1900) in Glamorgan together with Barker (1905) in Carmarthenshire all agree in their view as to the scarcity of the Redpoll as a breeding species at the turn of the century. Forrest, however, accurately observed that it was an expanding species at his time of writing. Ingram *et al.* (1966) recorded the Redpoll as still being scarce and local until the 1920s in Cardiganshire, but showed that by the late 1950s they were breeding extensively in the new conifer forests of mid and north Wales up to altitudes approaching 610 m. The rapid expansion of upland afforestation in the 1950s, 1960s and 1970s provided ideal conditions for the expansion of the species and only the Siskin can rival Redpoll in the extent to which it has profited from this new habitat in Wales. By 1977 the CBC index for Redpolls nationally had risen to four times the level of ten years earlier (Marchant *et al.* 1990), although Redpolls feature too sparingly in the few Welsh CBC sites for the accuracy of this increase to be tested for Wales. Subjective observations tend to support the national trend.

Nationally, the high populations of the mid-1970s dropped sharply from around 1977 and by the mid-1980s they were back to the pre-expansion levels of the early 1960s. The reasons for this fall are unclear but may possibly be related to the increasing unsuitability of many conifer forest plantations as they mature. Redpolls were noticeably fewer in most parts of Wales, although numbers had evidently picked up again by the end of the 1980s and early 1990s.

Although Redpolls are characteristically birds of birch scrub, they also breed in alders, sallows, lowland scrub (e.g. Mawddach Valley, Cors Caron, Cors Fochno) and similar overgrown sites. However, nowadays the Redpoll's stronghold in Wales is associated with the extensive conifer plantations, especially those in the upland core of the country. Although the loose colonies of breeding pairs are mobile in such areas, moving on once the young trees reach c6 m or so and the canopy closes, Redpolls are common breeders in all the mainland counties of north and mid-Wales (least so on Llŷn). On Anglesey they breed locally in the woodland bordering the Menai Straits and in east Anglesey, but only sparsely and irregularly elsewhere on the island (K. Croft in litt.) although the *New Atlas* indicates presence in all 10-km squares, with proved breeding in most areas. Breeding is extensive in mid-Wales as far south as the coalfield valleys. In Monmouthshire, Redpolls breed in some 20% of the tetrads, mainly on the high ground of the western valleys. In west Glamorgan they are absent from Gower but a strong correlation between conifer forests and the occurrence of Redpolls is demonstrated in other parts of the area. In Pembrokeshire – where breeding Redpolls are scarcer than in any other Welsh county – only 6% of tetrads are occupied, mainly in the north of the county on and around the flanks of the Preseli Mountains. Peers and Shrubb (1990) consider the population in Breconshire to be around 400 pairs in 24% of tetrads.

After the breeding season, many British Redpolls move away from the breeding forests and there is a well-documented southward movement, with birds crossing into northern France and Belgium by the short sea crossing. Evidence that Welsh birds follow this south-east movement is to be found in the small handful of ringing recoveries: a young bird ringed at Newborough Warren NNR in August 1964 was recovered in Belgium three months later. An adult bird from Holyhead was retrapped, also in Belgium, the following autumn, and another adult ringed in the Mawddach estuary in May 1984 (presumed breeding) was retrapped at Beachy Head (Sussex) in October 1985. A bird ringed in Cumbria in May 1981 was recorded wintering in Monmouthshire the following December, giving evidence of some degree of immigration of northern birds in winter.

The offshore islands provide regular evidence of spring and autumn passage, although numbers involved are usually very small. On Bardsey spring passage is usually negligible (mostly in late April) and 10–20 daily are the usual maxima in October–November any year. On the Pembrokeshire islands, conversely, Redpolls are slightly more numerous in spring than autumn, although again totals in any season are usually small.

Redpolls are circumpolar species with a northern distribution, mainly in arctic and sub-arctic regions, to which the isolated British population is one of the most southerly outposts. They are a complex group, but modern taxonomists prefer to divide them into two highly variable species, *C. flammea* (Common Redpoll) of the open northern forests, and the larger paler *C. hornemanni* (Arctic or Hornemann's Redpoll) of high arctic tundras (see next species). Within the Palearctic population of *C. flammea* – the British race is *C. f. cabaret* – there is much variation, with individuals from farther north tending to be paler and slightly larger. The subspecies *C. f. flammea*

(Mealy Redpoll) occurring from Finland eastwards, is the most clearly defined in this latter respect and occurs in Britain in small numbers in winter, mainly in counties bordering the North Sea. It is rare and irregular in Wales, with only a handful of confirmed records. One is alluded to in the Cardiff area " in the second half of the nineteenth century" (no date given). Walpole-Bond watched several near Builth Wells (Breconshire) in March 1903, and Forrest (1919) mentions alleged records from Anglesey and Montgomeryshire. In Carmarthenshire Barker (1905) claimed that they were "occasionally taken by bird catchers" but such indication of regularity is improbable. Two records are claimed in Monmouthshire in 1976 but otherwise confirmed records are limited to only a handful, all in the 1980s, as follows: Breconshire (2), Glamorgan (2), Pembrokeshire (1), Flintshire (1), Merioneth (1), Caernarfonshire (1).

Arctic Redpoll *Carduelis hornemanni*

Llinos Bengoch yr Arctig

A vagrant.

There is only one record in Wales:

3–4 May 1987 – One, Bardsey

This species breeds in the circumpolar Arctic and makes a variable limited movement southwards in winter. Up to 1991 there had been 259 records in Great Britain, chiefly from the east coast.

Two-barred Crossbill *Loxia leucoptera*

Croesbig Wenaden

A vagrant.

There is only one record in Wales:

November 1912 – Male picked up dead, Llandrindod Wells (Radnorshire)

This species breeds from northern Fenno-Scandia east to Amurland, also northern North America and West Indies. It is chiefly resident but some eruptive and dispersive movement takes place when the conifer seed crop fails.

Crossbill *Loxia curvirostra*

Gylfin Groes

An increasing resident species although numbers fluctuate; occasional irruptive immigrant.

The Crossbill is now a well-established resident bird in Wales, having become one of the principal species to benefit from the maturing of the post-war conifer plantations. It is a spruce specialist with its strongholds in northern Europe, and as the sitka spruce, although an introduced tree from the west coast of north America, is the predominant species in these new British forests, Crossbills have been quick to capitalise on the developing cone crops that have accompanied the maturing of these forests. Crossbills have now been proved to breed in all Welsh counties and are reasonably widespread and common, bearing in mind the fact that numbers fluctuate widely from year to year, being much influenced by cone crop and the scale of sporadic "invasions". For example, there were high numbers in 1991 after a large irruption, but they then departed and the species was relatively scarce in 1992. It is a species which is almost entirely restricted to a conifer habitat and although, strictly speaking, is normally resident, populations move from area to area with the shifting abundance of cone cropping. Although predominantly associated with spruce, Crossbills will also occur in other conifer species, pines and larches, e.g. Gwydir forest, Betws-y-Coed (Caernarfonshire).

As shown in the *Breeding Atlas* it was virtually unknown twenty years ago as a breeding species and was a relatively scarce bird. It was the large invasion of 1910 from which the original breeding population in eastern England originated. In the latter half of the present century the principal irruptions have occurred in 1953, 1956, 1958, 1959, 1962, 1966, 1972, 1983, 1985, and 1990. Such irruptions, caused by failure of cone crops in the breeding areas or a particularly successful breeding season, or a combination of both, usually occur from mid-summer onwards and last through the subsequent winter. Some birds not infrequently remain to breed in the months of late winter and early spring, thus producing the years when nesting in Wales is most noticeable. The case of the 1985 influx as it affected Glamorgan exemplifies a typical sequence of occurrence. The first birds were seen on 1 July, with an influx of birds building to a peak in October, with visible migration clear on the coast at the end of that month, the largest single group being 69 birds. Many more records occurred in November, December and the early months of 1986 although breeding was not proved that year (Harrop 1985). However, even in years when invasions of continental birds have taken place, little is documented in relation to its occurrence in Wales. The *New Atlas* (1988–91) records its presence in 82 10-km squares, with breeding established in 45. A summary of published records, county by county, is given below.

Anglesey

Forrest lists the Crossbill as "accidental" with only one record around 1880. There is a skin in the National

Museum of a male collected at Menai Bridge in 1899 and a party of 35 was seen at Penmon in June 1935. After the 1962 irruption (see below) several parties of Crossbills were seen between July and December. Crossbills probably breed in the Newborough Forest (M. Gould, pers. comm.) and are not infrequently seen at several other Anglesey sites. The *New Atlas* confirmed breeding at two sites in the south of the island.

Caernarfonshire

Large numbers were recorded in 1830 (Forrest 1907) and 1865 and pairs are known to have bred in 1890 and 1891. Small numbers wintered in 1898 and bred the following year. One was recorded in August 1898, otherwise it is referred to by Forrest as spasmodic, numerous in some winters and rare at other times. Present in the Gwydir Forest from 1957 (probably bred 1963). A few parties were present after the 1962 irruption and it now breeds, probably annually, at least in the Betws-y-Coed and Beddgelert areas.

Merioneth

In 1897 it was seen every month near Corwen and believed to have bred. Seen variously at the same place in 1903 from April (flock) to October. Numerous 1909 (all Forrest). Thirty were recorded at Dinas Mawddwy in August 1910. Sizeable flock (30+) Corwen 1929 and parties of 30+ Corris 1958. Small parties at two localities during the July 1962 irruption. Nowadays usually breeds each year but numbers are very variable: it was abundant in 1991.

Denbighshire and Flintshire

First recorded 1830. Recorded also 1839. Small numbers 1894. Two were shot in 1900 and one seen in 1902, otherwise sporadic except for large numbers 1909–10. Bred 1928. Flocks up to 100 strong in 1930. Several during 1962 irruption. Bred again 1978 and present annually, with breeding almost certainly regular since then in Clocaenog Forest and probably elsewhere. Newly fledged young were watched being fed near Nercwys in Flintshire in 1988 – the only suggestion of breeding in the county to date.

Montgomeryshire

Recorded sporadically in a number of years from 1868 onwards, invariably from the east of the county (there being no observers making records elsewhere). Proved to nest at Llanyblodwel in 1890. Newly fledged young were watched being fed near Carno in 1967, which represented the next firm evidence of breeding, and Crossbills were subsequently found breeding in the 1970s in Dyfnant Forest, Kerry Forest, Breidden Hills and Hafren Forest. More recently it has occasionally been plentiful (e.g. 1983, 1991) but in other years it has been difficult to find.

Cardiganshire

Unusually large numbers recorded for 1868 and 1879 and small numbers sporadically in subsequent years later that century and the early decades of this century although breeding was not proved until 1977. Since then, there have been numerous scattered records from many parts of the county with breeding substantiated or assumed, particularly in the north and east of the county, in forest areas such as Myherin, Rheidol, Tywi, Cwm Symlog. Numbers were notably high in 1985.

Radnorshire

Salmon and Ingram (1955) suspected breeding had taken place in 1910 although doubt was expressed at the time (*The Field*, 1911, p. 444). With the wisdom of hindsight and later knowledge there seems no particular reason to doubt the record. Breeding is now regarded as regular in Radnor Forest and Abbeycwmhir.

Breconshire

Cambridge Phillips (1899) found Crossbills generally rare but recorded in good numbers in some years, notably 1866 when they were apparently "abundant all over the county" and in 1887–88. A. T. L. Wilson recorded breeding in 1931 near Garth. Since at least 1975 breeding has almost certainly been annual in Crychan Forest and they were found in 11 tetrads in 1988–90 widely separated although, as elsewhere, numbers clearly fluctuate a great deal from year to year depending principally on the scale of irruptions the previous summer. Breeding is presumed to be firmly established.

Carmarthenshire

Barker (1905) could mention only two records, one shot in 1900, and an alleged family party in 1904. Otherwise records were very scarce through succeeding decades until a sizeable influx after the irruption in 1958. Since 1965 small numbers have been recorded in most years although breeding was not proved until 1975 on the Carmarthenshire side of the Crychan Forest (see also Breconshire). In subsequent years further evidence of breeding has been forthcoming from Crychan and also Brechfa (since 1978). It is regular and well established in Pembrey Forest where the predominant tree species is Corsican Pine.

Pembrokeshire

Mathew (1894) recorded it as an irregular visitor, substantiating this comment with mention of only two known records. Between 1858 and 1990 Donovan and Rees (in press) could trace records in only 22 years, commenting that 17 of these years were subsequent to 1950 thus indicating the likelihood that this represents a genuine increase in occurrence. Breeding has been strongly suspected in the north of the county since 1991.

Glamorgan

Breeding was suggested (at Margam) as long ago as 1898 after a particularly heavy irruption, although Heathcote *et al.* (1967) express doubt as to its validity. Years of high numbers in Glamorgan were 1930–31 and 1963–64. Records were sporadic in the 1960s but almost annual, although in fluctuating numbers, in the 1970s and 1980s, particularly 1972 and 1985. Breeding was first proved in 1964 and after that not again until 1977. However, in view of the regularity of records from some other localities, e.g. Welsh St. Donat's and Llanwonno, it seems extremely likely that breeding might have taken place earlier and indeed more regularly than has been reported. Thomas *et al.* (1992) regard Crossbills as "well established" in many of the large conifer plantations in west Glamorgan.

Monmouthshire

Up to 1930 there were only some ten records for the county, which nonetheless included anecdotal reports of breeding in 1901 and 1909. The Crossbill has now been regarded as a resident breeding species in Monmouthshire since at least 1971 with annual occurrences since 1961. Its strongholds are in the Wye Valley and Wentwood Forest.

Common Rosefinch *Carpodacus erythrinus*

Llinos Goch

A scarce migrant.

It breeds from Germany and southern Sweden eastwards to Kamchatka and from Georgia eastwards to central China and is recorded with any regularity in Wales only from Bardsey. The first Welsh record was a male in full plumage shot near Glascwm (Radnorshire) in about 1875 and there were no further records until one on Skokholm on 26 June 1949.

There are 43 dated records involving 45 individuals, of which all but five are from Bardsey and the Pembrokeshire islands; the other five are from mainland Caernarfonshire (2), mainland Pembrokeshire (2) and Anglesey. There are 24 spring records in the period, 3 May to 30 June, including a second-year male singing in the Bardsey Observatory garden from 30 May to 6 June 1980 and 16 autumn records in the period 17 August to 14 October. The other three records are in late June, late July and December.

It is notable that the species is being recorded with increasing regularity from Bardsey, where it has been seen annually since 1979.

Bullfinch *Pyrrhula pyrrhula*

Coch y Berllan

A fairly numerous resident in the lowlands, patchier in its distribution in the far west and scarce in the uplands, other than in some areas of recent afforestation.

The Bullfinch is a numerous species in lowland areas of Wales, where it is found in a variety of woodland edge and scrub habitats. It is particularly associated with tangled overgrown hedgerows, thickets, garden shrubbery and woodland understorey. Although the relatively high population is evidence of abundant suitable habitat, it should be remembered that many of the Welsh woodlands, unlike their English counterparts, are virtually devoid of understorey because of intensive sheep grazing and are completely unsuitable for occupation by Bullfinches. Accordingly the species is not as evenly and uniformly distributed as some other common finch species. Furthermore it is a somewhat skulking bird and despite its distinctive plumage and the relative brightness of the male's colours it is not always evident even where it occurs regularly; it is a notoriously elusive species on CBC and similar surveys.

The Breeding Atlases demonstrate the species' avoidance of the upland areas in the mountain spine of Wales and its patchiness in Anglesey and the coastal areas of the western seaboard. In the uplands, however, a new opportunity has been presented by the modern generation of conifer forests and many of these have been colonised by Bullfinches. Once the trees are 2 m high and until they reach mature-thicket stage (15–20 years), they can provide the necessary combination of nest sites and food supply. In conifer plantations in north Wales, Bibby *et al.* (1985) estimated densities of Bullfinches at 0.5 *individuals* per 10 ha; in restocked plantations they estimated densities at 2.1 *individuals* per 10 ha (Bibby *et al.* 1989). It may be noted that in northern Europe Bullfinches are predominantly birds of the spruce forests, where they have been recorded at densities of up to 15 pairs per km^2 (Palmgren 1930) (i.e. 1.5 pairs per 10 ha) although his evidence is not wholly convincing.

The spring habit of feeding on fruit-tree buds when wild fruits and seeds are exhausted, has resulted in the species being added to the list of pest species in parts of England, but is of no commercial consequence in Wales so birds are rarely persecuted on this account.

A general expansion of Bullfinch numbers in southern Britain from the mid-1950s is well documented and has been attributed (Marchant *et al.* 1990) at least in part to an expansion in the range of habitats which the species was prepared to occupy. It moved into a wider range of secondary habitats, e.g. open farmland and town suburbs, possibly (Newton 1972) aided by the great reduction in Sparrowhawk numbers at the time. Although this general increase is echoed in some of the Welsh county avifaunas and bird reports, there is no strong evidence that numbers have increased substantially here in the past 30 years. In fact, since the mid-1970s numbers fell back (interestingly coincidental with a rise in Sparrowhawk populations!). In the longer term there has been no clear evidence of any major change in status over the past 150 years or so. Locally, numbers have been affected by agricultural land clearance and improvement in the lowlands, e.g. Pembrokeshire, and to some extent by the spread of industrialisation and urbanisation.

Bullfinches have expanded their range by colonising many of the conifer plantations in the new generation of post-war upland forests. They are able to take advantage of the seral stages of recolonising areas of scrub and restocked conifer plots. In south Wales they are similarly able to exploit recolonising areas on the extensive tracts of derelict land which are the legacy of coal mining and other heavy industries.

The *Breeding Atlas* suggests that breeding densities over 20 pairs per km^2 are rare anywhere in Britain (although up to 50 pairs per km^2 have occasionally been found) (Newton 1972). In Monmouthshire, where Bullfinches are found in 85% of tetrads, CBC data (but again on rather slender evidence) suggest an average of 7.5 pairs per km^2 on farmland. In Pembrokeshire Donovan and Rees (in press) suggest a breeding population of 4000–5000 pairs with occupancy of 72% of tetrads and in west Glamorgan they occur in 75% of tetrads (no numbers given).

The endemic British and Irish race of Bullfinch (*P. p. pileata*) is given subspecific status on the basis of being smaller, less bright and generally darker than *P. p. pyrrhula* from northern Europe. The British population is notably sedentary, a fact borne out by the recoveries of Welsh-ringed birds, only two of which (out of a total of 28) have been recovered more than 10 km away from the place of ringing.

The continental birds are more prone to winter movements, usually determined by lack of food supply and it is suggested (Newton 1972) that there is a cyclic pattern (three to four years) to such emigrations. Birds of the northern race have occasionally been reported in Wales, although firm confirmation of sight records is usually lacking. Such continental birds may also be responsible for the occasional visible passage which occurs in autumn and spring. A party of 16 birds was watched flying south-east over the Caernarfon Bay light vessel in October 1885; some 34 records have occurred on Bardsey (March–May and October-November) since 1954 but on Skokholm the only record is for two birds in July 1972 and on Skomer the species has been recorded on only six occasions, although it has bred on Caldey Is. for over 50 years. Resident Bullfinches tend to be more visible in winter when they forage farther from dense scrub and woodland. Parties are usually small; a group of 30 in Pembrokeshire in January 1988 was notable. One habit that is not uncommon is that of feeding well up in the hills in winter, sometimes associated with conifer plantations, most often on heather seeds, occasionally using the surface of snow as a platform for feeding on protruding shoots of rank heather. In this context Bullfinches are sometimes found as high as 487 m even in the mid-winter months.

Hawfinch *Coccothraustes coccothraustes*

Gylfinbraff

Resident in small numbers in lowland areas, principally in east Wales. Formerly scarcer. Some evidence of small-scale immigration in winter.

The Hawfinch is one of the least well known and possibly most under-recorded birds in Wales. It is an inhabitant of deciduous woodland but is one of the few species which does not favour oak; it feeds on the large fruits of trees such as wild cherry, hornbeam, beech, field maple, sycamore, wych elm and crab apple, and so is limited in its local distribution partly by the occurrence of these tree species. Unknown in much of Wales through the 19th century, its strongholds in the 20th century have been the lowlands of Denbighshire and Flintshire and the Severn Valley in Montgomeryshire, although one suspects that it may always have occurred in the Wye Valley of Monmouthshire where there is now known to be a strong population and an abundance of suitable woodland. RSPB surveys in 1986 found Hawfinches in 17 separate woods there. Similarly, in Breconshire it was known in the Wye and Usk Valleys in the early years of the 20th century.

As well as being found in mature woodland, the Hawfinch also thrives in other less heavily wooded areas, such as parkland, churchyards and cemeteries and the grounds of country houses. Although its numbers are not high and it is neither universally nor evenly distributed (many suitable areas are apparently untenanted), it can be very faithful to favoured sites over many years, e.g. Powis Castle Park, Welshpool (Montgomeryshire) and Gwyrch Castle, Abergele (Denbighshire). The Hawfinch is a secretive and elusive species that probably awaits discovery in many other places in Wales in addition to those in which it is currently known.

It is suggested (Alexander & Lack 1944) that the Hawfinch may have begun to breed in England only as recently as the early 19th century (although summer depredations on pea crops were certainly known in East Anglia 200 years before that, suggesting the likelihood of breeding). Up to about 1850 it was confined to south-east England and the Midlands, around which time a period of

Status of the Hawfinch in Wales

County	*First recorded*	*First proved breeding*	*Present status*
Anglesey	1906	1930	Dubious occurence in south-east
Caernarfonshire	c1900	by 1919	Very local, coastal strip and lower Conwy Valley
Denbighshire	long established		Widespread but local. Abergele birds robbed by egg collectors annually 1930s–1950s
Flintshire	long established		Breeds sparingly
Merioneth	1888	post–1888	Very local. Harlech, Bala (eggs collected regularly 1950s). Happy Valley, bred in 1980s
Montgomeryshire	long established		Relatively common Severn Valley upstream to Newtown
Cardiganshire	Two shot 1868	1972 (possibly 1938)	Very local: 15 20th century records only
Radnorshire	?late 19th century	pre-1910	Scarce and local resident, notably Wye Valley near Newbridge
Breconshire	?	1890	Rare resident. One breeding site known; others possible; more frequent in winter
Pembrokeshire	1854	None	Only 16 occasional sightings
Carmarthenshire	Pair 1897	1926	A scarce resident
Glamorgan	listed by Dilwyn	1899	Well established in at least five localities – Cardiff, Tongwynlais, Penarth, Draethen, Dyffryn Gardens
Monmouthshire	Not documented but possibly long established		Found in 22 tetrads in 1981–85. Suggested 50–100 pairs

rapid increase and expansion began, bringing the first records of birds to Wales – Llanstinan (Pembrokeshire) 1854, Glandyfi (Cardiganshire) 1868 – although Dilwyn (1848) lists it as occurring in the Swansea area much earlier than this. The westward and northward expansion continued until about 1920, around which time it slowed down without completely abating. By this time, however, the colonisation of parts of Wales had taken place, although the numbers known to be present were very small and in individual counties birds were usually only found in one or two localities. By the time of the *Breeding Atlas* (1968-72) 28 10-km squares were shown to have Hawfinches present and, although this was undoubtedly an under-estimate of the true picture, the species has continued to strengthen its toehold since then (or is better recorded). It now breeds regularly in several counties and sporadically in others with the exception of Pembrokeshire and Anglesey. There is local evidence that the Hawfinch is still continuing to strengthen its numbers through the 1980s and 1990s.

Little is known about the movements of British Hawfinches, although it is assumed that the resident population is essentially sedentary with only local movements taking place in winter. Continental birds are known to be a little more migratory, although the extent to which this may bring birds to Britain probably varies depending on weather and available food. The slight bias towards the winter months which occurs in the Welsh records over the years may well be attributable to such ex-continental movements. Passage is only exceptionally recorded, including very occasional birds on the offshore islands (Bardsey – ten records mainly October–November; Skokholm – three; Ramsey – one, a November bird suffering the unlikely fate of being caught in a rabbit trap). At Lavernock Point (Glamorgan) 22 birds were watched flying west on 5 October 1967 and a party of eight similarly on 13 October 1962.

Black-and-white Warbler *Mniotilta varia*

Telor Brith

A vagrant.

This passerine, which breeds in North America and winters from southern USA to Ecuador, has been seen once in Wales:

10 September 1980 – One, Skomer

Yellow Warbler *Dendroica petechia*

Telor Melyn

A vagrant.

This species breeds in North and Central America and winters south to Peru. The first record from Britain and Ireland was a first-year bird trapped on Bardsey on 29 August 1964. It unfortunately died the following day. Full details are given in *British Birds* vol. 58, pp. 457–461.

Blackburnian Warbler *Dendroica fusca*
Telor Blackburn

A vagrant.

The first British record of this Nearctic warbler was of one seen on Skomer on 5 October 1961. There has only been one subsequent record in Britain, on Fair Isle in October 1988. Although the Skomer record was originally referred to as probable only, it has since been re-assessed and there are no doubts as to its true identity.

Blackpoll Warbler *Dendroica striata*
Telor Tinwen

A vagrant.

It breeds in Canada and winters in South America south to Brazil. It merits a place in the Welsh list by virtue of two records:

22–23 October 1968 – One trapped, Bardsey (second British record)
7–9 October 1976 – One trapped, Bardsey

Summer Tanager *Piranga rubra*
Euryn yr Haf

A vagrant.

Another Bardsey special! This species, which breeds in USA and northern Mexico and winters in Central America south to Ecuador and Brazil owes its inclusion on the British list to a first-winter male present on Bardsey from 11 to 25 September 1957; it was trapped on the first date. No further records have been obtained.

Song Sparrow *Zonotrichia melodia*
Llwyd Persain

A vagrant.

There is one record in Wales.

5–8 May 1970 – One, Bardsey

Unusually for most trans-Atlantic vagrants, the other five British records are also in the period April–May, but American buntings and sparrows are the exception to the rule and many of them may have had a ship-assisted passage.

White-throated Sparrow *Zonotrichia albicollis*
Llwyd Gyddfwyn

A vagrant.

From a total of 17 British and Irish records of this North American species, there is one from Wales:

15 October–7 November 1967 – First-year bird, Bardsey (trapped on 21 October)

Most of the other records (12) were in spring.

Dark-eyed Junco *Junco hyemalis*
Jynco Llygeitu

A vagrant.

Fifteen British records of this trans-Atlantic vagrant include one from Wales:

25 April–3 May 1975 – One, Bardsey

It is notable that all but three other records are from the period April to May.

Lapland Bunting *Calcarius lapponicus*

Bras y Gogledd

An uncommon but regular passage migrant in small numbers, especially in autumn, and an occasional winter resident.

The first record for Wales was two immature birds on Skokholm from 5 to 12 September 1936, after which there were very few records until the mid-1950s; from 1960 onwards it has been noted in Wales every year, particularly from coastal localities. The apparent recent increase may be a reflection of improved observer experience rather than a real increase, because it is a shy, easily disturbed bird with few easily-recognisable identification features.

There are records from all Welsh counties except Radnorshire and Montgomeryshire and from every month of the year, but mainly in the period 10 September to 21 October when autumn passage has been closely monitored at favoured sites, especially Skokholm, Bardsey and, in recent years, the Great Orme and Strumble Head. It is not known for certain where these migrants originate but the evidence suggests that small, regular numbers are from the Scandinavian population supplemented, in some years, by larger influxes from Greenland. Spring passage is not nearly so marked, with only 13 records in the period from 2 April to 10 June.

Most sightings are of single birds but flocks of 10 or more have included:

17 September 1956	– 10 on Skokholm
31 January 1961	– 18 near Newborough Forest (Anglesey)
3–4 October 1961	– At least 10 on Bardsey (3rd) and up to 20 on the 4th
19 October 1962	– 10 on Skomer
26 November1962 to the end of the year	– Up to 30, Bardsey
9 December 1966 *et seq.*	– Up to 15, Llangeinwen (Anglesey)
5 October 1987	– 26, Strumble Head
24 September 1988	– 26, Strumble Head

Snow Bunting *Plectrophenax nivalis*

Bras yr Eira

Regular autumn passage migrant and winter visitor in small numbers.

Small numbers of Snow Buntings have been known to winter regularly on parts of the north Wales coast for many years; Forrest first referred to them as a well-established feature at the Point of Air in 1907. In addition, parties are not infrequently encountered on the summits of high hills in winter and they are by no means rare on other coasts, e.g. Cardigan Bay and the south coast of Wales. Autumn passage at coastal sites such as Holyhead Mountain, Great Orme, Strumble Head and especially on Bardsey is quite well marked although numbers are modest, with the peak of occurrences in October and early November.

At this season small parties pass through coastal areas and may even turn up far offshore. S. L. White, spending several days on Grassholm at migration time in October 1982, encountered several small groups of Snow Buntings, some of which stopped off and others overflew the island, all heading north; there is a record of birds on the Smalls lighthouse 35 km offshore, as long ago as October 1884. Spring passage is less discernible, with only a few records between March and June (mainly April) (only 12 on Bardsey 1954–85).

In winter the most regular coastal site is the Point of Air/Gronant area where a flock of up to 30 or so has been fairly regularly recorded; smaller parties are also fairly regular at other sites as far west as the Conwy estuary on the north coast. On Anglesey they are most regularly recorded at north-coast sites such as Port Lynas, Holyhead Mountain and Cemlyn Lagoon although other records show that they are likely to occur sporadically in any number of coastal areas, from Penmon in the east to Newborough Warren in the south-west.

On the shores of Cardigan Bay small numbers occur irregularly both on passage and during the winter months. Records for the Merioneth stretches are rather scarce but more plentiful for Cardiganshire, especially in the vicinity of the Dyfi estuary, where birds occur in most years. Heavy snowfalls inland, or severe frost, can increase the likelihood of occurrences in these areas as in 1892, 1893, 1895 (Ingram *et al.* 1966) and 1982 when numbers were especially high. Although the north coast is the principal area for winter Snow Buntings in Wales, there are numerous sporadic records from the south coasts of Glamorgan and Carmarthenshire between October and February (occasionally March) with occasional passage birds turning up as early as September. Numbers are usually very small, often only individual birds or twos and threes, and a flock of 50 at Whitford Point (Gower) on 1 January 1960 was exceptional.

As well as this coastal distribution, Snow Buntings clearly use the high hills during winter, albeit sparingly, when weather conditions permit. When encountered they are almost invariably on or near the very tops of the hills, whatever the height may be, ranging from 468 m on the Preseli Hills to 1066 m in Snowdonia. In view of the obvious paucity of ornithologists visiting these habitats in mid-winter, the habit may well be more widespread and frequent than published records indicate. The number of hills and mountains on which Snow Buntings have been recorded in mid-winter (as opposed to autumn passage dates) is a long one and the following examples are not by any means exhaustive: Carneddau, Glyders and Snowdon in Caernarfonshire (probably regular), Moel Hebog and Siabod in the same county; Arddu – a large flock in 1904–05 – Berwyn Mountains, Arans (where a party of c60 was seen on 3 January 1977 (R. Cartwright, pers. comm.)), Arenigs (probably regular), all in Merioneth; Kerry Hills and Pumlumon (Montgomeryshire/ Cardiganshire); Radnor Forest; Brecon Beacons; Preseli Hills (Pembrokeshire).

Yellowhammer *Emberiza citrinella*

Melyn yr Eithin

A well distributed and fairly common resident throughout Wales; regular up to about 365 m above sea level in hill areas.

This, our commonest bunting, is well distributed throughout Wales and is found in a variety of farmland and other habitats. However, nowhere in Wales is it as numerous as in some of the arable and mixed farming areas of England; O'Connor and Shrubb (1986) have shown that the mean density on farmland in Wales (7.5 pairs/km^2) is little more than half that in the contiguous areas of the west Midlands. Although it favours farmland which has an arable component, the Yellowhammer readily occupies other habitats which can offer open ground with trees, bushes or other elevated song posts and rough vegetation. Accordingly it is found in Wales on roadside verges, railway lines and similar linear habitats, young conifer plantations, ffridd, overgrown coal tips and cliffs, gorse commons and uncultivated lowland areas.

The Common Birds Census shows that Yellowhammer populations vary very little from year to year and no long-term changes in numbers or distribution are known to have occurred. However, were CBC data to be available for more parts of rural Wales there is little doubt that considerable declines would be demonstrated as it is generally accepted that Yellowhammer numbers have reduced substantially in inland areas in the recent two decades or so. In the lowland agricultural areas of Wales, below approximately 185 m, its numbers are lowest in those areas of intensive sheep rearing where grassland is the universal crop and grazing intensity prevents the retention of many rough or uncultivated patches or rank hedgebanks; conversely, numbers are probably highest on bracken/scrub-covered ffridd and on the coastal headlands of the western peninsulas of Pembrokeshire, Gower, Llŷn and Anglesey where the terrain is more broken, rough areas abound and there is somewhat greater agricultural diversity. In the last two areas RSPB surveys found them in 251 and 193 1-km^2, respectively, in 1986.

On the offshore islands, the Yellowhammer has only a tenuous hold with no regular breeding now occurring, although three to four pairs nested regularly on Bardsey from the early 1950s until 1970. On most of the Pembrokeshire islands Yellowhammers are no more than scarce visitors; only five occurrences in the last 19 years on Skokholm, while records on ten individual days in 1988 on Skomer, mainly in April, were exceptional. They breed however on Caldey Is.

In the various breeding surveys which were carried out in south Wales counties during the 1980s, the Yellowhammer was found to be still widely distributed. In Monmouthshire they were present in 78% of tetrads (estimated population possilbly 1000 pairs), in west Glamorgan 42.5% and Breconshire 33% (with an estimated population of 450 pairs). Nowhere in Wales are Yellowhammers as numerous as in Pembrokeshire where they are shown to occur in 81% of tetrads and a breeding population of 14 000–15 000 pairs is estimated (Donovan & Rees, in press).

Yellowhammers are sedentary birds and O'Connor and Shrubb (1986) have shown an interesting relationship between their numbers in stock-rearing areas and the availability of winter food where stock are fed in stockyards or at sites in the open fields. The provision of loose hay with the grass seed which invariably falls from it forms an important food source as can be seen from the mixed flocks of finches and buntings which frequent such feeding sites daily. Furthermore, the Yellowhammer, as a grain feeder, is able to profit particularly from the gleanings of cattle feed in stockyards now that much of the feed is made from whole grains rather than ground meal (O'Connor & Shrubb 1986).

Yellowhammers can be found breeding up to about 365 m but leave the higher altitudes in autumn to congregate in small parties and flocks and forage on the lower ground. Winter flocks are not usually large, 10–30 being frequent but groups as large as 100 are occasionally recorded.

Some evidence of the decline in numbers may be gleaned from the fact that on Skokholm, where they were formerly of annual occurrence, they were recorded in only five years since 1971. On Bardsey, too, numbers have been "even lower" than previously, since 1970.

Few Yellowhammers in Wales have been ringed and subsequently recovered. Those that have (4) indicate no movement with recoveries being from the places of ringing.

Cirl Bunting *Emberiza cirlus*

Bras Ffrainc

Formerly a widespread breeding bird; now extinct as a breeder, occurring only as a rare and accidental visitor.

There has been much in common between the vacillating fortunes of the Cirl Bunting and the Woodlark (q.v.) over the past 150 years or so. The Cirl Bunting, which has always been at the north-west edge of its range in southern Britain (it was first recorded breeding, in Devon, only in 1880), is a bird which is more characteristic of warmer Mediterranean regions. Its zenith in Britain coincided with the period of general climatic warming that continued until the 1930s and once that trend reversed the Cirl Bunting decreased and contracted its range fairly rapidly, being lost as a breeding bird county by county in Wales from about 1935 onwards. The British range has continued to decline since then and at present the Cirl Bunting is restricted to approximately 330 pairs (1992), almost wholly restricted to south Devon.

O. V. Aplin (1892), in his review of the Cirl Bunting in Britain, recognised its increasing range into Wales and listed breeding in Glamorgan, Breconshire and Cardiganshire. He suggested that it was little more than a casual visitor in Pembrokeshire, a view shared by Mathew (1894), and he knew of only one record in Merioneth; however, it is inevitable in a country where observers were very few and far between, that an easily confused species such as this was probably a lot more widespread than these few records indicate. C. G. Beale, an annual visitor to the Ceiriog Valley (Denbighshire) had found a well-established colony there since 1875 and it was well known in a dozen or more sites in the lowlands of Denbighshire and Flintshire before the end of the century. Barker (1905) did not include Cirl Bunting in his list of Carmarthenshire birds but subsequently added one record at Abergwili in 1910. Cummings (1908) further reviewed the species in north Wales, adding a handful of regular breeding sites along the north coast as far west as Llandudno and the Conwy Valley with outposts as far west as Llanbedrog on Llŷn, He knew of no records on Anglesey although one was apparently shot there in either 1878 or 1879.

E. A. Swainson, a sportsman and collector, made the first Cirl Bunting record for Cardiganshire. In 1891 he heard two near Aberystwyth and shot one; it was the first his taxidermist had heard of in his 28 years in the area. In the years that immediately followed, Cirl Buntings were increasingly recorded as breeding from numerous localities throughout the coastal plain of the county, principally by J. H. Salter who left detailed diaries between 1891 and the onset of the Second World War (with a long break between 1908 and 1922). By the end of this time Salter was already reporting it as a decreasing bird and between 1936 and 1938 only single birds were noted at four sites; last known breeding was at Aberystwyth during the Second World War. Occasional birds were recorded in the winters of 1949 and 1951 and then none until three were seen with other finches and buntings near Tregaron in January 1960.

In Pembrokeshire breeding was first recorded near Solva and (possibly) Haverfordwest in 1895 but Lockley *et al.* (1949) knew only the small colony at Solva, which had apparently ceased to exist long before 1948. Radnorshire sites existed at Builth Wells (1902) and Clyro (1928) and a pair was watched near Michaelchurch-on-Arrow as recently as June 1960. Cambridge Phillips first noted Cirl Buntings in Breconshire in 1888 – one of which was immediately procured by one of his sons with a catapult – and ten years later it was breeding regularly in five or six localities, mainly in the Usk Valley; the last record was in 1915 near Hay-on-Wye.

Through the first two decades of the present century the Cirl Bunting prospered modestly in Wales, although some writers were already remarking on its growing scarcity. It was still breeding sparingly in all coastal counties (except Anglesey and possibly Merioneth) although it was very scarce in Pembrokeshire (where Salter (*Zoologist* 1896 pp. 24–25) met Cirl Buntings at Newport and Fishguard and Howard Saunders saw one at Dinas Island in 1893), Radnorshire and possibly Montgomeryshire. Once again it should be remembered that many of these areas were seriously under-recorded or even virtually unknown ornithologically.

In south Wales the Cirl Bunting was apparently not particularly common, although widely but thinly distributed in the lowlands of Monmouthshire (up to 1925, with an isolated pair reported in 1968) and in the Vale of Glamorgan. In the latter county it was first recorded in 1876 and then proved breeding (at Porthkerry) in 1891. It bred in scattered colonies along the coast, including the south side of the Gower peninsula up to about 1936. After this only two records exist, that of a pair near Marcross in 1934 and four seen at Whiteford Point on 20 September 1968 and again near Llangennith the next day. In Carmarthenshire Ingram and Salmon (1954) could record only three occurrences up to 1942, after which it went unrecorded until two were seen at Ferryside in 1960. Only occasional birds have been seen since then all from the lowlands in the south of the county, but including a party of seven in December 1981 and then again in the early days of the following February. One or two birds were found among other buntings and finches in stubble fields near Milford Haven (Pembrokeshire) in winter in the late 1960s and early 1970s.

Rock Bunting *Emberiza cia*

Bras y Graig

A vagrant.

There are only five British records (of six birds) of this species, which breeds from Iberia and north-west Africa eastwards to China. The small number of records is presumably a reflection of the fact that it is mainly resident but it does undertake some altitudinal movement.

Two of the records are from Wales:

15 August 1958 – One, Dale Fort (Pembrokeshire)
1 June 1967 – Male, Bardsey

Ortolan Bunting *Emberiza hortulana*

Bras y Gerddi

A scarce and irregular passage migrant to Wales.

It breeds in continental Europe and Fenno-Scandia eastwards to northern Iran and Mongolia, wintering in south Arabia and the sahel zone of the south Sahara.

The first record was of one on the Berwyn Mountains near Llanrhaeadr ym Mochnant on 20 May 1907. Up to the end of 1991 there has been a total of 88 records involving 113 individuals. Most of these (88%) have been in the autumn but there has also been a clearly marked spring passage (see histogram).

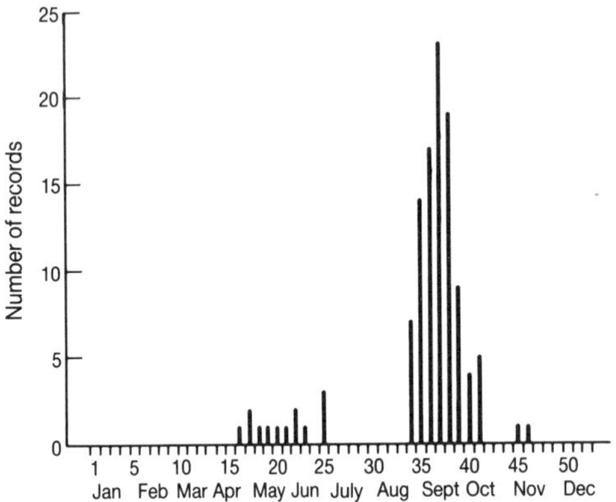

Weekly records of Ortolan Bunting to 1991

Most of the sightings are from the Pembrokeshire islands and Bardsey. Three or more have been seen on six occasions:

18–21 September 1941	– Three, Ramsey
11 September 1956	– Four, Skokholm (two on 9th; three on 12th)
29 September 1966	– Three trapped, Malltraeth (Anglesey)
22 September 1967	– Three, Skokholm
18 September 1968	– Five, near Brynsiencyn (Anglesey)
27 August–6 September 1970	– Up to three, Skokholm

Rustic Bunting *Emberiza rustica*

Bras Gwledig

A vagrant with four accepted records from Wales.

These are as follows:

8 June 1953	– Adult female trapped, Skokholm
6–7 June 1975	– Male trapped, Skokholm
29 March 1981	– Male, Bardsey
31 May 1987	– Male, Skomer

This species breeds from Fenno-Scandia eastwards to Kamchatka and winters from Turkestan to Manchuria and eastern China.

Little Bunting *Emberiza pusilla*

Bras Lleiaf

A rare visitor to Wales with 11 accepted records.

The records are chiefly from Bardsey:

8 January 1957	– One found injured, Llanddeusant (Anglesey)
28 September 1957	– One, Oxwich
11–16 October 1959	– One, Bardsey
25 September 1961	– One, The Skerries (Anglesey)
19 November 1967	– One, Skokholm
28–29 April 1984	– One in song, Bardsey
31 October–7 November 1984	– One, Bardsey
10 February–26 April 1986	– One, Bardsey
13 October 1986	– One, Bardsey
9 May 1987	One, Bardsey
1 October 1987	– One, Bardsey

The species breeds in north-east Europe and north Asia.

Yellow-breasted Bunting *Emberiza aureola*
Bras Bronfelen

A vagrant.

There is only one record from Wales

4–5 September 1973 – Female trapped, Bardsey

This species breeds from Finland eastwards to Kamchatka and Japan, wintering in India and south-east Asia.

Reed Bunting *Emberiza schoeniclus*
Bras y Cyrs

A fairly common resident breeding species, mainly in waterside habitats; evidence of a small autumn emigration.

The Reed Bunting is a familiar bird, inhabiting a variety of waterside habitats and other wet areas that have bushes or rank vegetation to provide nest sites and song posts. Up until about 1950 the status of the Reed Bunting had probably not changed radically in historical times. After 1950, however, the species showed a marked increase in numbers nationally and a corresponding expansion into sub-optimal breeding areas, often on drier sites, for example conifer plantations (Merioneth and Breconshire) and rough farmland hedgerows (Montgomeryshire). During the late 1950s the population was probably at an abnormally high level and a retraction occurred thereafter, due in part to losses caused by the severe winter of 1962–63 and later exacerbated by other winters, notably 1978–79 and 1981–82; as ground feeders, Reed Buntings are especially susceptible to hard weather, the moreso if snowfalls cover the food supply for protracted periods. The steepest decline in population occurred between 1975 and 1983 (Marchant *et al.* 1990) and it now appears to have stabilised at the lower level, probably also reflecting the continued intensification of agricultural land use in the 1980s, which entailed the further reduction of rough vegetation on marginal land, a feature of the boom in grassland improvement schemes in Wales.

Reed Bunting pairs are most numerous on extensive areas of wetland with rough vegetation. Oxwich NNR on Gower supported 66 pairs in 1977 declining to 25 in 1992 (Thomas *et al.* 1992); Kenfig NNR (c480 ha) usually has c30 pairs and 200 ha of bog in Breconshire supported 12 pairs in 1992 (M. Shrubb, in litt.). The distribution throughout the Welsh counties is strangely uneven, even allowing for the effect of hill areas. There are unexplained absences from the lower Wye Valley in Monmouthshire, south Cardiganshire and areas of Denbighshire and Flintshire; it is only thinly distributed in parts of Carmarthenshire although breeding occurs in every 10-km square except one. In Pembrokeshire only 21% of tetrads are occupied and in Monmouthshire (where a population estimate of c1000 pairs (Tyler *et al.* 1987) is almost certainly too high) there was a 27% tetrad occupation. Peers and Shrubb (1990) estimated a population of about 200 pairs in Breconshire, where there was 18% occupation of tetrads. RSPB surveys of 450 sites on Anglesey in 1986 found a minimum of 680 singing males/pairs and, simultaneously, work on Llŷn found a minimum of 384 pairs; both areas evidently offer much prime habitat for the species and as early as 1907 Forrest remarked on the abundance of Reed Buntings on Anglesey, particularly in the west. Clearly the population both there and in Llŷn is still healthy, albeit the relative interpretations of abundance since Forrest's days may have changed.

Although the Reed Bunting is classically associated with riversides, lake edges, canal banks and lowland marshes, it is an adaptable species which can exploit a fairly wide range of habitats. Reed Buntings will frequent willow scrub on wet dune slacks, overgrown rushy meadows and boggy hollows in the hills – often only pocket-handkerchief size – as high as 460 m (Merioneth, Caernarfonshire, Radnorshire). In winter there is a tendency for birds to group together in small parties and move farther away from the breeding sites although in some areas the breeding birds appear to remain sedentary and are even joined by others, e.g. Crychan Forest (Breconshire). Although parties are not usually large (and they will often mix with other buntings and finches) up to 100 have been recorded on occasion, e.g. Eglwys Nunydd (Glamorgan) in October 1966. When weather is particularly harsh, individuals may take advantage of feeding stations in suburban gardens.

There is rather scant evidence of post-breeding movement of Welsh birds but stronger evidence of a modest winter immigration of individuals from northern England. Ten individuals ringed in Wales in the winter months from November to February have later been recovered between 32 km and 328 km to the north or north-east. Five birds ringed in Lancashire showed south-westerly autumn movement when subsequently recovered in Wales, the farthest travelled was 244 km from Warrington to St. David's (Pembrokeshire). This autumn passage is poorly marked on the offshore islands. Reed Buntings are irregular autumn birds from year to year on Bardsey (where they bred from 1956 to 1977) and also on the Pembrokeshire islands (five or six pairs breed on Skomer and up to five pairs have bred sporadically on Skokholm), where a return spring movement is even less well marked.

Black-headed Bunting *Emberiza melanocephala*

Bras Penddu

A rare visitor.

There are 11 accepted records, predominantly in the second half of May:

27–29 May 1963	– One, Bardsey
16 May 1964	– One, Bardsey
11 May 1965	– One, Skokholm
3 June 1965	– One, Skokholm
27 May 1966	– One, Bardsey
18 May 1986	– One, Carmel Head (Anglesey)
26 May 1987	– One, Skokholm
about 30 May 1987	– One, near Aberystwyth (Cardiganshire)
31 May 1988	– One, Skomer
2 October 1988	– One, Bardsey
28–30 May 1990	– One, Marloes (Pembrokeshire)

All the records are of males. Though this is a common cagebird, the distribution of the above records both in time and location suggest that at least some of these are genuine vagrants rather than escapees.

Corn Bunting *Miliaria calandra*

Bras yr Ŷd

Formerly bred in all counties, plentifully in some coastal areas. Currently confined to a few pairs at one site in north-east Wales.

Few species have mirrored the spread of arable agriculture in Britain more closely than the Corn Bunting. With the decline of agriculture from the end of the 19th century, the fortunes of the Corn Bunting similarly declined, reaching a nadir between the two World Wars. Thus by 1940 it was a much scarcer bird in Wales and had withdrawn from many areas where it had been a familiar and plentiful species up to the 1920s. Although numbers nationally picked up again with the revival of agriculture after the Second World War, the resurgence was barely perceptible in Wales, where the Corn Bunting remains rare and irregular. Only in the extreme north-east, in the coastal strip of Flintshire around the Dee estuary, do a few residual pairs remain as outliers of the larger population in the lowlands of Cheshire and south Lancashire. The current absence of Corn Buntings elsewhere in Wales, nonetheless remains somewhat anomalous.

The Corn Bunting is a lowland species and its distribution in Wales was very strongly focused on the coastal strip as elsewhere in the west of Britain. Here, Forrest (1907) regarded it as abundant on western coasts, especially so in Anglesey, with at least 90% of the breeding population there being within 1.6 km of the coastline. In Pembrokeshire Mathew (1894) recorded it as a localised resident in coastal areas, notably the St. David's peninsula, and claimed that it was never to be found more than five miles (8 km) from the coast. By 1947, Lockley *et al.* saw it as a declining species, still most numerous in the St. David's area and it decreased rapidly in the 1950s to its final breeding in 1963. The pattern was similar in other coastal counties. In Glamorgan it bred on the coastal strip from Lavernock Point to Rhossili up to the early years of the present century but had gone by 1935; in Carmarthenshire it was local near the coast but had disappeared completely by 1950, since when there have been only a handful of records of individual birds in widely scattered coastal locations. It was "still numerous" (Salter) in north Cardiganshire in 1914 but had ceased to occur by 1934–35, with only a single record after that near Mwnt. At the turn of the century inland flocks were a feature of many farmyards on Llŷn during the winters and the demise of the species – almost totally unrecorded for Caernarfonshire – occurred between 1920 and 1950; there has been no definite record since 1956. On Bardsey it was "one of the most noticeable birds" in 1901 and probably continued as a breeding bird there until 1949. In Merioneth, birds remained only in the Tywyn area by 1929 and had disappeared even from there by 1935. In Monmouthshire only two breeding records were known (1901 and 1970) and in the inland counties of central Wales it was, at best, very uncommon. In Breconshire there have been only seven records this century, all since 1949.

Marchant *et al.* (1990) have shown that the decline of Corn Bunting numbers nationally accelerated sharply after 1981 and a close association has been suggested between the fortunes of the species and the distribution of lowland barley crops. This association has been identified by O'Connor and Shrubb (1986) even down to individual farm units. However, it is likely that other, more complex factors are also involved, for example easier harvesting (winter instead of spring sowing leading to increased nest loss), climatic change and the reduction of other arable crops including clover (Shrubb in litt.)

At present the Corn Bunting has no more than a toe-hold in Wales, represented by no more than 15 pairs in the area around the Dee estuary (Flintshire). Elsewhere it is only an occasional vagrant. A solitary individual, seen and heard by many hundreds of visitors *en route* for Skomer Island, remained at Martin's Haven (Pembrokeshire) for several years in the 1980s and produced a song claimed to be indistinguishable from that of a Yellowhammer.

In the late 1960s and early 1970s up to 30 Corn Buntings occurred in mixed winter flocks of buntings – including one or two Cirl Buntings (q.v.) – and finches in cliff-top stubble fields around Gellyswick (Pembrokeshire).

Rose-breasted Grosbeak *Pheucticus ludovicianus*
Gylfindew Brongoch

A vagrant.

Three of the 21 records of this trans-Atlantic vagrant are from Wales:

5 October 1967	– Female, Skokholm
14 October 1983	– First-year male trapped, Bardsey
29 September–11 October 1988	– First-winter female, Skomer

The dates coincide with the peak time for most Nearctic landbirds in Britain and Ireland.

Northern Oriole *Icterus galbula*
Euryn y Gogledd

A vagrant.

Three of the 17 British records of this trans-Atlantic vagrant are from Wales:

5-10 October 1967	– Female, Skokholm
6–7 May 1970	– Male, Hook (Pembrokeshire)
2 January–23 April 1989	– First-winter female, Roch (Pembrokeshire)

The last record was the first time that this species has been proved to have overwintered in Britain.

Appendices

APPENDIX 1: Scientific names of species mentioned in the text.

Trees

Alder	*Alnus glutinosa*
Arolla pine	*Pinus cembra*
Ash	*Fraxinus excelsior*
Beech	*Fagus sylvatica*
Birch species	*Betula* spp.
Blackthorn	*Prunus spinosa*
Corsican pine	*Pinus nigra*
Crab apple	*Malus sylvestris*
Douglas fir	*Pseudotsuga menziesii*
Elder	*Sambucus nigra*
European larch	*Larix decidua*
Field maple	*Acer campestre*
Hawthorn	*Crataegus monogyna* and *C. oxyacanthoides*
Hazel	*Corylus avellana*
Hornbeam	*Carpinus betulus*
'Mediterranean' Juniper	*Juniperus phoenicurus*
Norway spruce	*Picea abies*
Oak	*Quercus robur* and *Q. petrea*
Redwood	*Sequoia sequoia*
Rowan	*Sorbus aucuparia*
Scots pine	*Pinus sylvestris*
Sitka spruce	*Pices sitchensis*
Sycamore	*Acer pseudoplatanus*
Wild cherry	*Prunus avium*
Willow	*Salix* spp.
Wych elm	*Ulmus glabra*

Flowering plants

Bilberry	*Vaccinium myrtillus*
Bramble	*Rubus* spp.
Bracken	*Pteridium aquilinum*
Common reed	*Phragmites australis*
Cotton grass	*Eriophorum* spp.
Dandelion	*Taraxacum officinale*
Deergrass	*Scirpus cespitosus*
European gorse	*Ulex europaeus*
Heather	*Calluna vulgaris*
Heath rush	*Juncus squarrosus*
Western gorse	*Ulex gallii*
Knapweed	*Centaurea scabiosa*

Flowering plants – contd

Marram	*Ammophila arenaria*
Mat-grass	*Nardus stricta*
Red fescue	*Festuca rubra*
Sea beet	*Beta vulgaris*
Sea purslane	*Halimione portulacoides*
Sheep's-fescue	*Festuca ovina*
Sharp rush	*Juncus acutus*
Soft rush	*Juncus effusus*
Sorrels	*Rumex* spp.
Stinging nettle	*Urtica dioica*
Thrift	*Armeria maritima*
Tree-mallow	*Lavatera arborea*

Mammals

Brown rat	*Rattus norvegicus*
Common dolphin	*Delphinus delphis*
Field vole	*Microtus agrestis*
Fox	*Vulpes vulpes*
Grey squirrel	*Sciurus carolinensis*
Mink	*Mustela visor*
Pigmy shrew	*Sorex minutus*
Rabbit	*Oryctolagus cuniculus*
Skomer vole	*Clethrionomys glareolus skomerii*

Fish

Herring	*Clupea harengus*
Poor cod	*Gadus minutus*
Whiting	*Gadus merlangus*

Invertebrates

Cockle	*Cardium edule*
Earthworms	*Lumbricus* spp.
Mussel	*Mytilus edulis*

Birds

Sarus Crane	*Grus antigone*
Soft-plumaged Petrel	*Pterodroma mollis*
South Polar Skua	*Catharacta maccormicki*
Wandering Tattler	*Tringa incana*

APPENDIX 2: List of statutory and non-statutory sites of especial significance for birds in Wales.

1. Internationally designated sites.

Sites designated under the Ramsar Convention as internationally important wetlands (RAMSAR sites) and the EEC Wild Birds Directive (79/409) as Special Protection Areas (SPAs) (accurate to April 1993).

Aberdaron Coast and Bardsey	SPA
Burry Inlet	RAMSAR and SPA
Cors Caron (Tregaron Bog)	RAMSAR
Cors Fochno and Dyfi estuary	RAMSAR
Crymlyn Bog	RAMSAR
Dee estuary	RAMSAR and SPA
Grassholm	SPA
Holyhead Coast	SPA
Llyn Idwal	RAMSAR
Llyn Tegid (Bala Lake)	RAMSAR
Skokholm and Skomer	SPA
Traeth Lafan	SPA
Ynys Feurig, Cemlyn Bay, Skerries	SPA

In addition to the above sites the Severn estuary has awaited ratification by UK government since 1990 under both SPA and RAMSAR designations. Ten more sites are proposed as RAMSAR sites and 15 as SPA.

Sites identified by Bird Life International as Important Bird Areas (IBAs) in Wales.

Dee estuary; Clwyd/Cheshire/Merseyside
Berwyn; Clwyd/Powys/Gwynedd
Traeth Lafan; Gwynedd
Anglesey Islands and Cemlyn Bay; Gwynedd
Holy Island Coast; Gwynedd
Bardsey Island and Aberdaron Coast; Gwynedd
Cors Fochno and Dyfi; Dyfed/Gwynedd/Powys
Elenydd – Mallaen; Powys
Pembrokeshire Cliffs; Dyfed
Skokholm and Skomer; Dyfed
Grassholm; Dyfed
Carmarthen Bay; Dyfed
Burry Inlet; West Glamorgan/Dyfed
Swansea Bay – Blackpill; West Glamorgan
Severn estuary; Gloucestershire/Avon/Somerset/South Glamorgan/Gwent
Great and Little Orme; Gwynedd
Puffin Island; Gwynedd
St. Margaret's Island; Dyfed

2. National Nature Reserves.

This is a complete list of NNRs in Wales, many of which do not have a specific ornithological importance; those with particular bird interest are marked *.

Allt Rhyd y Groes
Cadair Idris*
Ceunant Llennyrch
Coed Camlyn
Coed Cymerau
Coed Dolgarrog
Coed Ganllwyd
Coed Gorswen
Coed Rheidol*
Coed Tremadog
Coed y Rhygen
Coed-y-Cerrig
Coedmor
Coedydd Aber*
Coedydd Maentwrog*
Cors Bodeilio
Cors Caron*
Cors Erddreiniog
Cors Geirch
Cors y Llyn
Corsydd Llangloffan
Craig Cerrig Gleisiad
Craig y Cilau
Crymlyn Bog*
Cwm Clydach
Cwm Glas Crafnant
Cwm Idwal
Dyfi*
Gower Coast*
Hafod Garregog
Kenfig Pool and Dunes*
Lady Park Wood
Morfa Dyffryn
Morfa Harlech
Nant Irfon*
Newborough Warren*
Ogof Ffynnon Ddu
Oxwich*
Penhow Woodlands
Rhinog*
Rhos Goch
Rhos Llawr Cwrt
Skomer*
Stackpole*
Stanner Rocks
Ty Canol
Whiteford*
Y Wyddfa*
Ynys Enlli*

3. Local Nature Reserves (declared to 31 March 1992).

The following nature reserves have been declared by local authorities under Section 21 of the National Parks and Access to the Countryside Act 1949.
BC – Borough Council, CC – County Council, NPA – National Park Authority

Reserve	*Local Authority Responsible*
Cliff Wood	Vale of Glamorgan BC
Cleddon Bog	Gwent CC
Taf Fechan	Merthyr Tydfil BC
Talybont Reservoir	Powys CC
Carreg Cennen Woodlands	Dyfed CC
Bishops Wood	Swansea City Council
Coed Dinorwig	Gwynedd CC
Coed y Cerrig	Brecon Beacons NPA
Flatholm	South Glamorgan CC
Kenfig Pool and Dunes	Mid Glamorgan CC
Traeth Lafan	Gwynedd CC
Cwmllwyd Wood	West Glamorgan CC
Glamorgan Canal	Cardiff City Council

Great Ormes Head	Aberconwy BC
Pant-y-Sais	Neath BC
Hermit Wood	South Glamorgan CC
Swansea Canal	West Glamorgan CC
Mumbles Hill	Swansea City Council
Howardian	Cardiff City Council

4. Non-statutory sites.

4.1 RSPB reserves.

Skerries, Anglesey: seabirds.
South Stack cliffs, Anglesey: seabirds and other cliff species.
Valley lakes, Anglesey: waterfowl.
Ynys Feurig, Anglesey: seabirds.
Dee estuary, Point of Air and Oakenholt Marsh, Flintshire: estuary birds.
Lake Vyrnwy, Berwyn Mountains, Montgomeryshire: upland and woodland birds.
Mawddach Valley, Merioneth: woodland communities.
Ynys-hir, Cardiganshire: woodland communities and estuary birds.
Elan woodlands and Corngafallt, Breconshire/Radnorshire: woodland communities.
Gwenffrwd and Dinas, Carmarthenshire: woodland communities.
Cwm Clydach, Glamorgan: woodland communities.
Ramsey Island, Pembrokeshire: seabirds and other cliff species.
Grassholm Island, Pembrokeshire: seabirds.

4.2 Other major bird reserves or sites with similar designations.

Cemlyn Lagoon, Anglesey (National Trust): seabirds.
Llyn Alaw, Anglesey (Welsh Water): waterfowl.
Connah's Quay Lagoons, Flintshire (Deeside Naturalists Society).
Glaslyn moorlands, Pumlumon (Montgomeryshire Wildlife Trust): upland birds.
Dolydd Hafren (Montgomeryshire Wildlife Trust): wetland birds.
Abergwesyn Common (National Trust): upland birds.
Elan Estate (Elan Estate Trust): upland birds.
Cardigan Island (Dyfed Wildlife Trust): seabirds.
Mynydd Du (BBNP)
Skokholm (Dyfed Wildlife Trust): seabirds.

APPENDIX 3: List of acronyms and initials used in the text.

BBNP – Brecon Beacons National Park
BoEE – Birds of Estuaries Enquiry
BTO – British Trust for Ornithology
CBC – Common Bird Census
CCW – Countryside Council for Wales
IBA – Important Bird Areas
NNR – National Nature Reserve
RSPB – Royal Society for the Protection of Birds
SCAN – Shropshire, Conwy and Anglesey (Ringing Group)
SPA – Special Protection Area (under the EEC Birds' Directive)
SSSI – Site of Special Scientific Interest
WWT – Wildfowl and Wetlands Trust

APPENDIX 4: Escapes and species which have not been admitted to the British and Irish list.

The list of birds recorded in Britain and Ireland is maintained by the British Ornithologists' Union through its Records Committee (British Ornithologists' Union 1992). Each species is categorised depending on the criteria for its admission to the British and Irish list and the category A, B or C is given to each species, defined as follows:

Category A. Species which have been recorded in an apparently wild state in Britain or Ireland at least once since 1 January 1958.

Category B. Species which were recorded in an apparently wild state in Britain or Ireland at least once up to 31 December 1957, but have not been recorded subsequently.

Category C. Species which although originally introduced by man have now established a regular feral breeding stock which apparently maintains itself without necessary recourse to further introduction.

A further category D which is defined as species which would otherwise appear in categories A or B except that:

(i) there is reasonable doubt that they have ever occurred in a wild state, or
(ii) they have certainly arrived with a combination of ship and human assistance, including provision of food and shelter, or
(iii) they have only ever been found dead on the tideline; also
(iv) species that would otherwise appear in Category C except that their feral populations may or may not be self supporting.

For Category D species it is therefore not appropriate to incorporate them in the systematic list although there is, of course, the possibility that the species concerned may be admitted to the British and Irish list at some future time. The Category D species recorded in Wales are as follows:

White Pelican	*Pelecanus onocrotalus*
Greater Flamingo	*Phoenicopterus ruber*
Wood Duck	*Aix sponsa*
Bald Eagle	*Haliaeetus leucocephalus*
Blue Rock Thrush	*Monticola solitarius*
Chestnut Bunting	*Emberiza rutila*
Red-headed Bunting	*Emberiza bruniceps*
Painted Bunting	*Passerina ciris*

Welsh records of the above include an adult Bald Eagle at Llyn Coron (Anglesey) on 17 October 1978, a Blue Rock Thrush at Moel y Gest (Caernarfonshire) on 4 June 1988 and a Chestnut Bunting at Bardsey on 18–19 June 1986.

In addition there are a number of vagrants that have been found in Wales but are also excluded from the systematic list for the reasons given. These are:

Cape Pigeon *Daption capense*

One, formerly in the collection at Gogerddan, was shot in the Dyfi estuary in 1879. This species, abundant in the Antarctic and sub-Antarctic, is not admitted to the British list beause individual birds are known to have been brought to European waters in ships and released.

Collared Petrel *Pterodroma leucoptera*

One was shot by a fisherman on the Cardiganshire coast near Aberystwyth in November or December 1889. This is the only British record but is not now accepted in view of the improbability of this Pacific Ocean species occurring naturally.

Snow Goose *Anser caerulescens*

There is no evidence to suggest that Welsh records of this species are of anything but captive origin, many are kept in waterfowl collections and escapes are quite frequent. Snow Geese breed in north-east Siberia eastward across arctic North America to north-west Greenland.

Egyptian Goose *Alopochen aegyptiacus*

In England this species, introduced in the 18th century, has long bred in a feral state in several areas. There are occasional records from different parts of Wales (e.g. Carmarthenshire, Glamorgan, Monmouthshire and Pembrokeshire) but it does not appear to have colonised Wales to any extent.

Black Vulture *Aegypius monachus*

This species breeds in small numbers as close to Wales as southern Europe. An immature bird first seen in the Llyn Heilyn area of Radnorshire in January 1978 stayed until early February when it moved to the upper Tywi Valley (Carmarthenshire). The species has not been admitted to the British list even as Category D.

Purple Gallinule *Porphyrula* sp.

In Carmarthenshire Ingram and Salmon (1954) record in a footnote that a Purple Gallinule was shot at Llandeilo Abercywyn and that it had been seen for several days before it was killed. It was seen in Jeffrey's taxidermy shop in September 1893. There is some doubt as to the origin of this bird and the possibility that it escaped from a collection cannot be overlooked although it could be an American Purple Gallimule which has been admitted to the British list on the strength of a record from the Isles of Scilly in November 1958.

Black Woodpecker *Dryocopus martius*

The case for the inclusion of the Black Woodpecker in the British list is argued in Snow (ed.) (1992), where no less than 82 records are referred to. In spite of this, the Black Woodpecker, which breeds as close as the Netherlands, has not been admitted to the British list. There is no doubt that some of the early records were fraudulent but one of the most reliable sight records is from Breconshire on 19 April 1903 near Builth Wells, seen by an eminent field ornithologist, J. Walpole Bond; he, himself, considered it had probably escaped from captivity. There is, however, a suggestive aggregation of records in the Welsh border area of Herefordshire, Forest of Dean and Breconshire in the years from about 1874 to 1903 where there were nine or ten occurrences.

Red-winged Blackbird *Agelaius phoeniceus*

One was recorded from the Nash Lighthouse (Glamorgan) on 27 October 1886. Alexander and Fitter (1955), who reviewed the known occurrences of American land-birds in Europe, suggest that this one was rendered suspect by association with known or possible releases in Britain. This species is a favourite cage-bird.

Rusty Blackbird *Euphagus carolinus*

One was shot on 4 October 1881 near Cardiff (Glamorgan). This is another American species and Alexander and Fitter (1955) suggest that it was more likely a genuine drift-migrant than an escaped cage-bird. However, the species has never been accepted on the British list and doubts remain as to how it arrived at Cardiff.

References

Aikin, A. 1797. *Journal of a tour in north Wales.*

Alexander, W. B. 1945–47. The Woodcock in the British Isles. *Ibis* 87:512–550; 88:1–24, 159–179, 271–286, 427–444; 89:1–28.

Alexander, W. B. & Fitter R. S. R. 1955. American Land Birds in Western Europe. *British Birds* 48:1–14.

Alexander, W. B. & Lack, D. 1944. Changes in status among British breeding birds. *British Birds* 38:42–45, 62–69, 82–88.

Allen, R. H. & Rutter, G. E. 1956. The moult migration of the Shelduck in Cheshire in 1955. *British Birds* 49:221–224.

Allport, G., O'Brien, M. & Cadbury, C. J. 1986. Survey of Redshank and other Breeding Birds on Salt Marshes in Britain, 1985. Unpublished RSPB Report.

Alsop, D. 1975. The Ring Ouzel in Derbyshire. *Derbyshire Bird Report* 1974, 10–14.

Andrews, J. H. & Bellamy, P. 1988. The Birds of Mynydd Du. Unpublished RSPB Report.

Anon. 1989. Goshawk breeding habitat in lowland Britain. *British Birds* 82:56–67.

Anon. 1990. Breeding biology of Goshawk in lowland Britain. *British Birds* 83:527–540.

Aplin, O. V. 1892. On the distribution of Cirl Bunting in Great Britain. *Zoologist* (3)16:121–181.

Aplin, O. V. 1990. The birds of Lleyn, west Caernarvonshire. *Zoologist* (4)4:489–505.

Aplin, O. V. 1902. The birds of Bardsey Island, with additional notes on the birds of Lleyn. *Zoologist* (4)6:8–17, 107–110.

Aplin, O. V. 1903. Additional notes on the birds of Lleyn. *Zoologist* (4)2:201–213.

Aplin, O. V. 1910. Summers in Lleyn with some notes on birds. *Zoologist* (4)14:41–50 & 99–108.

Aspden, W. 1933. Newborough Warren; some notes on its wildlife. *Anglesey Antiquarian Society Transactions* 118–122.

Aspden, W. 1933. Notes on Puffin I. *Anglesey Antiquarian Society Transactions* 63–106.

Aspden, W. 1939. Hen Harriers breeding in Anglesey. *British Birds* 32:326–328.

Atkinson, N. K., Davies, M. & Prater, A. J. 1978. The winter distribution of Purple Sandpipers in Britain. *Bird Study* 25:223–228.

Atkinson-Willes, G. L. 1956. *National Wildfowl Counts 1954–55.* The Wildfowl Trust, Slimbridge.

Atkinson-Willes, G. L. 1963. *Wildfowl in Great Britain.* London.

Avery, M. & Leslie, R. 1990. *Birds and Forestry.* Poyser.

Axell, H. E. 1966. Eruptions of Bearded Tits 1959–65. *British Birds* 59:513–543.

Baggott, G. K. 1986. The fat contents and flight ranges of four warbler species on migration in north Wales. *Ringing and Migration* 7:25–36.

Bain, C. 1987. Breeding Wader Habitats in an Upland Area of North Wales (Hiraethog). Unpublished RSPB Report.

Bainbridge, I. P. & Minton, C. D. T. 1978. The migration and mortality of the Curlew in Britain and Ireland. *Bird Study* 25:39–50.

Baker Gabb, W. Diaries in the National Museum of Wales.

Bannerman, D. A. 1955. *The Birds of the British Isles, IV.* Edinburgh & London.

Bark Jones, R. 1954. Nesting birds of the north Wales coast. *Merseyside Naturalists' Association Report.*

Barker, T. W. 1905. *Handbook to the natural history of Carmarthenshire.* Carmarthen.

Barnes, J. A. G. 1961. The Winter Status of the Lesser Black-backed Gull 1959–60. *Bird Study* 8:127–147.

Barrett, J. H. & Davis, T. A. W. 1958. The winter status of the Greenshank in west Wales. *British Birds* 51:525–526.

Barrett, M. 1992, Moussier's Redstart: new to Britain and Ireland. *British Birds* 85:108–111.

Batten, L. A. 1973. Population dynamics of suburban Blackbirds. *Bird Study* 20:251–258.

Baxter, E. V. & Rintoul, L. J. 1953. *Birds of Scotland.* Oliver & Boyd.

Bell, B. D., Catchpole, C. K., Corbett, K. J. 1968. Problems of censusing Reed Buntings, Sedge Warblers and Reed Warblers. *Bird Study* 15:16–21.

Bennett, T. 1982. *Welsh Shipwrecks* (Vol. 2).

Betts, M. 1992. *Birds of Skokholm.* Cardiff.

Bibby, C. J. 1981. Wintering Bitterns in Britain. *British Birds* 74:1–10.

Bibby, C. J. 1987. Food of Breeding Merlins in Wales. *Bird Study* 34:64–70.

Bibby, C. J. & Lunn, J. 1982. Conservation of reed beds and their avifauna in England and Wales. *Biological Conservation* 23:167–186.

Bibby, C. J. & Thomas, D. K. 1985. Breeding and diets of Reed Warblers at a rich and a poor site. *Bird Study* 32:19–31.

Bibby, C. J., Aston, N., Bellamy, P., 1989. The effect of broadleafed trees on birds of upland conifer plantations in North Wales. *Biological Conservation* 49:17–29.

Bibby, C. J., Phillips, B. N., Seddon, A. J. E. 1985. The Birds of Restocked Conifer Plantations in Wales. *Journal of Applied Ecology* 22:215–230.

Bingley, W. 1814. Bird Notes from N. Wales.

Birch, J. E., Birch, R. R., Birtwell, J. M., Done, C., Stokes, E. J. & Walton, G. F. 1968. *The Birds of Flintshire.*

Birkhead, T. R. 1974. Movement and mortality rates of British Guillemots. *Bird Study* 21:241–254.

Blackett, A. & Ord, W. 1962. Lesser Whitethroat breeding in Northern Ireland. *British Birds* 55:445.

Blaker, G. B. 1933. The Barn Owl in England – Results of the census. *Bird Notes & News* 15:169–172, 207–211.

Blurton Jones, N. G. 1956. Census of breeding Canada Geese 1953. *Bird Study* 3:153–170.

Bolam, G. 1913. *Wildlife in Wales.* London. Frank Palmer.

Bowen, D. A. 1981. *Geology – National Atlas of Wales.* University of Wales Press.

Bowes, A., Lack, P. C. & Fletcher, M. R. 1984. Wintering gulls in Britain, January 1983. *Bird Study* 31:161–170.

Bowman, N. 1980. The food of the Barn Owl in mid-Wales. *Nature in Wales* 17:84–88, 106–108.

Boyd, A. W. Diaries held at Merseyside County Council museums.

Boyd, H. & Ogilvie, M. A. 1961. The distribution of Mallard ringed in southern England. *Wildfowl Trust Annual Report* 12:125–136.

Branson, N. J. B. A., Ponting, E. D. & Minton, C. D. T. 1978.

Turnstone migrations in Britain and Europe. *Bird Study* 25:181–187.

Bray, R. P. 1968. Mortality rates of released pheasants. *Game Conservancy Annual Review* for 1967:14-33.

British Ornithologists' Union 1992. *Checklist of Birds of Britain and Ireland.* 6th edition. Tring, Herts. British Ornithologists' Union.

Brooke, M. 1990. *The Manx Shearwater.* Poyser.

Brooke, M. de L. & Davies, N. B. 1987. Recent changes in host usage by Cuckoos (*Cuculus canorus*) in Britain. *Journal of Animal Ecology* 56:873–883.

Brown, D. J. 1981. Seasonal variations in the prey of some Barn Owls in Gwynedd. *Bird Study* 28: 139–146.

Brown, R. G. B. 1955. The migration of the Coot in relation to Britain. *Bird Study* 2:135–142.

Bullock, I. D., Drewett, D. R. & Mickleburgh, S. P. 1983. The Chough in Wales. *Nature in Wales* 4:46–57.

Burton, J. F. 1956. Report on the national census of heronries, 1954. *Bird Study* 3:42–73.

Burton, J. F. 1957. Census of heronries, 1954: additions and corrections. *Bird Study* 4:50–52.

Buxton, A. & Lockley, R. M. 1950. *The Island of Skomer.* Staples London.

Buxton, E. J. M. 1961. The inland breeding of the Oystercatcher in Great Britain 1958–59. *Bird Study* 8:194–209.

Bye-gones. Relating to Wales & the border counties. 1871–1939. National Library of Wales.

Cadman, W. A. 1949. Distribution of Black Grouse in north Wales forests. *British Birds* 42:365–367.

Cadman, W. A. 1952. Red-breasted Goose on Montgomeryshire–Shropshire border. *British Birds* 45:293.

Cadman, W. A. 1955. Lesser White-fronted Goose in Cardiganshire. *British Birds* 48:325.

Cambridge, P. & Parr, C. 1992. Influx of Little Egrets in Britain and Ireland in 1989. *British Birds* 85:16–21.

Campbell, B. 1954–55. The breeding distribution and habitats of Pied Flycatcher in Britain. *Bird Study* 1:81–101; 2:24–32, 179–191.

Campbell, B. 1960. The Mute Swan census in England and Wales, 1955–56. *Bird Study* 7:208–223.

Campbell, J. W. 1957. The rarer birds of prey. Their present status in the British Isles. Hen Harrier. *British Birds* 50:143–146.

Caradoc & Severn Valley Field Club. Record of bare facts 1892–1956.

Cardiff Naturalist Society. 1900. *The Birds of Glamorgan.*

Cayford, J. T., Tyler, G. & Mackintosh-Williams, L. 1989. The ecology and management of Black Grouse in conifer forests in Wales. Unpublished RSPB Report.

Chance, E. P. Egg collection and catalogue held at the British Museum.

Chandler, R. J. 1981. Influxes into Britain and Ireland of Red-necked grebes and other waterfowl during winter 1978–79. *British Birds* 74:55–81.

Chater, A. O. M. & Chater, E. H. 1974. Rookeries in Cardiganshire. *Nature in Wales* 14(2):69–75.

Clarke, R. & Watson, D. 1990. The Hen Harrier *Circus cyaneus* winter roost survey in Britain and Ireland. *Bird Study* 37:84–100.

Colling, A. W. & Brown, E. B. 1946. The breeding of Marsh and Montagu's Harriers in North Wales in 1945. *British Birds* 39:233–243.

Conder, P. J. 1979. Britain's first Olive-backed Pipit. *British Birds* 72:2–4.

Condry, W. M. 1951. Woodlark in West Wales. 13th annual report of West Wales Field Society.

Condry, W. M. 1955. Breeding birds of the Welsh mountains. *Nature in Wales* 1:25–27.

Condry, W. M. 1966. *The Snowdonia National Park.* London.

Condry, W. M. 1970. Marsh & Willow Tits in Wales. *Nature in Wales* 12(1):1–6.

Congreve, W. Egg collection. Western Foundation of Vertebrate Zoology.

Coombs, C. J. F. 1978. *The Crows.* Batsford.

Corkhill, P. 1973. Manx Shearwater numbers on Skomer: population and mortality due to gull predation. *British Birds* 66:136–143.

Cornwallis, R. K. 1961. Four Waxwing invasions 1956–60. *British Birds* 54:1–30.

Cornwallis, R. K. & Townsend, A. D. 1968. Waxwings in Britain and Europe 1965–66. *British Birds* 61:97–118.

Coulson, J. C. 1963. Status of the Kittiwake in British Isles. *Bird Study* 10:147–179.

Coulson, J. C. 1983. Changing status of the Kittiwake in British Isles 1969–79. *Bird Study* 30:9–16.

Coward, T. A. Diaries held at the Edward Grey Institute.

Coward, T. A. 1893. Lesser Whitethroat breeding in Caernarfonshire. *Zoologist* (3)17:395.

Coward, T. A. 1895. Manx Shearwaters breeding on the coast of Caernarfonshire. *Zoologist* (3)19:72.

Coward, T. A. 1905. Supplementary notes on the birds of Anglesea. *Zoologist* (4)9:377–386, 423–426.

Coward, T. A. 1910 *Vertebrate fauna of Cheshire & Liverpool Bay.* 2 vol.

Coward, T. A. 1922. *Bird haunts and Nature Memories.* London.

Coward, T. A. & Oldham, C. 1902. Notes on the birds of Anglesea. *Zoologist* (4)6:401–415.

Coward, T. A. & Oldham, C. 1904. Notes on the birds of Anglesea. *Zoologist* (4)8:7–29.

Coward, T. A. & Oldham, C. 1905. Notes on the birds of Anglesea. *Zoologist* (4)9:213–230.

Cramp, S. 1963. Toxic chemicals and birds of prey. *British Birds* 56:124–138.

Cramp, S. & Olney, P. J. S. 1967. The sixth report of the Joint Committee of the British Trust for Ornithology and the Royal Society for the Protection of Birds on Toxic Chemicals.

Cramp, S., Bourne, W. R. P. & Saunders, D. R. 1974. *The Seabirds of Britain and Ireland.* Collins.

Cramp S. *et al.* 1983. *Handbook of the Birds of Europe, the Middle East and North Africa.* Oxford.

Crooks, J. & McLaughlin, K. 1989. Hen Harriers in Wales, 1989. Unpublished RSPB Report.

Cummings, S. G. & Oldham, C. 1906. Anglesea bird notes. *Zoologist* (4)10:94–104.

Cummings, S. G. 1908. Notes on the habits and distribution of the Cirl Bunting in North Wales. *British Birds* 1:275–279.

Currie, F. C. & Bamford, R. 1981. Bird populations of sample pre-thicket forest plantations. *Quarterly Journal of Forestry* 75:75–82.

Currie, F. C. & Bamford, R. 1982. The value to wildlife of retaining small conifer stands beyond normal felling age within forests. *Quarterly Journal of Forestry* 76:153–160.

Cusa, N. W. 1951. Ivory Gull in Pembrokeshire. *British Birds* 41:354–355.

da Prato, S. R. D. & da Prato, E. S. 1983. Movements of Whitethroat ringed in the British Isles. *Ringing and Migration* 4:193–209.

Dare, P. J. 1986a. Notes on the Kestrel population of Snowdonia, north Wales. *Naturalist* 111:49–54.

Dare, P. J. 1986b. Raven *Corvus corax* populations in two upland regions of north Wales. *Bird Study* 33:179–189.

Dare, P. J. & Schofield, P. 1976. An ecological survey of the Lavan Sands; Ornithological Survey, 1962–74. Cambrian Ornithological Society.

Davies, A. 1985. The British Mandarins – outstripping the ancestors. *BTO News* 136:12.

Davies, J. L. 1949. The Starling in Cardiganshire. *British Birds* 42:369–375.

Davies, M. 1978. Upland breeding bird surveys of the Brecon Beacons National Park. Unpublished RSPB Report.

Davies, M. 1988. The importance of Britain's Twites. *RSPB Conservation Review* 2:91–94.

Davies, W. 1858. *Llandeilo Fawr and its neighbourhood: past and present.*

Davis, P. E. 1993. The Red Kite in Wales: setting the record straight. *British Birds* 86:295–298.

Davis, P. E. & Davis, J. E. 1984. The breeding biology of a raven population in central Wales. *Nature in Wales (New Series)* 3:44–54.

Davis, P. E. & Davis, J. E. 1992. Dispersal and age of first breeding of Buzzards in central Wales. *British Birds* 85:578–587.

Davis, T. A. W. 1958. Breeding distribution of Great Black-backed Gulls in England and Wales in 1956. *Bird Study* 5:191–215.

Davis, T. A. W. & Saunders, D. R. 1965. Buzzards on Skomer Island 1954–64. *Nature in Wales* 9:116–124.

Dawkins, W. B. 1874. *Cave Hunting; caves and the early inhabitants of Europe.* London.

Day, J. C. U. 1984. Population and breeding biology of Marsh Harriers in Britain since 1900. *Journal of Applied Ecology* 21:773–787.

Day, J. C. U. & Wilson, J. 1978. Breeding Bitterns in Britain. *British Birds* 71:285–300.

Delany, S., Greenwood, J. J. D. & Kirby, J. 1992. National Mute Swan Survey. Report to the Joint Nature Conservation Committee.

Delgety, C. T. & Scott, P. 1948. A new race of the White-fronted Goose. *Bulletin of the B.O.C.* 68:109–121.

Dick, W. J. A., Pienkowski, M. W., Walter, M. & Minton, C. D. T. 1976. Distribution and Geographical Origins of Knot *Calidris canutus* wintering in Europe and Africa. *Ardea* 64:22–47.

Dickens, R. F. 1964. The north Atlantic population of the Great Skua. *British Birds* 57:209–210.

Dickinson, H. & Howells, R. J. 1962. Divers (Genus *Gavia*) in Wales. *Nature in Wales* 8:47–53.

Dillwyn, L. W. 1840. *Contributions towards a history of Swansea.*

Dillwyn, L. W. 1848. *Fauna and flora of Swansea and its neighbourhood.*

Dillwyn, L. W. 1849. *Catalogue & notes of the fauna of the neighbourhood of Swansea.* Commenced June 1849.

Dix, T. 1865. Notes on birds in Carmarthenshire. *Zoologist* 23:9663–4.

Dix, T. 1866. A list of birds observed in Pembrokeshire. *Zoologist* 2(1):132–140.

Dix, T. 1869. Ornithological notes from Pembrokeshire. *Zoologist* 2(4):1670–1681.

Dix, T. 1870. Quails in Pembrokeshire. *Zoologist* (2)5:2394–96.

Dobie, W.H. 1893. Birds of West Cheshire, Denbighshire and Flintshire. *Proceedings of Chester Society of Natural Science and Literature.* 282–351.

Dodd, S. G. 1987. Movements of SCAN Curlew. *SCAN Ringing Group Report* 1985 and 1986:30–31.

Dodd, S. & Moss, D. 1983. Turnstone population at Rhos-on-Sea. *SCAN Ringing Report* for *1981 and 1982*:56–57.

Dodderidge-Knight, E. 1853. An account of Newton Nottage. *Archaeologia Cambrensis.*

Doe, M. 1976. Census of nesting House Martins in Cardiff in 1976. *Glamorgan Bird Report.*

Donovan, E. 1805. *Descriptive excursions in Pembrokeshire.*

Donovan, E, 1805. *Descriptive excursions through South Wales & Monmouthshire in the year 1804, and the four preceding summers.* London.

Donovan, J. W. 1972. The Chough in Pembrokeshire. *Nature in Wales* 12:21–23.

Donovan, J. W. & Rees, G. H. In press. *Status and atlas of Pembrokeshire birds.* Dyfed Wildlife Trust.

Dovaston, J. F. M. 1832. *London's Magazine of Natural History* 5:83.

Drane, R. 1893. Natural history notes from Grassholm. *Transactions of the Cardiff Naturalists Society* 1893–4:1–13.

Drinnan, R. E. & Cole, H. A. 1957. Oystercatchers as pests of cockle and mussel beds. *Nature in Wales* 3:499–503.

Dymond, J. N., Fraser, P. A. & Gantlett, S. J. M. 1989. *Rare Birds in Britain and Ireland.* Poyser.

Eades, R. A. 1974. Monthly variations in foreign-ringed Dunlins on the Dee estuary. *Bird Study* 21:155–157.

Eades, R. A. 1982. Notes on the distribution and feeding of Little Gulls at sea in Liverpool Bay. *Seabird Report* 6:115–121.

Eades, R. A. & Okill, J. D. 1976. Weight variations of ringed Plovers on the Dee estuary. *Ringing and Migration* 1:92–97.

Eades, R. A. & Okill, J. D. 1977. Weight changes of Dunlins on the Dee estuary in May. *Bird Study* 24:62–63.

Ecroyd Smith, H. 1866. Nesting of the Little Tern. *Zoologist* (2)1:100.

Eltringham, S. K. 1963. The British population of the Mute Swan in 1961. *Bird Study* 10:10–28.

Elwes, H. J. Catalogue of eggs in the British Museum.

Emery, F. V. 1986. Edward Lhuyd and Snowdonia. *Nature in Wales* (New Series) 4:3–11.

Evans, D. 1990. Collation of Anglesey wildfowl data, 1980–90. NCC Report.

Evans, P. R. 1966. Some results from the ringing of Rock Pipits on Skokholm 1952–65. *Report of the Skokholm Bird Observatory*, 1966:22–27.

Extracts from correspondence with Keepers on Skerries (RSPB archives).

Eyton, T. C. 1839. An attempt to ascertain the fauna of Shropshire and north Wales. *Ann. Nat. Hist.* 1 (1838):285–292; 2 (1839):52–56.

Farrar, G. B. 1938. *The feathered folk of an estuary.* London.

Fenton, R. *Tours in Wales* 1804–1813.

Fenton, R. 1811. *A Historical Tour through Pembrokeshire.* London.

Ferguson-Lees, I. J. & Williamson, K. 1960. Recent Reports and News. *British Birds* 53:529–544.

Ferns, P. N. 1978. Spring passage of Dunlins, Sanderlings, Ringed Plovers and Turnstone through Britain. *Wader Study Group Bulletins* 24:7–9.

Ferns, P. N. 1979. Spring passage of Dunlins, Sanderlings, Ringed Plovers and Turnstone through Britain. Progress Report . *Wader Study Group Bulletins* 26:6–7.

Ferns, P. N. 1979. Spring passage of Dunlins, Sanderlings, Ringed Plovers and Turnstone through Britain. A Further Progress Report . *Wader Study Group Bulletins* 27:7.

Ferns, P. N. 1980. The Spring migration of Ringed Plovers through Britain in 1979. *Wader Study Group Bulletins* 29:10–13.

Ferns, P. N. 1981. Final Comments on the Spring migration of waders through Britain in 1979. *Wader Study Group Bulletins* 33:6–10.

Ferns, P. N. 1981. The spring migration of Dunlins through Britain in 1979 . *Wader Study Group Bulletins* 32:14–19.

Ferns, P. N. 1981. The Spring migration of Turnstone through Britain in 1979 . *Wader Study Group Bulletins* 31:36–40.

Ferns, P. N. & Green, G. H. 1979. Observations on the breeding plumage and prenuptial moult of Dunlins, *Calidris alpina* captured in Britain. *Le Gerfaut* 69:286–303.

Ferns, P. N. & Mudge, G. P. 1981. Accuracy of nest counts at a mixed colony of Herring and Lesser Black-backed Gulls. *Bird Study* 28:244–246.

Ferns, P. N., Green, G.H. & Round, P. D. 1979. Significance of the Somerset and Gwent Levels in Britain as feeding areas for migrant Whimbrels *Numenius phaeopus*. *Biological Conservation* 16:7-22.

Ferns, P. N., Humphreys, P. N., Sarson, E. T., Walker, I. R., Hamar, H. W., Kelsey, F. D. & Venables, W. A. 1977. *The Birds of Gwent*. Pontypool.

Fischer, J. (ed.) 1917. *Tours in Wales* (1804–1813) by Richard Fenton, edited from his MS journals in the Cardiff Free Library. London.

Fisher, J. 1948. Rook investigation. *Agriculture* 55:20-23.

Fisher, J. 1952. *The Fulmar*. Collins.

Fisher, J. 1953. The Collared Dove in Europe. *British Birds* 46: 153–181.

Fisher, J. 1966. Fulmar population of Britain and Ireland. *Bird Study* 13:5–76.

Fisher, J. 1966. *The Shell bird book*. Ebury & Michael Joseph. London.

Fitter, R. S. R. 1945. *London's Natural History*. London.

Flegg, J. J. M. 1973. A study of treecreepers. *Bird Study* 20:287–302.

Flegg, J. J. M. & Cox, C. J. 1972. Movement of Black-headed Gulls from colonies in England and Wales. *Bird Study* 19:228–240.

Foley, H. T. H. 1935. Red-breasted Goose in Pembrokeshire. *British Birds* 28:311-312.

Forrest, H. E. 1907. *The Vertebrate Fauna of North Wales*.

Forrest, H. E. 1919. *A Handbook to the Vertebrate Fauna of North Wales*.

Fox, A. D. 1986a. The breeding Teal *Anas crecca* of a coastal raised mire in central Wales. *Bird Study* 33:18–23.

Fox, A. D. 1986b. The Little Gull *Larus minutus* in Ceredigion, West Wales. *Seabird* 9:26–31.

Fox, A. D. 1988. Breeding status of the Gadwall in Britain and Ireland. *British Birds* 81:51–65.

Fox, A. D. 1991. History of the Pochard breeding in Britain. *British Birds* 84:83–97.

Fox, A. D. & Mitchell, C. R. 1988. Migration and seasonal distribution of Gadwall from Britain and Ireland. *Wildfowl* 39:145–152.

Fox, A. D. & Roderick, H. W. 1989. Wintering divers and grebes in Welsh coastal waters. *Welsh Bird Report* 1989:53–60. Welsh Ornithological Society.

Fox, A. D. & Salmon, D. G. 1988. Changes in the non-breeding distribution and habitat of the Pochard. *Biological Conservation* 46:303–316.

Fox, A.D. & Salmon, D. G. 1989. The winter status and distribution of Gadwall in Britain and Ireland. *Bird Study* 36:37–44.

Fox, A. D. & Salmon, D. G. In press. The breeding and moulting Shelduck *Tadorna tadorna* of the Severn Estuary. *Biological Journal of the Linnean Society*.

Fox, A. D. & Stroud, D. A. 1985. The Greenland White-fronted Goose in Wales. *Nature in Wales* (New Series) 4:20–27.

Fox, A. D., Stroud, D. A. & Francis, I. S. 1990. Uprooted cotton-grass *Eriophorum angustifolium* as evidence of goose feeding in Britain and Ireland. *Bird Study* 37:210–212.

Fox, A. D., Gitay, H., Owen, M., Salmon, D. G. & Ogilvie, M. A. 1989. Population dynamics of Icelandic-nesting geese, 1960–1987. *Ornis Scandinavica* 20:289–297.

Francis, I. 1992. The Birds of Llangorse Lake. Factors affecting population trends of wetland species. Unpublished RSPB Report.

Friese, J. 1988. Sand Martin colonies on the River Tywi. *Carmarthenshire Bird Report* 1988:27–28.

Friese, J. 1992. Birds of the Tywi valley. *Llanelli Naturalists' Newsletter*, summer 1992:17–21.

Fuller, R. J. 1982. *Bird Habitats in Britain*. Poyser.

Fuller, R. J. & Lloyd, D. 1981. The distribution and habitats of wintering Golden Plovers in Britain, 1977–1978. *Bird Study* 28:169–185.

Garnett, M. C. In prep. Song bird communities on ffridd in Wales.

Gibb, J. 1956. Food, feeding habits and territory of Rock Pipit. *Ibis* 98:506–530.

Gibbons, D. W., Reed, J. B. & Chapman, R. 1993. *The New Atlas of Breeding Birds in Britain and Ireland 1988–91*. Poyser.

Gibbs, R. G. & Wiggington, M. J. 1973. A breeding bird census in a sessile oak-wood at Aber, Caernarvonshire. *Nature in Wales* 13:158–162.

Gibbs, R. G. & Wood, J. B. 1974. A 1968 survey of Stonechat in Wales. *Nature in Wales* 14:7–12.

Gilbert, H. A. 1946. Gatherings of Ravens in Breconshire. *British Birds* 39:52.

Gilbert, H. A. & Brook, A. 1931. *Watchings and wanderings*.

Giraldus Cambrensis. *The journey through Wales*. 1188 – a diary of a preaching tour with Archbishop Giraldus and The Description of Wales c. 1192. Trans. Lewis Thorpe, 1978, Penguin Classics.

Gladstone, H. 1921. A sixteenth century portrait of the Pheasant. *British Birds* 15:67–69.

Glazebrook, F. H. 1926. The bird-life of the coast of Anglesey. *Anglesey Antiquarian Society Transactions* 106–109.

Glazebrook, F. H. 1928. The bird-life of the coast of Anglesey, Saint Dwynwen's Isle (winter). *Anglesey Antiquarian Society Transactions* 87–99.

Glazebrook, F. H. 1929. The bird-life of the coast of Anglesey, Saint Dwynwen's Isle (spring). *Anglesey Antiquarian Society Transactions* 90–94.

Glazebrook, F. H. 1931. The bird-life of the coast of Anglesey, 3. Ynys Cybi. *Anglesey Antiquarian Society Transactions* 109–112.

Glazebrook, F. H. 1934. The bird-life of the coast of Anglesey, 4. The terns. *Anglesey Antiquarian Society Transactions* 140–146.

Glue, D. & Morgan, R. 1972. Cuckoo hosts in British habitats. *Bird Study* 19:187–192.

Godman & Salvin. Manuscripts relating to egg collections. British Museum.

Gooch, S., Baillie, S. R. & Birkhead, T. R. 1991. Magpie and songbird populations. *Journal of Applied Ecology* 28:1068–1086.

Grant, P. J. 1983. Yellow-legged Herring Gulls in Britain. *British Birds* 76:192–194.

Green, M. 1992. Migration of the Dotterel *Charadrius morinellus* through Wales. *Welsh Bird Report* No 5:70–73.

Green, M & Williams, I. T. 1992. Status of Chough in Wales. Unpublished RSPB Report.

Gribble, F. C. 1962. Census of Black-headed gull colonies in England and Wales, 1958. *Bird Study* 9:56–71.

Gribble, F. C. 1976. Census of Black-headed gull colonies in England and Wales in 1973. *Bird Study* 23:139–149.

Gribble, F. C. 1983. Nightjars in Britain and Ireland in 1981. *Bird Study* 30:165–176.

Griffin, B. 1990. Breeding sawbills on Welsh rivers. Unpublished RSPB Report.

Griffin, B. & Rowland, G. 1992. Hen Harriers in Wales 1992. Unpublished RSPB Report.

Griffin, B., Saxton, N. & Williams, I. 1992. Breeding Redshank in Wales, 1991. RSPB Report.

Griffiths, J. 1963. Birdlife in an industrial valley in south Wales. *Nature in Wales* 8:179–183.

Griffiths, J. 1966. The migration of waders and terns through central Wales. *Nature in Wales* 10:17–21.

Griffiths, J. 1967. Summer flocks of Mute Swans in Breconshire. *Nature in Wales* 10:103–105.

Griffiths, J. 1968. Bird Life in Breconshire. Re-printed from *Brecon & Radnor Express* Nov. 1968. Brecknock Museum 1969.

Griffiths, J. 1971. Birdlife in Breconshire. Unpublished Report.

Griffiths, J. & Griffiths, G. 1964. Urban rooks in Merthyr Tydfil. *Nature in Wales* 9:14–16.

Ground, T, 1925. Notes on some birds of North Pembrokeshire 1894–1914. *British Birds* 18:231–236.

Grove, S. J., Hope-Jones, P., Malkinson, A. R., Thomas, D. & Williams, I. T. 1988. Black Grouse in Wales, spring 1986. *British Birds* 81:2–9.

Gurney, J. H. 1913. *The Gannet*. London.

Gurney, R. 1919. Breeding stations of the Black-headed gull in the British Isles. *Transactions of the Norfolk and Norwich Naturalists' Society* 10:416–447.

Hack, P. 1991. A Survey of the Breconshire and South Cardiganshire Section of the Elenydd SSSI for breeding Golden Plover and Dunlin (1991). Unpublished report for the Countryside Council for Wales.

Haines, G. Diaries in the National Museum of Wales.

Hardy, E. 1941. *The Birds of the Liverpool area*. Arbroath.

Harris, M. P. 1959. The status of the Eider in Wales. *Nature in Wales* 5:849–852.

Harris, M. P. 1963. The breeding biology of Larus gulls. PhD thesis, Swansea Univerity.

Harris, M. P. 1970. Rates and causes of increases of some British gull populations. *Bird Study* 17:325–335.

Harris, M. P. 1984. *The Puffin*. Poyser.

Harrison, C. J. O. 1977. Non-passerine birds of the Ipswichian Interglacial from the Gower caves. *Transactions British Cave Research Association* 4:441–442.

Harrison, R. 1970. Breeding of Bearded Tits in Anglesey. *British Birds* 63:83–84.

Harrison, T. H. & Hollom, P. A. D. 1932. The Great Crested Grebe enquiry, 1931. *British Birds* 26: 62–92, 102–131, 142–155, 174–195.

Harrisson, T. H. & Hurrell, H. G. 1933. Numerical fluctuations of the Great Black-backed Gull in England and Wales. *Proceedings of the Zoological Society London*, 1933:191–209.

Harrop, H. R. 1985. Influx of Crossbills 1985. *Glamorgan Bird Report*.

Harrop, J. M. 1961. The Woodcock in Denbighshire. *Nature in Wales* 7:79–82.

Harrop, J. M. 1970. The Chough in north Wales. *Nature in Wales* 12:65–69.

Hartham, A. J. 1958. Thirty years ago: an ornithological companion. *Nature in Wales* 4:578–580.

Hawthorn, I. & Mead, C. J. 1975. Wren movements and survival. *British Birds* 68: 349–358.

Heathcote, A., Griffin, D. & Morrey Salmon, H. 1967. The Birds of Glamorgan. *Cardiff Naturalists' Society*.

Hickling, R. A. O. 1954. The wintering of gulls in Britain. *Bird Study* 1:129–148.

Hickling, R. A. O. 1960. The coastal roosting of gulls in England and Wales 1955–56. *Bird Study* 7:32–52.

Hickling, R. A. O. 1977. Inland wintering gulls in England and Wales 1973. *Bird Study* 24:19–88.

Hoare, R. C. *The Journeys of Sir Richard Colt Hoare*, 1793–1810. Edited Diaries by M. W. Thompson, 1983.

Holdgate, M. W. 1971. The seabird wreck of 1969 in the Irish Sea. NERC Report.

Hollom, P. A. D. 1940. Report on the 1938 survey of Black-headed Gull colonies. *British Birds* 33:202–221, 230–244.

Hollom, P. A. D. 1957. The rarer birds of prey: their present status in the British Isles. Honey Buzzard. *British Birds* 50:141–142.

Hollyer, J. N. 1970. The invasion of Nutcrackers in autumn 1968. *British Birds* 63:353–373.

Holyoak, D. T. & Ratcliffe, D. A. 1968. The distribution of the Raven in Britain and Ireland. *Bird Study* 15:191–197.

Hope Jones, P. 1965. Birds recorded at Newborough Warren, Anglesey, June 1960–May 1965. *Nature in Wales* 9:197–215.

Hope Jones, P. 1974. *Birds of Merioneth*. Colwyn Bay.

Hope Jones, P. 1974. Wildlife Records from Merioneth Parish Documents. *Nature in Wales* 14:35–43.

Hope Jones, P. 1975. Migration of Redstarts through and from Britain. *Ringing and Migration* 1:12–17.

Hope Jones, P. 1976. Larger gulls nesting at inland waters in north Wales. *Cambrian Bird Report* 1976.

Hope Jones, P. 1980. Ring Ouzel territories in the Rhinog Hills of Gwynedd. *Nature in Wales* 17:267–269.

Hope Jones, P. 1987. A history of the Black Grouse in Wales. RSPB Unpublished Report.

Hope Jones, P. 1989. The chequered history of the Black Grouse in Wales. *Welsh Bird Report 1989*: 70–78. Welsh Ornithological Society.

Hope Jones, P. & Colling, A. W. 1984. Breeding and protection of Montagu's Harriers in Anglesey, 1955–64. *British Birds* 77:41–46.

Hope Jones, P. & Dare, P. 1976. *Birds of Caernarvonshire*. Colwyn Bay.

Hope Jones, P. & Roberts, J. L. 1983. Birds of Denbighshire. *Nature in Wales* 1(2):56–65.

Horton, N., Brough, T., Fletcher, M. R. Rochard, J. B. A. & Stanley, P. I. 1984. The winter distribution of foreign Black-headed Gulls in the British Isles. *Bird Study* 31:171–186.

Horwood, J. W. & Goss-Custard, J. D. 1977. Predation by the Oystercatcher, *Haematopus ostralegus* in relation to the cockle *Cerastoderma edule* fishing in the Burry Inlet, South Wales. *Journal of Applied Ecology* 14:139–158.

Hose, T. A. 1977. Cited Bird Specimens from North Wales extant in the Collections of the Grosvenor Museum Chester. *Nature in Wales* 15(4):224–225.

Houston, D. 1977. The effect of Hooded Crows on hill sheep farming in Argyll. *Journal of Applied Ecology* 14:1–15, 17–29.

Howe, G. M. & Thomas, P. 1963. *Welsh Landforms and Scenery*. London, Macmillan.

Howells, G. 1963. The status of the Red-legged Partridge in Britain. *Game Research Association Annual Report 2*:46–51.

Howells, R. 1968. *The Sounds Between*. H. H. Walters. Tenby.

Hudson, R. 1965. The spread of the Collared Dove in Britain and Ireland. *British Birds* 58:105–139.

Hudson, R. 1972. Collared Doves in Britain and Ireland 1965–70. *British Birds* 65:139–155.

Hughes, B. & Williamson, I. 1991. Hen Harriers in Wales 1991. Unpublished RSPB Report.

Hughes, S. W. M., Bacon, P. & Flegg, J. J. M. 1979. The 1975 census of the Great Crested Grebe in Britain. *Bird Study* 26:213–226.

Hume, R. A. 1973. Ring-billed Gull in Glamorgan: a species new to Britain and Ireland. *British Birds* 66:509–512.

Hume, R. A. 1975. Successful breeding of Ravens on city building. *British Birds* 68:515–516.

Hume, R. A. 1976. The pattern of Mediterranean Gull records at Blackpill, West Glamorgan. *British Birds* 69:503–505.

Humphreys, G. A. 1938. Some birds of Anglesey. *Anglesey Antiquarian Society Transactions* 84–89.

Humphreys, P. N. 1963. *The Birds of Monmouthshire*. Newport Museum.

Hurford, C. 1985. Gull trends in Glamorgan. *Glamorgan Bird Report for 1985*:5–31.

Hutcheson, M. 1986. *Cumbrian birds: a review of status and distribution 1964–84*. Kendal.

Hutchinson, C. D. 1989. *Birds in Ireland*. Poyser.

Hutchinson, C. D. & Neath, B. 1978. Little Gulls in Britain and Ireland. *British Birds* 71:563–581.

Ingram, G. C. S. & Salmon, H. M. *A list of the Birds of Roath Park 1920*. Plus addendum 1924.

Ingram, G. C. S. & Salmon, H. M. 1939. *The Birds of Monmouthshire*. Trans. Cardiff Nat. Soc. 70:93–127.

Ingram, G. C. S. & Salmon, H. M. 1954. *A hand list of the Birds of Carmarthenshire*. Haverfordwest.

Ingram, G. C. S. & Salmon, H. M. 1955. *The Birds of Radnorshire*. Leominster.

Ingram, G. C. S., Salmon, H. M. & Condry, W. M. 1966. *The Birds of Cardiganshire*.

Insley, H. & Wood, J. B. 1973. The bird community of a young conifer plantation at Plas Coed Mor, Anglesey. *Nature in Wales* 13:165–173.

James, P. C. & Rawlings, F. C. 1986. Little Shearwaters in Britain and Ireland. *British Birds* 79:28–33.

Jenkins, D. & Watson, A. 1967. Population control in Red Grouse and Rock Ptarmigan in Scotland. *Finn. Game Res.* 30:121–141.

John, A. N. G. & Roskell, J. 1985. Jay movements in autumn 1983. *British Birds* 78:611–637.

Johnson, C. 1983. County Bird Recording in Wales. *Nature in Wales*, New Series 2:57–62.

Johnson, G. 1983. Seasonal variation in wader weights in North Wales. *SCAN Ringing Group Report* 1981/82:45–57.

Johnson, G. 1985. The breeding areas of the Lavan Sands Redshank. *SCAN Ringing Group Report* 1983/84:32–34.

Jones, E. R. 1909. *A history of Barmouth and its vicinity*.

Jones, G. 1930. *Welsh folklore and folk customs*. London.

Jones, J. A. A. & Taylor, J. A. 1983. *Climate – National Atlas of Wales*. University of Wales Press.

Jones, W. E. 1968. *Natural history of Anglesey*.

Kalela, O. 1949. Changes in geographical ranges in the avifauna of northern and central Europe in relation to recent changes in climate. *Bird Banding* 20:77–103.

Kent, A. K. 1957. Breeding birds of a north Pembrokeshire farm. *Nature in Wales* 3:421–425.

Kirby, J. S., Rees, E. C., Merne, O. J. & Gardarrson, A. 1992. International census of Whooper Swans (*Cygnus cygnus*) in Britain, Ireland and Iceland: January 1991. *Wildfowl* 43:20–26.

Kirby, J. S., Waters, R. J. & Prys-Jones, R. P. 1990. *Wildfowl and Wader Counts 1989–90*. Slimbridge.

Kirby, J. S., Evans, R. J. & Fox, A.D. 1993. Wintering seaducks in Britain and Ireland: populations, threats, conservation and research priorities. *Aquatic Conservation* 3:105–137.

Lack, D. W. 1954. Habitats of Eurasian Tits. *Ibis* 96:565–573.

Lack, P. C. 1986. *The Atlas of Wintering Birds in Britain and Ireland*. BTO/IWC.

Langham, N. P. E. 1971. Seasonal movements of British terns in the Atlantic Ocean. *Bird Study* 18:155–175.

Langslow, D. R. 1977. Movements of Black Redstarts between Britain and Europe. *Bird Study* 24:169–178.

Lansdown, P. G. 1984. Ring-necked Parakeets in Glamorgan. *Glamorgan bird report*.

Leland, J. (ed. 1907–10). *Itinerary 1535–1543*. London.

Leversley, P., Williams, I., Griffin, B. & Brookes, D. 1990. Breeding Dunlin and Golden Plover on the Elan Uplands 1990. Unpublished RSPB Report.

Lewin, J. 1980. *Relief and Hydrology – National Atlas of Wales*. University of Wales Press.

Lindley, P. J. & Jenkins, Z. 1991. Status, distribution and prey selection of the Goshawk *Accipiter gentilis* in commercial upland forests in north Wales. Unpublished RSPB Report.

Linnard, W. 1982. *Welsh Woods and Forests: History and Utilization*. National Museum of Wales. Cardiff 1982.

Linton, E. & Fox, A. D. 1991. Inland breeding of Shelduck *Tadorna tadorna* in Britain. *Bird Study* 38(2):123–127.

Little, B. & Furness, R. W. 1985. Long distance moult migration by British Goosanders. *Ringing and Migration* 6:77–82.

Little, I. & Morris, S. 1990. Hen Harriers in Wales 1990. Unpublished RSPB Report.

Lloyd, B. Diaries.

Lloyd, B. 1934. Birds of Bardsey Island. *The North Wales Naturalist:* 331–334.

Lloyd, C. S. 1974. Movement and survival of British Razorbills. *Bird Study* 21:102–116.

Lloyd, C. S., Tasker, M. L. & Partridge, K. 1991. *The status of seabirds in Britain and Ireland*. Poyser.

Lockie, J. D. 1955a. The breeding and feeding of Jackdaws and Rooks. *Ibis* 97:341–369.

Lockie, J. D. 1955b. The breeding habits of Short-eared Owls after a vole plague. *Bird Study* 2:53–69.

Lockley, R. M. 1934. Unusual birds in Pembrokeshire, 1933. *British Birds* 27: 201.

Lockley, R. M. 1938. *Skokholm Observatory bird report for 1937*.

Lockley, R. M. 1942. *Shearwaters*. Dent.

Lockley, R. M. 1953a. On the movements of the Manx Shearwater at sea during the breeding season. *British Birds* 46 (supplement).

Lockley, R. M. 1953b. *Puffins*. Dent.

Lockley, R. M. 1956. The outermost rocks of Wales. *Nature in Wales* 2:257–266.

Lockley, R. M. 1961. Birds of the south-west peninsula of Wales. *Nature in Wales* 7:124–133.

Lockley, R. M. & Saunders, D. R. 1967. Middleholm, Pembrokeshire. *Nature in Wales* 10:146–151.

Lockley, R. M., Ingram, G. C. S. & Salmon, H. M. 1949. *The Birds of Pembrokeshire*. Haverfordwest.

Lovegrove, R. R. 1976. Scoter in Carmarthen Bay. Report to the Nature Conservancy Council.

Lovegrove, R. R. 1978(a). Breeding status of Goosanders in Wales. *British Birds* 71:214–216.

Lovegrove, R. R. 1978(b). Scoter in Carmarthen Bay. Second report to the Nature Conservancy Council.

Lovegrove, R. R., Hume, R. A. & McLean, I. 1980. The Status of Breeding Wildfowl in Wales. *Nature in Wales* 17:4–10.

Ludescher, F. B. 1973. Sumpfmeise und Weidenmeise als sympatrische Zwillingsarten. *J. Orn.* 114:3–56.

Mackworth-Praed, C. W. 1941. Orielton Decoy, Pembrokeshire – in Berry, J. Factors affecting the general status of wild geese and wild duck. *International Wildfowl Enquiry*: Volume 1: 67–83. Cambridge University Press.

Magee, J. D. 1965. The breeding distribution of Stonechat in Britain. *Bird Study* 12:83–89.

Marchant, J. H., Hudson, R., Carter, S. P. & Whittington, P. 1990. *Population trends in British breeding birds* BTO/NCC.

Marquiss, M., Newton, I. & Ratcliffe, D. A. 1978. The decline of the Raven in relation to afforestation in southern Scotland and northern England. *Journal of Applied Ecology* 15:129–144.

Marquiss, M. & Newton, I. 1982. A radio-tracking study of the ranging behaviour and dispersion of European Sparrowhawks *Accipiter nisus*. Journal of Animal Ecology 51:111–133.

Marquiss, M. & Reynolds, C. 1986. How many herons? *BTO News No. 143*: 12. March–April 1986.

Mason C. F. & Hussey, A. 1984. Bird population trends as shown by chick ringing data. *Ringing and Migration* 5:113–120.

Massey, M. 1974. Effects of woodland structure on breeding

bird communities in sample woods in south-central Wales. *Nature in Wales* 14(2):95–105.
Massey, M. 1976. Winter population of waterfowl in Brecon Beacons National Park. *Nature in Wales* 15(1):14–21.
Massey, M. 1976(b). *Birds of Breconshire*. Brecknock Naturalists' Trust.
Matheson, C. 1932. *Changes in the fauna of Wales within historical times*. Cardiff.
Matheson, C. 1953. The Partridge in Wales: a survey of gamebook records. *British Birds* 46:57–64.
Matheson, C. 1957. Further Partridge records from Wales. *British Birds* 50:534–536.
Matheson, C. 1960. Additional gamebook records of Partridges in Wales. *British Birds* 53:81–84.
Matheson, C. 1963. The Pheasant in Wales. *British Birds* 56:452–456.
Mathew, Rev. Murray A. 1894. *The Birds of Pembrokeshire and its islands*. London.
Matthews, G. V. T. 1953. Navigation in the Manx Shearwater. *Journal of Experimental Biology* 30:370–396.
McCarten L. 1958. The wreck of Kittiwakes in early 1957. *British Birds* 51:253–266.
McFadzean, S. 1988. Breeding bird communities on the semi-natural rough grazings of the Cambrian Mountains Environmentally Sensitive Area. 1988 RSPB unpublished report.
McFadzean, S. & Tyler, S. J. 1987. Lapwing and Curlew in north Powys: An evaluation of breeding sites. RSPB unpublished report.
Mead, C. J. 1973. Movement of British raptors. *Bird Study* 20:259–286.
Mead, C. J. 1984. Sand Martin slump. *BTO News* 133:1.
Meek, E. R. & Little, B. 1977. Ringing studies of Goosanders in Northumberland. *British Birds* 70:273–283.
Mercer, A. J. 1963. Migration studies on the Skerries, Anglesey, September 1961. *Nature in Wales* 8:109–115.
Metcalfe, N. B. & Furness, R. W. 1985. Survival, winter population stability and site fidelity in the Turnstone *Arenaria interpres*. *Bird Study* 32:207–214.
Monk, J. F. 1963. The past and present status of the Wryneck in the British Isles. *Bird Study* 10:112–132.
Montagu, George 1802. *Montagu's ornithological dictionary*. First edition.
Moore, J. W. 1984. Possible colony of Manx Shearwaters on Welsh mainland. *Seabird Group Newsletter* October 1984.
Moore, N. W. 1957. The past and present status of the Buzzard in the British Isles. *British Birds* 50:173–197.
Moralee, A. & Morall, S. Wintering Chough numbers in Anglesey RSPB unpublished report 1991.
More, A. G. 1865. On the distribution of birds in Great Britain during the nesting season. *Ibis* (2)1:1–27, 119–142, 425–458.
Moreau, R. E. 1956. Quail in the British Isles 1950–53. *British Birds* 49(5):161–168.
Morgan, I. K. 1982. Birds of the Carmarthenshire Coast. *The Carmarthenshire Bird Report* 1982:18–23.
Morgan, R. A. & Glue, D. E. 1981. The breeding of Black Redstarts in Britain in 1977. *Bird Study* 28:163–168.
Morris, G. E. 1990. Recent increases in wintering Black-tailed Godwit, Knot and Dunlin in the Flint Sands/Oakenholt Marsh/Connah's Quay area of the Dee Estuary. *Clwyd Bird Report* 1989:46–47.
Moser, M. E. 1987. A revision of Population Estimates for Waders (Charadrii) Wintering on the Coastline of Britain. *Biological Conservation* 39:153–164.
Moser, M. E. & Summers, R. W. 1987. Wader populations on the non-estuarine coasts of Britain and Northern Ireland: results of the 1984–85 Winter Shorebird Count. *Bird Study* 34:71–81.
Moss, D. 1985. Analysis of Redshank retraps. *SCAN Ringing Group Report* 1983/4:35–38.
Mulligan, W. & Venables, L. S. V. 1958. Reed Warblers in Anglesey. *British Birds* 51: 124–125.
Murton, R. K. 1965. *The Woodpigeon*. Collins.
Murton, R. K. *et al.* 1964. The feeding habits of Woodpigeon, Stock Dove and Turtle Dove. *Ibis* 106:107–188.
National Museum Of Wales Accession Lists (covers all skins and eggs).
National Rivers Authority. 1991. *The Quality of Rivers, Canals and Estuaries in England and Wales*. Water Quality Series No. 4.
Nelson, B. 1978. *The Gannet*. Poyser.
Nethersole-Thompson, D. 1973. *The Dotterel*. Collins.
Newton, I. 1972. *Finches*. Collins.
Newton, I. 1973. Egg breakage and breeding failure in British Merlins. *Bird Study* 20:241–244.
Newton, I. 1986. *The Sparrowhawk*. Poyser.
Newton, I., Davis, P. E. & Davis, J. E. 1982. Ravens and Buzzards in relation to sheep farming and forestry in Wales. *Journal of Applied Ecology* 19:681–706.
Nicholl, D. S. W. 1889. Notes on the rarer birds of Glamorgan. *Zoologist* 3(8):166–174.
Nicholson, E. M. 1929. Report on the "British Birds" census of Heronries, 1928. *British Birds* 22:270–323 and 334–372.
Nicholson, E. M. 1951. *Birds and Men*. Collins.
Nicholson, E. M. 1957. The rarer birds of prey. Their present status in the British Isles. Montagu's Harrier. *British Birds* 50:146–147.
Norman, D. 1987. Are Common Terns successful at a man-made nesting site? *Ringing and Migration* 8:7–10.
Norman, S. C. & Norman, W. 1985. Autumn movements of Willow Warblers ringed in the British Isles. *Ringing and Migration* 6:7–18.
Norris, C. A. 1953. The birds of Bardsey Island in 1952. *British Birds* 46:131–137.
Norris, C. A. 1954. Further notes on the birds of Bardsey Island. *British Birds* 47:206–207.
Norris, C. A. 1960. The breeding distribution of thirty bird species in 1952. *Bird Study* 7:129–184.
North, F. J., Campbell, B. & Scott, R. 1949. *Snowdonia*. London.
O'Connor, R. J. & Mead, C. J. 1984. The Stock Dove in Britain 1930–1980. *British Birds* 77:181–201.
O'Connor, R. J. & Shrubb, M. 1986. *Farming and Birds*. Cambridge University Press.
Ogilvie, M. A. 1968. The numbers and distribution of the European White-fronted Goose in Britain. *Bird Study* 15:2–15.
Ogilvie, M. A. 1981. The Mute Swan in Britain, 1978. *Bird Study* 28:87–106.
Ogilvie, M. A. 1986. The Mute Swan *Cygnus olor* in Britain 1983. *Bird Study* 33:121–137.
Ogilvie, M. A. 1987. Movements of Tufted Duck ringed in Britain: a preliminary assessment. *Wildfowl* 38: 28–36.
Oldham, C. Diaries held at the Edward Grey Institute.
Oldisworth, J. 1802 and 1832. *The Swansea Guide*.
Ormerod, S. J. & Tyler, S. J. 1987a. Aspects of the breeding biology of Welsh Grey Wagtails. *Bird Study* 34:43–51.
Ormerod, S. J. & Tyler, S. J. 1987b. Dippers and Grey Wagtails as indicators of stream acidity in upland Wales. ICBP technical publications 6:191–208.
Ormerod, S. J. & Tyler, S. J. 1990. Population characteristics of Dipper *Cinclus cinclus* roosts in mid and south Wales. *Bird Study* 37:165–170.
Ormerod, S. J., Tyler, S. J. & Lewis, J. M. S. 1985. Is the breeding distribution of Dippers influenced by stream acidity? *Bird Study* 32:32–39.
O'Sullivan, J. M. 1976. Bearded Tits in Britain and Ireland 1966–74. *British Birds* 69:473–489.

Owen, G. 1603. *The description of Pembrokeshire*. Part 1 1892 edition. London.

Owen, M. & Mitchell, C. R. 1988. Movements and migrations of Wigeon *Anas penelope* wintering in Britain and Ireland. *Bird Study* 35:47–60.

Owen, M. & Williams, G. M. 1976. Winter distribution and habitat requirements of Wigeon in Britain. *Wildfowl* 27:83–90.

Owen, M., Atkinson-Willes, G. L. & Salmon, D. G. 1986. *Wildfowl in Great Britain*. 2nd Edition. Cambridge University Press.

Owen, T. 1912. A season with the birds of Anglesey and north Caernarvonshire. *Zoologist* (4)16:304–313, 342–348.

Palmgren, P. 1930. Quantitative Untersuchungen uber die Vogelfauna in den Walden Sudfinnlands. *Acta. Zool. Fenn.* 7:1–219.

Pamplin, W. *Diary and Botanical Record*. 1866–94.

Parker, A. J. 1988. The birds of Roman Britain. *Oxford Journal of Archaeology* 7:197–226.

Parr, S. J. 1991. Occupation of new conifer plantations by Merlins in Wales. *Bird Study* 38:103–111.

Parrinder, E. R. 1989. Little Ringed Plovers in Britain in 1984. *Bird Study* 36:147–153.

Parrinder, E. R. & E. D. 1975. Little Ringed Plovers in Britain in 1968–73. *British Birds* 68:359–368.

Parslow, J. L. F. 1973. *Breeding Birds of Britain and Ireland*. Berkhamsted, Poyser.

Paterson, *et al.* 1990. *The Birds of Glamorgan*. Cardiff Natural History Society.

Paynter, D. B. Razorbill and Guillemot survey in north Pembrokeshire 1979. EGI report.

Peach, W. S. & Miles, P. M. 1961. An annotated list of some birds seen in the Aberystwyth district, 1949–56. *Nature in Wales* 7:11–20.

Peakall, D. B. 1962. The past and present status of the Red-backed Shrike in Great Britain. *Bird Study* 9:198–216.

Peal, R. E. F. 1968. The distribution of the Wryneck in the British Isles 1964–66. *Bird Study* 15:111–126.

Peers M. 1985. *Birds of Radnorshire and mid Powys*. Llandrindod Wells.

Peers, M. & Shrubb, M. 1990. *Birds of Breconshire*.

Pennant, T. 1778 and 1781. *Tours of Wales*.

Pennant, T. 1883 edition *Tours in Wales* Vol. III. Caernarvon.

Perrins, C. M. 1968. The numbers of Manx Shearwaters on Skokholm. *Skokholm Bird Observatory Report* 1967: 23–29.

Perrins, C. M. 1979. *British Tits*. Collins.

Phillips, E. Cambridge. 1882. *The Birds of Breconshire*. London.

Phillips, E. Cambridge. 1899. *The Birds of Breconshire*. Brecon.

Phillips, J. A. 1953. MAFF Infestation Control Division Report No. 36. The Raven in Pembrokeshire.

Picozzi, N. & Weir, D. 1976. Dispersal and causes of death in Buzzards. *British Birds* 69:193–201.

Plant, C. W. 1976. Some observations on the winter diet of the Barn Owl *Tyto alba* on Skomer Island, Dyfed, Wales. *Nature in Wales* 15:54–59.

Poole, J. & Sutcliffe, S. J. 1992. Seabird studies on Skomer Island in 1992. JNCC Report No. 139.

Powell, M. C. 1989. Leach's Petrel in Wales, December 1989. *Welsh Bird Report 1989* 4:36–37.

Prater, A. J. 1973. The Wintering Population of Ruffs in Britain and Ireland. *Bird Study* 20:245–250.

Prater, A. J. 1976a. Breeding population of the Ringed Plover in Britain. *Bird Study* 23:155–161.

Prater, A. J. 1976b. The midwinter estuarine population of Waders in Wales, 1971–74. *Nature in Wales* 15(1):2–7.

Prater, A. J. 1981. *Estuary birds of Britain*. Poyser.

Prater, A. J. 1989. Ringed Plover *Charadrius hiaticula* breeding population of the United Kingdom in 1984. *Bird Study* 36:154–159.

Prater, A. J. & Davies, M. 1978. Wintering Sanderlings in Britain. *Bird Study* 25:33–38.

Prestt, I. 1965. An enquiry into the recent breeding status of some of the smaller birds of prey and crows in Britain. *Bird Study* 12:196–221.

Prestt, I. & Mills, D. H. 1966. A census of the Great Crested Grebe in Britain 1965. *Bird Study* 13:163–203.

Proceedings of the Chester Society of Natural Science 1947.

Proceedings of the Llandudno and District Field Club 1907–1939.

Prys-Jones, R. P. & Davis, P. E. 1990. The Abundance and Distribution of Wildfowl and Waders on Carmarthen Bay (Taf/Tywi/Gwendraeth). Report from British Trust for Ornithology to the Nature Conservancy Council.

Raines, R. J. 1961–1963. *Birds of the Wirral Peninsula*. Liverpool Ornithologists' Club 1960, 1961, 1962.

Ratcliffe, D. A. 1962. Breeding Density in the Peregrine *Falco peregrinus* and Raven *Corvus corax*. *Ibis* 104:13–39.

Ratcliffe, D. A. 1963. The status of the Peregrine in Great Britain. *Bird Study* 10:56–90.

Ratcliffe, D. A. 1965. The Peregrine situation in Great Britain. 1963–64. *Bird Study* 12:66–82.

Ratcliffe, D. A. 1972. The Peregrine population of Great Britain in 1971. *Bird Study* 19:117–156.

Ratcliffe, D. A. 1976. Observations on the Breeding of the Golden Plover in Great Britain. *Bird Study* 23:63–116.

Ratcliffe, D. A. 1980. *The Peregrine Falcon*. Poyser.

Ratcliffe, D. A. 1984. The Peregrine breeding population of the United Kingdom in 1981. *Bird Study* 31:1–18.

Ray. 1713. *Synopsis Methodica Avium*.

Rees, E. I. S. 1969. An attempt to census the Puffin Island Herring Gulls.

Rees, G. H. 1985. Common Scoter movements at Strumble Head. *Pembrokeshire Bird Report* 1985.

Reynolds, C. M. 1974. The census of heronries 1969–73. *Bird Study* 21:129–134.

Richards, M. 1926. Birdlife of the neighbourhood. In: *The story of two parishes – Dolgellau and Llanelltyd*.

Richford, A. S. 1978. Unpublished PhD thesis.

Richford, A. S. 1978. The effect of Jackdaws on the breeding of auks on Skomer Island. *Nature in Wales* 16:32–36.

Richford, A. S. & Lawman, D. M. 1978. The breeding of Jackdaws on Skokholm. *Nature in Wales* 16:106–110.

Ridgill, S. C. & Fox, A. D. 1990. Cold weather movements of waterfowl in western Europe. IWRB Special Publication No. 13. IWRB, Slimbridge.

Riviere, B. B. 1930. *History of the birds of Norfolk*. London.

Roberts, D. H. V. 1983. Heronries in Carmarthenshire. *Carmarthenshire Bird Report* 1983.

Roberts, J. L. 1983. Whitethroats breeding on a Welsh heather moor. *British Birds* 76:456.

Roberts, J. L. & Bowman, N. 1986. Diet and ecology of Short-eared Owls *Asio flammeus* breeding on heather moor. *Bird Study* 33:12–17.

Roberts, J. L. & Green, D. 1983. Breeding failure and decline of Merlins on a north Wales moor. *Bird Study* 30:193–200.

Roberts, P. 1985. *The Birds of Bardsey*. Reading.

Robertson, I. S. 1977. Identification and European status of Eastern Stonechats. *British Birds* 70: 237–245.

Roderick, H. W. 1978. The Chough in Ceredigion. *Nature in Wales* 16(2): 127–128.

Roderick, H. W. & Cross, A. V. 1991. Past and present status of the Chough in Ceredigion. *Ceredigion Bird Report* 1991.

Rogers, E. F. & Gault, L. N. 1968. Sand Martins on the river Usk. *Nature in Wales* 11:15–19.

Rogers, M. J. 1982. Ruddy Shelducks in Britain in 1965–79. *British Birds* 75:446–455.

Rolfe, R. 1966. The status of Chough in the British Isles. *Bird Study* 13(3):221–236.

Rose, L. N. 1982. Breeding ecology of British pipits and their Cuckoo parasite. *Bird Study* 29:27–40.

Round, P. D. & Moss, M. 1984. The waterbird population of three Welsh rivers. *Bird Study* 31:62–68.

RSPB 1977. Breeding birds of the River Wye. Unpublished report.

RSPB 1977. Upland Bird Communities in Wales. Unpublished report.

RSPB 1978. Breeding birds of the River Severn. Unpublished report.

RSPB 1981. Sawbills in Wales – a short survey of selected rivers. Unpublished report.

RSPB 1986 and 1987. Breeding bird surveys on the Llŷn peninsula and Anglesey in 1986 and 1987. Unpublished report.

RSPB 1988. Ornithological studies at Trawsfynydd. Unpublished report to CEGB.

RSPB Watchers' Reports (Wales) 1918 *et seq.*

Sage, B. L. 1956. Notes on the birds of Caldey and St. Margaret's Islands, Pembrokeshire. *Nature in Wales* 2(4):333–340.

Sage, B. L. & King, B. 1959. The influx of Phalaropes in Autumn 1957. *British Birds* 52:33–42.

Sage, B. L. & Vernon, J. D. R. 1978. 1975 National survey of rookeries. *Bird Study* 25:64–86.

Sage, B. L. & Wittington, P. A. 1985. The 1980 sample census of rookeries. *Bird Study* 32:77–81.

Salmon, D. G., Prys-Jones, R. P. & Kirby, J. S. 1987. *Wildfowl and Wader counts 1986–87.* Slimbridge.

Salmon, D. G., Prys-Jones, R. P. & Kirby, J. S. 1988. *Wildfowl and Wader Counts* 1987–88. Slimbridge.

Salmon, H. M. 1971. A History of the Kite *Milvus milvus* in Wales. Unpublished MS.

Salmon, H. M. 1974. *A supplement to the Birds of Glamorgan.* 1967. Cardiff.

Salmon, H. M. & Ingram, G. C. S. 1957. *The Birds of Brecknock.*

Salomonsen, F. 1947. Forste forelobige Liste over genfunde gronlandske Ringfugel. *Dansk Ornithologisk Forenings Tidsskrift* 41:141–143.

Salter, J. H. Natural History Diaries in the National Library of Wales.

Salter, J. H. 1895. Observations on birds in mid-Wales. *Zoologist* (3)19:179.

Salter, J. H. 1900. List of the birds of Aberystwyth and neighbourhood. UCW Scientific Society.

Saunders, D. R. 1962. *Nature in Wales* 8:59–66.

Saunders, D. R. 1974. *A Guide to the Birds of Wales.* London, Constable.

Saunders, D. R. 1976. *A Brief Guide to the Birds of Pembrokeshire.* H. G. Walters (Tenby).

Schlapfer, A. 1988. Populationskologie der feldlerche in der intensiv gunetzten Agrarlandschaft. *Orn. Beob.* 85:309–371.

Scott, R. E. 1968. Rough-legged Buzzards in Britain in the winter of 1966/67. *British Birds* 61:449–455.

Scott, R. E. 1978. Rough-legged Buzzards in Britain in 1973/74 and 1974/75. *British Birds* 71:325–338.

Seabird Group report 1969.

Seel, D. C. & Walton, K. C. 1979. Numbers of Meadow Pipits on mountain farm grassland in N. Wales. *Ibis* 121:147–164.

Seel, D. C., Thomson, A. G. & Turner, J. C. E. 1986. A conservation assessment for the Barn Owl *Tyto alba* on Anglesey, north Wales. Occasional Paper No. 16. Bangor: Institute of Terrestrial Ecology.

Sellers, R. M. 1984. Movements of Coal, Marsh and Willow Tits in Britain. *Ringing and Migration* 5:79–89.

Sergeant, D. E. 1952. Little Auks in Britain, 1948–1951. *British Birds* 45:122–133.

Shankland, T. 1891. *A list of the birds of N. Wales.* Compiled for use of N. Wales Dialect Society in connection with N. Wales University College. Mold.

Shawyer, C. R. 1987. *The Barn Owl in the British Isles. Its past, present and future.* London: The Hawk Trust.

Shrubb, M. 1985. Breeding Sparrowhawks (*Accipiter nisus*) and organo-chlorine pesticides in Sussex and Kent. *Bird Study* 32:155–163.

Shrubb, M. 1988. The status of the Lapwing in Breconshire in 1987. *Breconshire Birds Annual Report* 1987.

Shrubb, M. & Lack, P. C. 1991. The numbers and distribution of Lapwings *Vanellus vanellus* nesting in England and Wales in 1987. *Bird Study* 38:20–37.

Simms, E. 1985. *British warblers.* Collins.

Simson, C. 1966. *A bird overhead.* Witherby.

Sitters, H. P. 1986. Woodlarks in Britain 1968–83. *British Birds* 79:105–116.

Smith, K. W. 1983. The status and distribution of waders breeding on wet lowland grasslands in England and Wales. *Bird Study* 30: 177–192.

Smith, L. T. 1906. *The itinerary in Wales of John Leland in or about the years 1536–1539.* London.

Smith, M. E. 1969. Kingfisher in Wales: effects of severe weather. *Nature in Wales* 11:109–115.

Smith, P. H. 1987. The changing status of Little Gulls *Larus minutus* in north Merseyside, England. *Seabird* 10:12–21.

Snow, D. W. 1954. The habitats of Eurasian Tits. *Ibis* 96:565–585.

Snow, D. W. 1968. Movements and mortality of British Kestrels. *Bird Study* 15:65–83.

Snow, D. W. (ed.) 1992. *Birds, Discovery and Conservation: 100 years of the British Ornithologists Club.* Article on the Black Woodpecker in Britain, pp. 138–147. Helm.

Southern, H. N. 1970. The natural control of a population of Tawny Owls *Strix aluco. Journal of Zoology, London* 162:197–285.

Southern, H. N. & Morley, A. 1950. Marsh Tit territories over six years. *British Birds* 43:33–47.

Stafford, J. 1962. Nightjar enquiry, 1957–58. *Bird Study* 9:104–115.

Stafford, J. 1969. The census of heronries 1962–63. *Bird Study* 16:83–88.

Stafford, J. 1971. The Heron population of England and Wales, 1928–70. *Bird Study* 18:218–221.

Stafford, J. 1979. The national census of heronries in England and Wales in 1964. *Bird Study* 26:3–6.

Stanley, P. I. & Minton, C. D. T. 1972. The unprecedented westward migration of Curlew Sandpipers in autumn 1969. *British Birds* 65:365–380.

Stowe, T. J. 1982. Recent population trends in cliff-breeding seabirds in Britain and Ireland. *Ibis* 124:502–510.

Stowe, T. J. 1987. The habitat requirements of some insectivorous birds and the management of sessile oakwoods. PhD thesis (unpublished).

Stowe, T. J. & Harris, M. P. 1984. Status of Guillemots and Razorbills in Britain and Ireland. *Seabird* 7:5–18.

Stowe, T. J., Newton, A., Green, R. & Mayes, E. 1993. Decline in Corncrake in Britain and Ireland in relation to habitat. *Journal of Applied Ecology* 30:53–62.

Stresemann, E. & Nowak, E. 1958. Die Ausbreitung der Turkentaube in Asien und Europa. *J. Orn.* 99:243–296.

Summers-Smith, J. D. 1989. *The Sparrows.* Poyser.

Sutcliffe, S. J. 1963. A survey of the breeding birds of St. Margaret's Island. *Nature in Wales* 8:126–127.

Sutcliffe, S. J. 1975. Common Scoter in Carmarthen Bay – an oiling incident. *Nature in Wales* 14(4):243–249.

Tallack, R. E. 1982. The breeding population of House Martins in Gower. *Gower Birds* 3(5):43–47.

Tatner, P. 1978. A review of House Martins in part of south Manchester, 1975. *Naturalist* 103: 59–68.

Taverner, J. H. 1959. The spread of the Eider in Great Britain. *British Birds* 52:245–258.

Taverner, J. H. 1963. Further notes on the spread of the Eider in Great Britain. *British Birds* 56: 273–285.

Taylor, J. A. 1981. *Vegetation – National Atlas of Wales.* University of Wales Press.

Tew, I. 1989. Quail records in Wales in summer 1989. *Welsh Bird Report* 38–39. Welsh Ornithological Society.

Thearle, R., Hobbs, J. T. & Fisher, J. 1953. Birds of St. Tudwal Islands. *British Birds* 46:182–188.

Thom, V. M. 1986. *Birds in Scotland.* Poyser.

Thomas, C. J. & Hack, P. C. 1984. *Breeding Biology of Dunlins and Golden Plovers in Central Wales in 1983.* Unpublished report to the RSPB and MSC.

Thomas, C. J., Hack, P. C. & Ferns, P. N. 1983. *Breeding Biology of Dunlins and Golden Plovers in Central Wales in 1982.* Unpublished report to the RSPB and MSC.

Thomas, D. K. 1984(a). Aspects of habitat selection in Sedge Warbler. *Bird Study* 31:187–194.

Thomas, D. K. 1984(b). Feeding ecology and habitat selection in the Reed Warbler. *Ibis* 126:454.

Thomas, D. K. 1992. A*n Atlas of Breeding Birds in West Glamorgan*. Neath.

Thomas, H. J. 1890–91. Visit to the Gannet settlement upon the island of Grassholm. *Transactions of the Cardiff Naturalists' Society* 1890–91:57–64.

Thomas, J. F. 1955. The Laugharne – Pendine Burrows. *Nature in Wales* 1:72–74.

Thomas, W. B. 1935. Some sea birds. *Anglesey Antiquarian Society Transactions* 182–188.

Thorpe, L. (trans.) 1978. *Gerald of Wales: The Journey through Wales/The Description of Wales.* Penguin Classics.

Thorpe, R. 1992. Counts of selected species in the northern section of Cardigan Bay, winter 1991/1992. Unpublished RSPB Report.

Ticehurst, N. F. 1919–1920. The birds of Bardsey Island. *British Birds* 13: 42–51, 66–75, 101–106, 129–134, 182–193, 214–216.

Ticehurst, N. F. & Jourdain, F. C. R. 1911. On the distribution of the Nightingale during the breeding season in Great Britain. *British Birds* 5:2–21.

Todd, P. 1987. Population dynamics and feeding ecology of Herring Gull and Lesser Black-backed Gull. Unpublished report to NERC.

Tomkinson, G. Egg collection and catalogue. Wildfowl and Wetlands Trust, Slimbridge.

Tong, M. 1967. Winter breeding of Shags. *British Birds* 60: 214–215.

Tubbs, C. R. 1972. Analysis of nest record cards for the Buzzard. *Bird Study* 19:96–104.

Tubbs, C. R. 1974. *The Buzzard.* David & Charles.

Tuite, C. H., Owen, M. & Paynter, D. 1983. Interaction between wildfowl and recreation at Llangorse Lake and Talybont Reservoir, South Wales. *Wildfowl* 34:48–63.

Tyler, S. J. 1985. Sawbills in Wales 1985. Unpublished RSPB Report.

Tyler, S. J. 1990. The breeding bird community at Penallt, Gwent 1978–1989. *Welsh Bird Report* 4:58–69.

Tyler, S. J. 1992a. A review of the status of breeding waders in Wales. *Welsh Bird Report* 5:74–86.

Tyler, S. J. 1992b. Little Ringed Plovers in Wales in 1991. Report to National Rivers Authority by RSPB.

Tyler, S. J. & Geach, J. 1987. Birds of the south Wales coalfield. Unpublished RSPB Report.

Tyler, S., Lewis, J., Venables, A. & Walton, J. 1987. *The Gwent Atlas of Breeding Birds.* Newport.

Tyler, S. J., Stratford J. O. & Lucas, R. M. 1988. Goosanders in Wales. *Welsh Bird Report* 1988.

Universal British Traveller 1779:661.

Venables, L. S. V. & U. M. 1969. The sequence of birds in a young conifer plantation, 1959–1967, Newborough Warren, Anglesey. *Nature in Wales* 11:176–182.

Venables, L. S. V. & U. M. 1972. Our vanishing swans. *Nature in Wales* 13:128–131.

Vernon, J. D. R. 1963. Icelandic Black-tailed Godwits in the British Isles. *British Birds* 56:233–237.

Vernon, J. D. R. 1963. The distribution of Starling roosts in Glamorgan and Monmouthshire. *Nature in Wales* 8:165–174.

Vernon, J. D. R. 1969. Spring migration of the Common Gull in Britain and Ireland. *Bird Study* 16:101–107.

Vernon, R. L. 1955. Red-breasted Mergansers breeding in Anglesey. *British Birds* 48:135–136.

Village, A. 1990. *The Kestrel.* Poyser.

Vinicombe, K. E. 1973. A second Ring-billed Gull in Glamorgan. *British Birds* 66:513–517.

Vinicombe, K. E. 1975. Ring-billed Gulls at Blackpill. *Gower Birds* 2:157–164.

Vinicombe, K. E. 1977. Bird Surveys in Gwynedd, *1976–77.* Unpublished RSPB Report.

Vinicombe, K. E. 1982. Breeding and population fluctuations of the Little Grebe. *British Birds* 75:204–218.

Vinicombe, K. E. 1985. Ring-billed Gulls in Britain and Ireland. *British Birds* 78:327–337.

Voous, K. H. 1960. *Atlas of European birds.* London.

Walford, T. 1818. *The scientific traveller through England, Wales and Scotland: Vol. II.*

Walker, T. G. 1956. *Birds of the Welsh Coast.* Cardiff University of Wales Press.

Walpole-Bond, J. A. 1904. *Birdlife in wild Wales* (with annotations by H. M. Salmon).

Walpole-Bond, J. 1914. *Field-studies of some rarer British Birds.* London.

Waters, W. E. 1968. Visible autumn migration at St. David's Head. *Nature in Wales* 11:20–27.

Watson, D. 1977. *The Hen Harrier.* Poyser.

Webb, A. *et al.* 1990. Seabird distribution west of Britain. NCC report.

Wells, C. H. Extracts from ornithological notebooks (Brecknockshire) 1920–1970.

Welsh Office. 1990. Welsh Agricultural Statistics No. 12.

Westlake, J. 1991. Nightjar species account. *Welsh Bird Report* No. 4 1990. Welsh Ornithological Society.

Whalley, P. E. S. 1954. List of birds seen in Anglesey and Caernarvonshire with notes on their distribution and status. *The North West Naturalist* 604–618.

Whalley, P. E. S. 1957. Terns nesting underground. *British Birds* 50:121–122.

Whitaker, A. Diaries in the Edward Grey Institute.

Williams 1863. *Llandudno guide.*

Williams, B. 1835. *The history and antiquities of the town of Aberconwy and its neighbourhood.*

Williams, G. A. 1977. Status and distribution of wildfowl in the Dee estuary. *Nature in Wales* 15(4):166–179.

Williams, G. A. 1978. Notes on the birds of Grassholm. *Nature in Wales* 16(1):2–15.

Williams, G. A. 1981. The Merlin in Wales: breeding numbers, habitat and success. *British Birds* 74: 205–214.

Williams, I. T. 1988. Breeding Short-eared Owls *Asio flammeus* in Wales. *Welsh Bird Report 1988.* Welsh Ornithological Society.

Williams, I. T. 1989. Short-eared Owl species account. *Welsh Bird Report 1989*. Welsh Ornithological Society.

Williams, I. T. 1992. The breeding population of the Peregrine in Wales in 1991. *Welsh Bird Report* 5:62–69.

Williams, I. T. In press. Honey Buzzard breeding in 1992, the first confirmed record for Wales. *Welsh Bird Report*.

Williams, I. T. & McLaughlin, K. 1988. Hen Harriers in Wales, 1988. Unpublished RSPB Report.

Williams, I. T., Anderson C. & Holloway, D. 1992. Black Grouse in Wales, spring 1992. Unpublished RSPB Report.

Williams, I., Griffin, B. & Holloway, D. 1991. The Red Grouse population of Wales based on winter counts 1990/91. Unpublished RSPB Report.

Williams, I. T., Griffin, B. M., Little, I. D. & Morris, S. J. 1990. Status of the Chough (*Pyrrhocorax pyrrhocorax*) on the Llŷn peninsula 1990. Unpublished RSPB Report.

Williams, J. 1830. *Faunula Grustensis*. Llanrwst.

Williams, J. 1957. The changed status of birds on St. Tudwal's Island, north Wales, 13–15 July 1956. *Merseyside Naturalists' Association Annual Report*.

Williamson, K. 1969. Habitat preferences of Wrens on farmland. *Bird Study* 16:53–59.

Williamson, K. 1971. Breeding birds in Redwood Grove. *Quarterly Journal of Forestry* 109–121.

Wilson, W. 1931. Some further notes on the birds of Bardsey Island. *British Birds* 24:121–123.

Wintle, W. J. 1925. Some Caldey birds. *Pax* (Quarterly Review of Benedictines of Caldy) 133–139.

Witherby, H. F. & Ticehurst, N. F. 1907–1908. On the more important additions to our knowledge of British birds since 1899. *British Birds* 1(6):52-56, 81–85, 109–114, 147–152, 178–184, 246–256, 280–284, 314–322, 347–350.

Witherby, H. F., Jourdain, F. C. R., Ticehurst, N. F. & Tucker, B. W. 1940–1941. *The Handbook of British Birds*. Witherby. 5 vols.

Worrall, D. H. (in litt.). The breeding birds of Flatholm.

Yapp, W. B. 1955. The birds of Welsh high level oakwood. *Nature in Wales* 1(4):161–166.

Yarker, B. & Atkinson-Willes, G. L. 1971. The numerical distribution of some British breeding ducks. *Wildfowl* 22:63–70.

Zienkiewicz, J. D. 1986. Excavations in the Scamnum Tribunarum at Caerleon: The Legionary museum site, 1983–85. *Britannica* 24.

Species Index

Page references of species accounts in bold numerals